The Handbook of Metabonomics and Metabolomics

The Handbook of Metabonomics and Metabolomics

Edited by

John C. Lindon
Jeremy K. Nicholson
and
Elaine Holmes
Imperial College London

ELSEVIER

Amsterdam • Boston • Heidelberg • London • New York • Oxford
Paris • San Diego • San Francisco • Singapore • Sydney • Tokyo

Elsevier
Radarweg 29, PO Box 211, 1000 AE Amsterdam, The Netherlands
The Boulevard, Langford Lane, Kidlington, Oxford OX5 1GB, UK

First edition 2007

Library of Congress Cataloging-in-Publication Data
A catalog record for this book is available from the Library of Congress

British Library Cataloguing in Publication Data
A catalogue record for this book is available from the British Library

ISBN-13: 978-0-44-452841-4
ISBN-10: 0-44-452841-5

For information on all Elsevier publications visit our website at books.elsevier.com

Printed and bound in the United Kingdom
Transfered to Digital Printing, 2011

Contents

Foreword vii
Sir Richard B. Sykes FRS

Preface ix
John C. Lindon, Jeremy K. Nicholson and Elaine Holmes

1 Metabonomics and Metabolomics Techniques and Their Applications in Mammalian Systems 1
Jeremy K. Nicholson, Elaine Holmes and John C. Lindon

2 Cellular Metabolomics: The Quest for Pathway Structure 35
Oliver Fiehn

3 NMR Spectroscopy Techniques for Application to Metabonomics 55
Hans Senn and Goetz Schlotterbeck

4 High-Resolution Magic Angle Spinning NMR Spectroscopy 113
Jin-Hong Chen and Samuel Singer

5 Chromatographic and Electrophoretic Separations Combined with Mass Spectrometry for Metabonomics 149
Ian D. Wilson

6 Chemometrics Techniques for Metabonomics 171
Johan Trygg and Torbjörn Lundstedt

7 Non-linear Methods for the Analysis of Metabolic Profiles 201
Timothy M.D. Ebbels

8 Databases and Standardisation of Reporting Methods for Metabolic Studies 227
Susanna A. Sansone, Michael D. Waters and Mark R. Viant

9 Metabonomics in Preclinical Pharmaceutical Discovery and Development 241
Donald G. Robertson, Michael D. Reily and Glenn H. Cantor

10 Applications of Metabonomics in Clinical Pharmaceutical R&D 279
Jeremy R. Everett

11 Exploiting the Potential of Metabonomics in Large Population Studies: Three Venues 289
Burton H. Singer, Jürg Utzinger, Carol D. Ryff, Yulan Wang and Elaine Holmes

12 Metabolite Profiling and Cardiovascular Disease 327
David J. Grainger

13 The Role of NMR-based Metabolomics in Cancer 345
Leo L. Cheng and Ute Pohl

14 NMR Spectroscopy of Body Fluids as a Metabolomics Approach to Inborn Errors of Metabolism 375
Udo F. H. Engelke, Marlies Oostendorp and Ron A. Wevers

15 A Survey of Metabonomics Approaches for Disease Characterisation 413
John C. Lindon and Elaine Holmes

16 Metabolic Profiling: Applications in Plant Science 443
Richard N. Trethewey and Arno J. Krotzky

17 *In vivo* NMR Applications of Metabonomics 489
Dieter Leibfritz, Wolfgang Dreher and Wieland Willker

18 Applications of Metabonomics Within Environmental Toxicology 517
Julian L. Griffin and Richard F. Shore

19 Global Systems Biology Through Integration of "Omics" Results 533
John C. Lindon, Elaine Holmes and Jeremy K. Nicholson

Index 557

Foreword

Biological systems which underpin life itself are continuously yielding their secrets as our ability to isolate processes and analyse their component parts improves. Recent advances have tended to focus on macromolecules – DNA, RNA and proteins – and there is considerable optimism that understanding the make up of our chromosomes and the differential expression of genes will lead to much better understanding of disease. It is also believed that we will soon be able to predict a person's likelihood of developing a disease during their lifetime and tailor curative or preventative treatment more effectively.

Whilst I believe that much of this optimism is well founded, it is clear that study of the behaviour of macromolecules will only give part of the information we need to observe the response of a complex organism, such as man, to a changing environment. To give a more complete picture, we need to be able to observe dynamic markers of biological status – real-time signals which reflect the integrated function of the organism in ways which allow diagnosis and prediction. This is where metabonomics is now demonstrating enormous potential.

Metabonomics was pioneered by a group of scientists now based at Imperial and headed by Jeremy Nicholson. Whilst the technological platform is only now starting to be recognised as a tool of major importance, the foundation studies involving NMR spectroscopy of biological fluids dates back to the mid-1980s. Metabonomics is based on the demonstration that correlation of changes in metabolite pool patterns with changes in the function of integrated biological pathways can lead to ways of predicting outcomes such as the magnitude of response to a drug, potential toxicity or even the path of a disease process.

That the concepts and mathematical algorithms being developed are of major significance is well evidenced by the keen interest being shown by the biopharmaceutical industry and regulatory agencies. Both sectors share the goal of identifying surrogate markers of biological or clinical outcome which are well validated and lead to speedier, cheaper ways of identifying which drug candidates may be efficacious or toxic and in which populations. In fact, the management of attrition in the drug discovery–drug development pipeline is one of the most challenging for the industry as development costs and our ability to produce candidate molecules escalate.

The regulators are caught in a difficult position – needing to help important new medicines reach needy patients quickly and also protecting the public at large. Any tools which can help to predict potential toxicity in particular populations, in speedy cost-effective ways, will undoubtedly be scrutinised closely to see whether they should become a standard part of the drug approval process.

Economic considerations will also raise the potential importance of metabonomic tests in the clinical setting. Huge opportunities exist wherever non-invasive tests costing just a few dollars can be shown to replace invasive tests which are nearly always more expensive.

It would be wrong to leave the impression that metabonomics is only of value in the field of human healthcare. In fact, the technologies are applicable to plant science, animal health and development of model organisms for research (yeast, bacteria or *in vitro* cell systems). The state of the art of metabonomics is now such that some of these possibilities – and more – are verging on being realities. It is thus timely for a comprehensive book on the subject to appear in print. For those closely engaged in the subject, the book will serve as a definitive reference; for those seeking to learn for the first time, the book will open a fascinating new world containing a plethora of possibilities for application.

Richard B. Sykes FRS
London

Preface

Since the 1990s and particularly since the determination of the human genome, there have been dramatic changes in the scientific techniques and approaches used in molecular biology and biochemistry. The accompanying change in mindset has appeared to lead to a pervasive attitude that genetic differences might be able to account for all disease processes, and that this would lead to new diagnostic approaches and thence to much better targeted therapies. The revolution in molecular biology really took off with the availability of automated micro-array methods for detecting changes in gene expression, leading to the new discipline of transcriptomics. The subsequent expansion in the ability to assay and then identify, using mostly mass spectrometry–based methods, the proteins in a system has led to the term "proteomics" being coined. However, during the past few years the full complexity of molecular biology has been realised and the complex interactions between genetic make-up and environmental factors have now been recognised. It is now accepted that understanding of these interactions is impossible at the transcriptomic level and difficult at the proteomic level.

The reality is that the small molecules involved in biochemical processes provide a great deal of information on the status and functioning of a living system under study both from effects caused by changes in gene expression, and also by differences in life style and diet in humans and other mammals. The process of monitoring and evaluating such changes is termed "metabonomics". A parallel approach has also been under way, mostly in model organisms and in plant systems, and that has led to the term "metabolomics". Metabonomics really grew out of work using NMR spectroscopy of biofluids going back to the mid-1980s, and which was subsequently combined with the use of pattern recognition and multivariate statistics investigation of the complex data sets. The term was not coined until much later and was formally defined in 1999 by Jeremy Nicholson and colleagues as "the quantitative measurement of the dynamic multiparametric metabolic response of living systems to pathophysiological stimuli or genetic modification". A little later, in 2001, the term "metabolomics" was introduced by Oliver Fiehn and defined somewhat differently as "a comprehensive and quantitative analysis of all metabolites" in a system. Although there remain some differences in concept, there is now a great deal of overlap in the philosophies and methodologies, and the two terms are often used interchangeably by scientists and organisations. In this volume, we have allowed authors to use their term of preference.

Metabolic analyses provide a data-dense approach to biochemistry by monitoring simultaneously the changes in concentrations (and in some cases, molecular dynamics and compartmentation) of a wide range of molecules. To aid the observation and understanding of those differences that are really significant between classes, it has been necessary to use multivariate statistics extensively and interpretation of such calculations leads directly to the real biochemical differences between sample classes and hence identification of biomarkers of the process under study.

The main aim of this book is to provide a state-of-the-art picture of where metabonomics and metabolomics stand today, to give authoritative education and guidance on the analytical and statistical techniques used, and to identify and review the main current areas of application. The main analytical techniques of NMR spectroscopy and chromatography linked to mass spectrometry are explained and reviewed in detail, as are the various chemometrics and statistics approaches that are widely used. In addition we have taken the opportunity to provide information on the recent attempts at setting standards in designing and reporting metabolic experimental studies. We have invited articles to cover a wide range of biological applications. These range from plants and model organisms through to pre-clinical and clinical pharmaceutical studies and human disease and epidemiological investigations.

We greatly appreciate the efforts of all of the authors of the chapters who have consistently provided excellent articles. These have a high educational content as well as providing reviews of very current and cutting-edge literature. We hope and believe that this provides a balanced view of the subject with both the advantages and the shortcomings of the various methodologies being explored. Each chapter also includes a relevant and comprehensive set of references for further reading.

The field of endeavour is expanding rapidly, but we believe that now is a good time to review the achievements in the area of metabonomics and metabolomics and for the scientific community to set standards for the future. Our intention is that this book will serve as the authoritative reference work for scientists entering the field and for those already conducting metabolic studies. We believe that it also has a substantial educational role and will be of use to postgraduate students. It should be of interest to analytical scientists and to those working in application areas as diverse as plant science, environmental science and human clinical medicine. In summary we believe that this volume should prove educational, informative, critical and thought-provoking.

John C. Lindon
Jeremy K. Nicholson
Elaine Holmes
London, UK
May 2006

The Handbook of Metabonomics and Metabolomics
John C. Lindon, Jeremy K. Nicholson and Elaine Holmes (Editors)

Chapter 1

Metabonomics and Metabolomics Techniques and Their Applications in Mammalian Systems

Jeremy K. Nicholson, Elaine Holmes and John C. Lindon

Department of Biomolecular Medicine, Faculty of Medicine, Imperial College London, Sir Alexander Fleming Building, South Kensington, London, SW7 2AZ UK

1.1. Metabonomics and metabolomics in relation to other "omics" approaches

Since the 1990s, there has been a revolution in the techniques and approaches used in molecular biology. This followed from the decoding of the human and other genomes, and as a consequence, the principal emphasis in biomedical studies has now largely switched to simultaneous determination of gene expression changes between subjects mainly carried out using micro-array technology [1], and this type of study has been given the name of transcriptomics. An equivalent impetus to map out all protein expression changes in a cell or tissue has also subsequently emerged and has been termed "proteomics". Nowadays, there are nearly 200 different named "omics", most of which are listed in Table 1.1 with the number of hits returned by the PubMed scientific paper search engine (as of February 2006). As can be seen, very many of these terms, which incidentally produce quite a large number of hits on the Web search engine Google because they relate to names of commercial organisations, do not arise in legitimate scientific publications. Many of these terms therefore will not survive because of their very specialist application, and indeed many of these terms are not necessary since they serve only to describe a methodology that already has a perfectly valid name. Table 1.1 shows that the terms "metabolomics" and "metabonomics" produce a significant number of returns and whilst not yet of the magnitude produced by genomics and proteomics (and

Table 1.1
A partial list of "-omic" terminologies, with, in brackets, the number of hits returned by the PubMed search engine for scientific publications, using each term for a search in titles and abstracts (February 2006). Those terms in italics had no publications detected.

agrigenomics, agronomics (5), *antigenomics*, *aquagenomics*, *bacteriomics*, *behaviouromics*, bibliomics (1), *biogenomics*, biomics (7), bionomics (16104), *cardiomics*, *cardioproteomics*, chemoproteomics (3), *chomics*, *choomics*, *cardiogenomics*, cellomics (37), chemogenomics (38), chromonomics (1), *chondriomics*, chronomics (32), clinomics (1), *complexomics*, *cryptomics*, crystallomics (1), *crystalomics*, cytomics (27), degradomics (4), *diagnomics*, embryogenomics (3), economics (405228), *enzymomics*, epigenomics (44), epitomics (2), *expressomics*, fluxomics (6), *foldomics*, *fragmentomics*, *functomics*, gastrogenomics (1), genomics (13819), glycomics (87), glycoproteomics (17), *hybridomics*, immunomics (12), immunoproteomics (22), *inomics*, integromics (12), ionomics (3), interactomics (7), kinomics (4), ligandomics (1), *liganomics*, *linkomics*, lipidomics (61), lipoproteomics (2), *localizomics*, metabolomics (267), metabonomics (123), metallomics (4), metalloproteomics (6), methylomics (2), microbiomics (1), microgenomics (5), mitochondriomics (1), *neuropharmacogenomics*, neuroproteomics (23), *nucleomics*, nutragenomics (2), nutrigenomics (61), *oncogenomics*, *oncopharmacogenomics*, operomics (4), orfeomics (1), *parasitomics*, pathogenomics (13), peptidomics (64), pharmacogenomics (3416), pharmacometabolomics (96), pharmacometabonomics (30), *pharmacomethylomics*, pharmacophylogenomics (1), pharmacoproteomics (12), phenomics (35), phosphatomics (5), phosphoproteomics (35), phylogenomics (43), phyloproteomics (1), *physiogenomics*, physiomics (5), *phytogenomics*, *phytoproteomics*, *postgenomics*, *predictomics*, *promoteromics*, proteogenomics (5) proteomics (6954), *pseudogenomics*, regulomics (1), *resistomics*, ribonomics (4), riboproteomics (2), Rnomics (19), *saccharomics*, secretomics (1), separomics (1), sialomics (1), *signalomics*, *somatonomics*, *stereomics*, *systeomics*, toponomics (1), toxicogenomics (256), *toxicomics*, *toxiconomics*, toxicoproteomics (9), transcriptomics (198), transgenomics (4), translatomics (1), *transportomics*, *unknomics*, *vaccinomics*, *variomics*, virogenomics (3), viromics (1).

economics, and bionomics which is related to economics and not other – omics as understood here), the values are larger than all of the rest. Furthermore, it is very clear that many of the "omics" disciplines have emerged because of financial rather than scientific drivers and now in a world dominated by systems biology, the need is to integrate "omics" sciences rather than to constantly dilute molecular biology by creation of ever more diverse and narrow sub-omics fields. Hence the time is probably opportune for a book summarising the state of the art in these disciplines.

During the 1980s prior to the development of the various "omics" approaches, the simultaneous analysis of the plethora of metabolites seen in biological fluids had been carried out largely using ^1H NMR spectroscopy [2], and the basic idea of determining multiple analytes simultaneously emerged and the concept of metabonomics was born even though the name was not coined then. These complex data sets were subsequently interpreted using multivariate statistics [3, 4] to classify the samples according to their biological status. Thus, metabonomics encompasses the comprehensive and simultaneous systematic profiling of multiple metabolite levels and their systematic and temporal changes caused by factors such as diet, lifestyle, environment, genetic effects and pharmaceutical effects both beneficial and adverse, in whole organisms. This is achieved by the study of biofluids and tissues with the data being interpreted using chemometrics techniques [5, 6]. A parallel approach mainly from plant science and from the study of *in vitro* and microbiological systems and using largely chromatographic–mass spectrometry techniques has led to the term "metabolomics" also being coined and defined [7] and the methods and approaches used in the two disciplines are now highly convergent. Indeed, Chapter 2 of this book covers specifically the approach known as metabolomics.

This multivariate approach holds out much promise in many biological areas of application, because one of the problems with transcriptomics, in particular, is the difficulty in some cases of relating observed gene expression fold changes to conventional end-points such as those used in disease diagnosis or for pharmaceutical evaluation. This does not apply so much to proteomics but this latter technology, usually based on mass spectrometry, is still slow and labour-intensive and technological advances are required before it can be made high-throughput. Conversely, metabolic profiling observes the biochemical effects in an organism and as such this represents a closer approach to real-world end-points.

In this book, we have set out to attempt to describe the broad approaches that have been termed "metabolomics" and "metabonomics" and to describe the main techniques and technologies used. We have tried to be as comprehensive as possible in coverage of applications, but we recognise that there will be omissions in such a rapidly developing subject.

1.2. The concept of global systems biology

Human populations face diverse health-related biological challenges including new infectious agents, antibiotic resistance, the increased incidences of cancer and age-related neurodegenerative conditions and the rapid and insidious rise in insulin resistance and obesity. All these problems involve interactions between multiple gene processes and environmental factors, and in many cases also interacting non-human genomes. In the quest to improve understanding of such disease mechanisms, advanced analytical platforms have been applied to generate new molecular information to complement data supplied by modern genomics and transcriptomics. The growth of a wide range of "omics" sciences has enabled the measurement of multiple features of complex systems at various levels of biomolecular organisation from the cell to whole organism level. However, these technologies generate massive amounts of data and it is a major task to model these robustly in a way that allows predictive modelling for disease classification.

The total mammalian system is complex because it has many spatially heterogeneous arrays of disparate cell types and thus it is not obvious what needs to be measured and modelled in order to describe the integrated function of the system in a way that can be used to predict modes of failure accurately. Hopefully by appropriate use of genomic knowledge within a framework of physiology and metabolism, improvements in the health of individuals via personalised healthcare solutions and the health of whole populations will be possible.

In addition, environmental and lifestyle effects have a large effect at all levels of biomolecular organisation. Gene and protein expression effects and metabolite levels can be altered by such factors and this variation has to be incorporated into any analysis as part of inter-sample and inter-individual variation. Even healthy animals and man can be considered as "superorganisms" with an internal ecosystem of diverse symbiotic gut microflora that have metabolic processes that interact with the host and for which, in most cases, the genome is not known. The complexity of mammalian biological systems and the diverse features that need to be measured to allow "omics" data to be fully interpreted have been reviewed recently [8] and it has been argued that novel approaches will continue to be required to measure and model metabolic processes in various compartments from such global systems with different interacting cell types, and with various genomes, connected by co-metabolic processes [9]. Interpretation of genomic data, in terms of real biological end-points, is a major challenge because of the conditional interactions of specific genetic complements with environmental factors that non-linearly change disease risks.

"Global Systems Biology" is a term coined [8] to cover attempts to integrate multivariate biological information to better understand gene–environment interactions. The measurement and modelling of such diverse information sets pose significant challenges at the analytical and bioinformatic modelling levels.

It is also important to try to measure biological system responses through time as this can show the development of a lesion with multiple organ effects occurring at different times, and also the recovery process [10]. This is particularly important where processes involve multiple cell or tissue systems that vary in metabolic composition and biosynthetic activity. For example, when a toxic stimulus occurs, there will be site- and cell-specific effects and downstream consequences in terms of changed interactions with other tissue and organ systems. In a single cell system such as a bacterial or plant cell culture or in a primary tissue cell line, the systems biology challenge is to understand the upstream and downstream relationships between the transcripts, proteins and intracellular and extracellular metabolites. The metabolite concentrations are influenced by substrate availability in the extracellular milieu, the enzyme activities and presence of co-factors within the cell and the activities of membrane transporters. In multicellular organisms the system is regulated at both the cellular level and also at a higher level and is under neurohormonal control. Here it is feasible to consider metabolic linkages of disparate nature that are dispersed through the medium of the extracellular fluids, for example blood plasma and lymph, but with further complication because there are multiple discrete secretory and excretory drains on the metabolite pool, for example urine, bile and sweat. Sampling these fluids also gives many clues as to what is happening in the integrated system, but can be more complex to map against specific pathway activity because analytical data represent the weighted average of the whole system. However, such measurements provide a means of studying the whole system responses collectively [5, 6].

1.3. Interactions between host and other genomes

1.3.1. Gut microbes

There is much evidence to support the supposition that most major disease classes have significant environmental and genetic components and that the incidence of disease in a population or an individual is a complex product of the conditional probabilities of certain gene combinations interacting with a diverse range of environmental triggers. Such interactions are well known for cancers, and other major diseases such as obesity and type-2 diabetes are undoubtedly influenced by environmental factors. Diet clearly has a major influence on many diseases and diet also modulates the complex internal community of gut micro-organisms [11, 12] usually referred to as the "microbiome" [13, 14]. These micro organisms, weighing up to 1.5 kg in a normal adult human may total ca. 100 trillion cells [15]. This means that the 1000 plus known species of symbionts probably contain more than 100 times as many genes than exists in the host [16]. Together these interacting genomes can be considered to operate as a "superorganism", with extensive coordination of metabolic and physiological responses, particularly at the gut-liver and the gut-immune system levels. The microbiome–host relationship has been discussed with respect to drug metabolism and toxicity and a new type of probabilistic model based on conditional mammalian genome–microfloral interactions that may account for some aspects of individual variation in drug responses as well as the idiosyncratic toxicity from certain compounds has been proposed [8]. The gut microbiotia influence cytochrome P450 levels in the host and have intrinsic drug-metabolising capabilities [17, 18]. They can influence immune status and such factors as PPAR-γ nuclear cytoplasmic shuttling in the host once thought to be purely under mammalian genome control [19]. The fact that modulation of the gut microbial populations results in significant changes at the macroscopic level of metabolism can be easily demonstrated by observing fluctuation in the levels of co-metabolised substrates, such as hippuric acid and hydroxyphenylpropionic acids and their conjugates present in urine at millimolar concentrations [20–22].

1.3.2. Parasites

The biocomplexity of mammalian systems is further increased by multi-way interactions between the mammalian host, parasites and microbes. Each of these organisms possesses considerable capability for metabolism and modification of substrates interactively. Several species of parasite can increase the host cytochrome P450 activity and alter the activity of phase I drug metabolising enzymes in the liver of the host, and this might impact on the capacity of the liver to activate or detoxify both endogenous and xenobiotic compounds [23–25]. Recently, a three-way interaction was proposed in a laboratory mouse model for *Schistosoma mansoni* infection whereby the introduction of the parasite into the host caused a depletion of microfloral products such as *n*-butyrate, propionate and hippurate followed by increased excretion of 4-cresol together with its ether glucuronide and sulphate [26]. Cresol metabolites are known to be antimicrobial products of *Clostridium difficile*, a mildly pathogenic obligate anaerobe which possesses 4-hydroxyphenylacetate decarboxylase activity [27]. Thus, the parasitic infection resulted in a perturbation of the intestinal microbiotia, allowing colonisation of *C. difficile* and resulting in a change in the metabolite signature of the host. This interaction is mediated by an unknown mechanism but probably involves selective bacteriocidal secretions of the parasite implying another level of biological selection and interaction exerted on the gut microbiome. The intestinal environment must therefore be visualised as an entire ecosystem, where chemical interactions occur at multiple organisational levels with cross-talk between the mammalian, parasite and microbial systems.

1.3.3. Impact of multi-genomic interactions on therapy and future drug design

As implied above, there is arguably a great deal of the information exchange between symbionts, parasites and their hosts via co-metabolism of substrates, some of which are pharmacologically active. Because of the conditional nature of these interactions the exact nature of the mammalian host–microbe interaction could influence certain aspects of drug metabolism and drug toxicity. The possible interactions that could affect drug toxicity and efficacy are summarised in Figure 1.1. From this, it can be can concluded that pharmacogenomic approaches alone, relying only on the mammalian genetic information (polymorphisms or transcripts), are unlikely to provide a general solution to the prediction of drug activity in individuals (something which would be needed for personalised healthcare) as these are only measuring a small part of the relevant biological interactions. Such a conclusion may give rise to profound changes in the way that basic drug research is conducted in the future.

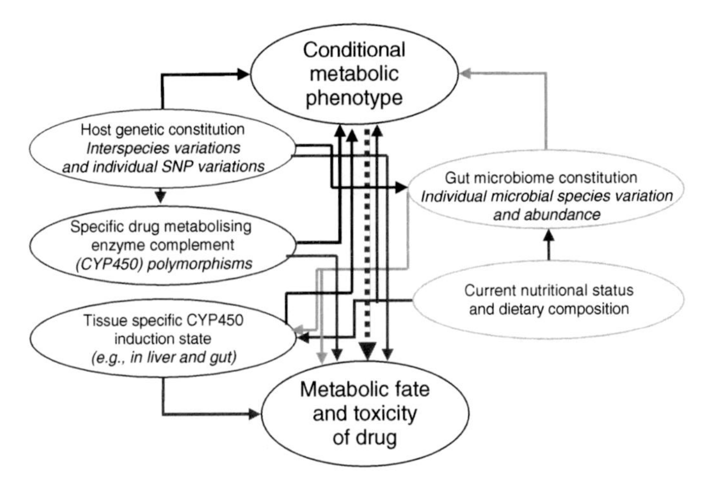

Figure 1.1. Mammalian–microbe interactions that can impact on drug metabolism and toxicity.

1.4. Time-scales of "-omics" events

It will be crucial to be able to integrate information at the transcriptomic, proteomic and metabonomic levels despite these different levels of biological control showing very different time scales of change.

A major problem in measuring and understanding a system, even at the single cell level, is the displacements in time between coupled gene, protein, metabolic and physiological events and their end-points [6]. This is one of the confounding issues to be gauged when attempting to relate gene expression data with, say, proteomic data using classical correlation methods or multivariate statistics. Attempts to achieve correlative "omics" have often proved to be unsatisfactory even for simple systems such as yeasts [28]. Thus, some time courses can be very rapid, such as gene switching, some require much longer time scales, for example protein synthesis, or, in the case of metabolic changes, can encompass enormous ranges of time scales. Counter-intuitively, biochemical changes do not always occur in the order, transcriptomic, proteomic, metabolic, because, for example, pharmacological or toxicological effects at the metabolic level can induce subsequent adaptation effects at the proteomic or transcriptomic levels. One important potential role for high-throughput and highly automated metabonomics methods therefore could be to direct the timing of more expensive or labour-intensive proteomic and transcriptomic analyses in order to maximise the probability of observing meaningful and relevant biochemical changes using those techniques.

Figure 1.2 illustrates a theoretical view of the problems that can occur when trying to cross-correlate protein levels with rises in gene products. Also it should

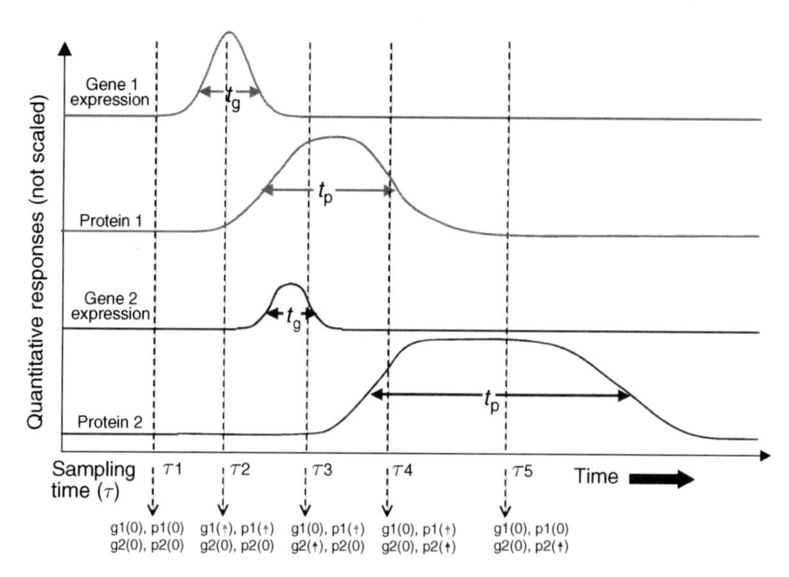

Figure 1.2. Time courses for two hypothetical gene–protein couples (relating transcription activity to protein level) that are up-regulated following a system stimulus such as a drug intervention. It can be seen that if the action of the stimulus is at the genetic level in the first instance it will take a finite amount of time in a cell for the associated protein synthesis (or post-translational modification) to occur and that the duration of the gene events (t_g) and protein events (t_p) may be very different. The practical consequence of this is that the observed co-variance of the gene and protein events is highly dependent on sampling time point and frequency and in some instances a single sampling point, say $\tau 3$ might lead to the incorrect assumption that gene 2 co-varied with protein 1. The differential displacement of the times of maximal activity or expression cannot be assumed to be constant and the variation in possible values for t_p and related turnover times is known to be large. Key: g1(0), p1(0) and so on describe the relative condition with respect to up regulation of each gene and protein at a given time-point (e.g. $\tau 1$, $\tau 2$ etc.), where (0) indicates baseline level and (↑) indicates up-regulation. Thus the post-intervention observation of relative state can be seen to be dependent on sampling time-point.

be remembered that the levels of mRNA (which have highly variable half-lives) are indirect measures of genome activity that in turn relate to gene switching events that operate on other undetermined time-scales. At the single cell level, gene events can be considered to be quantised (i.e. either "on" or "off" at any given time).

Understanding the true quantitative relationship between the variation in activity of every one of the thousands of hypothetical gene–protein couples in a cellular system is complicated by the time displacement of the genetic and protein synthetic and post-translational events, their different timescales and their half lives, such that the frequency and times of measurement of the transcripts and the proteins can markedly alter the modelled statistical relationships between these variables and, therefore, the conclusions to be drawn. Given that the time-scale of a gene switching "event" itself is short and difficult to measure and the half-life of a cytosolic protein can vary from a few to hundreds of hours [29], it is easy to understand why transcriptomic and proteomic data from the same system do not always agree. Another weakness when attempting to correlate proteomic with, say, metabolomic data is the fact that currently proteomic datasets contain no information on the activity of specific proteins, which is dependent on its exact location in the cell and the presence of co-factors or inhibitors.

An example of this problem is provided by work on the effects of orotic acid, an endogenous metabolite that when administered to mammals causes profound fatty changes in the liver [30]. Both transcript and metabolite levels were measured and cross-correlated in animals treated with orotic acid. At the statistical level there was little agreement between the gene expression and the liver, urinary or plasma metabolites. However, there were clear connectivities and patterns of behaviour of particular lipid metabolism pathways in the system as indicated by changes in gene regulation and metabolite levels that could be rationalised in biochemical terms. In simpler systems it has been shown that metabolic data can be effectively used to generate functional genomic information, and to uncover the phenotype of silent mutations, thus truly integrating the omics sciences involved [31]. This is highly appropriate for relatively static systems such as cells in culture (assuming a stationary growth phase), but hypercomplex systems such as mammals functioning with many interacting and spatially dispersed cell types and showing constant time-related variations require more sophisticated approaches where detailed time-responses must be measured.

For metabolic studies, it is usually crucial to measure the timed responses of the system to obtain a complete and evolving picture of metabolic injury following toxic insult or during a disease process [10, 32–35]. For practical reasons this may currently be uneconomic using gene chip technology or proteomics, although this might change in the future. There are also significant issues relating to comparisons and relative scaling of time-related metabolic phenomena because of differences in magnitude of effects in, say, comparative species studies in toxicology [35] but these problems are not confined to metabolic studies. Overall, -omics approaches should lead to the formulation of new hypotheses about pathway control and dysfunction, but they do not actually *prove* that a given data model provides a complete or even a correct biological explanation of the observed condition. Omics-generated hypotheses should of course be tested to make sure that the biological understanding is complete; however, this is infrequently done. When omic-generated hypotheses are tested properly, the results can be rewarding, for example it was possible to deduce the mechanism of toxicity of a failed drug using a metabonomic approach that identified key points in mitochondrial metabolism that were disrupted *in vivo* via timed post-dose biofluid analysis [36]. This was then tested *in vitro* proving that the hypothesis of the toxic mechanism generated by the exploratory metabolic studies was correct.

In future, as both transcriptomic and proteomic technologies become faster and cheaper, it should be possible to measure time-related fluctuations in much more detail leading to new levels of understanding. Indeed, one can already consider what such data might look like and how they could be analysed. Thus, at another level of time-related complexity in cellular responses to a stressor, the geometric shape of the time response could be considered as a descriptor (rather than just its magnitude or fold-change), and the mean time to recovery of gene expression in a reversibly perturbed system, for example after a low dose of a drug, would also be new parameters that could be modelled. This is an interesting concept, introducing a new set of time-related parameters that would describe the overall perturbation based not only on level but also on time to reach equilibrium following the intervention. This can be termed the "genetic (or proteomic) relaxation time" (which could have not only individual values for each gene or protein, but also global values for recovery of the system). The time-related relaxation patterns (with values in seconds or hours) would then provide an alternative signature of response to a specific intervention that did not rely on measurement of magnitudes of change.

Such differential geometric time responses can already be observed in real metabolic data showing responses to drug or toxin treatments. As in many cases there must be time-displaced relationships between metabolite level changes and transcript and protein levels, it is possible to surmise that similar responses would be observed in transcriptomic and proteomic data. These vary from time responses of simple mathematical form, where a particular metabolite closely follows the onset and recovery from the lesion, or can have complex forms such as damped oscillations as multiple organ systems interact to establish post-traumatic homeostatic control.

1.5. Samples for metabonomics and metabolomics

Metabonomics studies of biomedical relevance generally use biofluids or cell or tissue extracts. These are often easy to obtain and mammalian biofluids can provide an integrated view of the whole systems biology. Urine and plasma are obtained essentially non-invasively, and hence can easily be used for disease diagnosis and in a clinical trials setting for monitoring drug therapy. However, there is a wide range of fluids that have been studied, including seminal fluids, amniotic fluid, cerebrospinal fluid, synovial fluid, digestive fluids, blister and cyst fluids, lung aspirates and dialysis fluids [37]. In addition, a number of metabonomics studies have used NMR analysis of tissue biopsy samples and their lipid and aqueous extracts, such as from vascular tissue in studies of atherosclerosis [38]. The approach can also be used to characterise *in vitro* cell systems such as Caco-2 cells that are in widespread use for cell uptake studies [39], other model systems such as yeast [40], tumour cells

[41] as well as tissue spheroids that are used as model systems for liver or tumour investigations, for example [42]. In other fields of biology such as plant science, tissue extracts are usually employed (see Chapter 16 for a detailed review). In the environmental science field (see Chapter 18) both whole tissues and their extracts have been studied along with biofluids where they are available.

1.6. Analytical technologies

1.6.1. Introduction

The main analytical techniques that are employed for metabonomic studies are based on nuclear magnetic resonance (NMR) spectroscopy and mass spectrometry (MS). The latter technique requires a pre-separation of the metabolic components using either gas chromatography (GC) after chemical derivatisation, or liquid chromatography (LC), with the newer method of ultra-high-pressure LC (UPLC) being used increasingly. The use of capillary electrophoresis (CE) coupled to MS has also shown some promise. Other more specialised techniques such as Fourier transform infra-red (FTIR) spectroscopy and arrayed electrochemical detection have been used in some cases [43, 44]. The main limitation of the use of FTIR is the low level of detailed molecular identification that can be achieved, and indeed in the case quoted above, MS was also employed for metabolite identification. Similarly, although an array of coulometric detectors following high pressure LC (HPLC) separation does not identify compounds directly, the combination of retention time and redox properties can serve as a basis for database searching of libraries of standard compounds. The separation output can also be directed to a mass spectrometer for additional identification experiments [44].

All metabonomics studies result in complex multivariate data sets that require visualisation software and chemometric and bioinformatic methods for interpretation. The aim of these procedures is to produce biochemically based fingerprints that are of diagnostic or other classification value. A second stage, crucial in such studies, is to identify the substances causing the diagnosis or classification, and these become the combination of biomarkers that define the biological or clinical context.

1.6.2. NMR Spectroscopy

NMR spectroscopy is a non-destructive technique, widely used in chemistry, that provides detailed information on molecular structure, both for pure compounds and in complex mixtures as well as information on absolute or relative concentrations [45]. The NMR spectroscopic methods can also be used to probe metabolite molecular dynamics and mobility as well as substance concentrations through the interpretation of NMR spin relaxation times and by the determination of molecular diffusion coefficients [46]. The appropriate NMR theory and practical details necessary to conduct a metabonomic study are given in Chapter 3 and thus only a brief non-specialist overview is given here.

Automatic sample preparation is possible for NMR spectroscopy involving buffering and addition of D_2O as a magnetic field lock signal for the spectrometer and standard NMR spectra typically take only a few minutes to acquire using robotic flow-injection methods. For large-scale studies, bar-coded vials containing the biofluid can be used and the contents of these can be transferred and prepared for analysis using robotic liquid-handling technology into 96-well plates under LIMS system control. Currently, using such approaches, well over 100 samples per day can be measured on one spectrometer, each taking a total data acquisition time of only around 5 min. Alternatively, for more precious samples or for those of limited volume, conventional 5 mm or capillary NMR tubes are usually used, either individually or using a commercial sample tube changer and automatic data acquisition.

A typical ^{1}H NMR spectrum of urine contains thousands of sharp lines from predominantly low molecular weight metabolites. The large interfering NMR signal arising from water in all biofluids is easily eliminated by use of appropriate standard NMR solvent suppression methods, either by secondary RF irradiation at the water peak chemical shift or by use of a specialised NMR pulse sequence that does not excite the water resonance. The position of each spectral band (known as its chemical shift and measured in frequency terms, in ppm, from that of an added standard reference substance) gives information on molecular group identity and its molecular environment. The reference compound used in aqueous media is usually the sodium salt of 3-trimethylsilylpropionic acid (TSP) with the methylene groups deuterated to avoid giving rise to peaks in the ^{1}H NMR spectrum. The multiplicity of the splitting pattern on each NMR band and the magnitudes of the splittings (caused by nuclear spin–spin interactions mediated through the electrons of the chemical bonds, and known as J-coupling) provide knowledge about nearby protons, their through-bond connectivities, the relative orientation of nearby C–H bonds and hence also molecular conformations. The band areas relate directly to the number of protons giving rise to the peak and hence to the relative concentrations of the substances in the sample. Absolute concentrations can be obtained if the sample contains an added internal standard of known concentration, or if a standard addition of the analyte of interest is added to the sample, or if the concentration of a substance is known by independent means (e.g. glucose in plasma can be quantified by a conventional biochemical assay).

Blood plasma and serum contain both low and high molecular weight components, and these give a wide range of signal line widths. Broad bands from protein and lipoprotein signals contribute strongly to the ^{1}H NMR spectra, with sharp peaks from small molecules superimposed on them [47]. Standard NMR pulse sequences, where the observed peak intensities are edited on the basis of molecular diffusion coefficients or on NMR relaxation times, can be used to select only the contributions from macromolecules, or alternatively to select only the signals from the small molecule metabolites, respectively. It is also possible to use these approaches to investigate molecular mobility and flexibility, and to study inter-molecular interactions such as the reversible binding between small molecules and proteins [46].

Identification of biomarkers can involve the application of a range of techniques including two-dimensional NMR experiments [45] and as described in Chapter 3. The ^{1}H NMR spectra of urine and other biofluids, even though they are very complex, allow many resonances to be assigned directly based on their chemical shifts, signal multiplicities, and by adding authentic material, further information can be obtained by using spectral editing and two-dimensional techniques.

Two-dimensional NMR spectroscopy can be useful for increasing signal dispersion and for elucidating the connectivities between signals, thereby enhancing the information content and helping to identify biochemical substances. These include the ^{1}H–^{1}H 2-D J-resolved experiment, which attenuates the peaks from macromolecules and yields information on the multiplicity and coupling patterns of resonances, a good aid to molecule identification. The appropriate projection of such a spectrum on to the chemical shift axis yields a fingerprint of peaks from only the most highly mobile small molecules, with the added benefit that all of the spin-coupling peak multiplicities have been removed. Other 2-D experiments such as correlation spectroscopy (COSY) and total correlation spectroscopy (TOCSY) provide ^{1}H–^{1}H spin–spin coupling connectivities, giving information on which hydrogens in a molecule are close in chemical bond terms. Use of other types of nuclei, such as naturally abundant ^{13}C or ^{15}N, or where present ^{31}P, can be important to help assign NMR peaks and here such heteronuclear correlation NMR experiments are achievable. These now benefit from the use of so-called inverse detection, where the lower sensitivity or less abundant nucleus NMR spectrum (such as ^{13}C) is detected indirectly using the more sensitive/abundant nucleus (^{1}H) by making use of spin–spin interactions such as the one-bond ^{13}C–^{1}H spin–spin coupling between the nuclei to effect the connection. These yield both ^{1}H and ^{13}C NMR chemical shifts of CH, CH_2 and CH_3 groups, useful again for identification purposes. There is also a sequence that allows correlation of protons to quaternary carbons based on long-range ^{13}C–^{1}H spin–spin coupling between the nuclei.

A very useful recent advance in NMR technology has been the development of cryogenic probes where the detector coil and pre-amplifier (but not the samples) are cooled to around 20 K. This has provided an improvement in spectral signal–noise ratios of up to a factor of 5 by reducing the thermal noise in the electronics of the spectrometer. Conversely, because the NMR signal-to-noise ratio is proportional to the square root of the number of co-added scans, shorter data acquisition times by up to a factor of 25 become possible for the same amount of sample. The NMR spectroscopy of biofluids detecting the much less sensitive ^{13}C nuclei which only have a natural abundance (1.1%) also becomes possible because of this reduction in spectrometer noise [48]. This technology also makes the use of tissue-specific microdialysis samples more feasible [49].

Within the last few years, the development of a technique called high resolution ^{1}H magic angle spinning (MAS) NMR spectroscopy has made feasible the acquisition of high resolution NMR data on small pieces of intact tissues with no pretreatment [50–52]. Rapid spinning of the sample (typically at ~4–6 kHz) at an angle of 54.7° relative to the applied magnetic field serves to reduce the loss of information caused by line broadening effects seen in non-liquid samples such as tissues. These broadenings are caused by sample heterogeneity, and residual anisotropic NMR parameters that are normally averaged out in free solution where molecules can tumble isotropically and rapidly. MAS NMR spectroscopy has straightforward, but manual, sample preparation. NMR spectroscopy on a tissue sample in an MAS experiment is the same as solution state NMR and all common pulse techniques can be employed in order to study metabolic changes and to perform molecular structure elucidation and molecular dynamics studies. This topic is covered in more detail in Chapter 4.

Some typical ^{1}H NMR spectra are given in Figure 1.3 showing the varied profiles from mouse liver tissue, lipid and aqueous extracts of liver tissue and blood plasma.

Finally, it should not be forgotten that NMR spectroscopy can be conducted *in vivo* on whole live subjects and this approach has been used extensively for metabolic profiling, mainly for studies of human disease. This area is reviewed in Chapter 17.

1.6.3. Mass spectrometry

Mass spectrometry has also been widely used in metabolic fingerprinting and metabolite identification as well as being a mainstay technique in the pharmaceutical industry for identification and quantitation of drug metabolites. Although most MS-based studies to date have been on plant extracts and model cell system extracts, its application to mammalian studies is increasing. In general, a prior separation of the complex mixture sample using chromatography is required. MS is inherently considerably more sensitive than NMR spectroscopy, but it is necessary generally to employ different separation techniques (e.g. different LC column packings) for different classes of substances. MS is also a major technique for molecular identification purposes, especially through the use of tandem MS methods for fragment ion studies or using Fourier transform MS for very accurate mass determination. Analyte quantitation by MS in complex mixtures of highly variable composition can be impaired by variable ionisation and ion suppression effects. For plant metabolic studies, most investigations have used chemical derivatisation to ensure volatility

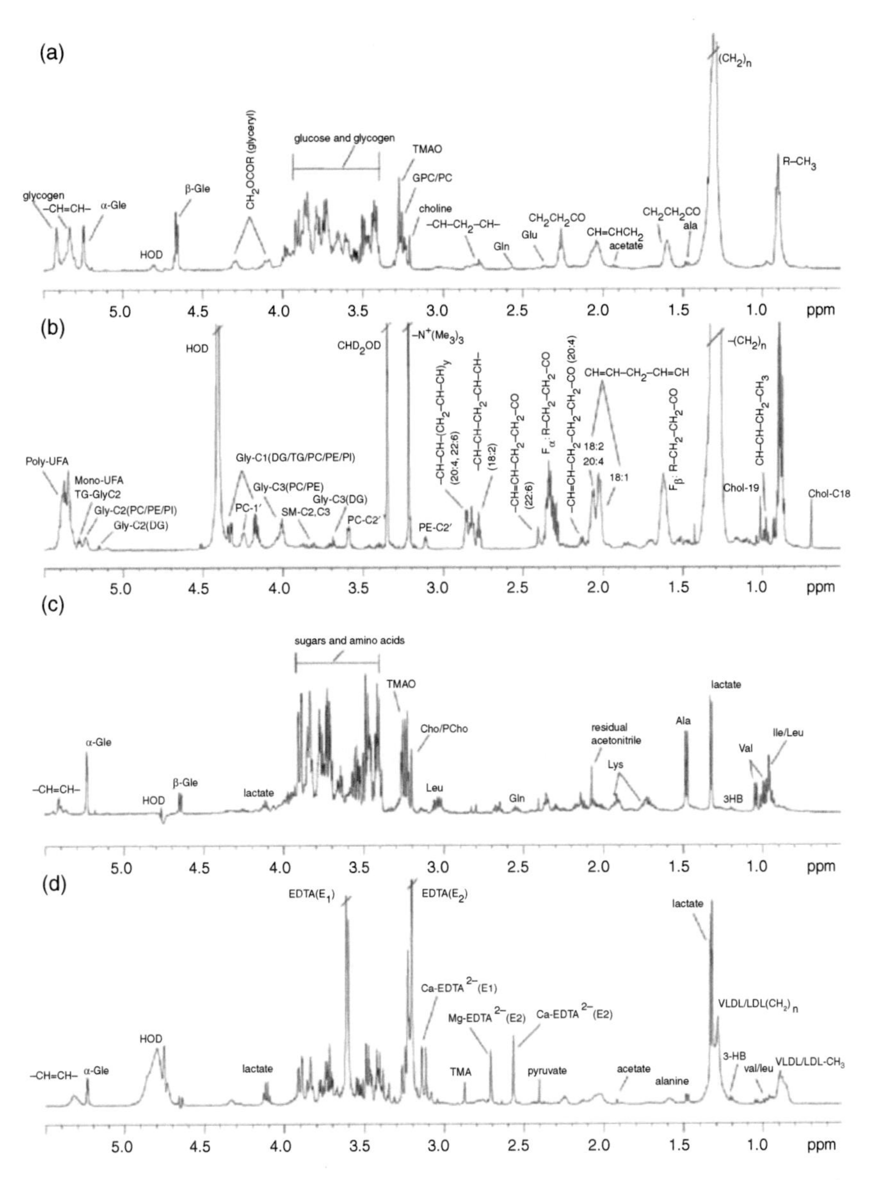

Figure 1.3. (a) ^{1}H MAS NMR CPMG spectrum (600 MHz) of intact control liver tissue, (b) ^{1}H NMR (600 MHz) spectrum of a control lipid soluble liver tissue extract, (c) solvent-presaturation ^{1}H NMR spectrum (600 MHz) of a control aqueous soluble liver tissue extract and (d) ^{1}H NMR CPMG spectrum (500 MHz) of control blood plasma. Key: 3HB, 3-D-hydroxybutyrate; Cho, choline; Chol, cholesterol; Glu, glucose; GPC, glycerophosphorylcholine; Gly, glycerol; LDL, low-density lipoprotein; PCho, phosphocholine; TMAO, trimethylamine-*N*-oxide; VLDL, very low density lipoprotein.

and analytical reproducibility, followed by GC-MS analysis. More recently, the use of GC-MS and GC-GC-MS has been exploited for mammalian metabonomics applications [53]. Some approaches using MS rely on more targeted studies, for example by detailed analysis of lipids [54]. A more comprehensive account of separation techniques coupled to MS for biomarker identification is provided in Chapter 5.

For metabonomics applications on biofluids such as urine, an HPLC chromatogram is generated with MS detection, usually using electrospray ionisation, and both positive and negative ion chromatograms can be measured. At each sampling point in the chromatgram, there is a full mass spectrum and so the data is three-dimensional in nature, that is retention time, mass and intensity. Given this very high resolution, it is possible to cut out any mass peaks from interfering substances such as drug metabolites, without unduly affecting the integrity of the data set.

Recently introduced, UPLC is a combination of a 1.7 μm reversed-phase packing material, and a chromatographic system, operating at around 12,000 psi. This has enabled better chromatographic peak resolution and increased speed and sensitivity to be obtained for complex mixture separation. UPLC provides around a 10-fold increase in speed and a threefold to fivefold increase in sensitivity compared to a conventional stationary phase. Because of the much improved chromatographic resolution of UPLC, the problem of ion suppression from co-eluting peaks is greatly reduced. UPLC-MS has already been used for metabolic profiling of urines from males and females of two groups of phenotypically normal mouse strains and a nude mouse strain [55]. A comparison of MS-detected HPLC and UPLC chromatograms from a mouse urine sample is shown in Figure 1.4.

Recently, CE coupled to mass spectrometry has also been explored as a suitable technology for metabonomics studies [56]. Metabolites are first separated by CE based on their charge and size and then selectively detected using MS monitoring. This method has been used to measure 352 metabolic standards and then employed for the analysis of 1692 metabolites from *Bacillus subtilis* extracts, revealing changes in metabolite levels during the bacterial growth.

For biomarker identification, it is also possible to separate out substances of interest on a larger scale from a complex biofluid sample using techniques such as solid-phase-extraction or HPLC. For metabolite identification, directly coupled chromatography-NMR spectroscopy methods can also be used. The most general of these "hyphenated" approaches is HPLC-NMR-MS [57] in which the eluting HPLC peak is split, with parallel analysis by directly coupled NMR and MS techniques. This can be operated in on-flow, stopped-flow and loop-storage modes and thus can provide the full array of NMR and MS-based molecular identification tools. These include two-dimensional NMR spectroscopy as well as MS-MS for identification of fragment ions and Fourier transform-MS (FT-MS) or time-of-flight-MS (TOF-MS) for accurate mass measurement and hence derivation of molecular empirical formulae.

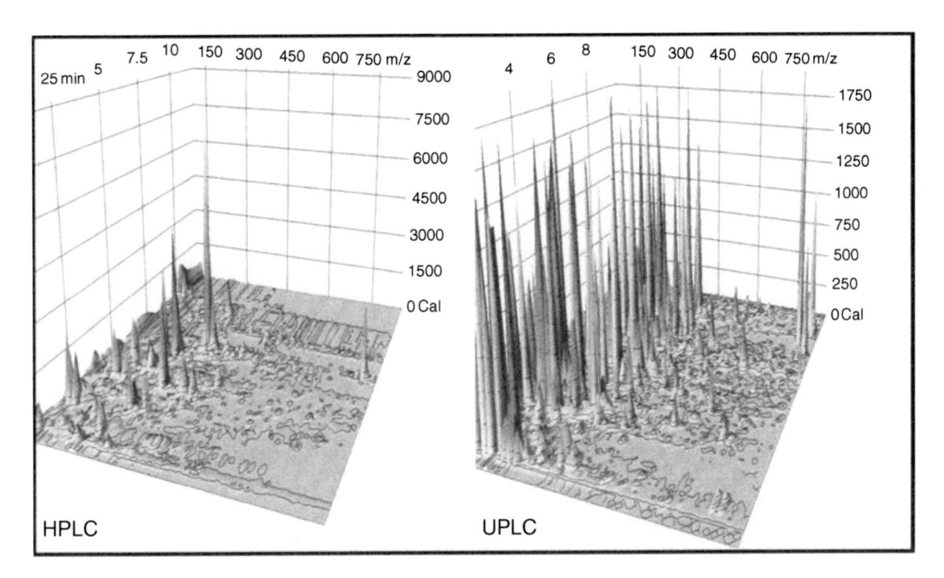

Figure 1.4. Three-dimensional plots of retention time, m/z and intensity from control white male mouse urine using (left) HPLC-MS with a 2.1 cm × 100 mm Waters Symmetry 3.5 μm C18 column, eluted with 0–95% linear gradient of water with 0.1% formic acid: acetonitrile with 0.1% formic acid over 10 min at a flow rate of 0.6 mL/min and (right) UPLC-MS with 2.1 cm × 100 mm Waters ACQUITY 1.7 μm C18 column, eluted with the same solvents at a flow rate of 0.5 mL/min. In both cases, the column eluent was monitored by ESI oa-TOF-MS from 50 to 850 m/z in positive ion mode. Reproduced with permission from Wilson *et al.* [55].

In summary, NMR and MS approaches are highly complementary, and use of both is often necessary for full molecular characterisation. MS can be more sensitive with lower detection limits provided the substance of interest can be ionised, but NMR spectroscopy is particularly useful for distinguishing isomers, for obtaining molecular conformation information and for studies of molecular dynamics and compartmentation, and given the now increasing use of cryoprobes, it is becoming ever more sensitive.

1.7. Chemometric methods

An NMR spectrum or a mass spectrum of a biofluid sample can be thought of as an object with a multi-dimensional set of metabolic coordinates, the values of which are the spectral intensities at each data point and the spectrum is therefore a point in a multi-dimensional metabolic hyperspace. The initial objective in metabonomics is to classify a spectrum based on identification of its inherent patterns of peaks and secondly to identify those spectral features responsible for the classification. The approach can also be used for reducing the dimensionality of complex data sets, for example by two-dimensional or three-dimensional mapping procedures, to enable easy visualisation of any clustering or similarity of the various samples. Alternatively, in what are known as "supervised" methods, multiparametric data sets can be modelled so that the class of separate samples (a "validation set") can be predicted based on a series of mathematical models derived from the original data or "training set" [58]. Two chapters in this book (Chapters 6 and 7) provide more comprehensive explanations of the various chemometrics techniques.

One of the simplest techniques that has been used extensively in metabonomics is principal components analysis (PCA). This technique expresses most of the variance within a data set using a smaller number of factors or principal components (PCs). Each PC is a linear combination of the original data parameters whereby each successive PC explains the maximum amount of variance possible, not accounted for by the previous PCs. Each PC is orthogonal and therefore independent of the other PCs and so the variation in the spectral set is usually described by many fewer PCs that comprise the number of original data point values because the less important PCs simply describe the noise variation in the spectra. Conversion of the data matrix to PCs results in two matrices known as scores and loadings. Scores, the linear combinations of the original variables, are the coordinates for the samples in the established model and may be regarded as the new variables. In a scores plot, each point represents a single sample spectrum. The PC loadings define the way in which the old variables are linearly combined to form the new variables and indicate those variables carrying the greatest weight in transforming the position of the original samples from the data matrix into their new position in the scores matrix. In the loadings plot, each point represents a different spectral intensity. Thus the cause of any spectral clustering observed in a PC scores plot is interpreted by examination of the loadings that cause any cluster separation. In addition, there are many other visualisation (or unsupervised) methods such as non-linear mapping and hierarchical cluster analysis.

One widely used supervised method (i.e. using a training set of data with known outcomes) is partial least squares (PLS) [59]. This is a method which relates a data matrix containing independent variables from samples, such as spectral intensity values (an **X** matrix), to a matrix containing dependent variables (e.g. measurements of response, such as toxicity scores) for those samples (a **Y** matrix). PLS can also be used to examine the influence of time on a data set, which is particularly useful for biofluid NMR data collected from samples taken over a time course of the progression of a pathological effect. PLS can also be combined with discriminant analysis (DA) to establish the optimal position to place a discriminant surface which best separates classes. It is possible to use such supervised models to provide classification probabilities and quantitative response factors for a wide range of sample types, but given the strong possibility of chance correlations when the number of descriptors is large, it is important to build and test such chemometric models using independent training data and validation data sets. Extensions of this approach allow the evaluation of those descriptors that are completely independent (orthogonal) to the **Y** matrix of end-point data. This orthogonal signal correction (OSC) can thus be used to remove irrelevant and confusing parameters and has been integrated into the PLS algorithm for optimum use [60]. The reader is directed to Chapter 6 for an expanded explanation of these techniques.

Apart from the methods described above that use linear combinations of parameters for dimension reduction or classification, other methods exist that are not limited in this way. These include methods that rely on comparison of distances in the sample point metabolic hyperspace, such as hierarchical cluster analysis. Other widely used techniques include artificial neural networks. Here, a training set of data is used to develop algorithms, which "learn" the structure of the data and can cope with complex functions. The basic software network consists of three or more layers including an input level of neurons (spectral descriptors or other variables), one or more hidden layers of neurons which adjust the weighting functions for each variable and an output layer which designates the class of the object or sample. Recently, probabilistic neural networks, which represent an extension to the approach, have shown promise for metabonomics applications in toxicity [61]. Other approaches that are currently being tested include genetic algorithms, machine learning and Bayesian modeling [62]. These methods are covered in Chapter 7.

1.8. New approaches to biomarker identification using chemometrics

Recently, a new method for identifying multiple NMR peaks from the same molecule in a complex mixture, hence providing a new approach to molecular identification, has been introduced. This is based on the concept of statistical total correlation spectroscopy and has been termed STOCSY [63]. This takes advantage of the multi-colinearity of the intensity variables in a set of spectra (e.g. ^{1}H NMR spectra) to generate a pseudo-two-dimensional NMR spectrum that displays the correlation among the intensities of the various peaks across the whole sample. This method is not limited to the usual connectivities that are deducible from more standard two-dimensional NMR spectroscopic methods, such as TOCSY. Added information is available by examining lower correlation coefficients or even negative correlations, since this leads to connection between two or more molecules involved in the same biochemical pathway. In an extension of the method, the combination of STOCSY with supervised chemometrics methods offers a new framework for analysis of metabonomic data.

In the first step, a supervised multivariate discriminant analysis can be used to extract the parts of NMR spectra related to discrimination between two sample classes. This information is then combined with the STOCSY results to help identify the molecules responsible for the metabolic variation. To illustrate the applicability of the method, it has been applied to ^{1}H NMR spectra of urine from a metabonomic

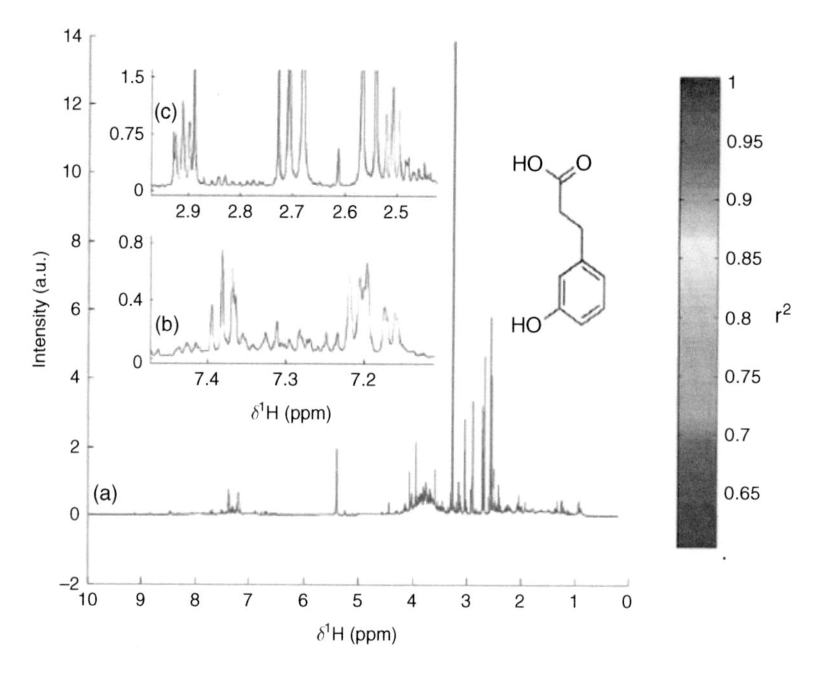

Figure 1.5. One-dimensional STOCSY analysis to identify peaks correlated to that at the chemical shift, δ 2.51. The degree of correlation across the spectrum has been colour-coded and projected on the spectrum. (a) Full spectrum; (b) partial spectrum between δ 7.1 and 7.5; (c) partial spectrum between δ 2.4 and 3.0. The STOCSY procedure enabled the assignment of this metabolite as 3-hydroxyphenylpropionic acid. Reproduced with permission from Cloarec *et al.* [63].

study of a model of insulin resistance based on the administration of a carbohydrate diet to three different mice strains in which a series of metabolites of biological importance could be conclusively assigned and identified by use of the STOCSY approach [63]. This is illustrated in Figure 1.5 where the approach has been used to identify the metabolite 3-hydroxyphenylpropionic acid.

The approach is not limited to NMR spectra alone and has been extended to other forms of data. It has recently been applied to co-analysis of both NMR and mass spectra from a metabonomic toxicity study [64]. This method, known as statistical heterospectroscopy (SHY), allowed better assignment of biomarkers of the toxin effect by using the correlated but complementary information available from the NMR and mass spectra taken on a whole sample cohort.

1.9. Standardisation of metabolic experiments and their reporting

A major initiative has been under way to investigate the reporting needs and to consider recommendations for standardising reporting arrangements for metabonomics studies, and to this end a Standard Metabolic Reporting Structures (SMRS) group has been formed (www.smrsgroup.org). This has produced a draft policy document that covers all of those aspects of a metabolic study that should be recorded, from the origin of a biological sample, the analysis of material from that sample and chemometric and statistical approaches to retrieve information from the sample data. A summary publication has also been produced [65]. The various levels and consequent detail for reporting needs, including journal submissions, public databases and regulatory submissions, have also been addressed. In parallel, a scheme called ArMet for capturing data and meta-data from metabolic studies has been proposed and developed [66]. This has been followed up with a workshop and discussion meeting sponsored by the US National Institutes of Health, from which firm plans are being developed to define standards in a number of areas relevant to metabonomics and metabolomics, including characterisation of sample-related meta-data, technical standards, and related data, parameters and QC matters for the analytical instrumentation, data transfer methodologies and schema for implementation of such activities, and development of standard vocabularies to enable transparent exchange of data [67].

1.10. Overview of applications of metabonomics and metabolomics to mammalian systems

1.10.1. Phenotypic and physiological effects

In order to relate therapeutic or toxic effects to normality or to understand the biochemical alterations caused by disease, it is necessary to have a good comprehension of what constitutes a normal biochemical profile. A number of studies have used metabonomics in this type of application to identify metabolic differences, in experimental animals such as mice and rats, caused by a range of inherent and external factors [68]. Such differences may help explain differential toxicity of drugs between strains and inter-animal variation within a study. Many effects can be distinguished using NMR-based metabonomics, including male/female differences, age-related changes, estrus cycle effects in females, diet, diurnal effects, and interspecies differences and similarities. Similarly some preliminary results have been obtained using the UPLC-MS method on normal and obese Zucker rats and on black, white and nude mice [69]. Considerable effort is being spent trying to elucidate the complex interactions between diet, health and therapy [70]. A number of these factors are touched upon in Chapter 11.

Metabonomics has also been used for the phenotyping of mutant and transgenic animals and the investigation of the consequences of transgenesis such as the transfection process used to introduce a new gene [71]. The development of a genetically engineered animal is often made using such transfection procedures and it is important to differentiate unintended consequences of this process from the intended result. Metabonomic approaches can give insight into the metabolic similarities or differences between mutant or transgenic animals and the human disease processes that they are intended to simulate. This leads to a better evaluation of their appropriateness for use as disease models and for drug efficacy studies.

The importance of the symbiotic relationship between mammals and their gut microfloral populations has been recognised [8] and highlighted by a study in which axenic (germ free) rats were allowed to acclimatise in normal laboratory conditions and their urine biochemical profiles were monitored for 21 days using ^{1}H NMR spectroscopy [21]. An interesting example of the phenotypic differences caused by variations in gut microflora has been highlighted by the study of the same strain of rat from the same supplier but housed in separate colonies at the supplier [72]. It was commented that the effect on drug metabolism and drug safety assessment of having different microfloral populations, in what would otherwise seem to be a homogenous population, is still unknown. Furthermore, the situation can be complicated by infections or pathological agents and the combined influence of gut microflora and parasitic infections on urinary metabolite profiles has also been elucidated [25].

1.10.2. Pre-clinical drug candidate safety assessment

The selection of robust candidate drugs for development based upon minimisation of the occurrence of drug adverse effects is one of the most important aims of pharmaceutical R&D, and the pharmaceutical industry is now embracing metabonomics for evaluating the adverse effects of candidate drugs and this is covered in more detail in Chapter 9.

Metabonomics can be used for definition of the metabolic hyperspace occupied by normal animals and the consequential rapid classification of a biofluid sample as normal or abnormal. If the sample is regarded as abnormal, then classification of the target organ or region of toxicity, the biochemical mechanism of that toxin, the identification of combination biomarkers of toxic effect and evaluation of the time-course of the effect, for example the onset, evolution and regression of toxicity, can all be determined. There have been many studies using ^{1}H NMR spectroscopy of biofluids to characterise drug toxicity going back to the 1980s [2], and the role of metabonomics in particular and magnetic resonance in general in toxicological evaluation of drugs has been comprehensively reviewed recently [73]. However, the combined use of NMR spectroscopy and HPLC-MS is beginning to be used for toxicity studies and this has been exemplified by a study on the nephrotoxin gentamycin [74].

The usefulness of metabonomics for the evaluation of xenobiotic toxicity effects has recently been comprehensively explored by the successful Consortium for Metabonomic Toxicology (COMET). This was formed between five pharmaceutical companies and Imperial College, London, UK [75] with the aim of developing methodologies for the acquisition and evaluation of metabonomic data generated using ^{1}H NMR spectroscopy of urine and blood serum from rats and mice for pre-clinical toxicological screening of candidate drugs.

A study was carried out at the start of the project, using the same detailed protocol and using the same model toxin, over seven sites in the companies and their appointed contract research organisations. This was used to evaluate the levels of analytical and biological variation that could both arise through the use of metabonomics on a multi-site basis. The inter-site NMR analytical reproducibility revealed the high degree of robustness expected for this technique when the same samples were analysed both at Imperial College and at various company sites. This gave a multi-variate coefficient of regression between paired samples of only about 1.6% [76]. Additionally, the biological variability was evaluated by a detailed comparison of the ability of the companies to provide consistent urine and serum samples for an in-life study of the same toxin, with all samples measured at Imperial College. There was a high degree of consistency between samples from the various companies and dose-related effects could be distinguished from inter-site variation.

To achieve the project goals, new methodologies for analysing and classifying the complex data sets were developed. For example, since the predictive expert system that was developed takes into account the metabolic trajectory over time (see Section 1.4), a new way of comparing and scaling these multivariate trajectories was developed (called SMART) [77]. Additionally, a novel classification method for identifying the class of toxicity based on all of the NMR data for a given study has been generated. This has been termed "Classification of Unknowns by Density Superposition (CLOUDS)" and is a novel non-neural implementation of a classification technique developed from probabilistic neural networks [78].

This consortium showed that it is possible to construct predictive and informative models of toxicity using NMR-based metabonomic data, delineating the whole time course of toxicity. The successful outcome is evidenced by the generated databases of spectral and conventional results for a wide range of model toxins (147 in total) that served as the basis for computer-based expert systems for toxicity prediction. The project goals of the generation of comprehensive metabonomic databases (now around 35,000 NMR spectra) and successful and robust multivariate statistical models (expert systems) for prediction of toxicity, initially for liver and kidney toxicity in the rat and mouse, have now been achieved and the predictive systems and databases have been transferred to the sponsoring companies [79]. In addition, interesting species differences (rat and mouse) in the toxicity of one compound have been published [80]. A follow-up project, COMET-2, has now commenced operation with the aim of improved understanding of biochemical mechanisms of toxicity relevant to pharmaceutical development.

1.10.3. Disease diagnosis and therapeutic efficacy

Many examples exist in the literature on the use of NMR-based metabolic profiling to aid human disease diagnosis, such as the use of plasma to study diabetes, CSF for investigating Alzheimer's disease, synovial fluid for osteoarthritis, seminal fluid for male infertility and urine in the investigation of drug overdose, renal transplantation and various renal diseases. A promising use of NMR spectroscopy of urine and plasma, as evidenced by the number of publications on the subject, is in the diagnosis of inborn errors of metabolism in children [81] as discussed in more detail in Chapter 14. Most of the earlier studies using NMR spectroscopy have been reviewed previously [37]. The increasing use of metabonomics in clinical pharmaceutical R&D studies is also reported in Chapter 10.

More recently, CSF sample analysis using NMR spectroscopy has been used to distinguish control subjects from those with meningitis and the various types of infection (bacterial, viral and fungal) could also be differentiated [82]. In another study, CSF analysis was used to investigate aneurismal subarachnoid haemorrhage and it was shown that metabolic profiles derived using NMR spectroscopy correlated with vasospasm and clinical outcome [83].

Tissues themselves can be studied by metabonomics through the magic-angle-spinning technique and published examples include prostate cancer [84], renal cell carcinoma [85], breast cancer [86] and various brain tumours [52, 87] as shown in Chapter 13. A number of mouse models of cardiac disease, including Duchenne muscular dystrophy, cardiac arrhythmia and cardiac hypertrophy, have been investigated using cardiac tissue MAS NMR spectroscopy [88]. It was shown that although the mouse strain was a major component of the mouse phenotype, it was possible to discover underlying profiles characteristic of each abnormality.

One area of disease where progress is being made using NMR-based metabonomics studies of biofluids is cancer. This is highlighted by a publication on the diagnosis of epithelial ovarian cancer based on analysis of serum [89].

Metabonomics using NMR spectroscopy has been used to develop a method for diagnosis of coronary artery disease non-invasively through analysis of a blood serum sample [90]. Based on angiography, patients were classified into two groups, those with normal coronary arteries and those with triple coronary vessel disease. Around 80% of the NMR spectra were used as a training set to provide a two-class model after data filtering techniques had been applied and the samples from the two classes were easily distinguished. The remaining 20% of the samples were used a test set and their class was then predicted based on the derived model with a sensitivity of 92% and a specificity of 93%. It was also possible to diagnose the severity of the disease that was present by employing serum samples from patients with stenosis of one, two or three of the coronary arteries. Although this is a simplistic indicator of disease severity, separation of the three sample classes was evident even though none of the wide range of conventional clinical risk factors that had been measured was significantly different between the classes.

1.10.4. Pharmacometabonomics

One of the long-term goals of using pharmacogenomic approaches is to understand the genetic make-up of different individuals (their genetic polymorphisms) and their varying abilities to handle pharmaceuticals both for their beneficial effects and for identifying adverse effects. If personalised healthcare is to become a reality, an individual's drug treatments must be tailored so as to achieve maximal efficacy and avoid adverse drug reactions. Very recently, an alternative approach to understanding inter-subject variability in response to drug treatment using a combination of multivariate metabolic profiling and chemometrics to predict the metabolism and toxicity of a dosed substance, based solely on the analysis and modeling of a pre-dose metabolic profile, has been developed [91]. Unlike pharmacogenomics, this approach, which has been termed "pharmacometabonomics", is sensitive to both the genetic and the modifying environmental influences that determine the basal metabolic fingerprint of an individual, since these will also influence the outcome of a chemical intervention. This new approach has been illustrated with studies of the toxicity and metabolism of compounds with very different modes of action, allyl alcohol, galactosamine and acetaminophen (paracetamol) administered to rats.

1.10.5. Other applications

There is a wide range of current and emerging applications of metabonomics and metabolomics. These include the study of invertebrate tissue extracts and biofluids as monitors of environmental toxicity and these applications are covered in more detail in Chapter 18. This Chapter also includes examples of the use of biofluids from wild rodents trapped from polluted ground sites.

Plant science has been investigated widely using metabolomics for many years using mainly GC-MS techniques. This field of application is covered in Chapter 16 of this book.

The important study of the functions of changes in gene expression as they relate to altered biochemical process (functional genomics) has seen many studies in model organisms such as yeast and this is covered in more detail in Chapter 2.

Finally, the availability of large cohorts of biofluids samples from epidemiological studies has allowed metabonomics investigations of differences in metabolism of various populations across the globe and this topic has been described in more detail in Chapter 11.

1.11. Concluding remarks

NMR- and MS-based metabonomics and metabolomics are now recognised as widely used techniques for evaluating the biochemical consequences of drug action, and they have been adopted by a number of pharmaceutical companies into their drug development protocols. For drug safety studies, it is possible to identify the target organ of toxicity, derive the biochemical mechanism of the toxicity and determine the combination of biochemical biomarkers for the onset, progression and regression of the lesion. Additionally, the technique has been shown to be able to provide a metabolic fingerprint of an organism ("metabotyping") as an adjunct to functional genomics and hence has applications in design of drug clinical trials and for evaluation of genetically modified animals as disease models.

Using the approach, it has proved possible to derive new biochemically based assays for disease diagnosis and to identify combination biomarkers for disease, which can then be used to monitor the efficacy of drugs in clinical trials. Thus, based on differences observed in metabonomic databases from control animals and from animal models of disease, diagnostic methods and biomarker combinations might be derivable in a pre-clinical setting. Similarly, the use of databases to derive predictive expert systems for human disease diagnosis and the effects of therapy requires compilations from both normal human populations and patients before, during and after therapy.

In summary, it is clear that metabonomics will have an impact in pharmaceutical R&D and Figure 1.6 summarises a SWOT (strengths, weaknesses, opportunities, threats) analysis of the discipline.

The analytical procedures used are stable and robust, and have a high degree of reproducibility, and although advances will obviously be made in the future, current data will always be readable and interpretable. In contrast to other -omics, metabonomics enjoys a good level of biological reproducibility and the cost per sample and per analyte is low. It has the advantage of not having to preselect analytes, and through use of biofluids it is minimally invasive with hypothesis generation studies being easily possible. Metabolic biomarkers are closely identifiable with real biological end-points and provide a global systems interpretation of biological effects, including the interactions between multiple genomes such as humans and their gut microflora. One major potential strength of metabonomics is the possibility

Strengths	**Weaknesses**
Robust and stable analytical platforms Excellent analytical/biological reproducibility No pre-selection of analytes Minimally invasive Exploratory studies possible Real biological end-points Whole system integration Multi-genomic Cross-species biomarkers Low cost per sample/analyte	Multiple analytical platforms Analytical sensitivity Analytical dynamic range Complexity of data sets Over-fitting of data possible No current standardisation of methods Regulatory body training needed High capital cost
Opportunities	**Threats**
Use of marker species in environmental studies Much experience from mammalian system studies (e.g. pathways) Potential of multi-omics integration Benefits of a central laboratory approach Web-based diagnostics	Scepticism of non-hypothesis led studies Conservatism (one significant analyte per test approach) Lack of well-trained scientists Disadvantages of a central laboratory approach

Figure 1.6. Strengths, weaknesses, opportunities, threats (SWOT) analysis of metabonomics.

that metabolic biomarkers will be more easily used across species than transcriptomic or proteomic biomarkers and this should be important for pharmaceutical studies.

On the other hand, metabonomics suffers from the use of multiple analytical technologies, there are questions of the sensitivity and dynamic range of the technologies used and the data sets are complex. Through the use of chemometrics, it is possible to over-interpret the data, but this is easily avoided by correct statistical rigour. At present, the groups using metabonomics are moving towards defining standards for data and for operations and a good start has been made, but there remains a need for the regulatory agencies to be trained in the data interpretation and for more well-trained practitioners.

There is an inherent conservatism that would like to be able to use a single biomarker or analyte for each diagnostic test. However, the reality of the complexity of disease and drug effects means that biomarker combinations will be more usual and thus there will be many opportunities for metabonomics that are as yet under-explored, such as its use in environmental toxicity studies, its use in directing the timing of transcriptomic and proteomic experiments and its use for deriving theranostic biomarkers. It will surely be an integral part of any multi-omics study where all the data sets are combined in order to derive an optimum set of biomarkers.

The ultimate goal of systems biology must be the integration of data acquired from living organisms at the genomic, protein and metabolite levels. In this respect, transcriptomics, proteomics and metabonomics will all play an important role. Through the combination of these, and related approaches, will come an improved understanding of an organism's total biology, and with this, better understanding of the causes and progression of human diseases and, given the 21st-century goal of personalised healthcare, the improved design and development of new and better targeted pharmaceuticals.

References

[1] P. Baldi, and G.W. Hatfield. *DNA Microarrays and Gene Expression*. pp. 230. Cambridge University Press, Cambridge, UK. ISBN: 0521800226 (2002).

[2] J.K. Nicholson, and I.D. Wilson. High-resolution proton magnetic resonance spectroscopy of biological fluids. *Prog. NMR Spectrosc.*, 21, 449–501 (1989).

[3] K.P.R. Gartland, S.M. Sanins, J.K. Nicholson, B.C. Sweatman, C.R. Beddell, and J.C. Lindon. Pattern recognition analysis of high resolution ^{1}H NMR spectra of urine. A nonlinear mapping approach to the classification of toxicological data. *NMR Biomed.*, 3, 166–172 (1990).

[4] K.P.R. Gartland, C.R. Beddell, J.C. Lindon, and J.K. Nicholson. The application of pattern recognition methods to the analysis and classification of toxicological data derived from proton NMR spectroscopy of urine. *Mol. Pharmacol.*, 39, 629–642 (1991).

[5] J.K. Nicholson, J.C. Lindon, and E. Holmes. 'Metabonomics': Understanding the metabolic responses of living systems to pathophysiological stimuli via multivariate statistical analysis of biological NMR spectroscopic data. *Xenobiotica*, 29, 1181–1189 (1999).

[6] J.K. Nicholson, J. Connelly, J.C. Lindon, and E. Holmes. Metabonomics: A platform for studying drug toxicity and gene function. *Nat. Rev. Drug Disc.* 1, 153–162 (2002).

[7] O. Fiehn. Metabolomics – the link between genotypes and phenotypes. *Plant Mol. Biol.*, 48, 155–171 (2002).

[8] J.K. Nicholson, and I.D. Wilson. Understanding 'global' systems biology: Metabonomics and the continuum of metabolism. *Nature Rev. Drug Disc.*, 2, 668–676 (2003).

[9] J.K. Nicholson, E. Holmes, J.C. Lindon, and I.D. Wilson. The challenges of modeling mammalian biocomplexity. *Nature Biotech.*, 22,1268–1274 (2004).

[10] E. Holmes, F.W. Bonner, B.C. Sweatman, J.C. Lindon, C.R. Beddell, E. Rahr, and J.K. Nicholson. NMR spectroscopy and pattern recognition analysis of the biochemical processes associated with the progression and recovery from nephrotoxic lesions in the rat induced by mercury II chloride and 2-bromoethanamine. *Mol. Pharmacol.*, 42, 922–930 (1992).

[11] F. Guarner, and J.R. Malagelada. Gut flora in health and disease. *The Lancet*, 360, 512–519 (2003).

[12] G.W. Tannock. *Normal Microflora*. Chapman and Hall (1995).

[13] J. Xu, M.K. Bjursell, J. Himrod, S. Deng, L.K. Carmichael, H.C. Chaing, L.V. Hooper, and J.I. Gordon. A genomic view of the human-*Bacteroides thetaiotaomicron* symbiosis. *Science*, 299, 2074–2076 (2003).

[14] M.S. Gilmore, and J.J. Ferretti. The thin line between gut commensal and pathogen. *Science*, 299, 1999–2002 (2003).

[15] R.D. Berg. The indigenous gastrointestinal microflora. *Trends Microbiol.*, 4, 430–435 (1996).

[16] D.A. Relman, and S. Falkow. The meaning and impact of the human genome sequence for microbiology. *Trends Microbiol.*, 9, 206–208 (2001).

[17] R. Gingell, J.W. Bridges, and R.T. Williams. The role of the intestinal flora in the metabolism of prontosil and neoprontosil in the rat. *Xenobiotica*, 1, 143–156 (1971).

[18] M.A. Peppercorn, and P. Goldman. Caffeic acid metabolism by bacteria of the human gastrointestinal tract. *Proc. Nat. Acad. Sci. USA*, 69, 1413–1415 (1972).

[19] D. Kelly, J.I. Campbell, T.P. King, G. Grant, E.A. Jansson, A.G.P. Coutts, S. Pettersson, and S. Conway. Commensal anaerobic gut bacteria attenuate inflammation by regulating nuclear-cytoplasmic shuttling of PPAR-γ and RelA. *Nature, Immunol.*, 5, 104–112 (2003).

[20] A.N. Phipps, J. Stewart, B. Wright, and I.D. Wilson. Effect of diet on the urinary excretion of hippuric acid and other dietary derived aromatics in the rat. A complex interaction between the diet, intestinal microflora and substrate specificity. *Xenobiotica*, 28, 527–537 (1998).

[21] R.E. Williams, H.W. Eyton-Jones, M.J. Farnworth, R. Gallagher, and W.M. Provan. Effect of intestinal microflora on the urinary metabolite profile of rats: a ^{1}H nuclear magnetic resonance spectroscopy study. *Xenobiotica*, 32, 783–794 (2002).

[22] A.W. Nicholls, R.J. Mortishire-Smith, and J.K. Nicholson. NMR spectroscopic based metabonomic studies of urinary metabolite variation in acclimatizing germ-free rats. *Chem. Res. Toxicol.*, 16, 1395–1404 (2003).

[23] S.A. Sheweita, S.A. Mangoura, and A.G. El Shemi. Different levels of Schistosoma mansoni infection induce changes in drug-metabolizing enzymes *J. Helminthol.*, 72, 71–77 (1998).

[24] S.A. Sheweita, J. Mubark, M.J. Doenhofe, M.H. Mostafa, G.P. Margison, P.J. O'Connor, and R.H. Elder. Changes in the expression of cytochrome P450 isoenzymes and related carcinogen metabolizing enzyme activities in Schistosoma mansoni-infected mice. *J. Helminthol.*, 76, 71–78 (2002).

[25] S. Satarug, M.A. Lang, P. Yongvanit, P. Sithithaworn, E. Mairiang, P. Mairiang, P. Pelkonen, H. Bartsch, and M.R. Haswell-Elkins. Induction of cytochrome P450 2A6 expression in humans by the carcinogenic parasite infection, opisthorchiasis viverrini. *Cancer Epidemiol Biomarkers Prev.*, 5, 795–800 (1996).

[26] Y. Wang, E. Holmes, J.K. Nicholson, O. Cloarec, J. Chollet, M. Tanner, B.H. Singer, and J. Utzinger. Metabonomic investigations in mice infected with Schistosoma mansoni: an approach for biomarker identification. *Proc. Nat. Acad. Sci. USA*, 101, 12676–12681 (2004).

[27] T. Selmer and P.I. Andrei. p-hydroxyphenylacetate decarboxylase from Clostridium difficile: A novel glycyl radical enzyme catalysing the formation of p-cresol. *Eur. J. Biochem.*, 268, 1363–1372 (2001).

[28] S.P. Gygi, Y. Rochon, B.R. Franza, and R. Aebersold. Correlation between protein and mRNA Abundance in Yeast. *Molec. Cell. Biol.*, 19, 1720–1730 (1999).

[29] R. Schoenheimer. *The Dynamic State of Body Constituents* (Harvard University Press, Boston) (1942).

[30] J.L. Griffin, J.L. Bonney, C. Mann, A.M. Hebbachi, G.F. Gibbons, J.K. Nicholson, C.C. Shoulders, and J. Scott. An integrated reverse functional genomic and metabolic approach to understanding orotic acid-induced fatty liver. *Physiol. Genomics.*, 17, 140–149 (2004).

[31] L.M. Raamsdonk, B. Teusink, D. Broadhurst, N. Zhang, A. Hayes, M.C. Walsh, J.A. Berden, K.M. Brindle, D.B. Kell, J.J. Rowland, H.V. Westerhoff, K. van Dam, and S.G. Oliver. A functional genomics strategy that uses metabolome data to reveal the phenotype of silent mutations. *Nature Biotech.*, 19, 45–50 (2001).

[32] M.E. Bollard, E. Holmes, J.C. Lindon, S.C. Mitchell, D. Branstetter, W. Zhang, and J.K. Nicholson. Investigations into biochemical changes due to diurnal variation and estrus cycle in female rats using high resolution ^{1}H NMR spectroscopy and pattern recognition. *Anal. Biochem.*, 295, 194–202 (2001).

[33] N.J. Waters, E. Holmes, A. Williams, C.J. Waterfield, R.D. Farrant, and J.K. Nicholson. NMR and pattern recognition studies on the time-related metabolic effects of α-naphthylisothiocyanate on liver, urine, and plasma in the rat: An integrative metabonomic approach. *Chem. Res. Toxicol.*, 14, 1401–1412 (2001).

[34] J. Azmi, J. Griffin, H. Antti, R.F. Shore, E. Johansson, J.K. Nicholson, and E. Holmes. Metabolic trajectory characterisation of xenobiotic-induced hepatotoxic lesions using statistical batch processing of NMR data. *The Analyst*, 127, 271–276 (2002).

[35] H.C. Keun, M. Bollard, O. Beckonert, E. Holmes, J.C. Lindon, and J.K. Nicholson. Geometric trajectory analysis of metabolic responses to toxicity can define treatment-specific profiles. *Chem. Res. Toxicol.*, 17, 579–587 (2004).

[36] R.J. Mortishire-Smith, G.L. Skiles, J.W. Lawrence, S. Spence, A.W. Nicholls, B.A. Johnson, and J.K. Nicholson. Use of metabonomics to identify impaired fatty acid metabolism as the mechanism of drug-induced toxicity. *Chem. Res. Toxicol.*, 17, 165–173 (2004).

[37] J.C. Lindon, J.K. Nicholson, and J.R. Everett. NMR spectroscopy of biofluids. Ann. Reports on NMR Spectrosc. Webb G.A. (Ed.), Academic Press, Oxford, UK, 38, 1–88 (1999).

[38] M. Mayr, Y.L. Chung, U. Mayr, X.K. Yin, L. Ly, H. Troy, S. Fredericks, Y.H. Yu, J.R. Griffiths, and Q.B. Xu. Proteomic and metabolomic analyses of atherosclerotic vessels from apolipoprotein E-deficient mice reveal alterations in inflammation, oxidative stress, and energy metabolism. *Arterioscl. Thromb. Vasc. Biol.*, 25, 2135–2142 (2005).

[39] R.J.A.N. Lamers, E.C.H.H. Wessels, J.J.M. van der Sandt, K. Venema, G. Schaafsma, J. van der Greef, and J.H.J. van Nesselrooij. A pilot study to investigate effects of inulin on Caco-2 cells through in vitro metabolic fingerprinting. *J. Nutrition*, 133, 3080–3084 (2003).

[40] S.G. Villas-Boas, J. Hojer-Pedersen, M. Akesson, J. Smedsgaard, and J. Nielsen. Global metabolite analysis of yeast: Evaluation of sample preparation methods. *Yeast*, 22, 1155–1169 (2005).

[41] J.E. Ippolito, J. Xu, S.J. Jain, K. Moulder, S. Mennerick, J.R. Crowley, R.R. Townsend, and J.I. Gordon. An integrated functional genomics and metabolomics approach for defining poor prognosis in human neuroendocrine cancers. *Proc. Nat. Acad. Sci. USA*, 102, 9901–9906 (2005).

[42] M.E. Bollard, J.S Xu, W. Purcell, J.L. Griffin, C. Quirk, E. Holmes, and J.K. Nicholson. Metabolic profiling of the effects of D-galactosamine in liver spheroids using ^{1}H NMR and MAS NMR spectroscopy. *Chem. Res. Toxicol.*, 15, 1351–1359 (2002).

[43] N.N. Kaderbhai, D.I. Broadhurst, D.I. Ellis, and D.B. Kell. Functional genomics via metabolic footprinting: Monitoring metabolite secretion by Escherichia coli tryptophan metabolism mutants using FT-IR and direct injection electrospray mass spectrometry. *Comp. Funct. Genom.*, 4, 376–391 (2003).

[44] P.H. Gamache, D.F. Meyer, M.C. Granger, and I.N. Acworth. Metabolomic applications of electrochemistry/mass spectrometry. *J. Am. Soc. Mass Spectrom.*, 15, 1717–1726 (2004).

[45] T.D.W. Claridge, *High-Resolution NMR Techniques in Organic Chemistry*. Elsevier Science, Oxford, UK, pp. 384. ISBN: 0080427995.

[46] M. Liu, J.K. Nicholson, and J.C. Lindon. High resolution diffusion and relaxation edited one- and two-dimensional ^{1}H NMR spectroscopy of biological fluids. *Anal. Chem.*, 68, 3370–3376 (1996).

[47] J.K. Nicholson, P.J.D. Foxall, M. Spraul, R.D. Farrant, and J.C. Lindon. 750 MHz ^{1}H and ^{1}H-^{13}C NMR spectroscopy of human blood plasma. *Anal. Chem.*, 67, 793–811 (1995).

[48] H.C Keun, O. Beckonert, J.L. Griffin, C. Richter, D. Moskau, J.C. Lindon, and J.K. Nicholson. Cryogenic probe ^{13}C NMR spectroscopy of urine for metabonomic studies. *Anal. Chem.*, 74, 4588–4593 (2002).

[49] K.E. Price, S.S. Vandaveer, C.E. Lunte, and C.K. Larive. Tissue targeted metabonomics: Metabolic profiling by microdialysis sampling and microcoil NMR. *J. Pharmaceut. Biomed. Anal.*, 38, 904–909 (2005).

[50] A. Tomlins, P.J.D. Foxall, J.C. Lindon, M.J. Lynch, M. Spraul, J.R. Everett, and J.K. Nicholson. High resolution magic angle spinning ^{1}H nuclear magnetic resonance analysis of intact prostatic hyperplastic and tumour tissues. *Anal. Commun.*, 35, 113–115 (1998).

[51] S.L. Garrod, E. Humpfer, M. Spraul, S.C. Connor, S. Polley, J. Connelly, J.C. Lindon, J.K. Nicholson, and E. Holmes. High-resolution magic angle spinning ^{1}H NMR spectroscopic studies on intact rat renal cortex and medulla. *Magn. Reson. Med.*, 41, 1108–1118 (1999).

[52] L.L. Cheng, I.W. Chang, D.N. Louis, and R.G. Gonzalez. Correlation of high-resolution magic angle spinning proton magnetic resonance spectroscopy with histopathology of intact human brain tumor specimens. *Cancer Res.*, 58, 1825–1832 (1998).

[53] J. Yang, G. Xu, Y. Zheng, H. Kong, T. Pang, S. Lv, Q. Yang. Diagnosis of liver cancer using HPLC-based metabonomics avoiding false-positive result from hepatitis and hepatocirrhosis diseases. *J. Chromatogr.*, B, 813, 59–65 (2004).

[54] M. Morris, and S.M. Watkins. Focused metabolomic profiling in the drug development process: Advances from lipid profiling. *Curr. Op. Chem. Biol.*, 9, 407–412 (2005).

[55] I.D. Wilson, J.K. Nicholson, J. Castro-Perez, J.H. Granger, K.A. Johnson, B.W. Smith, and R.S. Plumb. High resolution "ultra performance" liquid chromatography coupled to oa-TOF mass spectrometry as a tool for differential metabolic pathway profiling in functional genomic studies. *J. Proteome Res.*, 4, 591–598 (2005).

[56] T. Soga, Y. Ohashi, Y. Ueno, H. Nasaoka, M. Tomita, and T. Nishioka. Quantitative metabolome analysis using capillary electrophoresis mass spectrometry. *J. Proteome Res.*, 2, 488–494 (2003).

[57] J.C. Lindon, J.K. Nicholson, and I.D. Wilson. Directly-coupled HPLC-NMR and HPLC-NMR-MS in pharmaceutical research and development. *J. Chromatog.*, B, 748, 233–258 (2000).

[58] J.C. Lindon, E. Holmes, and J.K. Nicholson. Pattern recognition methods and applications in biomedical magnetic resonance. *Prog. NMR Spectrosc.*, 39, 1–40 (2001).

[59] H. Wold. "Partial Least Squares," in S. Kotz, and N. L. Johnson (Eds), *Encyclopedia of Statistical Sciences*, 6, 581–591. John Wiley & Sons, New York, USA. (1985).

[60] J. Trygg, and S. Wold. Orthogonal projections to latent structures (O-PLS). *J. Chemomet.*, 16, 119–128 (2002).

[61] E. Holmes, J.K. Nicholson, and G. Tranter. Metabonomic characterization of genetic variations in toxicological and metabolic responses using probabilistic neural networks. *Chem. Res. Toxicol.*, 14, 182–191 (2001).

[62] D.B. Kell. Metabolomics and machine learning: explanatory analysis of complex metabolome data using genetic programming to produce simple, robust rules. *Mol. Biol. Rep.*, 29, 237–241 (2002).

[63] O. Cloarec, M.E. Dumas, A. Craig, R.H. Barton, J. Trygg, J. Hudson, C. Blancher, D. Gauguier, J.C. Lindon, E. Holmes, and J. Nicholson. Statistical total correlation spectroscopy: An exploratory approach for latent biomarker identification from metabolic ^{1}H NMR data sets. *Anal. Chem.*, 77, 1282–1289 (2005).

[64] D.J. Crockford, E. Holmes, J.C. Lindon, R.S. Plumb, S. Zirah, S.J. Bruce, P. Rainville, C.L. Stumpf, and J.K. Nicholson. Statistical HeterospectroscopY (SHY), a new approach to the integrated analysis of NMR and UPLC-MS datasets: Application in metabonomic toxicology studies. *Anal. Chem.*, 78, 363–371 (2006).

[65] J.C. Lindon, J.K. Nicholson, E. Holmes, H.C. Keun, A. Craig, J.T.M. Pearce, S.J. Bruce, N. Hardy, S.-A. Sansone, H. Antti, P. Jonsson, C. Daykin, M. Navarange, R.D. Beger, E.R. Verheij, A. Amberg, D. Baunsgaard, G.H. Cantor, L. Lehman-McKeeman, M. Earll, S. Wold, E. Johansson, J.N. Haselden, K. Kramer, C. Thomas, J. Lindberg, I. Schuppe-Koistinen, I.D. Wilson, M.D. Reily, D.G. Robertson, H. Senn, A. Krotzky, S. Kochhar, J. Powell, F. van der Ouderaa, R. Plumb, H. Schaefer, and M. Spraul. Summary recommendations for standardization and reporting of metabolic analyses. *Nature Biotech.*, 23, 833–838 (2005).

[66] H. Jenkins, N. Hardy, M. Beckmann, J. Draper, A.R. Smith, J. Taylor, O. Fiehn, R. Goodacre, R.J. Bino, R. Hall, J. Kopka, G.A. Lane, B.M. Lange, J.R. Liu, P. Mendes, B.J. Nikolau, S.G. Oliver, N.W. Paton, S. Rhee, U. Roessner-Tunali, K. Saito, J. Smedsgaard, L.W. Sumner,

T. Wang, S. Walsh, E.S. Wurtele, and D.B. Kell. A proposed framework for the description of plant metabolomics experiments and their results. *Nature Biotech.*, 22,1601–1606 (2004).

[67] A.L. Castle, O. Fiehn, R. Kaddurah-Daouk, and J.C. Lindon. Metabolomics standards workshop and the development of international standards for reporting metabolomics experimental results. *Briefings in Bioinformatics*, 7, 159–165 (2006).

[68] M.E. Bollard, E.G. Stanley, J.C. Lindon, J.K. Nicholson, and E. Holmes. NMR-based metabonomics approaches for evaluating physiological influences on biofluid composition. *NMR Biomed.*, 18, 143–162 (2005).

[69] R.S. Plumb, J.H. Granger, C.L. Stumpf, K.A. Johnson, B.W. Smith, S. Gaulitz, I.D. Wilson, and J. Castro-Perez. A rapid screening approach to metabonomics using UPLC and oa-TOF mass spectrometry: Application to age, gender and diurnal variation in normal/Zucker obese rats and black, white and nude mice. *Analyst*, 130, 844–849 (2005).

[70] J.B. German, M.A. Roberts, and S.M. Watkins. Genomics and metabolomics as markers for the interaction of diet and health: Lessons from lipids. *J. Nutrition*, 133, 2078S–2083S (2003).

[71] K.K. Lehtimaki, P.K. Valonen, J.L. Griffin, T.H. Vaisanen, O.H.J. Grohn, M.I. Kettunen, J. Vepsalainen, S. Yla-Herttuala, J. Nicholson, and R.A. Kauppinen. Metabolite changes in BT4C rat gliomas undergoing ganciclovir-thymidine kinase gene therapy-induced programmed cell death as studied by H–1 NMR spectroscopy in vivo, ex vivo, and in vitro. *J. Biol. Chem.*, 278, 45915–45923 (2003).

[72] L.C. Robosky, D.F. Wells, L.A. Egnash, M.L. Manning, M.D. Reily, and D.G. Robertson. Metabonomic identification of two distinct phenotypes in Sprague-Dawley (Crl:CD(SD)) rats. *Toxicol. Sci.*, 87, 277–284 (2005).

[73] J.C. Lindon, E. Holmes, and J.K. Nicholson. Toxicological applications of magnetic resonance. *Prog. NMR Spectrosc.*, 45, 109–143 (2004).

[74] E.M. Lenz, J. Bright, R. Knight, F.R. Westwood, D. Davies, H. Major, and I.D. Wilson. Metabonomics with ^{1}H-NMR spectroscopy and liquid chromatography-mass spectrometry applied to the investigation of metabolic changes caused by gentamycin-induced nephrotoxicity in the rat. *Biomarkers*, 10, 173–187 (2005).

[75] J.C. Lindon, J.K. Nicholson, E. Holmes, H. Antti, M.E. Bollard, H. Keun, O. Beckonert, T.M. Ebbels, M.D. Reily, D. Robertson, G.J. Stevens, P. Luke, A.P. Breau, G.H. Cantor, R.H. Bible, U. Niederhauser, H. Senn, G. Schlotterbeck, U.G. Sidelmann, S.M. Laursen, A. Tymiak, B.D. Car, L. Lehman-McKeeman, J.M. Colet, A. Loukaci, and C. Thomas. Contemporary issues in toxicology: The role of metabonomics in toxicology and its evaluation by the COMET project. *Toxicol. Appl. Pharmacol.*, 187, 137–146 (2003).

[76] H.C. Keun, T.M.D. Ebbels, H. Antti, M. Bollard, O. Beckonert, G. Schlotterbeck, H. Senn, U. Niederhauser, E. Holmes, J.C. Lindon, and J.K. Nicholson. Analytical reproducibility in ^{1}H NMR-based metabonomic urinalysis. *Chem. Res. Toxicol.*, 15, 1380–1386 (2002).

[77] H.C. Keun, T.M.D. Ebbels, M.E. Bollard, O. Beckonert, H. Antti, E. Holmes, J.C. Lindon, and J.K. Nicholson. Geometric trajectory analysis of metabolic responses to toxicity can define treatment specific profiles. *Chem. Res. Toxicol.*, 17, 579–587 (2004).

[78] T. Ebbels, H. Keun, O. Beckonert, H. Antti, M. Bollard, E. Holmes, J. Lindon, and J. Nicholson. Toxicity classification from metabonomic data using a density superposition approach: 'CLOUDS'. *Anal. Chim. Acta*, 490, 109–122 (2003).

[79] J.C. Lindon, H.C. Keun, T.M.D. Ebbels, J.M.T. Pearce, E. Holmes, and J.K. Nicholson. The Consortium for Metabonomic Toxicology (COMET): Aims, Activities and Achievements. *Pharmacogenomics*, 6, 691–699 (2005).

[80] M.E. Bollard, H.C. Keun, O. Beckonert, T.M.D. Ebbels, H. Antti, A.W. Nicholls, J.P. Shockcor, G.H. Cantor, G. Stevens, J.C. Lindon, E. Holmes, and J.K. Nicholson. Comparative

metabonomics of differential hydrazine toxicity in the rat and mouse. *Toxicol. Appl. Pharmacol.*, 204, 135–151 (2005).

[81] S.H. Moolenaar, U.F.H. Engelke, and R.A. Wevers. Proton nuclear magnetic resonance spectroscopy of body fluids in the field of inborn errors of metabolism. *Ann. Clin. Biochem.*, 40, 16–24 (2003).

[82] M. Coen, M. O'Sullivan, W.A. Bubb, P.W. Kuchel, and T. Sorrell. Proton nuclear magnetic resonance-based metabonomics for rapid diagnosis of meningitis and ventriculitis. *Clin. Infect. Disease*, 41, 1582–1590 (2005).

[83] V.G. Dunne, S. Bhattachayya, M. Besser, C. Rae, and J.L. Griffin. Metabolites from cerebrospinal fluid in aneurysmal subarachnoid haemorrhage correlate with vasospasm and clinical outcome: A pattern-recognition H-1 NMR study. *NMR Biomed.*, 18, 24–33 (2005).

[84] M.G. Swanson, D.B. Vigneron, Z.L. Tabatabai, R.G. Males, L. Schmitt, P.R. Carroll, J.K. James, R.E. Hurd, and J. Kurhanewicz. Proton HR-MAS spectroscopy and quantitative pathologic analysis of MRI/3D-MRSI-targeted postsurgical prostate tissues. *Magn. Reson. Med.*, 50, 944–954 (2003).

[85] D. Moka, R. Vorreuther, H. Schicha, M. Spraul, E. Humpfer, M. Lipinski, P.J.D. Foxall, J.K. Nicholson, and J.C. Lindon. Biochemical classification of kidney carcinoma biopsy samples using magic-angle-spinning ^{1}H nuclear magnetic resonance spectroscopy. *J. Pharmaceut. Biomed. Anal.*, 17, 125–132 (1998).

[86] B. Sitter, U. Sonnewald, M. Spraul, H.E. Fjosne, and I.S. Gribbestad. High-resolution magic angle spinning MRS of breast cancer tissue. *NMR Biomed.*, 15, 327–37 (2002).

[87] S.J. Barton, F.A. Howe, A.M. Tomlins, S.A. Cudlip, J.K. Nicholson, B.A. Bell, and J.R. Griffiths. Comparison of in vivo ^{1}H MRS of human brain tumors with ^{1}H HR-MAS spectroscopy of intact biopsy samples in vitro. *Magma*, 8, 121–128 (1999).

[88] G.L.A.H. Jones, E. Sang, C. Goddard, R.J. Mortishire-Smith, B.C. Sweatman, J.N. Haselden, K. Davies, A.A. Grace, K. Clarke, and J.L. Griffin. A functional analysis of mouse models of cardiac disease through metabolic profiling. *J. Biol Chem.*, 280, 7530–7539 (2005).

[89] K. Odunsi, R.M. Wollman, C.B. Ambrosone, A. Hutson, S.E. McCann, J. Tammela, J.P. Geisler, G. Miller, T. Sellers, W. Cliby, F. Qian, B. Keitz, M. Intengan, S. Lele, and J.L. Alderfer. Detection of epithelial ovarian cancer using H-1-NMR-based metabonomics. *Int. J. Cancer*, 113, 782–788 (2005).

[90] J.T. Brindle, H. Antti, E. Holmes, G. Tranter, J.K. Nicholson, H.W.L. Bethell, S. Clarke, P.M. Schofield, E. McKilligin, D.E. Mosedale, and D.J. Grainger. Rapid and noninvasive diagnosis of the presence and severity of coronary heart disease using H-1 NMR-based metabonomics. *Nature Med.*, 8, 1439–1445 (2002).

[91] T.A. Clayton, J.C. Lindon, H. Antti, C. Charuel, G. Hanton, J.-P. Provost, J.-L. Le Net, D. Baker, R.J. Walley, J.R. Everett, and J.K. Nicholson. Pharmaco-metabonomic phenotyping and personalised drug treatment. *Nature*, 440, 1535–1542 (2006).

The Handbook of Metabonomics and Metabolomics
John C. Lindon, Jeremy K. Nicholson and Elaine Holmes (Editors)

Chapter 2

Cellular Metabolomics: The Quest for Pathway Structure

Oliver Fiehn

University of California Davis Genome Center, 451 E Health Sci. Dr., Davis CA 95616, USA

2.1. Introduction

Metabolomics can be used for two major and very different purposes: the screening for differences between global metabolic fingerprints of cohorts of populations, or efforts to understand the regulatory structure of metabolic pathways, its connectivity, control of cellular concentrations and fluxes of metabolites, and partitioning of metabolic products between cellular compartments and excretion. Almost certainly, any biomarkers or major differences that have been identified in metabolic fingerprinting will lead to the next level of query, a quest for an in-depth mechanistic understanding, for example why certain biomarkers were specific for a given biological condition or perturbation. *Vice versa*, once metabolic events are understood in a comprehensive way for a given cellular system, the logic next step is to ask how the observed regulatory structure responds to external stimuli and how its metabolic responsiveness relates to the integration of the cellular regulation into larger systems, be these on the level of tissues, organs or the organism. Even for unicellular systems it is known that the excretion of metabolites and ultimately the growth rates of cell cultures depend on the competition and interaction between these cells, especially when these are present in a mixture with very many other cell types of unicellular organisms. Although the two major approaches differ by emphasis of "mechanism" versus "prediction", eventually a better understanding of cellular regulatory circuits is always sought. Global metabolic fingerprinting [1] clearly aims at the ability to distinguish different metabolic states and to make predictions on the eventual outcome and the fate of organisms in a given biological condition.

A similar kind of predictability is also key for cellular metabolomics, which can be used, for example, in bioengineering efforts [2] with the aim of increasing specific metabolic levels [3], or for pharmacological approaches in which certain drugs are employed to exert control effects without redirecting unrelated metabolic fluxes in an unwanted way. Concluding, both major aims of metabolomics, ascertaining differences between metabolic systems and understanding metabolic control, are two complementary efforts which rely on each other. The major focus of this chapter is on highlighting difficulties for cellular metabolomics in understanding metabolic networks. Why do we have so many unknown metabolites and how can we identify them in a rapid manner? Why do metabolic networks behave in a highly flexible manner, yet are supposed to have a rigid structure of biochemical properties? Why is it so hard to predict metabolic systems despite decades of work, and the relatively low number of enzymes involved in primary metabolism and energy regulation? These and related questions are needed to be answered in order to make full use of the prospects of metabolomics.

2.2. How large is the metabolome?

It is a long-standing observation in many metabolomic research projects that the number of detected compounds is far higher than predicted by standard metabolic pathways. For example, the number of metabolites that were expected in the bacterium *Bacillus subtilis* were estimated to account for a maximum of 800 compounds, but more likely even fewer, when metabolic pathways were reconstructed from the genome sequences and enzyme coding genes. However, a combination of three capillary electrophoresis (CE) methods coupled to mass spectrometry (MS) identified and quantified up to 1500 metabolites from *B. subtilis* [4, 5], far more than expected from the simple calculations given above. One can argue that not all of these detected signals arose from genuine intracellular metabolites but some also might be caused by partial degradation or artifact formation during the sample preparation procedure. On the other hand, many metabolites might still have been missed by the aforementioned analytical chemistry methods. The strength of these techniques is the detection of charged molecules, and arguably, most metabolic intermediates bear acidic or basic functional moieties that would enable good separation by CE/MS [6]. However, both CE and MS are notoriously poor in separating and ionizing nonpolar (hydrophobic) and uncharged metabolites which are very likely also to be present in plasma membranes and subcellular structures.

One may further look at mammalian cellular metabolomes. In a similar way, when calculating the number of likely metabolites that originate from the sequenced genome and a consecutive prediction of biochemical pathways, some 2400 compounds have been predicted to comprise the complement of internal human metabolites. Does this actually hold true? There is valid reasoning in arguing that mammals do not need to synthesize many compounds which can instead be digested through food sources, and this may especially be true for those compounds that would be energetically costly to produce. However, there is also growing evidence that many enzymes are far less specific than previously assumed and might thus accept a variety of substrates that can be used for oxidative purposes, or that can be used as building blocks in biosynthetic routes. Given the converging role of catabolism and the great variety of foods that can be used by omnivores, it is imminent that a superficial view of mammalian enzymes does not hold to be as selective as is suggested by (reconstructed) genomic pathways [7]. It is simply not very realistic to exclusively reflect on (theoretical) substrate preferences and disregard the reality of mammals as food-eating organisms, for which the distinction between exogenous and endogenous compounds is far more artificial than, say, for photoautotrophic systems like plants. In addition, compounds that are produced and excreted by the gut microflora add to the complexity of mammalian metabolomic systems [8]. In humans, the metabolome therefore consists of a high complexity that relies on multiple parameters, not only on the human genome and reconstructed biochemical networks but also on food preferences and the intestinal microflora. Recently, first attempts have been reported on the prediction of metabolites that originate from ingestion of xenobiotics [9], but too little is yet known about the dynamics of absorption, transport and metabolism of complex diets to directly infer the complete metabolome within the human body.

Therefore, a complete analysis of the metabolome of photoautotrophic organisms and unicellular systems should, in principle, be much simpler. Many of these organisms have a clear preference for the major source of carbon (carbon dioxide for photoautotrophs, simple small molecules like acetate or glucose for microbial systems). Why is the complete set of small molecules unknown even for these rather simple systems? One major reason is that still many genes have only been vaguely annotated with properties such as 'catalytic activity'. The elucidation of pathways in which the corresponding putative enzymes work is tedious and not always straightforwardly interpretable. Consequently, functional genomics in such organisms necessarily calls for *de novo* structural elucidation of 'unknown' signals that were detected using advanced analytical chemical instrumentation.

2.3. Metabolite identification

'The identification of metabolites' is a less clear term than commonly believed [10]. First, a clear distinction is needed between (a) annotating peaks as 'known compounds' from a series of chromatograms and (b) *de novo* structural elucidation. Secondly, both tasks face different challenges. For example, peaks in GC/MS may be annotated as 'L-aspartate', however, in certain cases, also D-aspartate might occur which would be indistinguishable from its L-isomer if not special precautions are not taken. Conversely, even famous and Nobel-prize winning work for in-depth *de novo* structural elucidations can later turn out to be wrong [11], despite the many steps and various spectroscopic techniques that were involved. Therefore, any metabolite identification, especially in metabolomics, must precisely define how compound names are supported by experimental data and which analytical method and algorithm was used, including threshold levels.

It is now clear that there are different levels of confidence in naming metabolites. Many peaks will not safely match the criteria for clear-cut annotations, and accurate *de novo* identifications are too costly and too slow for most studies. However, if confidence levels are clearly ascribed, it might be valuable to annotate an unknown analytical signal with a less precise but still biochemically meaningful name. Consequently, metabolite annotations may be structured into four major groups:

(i) compounds that are identified by data acquisition of at least two different physicochemical parameters and using authentic reference standards. Typical examples comprise the analyses of small molecules such as cholesterol by volatility (in gas chromatography) and mass spectra (ion fragmentation pattern). As stated above, confidence thresholds for the identification algorithm must be given (e.g. retention time window, mass spectral similarity).

(ii) Secondly, compounds could be tentatively identified by deferring putative chemical structures from physicochemical properties. This process is called dereplication, and a potential workflow is detailed in Figure 2.1. Further, comparison of spectral properties of unknown signals to spectral libraries may lead to tentatively identified annotations, if the spectral similarity score and matching physicochemical parameters are beyond reasonable doubt. Usually, this process would not distinguish between closely related isomers.

(iii) Next, the structural information that is gained along the analytical process may not be sufficient to derive exact chemical structures or a small list of tentatively identified compounds, but it might still enable a classification of unknown compounds to a certain chemical group such as 'carbohydrate'. Such classifications may be aided by rules obtained using supervised statistics carried out on larger mass spectrometric libraries or deduced by expert knowledge of mass fragmentation rules.

(iv) Last, in unbiased metabolomic surveys there will always be analytical signals that are of just too low an abundance, too uncommon or too poor in spectral information to be classified to a certain group of metabolites. These compounds can still be used for phenotyping approaches using relative quantification of signal intensities between genetic or treatment perturbations, but in order to derive biochemical or mechanistic insights, the isolation of microgram quantities of these metabolites would be necessary.

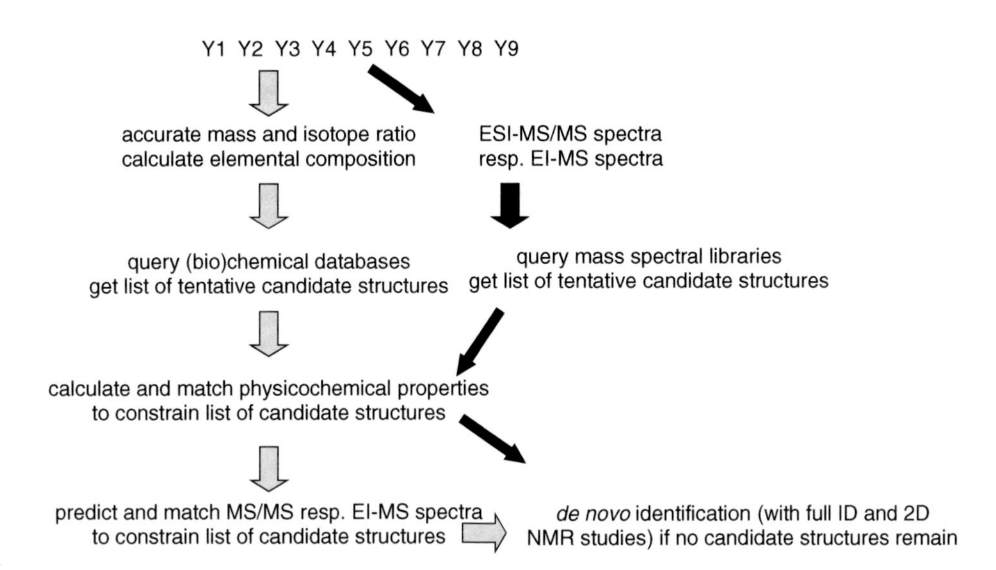

Figure 2.1. Proposed workflow for a rapid annotation of unknown analytical signals in LC/MS or GC/MS metabolite profiles. Such schema might evolve as a general database query tool similar to MASCOT or SEQUEST in proteomics research, returning a list of tentative structures with an overall similarity score. Y1–Y9 represent unknown metabolites.

Obviously, the preferred route of consistent, reliable and routine metabolite identification is to work with authentic reference compounds. Given the theoretically unlimited size of metabolomes, especially in heterotrophic multicellular organisms, only a fraction of the signals in metabolite profiles will generally be annotated by this means. It should become good metabolomic practice that at least the common set of conserved metabolites in primary metabolism are available as reference standards to verify certain signals. Publications should indicate which level of certainty is associated with metabolite naming, including referencing to authoritative chemical databases such as PubChem [12, 13], the (commercial) chemical abstracts service (CAS) or the InChI code [14]. In addition, if metabolite signals are reported that need yet to be structurally characterized, it is mandatory to label these signals in a way that enables tracking physicochemical characteristics (e.g. mass spectrum, quantification ion and chromatography retention times). Furthermore, it should be good reporting practice to label such unknown compounds in a way that is consistent for each specific laboratory in order to potentially learn about responses for this novel metabolite across biological studies, and hopefully even to exchange information about such compounds between laboratories. In 2005, an initiative was formed under the umbrella of the Metabolomics Society that is trying to foster such data exchange and reuse of metabolomic data by drafting and implementing ‘reporting standards’ on metabolomic studies [15]. The idea for this initiative is that metabolomic studies are so rich in data that novel conclusions may be derived if datasets are investigated from more than one point of view. Obviously, such efforts directly benefit from better strategies to structurally elucidate, annotate and report on metabolite identities.

As pointed out above, the identification of metabolic signals is best performed using authentic reference compounds. Where this is not possible, metabolite identity must be unraveled *de novo* as detailed as possible [16]. Classical methods have involved isolation of compounds followed by in-depth structural characterization using spectroscopic techniques. However, there are caveats. The parameter space for stereo configurations and positional isomers is tremendous, and even with elegant strategies and long time efforts, initial identifications have often proved wrong [11]. Furthermore, there is the risk that identifications are carried out on compounds that are later found to be already published in less well known journals. Such problems may be circumvented by applying a strategy that aims at tentatively annotating potential structural candidates from databases before performing more laborious work on isolation and *de novo* identification. Annotations would need to combine all available physicochemical information from a separation and spectral characterization of a specific compound without necessarily isolating it, and a workflow is outlined in Figure 2.1. The best way to approach such tentative identifications is by utilizing the combination of chromatography and mass spectrometry, calculating parameters from this information, and then matching these parameters with theoretical values that are calculated from molecular structures of database entries.

Two strategies may be distinguished: starting from elemental formulae and chemical databases or starting from mass fragment spectra and MS libraries. The first approach focuses on the elemental composition. Most recently, we have shown in a chemoinformatic approach that even 0.1 ppm mass accuracy would not enable unambiguous calculation of elemental formulae of low molecular weight compounds [17], whereas mass spectrometers with 3 ppm mass accuracy and 2% relative isotope ratio accuracy enables constraining compositional calculations in a most dramatic way to one or just a few candidate formulae. Such use of isotope ratio data can be applied to both LC- or GC-based methods [18, 19]. Calculated elemental compositions will be used to query chemical and biochemical databases, resulting in potentially a couple of thousand candidate structures. Good databases for these purposes are the publicly available *PubChem* effort (5 million entries as of January 2006) or the commercial *Dictionary of Natural Compounds* DNP (200,000 structures). For these structures, physicochemical parameters can be calculated, for example boiling points, lipophilicity (log K_{ow}) and pK_a values. In the next step, the same parameters can be calculated for the unidentified peaks using the experimental retention time information from liquid or gas chromatography. Matching the experimental to predicted parameters, along with the determination of elemental compositions, will constrain the candidate structure list to just a few structures, most of which will be positional isomers or closely related compounds. In principle, the commercial CAS could also be used for the identification of a list of potential structures. However, this service does not allow batch queries of structure searches and thus cannot be embedded into automatic algorithm routines because it needs to be handled in a manual way.

This constrained list of candidate structures can further be confined by matching tandem mass spectrometric information (MS/MS spectra) with theoretical fragmentations. Such theoretical MS/MS spectra can be predicted from commercially available software solutions based on proposed structures and known fragmentation rules (Mass Frontier) [20]. However, so far Mass Frontier has only implemented positive ionization rules, and it obviously relies on fully characterized fragmentation mechanisms published in the literature. Only if no structure from the initial list of tentative compounds remains, unidentified peaks must be regarded as novel metabolites which could be subjected to classical *de novo* identification, including isolation of the compound and full NMR analysis including two-dimensional methods.

A second approach would query mass spectral libraries. Once enough mass spectra have been annotated with structures, such as is the case for GC/MS spectra under electron impact ionization, further processes can be added to online structural annotations. For example, the large NIST 5.0 mass spectral database comprises some 5000 compounds with trimethylsilylated moieties, the most common derivatization technique used in metabolomics. From these compounds, substructures can be generated to be used as training data sets for supervised statistical tools. Such algorithms aim at learning rules to automatically annotate unknown mass spectra to belong to certain chemical groups such as ‘carbohydrate’, ‘primary amine’ and other metabolite families. The advantage of this approach is that multiple supervised methods may be tested simultaneously (such as partial least squares, linear discriminant analysis, tree-based models, feature selection, association rule models and others) which can then be investigated with respect to false discovery rates and robustness. In conjunction with calculating boiling points from retention indices, it will be further possible to assign the size of the molecules, for example mono-, di- and trisaccharides, sugar alcohols and other structural information such as the presence of furanoside and pyranoside rings. For any peak for which very high similarity or even virtual spectral identity is found, physicochemical properties (e.g. boiling points) could be calculated from the structures and matched with the experimentally determined parameters.

Only if no candidate structure remains after exploiting the techniques outlined above, is it reasonable to assume that an unidentified compound is a truly novel metabolite which would need to be identified using classical *de novo* compound identification, combining various structural characterization techniques including NMR.

2.4. Pathway identification

Once all biologically relevant metabolic signals are annotated in the workflow schema given above, these structures need further be associated to a biochemical pathway and a biological function in order to aid cellular interpretation of metabolomic findings.

The first step utilizes common pathways that are compiled in consensus biochemical maps such as BioCyc [21], or specific maps like AraCyc and comparable overview charts. A general layout of reconstructed pathways is outlined in Figure 2.2. These maps are certainly a good start for annotating compounds to pathways in which they are involved, but unfortunately, four major problems can be outlined: (1) for any given organism or tissue, these maps are usually sparse and do not detail the complement of already existing biochemical knowledge, and especially they do not contain information for many less common metabolites; (2) a number of metabolites that are believed to be involved in certain pathways cannot readily be determined by a given analytical-chemical technique (depicted as ‘empty boxes’ in Figure 2.2); (3) many novel metabolites may be detected (depicted as Y1–Y9 in Figure 2.2) for which no enzymes or biochemical reactions may be known; and (4) the pathway topology and the directionality of enzyme reactions are often deferred from homology to other organisms and might not reflect the actual *in vivo* function in the cellular system under study.

Therefore, pathways have to be elucidated using biochemical, genetic and molecular biology approaches which may be summed as ‘functional genomics’.

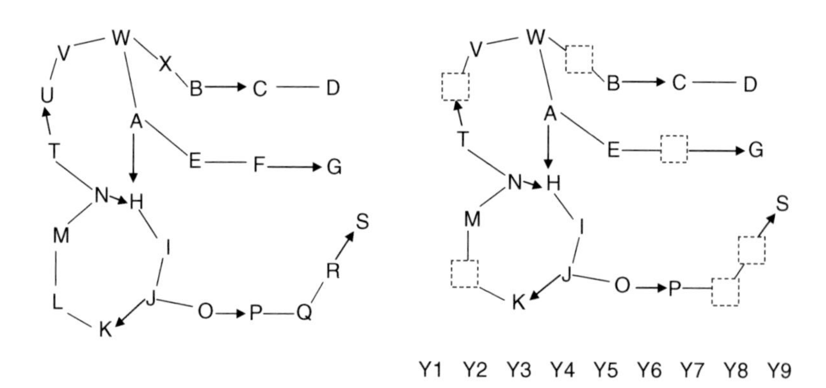

Figure 2.2. Left panel: Generalized metabolic pathway map (metabolites A–X) that may result from genomic reconstructions. Right panel: Result of a metabolite profiling study using a specific analytical technique. Six metabolites could not be detected (open squares) although these were supposed to be present by the reconstructed pathway map. In addition, nine novel metabolites were detected (Y1–Y9) which cannot immediately be mapped onto the pathway chart.

Three potential outcomes are possible for functional genomics approaches to pathway elucidations.

(1) Gene products may comprise specific catalytic activity, converting
  - already known substrates to known products in bypasses of classical reactions, or in cellular compartments that usually do not comprise these pathways.
  - known substrates to novel compounds, for example in anabolic reactions to fulfill specific biological roles such as communication and defense.
  - novel substrates to known compounds that may then merge into mainstream metabolism, for example in catabolic reactions to control turnover of compounds that were synthesized in the aforementioned process.
  - novel compounds to other novel compounds, which then define a completely new pathway.

(2) In addition, unspecific enzymatic activity must be considered, for example processes that convert a variety of known substrates to a plethora of products which can then be utilized for cellular communication or defense. Examples would be P450 mono-oxygenases or enzymes involved in release of plant volatiles. The evolutionary and biological roles of such unspecific broadband anabolic processes are poorly understood. Emission of a variety of different compounds instead of a single specific metabolite might aid in inter- and intraspecies communication, where signals are perceived in patterns rather than in activation of a single (specific) receptor.

(3) Last, novel compounds may be produced by non-enzymatic processes such as oxidation, hydrolysis or even cleavage and condensation. Molecular alterations of non-annotated genes may lead to changes of metabolite profiles; however, in order to conclude that enzymes encoded by these genes actually produce the final metabolic products, the exact reaction mechanisms must be worked out. Even more difficult to unravel are products that are entirely produced by physicochemical products but that are only present under certain physiological conditions, say heat stress. Generally, metabolic products are variable by interactions of *genotype* × *environment* × *time* × *spatial location*. This matrix of parameters complicates any simple relationship of using metabolic phenotypes to understand gene functions, so any pathway hypothesis that is developed by metabolomics approaches needs to be verified by molecular and biochemical studies.

Given these constraints, novel pathways might be unraveled for compounds like Y3 and Y9 in the generalized pathway given in Figure 2.3. Due to the reasons outlined above, such pathway elucidation is a rather tedious process that usually would not allow quick mapping of the other novel metabolites Y1,2,4–8 into the biochemical pathway structure.

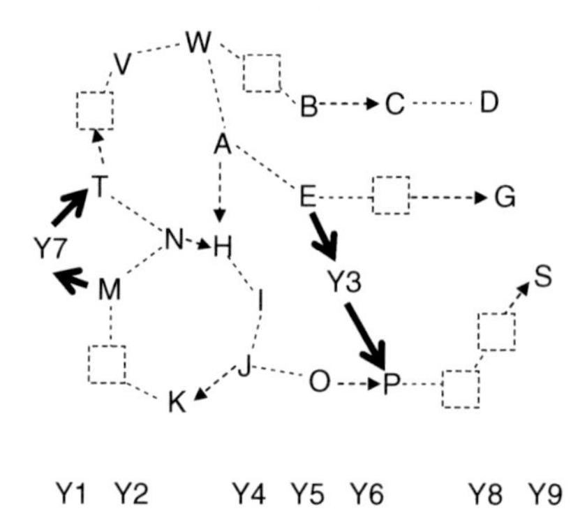

Figure 2.3. Functional annotation of novel metabolites onto biochemical pathways depicted in Figure 2.2. Using a variety of genomic, analytical and biochemical experiments, pathways might be unraveled for some of the new compounds. Such pathway elucidations involve laborious wet laboratory work and thus leave many other uncommon metabolites without biochemical annotation.

Furthermore, the situation may even be more complicated than outlined so far. We have considered linear 1:1 relationships of enzymes, substrates and products that can be unraveled using classical biochemical or modern molecular biology tools. However, there is still the question why there are seemingly more metabolites than enzymes, why there is so much diversity of metabolomes within a genus or between closely related species, and why crosses of these species will often reveal metabolites that are present in neither of the parents. This can hardly been explained by metabolite channeling or by kinetic parameters. This phenomenon might rather point to differences in substrate availability and transport between cellular compartments, organs or even within a cellular compartment. At least in eukaryote cells, but likely also in prokaryotes, the intraplasmic space (e.g. in the cytosol) cannot be regarded as an aqueous solution that allows free diffusion of substrates. We might need to consider the interaction between different protein complexes that carry enzymatic activities, in addition to allelic complementation of missing pathway links that may distinguish crosses from parental lines. Protein complexes are the focus of very active research in many areas of biology and biomedicine, but so far, enzymatic consequences have rarely been studied. It is known that protein folding, topology and protein complexes rely on post-translational modification as well as allosteric modification, both of which may largely change in response to genetic differences (e.g. in crosses) or environmental perturbation (e.g. stress). Consequently, the formation of novel compounds may also largely depend on the actual formation and disassembly of such protein complexes in a given intracellular environment. Such events would easily be missed by typical molecular biology techniques which focus on the identity and activity of a single enzyme, for example by over-expression of eukaryote enzymes into *Escherichia coli* for purification purposes. It may be due to these theoretical difficulties that just a few examples have been reported so far where use of metabolite profiling actually led to the discovery of new pathways, such as for a direct pathway from glycine to glyoxylate in yeast [22].

2.5. "Omics" data integration

The paradigm of (molecular) biology implies a more or less linear hierarchy from genome to phenotype. This linearity would start at gene expression, splicing, translation to encoded proteins, post-translational modifications, and eventually continue to metabolites as victims of the overall process, substances regarded as useful tools to monitor and predict the ultimate organismal phenotypes. In current research proposals, research panels and scientific boards, this view on the biological paradigm leads to demanding an integration of data across these levels of cellular organization. The integration of data from different levels of cellular regulation focuses on biochemical maps that are derived from genomic annotations and homology of genes and enzymes to well-studied organisms [23].

Theoretically, this is a compelling idea that might help understanding of cellular biology by using the data to model the dynamics of the organism in a Systems Biology approach. However, the linearity of the paradigm represents an overly simplistic view that does not lend a good framework for actual data integration beyond biomarker detection for classification purposes. Despite the number of years that Systems Biology has been announced as a valuable goal, the number of research papers and the quality of these yet do not justify the verve with which the demand for 'Omics data integration' is put forward.

The problem is that the level of complexity increases with each level of cellular organization. The genome itself can readily be used for in-depth studies and comparisons, but in many organisms, there is a $1:n$ relationship on subsequent levels of organization. One gene may be spliced and transcribed into more than one gene product, one mRNA may be translated and modified to more than one protein, and one enzyme may work on one or more substrates and may be involved in many pathways. Consequently, there is no easy way to infer cellular regulation from metabolite levels, at least not for higher organisms, and especially not for efforts integrating transcriptomics and metabolomics levels [24, 25].

Even for simple and well-studied unicellular models, the dynamics of metabolism can only be modeled for a couple of seconds after a certain perturbation [26]. In eukaryotes, cellular compartmentation and metabolic specialization of organs further complicate any reasonable biological interpretation of findings beyond simple statements such as 'the rate of glycolysis is increased'. Recently, a study on yeast metabolism under sulfur deprivation using a combined approach of proteomics and metabolomics revealed that predictions of use of pathways could not be made on transcriptomics or proteomics alone [27].

For higher organisms, it is a truism that these consist of many organs, each organ may include many tissue types and each tissue type may comprise various cell types. All published reports so far support the notion that different tissue types comprise varying metabolomes. Different biological roles of individual cell types support the further expectation of detecting striking differences on the low-level spatial resolution, for example between trichome and epidermis cells or between parenchyma and bundle sheath cells in plants [28]. Lastly, intracellular organization of metabolism is also highly structured into compartments, each of which serves specific functions which lead to large metabolic differences. For this reason, *in vivo* measurements are highly advantageous to study both the dynamics and the subcellular localization of metabolites in real time, such as with genetically encoded fluorescent nanosensors [29]. Another report on subcellular studies of metabolites focused on metabolite profiling of isolated chloroplasts and subfractions including the envelope, the stroma and the thylakoids in a study on the activity of three 13-lipoxygenases under stress conditions [30]. So far, the integration of metabolomics data with proteomic or transcriptomic data has not gone beyond simple correlation analysis or statistical discrimination of phenotypes or treatment parameters. This use of data is inadequate to fulfill the vision of Systems Biology which aims at a comprehensive understanding and regulatory modeling of the complex interrelationships of cellular organisms [31], based on intensive computer simulations [32] followed by subsequent experimental testing of hypotheses derived from such models. In order to enable metabolomics (or proteomics!) to be a useful tool in such endeavor, metabolites need to be analyzed at high temporal and spatial resolution under carefully designed experiments in response to a range of genetic or environmental perturbations (not just plus/minus type of experiments such as 'healthy vs. disease'). Today's analytical methods still seem to be inadequate with respect to acquiring the full complement of metabolites at ultimate sensitivity and for multiple biological snapshots. Instead of metabolomics approaches, hypothesis-driven approaches seem currently to be more feasible for integrating gene, protein and metabolite levels.

2.6. Metabolic fluxes

The result of metabolomic analyses is a series of measurements of metabolite levels: snapshots of metabolism. Recently, attempts have been reported to use stable isotope incorporation for better quantification of cellular metabolism [33] whereas most researcher seem to utilize mature technology, such as GC/MS for high throughput profiling of steady state metabolite levels in yeast [34], algae [35] or plant cells [36]. However, such measurements just represent one side of the coin to study pathway structures. For example, with isotope-labeled intermediates and isotopomer analysis, the different contributions of central carbon metabolic pathways can be unraveled for simple cell types [37]. Metabolic snapshot data, however, are usually not sufficient to directly derive enzyme activities and hierarchical structures of pathways, although metabolic changes caused by lack of enzyme activities are sometimes interpreted as alterations in metabolic fluxes [38]. Generally, changes of metabolite levels may be due to drastically different causes: activities of membrane transport proteins may have been altered, rates of catabolism or anabolic reactions may have shifted, or branch point enzyme activities might have changed. Even if metabolite levels are found unchanged between different experimental situations, the underlying flux differences and enzymatic activities might still have changed. For example, if both anabolic and catabolic rates change in the same way and intensity, steady state levels of substrates and products involved in this reaction should not change, although flux through the pathway would have clearly increased.

Therefore, metabolite snapshot data should be complemented by flux data, which has been proposed to be best accomplished by *in vivo* NMR measurements [39]. In principle, it should be also possible to derive enzyme activities and fluxes from snapshot measurements if we had the ability to measure true concentrations of all substrates and products in fast intervals, assuming we would know the total network structure. In practice, however, metabolomic methods miss important intermediates unless methods are tailored to meet these requirements, for example by using a range of different tools and technologies. Furthermore, even if snapshots were taken in time series, and even if all substrates and products of a pathway were covered, we are still unable to unravel the flux structure (i.e. the activity of reversible reactions, futile cycles or other back flows of products into the pathways via other routes through the metabolic network). Consequently, potential new side fluxes out of or into pathways by unforeseen additional enzymatic activities can only be detected by use of labeled compounds, either employing radioactive or stable isotope tracers.

Unfortunately, these techniques are restricted in use by the need to feed in labeled substrates which (a) may not been taken up quickly enough into the cells and (b) are subsequently quickly diluted within the metabolic network. Therefore, only short distances or small parts of the total metabolic network that have reasonably high metabolic turnover rates such as central C/N pathways in extremophiles [40] or that lead to and from strong carbon sinks such as starch in plant cells can be deduced. A potential outcome of such a flux study is depicted in Figure 2.4, with the compound 'A' used as labeled starting point from which relative fluxes $A \rightarrow D$ and $A \rightarrow P \rightarrow S$ can be followed. In this idealized map, the conclusion of such a flux study would be that the major flux of carbon is routed from $A \rightarrow S$ via the $H \rightarrow O$ pathway but much less via the novel $E \rightarrow Y3$ pathway. Significantly less carbon was partitioned here to the $A \rightarrow D$ pathway, and no net flux could be detected towards $A \rightarrow M$ in this artificial study result. Still, such a biochemical map would only include

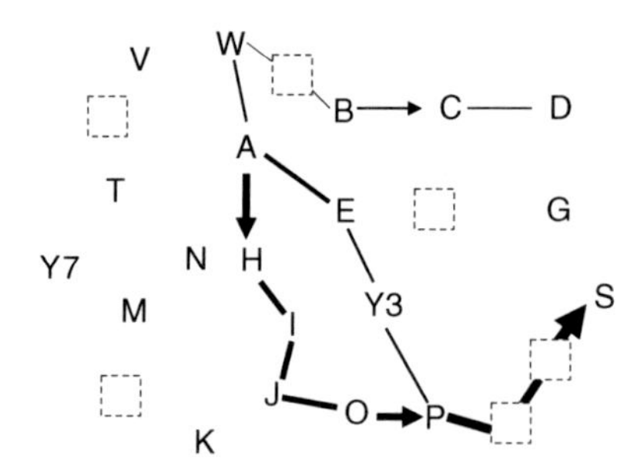

Figure 2.4. Potential outcome of a flux study using the labeled metabolite 'A' from the biochemical pathway depicted in Figure 2.2. Differences in use of alternative biochemical pathways result from flux studies, enzymatic activities can be calculated and the involvement of novel metabolites (Y3) is confirmed. Usually, flux data are not obtained for more distant pathways, or pathways with low overall metabolic turnover.

35 compounds, which comprises a realistic and achievable aim in today's flux studies. A global view on all metabolic fluxes (a 'fluxome' [41]) is still out of reach by current techniques, even if fluxes are inferred from other metabolic sinks such as proteins. Using current methods, a combination of metabolomic snapshot data at a high number of biological replicates (to get the breadth of metabolic networks at high statistical significance levels) and flux measurements (on select and important pathways [42]), therefore seems to present the most practical solution to reach a more complete picture of metabolic control and regulation.

2.7. Metabolic networks

The topology of metabolic networks, as being computed from genomic information, already comprises information about general systems properties [43]. Differences in metabolic levels can be used to interpret potential changes in pathway regulation using such known pathways. However, in general, it is not feasible to directly infer the global biochemical pathways structure or regulatory organization from 'omics' data or even labeled intermediates and flux data. Still, metabolomics data can actually be used for gathering information that is complementary to differences in metabolite levels or patterns between genotypes or treatments. Metabolomics workflows consist of one or many perturbations of a species with a number of biological replicates per group according to the question and study design, and subsequently, data are compared by univariate and multivariate statistical means focusing on differences in average values between these statistical groups. However, one may ask if there actually is an average mouse, an average plant or even an average cell in a microculture? Since such a perfect average individual does not physically exist, how much biological interpretation and meaning can we generate out of the analysis of differences between averages of statistical groups?

What these statistical methods usually ignore is the within-group variance, or the individuality of every given sample, for example by applying Bayesian likelihood calculations [44], Pearson's correlations [45] or partial correlation analysis [46] to construct undirected dependency graphs from metabolomics data. This is a complementary way of looking at the data, which allows constructing a snapshot of a metabolic network for a given biological situation reflecting its underlying regulatory structure. Two theoretical papers have outlined the origin and meaning of metabolite–metabolite covariance and correlation [47, 48], and some further reports have based biological conclusions from comparisons of metabolomic correlation networks [49, 50]. The general outline of metabolic correlation networks is depicted in Figure 2.5, from which it becomes clear that such networks hardly resemble the underlying biochemical network (Figure 2.1). However, when comparing correlation networks under different conditions (e.g. environmental perturbation I and II, left and right panel of Figure 2.5), such networks can still be utilized in several ways: (1) novel metabolites Y2–Y9 can be mapped to other compounds for which the biological role and cellular compartment is supposed to be known, thus hypotheses about these novel compounds can be generated in an easier way compared to flux studies; (2) Network topology differences emphasize differences in overall regulation and carbon partitioning. For example, compound *A* lost its dominant hub role under the hypothetical biological situation II, whereas metabolite *W* becomes

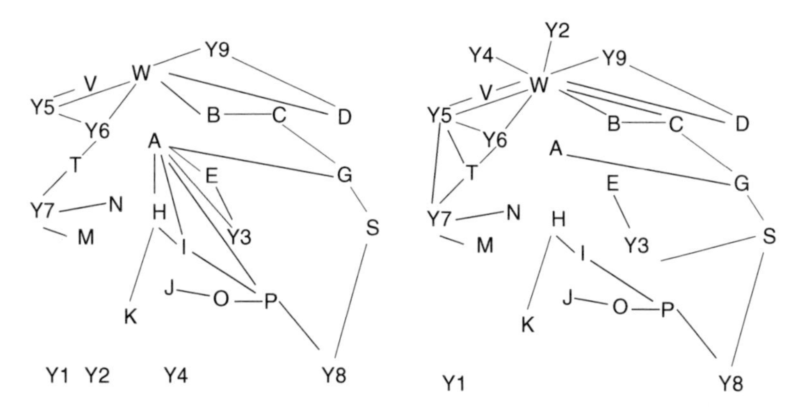

Figure 2.5. Metabolic network graphs resulting from correlation or linearity analysis of metabolite pairs under two different conditions (left and right panel). Some correlations will reflect the underlying biochemical pathway structure depicted in Figure 2.2, whereas other correlations refer to differences in overall metabolic regulation (e.g. by activation of transcription factors). Often, such network graphs enable the generation of improved hypotheses on the biological roles of pathways and the known and novel metabolites (Y2–Y9).

much more connected to other compounds. Such findings can best be interpreted if further biological data are added, such as specific enzyme or transport activities, or activation of transcription factors that would elicit a range of biological pathways and therefore impact biochemically unrelated metabolites in a similar direction and magnitude. More detailed analysis might also focus on the strength (the statistical power) of a given linear relationship, or the slope of the linear regression (which is equivalent to the ratios of pairs of metabolites). Biochemically, changes in metabolite ratios (such as ADP/ATP or Glu/Gln) can readily be interpreted as important physiological parameters. Correlation network analysis may add an overview on regulation of metabolite pairs, hence potentially bridging the analysis of metabolic snapshots ('steady state levels') to flux data by detailing the relative partitioning of metabolite pools under different biological conditions.

There are further theoretical considerations that support this notion of utilizing within-group variance as a surrogate for the actual dynamics of the intracellular metabolic network. All biological systems share network properties which are called 'robustness and flexibility'. Cells are hit in short time intervals by stochastic factors such as influx deviations of external transport metabolites, intensity differences in environmental parameters or subtle physical interferences such as 'wind'. Metabolic systems would become very unstable if each of these short-term pulses would be taken up in immediate responses. There are a number of regulatory steps that inhibit metabolic overreactions but instead introduce response lag times by using threshold systems, active transport steps or reversibility of reactions. In total, these delay steps render the system to become 'robust' which is an important property to maintain the system at a given steady state. Complementary to such robust regulation of the network structure is the necessity to quickly alter metabolite levels depending on certain stress conditions or developmental needs. The responsible general system property is called 'flexibility'. System flexibility is a prerequisite of the capability to 'control' or alter defined steady states without affecting other parts of the system, depending on external or internal stimuli. Any system needs capabilities to react in a fast and coordinated manner on immediate needs and threats, even if the triggering signals for such needs are of a low abundant and transient nature. Examples might be heat shock, wounding responses or herbivore attacks, among others. Very fruitful research has been reported on a combination of a calculation of the metabolic feasibility space of prokaryotes and confirmation of predictions of system responses using flux data [51]. One further step comprises use of metabolite concentrations to gain information on the feasible parameter space of enzyme kinetics in yeast [52]. It is this kind of model-based computation that needs further refinements and application for predicting metabolic responses in complex eukaryote systems and higher organisms.

It is interesting to note that there is a difference in terminology between theory of metabolism and molecular biology. David Fell has pointed out in his famous book

'*Understanding the Control of Metabolism*' (1997) [53], that the terms 'control' and 'regulation' point to biochemical properties that are rather different in their respective meanings. Regulation is the ability of a complex system to maintain its basic properties (e.g. metabolite levels) independent of external factors that continuously try to push the system out of balance, whereas control was defined as a system property that enables changes between different states of a system. In this respect, terminology of metabolic theory reflects the understanding of network properties (flexibility and robustness), whereas in molecular biology, regulation and control are used as synonyms describing only the ability to change a system, but not how to maintain it. Examples of 'control' are found in classic physiology. In terms of plant physiology, cold acclimation (by increased values in carbohydrates) or leaf senescence (altered ratios of catabolism versus anabolism) are examples of 'control' or 'system flexibility', whereas the tendency to keep metabolic fluxes in a narrow range under a given set of environmental parameters (the steady state) is an example for metabolic 'regulation' or system 'robustness'.

Apart from kinetics and flux rates, further properties of metabolic networks are the stoichiometric structure which may be used to define metabolic feasibility spaces for cellular growth, and connectivity [54, 55], which define the relative importance of metabolites as branching points to allow redirection and partitioning of the ratio of carbon, nitrogen, phosphor and sulfur between pathways, organs and compartments. Interestingly, metabolic correlation networks seem to reflect the different needs and partitioning between pathways and thus may enable bridging information that is derived from steady state levels and from flux information. However, so far information garnered from theoretical or experimental metabolic networks has not enabled probing biochemical pathway structure with the aim at detecting novel metabolic routes. At least, such work seems to be more feasible within the foreseeable future than meaningful integration with data from other omics approaches.

Acknowledgments

Continuing discussions with Wolfram Weckwerth (MPI-MP, Potsdam, Germany) and Tobias Kind (UC Davis) have been helpful to elaborate these considerations.

References

[1] Fiehn, O. (2001) Combining genomics, metabolome analysis, and biochemical modelling to understand metabolic networks. *Comp. Funct. Genom.* 2, 155–168.

[2] Forster, J., Gombert, A.K., and Nielsen, J. (2002) A functional genomics approach using metabolomics and in silico pathway analysis. *Biotechnol. Bioeng.* 79, 703–712.

[3] Levin, I., Lalazar, A., Bar, M., and Schaffer, A.A. (2004) Non GMO fruit factories strategies for modulating metabolic pathways in the tomato fruit. *Ind. Crops Prod.* 20, 29–36.

[4] Soga, T., Ohashi, Y., Ueno, Y., Naraoka, H., Tomita, M., and Nishioka, T. (2003) Quantitative metabolome analysis using capillary electrophoresis mass spectrometry. *J. Proteome Res.* 2, 488–494.

[5] Soga, T., Ueno, Y., Naraoka, H., Ohashi, Y., Tomita, M., and Nishioka, T. (2002) Simultaneous determination of anionic intermediates for Bacillus subtilis metabolic pathways by capillary electrophoresis electrospray ionization mass spectrometry. *Anal. Chem.* 74, 2233–2239.

[6] Sato, S., Soga, T., Nishioka, T., and Tomita, M. (2004) Simultaneous determination of the main metabolites in rice leaves using capillary electrophoresis mass spectrometry and capillary electrophoresis diode array detection. *Plant J.* 40, 151–163.

[7] Romero, P., Wagg, J., Green, M.L., Kaiser, D., Krummenacker, M., and Karp, P.D. (2005) Computational prediction of human metabolic pathways from the complete human genome. *Genome Biol.* 6: Art. No. R2 2005.

[8] Nicholson, J.K., Holmes, E., Lindon, J.C., and Wilson, I.D. (2004) The challenges of modeling mammalian biocomplexity. *Nat. Biotechnol.* 22, 1268–1274.

[9] Mayeno, A.N., Yang, R.S.H., and Reisfeld, B. (2005) Biochemical reaction network modeling: Predicting metabolism of organic chemical mixtures. *Environmental Sci. Technol.* 39, 5363–5371.

[10] Milman, B.L. (2005) Identification of chemical compounds. *Trends Anal. Chem.* 24, 493–508.

[11] Nicolaou, K.C. and Snyder, S.A. (2005) Chasing molecules that were never there: Misassigned natural products and the role of chemical synthesis in modern structure elucidation. *Angew. Chem. Int. Ed.* 44, 1012–1044 2005.

[12] Pubchem project. URL cited April 30, 2006 [http://pubchem.ncbi.nlm.nih.gov/].

[13] Feldman, H.J., Snyder, K.A., Ticoll, A., Pintilie, G., and Hogue, C.W.V. (2006) A complete small molecule dataset from the protein data bank. *FEBS Lett.* 580, 1649–1653.

[14] Murray-Rust, P., Rzepa, H.S., Stewart, J.J.P., and Zhang, Y. (2005) A global resource for computational chemistry. *J. Mol. Model.* 11, 532–541.

[15] Metabolomics Standards Initiative. URL cited April 30, 2006 [http://metabolomicssociety.org/mstandards.html].

[16] Steinbeck, C. (2004) Recent developments in automated structure elucidation of natural products. *Nat. Prod. Rep.* 21, 512–518.

[17] Kind, T. and Fiehn, O. (2006) Metabolomic database annotations via query of elemental compositions: Mass accuracy is insufficient even at less than 1 ppm. *BMC Bioinformatics* 7, 234.

[18] Fiehn, O., Kopka, J., Trethewey, R.N., and Willmitzer, L. (2000) Identification of uncommon plant metabolites based on calculation of elemental compositions using gas chromatography and quadrupole mass spectrometry. *Anal. Chem.* 72, 3573–3580.

[19] Fiehn, O. and Major, H. (2005) Exact Molecular Mass Determination of Polar Plant Metabolites Using GCT with Chemical Ionization. *Waters Application Note*, URL cited January 08, 2006 [http://www.waters.com/WatersDivision/SiteSearch/AppLibDetails.asp?LibNum=720001260EN].

[20] Tolstikov, V.V. and Fiehn, O. (2002) Analysis of highly polar compounds of plant origin: combination of hydrophilic interaction chromatography and electrospray ion trap mass spectrometry. *Anal Biochem.* 301, 298–307.

[21] BioCyc. URL cited April 30, 2006 [http://biocyc.org/].

[22] Villas-Boas, S.G., Akesson, M., and Nielsen, J. (2005) Biosynthesis of glyoxylate from glycine in Saccharomyces cerevisiae. *FEMS Yeast Res.* 5, 703–709.

[23] Lange, B.M. and Ghassemian, M. (2005) Comprehensive post-genomic data analysis approaches integrating biochemical pathway maps. *Phytochemistry* 66, 413–451.

[24] Griffin, J.L., Bonney, S.A., Mann, C., Hebbachi, A.M., Gibbons, G.F., Nicholson, J.K., Shoulders, C.C., and Scott, J. (2004) An integrated reverse functional genomic and metabolic approach to understanding orotic acid-induced fatty liver. *Physiol. Genomics* 17, 140–149.

[25] Kant, M.R., and Ament, K., Sabelis, M.W., Haring, M.A., and Schuurink, R.C. (2004) Differential timing of spidmite-induced direct and indirect defenses in tomato plants. *Plant Physiol.* 135, 483–495.

[26] Buchholz, A., Hurlebaus, J., Wandrey, C., and Takors, R. (2002) Metabolomics: Quantification of intracellular metabolite dynamics. *Biomol. Eng.* 19, 5–15.

[27] Lafaye, A., Junot, C., Pereira, Y., Lagniel, G., Tabet, J.C., Ezan, E., and Labarre, J. (2005) Combined proteome and metabolite-profiling analyses reveal surprising insights into yeast sulfur metabolism. *J. Biol. Chem.* 280, 24723–24730.

[28] Schad, M., Mungur, R., Fiehn, O., and Kehr, J. (2005) Metabolic profiling of laser microdissected vascular bundles of Arabidopsis thaliana. *Plant Methods* 1:2.

[29] Deuschle, K., Fehr, M., Hilpert, M., Lager, I., Lalonde, S., Looger, L.L., Okumoto, S., Persson, J., Schmidt, A., and Frommer, W.B. (2005) Genetically encoded sensors for metabolites. *Cytometry A* 64A, 3–9.

[30] Bachmann, A., Hause, B., Maucher, H., Garbe, E., Voros, K., Weichert, H., Wasternack, C., and Feussner, I. (2002) Jasmonate-induced lipid peroxidation in barley leaves initiated by distinct 13-LOX forms of chloroplasts. *Biological Chemistry* 383, 1645–1657.

[31] Yang, Y.T., Engin, L., Wurtele, E.S., Cruz-Neira, C., and Dickerson, J.A. (2005) Integration of metabolic networks and gene expression in virtual reality. *Bioinformatics* 21, 3645–3650.

[32] Ishii, N., Robert, M., Nakayama, Y., Kanai, A., and Tomita, M. (2004) Toward large-scale modeling of the microbial cell for computer simulation. *J. Biotechnology* 113, 281–294.

[33] Kim, J.K., Harada, K., Bamba, T., Fukusaki, E., and Kobayashi, A. (2005) Stable isotope dilution-based accurate comparative quantification of nitrogen-containing metabolites in Arabidopsis thaliana T87 cells using in vivo N-15-isotope enrichment. *Biosci. Biotechnol. Biochem.* 69, 1331–1340.

[34] Villas-Boas, S.G., Moxley, J.F., Akesson, M., Stephanopoulos, G., and Nielsen, J. (2005) High-throughput metabolic state analysis: The missing link in integrated functional genomics of yeasts. *Biochem. J.* 388, 669–677.

[35] Boelling, C. and Fiehn, O. (2005) Metabolite profiling of Chlamydomonas reinhardtii under nutrient deprivation. *Plant Physiol.* 139, 1995–2005.

[36] Broeckling, C.D., Huhman, D.V., Farag, M.A., Smith, J.T., May, G.D., Mendes, P., Dixon, R.A., and Sumner, L.W. (2005) Metabolic profiling of Medicago truncatula cell cultures reveals the effects of biotic and abiotic elicitors on metabolism. *J. Exp. Bot.* 56, 323–336.

[37] Marin, S., Chiang, K., Bassilian, S., Lee, W.N.P., Boros, L.G., Fernandez-Novell, J.M., Centelles, J.J., Medrano, A., Rodriguez-Gil, J.E., and Cascante, M. (2003) Metabolic strategy of boar spermatozoa revealed by a metabolomic characterization. *FEBS Lett.* 554, 342–346.

[38] Watkins, S.M., Zhu, XN., and Zeisel, S.H. (2003) Phosphatidylethanolamine-N-methyltransferaactivity and dietary choline regulate liver-plasma lipid flux and essential fatty acid metabolism in mice. *J. Nutrition* 133, 3386–3391.

[39] Mesnard, F. and Ratcliffe, R.G. (2005) NMR analysis of plant nitrogen metabolism. *Photosynth. Res.* 83, 163–180.

[40] Maskow, T. and Kleinsteuber, S. (2004) Carbon and energy fluxes during haloadaptation of Halomonas sp EF11 growing on phenol. *Extremophiles* 8, 133–141.

[41] Zamboni, N. and Sauer, U. (2004) Model-independent fluxome profiling from H-2 and C-13 experiments for metabolic variant discrimination. *Genome Biology* 5, 12, Art. No. R99.

[42] Fernie, A.R., Geigenberger, P., and Stitt, M. (2005) Flux an important, but neglected, component of functional genomics. *Curr. Opin. Plant Biol.* 8, 174–182.

[43] Giuliani, A., Zbilut, J.P., Conti, F., Manetti, C., and Miccheli, A. (2004) Invariant features of metabolic networks: A data analysis application on scaling properties of biochemical pathways. *Physica A – Stat. Mech. Appl.* 337, 157–170.

[44] Likelynet – software for exploring linear relationships in multidimensional data sets. URL cited April 30, 2006 [www.likelynet.com].

[45] Kose, F., Weckwerth, W., Linke, T., and Fiehn, O. (2001) Visualizing plant metabolomic correlation networks using clique-metabolite matrices. *Bioinformatics* 17, 1198–1208.

[46] de la Fuente, A., Bing, N., Hoeschele, I., and Mendes, P. Discovery of meaningful associations in genomic data using partial correlation coefficients. *Bioinformatics* 2004, 20, 3565–3574.

[47] Steuer, R., Kurth, J., Fiehn, O., and Weckwerth, W. (2003) Observing and interpreting correlations in metabolic networks. *Bioinformatics* 19, 1019–1026.

[48] Camacho, D., de la Fuente, A., and Mendes, P. (2005). The origin of correlations in metabolomics data. *Metabolomics* 1, 53–63.

[49] Fiehn, O. (2003) Metabolic networks of Cucurbita maxima phloem. *Phytochemistry* 62, 875–886.

[50] Weckwerth, W., Loureiro, M.E., Wenzel, K., and Fiehn, O. (2004) Metabolic networks unravel the effects of silent plant phenotypes. *Proc. Natl. Acad. Sci. USA* 101, 7809–7814.

[51] Wiback, S.J., Mahadevan, R., and Palsson, B.O. (2004) Using metabolic flux data to further constrain the metabolic solution space and predict internal flux patterns: The Escherichia coli spectrum. *Biotechnol. Bioeng.* 86, 317–331.

[52] Famili, I., Mahadevan, R., and Palsson, B.O. (2005) k-cone analysis: Determining all candidate values for kinetic parameters on a network scale. *Biophys. J.* 88, 1616–1625.

[53] Fell, D. (1997) *Understanding the Control of Metabolism*. Portland Press, London.

[54] Duarte, N.C., Palsson, B.O., and Fu, P.C. (2004) Integrated analysis of metabolic phenotypes in Saccharomyces cerevisiae. *BMC Genomics* 5: Art. No. 63.

[55] Dandekar, T., Moldenhauer, F., Bulik, S., Bertram, H., and Schuster, S. (2003) A method for classifying metabolites in topological pathway analyses based on minimization of pathway number. *Biosystems* 70, 255–270.

The Handbook of Metabonomics and Metabolomics
John C. Lindon, Jeremy K. Nicholson and Elaine Holmes (Editors)

Chapter 3

NMR Spectroscopy Techniques for Application to Metabonomics

Alfred Ross, Goetz Schlotterbeck, Frank Dieterle, and Hans Senn

Pharma Research, F. Hoffman La-Roche AG, Basel Switzerland

3.1. Introduction

Since its discovery in the 1940s, **N**uclear **M**agnetic **R**esonance (NMR) Spectroscopy has become a powerful, interdisciplinary method. A brief historical review would reveal as many as nine Nobel Prize laureates since the time when Isador I. Rabi developed resonance methods for recording the magnetic properties of atomic nuclei and was awarded the Nobel prize in physics (1944). The NMR phenomenon was soon later demonstrated for protons. After years of continuous development, Fourier Transform (FT) NMR entered the scene in the 1960s, followed by the evaluation of non-invasive preclinical and medical imaging in the early 1980s. At the same time, three-dimensional (3D) structure elucidation of proteins at atomic resolution in aqueous environment was developed [1]. This breakthrough was made possible also by the availability of superconducting materials and stable and robust electronic equipments.

NMR has since been used in an almost unlimited variety of ways in physics, chemistry and biology. For the investigation of biological systems it is convenient to distinguish between three types of applications [2]: (1) to study structure and function of macromolecules, (2) to study metabolism, and (3) to obtain *in vivo* images of anatomical structure and functional (physiological) states.

The use of ^{1}H NMR for metabolic studies was described as early as 1977 when it was shown that ^{1}H signals could be observed from a range of compounds in a suspension of red blood cells, including lactate, pyruvate, alanine and creatine [3]. A great deal of metabolic information can be derived from such metabolic studies and it was soon recognized that ^{1}H NMR of body fluids has a considerable role to play in areas of pharmacology, toxicology and the investigations of inborn errors of metabolism [4–6].

Since these early applications, ^{1}H NMR of biofluids and cell extracts including high-resolution magic angle spinning (HR-MAS) NMR of soft tissues have been successfully applied to investigate numerous diseases and toxic processes [7–9].

This chapter is meant to give an introduction and overview on NMR under the view point of its application in metabolic profiling (metabonomics). It is not meant to review all the different applications of NMR in this field but rather to concentrate on the essentials and prerequisites of its successful implementation as a metabolite profiling tool. To this end, specific NMR hardware requirements for metabolite profiling are reviewed and compared as well as automation and robotics control of the work flow, which are crucial for achieving high throughput and consistent quality of results. It is then shown that sample preparation and handling is of utmost importance for meaningful comparison of hundreds of samples and thousands of spectral variables in a metabolite profiling study. Therefore all aspects known to us which may change the sample property of biofluids are included and extensively discussed. The handling and preparation of urine samples is especially demanding in order to control and handle the wide variability of conditions that influence the ^{1}H NMR spectrum. These problems are less severe with blood serum, plasma or cerebrospinal fluid (CSF) where homeostasis ensures a much narrower range of sample variability. In the section on NMR Experiments and Processing, the information which is necessary to obtain high-quality one-dimensional (1D) and two-dimensional (2D) NMR spectra of biofluids is summarized. Finally, the state-of-the-art of data pre-processing, which is the intermediate step between recording NMR raw spectra and applying uni- or multivariate data analysis and modeling methods, is discussed in great detail. In metabonomics, this is an important step to make subsequent analysis and modeling easier, more robust and more accurate.

3.2. Principles of NMR

The theory of NMR is highly developed and the dynamics of nuclear spin-systems fully understood [10, 11]. A first-principle quantum mechanical description is available, but beyond the scope of this introduction. What is missing to complete the theoretical picture is a reliable and accurate prediction of the chemical shift of spins. Therefore, recourse to spectral databases is needed if a chemical interpretation of metabonomics data is envisaged.

We present a phenomenological description of magnetic resonance needed for the understanding of the metabonomic literature. It has to be stressed, however, that nearly any aspect of modern NMR spectroscopy is of importance for the acquisition of high-quality data needed for a reliable biological interpretation of results. After an introduction into the principles of NMR (magnetization, chemical shift, relaxation, J-coupling) we touch upon more advanced concepts involving chemical exchange, 2D and heteronuclear experiments. We will see that all effects are of relevance for metabonomic research. We do not seek for a complete review of literature of the topic. If possible we provide examples from our own work.

3.2.1. Magnetism

Magnetism and spin: Matter is composed of molecules built of atomic nuclei with a characteristic proton/neutron composition. Nuclei are surrounded by electronic “clouds”. Besides charge and mass, a further property of protons, neutrons and electrons is an angular momentum $\vec{I}$ known as spin. Because of the magnitude of this angular momentum $\vec{I} \cdot \vec{I} = \frac{h}{2\pi} \cdot \frac{3}{4}$ the aforementioned particles are known as spin-1/2 particles.[1] The total spin of a nucleus depends on its nucleon content. Some nuclei carry a total spin (^{1}H, ^{2}H, ^{13}C, ^{15}N, ^{19}F, ^{31}P, ...)[2] resulting in a magnetic moment $\vec{M} = \gamma_X \cdot \vec{I}$ of different magnitude and sign, others (e.g. ^{12}C, ^{14}C, ^{16}O, ...) do not; γ_X is the gyromagnetic ratio of the atomic nucleus X.

In NMR-spectroscopy samples of liquid or solid material[3] are exposed to an external static and homogeneous magnetic field referred to as B_0. The direction of this field is usually defined along z. The magnetic moments in the sample align along B_0 according to a Boltzmann distribution. In contrast to classical physics, quantum mechanics shows that magnetic moments due to spin-1/2 particles can only align parallel (called up) or anti-parallel (called down) with respect to this external field, these two states have a difference of energy given by $\Delta E = \gamma_X \cdot B_0$. Due to statistic averaging, the magnetic moment of any macroscopic sample can be treated in many respects like a classical macroscopic magnetic moment $\vec{M}$. In accordance with the definitions above, the thermal equilibrium magnetization of a sample aligned along z is called M_0. For the detection of the NMR signal, M_0 is flipped orthogonal to B_0 by use of a high-frequency magnetic field B_1 applied for a defined time period. B_1 is also applied orthogonal to z (see 3.2). This is called a B_1-pulse. Classical physics shows that $\vec{M}$, now aligned without loss of generality along the x-direction, will precess with a (resonance) frequency given by

$$f_0 = \frac{\gamma_X}{2\pi} \cdot B_0 \tag{3.1}$$

[1] h is known as Planck's constant.

[2] H is a spin-1 particle. To date there is no application for ^{2}H and other “higher spin” nuclei reported for metabonomic research.

[3] There are metabonomic applications for tissues. Tissues are not real solids, but more gel-like and soft structures where the spin-physics can be transformed to liquid-like behavior by use of the so-called HR-MAS technique.

This is the reason why NMR spectrometers are normally classified in frequencies. The magnetization of a sample of hydrogen atoms experiencing a magnetic field of 14.1 T will, for example, rotate with 600 MHz. Such a spectrometer is called a 600 MHz apparatus. Other nuclei will rotate in the same field with another frequency. A ^{13}C spin will precess in the same 600 MHz magnet at about 150 MHz.

Detection of the signal: The sample is positioned inside a detection coil (see Figure 3.1). According to the law of inductivity the precessing magnetization will induce a voltage U_{ind} modulated with f_0. The amplitude of this voltage is directly proportional to $\vec{M}$ and thus with the number of spins rotating with f_0 located inside the observe volume of the apparatus. In the NMR literature the signal detected is called a **F**ree **I**nduction **D**ecay or FID.

Due to diamagnetism of the electron clouds of the molecule the local magnetic field experienced by a spin is changed by a small amount compared to B_0. Thus the frequency f_A measured for spin A directly reports the electronic/chemical neighborhood of the nucleus observed. The value $\frac{(f_A - f_0)}{f_0} \cdot 10^6$, measured in parts per million (ppm) in respect to B_0, is called the chemical shift of this spin. This definition is independent of B_0 – thus data taken at different field strength can be compared easily. The chemical shift of different spin species covers different ranges: ^{1}H nuclei resonate in most cases within a width of 15 ppm; ^{13}C nuclei cover more than

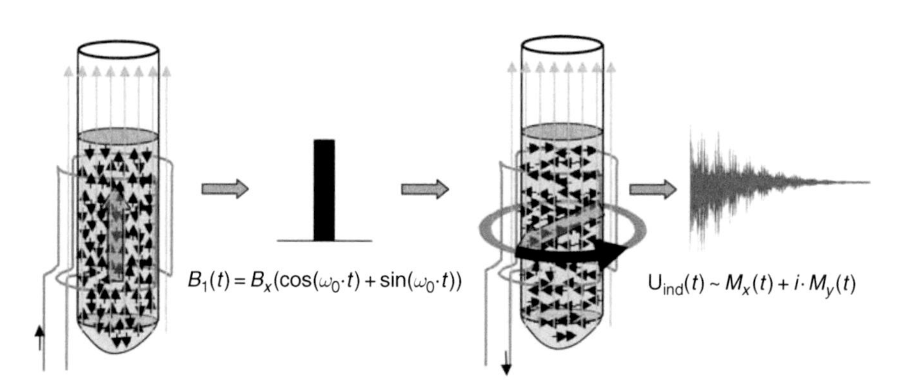

Figure 3.1. Detection of an NMR Signal: (left) a sample with many spins (>10^{12}) is placed inside a strong external magnetic field B_0 (orange) oriented along the z-direction. The sample is enclosed in a detection coil (shown in green). At thermal equilibrium about one spin in every 10,000 contributes to a macroscopic magnetization M_0 shown in transparent red. After short application of a high-frequency B_1 field (90° pulse: black bar and equation at the left panel) the magnetization is aligned along the x-axis of the rotating B_1 field, and starts to precess around B_0. The voltage, U_i, induced in the detection coil (dark-blue and equation at the right panel) can be described by the equation shown on the bottom of the right panel.

200 ppm. For the definition of f_0 normally a reference compound is added to the sample.[4]

Relaxation [12]: Each spin creates a small dipolar magnetic field spread over space. The magnitude of this additional magnetic field experienced by another spin species in the neighborhood (e.g. the same molecule) is dependent on the angle and the length of the vector connecting both spins with respect to B_0. As Brownian motion is active in liquids this angle is time-dependent. Thus besides a huge constant magnetic field each spin experiences an individual small time-dependent "jittery" magnetic field. As a consequence the frequency of rotation is time-dependent – each molecule experiences its own time-dependence. The signal measured is an ensemble average. This time-dependence will lead to an individual precession trajectory of the magnetization for any molecule of the sample reflecting itself in a "dephasing" of the measured macroscopic magnetization $\vec{M}$. The same happens with the induced voltage U_{ind}. The magnitude of the signal will decay over time. This phenomenon is known as dipolar relaxation. There are other sources of relaxation like quadrupolar relaxation. The effect of all relaxation mechanisms active for a certain spin species manifest themselves as an exponential decay of the signal characterized by a constant T_2 known as transverse relaxation.

If an unmagnetized sample is placed into a magnet the built up of M_0 will take a certain time. The time-constant necessary to approach this thermal equilibrium is known as longitudinal relaxation T_1. The physical mechanisms (e.g. dipolar interaction of spins) behind this phenomenon are essentially the same as those for T_2 but they are active in general with different rate constants here. For small molecules as a rule of thumb $T_1 \approx T_2$ is found. For macromolecules $T_1 \gg T_2$ is valid.[5] Stokes law $\tau_C \sim R_H^3$ $T_2^{-1} \sim \tau_C$ (τ_C, R_H being the rotational correlation time and the hydrodynamic radius of the molecule respectively) ensures that NMR signals of large molecules decay much faster than those of small molecules. This has been exploited in metabonomics for the suppression of protein-related signals [13] (see Section 3.5.1.2). For small molecular metabolites mostly investigated in metabonomics, the relaxation times are in the range 1–3 s. The magnitude of the "jittery field" and thus dipolar relaxation depends on the gyromagnetic ratios of the spins involved. Nuclei with a small gyromagnetic ratio (e.g. ^{13}C) relax more slowly and longer relaxation times between repetition of pulses is needed for their NMR detection (see Section 3.2.1.1).

---

[4] In metabonomics, the samples are usually aqueous solutions (e.g. urine) to which a small amount of partially deuterated TSP (**t**rimethyl**s**ilyl**p**ropionic acid-d_4) is added for reference purposes. It must be noted, however, that TSP does bind to proteins (e.g. human serum albumin, HSA). This has to be taken into account for biofluids like blood plasma or serum, as protein binding influences the chemical shift of TSP (see Section 3.2.2.1). The addition of TSP also allows the monitoring of the quality of the B_0 homogeneity across the sample, as the line width of this signal is expected to be identical for all samples as long as HSA binding can be excluded.

[5] T_1 and T_2 are dependent slightly on B_0, but this can be neglected for low-molecular-weight molecules investigated in metabonomics analysis.

Bloch equation: All what has been described above can be summarized phenomenological in the so-called Bloch equation: here the time evolution of any type of magnetization $\vec{M}$ can be described if direct spin–spin interactions (J-coupling, NOE, see Section 2.1.3) are neglected:

$$\frac{\partial}{\partial t}\begin{pmatrix} M_x \\ M_y \\ M_z \end{pmatrix} = \gamma_X \begin{pmatrix} M_x \\ M_y \\ M_z \end{pmatrix} \times \begin{pmatrix} B_x \\ B_y \\ B_z \end{pmatrix} - \frac{1}{T_2}\begin{pmatrix} M_x \\ M_y \\ 0 \end{pmatrix} - \frac{1}{T_1}\begin{pmatrix} 0 \\ 0 \\ M_z - M_0 \end{pmatrix} \quad (3.2)^6$$

As a consequence of Equation 3.2, z-magnetization rotates in the z-y plane by application of a x-pulse. Magnetization in the x-y plane will evolve as a superposition of cosine-modulated and T_2-damped contributions under the presence of B_0. Irradiation, long on the timescale of T_1 and T_2, will lead to a substantial reduction of the NMR signals. This phenomenon is known as saturation and does find its application for the purpose of solvent suppression in metabonomics (see Section 3.5.1.1).

The NMR signal: The amplitudes $M^i_{z,0}$ of the individual contributions originating from one molecular species in a mixture are given by the number of spins in a molecule which resonate at a certain frequency f_i multiplied by the concentration c_j of the respective molecular species.[7] Thus the signal detected can be described as:

$$\begin{aligned} \vec{M}_{\text{molecule}}(t) &\sim \sum_i M^i_{z,0} \cdot \left[\cos(2\pi \cdot f_i \cdot t) \cdot \vec{e}_x + \sin(2\pi \cdot f_i \cdot t) \cdot \vec{e}_y\right] \cdot e^{\frac{t}{T_2}} \\ \vec{M}_{\text{total}}(t) &= \sum_j c_j \cdot \vec{M}^j_{molecule}(t) \end{aligned} \quad (3.3)$$

Here the summation over "i" is done over the spin-types of a molecular species. The summation over j is performed over the different molecular species composing the sample.

Fourier transformation (see Section 3.2.1.2) of Equation 3.3 leads to the NMR spectrum of the sample (see Figure 3.1). The spectrum of a molecule is given

[6] All quantities in Equation 3.2 except relaxation times and the gyromagnetic ratio are in general time-dependent. Only time-independent magnetic fields are analytically solvable.

[7] $M^i_{z,0}$ is equal to the thermodynamic equilibrium value only, if the sample is in full equilibrium for each repetition of the experiment. (This is needed to achieve high signal-to-noise (see below).) Otherwise the simple assumption that a proton signal emerging from a methyl-group is three times higher compared to that of a methine group will break down. Equilibration is achieved to a good approximation by a waiting time between repetitions of the experiment in the order of 3 s. Experience shows that in this regime substantial falsification of amplitudes is prevented.

by resonance lines with defined frequencies and amplitudes.[8] The spectrum of a mixture is given by the concentration-weighted summation of the spectra of the molecular species constituting the mixture. This feature of perfect linearity makes the detection of NMR signals highly attractive for the quantitative determination of the concentrations of components of a mixture.

3.2.1.1. Noise, sensitivity, limit of detection

The detected signal is proportional to the total magnetic moment of a sample (law of induction). The latter depends linearly on the number of spins N (atomic magnets) in the sample and the difference between spins aligned in parallel with the B_0-field and those aligned anti-parallel. According to the high-temperature approximation of Boltzmann's law this difference increases linearly with B_0. There is an additional B_0 dependence in the detection process itself (the energy provided by any spin flip). Other important factors influencing the sensitivity of the measurement are the gyromagnetic ratio γ_x, the sensitivity of the detector and the noise created by the sample, the latter two summarized in S_D. The sensitivity defined as the signal-to-noise ratio (S/N) of a single repetition of the experiment is given by [14]:

$$\text{Sensitivity}_{\text{scan}} \sim \gamma_X^3 \cdot N \cdot B_o^2 \cdot S_D \qquad (3.4)^9$$

There exist two approaches to increase S_D (see Sections 3.1 and 3.3): (a) Noise in electric circuits is caused by thermal motion of electrons in wires. Consequently this noise can be reduced by cooling the wires of the detection coil and the preamplifier used to cryogenic temperatures of a few K. Such probeheads are called cryogenic or cold-probes [15, 16] (b) S_D can be increased by reducing the size of the detection coil [17, 18] – naively it is clear that shorter wires create less noise. It must be noted that for biological samples used in metabonomics the concentration of the sample is given by biology. Here the application of miniaturized detection devices is only recommended for volume-limited samples (e.g. CSF taken from a mouse).[10]

A sample of high conductivity (salt) reduces the sensitivity of the detection by induced eddy-currents. This situation is often encountered for biological samples having high ionic strength. The effect is more pronounced for cryogenic probeheads [19].

---

[8] Due to rotational averaging the chemical shifts of fast rotating structural moieties (e.g. methyl groups) will appear at the same frequency – they are called "degenerate". The same holds true for spins in fast chemical exchange between two sites (see Section 3.2.2.1).

[9] A more thorough calculation shows proportionality in $B_0^{7/4}$ and $\gamma_X^{11/4}$.

[10] This is true as the gain achieved for S_D by miniaturization is normally more than counterbalanced by the reduction of N if a smaller sample volume is used.

From Equation 3.4 it is also clear that a higher static magnetic field does increase the sensitivity of the NMR detection. This is one reason why magnets with higher B_0 field are attractive.

If the same experiment is repeated by adding the detected signal, it is important to realize that the signal sums coherently (proportional to the number of repetitions called scans (NS) or transients). Statistic averaging leads to an increase of noise linear in $\sqrt{\text{NS}}$. Thus a needed increase in signal-to-noise by two enforces a fourfold increase in experimental time.

Another important factor influencing the outcome is the rate of repetition of scans. A slow rate, that is long inter-scan delays, allows for full relaxation of the magnetization. This ensures that the NMR signal is perfectly linear to the concentration of the analytes. The trade-off in this regime is a long experimental time for a given number of repetitions. Analysis by use of Equation 3.2 shows that highest sensitivity under a loss of perfect linearity is obtained if the repetition rate is 1.26^*T_1. The compromise used in metabonomics is given by a repetition of scans every 3 to 4 sec.

3.2.1.2. Principles of spectral processing [20]

Modern NMR spectrometers detect magnetization along the x- and the y-direction over time. This process can be visualized as if the measurement of induced voltages were done with two detection coils arranged along the x- and y-direction respectively.[11] For any time-point, two values M_x and M_y are recorded – this technique is called phase-sensitive detection.[12] For the calculation of the NMR spectrum, a complex valued time domain signal, constructed as $M_x(t)+iM_y(t)$ (with $i:=\sqrt{-1}$), is used as input for a complex valued FT. The mathematics of FT tells that the resulting real part of this calculation performed on a time domain signal expressed in Equation 3.3 will show individual lines with positions given by the chemical shifts. The width of the lines is characterized by $(\pi^*T_2)^{-1}$ and shows a so-called absorptive lorenzian line-shape. This holds true only if all magnetization flipped in the x-y plane is perfectly aligned along the x-axis at the beginning of the detection. Due to limited strength of the B_1 field in use[13] and due to limitations of spectrometer electronics this does not hold true in practice. Magnetization is oriented under a small

[11] In reality the detection is done with a single coil using two-channel phase-shifted high-frequency mixing.

[12] Not only the magnitude of the transversal magnetization is measured, but also the orientation in the x-y plane, normally expressed as an angle (phase) with respect to the x-axis.

[13] Typically a B_1 pulse needed for flipping the magnetization by 90° lasts less than 7.5 μs for modern 100 W amplifiers. Thus a 360° rotation with the same pulse will last 30 μs. The reciprocal of this value defines the B_1 field strength in Hz. In this example, we have a B_1 field of 33 kHz. The excitation bandwidth of a pulse given in Hz is approximately 2^*B_1. A pulse of 33 kHz excites at a 600 MHz spectrometer, a bandwidth of 60 kHz or 100 ppm for protons, easily covering the typical chemical shift range found in biofluids. The situation changes if multiple pulses are applied.

initial and frequency-dependent phase with respect to the x-axis. As a consequence, the real part of the FT will also contain a frequency-dependent contribution of the imaginary part of a perfect signal. This imperfection can be removed after FT by a process called "phasing". Another artifact found in NMR spectra is a distortion of the baseline due to imperfections and non-linearities of the electronic detection process. This baseline distortion can be corrected by the subtraction of a tailored polynomial from the raw spectrum obtained.

Unfortunately, both processes, phasing and baseline correction, can be performed automatically only with limited reliability. Visual inspection of processed spectra for artifacts is of importance.

The signal-to-noise and line-shape of spectra can be tailored according to needs if the time-domain signal is multiplied with a so-called window function [21] prior to FT. In metabonomics, the window function used for this purpose is typically the so-called exponential-window given by $e^{-t \cdot \pi \cdot \mathrm{LB}}$ employed with a so-called line-broadening LB of 1 Hz (see Figure 3.2). This offers an acceptable compromise between signal-to-noise and spectral resolution. If only a short acquisition time (e.g. in 2D applications, see Section 3.5.2.) is allowed, a shifted sine or squared sine-shaped window ensures the absence of processing artifacts.

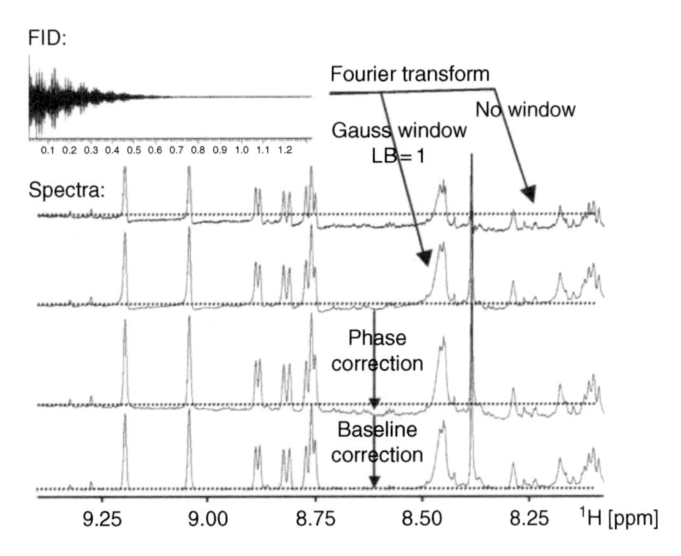

Figure 3.2. Processing of a recorded time-domain signal (top). The FT of the signal is shown second trace: a low signal-to-noise ratio, baseline-offset and dispersive contribution of the lines are seen here. Application of an exponential window function of 1 Hz LB prior to FT results in the third trace. Here signal-to-noise is substantially improved but the other artefacts remain. A spectrum suited for metabonomics interpretation is only obtained after phase correction (fourth trace) followed by baseline correction (fifth trace, bottom). The last two steps have to be inspected visually.

It is of importance that the integral of an NMR signal in the spectrum (not its amplitude) is linear in the number of NMR active nuclei present in a molecular moiety. The concentration of this molecule can be determined from this integral if the signal was assigned to one or a group of nuclei in the molecule. Comparison of integral values with a reference compound or an electronically created standard [22] allows an absolute value (e.g. mg/ml) determination of concentration. More details for spectral processing can be found in Section 3.5 along with the detailed description of NMR experiments.

3.2.1.3. Spin–Spin interaction

The description of NMR based on the Bloch Equation breaks down if spin–spin interaction beyond T_1 and T_2 relaxation is occuring. The detailed description of the physics behind these phenomena is beyond the scope of this introduction. Only a qualitative description of the phenomena will be presented.

J-coupling: Mediated through the spin-orbit coupling of the nuclear spins with the electronic system there is an interaction between two active A and B spins, if both spins share a common electronic orbital. This is often (but not necessarily) fulfilled if both spins are separated by less than 5 chemical bonds in a molecule. As a consequence, the magnetic field seen by spin A is dependent on the orientation of spin B and vice versa. Thus the signals of spins A and B appear as a doublet of 1:1 intensity ratio. The difference between both components (expressed in Hz) is a direct measure of the electronic overlap and thus of the nature and topology of the chemical bond(s) linking both spins.[14] This difference is B_0 independent and is called scalar coupling (constant) J_{AB}. Besides homonuclear scalar couplings (e.g. coupling from ^1H to another ^1H) there also exist heteronuclear couplings (e.g. ^1H to ^{13}C) [23].[15] Normally the number of bonds separating the two spins is provided as additional information: $^3J_{HC}$ refers to a three-bond coupling measured for a ^1H nucleus separated from a ^{13}C nucleus by three bonds. If a spin has a scalar coupling to more than one spin, complicated but highly informative coupling patterns are seen in the spectrum. These patterns help to classify spin systems: an AX spin system consists of two spins sharing one coupling and in the NMR spectrum two doublet signals (1:1 ratio) of equal intensity are observed. An AX_2 spin system has three spins: spin A corresponds to one nucleus (e.g. CH), spin X is given by

[14] For a given type of bond (e.g. sp^3 hybridization) the size of the coupling constant is related to the dihedral angle of the bond topology via the so-called Karplus Equation.

[15] Homo and heteronuclear J-couplings (multiplets) can be removed from the spectrum by decoupling. With this technique one or more spins of the spin system are selectively irradiated with a B_1 field (strong compared to J_{AB}); for example at the resonance frequency of spin B. This field will lead to a fast exchange of the up and down state of this spin. Thus the coupled spin A in A-B only 'sees' an averaged state of spin B leading to a collapse of the spin A signal to a singlet.

two nuclei (e.g. CH_2) having identical chemical shifts and coupling topology to spin A. The X spins are called magnetically equivalent and no coupling between the two X spins is seen in the NMR spectra. The spectrum of the AX_2 system is given by one triplet signal (1:2:1 ratio of the three components) with intensity 1 for spin A, and one doublet signal with intensity 2 for spins X_2. More complicated spin-topologies can occur in molecules. It must be noted that a simple interpretation of spin-systems based on their multiplet structure is possible only if the coupling constants involved are much smaller than the differences of their chemical shifts (both in Hz). This regime is called "weak coupling". If this assumption is not valid, the spin system is strongly coupled and only a numerical fit can resolve the individual coupling constants. It is clear that a strong coupling effect at a low B_0 field strength can become weak coupling at higher field strength.[16] An important example of a strongly coupled spin-system in metabonomics is citrate (see Figure 3.10).

Complex spin systems can be resolved by use of 2D NMR methods if individual components are overlapping in the 1D spectrum (see Sections 3.2.1.4 and 3.5.2.1).

Nuclear Overhauser Effect (NOE): A second spin–spin interaction of importance for the richness in information of NMR spectra is the so-called NOE. It rests on the motional average of through-space dipolar interaction between spins.[17] In contrast to J-coupling, NOE interacting spins do not have to share a common electronic orbital – the strength of this effect scales with the spatial distance between the interacting spins. If the NMR signal of a spin A is saturated (see above) by irradiation with a B_1-field, this distortion of signal amplitude will progress to other spins B_i having dipolar interactions with the saturated spin. As a consequence, the amplitudes of the signal detected for B_i spins will also change. This is due to double and zero quantum transitions in the interacting spin system. By bookkeeping of all populations of the different energy levels,[18] a phenomenological equation known as the Solomon equation [24] is derived. The Solomon equation describes the evolution of longitudinal non-equilibrium magnetization $\Delta M_z^{A/B}(t) := M_z^{A/B}(t) - M_0^{A/B}$ of two spins with dipolar interaction:

$$\frac{\partial}{\partial t}\begin{pmatrix}\Delta M_z^{A}\\ \Delta M_z^{B}\end{pmatrix} = -\begin{pmatrix}1/T_1^{A} & \sigma\\ \sigma & 1/T_1^{B}\end{pmatrix}\cdot\begin{pmatrix}\Delta M_z^{A}\\ \Delta M_z^{B}\end{pmatrix} \tag{3.5}$$

[16] On a 400 MHz spectrometer a chemical shift difference of 0.1 ppm corresponds to 40 Hz. This difference doubles at 800 MHz.

[17] If motional averaging is restricted, for example in tissues due to high viscosity, the dipolar interaction manifests itself in line-broadening of signals making their interpretation impossible. This effect can be reduced by application of the HR-MAS technique here the sample is rotated at high frequencies of several kHz. This introduces the motional averaging needed to narrow lines by reducing the strength of the dipole-dipole interaction. Applications of this technique are discussed in Section 2.1.4. A generalization is also possible for powder or crystalline samples.

[18] A two spin AB system has four energy levels, given by the four states up_A-up_B, up_A-$down_B$, $down_A$-up_B, $down_A$-$down_B$.

Details of the parameters T_1 and σ depend on the gyromagnetic moments and the rotational tumbling of the molecule involved. In general the effect is more pronounced for high-molecular-weight compounds. If the spins of interest are both protons, the saturation of spin A will result in a reduction of the signal observed for spin B. The NOE is used for the determination of the 3D fold of molecules, as the effect is proportional to r_{AB}^{-6} with r_{AB} being the distance between both spins. Formally an equation identical to Equation 3.5 is obtained if spin A is in chemical exchange with spin B. This is, for example, the case for protons of urea in exchange with those of water. Thus in full analogy with the NOE, the saturation of water (used for solvent suppression) will influence the amplitude of any proton (–OH, –NH, $-NH_2$) in chemical exchange with water.

3.2.1.4. Two-Dimensional methods

In 1D NMR the molecule is characterized by a spectral amplitude plotted versus a single frequency axis. This method allows the determination of the composition of a sample only if there is no severe overlap of signals occurring in the spectrum.

In the mid-1970s, a method called 2D NMR was developed [25]. Here a series of 1D NMR spectra is acquired under systematic variation of an experimental parameter of a sequence of B_1 pulses (see Figure 3.3). In most cases the parameter varied is an inter-pulse delay between two pulses (or sequences thereof). During this variable interpulse delay, named t_1, a magnetic coherence state[19] which was prepared at the beginning of the pulse sequence is evolving. Step-wise incrementation of the delay allows the monitoring of the evolution of the prepared spin state. If this evolution is, for example, a chemical shift precession, the chemical shift of the spins involved in this prepared state can be measured. Thereby, the evolving spin state is transferred at the end of t_1 to magnetization resting on a 2nd spin followed by the detection of the modulated NMR signal evolving under t_2. The 2D spectrum obtained after FT with respect to the step-wise increased t_1 value and the t_2 acquisition time will result in a 2D map.[20] Here the spectral amplitude is plotted as a function of two frequency dimensions. A cross-signal in this map will indicate a pair of spins between which a transfer of magnetization has occurred during the pulse sequence. Details of the architecture of the sequence determine which interaction and which types of spins are selected. If both spins are of the same type (e.g. hydrogen atoms) the experiment

[19] Coherences are states of the spin-system which cannot be described in classical terms. The quantum mechanical picture is needed here [10]. Important examples of coherences are the so-called anti-phase state used for *J*-coupling-based transfer of magnetization between spins (see Section 3.5.2.1) and multiple-quantum coherences (see Section 3.5.2.2) evolving with the sum or the difference of the chemical shift of the spins entangled in the state.

[20] Phase sensitive detection is realized along t_1 by always acquiring pairs of experiments for identical settings of t_1. In these pairs, x versus y direction coherence is the read out at the end of the evolution period.

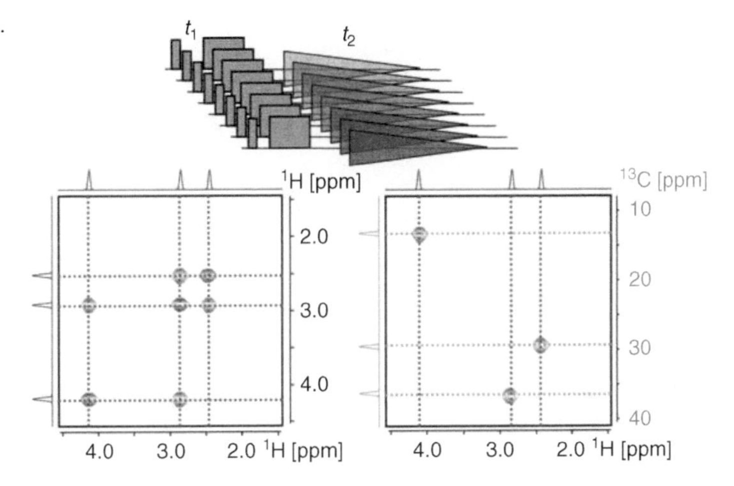

Figure 3.3. Principle of 2D NMR spectroscopy: A set of 1D spectra is acquired under incrementation of a t_1 evolution delay (top panel). FT is performed with respect to t_1 and t_2. A homonuclear shift-correlation spectrum is shown (left bottom). Diagonal signals (in blue) code for magnetization which was initially on a Spin A and was not transferred to a 2nd spin along the sequence. Thus both chemical shifts are identical. The red off-diagonal cross-signals code for magnetization which was transferred by a selected interaction mechanism to a 2nd spin B. It is clear that a heteronuclear shift correlation spectrum (bottom right) only contains non-diagonal cross signals. Here proton magnetization was transferred to carbons (evolution along t_1) and back to protons for detection along t_2. Here axes show the chemical shift of ^{13}C and ^{1}H respectively.

is called homonuclear otherwise it is termed heteronuclear.[21] The 2nd dimension can also code a physical relaxation or a diffusion parameter measured for a certain spin.

Popular examples of 2D NMR which are of some significance for metabolic profiling are discussed in Section 3.5.2. For all 2D methods the detection of the NMR signal has to be performed several hundred times under incrementation of the t_1 duration. Therefore the time needed for such experiments normally comprises several hours up to days. It is clear that this type of experiments normally cannot be done on large sample arrays. It is noteworthy that all that has been said for 1D NMR methods with respect to sensitivity, linearity of detection, methodology of spectral processing can be extended to the 2nd dimension here. Thus relative or absolute quantification of metabolites for comparison of different samples can also be achieved on signals of 2D NMR spectra. Absolute quantification (in mg/ml) is

[21] It is of importance to realize that the sensitivity of the experiment is given by the sensitivity of the nucleus detected during t_2 multiplied by the lowest natural abundance of all nuclei involved in the transfer. This offers the possibility that the chemical shift of carbon nuclei can be detected with a much higher sensitivity compared to the direct measurement of a ^{13}C spectrum, namely via indirect ^{1}H detection of its ^{13}C-coupled ^{1}H nuclei. It has to be kept in mind that the sensitivity of this experiment is still hampered by the low natural abundance of ^{13}C – only 1% of the sample will contribute to the measurement.

more difficult, as the transfer steps of the pulse-sequences involved may reduce the magnitude of the signals because of dependence on many molecular parameters (T_2, T_1, $J_{AB} \ldots$).

Multivariate analysis of 2D NMR data has been described in the literature [26]. Interestingly pseudo 2D spectral information was achieved in metabonomics recently as a spectral map of the correlation coefficient between data points calculated for 1D spectra of a large ensemble of samples of a metabonomic study. Analysis of this map allows for the identification of up- or down-regulated metabolites [27].

3.2.2. Special aspects for biological samples

3.2.2.1. Exchange between different states

The NMR parameters, such as chemical shift, T_1 or T_2, are influenced if molecules exchange between different chemical states. The molecule will have different resonance frequencies for its NMR signals in the different states which are separated by Δf [Hz]. The parameter $\tau = (2 \cdot \pi \cdot \Delta f)^{-1}$ defines the NMR timescale of exchange in seconds.[22] If the exchange time τ_{ex} is slow compared to τ, this will reflect itself in the NMR spectrum as individual lines with amplitudes reflecting the populations of the different chemical states. This situation is referred to as the slow-exchange regime. In the fast-exchange regime given by $\tau_{ex} \ll \tau$ the NMR spectrum will be a single line with a position given by the population weighted average of the chemical shifts of all chemical states contributing. In between, the situation is more complicated and can be described analytically only for the two-site exchange by the so-called McConnell equation [27, 28]. This equation allows the fit to kinetic and stochiometric parameters of the exchange process by detailed analysis of the line-shape. Another interesting case is the so-called coalescence point given by $\tau_{ex} \approx 1.1 \cdot \tau$ where the signals of the exchanging spins will just merge and can become very broad and become nearly invisible in the spectrum.

In this description different states can be represented by different molecules which are interconverting by spin exchange (e.g. protons), or different conformations (e.g. bound and free, or chair and boat forms) of the same molecule which are in dynamic exchange. Chemical exchange is, for example, seen for any signal in a molecule where protons (OH, NH, NH_2) exchange with water. The exchange can also occur between the free-state and a bound-state of a molecule (which is, e.g., seen for citrate forming a non-covalent complex with Ca^{2+} or Mg^{2+}, see Section 3.4.2.3) [30]. The referencing compound normally used in metabonomics TSP (trimethylsilylpropionic acid) has a certain binding affinity to the protein HSA (human serum albumin). As a consequence, the referencing of chemical shift (and

[22] τ is dependent on B_0.

therefore of all lines in the spectrum) will change if samples with different HSA/TSP ratios are compared. Another origin of exchange phenomena is interconversion between different conformations of the same molecule. In this context it is important to realize that the population of a protonated species of a molecule is shifted towards the deprotonated state if the pH of the sample is increased. Thus it is clear that the position of NMR signals of molecules having a titrateable group will depend on the pH of the sample (see Figure 3.10). This effect is compensated for in many applications by introduction of a buffering agent into the biological sample prior to the NMR measurement. It is also clear that any conceivable origin of exchange phenomena can be encountered in complex mixtures such as biological samples. The ideal situation to control the positions of all NMR lines of identical metabolites in biofluids across all samples of a biological study perfectly is therefore approached only asymptotically. Our approach to improve on this topic is discussed in Section 3.4.2.

3.2.2.2. Diffusion

It has been described in Section 3.2.1 that the tumbling of the molecules in mixtures can be used to edit spectra with respect to molecular size. If a so-called T_2-filtration is applied, only NMR signals of low-molecular-weight molecules will be seen in the spectrum (3.5.1.2). Another parameter sensitive to molecular size is translational diffusion characterized by the diffusion constant D_t. In a properly tailored NMR sequence, small molecules with fast Brownian motion can be filtered out of the spectrum [31]. Only macromolecules will be seen in the final data (see Figure 3.18). By systematic variation of a gradient, pulse mixtures can be characterized based on their propensity of different diffusion timescales. This is, for example, highly attractive for the classification of blood samples with respect to different sizes of lipoprotein fractions for which LDL and HDL values reported by classical clinical chemistry are only the most prominent species in a distribution of 14 differently sized fractions with different D_t [32, 33].

3.3. Hardware requirements and automation

The specific demands on the NMR technology and concomitantly on the NMR hardware have been changing over time, although the basic physical principles remain the same. Structural elucidation of large biomolecular structures require multi-dimensional experiments with long data acquisition times depending on complex pulse-sequences and spin physics. But rarely more than a handful specifically prepared samples, for example of a protein are involved [34]. This has changed significantly as the methodological developments has begun to include screening for ligand binding to target proteins [35], and, increasingly also for metabolic profiles of biofluids [8]. Metabonomic studies involve large sample arrays with several hundred biofluids or cell extracts. As a consequence, the NMR data acquisition needs to be fast. This requires relatively simple but highly robust 1D or simple 2D ^{1}H NMR experiments which are amenable for high sample throughput. Therefore a high level of automation is required starting from sample preparation, followed by automatic transfer of the sample in and out of the magnet, and ultimately also for data processing and evaluation including the build-up of biologically annotated spectroscopic databases. Another recent challenge in hardware development was dictated by the ever increasing need to optimally handle and measure mass- or volume-limited samples from biological sources or from combinatorial chemistry and HT-screening. This has led to the miniaturization of the probehead and the adaptation of the cryoprobe technology to reduced volume flow cells. In the following, the state of the NMR hardware for high-throughput applications in metabolite profiling is discussed.

3.3.1. Magnetic field strength

Most metabolite profiling studies reported in the literature are carried out at high magnetic field strengths ($\geq$600 MHz) to achieve a good spectral dispersion [8]. Nevertheless, the spectral overlap problem in ^{1}H NMR spectra of complex biofluids, especially of urine containing thousands of signals arising from hundreds of endogenous molecules, remains severe, and the spectra are comparable in complexity with spectra of large proteins, [1, 36]. In principle the application of higher magnetic fields (700–900 MHz) would be advantageous for resolving more individual metabolite signals. However, to control the data resolution across a set of many samples in a practical metabonomics study is a challenging problem. This is mainly due to the observed variation of chemical shifts between corresponding metabolite signals as a consequence of varying biofluid compositions, especially of urine samples (see Section 3.4.2). Thus it is often mandatory to reduce artificially the originally obtained spectral resolution of 16 or 32 k data points to a much smaller number of bins in order to gain control over the spectral variable for subsequent multivariate analysis (see Section 3.6). To make full use of the enormous dispersion power of very high field spectrometer, the data points of identical structures across all the biofluid samples of a metabolite profiling study must be related and aligned in one way or another. This remains a challenge for sample preparation (see Section 3.4.2) and post-measurement alignment methods [37–39].

3.3.2. Spectrometer console

The electronic hardware for metabonomics studies should ideally include two independent high frequency channels with full phase and pulse shape control for ^{1}H (proton) and for ^{13}C (carbon) excitation and acquisition. In addition, a gradient amplifier unit for generation of pulsed field gradients (z-gradient pulses) is mandatory in almost any pulse sequence used today. Pulsed field gradients are applied to select the magnetic spin state(s) (coherences) of interest or to destroy unwanted magnetization without the need of lengthy phase cycling. Pulse field gradient pulses are also important for efficient water suppression by purge pulses. Thus the experiment time can significantly be reduced and spectral quality dramatically increased. Further, signal detection with optimal dynamic range of the electronic receiver system is possible as only the signal of interest is detected.

In addition to higher magnetic fields, also technical means to control and improve the homogeneity of the magnetic field have been constantly improved over the years, for example by new field lock and shim technologies. Most importantly, also the sensitivity has been steadily increased over the years by improved low-noise receiver technology, digital filters, oversampling and notably boosted by new probehead technologies [16].

3.3.3. Probeheads and sample volumes

The probehead is the sensor positioned in the center of the magnet containing a coil that is used both to send radiofrequency (RF) pulses to the sample and to detect the NMR signals returning from the excited atomic nuclei of the sample. The sensitivity of the NMR spectrometer depends, besides the strength of the applied magnetic field, primarily on the inherent sensitivity of the probehead (see Section 3.2.1.1). There are different probeheads available to detect different nuclei with optimal signal-to-noise. For metabolite profiling studies where the endogenous molecules in the biofluids are mostly rather diluted, the probe should be optimized for ^{1}H detection and the NMR observe volume of the probe coil has to be filled completely with the sample to allow for highest sensitivity. In this case the filling factor of the probe is 1. The volume of the sample residing within the boundaries of the NMR active detection region of the coil is called the observe volume. It is always smaller than the overall volume of the sample which is needed to fill the probe such that the sample extends homogeneously over the two coil ends. There are probeheads available with different coil sizes, ranging from 10 to 1 mm in coil diameter and with corresponding sample volumes from a few ml to as low as 2 μl. The standard NMR probehead is usually equipped with a 5 mm coil and requires 550–600 μl sample volume (see Figure 3.4).

The sample volume depends critically on how the sample is transferred to the probehead. It can be provided in a discrete sample glass tube as shown above or by flow injection through a transfer capillary. In the latter case the probe must contain a flow detection cell which is linked to the transfer capillary. Such a probehead is called a flow probe (see Figure 3.5). The sample volume needed for filling

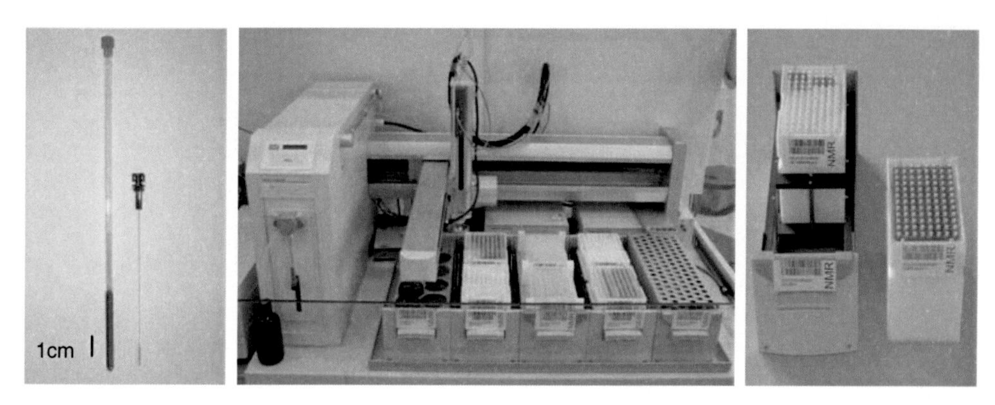

Figure 3.4. *Left:* Standard NMR sample tube with 5 mm diameter (left) and NMR capillary tube with 1 mm diameter (right). The sample volume needed is 550 μl and 5 μl, respectively. *Middle:* Liquid handling Robot (Gilson with NMR micro addition) which can automatically fill NMR sample tubes from 5 mm down to 1 mm in diameter. *Right:* NMR samples are organized in standard 96-well-plate-sized NMR tube racks. These racks can hold, for example, sample tubes and capillaries with 5 mm and 1 mm in diameter, respectively. Each rack is identified by a bar code and each sample by a dot code on the sample cap.

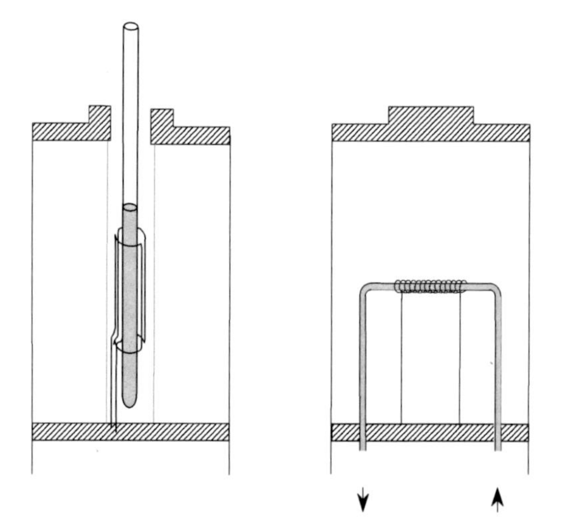

Figure 3.5. Schematic view of the inner part of an NMR probehead. The orientation of the static magnetic field B_0 is vertical. *Left:* NMR detection coil with Helmholtz design filled with a discrete NMR tube. Probeheads are commercially available with coils ranging from 10 to 1 mm in diameter with corresponding observe volumes as small as 2 μl. The Helmholz coil design lends itself also for flow application (direct injection or on-line LC-NMR). Flow probes with various cell volumes down to 30 μl are available. *Right:* NMR detection coil with solenoidal design [18]. Probeheads with this design can only be used for flow application (direct injection or online LC-NMR) as the cell axis is perpendicular to the opening of the magnet and the B_0 field. A probehead equipped with a 5 μl flow cell having 1.5 μl NMR active observe volume (observe volume) is commercially available.

an identical active volume is always larger for a flow cell than by using discrete sampling tubes. In case of the 1 mm coil, approximately two times more sample volume is needed to fill the flow cell (9 µl) than to fill a discrete 1 mm capillary (4.5 µl). The choice of the appropriate transfer method depends on sample properties such as volume, viscosity and solute concentration (see below). Also the sample exchange time varies between the two basic supply modes.

With decreasing diameter of the NMR detector coil, the mass sensitivity (S/N per mole) increases with 1/d to a first approximation [18, 40–42]. To acquire high-quality spectra of volume-limited biofluids, for example CSF from mice, where only a few µl are available, the measurement is ideally done in a 1 mm probehead having an NMR active volume of ~2.5 µl [17] with a fourfold to fivefold increased mass sensitivity when compared to a conventional 5 mm probe. Micro-probeheads are available for discrete tube and flow sampling [17, 43]. Alternatively, the CSF sample could be measured in a cryogenically cooled probe [44] with approximately the same mass sensitivity available for operation with 5 and 3 mm sample tubes (500–200 µl sample volume), either by using the same 1 mm capillary or, after dilution with buffer, in a 3 or 5 mm tube. The cryogenically cooled probehead technology enhances the signal-to-noise of larger sample volumes by a factor of approximately 4 compared to a conventional probe, or alternatively it allows for a 16-fold reduction in measuring time (see Figure 3.6). The high sensitivity of this technology has already notably stepped up the efficiency of drug metabolite identification and structure elucidation [44] . The use of a cryogenically cooled probe will be particularly beneficial in metabolite profiling and biomarker identification where ultimate sensitivity is important. It emerges to be the first choice for sample volumes >10 µl, especially since a convertible cryogenic flow probe with removable

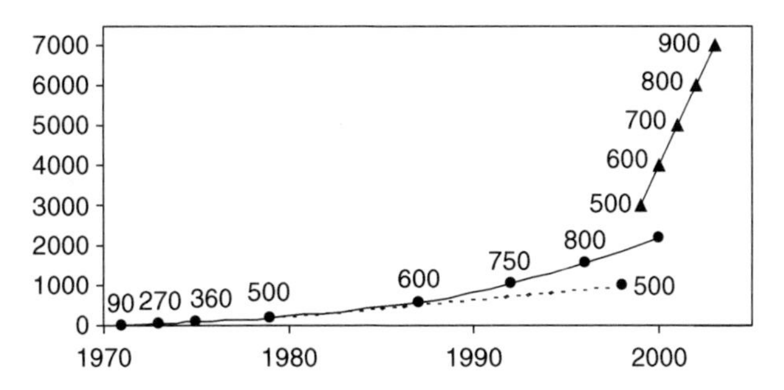

Figure 3.6. The specified signal-to-noise ratio of 0.1% ethylbenzene (EB) in $CDCl_3$ for ^1H-observe probes plotted as a function of time. The black dots denote the sensitivity of a conventional probe at the launch of a magnet operating at a particular field, and the triangles mark the launches of cryogenic probes at different fields (all data from Bruker BioSpin). The magnetic field (indicated by the ^1H operating frequency MHz) is given above the marker. The dashed line indicates the increase in specified sensitivity during two decades for a conventional probe operating at 500 MHz.

insert cells is becoming available [44]. Thus this probe allows switching between flow injection and discrete sample mode.

Important practical accessories to the probehead are a temperature control unit and a motor device which allows for automatic matching and tuning of the resonant circuit of the coil to the impedance of each individual sample.

3.3.4. Robotic sample changer

To monitor beneficial or toxic effects of drug candidates, or to recognize metabolic disease patterns, often demands the screening of hundreds of biofluid samples [8]. In case of small animals such as mice and rats, the biofluids are available only in limited volumes, CSF from 2 to 50 μl, blood from 50 to 200 μl, and urine from <1 ml to few ml. The fast supply of large sample arrays of different volumes to the spectrometer magnet is a critical issue in metabolic profiling of biofluids, in biomolecular NMR screening and in structural analysis of combinatorial chemistry products.

As described above there are two fundamentally different ways to supply and exchange the samples to the magnet. Conventionally, glass NMR tubes of 5 mm in diameter, also available as disposables, are filled and supplied to the magnet. This process has been in operation with a low-speed robotic sample changer over 20 years [45, 46]. It is very limited in efficiency for high sample throughput as the time for sample exchange can be much longer than the actual measuring time needed for data acquisition in the magnet. An alternative approach was introduced in 1997 [47, 48] in a study employing urine and samples from combinatorial chemistry. It uses a flow probe in which a direct transfer of samples is possible from a 96 well plate by a flow-injection device resulting in a significantly increased rate of sample throughput. If the washing step of the transfer capillary is not optimal, flow injection may suffer from sample spillover and contaminations or even from bacterial infections within the transfer tube. This could particularly be the case with viscous or concentrated samples, for example blood plasma. An injected sample may extend over 1 m in the transfer capillary and is thus exposed to a huge surface area. When it enters in the wider NMR flow cell, the sample lengths is dramatically 'contracted' to a few mm in length with corresponding changes in flow dynamics. Besides this fluidic problem, sample dilution will also occur if a system solvent is used between sample plugs. The recovery of the sample is thus further hampered. As previously discussed, flow injection needs a sample volume to fill the cell which is approximately two times larger than the flow cell volume. This is due to the large dead volume of the flow system. In the case of discrete sampling with tubes or capillaries, there are only minimal dead volumes involved and the samples are completely shielded from each other. This difference may become significant if only limited and small sample volumes of biofluids are available, for example from small animals.

In the past few years, both discrete and direct flow-injection sampling methods have been further developed and miniaturized. This development occurred in parallel with the miniaturization of the NMR detection coil of the probehead [17, 43]. Today, discrete 1 mm NMR sample capillaries can easily, quickly and reproducibly be filled with 5–8 μl sample by a liquid handling robot. Also automatic flow-injection systems are in place for handling sample volumes as small as a few microliters.

A very promising development has recently led to a new innovative sample changing system (Figure 3.7) called SampleJet (Bruker Biospin). It combines high speed and throughput with the advantages of handling discrete NMR sample tubes having sample volumes from 600 to 5 μl and sample diameters from 5 to 1 mm, respectively. The NMR samples are spatially organized in the standard well-plate format that lends itself ideally for high-speed, automated sample handling systems. Sample exchange times as short as 30 s can be achieved. There is room for even higher throughput using this platform. The exchange time is thus comparable to those achieved with fast flow-injection systems. Continuous NMR sample tube automation is available in the microtiter-plate format including automated liquid and tube handling coordinated by NMR automation software.

3.3.5. Connecting lab bench and NMR spectrometer

The NMR-based metabolite profiling of biological samples might need automation which includes in addition to the NMR measurement also the time-coordinated preparation of the sample. This needs to be the case if the biological sample has

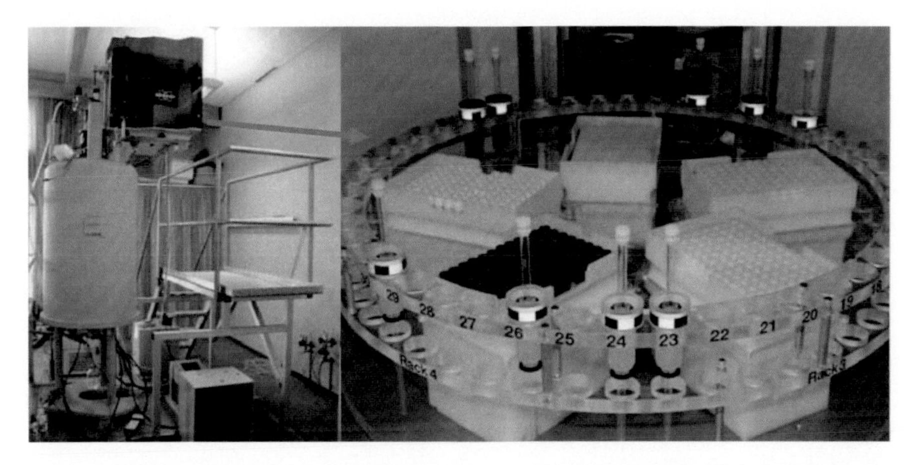

Figure 3.7. A sample changing robot (Sample Jet) allows fast sequential single tube submission under controlled conditions from five positions each holding a 96-well-plate-sized NMR tube rack. The system can thus efficiently handle batches with up to 480 samples. Also manual sample handling with standard NMR tubes and turbines is possible (outer circle).

to be prepared freshly, that is just-in-time for measurement and is not allowed to stand in a waiting queue in order not to deteriorate. In this case, automation should include the following individual steps:

- Just-in-time sample preparation.
- Transfer of the sample to the magnet.
- Setup of the NMR apparatus including locking of the field and shimming.
- Measurement of NMR experiments comprising a set of selected 1D and possibly 2D techniques.
- Back-transfer of the sample to a park position outside the magnet and storage.

Computerized book-keeping is absolutely compulsory in order not to lose track of hundreds or thousands of NMR experimental runs. The need for high throughput and reliability for all steps involved can hardly be over emphasized. A robotic system which has been described and successfully used on thousands of sample in biostructural NMR screening can also be employed in biofluid screening [49, 50]. It consists of a Genesis sample handling robot (Tecan) which prepares the sample by mixing the required components in an NMR tube immediately before it gets automatically transferred to the spectrometer by a Bruker SampleRail system.

3.4. Sample handling

NMR spectroscopy on native biological samples differs in many aspects from conventional NMR spectroscopy on material from synthetic sources. Many parameters must be considered to obtain relevant NMR spectra of good spectral quality. Potential sources of any sort of artifacts have to be consistent and to be avoided by all means during the whole process from collecting the biological sample until the analytical result is provided for further data processing.

3.4.1. Sample collection

Factors, such as sample type, time of collection, containers used, preservatives and other additives, transportation and length of transit time affect the quality of the samples and must be carefully considered before the initial collection stage [51]. In addition the design of the metabolic cages used for animal studies as well as the protocol for sample handling at the animal housing facility influence the sample quality. The metabolic cage for animal housing must prevent feces and food from entering the urine collection container. For urine, cooled sampling units attached to the metabolic cage are necessary. Microbial degradation[23] during sampling intervals of several hours can be dramatically reduced by such devices. Therefore it is absolutely essential to set up a proper study protocol with clear instructions for all aspects of the sample collection process to ensure a reproducible high quality of samples.

In the following, we focus on the factors affecting the quality of biological samples and some of the provisions that must be made during collection, processing and storage of samples, based on our experience with a focus to urine samples, which are with respect to their variability for obvious biological reasons most demanding.

3.4.1.1. Collection containers

The choice of the containers for sample collection is an extremely important prerequisite for a successful realization of metabolite profiling by any analytical techniques. A wide range of different types of containers for sample collection are available. Examples of spectra of different types of blood sample tubes leached with phosphate buffered water are depicted in Figure 3.8. It has to be pointed out that tubes of

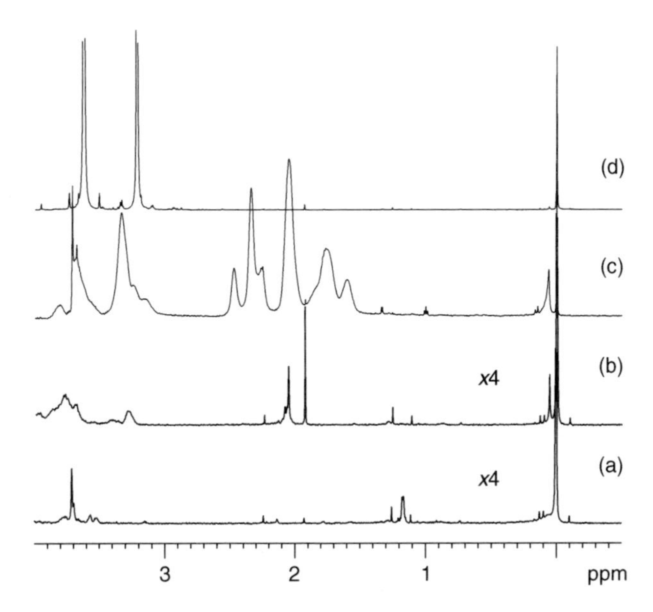

Figure 3.8. ^{1}H NMR spectra of a cross section of sample containers for biofluids of different manufactures. All tubes were leached with buffered water (pH = 7.4). Spectra were originally scaled to same TSP intensity and enlarged as indicated. (a) native PP tube, (b) Li-Heparin coated tube for plasma samples, (c) PP tube coated with a clot activator for serum samples, (d) EDTA coated tube for plasma samples. Spectra for the same tube type but from different manufacturer may vary significantly.

[23] In the collection containers of cages often 1 ml of a 1% solution of NaN_3 solution is provided. It is often overlooked that this sample collection scheme typically adds a concentration of 160 mM Na^+ to urine samples.

the same type, for example clot activated tubes for serum collection of different manufactures or different batches of the same supplier, can show markedly different additive profiles. It can be clearly seen that native tubes without any additives or stabilizers generally show least interfering signals in the ^{1}H NMR spectrum. To test the container (of the same batch that will be used for the study) for possible contaminants before sample collection is therefore strongly recommended. Greatest care has to be taken to select the right sample tube for metabolic profiling studies because analysis of data can fail if contamination by inadequate sample tubes is introduced.

Urine and CSF: For urine or CSF collection no special functionality is needed for the sample container. Here native tubes without any coating, made from glass or polypropylene (PP), are strongly recommended. This is of advantage as more functionality incorporated in the tube holds the risk to render the sample useless for metabolite profiling.

Serum and plasma: Typical sample tubes for blood collection often contain additives or stabilizers that simplify further sample handling or are needed for the preparation of the desired serum or plasma sample. Often the decision whether to collect plasma or serum samples is influenced by practical considerations at the sample collection site. More important for metabolic profiling investigations is the spectroscopic suitability of the finally obtained biofluid. For plasma and serum collection the type of sample container must be carefully evaluated (see Figure 3.8).

Plasma is a cell-free supernatant of anti-coagulated blood and can be obtained fast and in high yields. The risk of uncontrolled and incomplete clotting is low. Therefore plasma is often the favored blood derivative. Tubes containing unspecified anti-coagulants should be avoided (see Figure 3.8) unless they are completely free of any ^{1}H NMR signals. In most cases, Li-Heparin or EDTA (**E**thylene-**d**iamine-**t**etraacetic **a**cid) is used as anti-coagulant. Both types of tubes show signal overlap of endogenous metabolites with EDTA and Heparin resonances in the NMR spectrum. Ideally, per-deuterated EDTA is used as anti-coagulant as this compound is completely free of ^{1}H NMR signals. However, such tubes are not yet commercially available, but can easily be prepared. As a second choice, Li-Heparin tubes are recommended.

Serum is obtained from whole blood by centrifugation after clotting is completed. This process takes about 30 min and may not be exactly controlled at the collection site. The advantage of serum is that no additives are necessary to obtain this biofluid. Native glass tubes can be used for serum collection. The time samples are allowed to clot naturally at room temperature and the cooling chain during preparation and transportation has to be controlled and monitored carefully. Tubes coated with unknown clot activators generally are not recommended for metabolic profiling studies because contamination can be introduced (see Figure 3.8).

3.4.1.2. Stability and storage

The consistency of the metabolite profile across many samples in a study depends on the quality of the biological samples. Not only inappropriate sample containers may lead to unwanted variation but also unequal treatment or storage of individual samples of a study. For example, the study protocol must define maximum allowed storage periods for samples at room temperature, as many endogenous metabolites are sensitive to chemical or microbial degradation. A long total storage time between sample collection and analytical measurement can induce variation in the data. Investigations during the Consortium for Metabonomic Toxicology (COMET) project [52] revealed that biological samples like urine are stable for at least 9 months at −40°C. Slight biochemical changes in tricarboxylic acid (TCA) cycle intermediates were found in urine samples after an 18 months storage period at −40°C. For plasma samples the influence of short- and long-term storage at various temperatures was assessed in detail by Deprez *et al.* [53]. The authors reported no observable changes in plasma NMR spectra after 6 months storage at −80°C.

3.4.1.3. Microdialysis

Microdialysis is an established *in vivo* tool in neuroscience and has been used widely for pharmacological and metabolic profile analysis. The dialysate which is collected for analysis represents the local profile of the extracellular environment of a specific tissue or biofluid. For microdialysis a probe with a hollow fiber dialysis membrane is implanted into the organ or biological matrix. The perfusate, a solution that mimics the physiological composition of the extracellular fluid of the target, is slowly pumped through the probe. Small molecules can diffuse through the membrane and are carried by the dialysate. In most cases the microdialysate is free of large molecules like proteins, but also small molecules bound to proteins are excluded by the membrane [54]. Therefore ^{1}H NMR resonances of endogenous metabolites of microdialysate samples are not obscured by broad lipid or protein signals [55]. This facilitates data interpretation. In addition, microdialysates are metabolically more stable than native biofluids or tissues because enzymes are not able to pass the membrane. A drawback of microdialysis from an analytical point of view is the small volume of dialysates and the low concentration of analytes present in the dialysate. Therefore the solvent is completely removed by vacuum evaporation in most applications [56] and measurements have been performed in miniaturized NMR probes [17, 57].

3.4.2. Sample preparation

Sample preparation must fulfill the following criteria: (i) reproducibility, (ii) robustness, (iii) ease of use, (iv) non-discrimination between samples, (v) reduction of unwanted variation in the data, due to different pH of samples and so on, (vi) maintenance or improvement of spectral quality regarding resolution and sensitivity, (vii) freedom of analytically visible artifacts.

In contrast to other analytical techniques [58] or clinical chemistry methods, metabolite profiling of biofluids by NMR like urine, plasma or CSF requires in general less sophisticated sample preparation procedures [59]. In most cases, addition of water or buffer to account for pH variation or to reduce viscosity is sufficient as sample preparation before the NMR measurement [60]. This means that one potential source of error, for example due to sample extraction procedures, is absent. However, a homogenous data set is highly desirable for further data processing and data evaluation steps.

3.4.2.1. Concentration and lyophilization

Lyophilization and reconstitution in an appropriate solvent is frequently a practical sample preparation method for samples of biological origin to enhance the analytical sensitivity and spectral quality, or to stop the inherent enzymatic activity of biological samples. For CSF, significant sharpening of small molecule resonances was reported after lyophilization [61]. However, the loss of volatile or unstable components from biofluid samples may occur during the lyophilization process, for example for acetate or formate. Selective deuteration of acidic protons on reconstitution with D_2O may further complicate spectral interpretation. In Figure 3.9 ^1H NMR spectra of human urines are depicted where the effect of lyophilization and concentration was investigated. Aliquots of urine were directly measured after buffering (a), reconstituted in buffer to the original volume (b), and after lyophilizing the sample and concentrating by a factor of 4 (c). The native and lyophilized spectrum show minute but visible differences in metabolite compositions. For example, besides other subtle changes resonances of acetate, dimethylamine and succinate were reduced after lyophilization. Due to higher viscosity the fourfold concentrated sample reveals slightly increased line widths. If sensitivity is not an urgent issue, lyophilization and concentration should be considered with care. Every sample preparation step is a direct intervention into the sample constitution and bears a risk for both loss of metabolites and contamination. It is the power of NMR spectroscopy not to discriminate between analytes and to need only sparse sample preparation steps. This has to be critically compared to the advantages of lyophilization.

3.4.2.2. pH and buffering

The common sample preparation protocol for urine samples includes pH-adjustment with phosphate buffer (e.g. pH 7.4) and the addition of D_2O, TSP and sodium azide to final concentrations of about 5%, 0.3 mM and 1 mM respectively [8]. The pH of different urine samples may vary from 5 to 8, according to the physiological condition in the individual, but usually it lies between 6.5 and 7.5. The

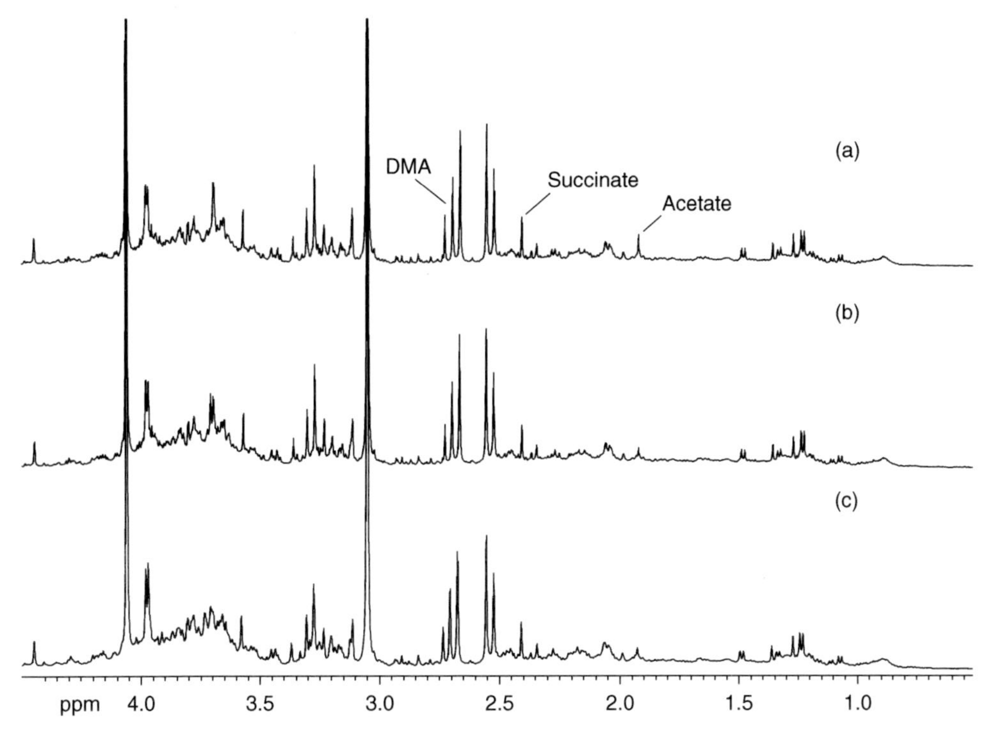

Figure 3.9. Comparison of ^{1}H NMR spectra of: (a) native human urine; (b) lyophilization and reconstitution to original concentration; (c) lyophilization and reconstituted with fourfold concentration. The spectra are scaled to the creatinine intensity.

addition of 100–200 mM phosphate buffer normalizes the pH in most cases to a range of 6.7–7.6, at least for several hours [59]. Nevertheless several small molecular endogenous metabolites present in biofluids show a pronounced pH-dependent chemical shift variation in ^{1}H NMR spectroscopy even after buffering [62]. This is, for example, the case for citrate and histidine. The dependency of the citrate resonances on pH is depicted in Figure 3.10. The remaining chemical shift variation (see also Section 3.2.2.1) between different samples has a direct impact on data reduction procedures and complicates further data evaluation steps. It may thus influence the data interpretation. Therefore, the binning of each spectrum into segments with sufficiently large spectral width was introduced [63] to scope with the shift variation (see also Section 3.6.2).

3.4.2.3. Salt concentration, EDTA

Adjustment of pH of biological fluids alone does not fully remove the chemical shift variation of organic acids like citrate [62], lactate, taurine or others. Further effects related to osmolality of the samples, ionic strength or metal ion composition

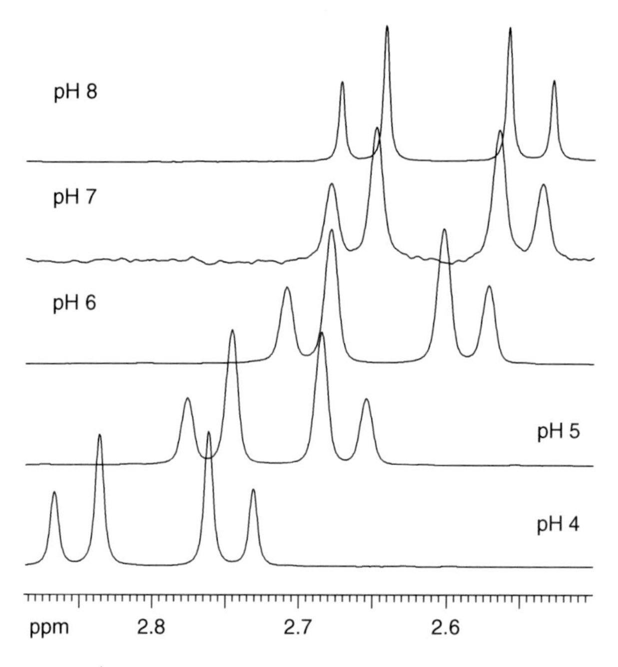

Figure 3.10. Variation of ^1H NMR citrate chemical shifts at different pHs ranging from pH 4–8.

significantly contribute to the pH-independent chemical shift variation (see also Section 3.2.2). This is especially the case not only for urine and plasma but also for CSF or saliva [30].

The inorganic metal ion composition of biological fluids, especially of urine, shows a great variance between different individuals and also between different time points of an individual. The major positively charged urinary inorganic constituents are the cations of sodium, potassium, calcium and magnesium. Typical concentrations for these alkali and earth alkali elements in human urine are 224 mmol/d, 70 mmol/d, 5.9 mmol/d and 5.4 mmol/d respectively [64]. Citric acid is known for its avid chemical propensity to bind metal ions like calcium and magnesium, and to a smaller extent also sodium. High-resolution ^1H NMR spectroscopy can readily distinguish the various complex/chelate states of endogenous metabolites via the assessments of precise frequency or line widths [30]. In Figure 3.11, the change of the chemical shifts of citric acid in the presence of different sodium concentrations is depicted.

Bivalent calcium and magnesium ions influence the citrate chemical shift more pronouncedly than the sodium cation (see Figure 3.12). Besides pH dependency, the different Ca^{2+} and Mg^{2+} concentrations of biofluids are the major source for the chemical shift variation of citrate and other organic acid anions observed in biofluids. Interestingly, only the chemical shift of one of the two protons of the AB CH_2 spin system of citrate is markedly influenced by added Mg^{2+}, whereas Ca^{2+}

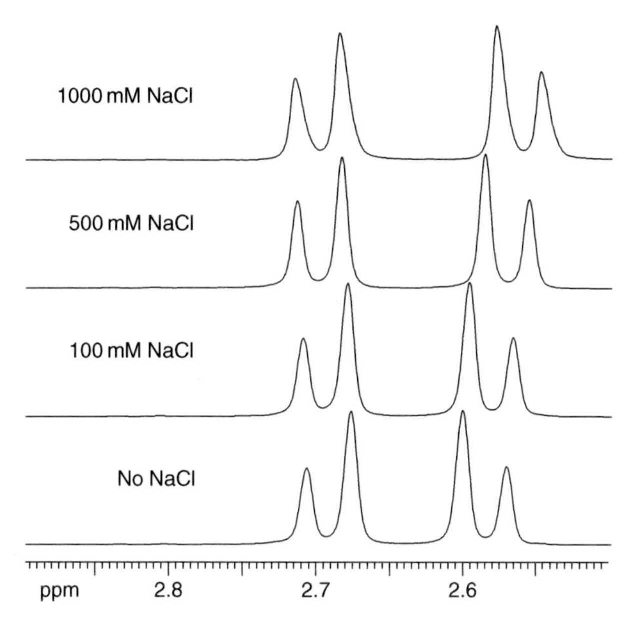

Figure 3.11. Variation of ^1H NMR citrate chemical shifts against change of the sodium chloride concentration (ionic strength) referenced to TSP.

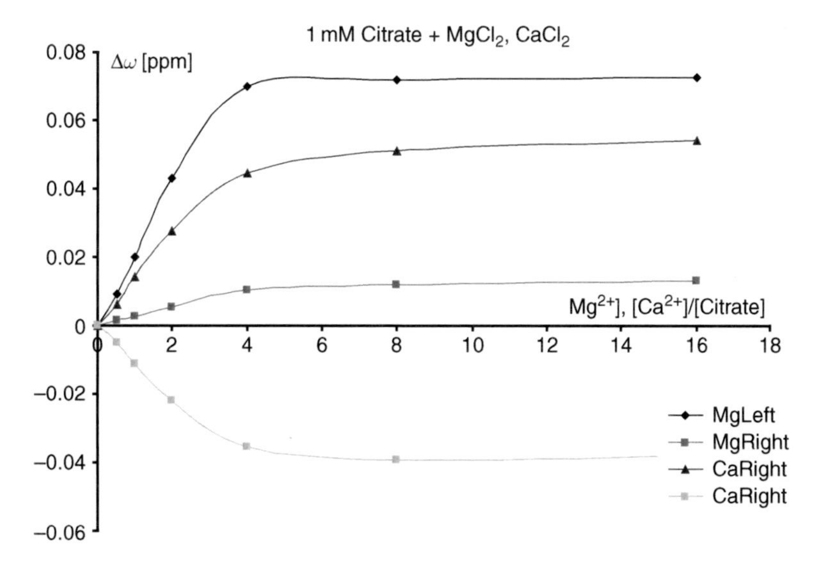

Figure 3.12. Plot of modification of citrate-CH_2 group chemical shift values referenced to TSP ($\Delta\omega$) versus added Ca^{2+} and Mg^{2+} concentrations. Left and Right respectively denote the deshielded and shielded resonances of the AB group of citrate.

changed both resonances of the AB spin system. This may be attributed to different stereochemistry of the Ca^{2+} and Mg^{2+} complexes due to differences in ionic radii (100 pm versus 72 pm [65] respectively).

From an analytical and data evaluation point of view, a homogenous sample set with respect to overall concentration, pH, metal ion composition and osmolality would be highly desirable, but for practical reasons the sample preparation can not account all these sources of variation in biofluids. For example, the adjustment of all samples of a study to the same overall concentration can only be done with respect to the most dilute sample in the series. This would lead to drastically increased experimental acquisition times. As a compromise we propose the addition of perdeuterated EDTA to account for different earth alkali cation composition of biofluids in addition to buffering. This is a simple modification of the sample preparation procedure that fulfills most of the criteria mentioned in Section 3.4.2. The two major sources (pH and bivalent metal ion composition) of variation in chemical shifts can be reduced. The span of chemical shifts for citrate can be dramatically decreased and peak overlap in different bins after data reduction can be avoided. This helps for statistical data analysis and simplifies the binning procedure (see Section 3.6.2). Figure 3.13 shows the effect of addition of 10 mM EDTA-d_{12} to 80 samples of phosphate-buffered control rat urines. It can be clearly seen that not only was the variation in citrate chemical shift minimized but also the potential overlap with a resonance of dimethylamine at $\delta = 2.706$ ppm was prevented. This modified procedure mainly affects the chemical shifts of chelating metabolites like citric acid, taurine or lactic acid. The remaining signal of not fully deuterated EDTA is negligible and does not interfere with further data analysis steps.

If creatinine is taken as a measure for urinary concentration[24] and citrate chemical shifts are plotted against urinary concentration (creatinine) in a regime free of interaction between endogenous components, there should be no correlation visible. However, the situation depicted in Figure 3.14 clearly shows a concentration dependency of citrate chemical shifts in both cases with and without addition of EDTA-d_{12}. The variation of citrate resonance is reduced by a factor of about 2.4 in standard deviation after addition of EDTA-d_{12}. The slope of the trend line indicates remaining interaction modes of the non-chelated citrate with urinary components which is concentration dependent, although less than in absence of EDTA.

Besides the reduction in citrate chemical shift variation, the overall explained variance of a Principal Components Analysis (PCA) model is a measure for the effect of EDTA-d_{12} addition. In Figure 3.15, the convergence of the explained variance of PCA models of 96 human urines with and without addition of EDTA-d_{12} is shown. The

[24] This is valid to a good approximation for samples, where a drug induced change of creatinine signal can be safely excluded.

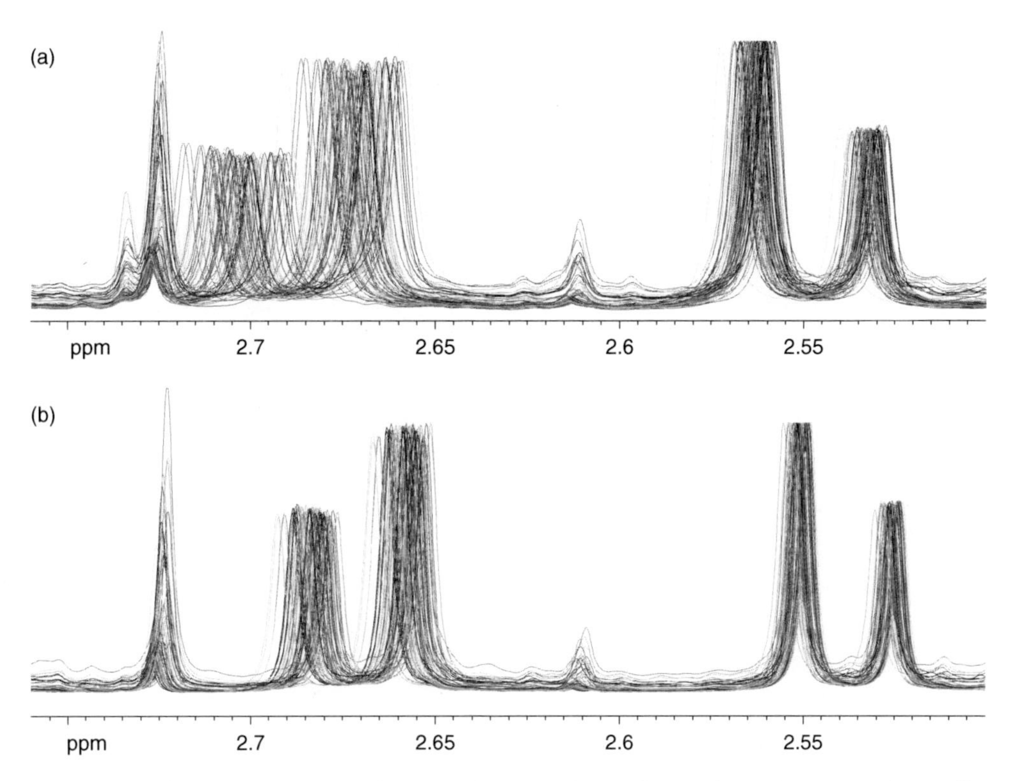

Figure 3.13. (a) Citrate chemical shift region of 80 rat urines buffered with 100 mM phosphate buffer (pH 7.4). (b) Citrate chemical shift region of the same 80 rat urines buffered with 100 mM phosphate buffer and addition of 10 mM EDTA-d_{12}.

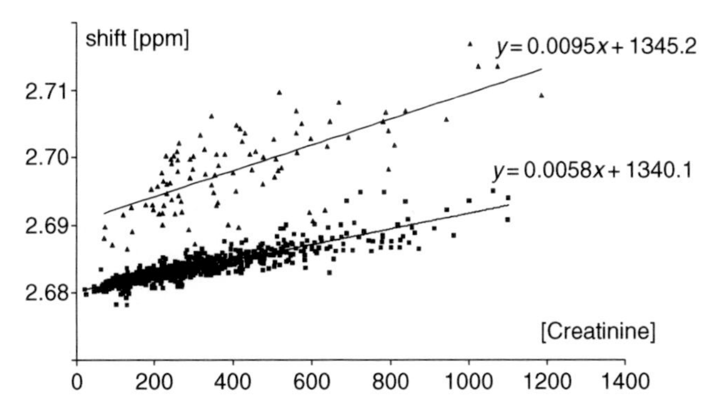

Figure 3.14. Correlation of citrate chemical shift (centre of the low-field doublet) with (lower trace) and without EDTA-d_{12} (upper trace) against urine concentration represented by creatinine intensity. The standard deviation for citrate chemical shifts is reduced from ±3.1 Hz to ±1.3 Hz after addition of EDTA-d_{12}, also the slope of the trend line is smaller.

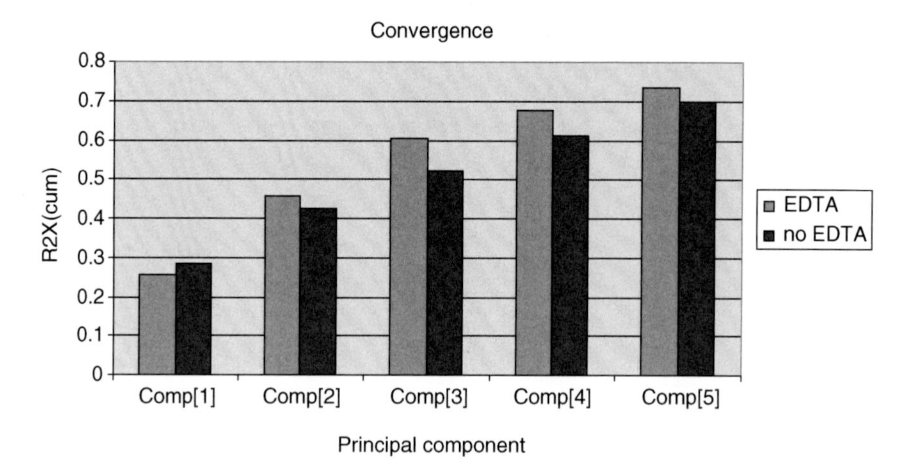

Figure 3.15. Convergence of explained variance of PCA with integral normalized samples with and without EDTA-d_{12} addition.

explained variance increases faster after addition of EDTA-d_{12}, therefore, the beneficial effect of decreased citrate variation is beneficial also for the complete data set. It can be seen that by removing one source of non-dose-dependent variation from the data, the PCA models converge faster and statistical analysis should be facilitated.

3.5. NMR Experiments and their processing

Compared to other applications of NMR, the pulse sequences applied in metabonomics require a lower level of sophistication and the application of all sequences is hampered by the following reason:

All NMR experiments which rely on the detection of hydrogen nuclei (protons) in biological samples face the problem that molecules have to be detected in an aqueous sample at μM or lower concentrations against a background concentration of 110 M water protons. Thus a dynamic range of the receiver and digitizer system of 10^8 would be needed. Modern state of the art receivers allow for 16-bit signal digitization, ensuring a dynamic range below 10^6. Thus a suppression of the water signal of at least 10^2 is required such that a signal of a compound at low concentration "fills at least a single bit". A number of water suppression techniques have been reviewed in the NMR literature [66]. In the following, solvent suppression techniques and related pulse sequences that are applied in metabonomics analysis will be described. The technique used for water suppression is of high importance for the comparability between NMR data. Results obtained from multivariate analysis of NMR spectra taken under different solvent suppression conditions can be erroneously interpreted as biological differences [62] if water suppression artifacts are not carefully excluded.

3.5.1. 1D NMR

Most applications of NMR for metabonomics research rely on 1D NMR experiments. This is dictated by long acquisition times of most multi-dimensional NMR experiments making their application a time-consuming task for studies with large sample arrays. This situation is normally encountered for time- and dose-resolved studies with preclinical or clinical samples. Typical 1D NMR spectra of serum, urine and CSF are shown in Figure 3.16.

3.5.1.1. Water supression

Presaturation: The most widespread solvent suppression in use is the so-called presaturation technique [67]. This technique is applied in an NMR pulse method known as noesy-presat [68]. This pulse sequence is identical to the 1st time increment of the 2D NOESY[25] experiment (Figure 3.17). The water resonance is saturated by

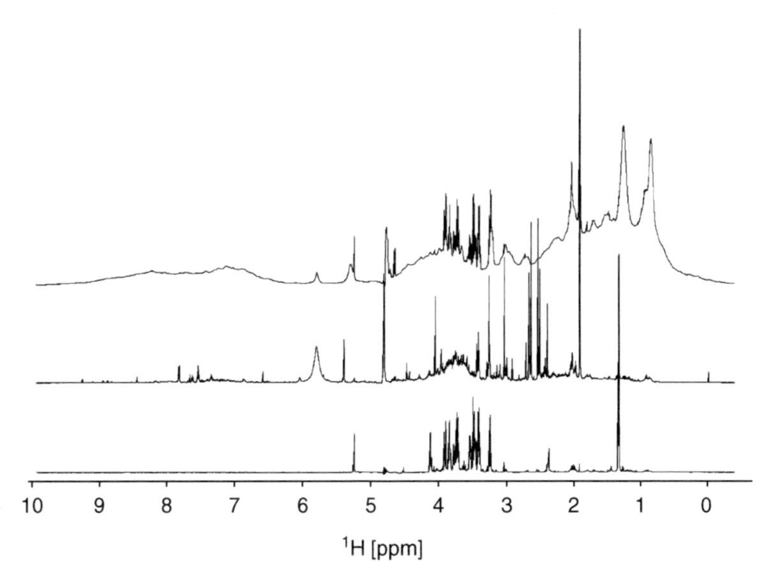

Figure 3.16. 1D ^{1}H NMR spectra from the top of serum, urine and CSF of rat. Very different spectral signatures of the body fluids are clearly visible. The overall shape of the serum spectrum is dominated by a broad signal background having its origin in high-molecular weight compounds (proteins, lipoproteins). The eye-striking feature of urine is a huge urea signal at around 5.9 ppm, whereas the CSF spectrum is dominated by high glucose (3–4 ppm) and lactate (1.2 ppm) content.

[25] **N**uclear **O**verhauser **S**pectroscop**y** (NOESY): Here a 2D spectrum is acquired allowing for the determination of distances between protons within a molecule. The NOESY is widely applied in structure determination of proteins by NMR[1]. The technique has found no application in metabonomics.

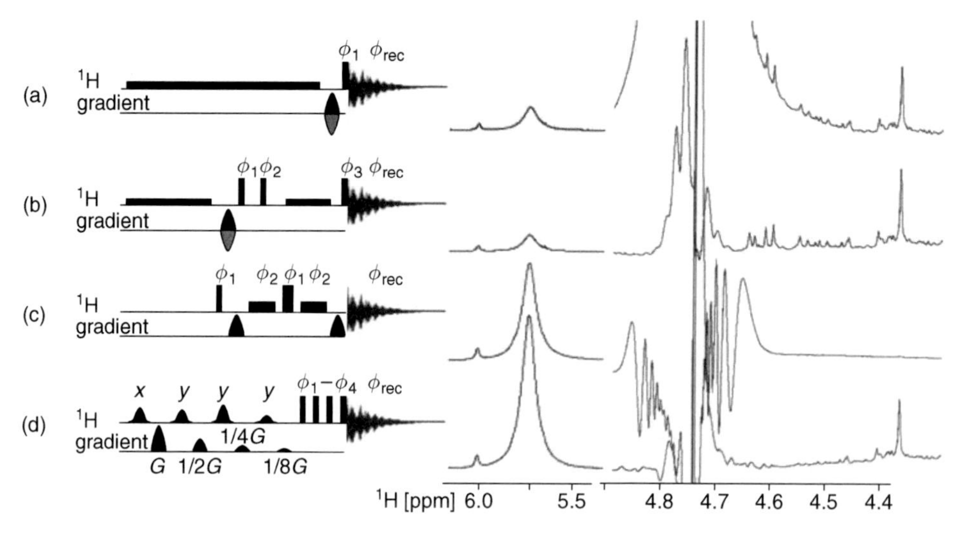

Figure 3.17. Different methods of water suppression: From top to bottom the following methods together with typical results obtained for a urine sample are shown: (a) *standard water presaturation* of 3 s with 50 Hz (phase cycling as follows: $\phi_1 = \phi_{rec} = \{x, -x, -x, x, y, -y, -y, y\}$); (b) *1D noesy-presat* during 3 s and 100 ms mixing time with 50 Hz (phase cycling: $\phi_1 = \{x, x, -x, -x\}$, $\phi_2 = \{16x, 16(-x)\}$, $\phi_3 = \{4x, 4(-x), 4y, 4(-y)\}$, $\phi_{rec} = \{2x, 4(-x), 2x, 2y, 4(-y), 2y, 2(-x), 4x, 2(-x), 2(-y), 4y, 2(-y)\}$); (c) *WATERGATE* with 1 ms selective pulses (phase cycling: $\phi_1 = \phi_{rec} = \{x, y, -x, -y\}$, $\phi_2 = \phi_1 + 180°$); (d) *WET* with 10 ms selective shaped pulses – power levels of pulses compared to a 90° shaped pulse are +0.87, −1.04, +2.27, −5.05 dB counting from left to right (phase cycling: $\phi_1 = \{y, 2(-y), y, -x, 2x, -x\}$, $\phi_2 = \{-x, x, -x, x, -y, y, -y, y\}$, $\phi_3 = \{-y, 2y, -y, x, 2(-x), x\}$, $\phi_4 = \{x, -x, x, -x, y, -y, y, -y\}$, $\phi_{rec} = \{2(x, -x), 2(y, -y)\}$). In the pulse schemes (a)–(d) shown to the left, pulses on the top trace represent proton pulses (proton frequency). The heights of the pulse go in parallel with the applied B_1 field strength. The bottom trace shows gradient pulses applied, whereby all gradients are applied with sine shape. The black/grey shaded gradient pulses indicate application with alternating sign of this pulse. The spectral region shown in the middle column shows the size of the urea signal. Urea protons experience chemical exchange with water. Therefore a substantial suppression of this signal is seen for pulse methods which are "slow" on the exchange timescale (presaturation for 3 s in (a) and 2 s in (b) versus selective pulses of a few milliseconds for WATERGATE and WET). Therefore quantification of urea is not possible and the signal is typically excluded from analysis. Note: We have included here the phase cycling scheme that was employed. The addition of scans is performed with different directions of applied pulses. This has to be done in concert with phase cycling of the receiver to obtain coherent addition of signals. Phase cycling is applied for these techniques to remove artifacts due to spectrometer imperfection and relaxation phenomena accruing during the course of the pulse sequence.

frequency selective irradiation during the inter-scan delay and the mixing time of the sequence thus leading to a reduction of the equilibrium population difference of the spin species resonating at the frequency of the weak (<50 Hz) B_1 field applied. The big advantage of this method is given by a chemical shift selectivity superior to any other technique. If a reliable quantification of signals resonating close to water (e.g. anomeric proton of glucose) is needed, this technique will be the choice. The disadvantage of presaturation comes from the long duration (typically 2s). Therefore hydrogen atoms in chemical exchange with water (OH, NH, NH_2) (see Section 3.2.2.1), protons having an NOE (see Section 3.2.1.3) with a proton under the water line, or which are exposed to a hydration shell will also be saturated. The degree of saturation is given by the rate of the chemical exchange and the strength of the NOE respectively.[26] Any interaction process slow on a 2s timescale can be neglected. The most prominent signal in urine exposed to saturation exchange is urea, which cannot be quantified if water presaturation is applied (see Figure 3.17). This spectral region is excluded normally from metabonomics analysis. The mixing time, together with the intense phase cycling used in noesy-presat, has a positive effect on a flat baseline. This can be explained by a small degree of spatial selectivity[27] encoded in the noesy-presat pulse sequence reducing the influence of protons entering or leaving the sensitive volume of the detection coil during the detection of the FID. It must be noted that a mixing time will falsify the absolute amplitudes of NMR signals detected due to differential T_1, NOE and chemical exchange effects.

Experience in our laboratory shows that for cryogenic probeheads offering higher sensitivity the quality of water suppression achievable by this technique is poor when measuring urine in 5 mm sample tubes. In this case we have successfully used a sign-alternating gradient pulse applied between the end of the presaturation and the first 90° pulse to significantly improve water suppression.

In summary, the noesy-presat technique provides a simple, highly reproducible and robust method for the acquisition of high-quality NMR spectra in aqueous solutions. It has to be kept in mind, however, that absolute value quantification is possible only with limited accuracy.

Gradient pulse based technique: This water suppression technique rests on the application of the so-called gradient pulses. A gradient pulse G_z (typically 1 ms) creates an inhomogeneous magnetic field along the z-axis. Magnetization of amplitude M_0 present in the transverse plane prior to the pulse will be dephased efficiently by G_z, as the frequency of the precession is z-dependent during G_z. Only the integral along the z-direction of the sample given by $M_0 \int_{-z/2}^{+z/2} e^{i \cdot \gamma_x \cdot G_z Z'} dZ' \approx 0$ is detected by the receiver. If the gradient is strong enough (>10 G/cm), the signal is

[26] This phenomenon is referred to as saturation transfer.

[27] Spatial selectivity of B_1 pulses is obtained, as the distribution of the B_1 field strength is not homogeneous over the sample. Therefore, long and/or many B_1 pulses will lead to a reduction of signal amplitude. The region of highest B_1 inhomogeneity is found at the upper and the lower rims of the sample. Therefore contributions of these regions (also prone to high B_0 inhomogeneity and diffusion effects) are suppressed most efficiently by this technique.

dephased by more than 99%.[28] For the WET (water suppression enhanced through T_1 effects) [69] sequence which is used in metabonomics the transverse magnetization of the solvent signal is thus suppressed by the application of a series of weak frequency selective B_1 pulses, interleaved by G_z pulses. The flip-angles (durations) of the selective pulses and the strengths of the gradient pulses were optimized to achieve high-frequency selectivity, robustness and a flat baseline. The performance of the sequence is further improved by the application of a composite 90° excitation pulse as the last pulse taking advantage of its higher spatial selectivity [27]. Effects from chemical exchange and NOE are less serious for WET, as the overall duration of the sequence is more than one magnitude of order shorter compared to the duration of presaturation pulses described in above. The drawback is a reduced performance in respect to frequency selectivity, having its origin in the higher B_1 field strength typically used. All other parameters are essentially the same as in noesy-presat. The absence of the mixing time allows for an accurate quantification here. As a consequence, data taken with WET and noesy-presat cannot be analyzed statistically in a combined data set but must be treated separately.

3.5.1.2. Editing techniques

Relaxation editing: Rigid molecules (e.g. proteins) with high molecular weight experience a fast transverse relaxation (see Section 3.2.1). This is exploited in metabonomics for the removal of signals of high-molecular-weight compounds. Thereby a 40–100 ms relaxation delay [70, 71] is included prior to data acquisition. J-coupling effects are most efficiently removed by use of the so-called CPMG (named after its inventors **C**arr **P**urcell **M**eiboom and **G**ill) [72] pulse sequence with a refocusing delay set to 1 ms. This technique is especially attractive for blood serum or plasma samples having high content of proteins (see Figure 3.18). For applications on soft tissues the technique was successfully applied using a HR-MAS probe-head [73].

Diffusion editing: The principle of gradient pulses was introduced in Section 3.5.1. The application of a pair of z-direction gradient pulses of opposite sign will result in a dephasing of transverse magnetization followed by a rephasing. No signal is lost. This only holds true if the magnetization in a certain z-plane of the sample has not moved to another z-plane between the application of the pair of gradient pulses. If a delay is inserted between the two gradients, rephasing is only partially possible due to Brownian motion and depending on the molecular weight of the solute molecule. This phenomenon is used to remove signals from "small" molecules

[28] Another important application of gradient pulses is the selection of coherence pathways. This technique allows, for example in HSQC type experiments, for the selection of signals of protons bound to ^{13}C, and artefacts due to protons bound to ^{12}C are removed. For details, the reader is referred to citations in Section 3.5.2.2.

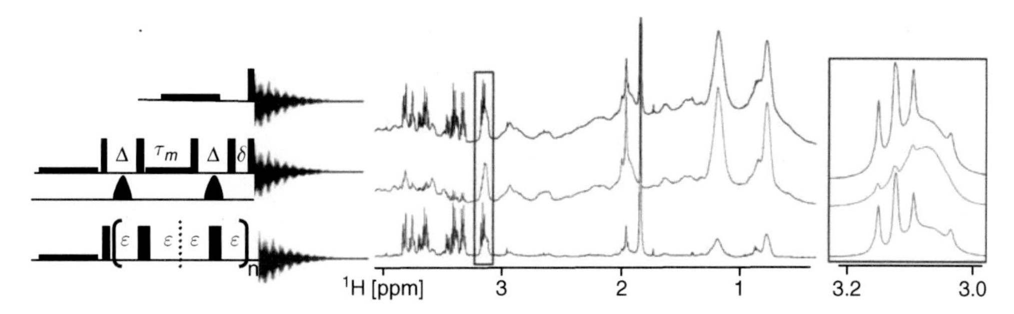

Figure 3.18. Diffusion and T_2 weighted filtration of a serum ^1H NMR spectrum: The top row shows a 1D presaturation spectrum together with a boxed expansion. The middle row shows a spectrum and expansion with diffusion-weighted editing. A delay τ_m of 300 ms serves to reduce signals from low-molecular-weight compounds (see triplet in expansion). The bottom row shows a CPMG-based T_2 weighted result. Here, n was set to 150 ($150^*4\varepsilon = 200$ ms) which reduces the broad lipoprotein hump in the background of the triplet substantially (see expansion).

which experience a high mobility [31]. To overcome fast signal decay due to T_2 relaxation, the magnetization, dephased by the first gradient, is aligned along the z-axis and flipped back to the transverse plane prior to the 2nd gradient after 50–300 ms. This technique is referred to as PGSE (**P**ulsed **G**radient **S**pin **E**cho) [74], whereby bipolar gradients [75] up to 100 G/cm are used. An extension to 2D by incrementation of the gradient strength is known by the acronym DOSY [76] (**D**iffusion **O**rdered **S**pectroscop**y**). This technique allows "the sorting" of the compounds in a mixture according to their translation diffusion; applications for metabonomics research including tissues can be found in the literature [32, 77].

3.5.1.3. ^{13}C spectroscopy

The chemical shift dispersion of ^{13}C spins is higher by approximately a factor of 20 when compared to protons. In addition, all signals in a ^{13}C spectrum appear as singlets if ^1H decoupling is applied[15]. Homonuclear ^{13}C–^{13}C scalar coupling is absent due to the high dilution of ^{13}C nuclei (only 1.11% of all carbon nuclei carry a spin). Thus the reduction in spectral complexity combined with substantially reduced spectral overlap would make the ^{13}C nucleus an ideal candidate for metabolic profiling. However, this is prohibited by the low intrinsic sensitivity of ^{13}C detection given by $S_{13C} = S_{1H} \cdot \gamma_{13C}^3 \cdot 0.011 \approx S_{1H} \cdot 3.4 \cdot 10^{-4}$. Thus the acquisition of a ^{13}C 1D NMR spectrum that is identical in signal-to-noise to a ^1H spectrum of the same sample will need $\sim 10^8$ times as long. Compromises in spectral amplitude are unpreventable and highest sensitivity equipment (^{13}C cryogenic probehead together with 900 MHz) is absolutely compulsory. An additional gain in signal-to-noise compared to ^1H is of course seen as no spin multiplicities reduce the amplitude. The situation changes slightly if sensitivity enhancement by polarization transfer from protons (INEPT [78]) is used (see also 3.5.2.2).

In the literature there is one example of a metabonomics analysis by use of ^{13}C spectra [79]. However, general use of ^{13}C detection can only be recommended if further substantial improvements in sensitivity of the detection of NMR signals will be achieved. What is of interest is the enrichment of ^{13}C in biofluids or tissues by use of ^{13}C-labeled tracer molecules in biological studies. With this labeling approach a new window into metabolic analysis and diagnostics may be opened also in higher organisms and metabolic pathway analysis including fluxes can be performed with acceptable sensitivity [80].

3.5.2. 2D NMR

It was mentioned before that a widespread use of 2D NMR in metabonomics is limited to small sample arrays by the long acquisition times needed. Nevertheless examples of these techniques can be found in the literature and are applied in our laboratory. The 2D techniques unfold their full power if a *de novo* structure determination of a newly found metabolite in a biological sample is needed. We will not go into the process of structure determination here. The reader is referred to the expert literature [81].

3.5.2.1. Homonuclear 2D NMR

J resolved (J-Res): The most simple homonuclear 2D NMR experiment is the so-called J-resolved experiment [82]. Here J-coupling information is separated into a 2nd dimension from the chemical shifts. The experiment was successfully applied to many body fluids including urine [83], CSF[61], seminal fluid [84] and blood plasma [71]. An attractive feature of the 2D data matrix obtained is the calculation of a so-called chemical shift spectrum. It is obtained as a tilted projection of the full 2D matrix. All signals (except those with strong coupling, e.g. citrate) appear as singlets positioned at the center of their respective multiplets in the standard 1D counterpart (see Figure 3.19). Thus the complexity of the spectrum is substantially reduced. The integral of signals in chemical shift spectra is strongly influenced by T_2 relaxation. A straightforward way to use this data for absolute quantification of concentration is not possible, but comparison with a spectral database of metabolites recorded under identical experimental conditions might be promising for relative quantification. Typical experimental settings require the acquisition of 4–8 k complex data points along t_2 for 32–128 t_1 increments. An interscan delay of 2 s allows for T_1 relaxation. Exponential or shifted sine-bell windowing is required along t_2. Along t_1 a sine or squared sine multiplication together with data summation can be recommended.

Correlation Spectroscopy (COSY) [85]: In this type of spectroscopy a cross-peak is obtained if two spins are connected by a homonuclear J-coupling (over

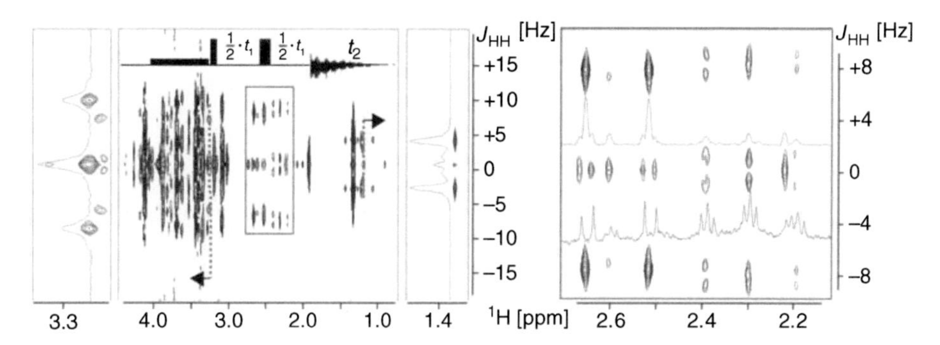

Figure 3.19. J-coupling resolved spectroscopy of rat CSF: Aliphatic region of a tilted 2D J-Res spectrum. On the panels left and right of the overview plot, expanded regions together with 1D traces are shown. The position where these 1D traces where taken are indicated in the overview plot by green dotted lines and black arrows. Along f_1 detailed information on J-coupling patterns can be extracted from the 1D traces. The red boxed region is shown in expansion on the right. Here, for comparison, the corresponding region of the 1D spectrum (black) together with the projection of the 2D dataset (red) is inserted. This demonstrates that only chemical shift information is retained in the 1D projection calculated, leading to a substantial simplification of the spectra. The pulse sequence for 2D *J*-Res is shown.

3–5 bonds). Thus information on spin-system topologies can be extracted. The cross-peaks contain fine-structure allowing for the determination of the values of active and passive J-couplings. The latter is physically not contributing to the origin of the cross-peak. These values are of importance for structure elucidation of unknown metabolites. In metabonomics also COSY45 [85] using a 45° mixing pulse and Double-Quantum COSY [86, 87] are applied. As the magnitude of the cross-peak is related to the maximum length of t_1 and t_2, the acquisition of at least 512 times 4 k data points along the respective dimensions can be recommended. Sine or squared sine window functions along both time-domain dimensions are standard. Body fluids for which this experiment has been applied together with suited references are listed in the *J*-Res section above.

Total Correlation Spectroscopy (TOCSY) [88]: In contrast to COSY described above, cross signals appear here also between pairs of spins which have a common coupling to a third or fourth (and so on) spin. This is achieved by multistep transfer of magnetization over many spins. Chemical shift information on the complete spin-system may thus be obtained for any of the members of the spin-system (see Figure 3.20). The much higher information content of the 2D TOCSY map can cause serious signal overlap on cross-peaks.[29] The size of the cross-peak is determined

[29] It must be noted that overlap is reduced compared to COSY due to more favourable line-shapes of TOCSY cross-peaks.

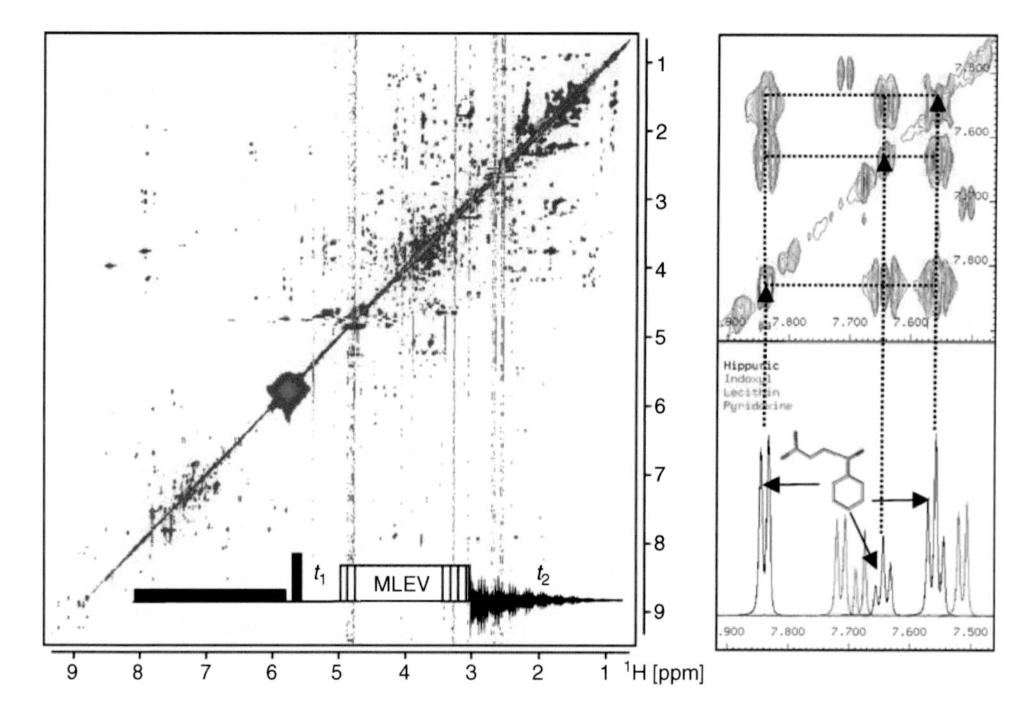

Figure 3.20. Total Correlation spectroscopy (TOCSY): The left panel shows an overview TOCSY spectrum taken from urine of one of the authors. Spin systems of metabolites can be identified as rows of cross-signals correlating the chemical shifts of all spins involved in a spin system topology. The insert (left) shows a schematic drawing of the pulse sequence. The MLEV mixing sequence serves to transfer magnetisation along J-coupled networks. The right panel shows an expanded region correlating protons of aromatic moieties. As an example the spin system of hippurate is identified (right). It is immediately clear which signals belong to the aromatic protons of this molecule. Another spin system also easily assigned in the same spectral region is that of indoxyl. Note: lecithin and pyridoxine, which only show singlet signals in the 1D, do not cause off-diagonal cross-signals in TOCSY. Accurate chemical shift measurements to distinguish both metabolites are mandatory. For comparison the 1D spectra of all four compounds are shown at the bottom (right).

by the length of a so-called mixing sequence. In metabonomics [89] the Malcolm Levitt's CPD sequence (MLEV) mixing sequence with duration of 80–300 ms is applied. Smaller t_1/t_2 data-matrices (e.g. 256*2 k data points) are possible here making the application of squared cosine shaped window functions attractive. Also applications to tissues employing HR-MAS are found in the literature [90, 73]. A combination of TOCSY with diffusion editing (300 ms diffusion delay) has also been reported [32].

3.5.2.2. Heteronuclear correlation experiments

In contrast to the direct 1D-based observation of ^{13}C information, the indirect measurement via J-coupled ^{1}H spins offers several advantages: (1) the sensitivity of these class of experiments depends on the higher gyromagnetic ratio of the protons. As a consequence, ^{13}C spins are observed with "only" 100 times reduced sensitivity compared to ^{1}H spectroscopy; (2) the chemical shift of the carbons is correlated with the chemical shift of J-coupled protons. Several experiments found in metabonomics make use of these advantages.

Heteronuclear Single Quantum Coherence (HSQC) [71, 91, 92]: The chemical shift of ^{13}C is correlated via the large one-bond coupling constant $^{1}J_{CH}$ to the ^{1}H chemical shift of the directly bound proton (see Figure 3.21). Typical experimental settings are a relaxation delay of 1 s and a refocusing delay of 4.2 ms. In total, 1024 data points with 200–300 scans for each of up to 512 t_1 increments are acquired. The spectral width is set to 140–180 ppm along the ^{13}C dimension [93]. The standard for processing is a squared cosine window function applied along both dimensions. It must be noted, however, that the acquisition time for such an experiment will be in the range of half a day making the application attractive only for selected samples.

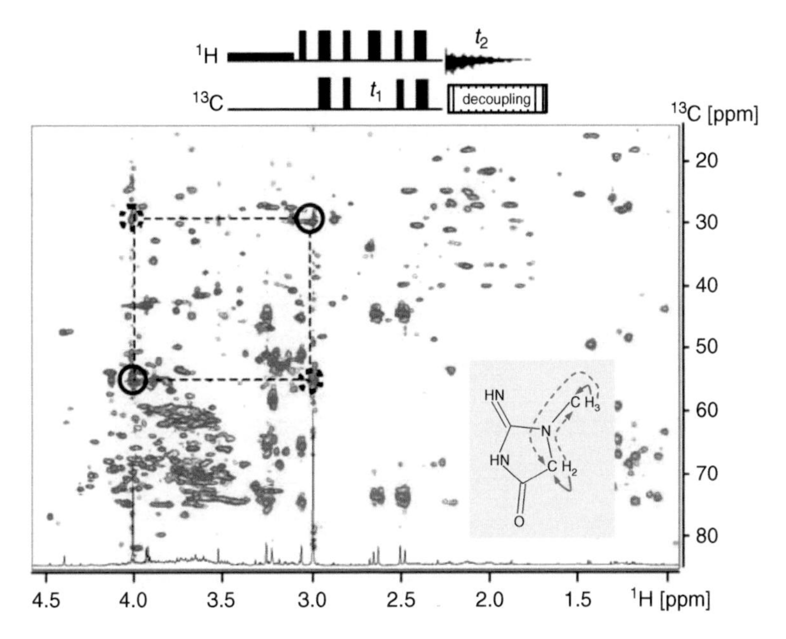

Figure 3.21. The schematic drawing of the HSQC sequence (HMBC not shown) is shown at the top. The 2D spectrum shows a superposition of a HSQC spectrum (blue) and a HMBC spectrum (red) of urine of one of the authors; only the aliphatic region is shown. Highlighted are the correlation signals for the CH_2 and CH_3 groups of creatinine. The full circles show the direct $^{1}J_{HC}$ cross-peaks of the CH_3 and CH_2 group, respectively, in the HSQC spectrum (blue); the dotted circles highlight the long range $^{3}J_{HC}$ cross-peak observed in the HMBC spectrum of the two groups. The direct and long range correlations are also indicated in the structure (insert) with full blue and dotted red arrows, respectively. Many other very informative correlations are seen in the two 2D spectra. The acquisition of these two data sets took one weekend of experimental measuring time.

Heteronuclear Multiple Quantum Coherence (HMQC) [94]: The information content obtained from this experiment is identical to that from HSQC. Applications in metabonomics [79] reported 800 scans and 128 t_1 increments. A short 500 ms interscan relaxation is described in literature [95]. The same holds true for tissue-based work employing HR-MAS [90, 73]. A single example of the application of ^{1}H-^{31}P HMQC-TOCSY [96] can also be found. Here the ^{1}H spin-system information, obtained via TOCSY transfer, is correlated with the chemical shift of phosphorous. This technique is not limited in sensitivity due to the high natural abundance of the NMR active phosphorous isotope.

Heteronuclear Multiple Bond Correlation (HMBC) [97]: With this experiment ^{13}C nuclei having a small multiple bond coupling $^nJ_{CH}$ are correlated via their chemical shifts to ^{1}H (see Figure 3.21). For each proton several cross signals related to coupled ^{13}C nuclei appear in the spectrum. The application in metabonomics [98] is hampered by the long experimental time. About 128–512 FIDs encompassing 220 ppm ^{13}C spectral width and up to 400 scans per increment are required. Along t_2, between 2 k and 32 k data points can be recorded. A squared sine-shaped window function is recommended for processing. Typically a delay of 70 ms was used for transfer of magnetization. The HR-MAS applications are described in literature [89, 90].

3.6. Data pre-processing

Data pre-processing can be seen as an intermediate step between recording of raw spectra and applying data analysis and modeling methods. Data pre-processing transforms the data in a way that subsequent analyses and modeling are easier, more robust and more accurate. For NMR spectra from metabonomic studies, pre-processing methods are usually used to reduce variances and influences, which are not of interest or which might interfere with data analysis. These variances can be caused by measurements, by biology or by combined effects. Among these effects, NMR typical effects such as incorrect phases of the spectrum or varying baselines, animal study specific effects such as varying overall concentrations of biofluids and combined effects such as shifting peak locations in the spectrum play a role. Data pre-processing methods, which account for effects typical for NMR measurement and which are routinely applied to NMR spectra in many fields, are discussed in Section 3.2.1.2. This section covers data pre-processing methods, which are more specific for metabonomics. Data pre-processing for metabonomic studies usually accounts for three effects discussed in the following sections, whereby not all pre-processing steps are mandatory.

3.6.1. Exclusion regions

Usually, the first step of data pre-processing is an exclusion of spectral regions, which contain non-reproducible information or which do not contain information about metabolites. Typically, the spectral width of recording NMR spectra is wider than chemical shifts populated by endogenous metabolites. On the one hand, a minimum spectral width is predefined by a symmetrical range around the water resonance when using water suppression techniques. On the other hand, a broad spectral width simplifies NMR pre-processing methods such as phasing and baseline corrections. For the data analysis, spectral regions not populated by endogenous metabolites are not useful. As these regions are sensitive to spectral artifacts such as inadequate phasing, exclusion is beneficial. Therefore, the spectrum outside the window of 0.2–10 ppm is usually excluded. Other parts of the spectrum, which are usually excluded, are resonances of the solvent water. Although special suppression techniques for water resonances are used, the remaining signal of water, which varies due to different pre-saturation conditions, dominates the spectrum between 4.6 ppm and 5.0 ppm rendering the analyses of signals of metabolites lying below the water resonances impossible. For spectra of urine, the signal of urea, which is very close to water, is the most dominating peak besides that of water, as urea is the most concentrated metabolite in urine. Yet, the urea peak is not quantitative, as the protons exchange with water and consequently the peak intensity varies with the quality of water suppression (see Figure 3.17). As this peak has a high intensity and a broad range between 5.4 and 6.0 ppm, and as this peak varies, it highly influences methods of multivariate data analysis unless the peak is excluded.

In many publications on the analysis of urine samples, the exclusion regions of water and of urea are combined and the complete spectral region between 4.50 and 5.98 ppm is excluded. Although this combination of exclusion areas also excludes spectral regions with signals of metabolites of interest (for example the resonance of the α-anomeric proton of glucose at 5.24 ppm), it is the *de facto* standard in the literature. Consequently, only the spectral range between 0.2 and 4.5 ppm and between 5.98 and 10 ppm is analyzed in the case of urine samples. For plasma samples and extracts of tissue the spectral ranges from 0.2 to 4.6 ppm and from 5.0 to 10 ppm are used, whereby additional exclusion regions might be necessary depending on resonances of solvents for tissue extracts.

3.6.2. Peak shifts and binning

One of the major challenges for automated data analyses and peak assignments for spectra of biofluids from metabonomic studies is the effect of shifting peaks (see Section 3.4.2). As the matrix of biofluids and, in particular, urine is highly varying, the local environment of protons also varies. Thereby many parameters play a role such as changes of the pH, changes of the salt concentration, overall dilution of the sample, relative concentrations of specific ions, relative concentration of specific metabolites and many more. All these parameters can influence the shifts of peaks, whereby not all peaks are affected and different peaks are affected to a different extent even when belonging to the same metabolite. In the left panel of Figure 3.22 an example with ^{1}H NMR spectra of 30 samples from a human metabonomic study is shown. The clipping between 3.35 ppm and 3.49 ppm shows one singlet belonging to para-hydroxyphenylacetate and one triplet belonging to taurine. The position of the singlet is very stable at 3.455 ppm whereas the position of the taurine triplet varies between 3.407 and 3.435 ppm. As the pH of the samples was constant due to buffer added, the shifts of the taurine triplet are caused by different concentrations of salts, of metal ions, of metabolites, or by varying dilutions of the samples.

In a recent publication [99], it has been shown that peak shifts can be beneficial for classifying groups of samples, if the locations of peaks are systematically different between the various groups of samples. Yet, it is not possible to trace these differences back to a single parameter such as systematic differences of pH, as many parameters interact with peak shifts. In addition, the interpretation of classifications and separations based on shifts of peaks is difficult, as the peak shifts are usually overlaid by changes of the intensity of peaks due to changes of concentrations of metabolites. Consequently, the loadings of decomposition methods (PCA, partial least squares (PLS), O-PLS) show distorted shapes of peaks, which need thorough manual interpretations. Therefore, a reduction of peak shifts either in silico or by special sample preparation methods (see Section 3.4.2.3) is most beneficial for the majority of applications. The mathematical approaches of reducing the influence of peak shifts can be classified into two categories. The first widespread category reduces the resolution of spectra (by the so-called binning or bucketing procedures), whereas methods belonging to the second category try to align peaks.

Equidistant binning: The most popular methods of reducing the influence of shifting peaks are the so-called binning or bucketing methods, which reduce the resolution of spectra.[30] Thereby the spectra are integrated within small spectral regions, which are called "bins" or "buckets". Subsequent data analysis procedures, which are applied to the binned spectra, are not influenced by peak shifts, as long as these shifts remain within the borders of the corresponding bins. The vast majority of ^{1}H-NMR metabonomics literature uses an equidistant binning of 0.04 ppm. This means that the spectrum is split into evenly spaced integral regions with a spectral width of 0.04 ppm. In addition, it is common practice to sum up bins covering the citrate doublets (two bins per doublet) into super-bins, as the shifts of the two citrate

[30] This reduction of resolution is not identical to that obtained by a lower B_0 field strength. A multiplet with a total width of 24 Hz covers 0.04 ppm at 600 MHz. The ppm-width doubles at 300 MHz.

doublets span more than one bin width each. With respect to exclusion areas the spectra are typically reduced to 206 bins visualized in Figure 3.23.

The major drawback of equidistant binning is the non-flexibility of the boundaries. If a peak crosses the border between two bins, this peak shift can significantly influence the data analysis. For example, when analyzing binned spectra by PCA or PLS, peak shifts result in loadings with opposite signs. If the spectra shown in Figure 3.22 are binned by the wide-spread equidistant binning method with a bin width of 0.04 ppm, the signal of the triplet of taurine is distributed among two bins (see Figure 3.22, left). The intensity assigned to each of the two bins highly depends on the exact location of the triplet. Thus, significant variance is added to the two bins, which is not related to the concentrations of taurine and para-hydroxyphenylacetate. Therefore the analysis of the concentrations of these metabolites on the basis of these bins is hampered.

Non-equidistant binning: To prevent peaks being cut by the boundaries of bins, binning methods have been proposed, which are based on non-equidistant spacing of bins. In the ideal case, the borders of the bins are adjusted in a way that bins cover only complete peaks including all possible locations of the peaks. This means that the bin width depends on the width of the peak shape and on the shift width of the peak. An example is shown on the right side of Figure 3.22. None of the peaks is cut by the

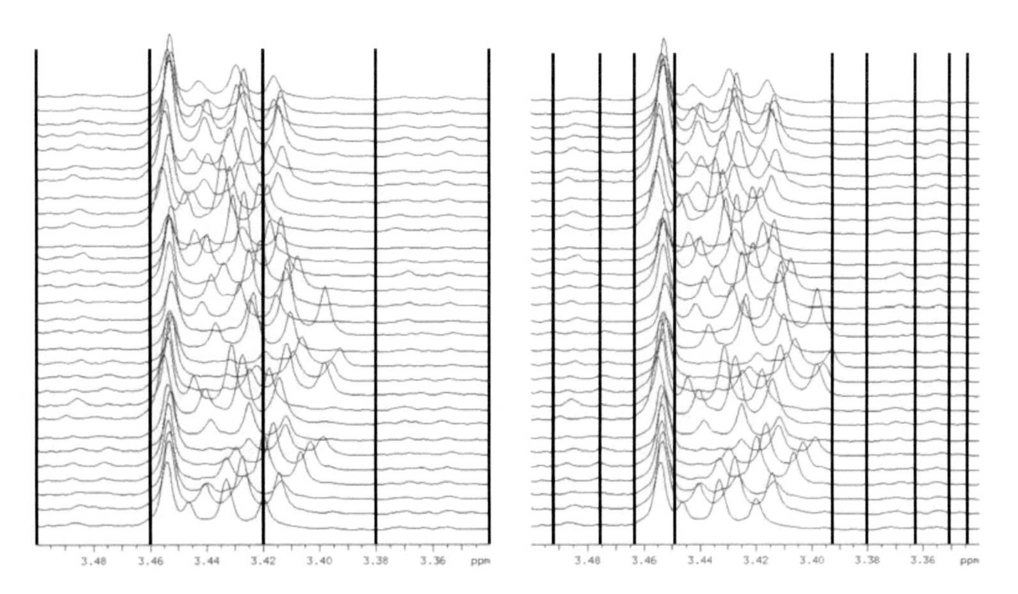

Figure 3.22. Two different binning methods applied to a set of spectra. On the left side the boundaries of the widely used equidistant binning with a bin width of 0.04 ppm are shown. The right side shows the boundaries of a non-equidistant binning method, which locates the boundaries of the bins into minima of mean spectra.

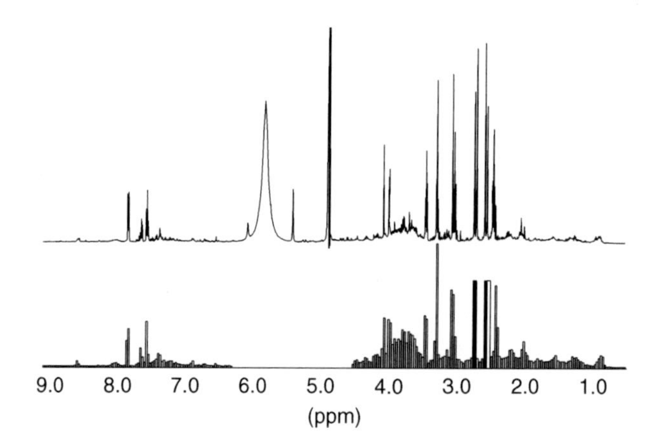

Figure 3.23. Spectrum before and after equidistant binning with a bin width of 0.04 ppm. In addition, two super bins for the citrate doublets were created and the spectral area between 4.5 and 5.98 ppm was excluded. A vast majority of ^{1}H NMR metabonomics literature uses this form of reduced spectra for multivariate data analyses.

bin boundaries. The width of the bin covering the peak of para-hydroxyphenylacetate is narrow, as the peak does not shift. On the other side, the triplet of taurine is quite wide and shows extensive shifts. Therefore, the corresponding bin is very wide.

The binning shown on the right side of Figure 3.22 is based on positioning the borders of the bins on local minima of the mean spectrum of the set of control spectra of the study. The considerable number of spectra results in a very smooth spectrum. Therefore, the local minima separate well between shifting peaks while conserving each peak in a bin. For static peaks, the hyperfine structure is split into separate bins. Overall, 604 bins were generated for the spectral range of 0–10 ppm with the water peak excluded.

The unequal bin width also has some fundamental consequences for the subsequent data analysis. For the example discussed above the bin width varied between 0.0014 and 0.46 ppm (broad peak of urea). If no further processing of the spectrum is performed, data points belonging to different bins influence the subsequent data analysis to a different extent. For example, a data point of a very small bin has a higher influence than a data point of a very wide bin. If the subsequent data analysis is to represent the data points of the raw spectra as adequate as possible, it might be appropriate to scale the integral of the bins by the inverse of the bin width. On the other side, each bin represents one single signal in the ideal case. Therefore the integral of a bin represents the "proton concentration" and with it the concentration of the metabolite behind the signal. Thus, no further scaling is needed, if the data analysis is to be based on the concentrations of metabolites and changes of concentrations of metabolites.

A combination of equidistant binning and non-equidistant binning has been proposed and applied in metabonomics studies recently [100–102]. This method starts with the traditional bucket size of 0.04 ppm and allows the adjustment of the borders by 50% resulting in a bin width between 0.02 and 0.06 ppm. The adjustment is based on finding local minima of a sum of spectra or of a skyline projection of spectra.

Peak alignment: The alignment of peaks is an alternative to reducing the spectrum to account for peak shifts. A very promising approach has been proposed recently [37]. Thereby the spectra are segmented and the segments are shifted, stretched and shrunken to maximize the correlation coefficient between the segments of a spectrum and a target spectrum. Overlapping connections after alignment are deleted and missing connections are interpolated. The optimal parameters are determined either by a genetic algorithm [37] or by a grid-search [103]. Although first results are very promising, further validation has to be performed as eliminations of peaks, changes of peak shapes and erroneous alignments cannot be excluded. Another approach for the alignment of NMR spectra of complex mixtures with the same strength and drawback has been proposed by Stoyanova [38]. In this approach peak shifts are identified by comparing a set of typical derivative shapes (e.g. derivatives of Lorentzian doublets with various coupling constants) with the loadings of the second principal component of the set of spectra. Identified regions with peak shifts are aligned to a certain average frequency, whereby gaps are filled with zeros.

3.6.3. Normalization

A crucial step in data pre-processing of spectra from metabonomic studies is the so-called normalization. The normalization step tries to account for variations of the overall concentrations of samples. Especially for urine, these variations of the overall concentrations of samples are very distinctive. Urine of animals and humans can typically be diluted by a factor of 4–5, whereby the dilution due to food deprivation or due to drug effects can exceed a factor of 10. In addition, sample preparation steps, such as addition of buffer, can further contribute to dilution of urine. In contrast to changes of the overall concentration of urine, metabonomic responses and fluxes mainly influence only a few analytes in urine. Thereby concentrations of these few metabolites change specifically. These changes are visible as relative changes of concentrations of the few metabolites related to the concentrations of all other metabolites, which represent the overall concentration of urine. Usually these relative specific changes are of interest in metabonomic studies. Therefore, a normalization step, which compensates for differences of the overall concentration, is crucial, as variations of overall concentrations obscure specific changes of metabolites. Besides compensating for changes of the overall concentration of samples, normalization can also be necessary due to technical reasons. If spectra are recorded using different number of scans or if spectra are recorded with different devices, the absolute values of the spectra are different rendering a joint analysis of the spectra without prior normalization impossible.

All normalization procedures scale complete spectra in a way that these spectra represent the same overall concentration. The scaling factor of each spectrum corresponds to the dilution factor of the corresponding sample. In general, most normalization methods (also in other scientific fields) are special formulations of the general equation:

$$I(i) = \frac{I^{\text{old}}(i)}{\left[\sum_k \int_{j_k^l}^{j_k^u} (I(x))^n dx\right]^{\frac{1}{n}}} \quad (3.6)$$

Hereby $I^{\text{old}}(i)$ and $I(i)$ are the intensities of the variable i (spectral feature, wavelength, bin, chemical shift) before and after normalization, k is an index of the spectral regions used for normalization, j_k^l and j_k^u are the lower and upper borders of the spectral region k, for which the power n of the intensities $I(x)$ are integrated.

For the integral normalization, the power n is set to 1. Special variants of integral normalization, such as creatinine normalization in metabonomics, use only parts of the spectral range. Thereby j_k^l and j_k^u mark the borders of the k parts (e.g. left and right borders of the 2 creatinine peaks). The vector length normalization, which has not been used in the field of metabonomics up to now, uses a power n of 2. In the next sections, the normalization procedures used in the field of metabonomics are discussed in more detail.

Integral normalization: This method is the *de facto* standard of normalizing NMR spectra of biofluids. Integral normalization assumes that the integrals of spectra are mainly a function of the overall concentrations of samples. A linear concentration series of urine should result in a linear series of integrals of the corresponding spectra. Influences of changes of individual concentrations of single analytes are assumed to be small compared to changes due to varying overall concentrations of urine. In addition, specific down-regulations of metabolites should balance up-regulations of metabolites to a certain extent. The integral normalization procedure divides each signal or bin of a spectrum by the integral of the spectrum or part of it. Virtually all publications dealing with ^{1}H-NMR metabonomics of urine samples exclude the spectral range of urea and water resonances from the integration and multiply each variable additionally by a factor of 100 ending up in a total integral of 100 for each spectrum [37, 104–108].

In Figure 3.24, two NMR spectra of rats from the same study are shown before normalization and after normalization. The spectrum shown in black is diluted

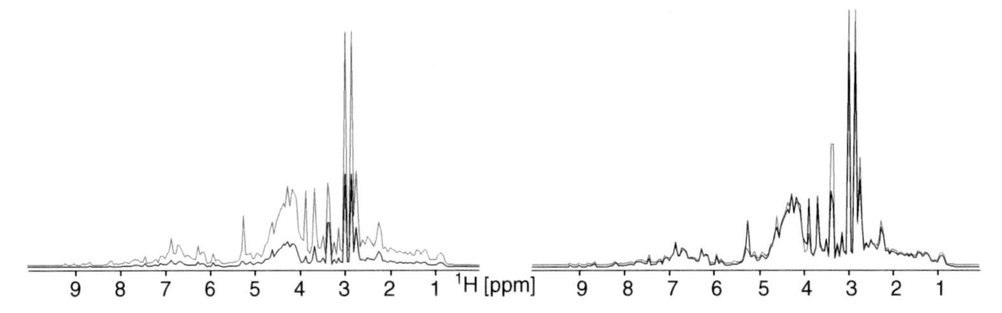

Figure 3.24. Two NMR urine spectra before normalization (left side) and after normalization (right side). The black spectrum is diluted by a factor of 3 compared to the grey spectrum. After normalization it is obvious that only very few differences of relative concentrations of metabolites between both spectra exist.

by a factor of 3 compared to the grey spectrum. Therefore the spectra look very different before normalization. The right side of the figure shows the two spectra after normalization. Only very few differences are visible. These differences are based on differences of relative concentrations of metabolites (specific differences of the metabolic profiles) and are typically of interest. Before normalization these differences were covered by the dilution effect.

Although integral normalization is the *de facto* standard for NMR measurements for metabolic profiling, the robustness, which is crucial for analysing highly varying biofluids, is the main weakness of the method. As soon as the integral of a spectrum is dominated not only by the overall concentration but also by specific changes of metabolites, the integral normalization does not scale the corresponding spectrum correctly. In Figure 3.25, an example of a sample is shown, which contains extreme

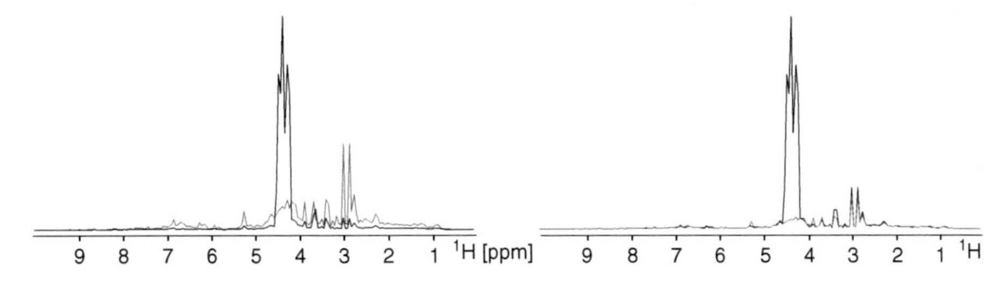

Figure 3.25. Two NMR urine spectra after integral normalization (left side) an optimal normalization (right side, quotient normalization). The black spectrum contains massive amounts of glucose due to a glucosuria of the animal. As the peaks of glucose account for 75% of the integral of the spectrum, the integral normalization method has incorrectly downscaled the spectrum by a factor of 0.25. An optimal normalization shown on the right side reveals that besides the extreme amounts of glucose hardly any differences between the two samples exist.

amounts of glucose due to glucosuria of the corresponding animal. As the peaks of glucose account for 75% of the integral of the spectrum, the integral normalization has downscaled the complete spectrum by a factor of 0.25 compared to an optimal normalization. Besides the presence of extreme amounts of endogenous metabolites, drug metabolites, for which the signals were not excluded or replaced, can also dramatically mislead the integral normalization.

Creatinine Normalisation: This method is a special version of the integral normalization originating from clinical chemistry. For the investigation of urine of humans and animals in clinical chemistry, it is a common procedure to normalize concentrations of analytes by the concentration of creatinine [111, 112–114]. The assumption is a constant excretion of creatinine into urine (often referred to as creatinine clearance). Thus, creatinine is an indicator of the concentration of urine. As the concentration of creatinine can be determined by its peaks in the spectra (peaks at 3.05 and 4.05 ppm), it has been proposed in early publications to normalize spectra by the concentration of creatinine [113].

Yet, the practical application of the creatinine normalization is faced by technical and biological difficulties. From the technical point of view, metabolites with overlapping peaks can interfere with the determination of the creatinine concentration (e.g. creatine at 3.04 ppm). In addition, the chemical shift of creatinine at 4.05 ppm depends on the pH of samples rendering a sophisticated peak picking algorithm necessary. Biological challenges for creatinine normalization are changes of the concentrations of creatinine due to metabonomic responses, which has been shown in several studies [114]. In that case, the normalization by creatinine is worthless. As at the step of normalization of spectra a possible increase of the creatinine level due to metabonomic responses is usually not yet known, a creatinine-based normalization is not of general use in metabonomics.

Quotient normalization: Recently, a new approach of normalization has been proposed [115], which is based on the assumption that changes in concentrations of single analytes only influence parts of the spectra, whereas changes of overall concentrations of samples influence the complete spectrum. Instead of the assumption of a constant total integral of the integral normalization, a most probable quotient between a spectrum and a reference spectrum is calculated as normalization factor. This most probable quotient for a specific spectrum can be derived from the distribution of each bin (variable, spectral value, data point) of a spectrum divided by the corresponding bin of a reference spectrum. In Figure 3.26, two histograms of the distributions for the quotients of two spectra each are shown. The histogram on the left side represents the quotients of the 205 bins of the black and grey spectrum shown in the left part of Figure 3.24. It is clear that the distribution of the quotients is rather sharp. The most probable quotient is located around 0.33, which corresponds to a threefold dilution of the black spectrum. The histogram on the right

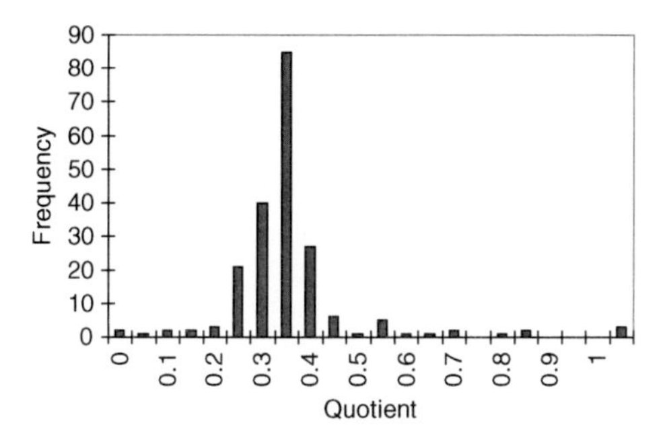

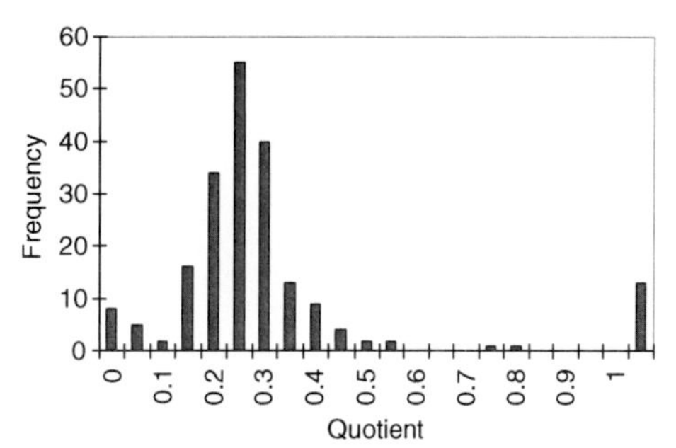

Figure 3.26. Histograms for the distributions of the quotient of two spectra each. The left histogram represents the quotients of the two spectra each shown in the left part of Figure 3.24. The dilution factor of 3 of the black spectrum is represented as most probable quotient around 0.33. The right histogram represents the quotients of the two spectra shown in the left part of Figure 3.25. The most probable quotient is around 0.25. The dominance of the glucose peaks is only visible as a rather small bar for quotients higher than 1 (last bar on the right side).

side of Figure 3.26 represents the quotients of the two spectra shown in the left part of Figure 3.25. The distribution shows that the most probable quotient is around 0.25. This corresponds to the incorrect downscaling of the black spectrum by the integral normalization, as the glucose peak contributes 75% to the total integral. The dominance of the glucose peak hardly influences the distribution and the most probable quotient. It is only visible as rather small bar on the right side, which represents quotients higher than 1.

The quotient normalization is based on the robustness of the distribution of quotients in contrast to integrals. The most probable quotient, which is needed for the normalization, can be estimated in several ways. Using the median as estimation of the most probable quotient has been proven to be a very robust and very exact method. For the calculation of the quotients, a reference spectrum is needed in contrast to integral normalization. The reference spectrum can be a single spectrum of the study, a "golden" reference spectrum from a database, or a calculated median or mean spectrum on the basis of all spectra of the study or on the basis of a subset of the study. It has been shown that the choice of reference spectrum is uncritical, and a median spectrum of control animals seems to be the most robust reference spectrum for studies with only few animals. It is also recommended to perform an integral normalization prior to quotient normalization to scale different studies to the same absolute magnitude. The algorithm of the quotient normalization can be summarized as follows:

1. Perform an integral normalization (typically a constant integral of 100 is used)
2. Choose/calculate a reference spectrum (best approach: calculate median spectrum of control samples).

3. Calculate the quotients of all variables of interest of the test spectrum with those of the reference spectrum.
4. Calculate the median of these quotients.
5. Divide all variables of the test spectrum by this median.

The quotient normalization can be applied to raw spectra or to binned spectra, but variables not containing signals should be excluded from calculating the most probable quotient if possible.

The robustness of the quotient normalization can be seen in Figure 3.25. The integral normalization is heavily hampered by massive amounts of glucose in the sample represented by the black spectrum. Therefore the spectrum shown in black is incorrectly downscaled (left part of the figure). On the right side of Figure 3.25, the results of the quotient normalization are shown. It is clear that the quotient normalization is not influenced by the glucose peak, but performs an ideal normalization: The quotient normalization reveals the high similarity of both spectra representing nearly identical relative concentration levels of most metabolites except glucose. This similarity remains hidden when using the integral normalization.

The quotient normalization performs better than the integral normalization not only in the case of very intense signals due to strong metabolic responses or due to drug metabolites, but also in the case of "normal" spectra of control animals showing only small normal metabolic variations. For example, 4023 spectra from control animals and pre-dosed animals were normalized by the total integral normalization procedure and by the quotient normalization procedure. As all animals had not been dosed, the creatinine levels of the samples were not influenced by metabonomic changes and consequently represent a good measure for the overall concentration of samples. The variations of the creatinine levels, which can be seen as performance indicator for the normalization, are 7.6% for the integral normalization and 6.7% for the quotient normalization. This means that even when looking at control animals, small specific variations due to metabonomic fluxes negatively impact the integral normalization.

A drawback of the quotient normalization compared to integral normalization is the absence of a fixed independent absolute reference value (e.g. a total integral of 100). This renders a re-normalization of a combined data set necessary, if two data sets which were normalized separately are merged or compared. This re-normalization can be prevented if a common reference spectrum is used for the separate normalizations of various data sets.

3.7. Outlook

The intrinsic properties of complex biological samples will continue to be a big challenge for NMR in metabolic analysis:

- Metabonomics fundamentally rests on large sample arrays. Perfect control of all spectral parameters (including chemical shifts) in mixtures cannot be achieved. As a consequence, assigning corresponding variables across different samples of one and the same study to the same molecular structure is of paramount importance. NMR will have to deal with "imperfect" samples although improvements in sample preparation are certainly possible.
- Currently there is a severe "assignment gap" – many NMR peaks in body fluids are not assigned to metabolites. Recourse to pattern-only analysis without knowing their most prominent molecular features is unsatisfactory for mechanistical biochemical interpretation.
- In principle, using NMR spectroscopy submicrogram quantities can be quantified. However, this is possible in practice only if signals of high amplitude do not overlap with small signals. This prerequisite is not fulfilled for biological samples, where concentration ranges of metabolites can span several orders of magnitude. Large changes of compounds that are low in concentration can only be detected if signal overlap is avoided. Here liquid chromatography (LC)-based separation techniques might come into play combined with NMR. It might well be that the fundamental "dogma" of metabonomics "no sample preparation needed" will soon have to be revisited.

Thus from the NMR spectroscopic point of view, complete profiling of biological mixtures will remain difficult for very fundamental technology reasons.

The ongoing development of NMR is addressing some of these problems:

- The trend to higher field-strength will improve the analysis of crowded spectra. More metabolites will become visible without overlap of lines. The 1 GHz magnet is within reach. Nevertheless the dynamic range problem in signal intensity combined with signal overlap will always cause difficulties in the detailed analysis of metabolites that are low in concentration.
- The introduction of cryogenic probeheads has improved the sensitivity of detection within the last 10 years by a factor of 4–5. Now it is a fact that for aqueous biological samples the "sample noise" is one limiting factor of NMR sensitivity. It seems unlikely that NMR devices will be available in the mid-term future which will allow the detection of compounds which are orders of magnitude lower in concentration than those analyzed today. We estimate that less than 100 metabolites in urine provide more than 95% of the total signal amplitude, and this will be true at any given sensitivity.
- The low concentration and the high range to be detected prevent the application of multi-dimensional NMR techniques, which could resolve the overlap problem for large sample arrays. Due to the extensive measuring times needed for diluted samples, these methods can be employed only for very focused questions like the *de novo* determination or verification of the structure of an unknown endogenous

or xenobiotic metabolite. There is no obvious gap in the availability of NMR pulse sequences. Nevertheless improvement in details (see Section 3.5.1.1) is needed to adjust the pulse sequence to the given sample and problem.

Having said that, we foresee the two following main developments in NMR methods that will open new avenues in metabonomic analysis:

- Currently different vendors of NMR spectrometer and industrial research groups are building up spectral databases [116] of metabolites known to be present in biological samples at concentrations visible in NMR. The aim of these efforts is to develop an analysis tool for *a priori* assignment of spectra of biofluids to individual metabolites. The analysis rests today almost exclusively on multivariate analyses of spectral features and patterns, in order to identify and extract significant peaks which are *a posteriori* assigned. The *a priori* assignment would open up a window to concentration values of hundreds of known metabolites [117]. Different software companies [116, 118, 119] are working on algorithms to allow extraction of this information by interactive analysis of 1D NMR spectra.
- A second development expected to have impact in the field is the concerted analysis of data acquired with different analytical methods on the same sample (NMR, GC-MS, LC-MS) [120]. The richness of information to be obtained by this approach cannot be overestimated.

We expect that both developments will have a substantial impact on the proliferation of metabolite profiling methods in wide fields of discovery and applied medical research.

References

[1] K. Wüthrich, *NMR of Proteins and Nucleic Acids*, Wiley and Sons, New York, 1986.
[2] D.G. Gadian, *NMR and its application to living systems* Oxford University Press, Oxford, 1995.
[3] F.F. Brown, I.D. Camplell, P.W. Kuchel, D.C. Rabenstein, *FEBS Lett.* 82 (1977) 12.
[4] J.R. Bales, D.P. Higham, I. Howe, J.K. Nicholson, P.J. Sadler, *Clin. Chem.* 30 (1984) 426.
[5] J.K. Nicholson, I.D. Wilson, *Prog. NMR Spectrosc.* 21 (1989) 449.
[6] Bell, J.D. *MRS in biology and medicine*, Eds J.D. de Certaines, W.M.J. Bovee, F. Bodo, Pergamon Press, Oxford.
[7] J.K. Nicholson, J. Conelly, J.C. Lindon, E. Holmes *Nature Reviews, Drug Disc.* 1 (2002) 153.
[8] J.C. Lindon, J.N. Nicholson, E. Holmes, H. Antii, E. Bollard, H. Keun, O. Beckonert, T. Ebbels, M. Reily, D. Robertson, G. Stevens, P. Luke, A. Breau, G. Cantor, R. Bible, U. Niederhauser, G. Schlotterbeck, H. Senn, U. Sidelmann, A. Laursen, A. Tymiak, B. Car, L. Lehman, J.-M. Colet, C. Thomas, *Tox. Appl. Pharm.* 187 (2003) 137.
[9] A.R. Fernie, R.N. Tethewey, A.J. Krotzky, L.W. Willmitzer, *Nature Reviews, Molec. Cell Biol.* 5 (2002) 1.

[10] R. Ernst, G. Bodenhausen, A. Wokaun, *Principles of Nuclear Magnetic Resonance in One and Two Dimensions*, Oxford University Press, Oxford, 1990.
[11] M. Goldman, *Quantum Description of High-Resolution NMR in Liquids*, Oxford University Press, Oxford, 1991.
[12] J.L. Sudmeier, S.E. Anderson, J.S. Frye, *Conc. Magn. Reson.* 2 (1990) 197.
[13] Q.N. Van, G.N. Chmurny, T.D. Veenstra, *Biochem. Biophys. Res. Comm.* 301 (2003) 952.
[14] P.T. Callaghan, *Principles of Nuclear Magnetic Resonance Microscopy*, Oxford University Press, Oxford, 1991.
[15] P.F. Flynn, D.L. Mattiello, H.D.W. Hill, A.J. Wand, *J. Am. Chem. Soc.* 122 (2000) 4824.
[16] D. Moskau, *Concepts Magn. Reson.* 15 (2002) 164.
[17] G. Schlotterbeck, A. Ross, R. Hochstrasser, H. Senn, T. Kühn, D. Markek, O. Schett, *Anal. Chem.* 74 (2002) 4464.
[18] D.L. Olson, T.L. Peck, A.G. Webb, R.L. Magin, J.V. Sweedler, *Science* 270, (1995) 1967.
[19] T. Horiuchi, M. Takahashi, J. Kikuchi, S. Yokoyama, H. Maeda, *J. Magn. Reson.* 174 (2005) 34.
[20] J.C. Hoch, A.S. Stern , *NMR Data Processing*, John Wiley & Sons Inc. New York, 1997.
[21] D.D. Traficante, M. Rajabzadeh, *Conc. Magn. Reson.* 12 (2000) 83.
[22] V. Silvestre, S. Goupry, M. Trierweiler, R. Robins, and S. Akoka, *Anal. Chem.* 73 (2001) 1862.
[23] R. Freeman, E. Kupce, *NMR in Biomed.* 10 (1997) 272.
[24] I. Solomon, *Phys. Rev.* 99 (1955) 559.
[25] W.P. Aue, E. Bartholdi, R.R. Ernst, *J. Chem. Phys.* 64 (1976) 2229.
[26] A. Ross, G. Schlotterbeck, W. Klaus, H. Senn, *J. Biomol. NMR* 16 (2000) 139.
[27] O. Cloarec, M.-E. Dumas, A. Craig, R.H. Barton, J. Trygg, J. Hudson, C. Blancher, D. Gauguier, J.C. Lindon, E. Holmes, J. Nicholson, *Anal. Chem.* 77 (2005) 1282.
[28] H.M. McConnell, *J. Chem. Phys* 28 (1958) 430.
[29] L.Y. Lian, G.C.K. Roberts, *NMR of Macromolecules – A Practical Approach*, IRL Press, Oxford, 1993.
[30] C.J.L. Silwood, M. Grootveld, E. Lynch, *J. Biol. Inorg. Chem.* 7 (2002) 46.
[31] E.O. Stejskal, J.E.J. Tanner, *Chem. Phys.* 42 (1965) 288.
[32] M. Liu, J.K. Nicholson, J.C. Lindon, *Anal. Chem.* 68 (1996) 3370.
[33] J.D. Otvos, E.J. Jeyarajah, D.W. Bennett, *Clin. Chem.* 37 (1991) 377.
[34] M. Pellechia, D.S. Sems, K. Wüthrich, *Nature Rev. Drug Disc* 1 (2002) 211.
[35] P.J. Hajduk, T. Gerfin, J.M. Boehlen, M. Häberli, D. Marek, S.W. Fesik, *J. Med. Chem.* 42 (1999) 2315.
[36] J.C. Lindon, E. Holmes, J.K. Nicholson, *Prog. in NMR Spectrosc.* 45 (2004) 109.
[37] J. Forshed, I. Schuppe-Koistinen, S.P. Jacobssen, *Anal. Chim. Acta* 487 (2003) 189.
[38] R. Stoyanova, A.W. Nicholls, J.K. Nicholson, J.C. Lindon, T.R. Brown, *J. Magn. Reson.* 170 (2004) 329.
[39] G.C. Lee, D.L. Woodruff, *Anal. Chim. Acta.* 513 (2004) 413.
[40] T.L. Peck, R.L. Magin, P.C. Lauterbur, *J. Magn. Reson.* 108 (1995) 114.
[41] G.E. Martin, R.C. Crouch, A.P. Zens, *Magn. Reson. Chem.* 36 (1998) 551.
[42] W.L. Fitch, G. Detrer, C.P. Holmes, J.N. Shoolery, P.A. Keifer, *J. Org. Chem.* 59 (1994) 7955.
[43] D.L. Olson, J.A. Norcross, M. O'Neil-Johnson, P.F. Molitor, D.J. Detlefsen, A.G. Wilson, T.L. Peck, *Anal. Chem.* 76 (2004) 2966.
[44] H. Kocacs, D. Moskau, M. Spraul, *Prog. NMR Spectrosc.* 46 (2005) 131.
[45] A. Geiger, M. Holz, *J. Phys. E: Sci. Instrum.* 13 (1980) 697.
[46] C.G. Wade, R.D. Johnson, S.B. Philson, J. Strouse, F.J. McEnroe, *Anal. Chem.* 61 (1989) 107A.

[47] M. Spraul, M. Hofmann, M. Ackermann, A.W. Nicholls, S.J.P. Damment, J. N. Halselden, J.P. Shockcor, J.N. Nicholson, J.C. Lindon, *Anal. Commun.* 34 (1997) 339.

[48] P.A. Keifer, *Drug Disc. Today* 2 (1997) 468.

[49] A. Ross, H. Senn, *Drug Disc. Today* 6 (2001) 583.

[50] A. Ross, H. Senn, *Curr. Top. Med. Chem.* 3 (2003) 55.

[51] N.T. Holland, M.T. Smith, B. Eskenazi, M. Bastaki, *Mutat. Res.* 543 (2003) 217.

[52] COMET Consortium internal presentation O. Beckonert, 2002.

[53] S. Deprez, B.C. Sweatman, S.C. Connor, J.N. Haselden, C.J. Waterfield, *J. Pharm. Biomed. Anal.* 30 (2002) 1297.

[54] D.K. Hansen, M.I. Davies, S.M. Lunte, C.E. Lunte, *J. Pharm. Sci.* 88 (1999) 14.

[55] K.E. Price, S.S. Vandaveer, C.E. Lunte, C.K. Larrive, *J. Pharm. Biomed. Anal.* 38 (2005) 904.

[56] P. Khandelwal, C.E. Beyer, Q. Lin, L.E. Schechter, A.C. Bach, *Anal. Chem.* 76 (2004) 4123.

[57] P. Khandelwal, C.E. Beyer, Q. Lin, P. McGonigle, L.E. Schechter, A.C. Bach, *J. Neurosci. Methods* 133 (2004) 181.

[58] O. Fiehn, *Plant Molec. Biol.* 48 (2002) 155.

[59] J.C. Lindon, J.K. Nicholson, E. Holmes, J.R. Everett, *Concepts Mag. Reson.* 289 (2000) 12.

[60] H.C. Keun, T.M.D. Ebbels, H. Antti, M.E. Bollard, O. Beckonert, G. Schlotterbeck, H. Senn, U. Niederhauser, E. Holmes, J.C. Lindon, J.K. Nicholson, *Chem. Res. Toxicol.* 15 (2002) 1380.

[61] B.C. Sweatman, R.D. Farrant, E. Holmes, F.Y. Ghauri, J.K. Nicholson, J.C. Lindon, *J. Pharm. Biomed. Anal.* 11 (1993) 651.

[62] B.C.M. Potts, A.J. Deese, G.J Stevens, M.D. Reily, D.G. Robertson, J. Theiss, *J. Pharm. Biomed. Anal.* 26 (2001) 463.

[63] R.D. Farrant, J.C. Lindon, E. Rahr, B.C. Sweatman, *J. Pharm. Biomed. Anal.* 10 (1992) 141.

[64] Ciba-Geigy, Eds, *Wissenschaftliche Tabellen Geigy*, vol. 1, 8th edn, Ciba-Geigy, Basel, 1980.

[65] N.N Greenwood, A. Earnshaw, *Chemistry of the elements*, Pergamon Press, Oxford, 1986.

[66] W.S Prince, *Ann. Rep. NMR Spectr.* 38 (1999) 289.

[67] D.I. Hoult , *J. Magn. Reson* 21 (1976) 337.

[68] A. Kumar, R.R. Ernst, K. Wüthrich, *Biochem. Biophys. Res. Commun.* 95 (1980) 1.

[69] R.J. Ogg, P.B. Kingsley, J.S. Taylor, *J. Magn. Reson.* 104B (1994) 1.

[70] E.M. Lenz, J. Bright, I.D. Wilson, S.R. Morgan, A.F.P. Nash, *J. Pharm. Biomed. Anal.* 33 (2003) 1103.

[71] J.K. Nicholson, P.J.D. Foxall, M. Spraul, R.D. Farrant, J.C. Lindon, *Anal. Chem.* 67 (1995) 793.

[72] S. Meiboom, and D. Gill, *Rev. Sci. Instrum.* 29 (1958) 688.

[73] M.E. Bollard, S. Garrod, E. Holmes, J.C. Lindon, E. Humpfer, M. Spraul, and J.K. Nicholson, *Magn. Reson. Med.* 42 (2000) 201.

[74] S.J. Gibbs, and C.S. Johnson Jr, *J. Magn. Reson.* 93 (1991) 395.

[75] G. Wider, V. Dötsch, K. Wüthrich, *J. Magn. Reson.* 108A (1994) 255.

[76] K.F. Morris, C.S. Johnson Jr, *J. Am. Chem. Soc.* 114 (1992) 3139.

[77] J.L. Griffin, H.J. Williams, E. Sang, and J.K. Nicholson, *Magn. Reson. Med.* 46 (2001) 249.

[78] G.A. Morris, R. Freeman, *J. Am. Chem. Soc.* 101 (1979) 770.

[79] H.C. Keun, O. Beckonert, J.L. Griffin, C. Richter, D. Moskau, J.C. Lindon, J.K. Nicholson, *Anal. Chem.* 74 (2002) 4588.

[80] J.G. Jones, M.A. Solomon, A.D. Sherry, F.M.H. Jeffrey, C.R. Malloy, *Am. J. Physiol.* 275 (1998) E843.

[81] H. Günther, *NMR Spectroscopy,* Wiley & Sons Inc. New York, 1992.

[82] W.P. Aue, J. Karhan, R.R. Ernst, *J. Chem. Phys.* 64 (1976) 4226.

[83] E. Holmes, P.J.D. Foxall, M. Spraul, R.D. Farrant, J.K. Nicholson, J.C. Lindon, *J. Pharm. Biomed. Anal.* 15 (1997) 1647.

[84] M.J. Lynch, J. Masters, J.P Pryor, J.C. Lindon, M. Spraul, P.J.D. Foxall, J.K. Nicholson, *J. Pharm. Biomed. Anal.* 12 (1994) 19.
[85] A. Bax, and R. Freeman, *J. Magn. Reson.* 44 (1981) 542.
[86] A. Derome, and M. Williamson, *J. Magn. Reson.* 88 (1990) 177.
[87] B. Ancian, I. Bourgeois, J.-F. Dauphin, and A.A. Shaw, *J. Magn. Reson.* 125A (1997) 348.
[88] A. Bax, and D.G. Davis, *J. Magn. Reson.* 65 (1985) 355.
[89] A.W. Nicholls, E. Holmes, J.C. Lindon, J.P. Shockcor, R.D. Farrant, J.N. Haselden, S.J.P. Damment, C.J. Waterfield, J.K. Nicholson, *Chem. Res. Toxicol.* 14 (2001) 975.
[90] S. Garrod, E. Humpfer, M. Spraul, S.C. Connor, S. Polley, J. Connelly, J.C. Lindon, J.K. Nicholson, E. Holmes, *Magn. Reson. Med.* 41 (1999) 1108.
[91] A.G. Palmer III, J. Cavanagh, P.E. Wright, M. Rance, *J. Magn. Reson.* 93 (1991) 151.
[92] L.E. Kay, P. Keifer, T. Saarinen, *J. Am. Chem. Soc.* 114 (1992) 10663.
[93] S. Ringeissen, S.C. Connor, H.R. Brown, B.C. Sweatman, M.P. Hodson, S.P. Kenny, R.I. Haworth, P. Mcgill, M.A. Price, M.C. Aylott, D.J. Nunez, J.N. Haselden, C.J. Waterfield, *Biomarkers* 8 (2003) 240.
[94] R.E. Hurd, B.K. John, *J. Magn. Reson* 91 (1991) 648.
[95] A.W. Nicholls, J.K. Nicholson, J.K. Haselden, C.J. Waterfield, *Biomarkers* 5 (2000) 410.
[96] M. Coen, S.U. Rüpp, J.C. Lindon, J.K. Nicholson, F. Pognan, E. Lenz, I.D. Wilson, *J. Pharm Biomed. Anal.* 35 (2004) 93.
[97] A. Bax, M.F. Summers, *J. Am. Chem. Soc.* 108 (1986) 2093.
[98] M.-E. Dumas, C. Canlet, F. André, J. Vercauteren, A. Paris, *Anal. Chem.* 74 (2002) 2261.
[99] O. Cloarec, M.E. Dumas, J. Trygg, A. Craig, R.H. Barton, J.C. Lindon, J.K. Nicholson E. Holmes, *Anal. Chem.* 77 (2005) 517.
[100] B. Lefebvre, S. Golotvin, L. Schoenbachler, R. Beger, P. Price, J. Megyesi, R. Safirstein, Intelligent Bucketing for Metabonomics – Part 1, Poster, http://www.acdlabs.com/download/publ/2004/enc04/intelbucket.pdf.
[101] B. Lefebvre, R. Sasaki, S. Golotvin and A. Nicholls, Intelligent Bucketing for Metabonomics – Part 2, Poster, http://www.acdlabs.com/download/publ/2004/intelbucket2.pdf.
[102] L. Schnackenberg, R.D. Beger, Y. Dragan, *Metabolomics*, 1 (2005) 87.
[103] G. Lee and D.L. Woodruff, *Anal. Chim. Acta* 513 (2004) 413.
[104] H.C. Keun, T.M.D. Ebbels, H. Antti, M.E. Bollard, O. Beckonert, E. Holmes, J.C. Lindon and J.K. Nicholson, *Anal. Chim. Acta* 490 (2003) 265.
[105] H. Antti, M.E. Bollard, T. Ebbels, H. Keun, J.C. Lindon, J.K. Nicholson and E. Holmes, *J. Chemometrics* 16 (2002) 461.
[106] J.T. Brindle, J.K. Nicholson, P.M. Schofield, D.J. Grainger, E. Holmes, *Analyst* 128 (2003) 32.
[107] O. Beckonert, M.E. Bollard, T.M.D. Ebbels, H.C. Keun, H. Antti, E. Holmes, J.C. Lindon, J.K. Nicholson, *Anal. Chim. Acta* 490 (2003) 3.
[108] T. Ebbels, H. Keun, O. Beckonert, H. Antti, M. Bollard, E. Holmes, J. Lindon, J. Nicholson, *Anal. Chim. Acta* 490 (2003) 109.
[109] The Merck Veterinary Manual, http://www.merckvetmanual.com/mvm/index.jsp?cfile=htm/bc/150204.htm.
[110] P. Jatlow, S. McKee, S.Ó. Malley, *Clin. Chem.* 49 (2003) 1932.
[111] Sterlin Reference Laboratories, http://www.sterlingreflabs.com/thccrearatios.html.
[112] G. Fauler, H.J. Leis, E. Huber, C. Schellauf, R. Kerbl, C. Urban and H. Gleispach, *J. Mass Spectrom.* 32 (1997) 507.
[113] E. Holmes, P.J.D. Foxall, J.K. Nicholson, G.H. Neild, S. M. Brown, C.R. Beddell, B.C. Sweatman, E. Rahr, J.C. Lindon, M. Spraul, P. Neidig, *Anal. Biochem.* 220 (1994) 284.
[114] J.P. Shockor and E. Holmes, *Curr. Top. Med. Chem.* 2 (2002) 35–51.

[115] F. Dieterle, A. Ross, G. Schlotterbeck, H. Senn, *European Patent Application* 05006476.5 (24.03.2005).
[116] AMIX Metabolite Database, Bruker BioSpin AG, Karlsruhe, Germany.
[117] F. Dieterle, A. Ross, G. Schlotterbeck, H. Senn, *Anal. Chem.* 78 (2006) 3551.
[118] S.W. Provencher, *NMR Biomed.* 14 (2001) 260.
[119] Chenomx Eclipse, Chenomx Inc. Edmonton, Canada.
[120] C.D. Eads, I. Noda, J. *Am. Chem. Soc.* 124 (2002) 1111.

The Handbook of Metabonomics and Metabolomics
John C. Lindon, Jeremy K. Nicholson and Elaine Holmes (Editors)

Chapter 4

High-Resolution Magic Angle Spinning NMR Spectroscopy

Jin-Hong Chen and Samuel Singer

Sarcoma Program, Memorial Sloan-Kettering Cancer Center, New York, NY 10021, USA

4.1. Introduction

Nuclear magnetic resonance (NMR) is an incredibly versatile technique. It emerged from physics [1, 2] in the mid-1940s and has since then achieved great success in chemistry, biochemistry, physiology and medicine. NMR techniques can be used to determine molecular composition, structure, dynamics and molecular reactions and provide information on all forms of matter such as solid, liquid, liquid-crystal and gas states. The determination of macromolecular structures serves as an excellent example of the power and elegance of NMR as multiple techniques are used for signal assignment and measurement of distance and angle constraints [3, 4]. Compared to crystal diffraction techniques for structure determination, NMR obtains the three-dimensional (3D) macromolecular structure of a protein in solution, which more realistically represents the structure and function of a protein in its natural state.

NMR applications include spectroscopy and magnetic resonance imaging (MRI). NMR spectroscopy techniques have been developed and applied to liquid-state systems, solid-state systems and intact organisms using local spectroscopy combined with MRI. The NMR techniques and experiments used for each of these systems are very different because the sample properties, especially the spin interactions that determine the NMR spectrum, are quite distinct and unique for each system. For example, coherence transfer is largely through J-coupling in liquid-state NMR but through cross-polarization utilizing the large dipolar coupling in solids. While both liquid- and solid-state NMR spectroscopy usually acquires signals from the whole of a small sample, localized spectroscopy acquires signal from a specific voxel determined by magnetic field gradients applied over a large region. In the application of NMR for samples in different phases, liquid-state NMR offers the highest resolution and sensitivity.

In biomedicine, MRI is the most successful magnetic resonance technique and has become a routine clinical tool. However, in most MRI applications only the large water signal can be used for imaging because of the technique's low sensitivity. This largely limits the power of MRI for detecting certain disease states and processes. The ability to use information on additional less abundant molecular species in the tissue would significantly increase the power of NMR techniques in medicine. *In vivo*, localized spectroscopy detects more of these weak signals but is still limited to the relatively abundant signals in the tissue such as *N*-acetylaspartate (NAA), lactate, choline, and so on.

The limitation of the low detection sensitivity and resolution of localized spectroscopy may be partially addressed by performing *ex vivo* experiments on spectrometers with substantially higher field strengths and more sensitive detection probes than conventional MRI instruments. Even at high field strengths the spectral sensitivity and resolution are still limited inherently by the multiple line-broadening mechanisms, most notably dipolar coupling and susceptibility heterogeneity, existing in biological samples such as tissue and cells.

An NMR technique, originally developed for solid-state NMR by Andrew [5] and Lowe [6], has the ability to average multiple line-broadening mechanisms and has been used to acquire high-resolution spectra of tissue and cell samples. This technique is high-resolution magic angle spinning (hr-MAS) NMR spectroscopy. In this chapter we will present the basic principles of this technique, outline the mechanisms of line-broadening in tissue and cell samples and discuss the averaging by MAS of these mechanisms. A formal description of the principles of MAS averaging can involve a copious amount of tedious mathematical treatments and multiple frame-transformations. However, in this chapter, the theoretical discussion will be focused only on the MAS coherence averaging of the two line-broadening mechanisms that are important in acquiring spectra of a tissue or cell sample, namely dipolar coupling and heterogeneous isotropic susceptibility. A special emphasis will be placed on MAS detection of molecules with rapid axial rotation, an important molecular motion for phospholipids and small proteins in membrane structures. The final part of this chapter will discuss the pulse sequence techniques used in hr-MAS NMR spectroscopy and show some application examples in cell and tissue studies.

4.2. Nuclear spin Hamiltonian

4.2.1. The nuclear spin Hamiltonian and line-broadening by the local field effect

The mechanism of line-broadening may be best illustrated using the localized field-dependent spin Hamiltonian. The spin Hamiltonian depends on the multiple interaction of spins with the external fields and with other internal spins. We only include the relevant terms that are important in our discussion. In the absence of an *rf* field, the spin Hamiltonian may be written as

$$H(r) = H_0 + H_{DD}(r) + H_{sus}(r) \tag{4.1}$$

in which $H_0 = -\sum_j \gamma^j B_0 I_z^j$ is the interaction of spins with the external static field B_0 where γ is the gyromagnetic ratio and I is spin angular momentum and the summation is over all the spins in the sample. In this term we assume the external field is homogeneous throughout the sample based on the assumptions that a very homogeneous static field can be achieved through shimming and that the inhomogeneity can be neglected especially compared to the remaining terms in the spin Hamiltonian. The second term in Equation 4.1 comes from the dipolar coupling. In a general treatment, this dipolar coupling can be intramolecular or intermolecular. The third term comes from the heterogeneous susceptibility. Magnetic materials have a finite magnetic susceptibility and when placed in an external field, this modifies the magnetic field both inside and outside the material. For a theoretical treatment of a homogeneous liquid sample, this effect may be included into the static field with a slight modification. However, in tissue and cells this term is the major line-broadening mechanism for small molecules dissolved in cell cytosol and therefore deserves careful consideration.

Both the dipolar and susceptibility terms are position dependent in the sample. This stems from the fact that the local magnetic field created by dipolar coupling and by heterogeneous susceptibility varies from position to position and the resonance frequency thus varies as well. The spreading of the frequency from this local field difference causes the line-broadening.

4.2.2. The Dipole–dipole interaction spin Hamiltonian

4.2.2.1. Dipolar–dipolar coupling of paired spins

Each spin has a magnetic moment $\mu = \gamma \hbar \vec{I}$ inducing a dipolar magnetic field around it. A second neighboring spin interacts with this field. This interaction is mutual and is called through-space dipole–dipole coupling. The dipole–dipole coupling of two spins depends on the magnitude and orientation of their magnetic moments and also on the distance and orientation of the vector describing their relative positions. We start with a pair of rigid spins at a fixed distance apart and the well-known dipolar coupling is

$$H_{\mathrm{DD}}^{ij} = \frac{q_{ij}}{r_{ij}^3}[\vec{I}_i\vec{I}_j - 3(\vec{I}_i \cdot \vec{e}_{ij})(\vec{I}_j \cdot \vec{e}_{ij})] \tag{4.2}$$

in which $q = \frac{\mu_0}{4\pi}\gamma_i\gamma_j$ and $\mu_0 = 4\pi \times 10^{-7}\mathrm{NA}^{-2}$ the induction constant or magnetic constant, $\vec{e}_{ij}$ is the unit vector from spin i to spin j and r_{ij} is the distance between them. The magnitude of the dipolar coupling is inversely proportional to the cubic distance. For example, two ^{1}H spins on a membrane phospholipid chain methylene group separated by a distance of 0.173 nm have a dipolar interaction of approximately 23.2 kHz. In a solid crystal, the sum of the dipolar couplings to a given spin may be much larger because there can be multiple neighbor protons in close proximity.

In high magnetic field applications, the non-secular part of the dipolar coupling in Equation 4.2 can be dropped. The secular part of dipolar coupling depends on whether the interacting spins belong to the same species. The secular approximation is based on the energy level difference which depends on whether the spin system is homonuclear or heteronuclear [7]. The secular homonuclear dipolar coupling is

$$H_{\mathrm{DD,homo}}^{ij} = \frac{q_{ij}}{r_{ij}^3}\frac{3\cos^2\theta_{ij} - 1}{2}(3I_{iz}I_{jz} - \vec{I}_i \cdot \vec{I}_j) \tag{4.3}$$

where θ_{ij} is the angle between the vector $\vec{e}_{ij}$ and the static magnetic field. The secular heteronuclear dipolar coupling is

$$H_{\mathrm{DD,hetero}}^{ij} = \frac{q_{ij}}{r_{ij}^3}\frac{3\cos^2\theta_{ij} - 1}{2}2I_{iz}I_{jz} \tag{4.4}$$

The spin Hamiltonian is given by the sum over each spin pair in the sample

$$H_{\mathrm{DD}} = \sum_{i \neq j} H_{\mathrm{DD}}^{ij} \tag{4.5}$$

In the homonuclear dipolar coupling spin Hamiltonian, the spin operator $\vec{I}_i \cdot \vec{I}_j$ includes an $\vec{I}_{i\pm}\vec{I}_{j\mp}$ term which generates fast spin-diffusion in the molecules. Fast spin-diffusion from net dipolar coupling is a very rapid magnetization transfer mechanism in tissue and cell samples [8].

4.2.2.2. Motional averaging

The dipolar coupling tensor depends on the direction and distance of the paired spins. Any changes in direction or distance of the paired spins would influence the dipolar coupling. As a result, the strength of dipolar coupling depends strongly on the type and the time scale of molecular motion. Molecular motion modes include rotational tumbling of individual molecules, relative translational motion of molecules and migration of atoms or group of atoms between molecules. Molecules can have internal motion as well, such as vibration, rotation of the internal groups and chemical isomerizations. For a molecule undergoing a rapid motion, the dipolar coupling fluctuates in time. The dipolar coupling Hamiltonian can be formally split into two parts, a time invariant component and a fluctuating component.

$$H_{\mathrm{DD}}(t) = \overline{H}_{\mathrm{DD}} + H^{1}_{\mathrm{DD}}(t) \tag{4.6}$$

with

$$\overline{H}_{\mathrm{DD}} = \tau^{-1} \int_{0}^{\tau} H_{\mathrm{DD}}(t)dt \tag{4.7}$$

and

$$H^{1}_{\mathrm{DD}}(t) = H_{\mathrm{DD}}(t) - \overline{H}_{\mathrm{DD}} \tag{4.8}$$

The time-fluctuating term $H^{1}_{\mathrm{DD}}(t)$ accounts for spin relaxation but the details of relaxation by this fluctuating Hamiltonian will not be discussed in this chapter. In a solid sample, the molecular motions may be less but still existent. The large un-averaged dipolar coupling is the primary source of the NMR spectral line shape and linewidth in a solid state NMR spectrum, whilst the time-dependent component of dipolar coupling is responsible for relaxation. In gases and isotropic liquids, the fast tumbling of molecules completely averages the dipolar coupling $\overline{H}_{\mathrm{DD}}$ to zero [7]; and hence the dipolar coupling is essentially a pure relaxation mechanism in these samples. For an anisotropic liquid, there is preferential molecular orientation. In this medium, the dipolar coupling is not completely averaged out. The un-averaged smaller residual dipolar couplings seen in anisotropic liquids are very useful for molecular structure determination. Currently, dissolving molecules in anisotropic liquid to determine the macromolecular structure using the residual dipolar coupling is one of the most promising and active fields of study in liquid-state NMR [9–11]. Cell membranes share similar NMR properties to anisotropic liquid crystals. Based on our interest in the detection of cell membrane phospholipids by MAS, we will now outline how molecular axial rotation results in partial averaging of the homonuclear dipolar coupling in cell and tissue systems.

In an anisotropic liquid, either the molecules have a preferential orientation or the molecular motion is confined in some direction. Figure 4.1a illustrates molecules in a small region of a membrane bilayer. These molecules orient along the membrane surface normal direction (perpendicular to the surface of the membrane shown using the broad arrow). These molecules have a much faster axial rotation than translation. In Figure 4.1a, the intramolecular dipolar couplings are illustrated by

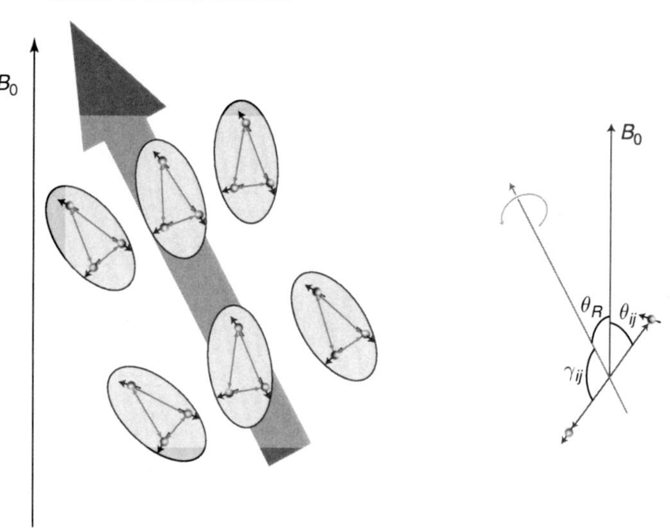

Figure 4.1. (a) Molecular motion in the anisotropic liquid crystal structure of a model membrane. Ellipses represent molecules in a bilayer structure. (b) Important angles in the description of dipolar coupling for a rotating molecule.

vectors between paired spins. The evaluation of rotation on dipolar coupling is expressed in terms of the angles shown in the schematic (Figure 4.1b), in which paired spins with angle θ_{ij} to the magnetic field rotate along the rotation axis. The rotation axis makes an angle θ_R with respect to the magnetic field and an angle γ_{ij} to the spin pairs. The evaluation of the time average of dipolar coupling in Equation 4.7 only requires calculating the angular average $\overline{3\cos^2\theta_{ij} - 1}$. This is done using the following relationship (p. 454 of [12] or p. 75 of [13])

$$\overline{3\cos^2\theta_{ij} - 1} = \frac{1}{2}(3\cos^2\theta_R - 1)\overline{(3\cos^2\gamma_{ij} - 1)} \tag{4.9}$$

If the molecule rotates sufficiently fast, Equation (4.9) becomes

$$\overline{3\cos^2\theta_{ij} - 1} = 3\cos^2\theta_R - 1 \tag{4.10}$$

This result suggests that for molecules under fast axial rotation the remaining dipolar coupling may still be significant but only along the molecular rotation axis, that is the membrane normal or the preferential direction of the anisotropic liquid. It is very important for the MAS NMR detection of molecules with residual homonuclear dipolar coupling to be aligned only along one direction as a result of rapid molecular axial rotation.

4.2.3. *The spin Hamiltonian from isotropic susceptibility heterogeneity*

To formulate the spin Hamiltonian from the interaction of spins with the magnetic field induced by the magnetic material, it is necessary to calculate the magnetic field at each individual spin. Classic treatment of the magnetic field generated in a magnetic material averages the collective effect of all the molecules in the medium over a distance scale much larger than a molecule. To address this macroscopic problem of the magnetic state, it is convenient to introduce the magnetization $\vec{M}$, defined as the average dipolar moment per unit volume. However, the microscopic magnetic field sensed by each spin is on the scale of the molecular size which is substantially smaller than the distance scale of the macroscopic field. Finding the exact magnetic field acting on an individual spin is a classical electrodynamics problem [14] and has also been extensively studied in NMR [15–18]. In the following, we outline a formal treatment for the interaction of a spin with the susceptibility heterogeneity, starting by calculating the macroscopic field induced by susceptibility and then using the classical approach to calculate the microscopic field [14].

4.2.3.1. *The macroscopic magnetic field in a magnetic medium*

When a magnetic material is placed into a static magnetic field B_0 (along the z direction with unit vector $\vec{e}_z$), the material is magnetized with the magnetization defined as the magnetic average dipolar moment per unit volume,

$$\vec{M} = \mu_0^{-1} \chi B_0 \vec{e}_z \tag{4.11}$$

in which χ is the magnetic susceptibility and is material dependent. In Equation 4.11, the susceptibility is assumed to be isotropic. Inhomogeneous susceptibility means that the magnitude of the susceptibility varies from position to position in the matter, whereas anisotropic susceptibility indicates that the susceptibility at each position is a tensor with multiple directions. Anisotropic susceptibility is more complicated and will be briefly discussed later.

The magnetization is determined by both the substance and the applied magnetic field. Here it is important to distinguish spin magnetization and electronic magnetization [17]. Spin magnetization is the summation of all the nuclear spin moments and provides all the NMR signals. The electronic magnetization originates in the perturbation of the orbital and spin motions of the electrons by the applied magnetic field. In most conditions, electronic magnetization dominates the overall magnetization. For water in a magnetic field of 14.09 T at 293 K, the spin magnetization and electronic magnetization are $4.92 \times 10^{-3}\,\mathrm{Am}^{-1}$ and $1.83 \times 10^{3}\,\mathrm{Am}^{-1}$, respectively, differing by five orders of magnitude [17]. The magnetic field in the material is modified by the magnetization and is no longer determined by B_0 alone. We can safely neglect the influence of spin magnetization and only consider the electronic magnetization when we calculate how the magnetic field in the material is modified by the magnetic material.

If the sample is homogeneous throughout space, the magnetic field induced by the magnetization in the material is $\vec{B}_M = \chi B_0 \vec{e}_z$. However, the susceptibility varies in natural samples such as cells or tissue and is even different across the cell membrane [19]. Therefore, the susceptibility constant $\chi(\vec{r})$ is position dependent. A mathematical description of the magnetic field induced by susceptibility heterogeneity requires a precise expression for the susceptibility in the material and a solution to the magnetostatics boundary-value problem. A comprehensive description and solution to this problem is not necessary for our discussion of line-broadening by susceptibility heterogeneity; instead, we provide below a phenomenological description of this effect. As shown in Figure 4.2, a heterogeneous sample may be split into many small regions and each region has a uniform susceptibility $\chi_i = \chi(r_i)$. These individual regions with volume $\Delta\nu$ are described using a magnetic moment

$$\vec{m}_i = \vec{m}(r_i) = \mu_0^{-1} \chi_i B_0 \Delta\nu \vec{e}_z \tag{4.12}$$

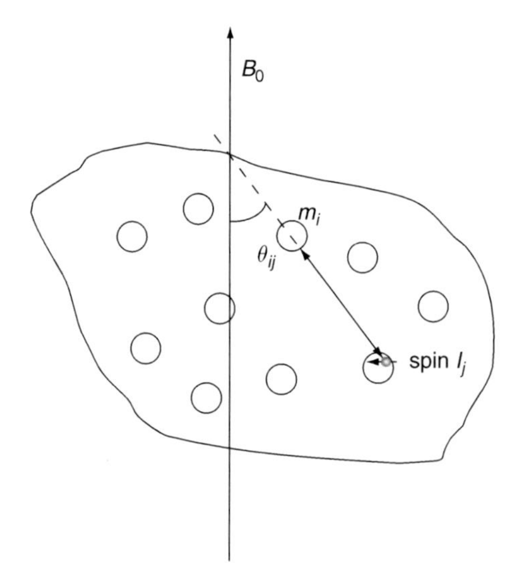

Figure 4.2. Illustration of a spin interaction with the heterogeneous isotropic susceptibility. The circles represent regions in which susceptibility can be viewed as homogeneous. The whole sample can be split into numerous small regions.

Each magnetic moment induces a homogeneous field inside the region and a dipolar field around it. The magnetic field can then be written as linear superposition of all the dipolar fields

$$\vec{B}_m(r) = \frac{2}{3}\mu_0\vec{m}(r) + \sum_i \frac{\mu_0}{4\pi r_i^3}[3(\vec{m}_i \cdot \vec{e}_i)\vec{e}_i - \vec{m}_i] \tag{4.13}$$

This is the macroscopic magnetic field.

4.2.3.2. The microscopic local magnetic field and spin Hamiltonian

The nuclear spins are magnetic dipoles with negligible spatial extent and so the exact magnetic field that a spin experiences is a microscopic field. Calculating the microscopic field uses the principle – "The local field on a spin is the same as it would be at the center of the small hole which would be left if the spin is taken out and all the neighbor dipole moments are the same" [14]. In NMR, calculation of the microscopic local field normally considers the demagnetization field of neighbor spins [17, 18]. The microscopic local field relates to the macroscopic field by

$$\vec{B}_{\text{sus}}(r) = \vec{B}_m(r) - \frac{2}{3}\mu_0\vec{m}(r) \tag{4.14}$$

In this equation, the local field $\vec{B}_{\text{sus}}(r)$ now depends on each spin's position r which is defined slightly differently from the macroscopic field in Equation 4.13, where r represents a macroscopic scale.

The secular term of the spin Hamiltonian of spin I_{jz} at position r_j from the interaction of spins with the electronic magnetization induced by susceptibility heterogeneity becomes

$$H_m(r) = \sum_i \frac{\mu_0}{4\pi r_i^3}(3\cos^2\theta_{ij} - 1)I_{jz} \tag{4.15}$$

And the overall spin Hamiltonian is

$$\hat{H}_m = \sum_{j\neq i}\sum_i \frac{\mu_0}{4\pi r_i^3}(3\cos^2\theta_{ij} - 1)I_{jz} \tag{4.16}$$

The subscript i is over all the homogeneous regions of the sample which was formally split and j is over all the spins in the sample. This phenomenological expression of the heterogeneous susceptibility spin Hamiltonian includes two important features: it is proportional to $(3\cos^2\theta - 1)$ and it only includes the spin operator I_z. Again note that we have ignored the anisotropic component of susceptibility that has been discussed by Samoson *et al.* [20].

4.3. Coherence averaging by magic angle spinning

In the static state, NMR peaks of molecules in cell and tissue samples are broadened by the residual dipolar coupling and heterogeneous susceptibility. MAS is introduced to remove the line-broadening caused by these mechanisms. As outlined above, the spin Hamiltonians of these interactions all are proportional to the well-known $(3\cos^2\theta - 1)$ term. Spinning the sample at the magic angle rapidly averages these interactions to zero and the high-resolution spectrum is readily achieved. For the applications of MAS on tissue and cell samples, it is necessary to maintain the sample integrity. Spinning the sample at an unnecessarily high speed could increase the sample degradation rate. The central question then is, how fast should we spin the sample in our experiment? To answer this question, we have to understand how MAS achieves the coherence averaging and how the interaction property influences the coherence averaging by MAS. In this section we first introduce the average Hamiltonian theory and then discuss the interplay of molecular motion and MAS.

4.3.1. The average Hamiltonian

4.3.1.1. Average Hamiltonian theory

For a discussion of coherence averaging by MAS on the sample interactions, it is convenient to use the average Hamiltonian. Spinning the sample physically along a fixed axis with a spin rate ω_r introduces a periodic modulation into the spin Hamiltonian, with period $\tau_r = 2\pi/\omega_r$. The basic idea of average Hamiltonian theory is, if the Hamiltonian is periodically time-dependent, the evolution of the spin system over one period can be described using an 'average Hamiltonian' [21–23], where the periodic time-dependent Hamiltonian is replaced by a time-independent average Hamiltonian, defined over one rotation period as follows:

$$\overline{H} = \sum_{\mu=0}^{\infty} H_\mu \tag{4.17}$$

$$H_0 = \frac{1}{\tau_r}\int_0^{\tau_r} H(t)dt \tag{4.18a}$$

$$H_1 = \frac{-i}{2\tau_r}\int_0^{\tau_r}\int_0^{t} [H(t), H(t')]dtdt' \tag{4.18b}$$

$$H_2 = \frac{-1}{6\tau_r}\int_0^{\tau_r}\int_0^{t}\int_0^{t'} [H(t), [H(t'), H(t'')]] + [H(t''), [H(t'), H(t)]]dtdt'dt'' \tag{4.18c}$$

and so on, in which $i = \sqrt{-1}$ and the commutator $[H(t), H(t')] = H(t)H(t') - H(t')H(t)$ involves the commutation of the Hamiltonian with itself at a different time. In Equation 4.18 the higher order converges as

$$H_{\mu+1} \sim \frac{\Delta\tau_r}{\mu+1} H_{\mu} \tag{4.19}$$

where Δ is the size of the static linewidth of the sample. As the rotation period τ_r shortens relative to the inverse of the static linewidth of a non-spinning sample $(1/\Delta)$, the higher order terms become less and less important. If $\Delta\tau_r \ll 1$, that is the spinning rate is much larger than the static linewidth, all the higher order terms can be dropped and the remaining term is only the zero order which does not result in line-broadening.

4.3.1.2. Homogeneous and inhomogeneous interactions

If the interaction Hamiltonian commutes with itself at different times, the commutator in Equation 4.18b vanishes. All the higher order terms will also equal zero since it depends on the same commutator. The average Hamiltonian then collapses into Equation 4.18a. On the other hand, if the spin Hamiltonian does not commute with itself, all the higher order terms in Equation 4.18 must be considered. According to Maricq and Waugh [23], there are two types of interactions, defined as inhomogeneous or homogeneous interactions. Specifically, an interaction is inhomogeneous if its Hamiltonian commutes with itself at different times. An interaction is homogenous if its Hamiltonian does not commute with itself at different times.

For inhomogeneous interactions, removal of line-broadening is independent of the spin rate because all the higher order terms in Equation 4.18 are zero. A slow spin rate would remove all the inhomogeneous interaction and result in an isotropic spectrum. Rapid spinning $\omega_r \gg \Delta$ does not further improve the spectral resolution for an inhomogeneous interaction. However, sample spinning results in spinning sidebands that spread over the frequency range comparable to the static linewidth. The distance between two successive sidebands equals the spin rate; therefore, there will be Δ/ω_r spinning sidebands. For example, if the static linewidth is 5 kHz and is purely caused by an inhomogeneous interaction, spinning the sample at 100 Hz will obtain an isotropic spectrum but with 50 sidebands. Sidebands from the resonances of different spin groups overlap with each other and can lead to great difficulty in spectral analysis and interpretation. A further drawback of slow spinning is that the detection sensitivity for the isotropic peak is low because the signal intensity spreads into all the sidebands. Thus, for peaks broadened by inhomogeneous interactions, rapid spinning may not increase the resolution of the isotropic MAS spectrum; however, it does simplify the spectrum and increase the detection sensitivity. For a homogeneous interaction, rapid spinning $\omega_r \gg \Delta$ is necessary to reduce the contribution of the higher order terms in Equation 4.18 so as to improve the spectral linewidths. In the following section we discuss a specific homogeneous interaction: the homonuclear dipolar coupling which is important for MAS NMR analysis of tissue and cell samples.

4.3.2. The effect of MAS on homonuclear dipolar coupling

The spin Hamiltonian of homonuclear dipolar coupling does not commute with itself at different times and thus is considered a homogeneous interaction. To achieve a significant improvement in line-broadening, the higher order terms in Equation 4.18 must be reduced by high speed spinning. The basic requirement is that the spin rate must be much larger than the static linewidth: $\omega_r \gg \Delta$. A more precise estimate of the optimal spin rate can be determined by analyzing the contribution of the higher order terms in Equation 4.18 to the linewidth for a given MAS rate. According to Haeberlen and Waugh, the contribution of the μth order term is not more than

$$\Delta_\mu \sim \frac{(\Delta/\omega_r)^\mu}{(\mu+1)!}\Delta \tag{4.20}$$

For fast spinning $\omega_r \gg \Delta$, the higher order decays rapidly with increasing order μ. For two methylene protons with a static dipolar coupling of 23.2 kHz, the static linewidth is of the same order $\sim$ 23.2 kHz. For an MAS rate of 23.2 kHz, the contribution of the first order average Hamiltonian H_1 remains 50%, which means MAS reduces the linewidth by half. If we increase the spinning rate by 10 times to 232 kHz (a spin rate that is not achievable using current MAS hardware), the first order still remains 5% and the resulting linewidth is more than 1.1 kHz. Adding the higher order terms will further increase the linewidth above this initial estimate. However, in the presence of molecular motion, the homonuclear dipolar coupling is largely reduced, but even with fast spinning it can still be extremely difficult to achieve very high resolution. For example, spinning the sample at 20 kHz for a static linewidth of 5 kHz would result in a spectrum with a peak linewidth of more than 250 Hz. Thus, even under MAS, a high-resolution proton spectrum is hard to achieve in the NMR analysis of solid or 'soft' solid samples.

For homonuclear dipolar coupling, Maricq and Waugh in their classic paper predicted a special case [23]: homonuclear dipolar coupling becomes inhomogeneous if all the nuclei align on a line. If molecules have a fast axial rotation and Equation 4.10 is satisfied, the residual dipolar couplings are then along one direction. As a result, molecules can be viewed as aligning on a line. In this special case, a high-resolution spectrum is obtainable with a slow MAS rate even though the static spectrum is broad.

4.3.3. *The effect of MAS on heterogeneous susceptibility and heteronuclear dipolar coupling*

The spin interaction associated with isotropic susceptibility is an inhomogeneous interaction; thus, the higher order terms of the average Hamiltonian in Equation 4.18 are zero. Isotropic susceptibility is thus averaged out under MAS and the ability of MAS to eliminate isotropic susceptibility has long been realized [16, 24]. The anisotropic susceptibility, however, cannot be eliminated by MAS and was suggested to be the determinant of the spectral linewidth under MAS in many cases [20, 25]. For isotropic susceptibility in the static state, the summation in Equation 4.16 is over the entire sample. Apparently, reducing sample size could reduce the susceptibility heterogeneity. The magnitude of the susceptibility heterogeneity also depends on the sample species. For example, a blood-cell-abundant tissue sample has significantly more susceptibility heterogeneity than a blood-free tissue sample.

In contrast to homonuclear dipolar coupling (Equation 4.3), heteronuclear dipolar coupling (Equation 4.4) is an inhomogeneous interaction which can be eliminated by a slow MAS. This is an important conclusion in practice. In the acquisition of a ^{13}C NMR spectrum, MAS alone will decouple dipolar couplings from ^{1}H nuclei. Furthermore, in the acquisition of two-dimensional (2D) heteronuclear spectroscopy, for example HSQC, we can focus our attention on magnetization transfer by scalar J-coupling without worrying about a fast magnetization transfer by heteronuclear dipolar coupling.

4.3.4. *MAS detection of the mobile molecules in cells*

With all the above theoretical preparation, it is time to consider the important spin interactions of molecules in a tissue or cell sample and the coherence averaging of MAS on these interactions. As discussed above, molecular motion modulates spin interactions. A natural sample such as cells or tissue consists of molecules with dramatically different mobilities. Therefore, the interactions, especially the dipolar couplings, are different from molecule to molecule.

In cells, the mobility of molecules is dependent on their size, their environment and their congregation pattern. For example, small molecules tumble faster than macromolecules and a small molecule dissolved in the cell cytosol has a more rapid mobility compared to a small molecule associated with a macromolecule. Congregated molecules may limit their motion in some direction. What makes it more complex in a natural sample is that a molecule may exist in different environments and thus possess different mobilities. For example, water is the most abundant molecular species in cells and constitutes the major component of the cell cytosol. A small portion of water molecules stay in the core of proteins and have a different mobility than the cytosolic water.

Based on the molecular mobility and the resulting proton dipolar coupling, we can roughly split the molecules in cells or tissues into three categories: small molecules with free tumbling, molecules congregating together with fast axial rotation and molecules with slow mobility.

Most free-tumbling molecules in cells or tissue are usually found as small metabolites in the cell cytosol. For these molecules, both intermolecular and intramolecular dipolar couplings are averaged to zero. From the point of view that the net dipolar coupling is zero, some small molecules associated with macromolecules can also be seen as free molecules. For example, water molecules in the core of proteins have a tumbling rate approximately of 1 ns [26], which is much slower than free water but fast enough to average out all the dipolar coupling. Therefore, there is no net proton dipolar coupling that causes line-broadening. (These associated waters with slower tumbling rate do possess faster transverse relaxation than free water.) For these molecules, the major line-broadening mechanism is heterogeneous susceptibility. An accurate estimate of the magnitude of isotropic susceptibility heterogeneity can be made from the linewidth of the mobile small molecules in the static state. The overall contribution is not expected to be more than 2 ppm. At 600 MHz, this is less than 1.2 kHz. As discussed above, line-broadening by interaction of the spin with isotropic susceptibility is inhomogeneous and can be completely eliminated by slow MAS. However, as shown above, a slow spin rate could lead to spinning sidebands in the spectral region of interest. For the detection of these free-tumbling molecules, the MAS rate is thus only limited by the sideband problem. If an appropriate approach can be used to suppress these sidebands, a hr-MAS NMR spectrum may be acquired using slow MAS rate. However, sideband suppression is difficult in the presence of molecular diffusion [27].

In cells, phospholipids are in a smectic liquid crystal phase. The motion of phospholipids is different in rotation and diffusion. In a model membrane, the axial rotation and lateral diffusion rate of phospholipids are $\tau_R \sim 10^{-10}$ sec and $\tau_L \sim 10^{-7}$ sec, respectively [28]. As a result, the intermolecular dipolar couplings are averaged to zero and the intramolecular dipolar couplings are reduced and are only seen along the membrane normal direction [29, 30]. This reduces the magnitude of proton dipolar coupling. In the static state in a model membrane, the linewidth of membrane ^{1}H NMR peaks is between 2 and 4 kHz depending on the preparation of the membrane and the hydration level [28, 31]. More importantly, rapid axial rotation of phospholipids results in a homonuclear dipolar coupling that produces an inhomogeneous spin Hamiltonian. Thus even a slow spin rate removes this interaction and detects an isotropic spectrum of phospholipids in cells.

Homonuclear dipolar coupling is more complex for protein protons. For small proteins with rapid axial rotation, the ^{1}H dipolar coupling spin Hamiltonian also becomes inhomogeneous and the proton spectrum is isotropic at a slow MAS rate. However, resolving protein protons for a natural cell or tissue sample is impossible because so many weak proton peaks are overlapped in relatively small spectral regions. In practice we may only observe envelopes of different protein proton groups in the acquired MAS proton spectrum. Observation of a specific protein may be implemented by employing ^{13}C and/or ^{15}N enrichment in this protein. If uniform enrichment can be achieved on a specific protein, we may be able to determine the protein structure in cells using MAS NMR with the same pulse sequence techniques developed for liquid-state NMR.

For large proteins with limited axial rotation rate, the proton dipolar coupling is a homogeneous interaction. For these proteins, moderate increases in spin rate will not increase the resolution of the proton spectrum. These proteins have more solid-like features and may be best studied using solid-state NMR techniques such as cross-polarization.

To summarize, hr-MAS NMR spectroscopy detects a high-resolution isotropic spectrum of small free-tumbling metabolites, membrane phospholipids and small proteins in cell and tissue samples. Of course, the acquired spectrum also strongly depends on the pulse sequence used in the NMR experiment.

4.4. MAS NMR experiments

4.4.1. Magic angle spinning of samples

MAS spins the sample at an angle $\theta_{mas} = 54.7^0$ (more accurately $3\cos^2\theta_{mas} - 1 = 0$) with respect to the magnetic field. Figure 4.3a illustrates a rotor spinning along the magic angle and Figure 4.3b shows a Bruker 4 mm rotor with an insert. Nitrogen gas is used to drive the rotor spinning. The sample is placed inside the rotor with the sample size varying from 12 to 80 μl.

4.4.2. Practical aspects of MAS NMR experiments

Spinning samples in a well-sealed rotor may result in different degrees of sample deterioration, depending on the experimental time and the tissue or cell types analyzed [32–34]. Under MAS at 600 MHz observation for ^1H NMR, a one-dimensional (1D) proton experiment with sufficient sensitivity for quantitative analysis takes approximately 10 min. Homonuclear 2D spectra can be acquired in an hour, while heteronuclear 2D experiments may take several hours. There have been studies to examine the effects of MAS on cell line samples. Studies have found 12% and 3% more trypan blue staining for F442A preadipocyte cells (3.5 kHz, 2 h spinning) and B104 neuroblastoma cells (5 kHz, 30 min), respectively, as compared to non-spun control cells [32, 35]. Light-microscopy revealed no signs of cell lysis or fragmentation for undifferentiated cells, but about 15–19% lysis for larger, lipid-abundant

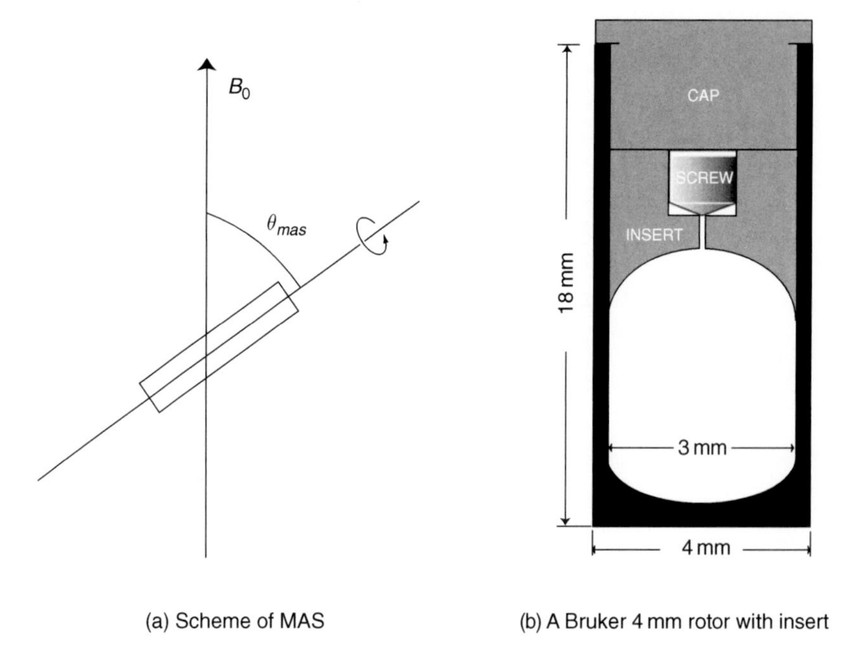

Figure 4.3. Illustration of (a) magic angle sample spinning and (b) a rotor with insert.

differentiated adipocyte cells as a result of the sample spinning [32]. Tissue samples tend to be more robust for MAS studies than isolated cell samples based on the presence of a structural interstitial matrix that supports individual cells within the tissue.

Spinning rates of 1–6 kHz are generally used to acquire spectra on tissue specimens or cell line samples. As discussed above, increasing the MAS spinning rate will not increase the resolution or sensitivity for tissue or cell samples. A practical spin rate may be chosen that places the spinning sidebands outside the spectral region of interest which is approximately 10 ppm in most tissue types and cell lines. On a 600 MHz instrument, therefore, the best spinning rate may be 5–6 kHz.

Tissue degradation caused by MAS may be alleviated by maintaining the sample at a relatively low temperature, 4 °C or lower, during the experiments. The real temperature of the sample inside the rotor may be slightly higher than the temperature measured by the thermocouple outside the rotor. The inside temperature can be calibrated using a routine protocol, for example measuring the chemical shift between the two proton peaks of 4% CH_3OH in CD_3OH [36].

4.4.3. Addition of a gradient coil to an MAS probe

Although MAS was adopted from solid-state NMR, most of the pulse sequence techniques used for the study of tissue and cell samples were from liquid-state

NMR. This is not surprising since the resolution of hr-MAS NMR on cell and tissue sample approaches the liquid-state spectrum. Therefore, experiments based on J-coupling connectivities, for example COSY, TOCSY, HSQC and so on, can be easily performed on tissue and cell samples under MAS. The use of pulsed field gradients for spectral acquisition has become a key component of modern pulse sequence techniques employed in high-resolution liquid NMR spectroscopy applications. The addition of a gradient capability to a high-resolution MAS probe significantly increases the power of hr-MAS NMR spectroscopy. For solid-state applications there has been limited benefit in using gradients to acquire spectra, and a gradient coil has been added to MAS probes only in very special applications. In solid imaging, adding a gradient was focused on creating an imaging reference frame tied to the spinning sample instead of the laboratory reference frame. It was suggested by Wind and Yannoni [37], and implemented by Cory and Veeman [38] and others [39, 40]. In these designs the gradient fields rotate with the sample by modulating the currents (through specially designed gradient coils) synchronously with the spinning. In high-resolution liquid-state experiments to suppress the demagnetization field, the gradient was designed to align with the magic angle direction [41], which is not compatible with conventional MAS probe designs.

A gradient for a high-resolution MAS probe was implemented by Mass, Laukin and Cory [42] and is aligned along the MAS direction in contrast to the gradients in a liquid-state NMR probe in which the gradients are generally along or perpendicular to the static magnetic field. The importance of aligning the gradient along the MAS direction is readily appreciated from Figure 4.4. The planes show the homogeneous field in the presence of a gradient. In Figure 4.4a, the gradient is along the z direction (the direction of static magic field). When the sample rotates, a nuclear spin in the sample enters into a different field by MAS. This imposes a partial averaging of the added gradient field

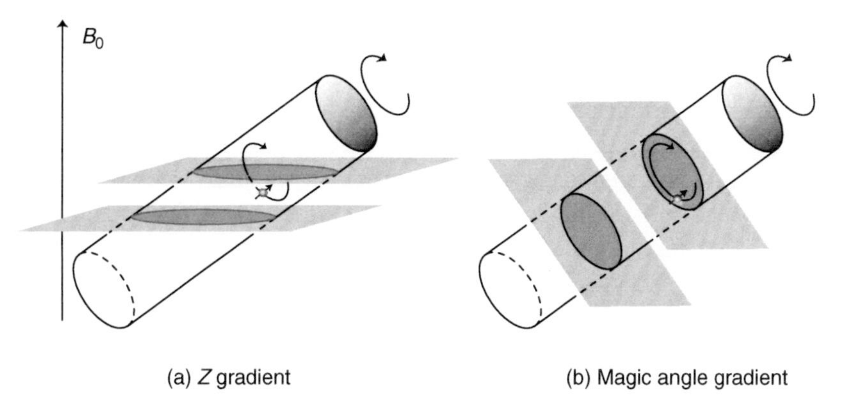

Figure 4.4. Under MAS a nuclear spin packet (a) spins into a different field when a Z gradient is applied and (b) remains in a uniform field when a magic angle gradient is applied (b).

and would require a strict synchronization of experiments with the spinning period in multiple-pulse experiments. If the gradient is aligned along the spinning direction as shown in Figure 4.4b, this added complexity can be avoided.

The three most important applications for gradients in hr-MAS NMR can be summarized as follows:

(1) Selection of a coherence pathway. In multi-dimensional experiments, pulsed gradients are able to select a coherence pathway in a cleaner fashion than traditional phase cycling. This is especially obvious in heteronuclear ^{1}H–^{13}C experiments for non-labeled samples in that 99% ^{1}H are bonded to ^{12}C. If phase cycling is used to select the coherence pathway, the large signals of protons bonded to ^{12}C are generally subtracted in the successive odd and even scans with pulses applied with opposite phases.

    However, the odd scan spectra will detect a large signal from these protons and this prevents the use of large receiver gain and thus limits the detection sensitivity. Selection of a coherence pathway using a gradient only allows the desired coherence to be detected at each scan. The large proton signal (that bonded to ^{12}C) is dephased by the pulsed gradient and cannot be detected. As a result, a large receiver gain can be used. Furthermore, a specially designed combination of gradient and phase cycling can further increase the detection sensitivity [43, 44].

(2) Suppression of unwanted signals using a "crusher" gradient. In many experiments, unwanted coherence or magnetization by pulse imperfection and/or relaxation leads to signal phase distortion in 1D experiments and t_1 noise in 2D experiments. These signals can be substantially suppressed by employing a pulsed gradient. Crushing these unwanted magnetizations is especially important in the presence of MAS. The sample spinning introduces two additional modulations into the experiment [42]. First, the finite wire size and discrete wire placement create significant *rf* field inhomogeneity and lead to signal modulations in the presence of sample rotation. Second, the spinner dynamics introduces an overall modulation of the sample coupled to the receiver (known as Q-modulation) and this appears as an incoherent modulation, since in hr-MAS experiments the pulsing is often not synchronized with the spinning frequency. The incoherent term results in a broad spectrum of t_1 noise and this effect is most commonly seen in heteronuclear 2D measurements.

(3) Measurement of diffusion coefficients of metabolites and selective detection of slowly diffusing molecules in the sample using a diffusion filter. Diffusion coefficients serve as important quantitative descriptors of molecular dynamics and can provide information about the association of small molecules with macromolecules in cell or tissue samples. NMR gradient echoes serve as straightforward approaches to measure these diffusion coefficients. A selective

analysis of slowly diffusing molecules can be achieved by using a gradient filter to remove all the signals arising from molecules undergoing fast diffusion (see the 1D pulse sequence in Section 4.5).

4.4.4. The slow MAS experiment

High-resolution ^{1}H spectra in organs and tissues have also been acquired using a slow MAS rate [45–48]. In these slow spinning MAS experiments, a clean suppression of spinning sidebands is critical for the acquisition of NMR spectra that can be readily interpreted and analyzed. A 2D phase-corrected magic angle turning (PHORMAT) sequence can be used to acquire the high-resolution NMR spectrum at a low spin rate and this has now been achieved in a live mouse [48]. These 2D protocols have the ability to detect the isotropic part of metabolites in one dimension and the anisotropic part in the second dimension with the isotropic spectrum achieving high-resolution even at spin rate of 1 Hz. These proposed techniques, with much lower sensitivity and longer experimental time, have not yet been widely applied to the NMR analysis of biological systems.

4.5. Pulse sequences

Hr-MAS NMR detects the molecular profile of a tissue or cell sample. In such a sample, molecules exist in heterogeneous micro-environments. As a result, the abundance, relaxation, diffusion, and the ability to exchange with water may differ by order of magnitudes for different metabolites. Whilst this complicates the selection of pulse sequences, it also provides an opportunity for spectral editing to selectively detect desired metabolites in the sample. Besides the sample metabolic profiles, hr-MAS NMR can also detect dynamics, compartmentalization, clustering and segmentation of metabolites by using different pulse techniques to estimate the relaxation time constants and diffusion coefficients of these metabolites.

The versatility of NMR pulse sequence techniques is one of the key features that makes NMR spectroscopy so useful in different fields from physics, chemistry and biology to medicine. Historically, NMR pulse sequences have been developed specifically for solid- and liquid-state NMR applications. Tissue specimens and culture cell lines have substantially different magnetization properties than pure solid or liquid samples. Some metabolites in cells or tissue samples such as molecular components of the cell lattice, including membrane phospholipids and proteins, have more solid-like features; however, the low molecular weight metabolites exhibit more liquid-like characteristics. The resolution achieved by hr-MAS NMR on tissue specimens and cultured cell line samples is comparable to the resolution of conventional high-resolution liquid-state experiments. As the result, most pulse sequences used in hr-MAS NMR are directly applied from liquid-state NMR without any modifications. Careful consideration of the solid-like properties of some metabolites in cell or tissue samples, such as short transverse relaxation time and slow diffusion, has to be taken into account when designing experiments since these features may significantly influence the acquired spectrum if a relaxation or diffusion filter is used in the pulse sequence. Furthermore, ^{1}H–^{1}H homonuclear dipolar coupling in large proteins with slow axial rotation cannot be averaged out by MAS. The fast spin-diffusion by this un-averaged dipolar coupling can easily transfer magnetization over long distances and to other small metabolites through a further-step chemical exchange or NOE. These magnetization transfer mechanisms could also significantly influence the acquired spectrum [8].

For the purpose of quantitative analysis of metabolites in cell line and tissue specimens, 1D ^{1}H spectra are the most popular and useful techniques. These spectra reflect the metabolite distribution and their relative abundance in the sample. However, various amounts of signals may be lost throughout the 1D spectrum when different pulse sequences are employed. In this section, 1D pulse sequences are reviewed with an emphasis on how the pulse sequence influences the detected metabolite profile and comparison of the spectra acquired with different 1D pulse sequences is illustrated using a solid tumor specimen. The 2D sequences are much more complex and are briefly explained in this chapter. A very detailed explanation of all the basic 2D experiments can be found in the book by Cavanagh *et al.* [4], and are directly applicable to hr-MAS NMR experiments in most cases.

4.5.1. *One-dimensional NMR pulse sequences*

Water is normally approximately 3 orders of magnitude more abundant than other metabolites in a tissue or cell samples. Therefore, water NMR peak suppression is necessary and generally applied to increase the detection sensitivity. In hr-MAS NMR applications, presaturation has been the most widely used water suppression technique. However, in a sample such as a tissue specimen or cultured cell line, water exchanges magnetization with the cell lattice as well as some small molecules [49–51]. Irradiation at the water resonance for several seconds as used in presaturation may result in significant signal losses in the NMR peaks of various other metabolites. A pulse sequence named SEEN (selective excitation to empirically null), which is modified from the combination of three chemical shift-selected (CHESS) sequences, was recently applied in hr-MAS NMR studies of cells and tissues. The SEEN sequence was found to acquire quantitative hr-MAS NMR spectra on tissue specimens and cultured cell lines without signal loss [51]. Both water suppression approaches can be easily combined with other pulse sequence techniques. The combination of water suppression with a series of spin-echo pulses acquires only low molecular weight metabolites by filtering out the macromolecules or molecule clusters with long correlation times. In contrast, pulsed gradients are applied to selectively detect macromolecules by eliminating the signals from low molecular weight metabolites with faster diffusion. The most widely used 1D pulse sequences are summarized in Figure 4.5 and explained below.

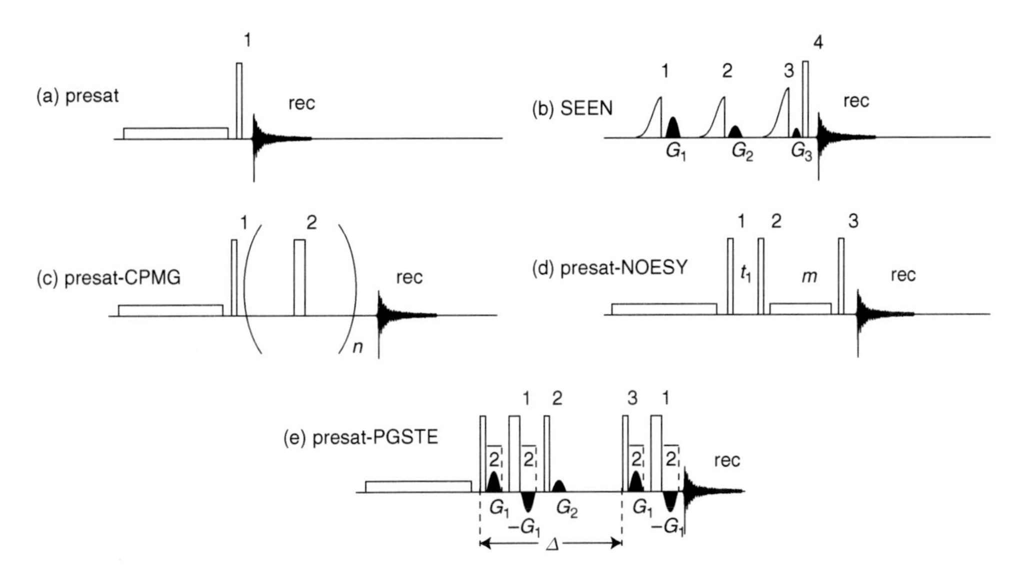

Figure 4.5. One-dimensional NMR pulse sequences frequently used in hr-MAS experiments. The symbols in the pulse sequences are: high narrow open blocks, 90° pulse; high broad open blocks, 180° pulse; low long open blocks, low power long irradiation on water; open curve, shaped selective pulse; filled curves, pulsed field gradient. (a) Water suppression using presaturation. The presaturation pulse is always on the x axis, and $\phi_1 = \phi_{rec} = xx - x - xyy - y - y$. The low power presaturation uses a field strength of about 100 Hz and a length of 2–5 sec. An extra delay time may be used before presaturation to allow the spin system a full recovery to the equilibrium state. (b) SEEN uses the combination of three selective pulse and gradients to suppress the water signal. This pulse sequence includes three selective pulses and one hard detecting pulse with three pulsed gradients. The first two selective pulses are at 90° and the third one needs to be empirically calibrated to achieve a null water signal. The fourth pulse is a hard 90° pulse for detection. The phases for the pulses are $\phi_1 = -x\ x\ x\ -x\ \ -y\ y\ y\ -y$; $\phi_2 = y\ y\ -y\ -y\ x\ x\ -x\ -x$; $\phi_3 = x\ -x\ y\ -y$; $\phi_4 = \phi_{rec} = x\ -x\ -x\ x\ y\ -y\ -y\ y$. (c) Combination of presaturation and series spin-echo using a CPMG phase cycling. The presaturation uses x phase and is the same as in (a). CPMG phase cycling is: $\phi_1 = \phi_{rec} = xx - x - xyy - y - y$, $\phi_2 = y - yy - yx - xx - x$. The best choice for the short delay τ may be the MAS period. The spectra acquired using this pulse sequence will result in signal loss of the broad metabolite peaks. (d) Combination of presaturation with the 1D NOE sequence. The phase of presaturation and this irradiation are on the x axis. $\phi_1 = x - x$, $\phi_2 = 8(x)8(-x)$, $\phi_3 = xx - x - xyy - y - y$, $\phi_{rec} = x - x - xxy - y - yy - xxx - x - yyy - y$. (e) Combination of presaturation with the pulse gradient stimulated echo sequence. The presaturation pulse is on the x axis and is the same as in (a). The rest of the phase cycling is: $\phi_1 = 4(x)4(-x)$, $\phi_2 = 8(x)8(-x)$, $\phi_3 = xy - x - y$, $\phi_{rec} = 2(x - y - xy)2(-xyx - y)$.

(a) The presaturation (presat) pulse sequence [52] is depicted in Figure 4.5a and the phase of the presaturation pulse remains constant. The 90° pulse and the receiver use a double CYCLOPS [53] phase cycling ($x\ x\ -x\ -x\ y\ y\ -y\ -y$) to suppress quadrature images. The strength of the weak irradiation used for presaturation is approximately 100 Hz and the length can vary from 1 to 5 sec. Typically, this pulse sequence is sufficient to obtain efficient water peak suppression and a flat baseline for a tissue or cell sample. The effectiveness of water suppression by presaturation in a cell or tissue sample is influenced by the homogeneity of the static field across the sample. Better shimming results in a more efficient water suppression and improves the shape of the residual water signal. Splitting the long irradiation pulse into two uneven parts with a 90° phase shift for the second part normally increases the efficiency of presaturation as well, but this usually is not required in hr-MAS NMR applications. An extra recovery delay before the low power irradiation may be added to allow the spin system a full recovery to the equilibrium state following each scan.

(b) The SEEN pulse sequence is depicted in Figure 4.5b [51]. This water suppression technique was developed to suppress the water peak whilst retaining all the signal intensities of other metabolites from tissue specimens and cell line samples. This pulse sequence includes three repetitions of CHESS with the last selective excitation to empirically null the water signal. Water exchanges magnetization with membrane phospholipids fairly quickly and therefore a 2 sec presaturation can result in a loss of more than 70% of the membrane lipid signals. Other metabolites exchanging magnetization with water include proteins and some small molecules containing amide protons. The signals from these protein- and amide-containing molecules are also retained in the spectrum when SEEN is used to suppress the water resonance. Because magnetization exchange still occurs during the selective pulse, the length of the shaped pulse needs to be as short as possible. It has been shown that a 10 ms half Gaussian for this selective excitation results in no significant change in signal intensities for lipid-, protein- or amide-containing metabolites.

(c) The presat-CPMG is the combination of a presaturation irradiation with a CPMG spin-echo sequence [54]. The phase cycling developed for the series spin-echo by Carr-Purcell-Meilboom-Gill has the ability to compensate for any small inaccuracy in the 180° pulse. This pulse sequence selectively acquires metabolites with sharp NMR peaks. These are compounds with high mobility or short correlation time. Two factors need to be considered for selecting the length of delay τ: the phase distortion during this delay by homonuclear scalar coupling and the synchronization with the MAS rate. The most straightforward approach is to select the length of delay based on a simple calculation of the inverse of the spin rate. For example, if the MAS rate is 5 kHz, the ideal delay τ to use is 200 μs. The overall time of $2n\tau$ varies from 30 ms to seconds depending on the desired extent of broad peak suppression. An estimate of the transverse relaxation time can also be made using this pulse sequence [36]. This pulse sequence has been widely used to acquire the metabolic profiles of tissue or cell samples with the major drawback largely related to the loss of information from low mobility metabolites that appear as broad peaks in the proton NMR spectrum.

d) The presat-Nuclear overhauser spectroscopy (presat-NOESY) pulse sequence acquires the first free induction decay (FID) of the 2D NOESY experiment. This is a widely used pulse sequence by several NMR groups. The two delays are normally chosen with t_1 of about 3 μs and τ_m of about 100 ms. The low power irradiation during the mixing time τ_m prevents the water signal recovery by longitudinal relaxation. This irradiation is not required if this delay is much shorter, for example around 10 ms. A short crusher pulsed gradient may be added at the beginning of mixing time to eliminate the residual transverse magnetization. The Presat-NOESY can obtain a better spectral baseline than presaturation alone and a 100 ms mixing time may not generate significant NOE for high-mobility metabolites. For macromolecules or molecular clusters such as membrane lipids, a mixing time of 100 ms does induce a significant intramolecular NOE effect and will redistribute the signal intensities between different resonances.

(e) The presat-PGSTE pulse sequence is the combination of presaturation with a pulse gradient stimulated echo sequence. Instead of using a single gradient, the pulse sequence shown here employs bipolar gradient pairs which have the ability to reduce the long eddy current found for the unshielded gradient coil used in a hr-MAS probe. With proper setting of the gradient intensities and diffusion time, this pulse sequence can only acquire only the slow diffusing metabolites. Generally, these metabolites include macromolecules such as membrane phospholipids, proteins and some low molecular weight metabolites that bind to macromolecules. Thus, this pulse sequence can also study the binding activity of low molecular weight molecules to macromolecules. The diffusion coefficients can also be measured using this pulse sequence by varying the gradient intensities. The drawback of this pulse sequence is that water peak presaturation already removes a significant amount of signal intensity of macromolecules because macromolecules in cells or tissue exchange magnetization with water. The combination of SEEN with PGSTE is expected to be more precise in reflecting the concentrations of macromolecules in a tissue or cell sample.

In Figure 4.6, spectra are shown that were acquired using the above pulse sequences on a Bruker Avance 600 MHz with a hr-MAS $^{1}H/^{13}C/^{15}N$ triple resonance probe. The sample is a snap-frozen gastrointestinal stromal tumor (obtained with the consent of the patient and approval of the institutional review board) of 21.6 mg including 3 μl D_2O/PBS/95 mM TSP (3-(trimethylsilyl) [2,2,3,3$-D_4$] propionate

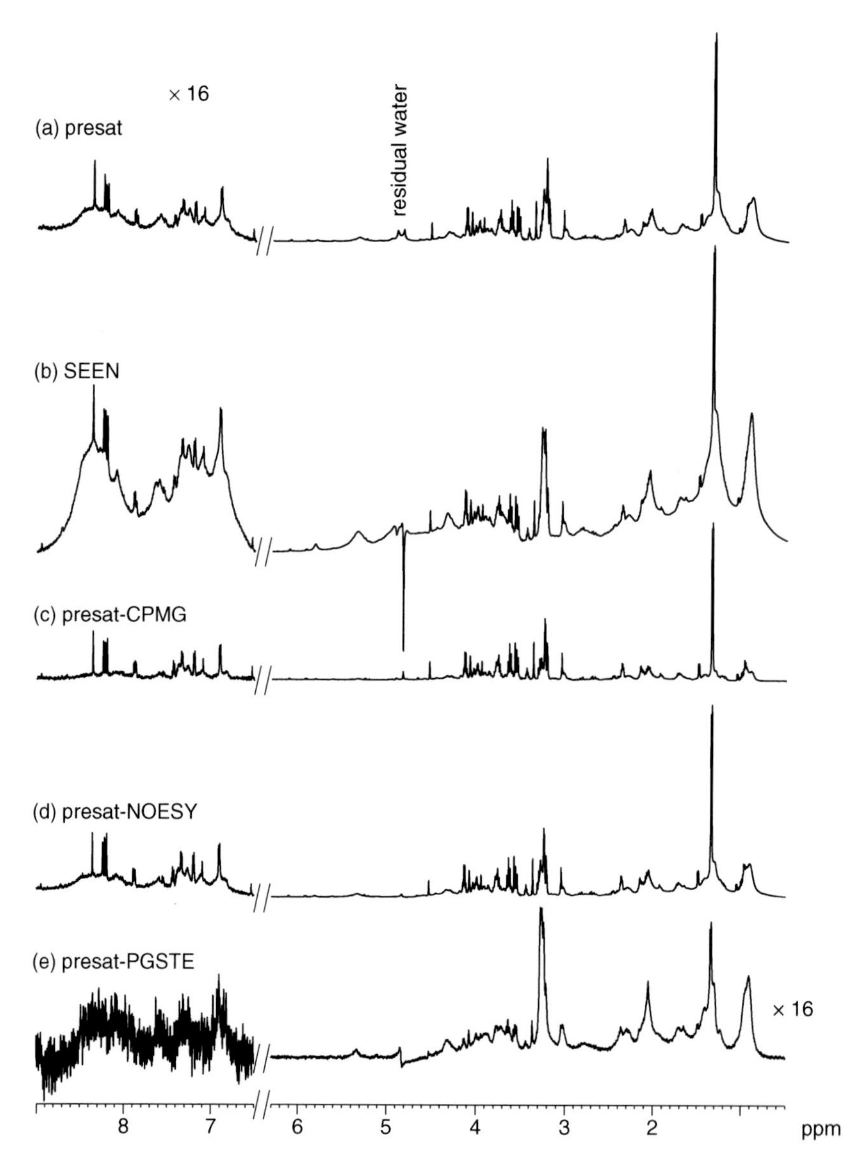

Figure 4.6. Comparison of the 1D spectra acquired using the pulse sequences in Figure 4.5 on a Bruker Avance 600 MHz spectrometer with a hr-MAS $^{1}H/^{13}C/^{15}N$ triple resonance probe. The plots of the left regions of all the spectra are enhanced by 16 times compared to the right region and spectrum (e) was enhanced 16 times compared to spectra (a) to (d). All the spectra were obtained using the same receiver gain, approximately the same acquisition time (13 min) and with the same following parameter settings: 16 k and 32 k data points in the time and frequency domain, respectively; 128 scans; repetition time of 6.14 sec including a 5 sec delay (3 sec delay in presaturation experiments) and 1.14 sec acquisition time for a single scan; a line-broadening of 1 Hz before Fourier transformation. (a) Acquired using 2 sec presaturation. (b) Acquired with SEEN. The three sine-shaped pulsed field gradients were

sodium salt) loaded in a 4 mm rotor with an insert. The MAS rate was 5 kHz. All the spectra were obtained using the same receiver gain, approximately the same acquisition time (13 min) and with the same parameters. The presaturation irradiation at the water resonance for 2 sec achieved good water suppression and a flat baseline (spectrum a). The SEEN (spectrum b) also had efficient water suppression and acquired many more signals than presaturation. In this spectrum, the protein aromatic ring protons and amide protons are shown as envelopes overlapped with some small metabolites between 6.5 and 8.8 ppm. In this solid tumor, a 60 ms spin-echo time almost removes all the broad peaks (spectrum c). A 100 ms mixing time used in the 1D NOESY sequence does not have a significant effect on the sharp peaks but does have observable effects on the broad peaks (see spectrum d) when compared to presaturation alone. In spectrum (e), some low molecular weight molecules survived the diffusion filter probably due to their binding to macromolecules.

4.5.2. Two-dimensional NMR pulse sequences

The 2D spectra of tissue or cell samples serve as the major assignment tool for resonances. Some metabolites that are not well resolved in the 1D spectrum are easily revealed in 2D spectra. It is especially useful for the study of subtle differences between different tumor types or the effects of chemotherapy on tumor or cell lines [55]. The 2D pulse sequences used in hr-MAS NMR are generally the same as used in liquid-state experiments. Processing of 2D data is identical to that used in liquid-state NMR as well. The commonly used 2D experiments are briefly reviewed below.

(a) Correlation Spectroscopy (COSY), the first 2D NMR experiment to be devised [56, 57], is commonly used in hr-MAS NMR. The COSY establishes coherence between coupled spins through scalar coupling. In COSY, cross peaks exhibit an antiphase lineshape. The diagonal and cross peaks also differ in phase by 90° and thus cannot be phased to an absorption lineshape simultaneously. When the cross peaks are phased to an absorptive lineshape, the dispersive tails of

---

Figure 4.6. *(Continued)* G_1: 1 ms × 0.19 T/m, G_2: 1 ms × 0.085 T/m, G_3: 1 ms × 0.055 T/m. The selective pulses had a half-Gaussian shape with length of 10 ms which excites a frequency band of 188 Hz. (c) Acquired using 2 sec presaturation and 60 ms spin-echo time with the delay $\tau = 200\,\mu s$. The 2 sec presaturation with 60 ms spin-echo almost completely removed the broad components in the spectrum. (d) Acquired using 2 sec presaturation and 100 mixing time in the presat-NOESY sequence. The 100 ms NOE does not have significant effects on the sharp peaks but does have observable effects on broad peaks when compared to presaturation alone. (e) Acquired using 2 sec presat-PGSTE. The two sine-shaped pulsed field gradients were G_1: 3 ms × 0.54 T/m, G_2: 1 ms × 0.10 T/m. $\Delta = 20$ ms.

the diagonal peaks prevent the resolution of nearby cross peaks. The double-quantum-filtered (DQF) COSY sequence [58, 59] can overcome some of the long tail problems found in the simple COSY experiment. In particular, all uncoupled singlet resonances are unable to generate double quantum signals and, therefore, do not appear in DQF–COSY spectra.

(b) Total Correlation Spectroscopy (TOCSY) [60] is the most widely used homonuclear 2D experiment acquired on a tissue or cell samples and has the advantage of requiring much shorter acquisition times than other 2D experiments. The TOCSY transfers in-phase magnetization between spins via the scalar couplings. The magnitude of the cross peaks in TOCSY depends on the topology of the spin system, the coupling constants between spins, the efficiency of the mixing sequence and the relaxation rate of the spins. Thus, special attention must be used when comparing the relative cross peak intensities between different metabolites if a quantitative analysis of metabolite levels is desired. The diagonal and cross peaks have the same phase and can always be adjusted to an absorption lineshape. The anti-phase zero-quantum coherence has not been found to be a problem for hr-MAS NMR applications in cell and tissue samples.

(c) Frequently used mixing sequences in TOCSY include WALTZ-16, MLEV-17 and DIPSI-2. The length of the mixing period varies between 40 and 150 ms. During the mixing period, the supercycle of the mixing sequence is executed for integer times. The supercycle number is normally automatically calculated by the spectrometer software based on the input mixing time. The real mixing time for spin mixing is typically slightly shorter than the input time to satisfy the integer supercycle requirement. The *rf* strength used for mixing is around 10 kHz. In TOCSY spectra the rotating frame Overhauser effect (ROE) may result in peaks of opposite sign and this may in turn reduce the intensity or even cancel the TOCSY peak. To account for this effect, it is best to acquire multiple TOCSY spectra with different mixing times so as to detect all the metabolite cross peaks.

(d) $^{13}C-^{1}H$ HMQC (heteronuclear multiple quantum coherence spectroscopy) [61] and $^{13}C-^{1}H$ HSQC [62] (heteronuclear single quantum coherence spectroscopy) are proton-detected $^{13}C-^{1}H$ correlation experiments using a multiple and single quantum transfer mechanism, respectively. The effective F1 linewidths of signals in the HMQC and HSQC spectra accordingly depend on multiple quantum coherence and single quantum coherence. Thus, HSQC has narrower F1 linewidths and better sensitivity than HMQC for signals from cell membrane phospholipids. The gradient-enhanced HSQC [63] experiment is now generally used in hr-MAS NMR applications. This pulse sequence uses a pulsed gradient to select coherence and enhance sensitivity. A complete description of this HSQC sequence and the efficiency of heteronuclear coherence transfer is described in Chapter 7 of Reference [4].

4.6. Applications

High-throughput screening of the alterations in DNA, RNA and protein levels in normal and malignant tissue or cell line samples has achieved tremendous success and has dramatically improved our understanding of disease processes and has enhanced our ability to identify new therapeutic targets. The correlation of metabolomics with these conventional screening techniques will provide a more complete picture of tumor biology. Metabolites of cultured cells or tissue specimens at a specific point in time are largely the collective result of all the biological activity prior to this time point. The change of a single metabolite in the tumor cells or tumor tissue may reflect the activity of a specific protein or the collective activity of several proteins and retrospectively the genome expression. Hr-MAS NMR is one of the key techniques to screen a large number of metabolites and has been used to evaluate drug toxicity; find biomedical markers for various diseases and cancers; distinguish tumor types; and study chemotherapy effects in cultured cell lines, animal models and patients. In addition to monitoring fluctuation in metabolite profiles, Hr-MAS NMR can be used to investigate the dynamics, compartmentalization, clustering and segmental motion of metabolites. Thus, this technique can be further extended to monitor the metabolic response of tissue or cells to different interventional techniques such as chemotherapy, radiotherapy, hormonal response, nutritional changes and so on.

4.6.1. Applications in tissue specimens

For the identification and rapid quantification of tissue components and determination of changes in disease states, metabolite profile analysis by hr-MAS NMR has been used to characterize different normal tissues, including intact human kidney [64], rat renal cortex and medulla [65], kidney of wild and laboratory rats [66], rat liver [67, 68], mice dystrophic cardiac tissues [69], rat testicular tissues [70], rat intestine [71] and rat brain tissue [72]. These studies reveal the potential of using the metabolite profile as a fingerprint for classification of tissue type [73].

One of the important advantages of the NMR study of tissue over other types of spectroscopy is that NMR spectroscopy is non-invasive to the sample. After NMR experiments, tissue can be maintained in formalin for histopathological analysis. The ability to morphologically examine samples after NMR experiments prevents errors of classification and allows better correlation of the NMR results to the pathological state of the tissue under investigation. Hr-MAS NMR has been used extensively to study a variety of disease processes including different tumor types such as human brain tumors [34, 74–77, 78], breast cancer [79, 80], prostate tumor [81–84], cervical cancer [85–88], kidney renal cell carcinoma [89], human renal cortex [90], the neuronal ceroid lipofuscinoses [91] and liposarcomas [92, 93]. Chapter 13 of this book is dedicated to the NMR applications to cancers.

Chemotherapy can have considerable adverse side effects in patients. An important application of hr-MAS NMR spectroscopy is to detect the metabolite changes induced by toxicants in animal models. This provides invaluable knowledge of how the patients may respond to a certain treatment. This kind of study also is important in drug discovery to rapidly elucidate the mechanisms contributing to drug toxicity. Most of these studies use conventional liquid-state NMR to study the urine and serum of animal models in addition to hr-MAS NMR examination of the animal liver [94–96] and kidney [97, 98] tissue samples.

4.6.2. Applications in cultured cell lines

Biological experiments on cell lines have been used to elucidate the biological mechanism and interactions involved in cellular pathways, to screen promising drugs for activity and to identify and validate new cellular targets. The use of hr-MAS NMR for cell studies has not been as widely applied compared to hr-MAS NMR analysis of tissue samples. Lipids are the major visible metabolites in cell line studies using hr-MAS NMR [55, 99, 100]. The lipid CH_2 moieties with an NMR resonance at 1.3 ppm are associated with the intracellular accumulation of fat droplets and may serve as a marker of apoptosis. In a study of BT4C glioma during ganciclovir-thymidine kinase gene therapy–induced apoptosis, it was further suggested that the ^1H NMR lipids detected during apoptosis arise from cell constituent breakdown products forming lipid vesicles in dying cells [100].

A diffusion- and relaxation-weighted hr-MAS NMR analysis was used to metabolically characterize Ishikawa cells, a human cell line derived from endometrial adenocarcinoma [101]. The changes in the metabolite profile of this cell line following exposure to tamoxifen, a selective estrogen receptor modulator (SERM), were investigated. A statistical regression technique of prediction to latent structures by partial least squares (PLS) was used to build a predictive model of the metabolite profile alterations following exposure to tamoxifen. These spectral changes were characterized by increased resonance intensities from ethanolamine, glucose, glutamate, tyrosine (7.24 ppm), uridine and adenosine, and a relative decrease in the contribution from myo-inositol resonances. The nucleotide changes suggest that tamoxifen affects RNA transcription, while the changes in ethanolamine and myo-inositol concentrations are indicative of cell membrane turnover.

Drug treatment to induce differentiation has been investigated by hr-MAS NMR [55]. Thiazolidinediones, a class of synthetic ligands to the peroxisome proliferator-activated receptor-gamma, which have already been used as alternative therapeutic agents for the treatment of liposarcoma in clinical trials, induce terminal adipocyte differentiation of 3T3 F442A cells. The biochemical changes occurring in the 3T3 F442A cell line and a well-differentiated liposarcoma following induction of adipocyte differentiation with the thiazolidinedione troglitazone were measured.

The 3T3 F442A cell differentiation was characterized by a large accumulation of intracellular triglyceride and withdrawal from the cell cycle. Phosphatidylcholine, phosphocholine, myo-inositol and glycerol were found to be possible biochemical markers for adipocyte differentiation induced by thiazolidenediones. The molar ratio of phosphatidylcholine to phosphocholine increased fourfold in differentiated 3T3 F442A cells compared to undifferentiated cells, suggesting a substantial increase in CTP:phosphocholine cytidylyltransferase activity with differentiation. This ratio was increased by 2.8-fold in the lipoma-like well-differentiated liposarcoma of three patients who were treated with troglitazone compared to liposarcoma from patients not treated with this drug. It was suggested this ratio may be an NMR-detectable marker of troglitazone efficacy and response to differentiation therapy for liposarcoma.

While most studies use proton NMR spectroscopy, the technique of ^{19}F NMR, with a wide range of chemical shifts spanning about 300 ppm, is very sensitive to changes in chemical structure. A recent study used ^{19}F MAS NMR spectroscopy to trace 2,2-difluoropentanedioic acid (DFPA) uptake and accumulation [102]. A strong ^{19}F NMR signal was measured in the alpha-ketoglutarate permease (KGTP) strain incubated with DFPA, whereas only a weak ^{19}F signal was recorded in wild-type cells treated under similar conditions. Neither the wild type nor the KGTP strain not treated with DFPA produced any ^{19}F signals. These results indicated that DFPA enters the cells preferentially via the 2-oxoglutarate permease. In addition, ^{19}F MAS NMR data demonstrated that this molecule is not metabolized into other components because the ^{19}F signal characteristic of the DFPA molecule in solution is identical to that detected in whole cells. As a consequence, DFPA added at 0.5 mM to the KGTP cell cultures accumulated in the cells and reached a plateau within 3–6 h of incubation.

Acknowledgements

The authors wish to thank Prof. D. Cory at MIT for initiating an outline of the content. J.H.C is especially in debt to Ms. DeCarolis and Ms. O'Connor for clarifications on many points in this chapter.

References

[1] Bloch, F., Hansen, W.W., Packard, M. Nuclear Induction. *Phys Rev.* 1946;69:127.
[2] Purcell, E.M., Torrey, H.C., Pound, R.V. Resonance Absorption by Nuclear Magnetic Moments in a Solid. *Phys Rev.* 1946;69:37–38.
[3] Wüthrich, K. *NMR of Proteins and Nuclear Acids* (John Wiley & Sons, New York) 1986.

[4] Cavanagh, J., Fairbrother, W.J., Palmer III, A.G., Skelton, N.J. *Protein NMR Spectroscopy* (Acedemic Press, San Diego) 1996.
[5] Andrew, E.R., Bradbury, A., Eades, R.G. Removal of dipolar broadening of NMR spectra of solids by specimen rotation. *Nature* 1959;183:1802–1803.
[6] Lowe, I.J. Free Induction Decays of Rotating Solids. *Phys Rev. Lett.* 1959;2:285–287.
[7] Levitt, M.H. Spin dynamics: Basics of nuclear magnetic resonance. (John Wiley & Sons, New York) 2001.
[8] Chen, J-H., Sambol, E.B., DeCarolis, P., O'Connor, R., Geha, R.C., Wu, Y.V., Singer, S. High-Resolution MAS NMR Spectroscopy Detection of the Spin Magnetization Exchange by Cross-Relaxation and Chemical Exchange in Intact Cell lines and Human Tissue Specimens. *Magn. Reson. Med.* 2006;55:1245–1256.
[9] Tjandra, N., Bax, A. Direct measurement of distances and angles in biomolecules by NMR in a dilute liquid crystalline medium (Vol. 278, p. 1111). *Science* 1997;278:1697–1697.
[10] Tjandra, N., Omichinski, J.G., Gronenborn, A.M., Clore, G.M., Bax, A. Use of dipolar H-1-N-15 and H-1-C-13 couplings in the structure determination of magnetically oriented macromolecules in solution. *Nature Structural Biology* 1997;4:732–738.
[11] Prestegard, J.H. New techniques in structural NMR – anisotropic interactions. *Nature Structural Biology* 1998;5:517–522.
[12] Abragam, A. Principles of Nuclear Magnetism (Oxford University Press, London) 1961.
[13] Slichter, C.P. *Principles of Magnetic Resonance.* (Springer-Verlag, New York) 1978.
[14] Feynman, R.P., Leighton, R.B., Sands, M. *The Feynman Lectures on Physics*, Vol. 2, (Addison-Wesley, New York) 1989, pp. 36.
[15] Bowtell, R. Indirect Detection Via the Dipolar Demagnetizing Field. *J. Magn. Reson.* 1992;100:1–17.
[16] Garroway, A.N. Magic-Angle Sample Spinning of Liquids. *J. Magn. Reson.* 1982;49:168–171.
[17] Levitt, M.H. Demagnetization field effects in two-dimensional solution NMR. *Concepts Magn. Reson.* 1996;8:77–103.
[18] Vlassenbroek, A., Jeener, J., Broekaert, P. Macroscopic and microscopic fields in high-resolution liquid NMR. *J. Magn. Reson. Series A* 1996;118:234–246.
[19] Peled, S., Cory, D.G., Raymond, S.A., Kirschner, D.A., Jolesz, F.A. Water diffusion, T(2), and compartmentation in frog sciatic nerve. *Magn. Reson. Med.* 1999;42:911–918.
[20] Samoson, A., Tuherm, T., Gan, Z. High-field high-speed MAS resolution enhancement in 1H NMR spectroscopy of solids. *Solid State Nucl. Magn. Reson.* 2001;20:130–136.
[21] Haeberle, U., Waugh, J.S. Coherent Averaging Effects in Magnetic Resonance. *Phys. Rev.* 1968;175:453–467.
[22] Haeberle. U., Waugh, J.S. Spin-Lattice Relaxation in Periodically Perturbed Systems. *Phys. Rev.* 1969;185:420–429.
[23] Maricq, M.M., Waugh, J.S. Nmr in Rotating Solids. *J. Chem. Phys.* 1979;70:3300–3316.
[24] Stoll, M.E., Majors, T.J. Elimination of Magnetic-Susceptibility Broadening in Nmr Using Magic-Angle Sample Spinning to Measure Chemical-Shifts in Nbhx. *Physical Review B* 1981;24:2859–2862.
[25] Alla, M., Lippmaa, E. Resolution Limits in Magic-Angle Rotation Nmr-Spectra of Polycrystalline Solids. *Chem. Phys. Lett.* 1982;87:30–33.
[26] Otting, G. NMR studies of water bound to biological molecules. *Progress in Nuclear Magnetic Resonance Spectroscopy* 1997;31:259–285.
[27] Liu, Y., Leu, G., Singer, S., Cory, D.G., Sen, P.N. Manipulation of phase and amplitude modulation of spin magnetization in magic angle spinning nuclear magnetic resonance in the presence of molecular diffusion. *J. Chem. Phys.* 2001;114:5729–5734.

[28] Feigenson, G.W., Chan, S.I. Nuclear magnetic relaxation behavior of lecithin multilayers. *J. Am. Chem. Soc.* 1974;96:1312–1319.

[29] Bloom, M., Burnell, E.E., MacKay, A.L., Nichol, C.P., Valic, M.I., Weeks, G. Fatty acyl chain order in lecithin model membranes determined from proton magnetic resonance. *Biochemistry* 1978;17:5750–5762.

[30] Forbes, J., Bowers, J., Shan, X., Moran, L., Oldfield, E., Moscarello, M.A. Some New Developments in Solid-State Nuclear Magnetic-Resonance Spectroscopic Studies of Lipids and Biological-Membranes, Including the Effects of Cholesterol in Model and Natural Systems. *J. Chem. Soc. Farad Trans* 1988;84:3821–3849.

[31] Davis, J.H., Auger, M., Hodges, R.S. High resolution 1H nuclear magnetic resonance of a transmembrane peptide. *Biophys. J.* 1995;69:1917–1932.

[32] Weybright, P., Millis, K., Campbell, N., Cory, D.G., Singer, S. Gradient, high-resolution, magic angle spinning 1H nuclear magnetic resonance spectroscopy of intact cells. *Magn. Reson. Med.* 1998;39:337–345.

[33] Waters, N.J., Garrod, S., Farrant, R.D., Haselden, J.N., Connor, S.C., Connelly, J., Lindon, J.C., Holmes, E., Nicholson, J.K. High-resolution magic angle spinning (1)H NMR spectroscopy of intact liver and kidney: optimization of sample preparation procedures and biochemical stability of tissue during spectral acquisition. *Anal. Biochem.* 2000;282:16–23.

[34] Martinez-Bisbal, M.C., Marti-Bonmati, L., Piquer, J., Revert, A., Ferrer, P., Llacer, J.L., Piotto, M., Assemat, O., Celda, B. 1H and (13)C HR-MAS spectroscopy of intact biopsy samples ex vivo and in vivo (1)H MRS study of human high grade gliomas. *NMR Biomed.* 2004;17:191–205.

[35] Griffin, J.L., Bollard, M., Nicholson, J.K., Bhakoo, K. Spectral profiles of cultured neuronal and glial cells derived from HRMAS (1)H NMR spectroscopy. *NMR Biomed.* 2002;15:375–384.

[36] Braun, S., Kalinowski, H-O., Berger, S. *150 and more basic NMR experiments* (Wiley-VCH, New York) 1998.

[37] Wind, R., Yannoni, C.S. U.S. Patent 4301410. 17 November 1981.

[38] Cory, D.G., Vanos, J.W.M., Veeman, W.S. Nmr Images of Rotating Solids. *J. Magn. Reson.* 1988;76: 543–547.

[39] Buszko, M., Maciel, G.E. Magnetic-Field-Gradient-Coil System for Solid-State MAS and Cramps NMR Imaging. *J. Magn. Reson.* 1994;A107:151–157.

[40] Schauss, G., Blumich, B., Spiess, H.W. Conditions for Generating Rotating Gradients in Mas Nmr Imaging. *J. Magn. Reson.* 1991;95:437–441.

[41] Bowtell, R., Peters, A. Magic-Angle Gradient-Coil Design. *J. Magn. Reson.* 1995;A115:55–59.

[42] Maas, W.E., Laukien, F.H., Cory, D.G. Gradient High Resolution, Magic Angle Sample Spinning NMR. *J. Am. Chem. Soc.* 1996;118:13085–13086.

[43] Cavanagh, J., Palmer, A.G., Wright, P.E., Rance, M. Sensitivity Improvement in Proton-Detected 2-Dimensional Heteronuclear Relay Spectroscopy. *J. Magn. Reson.* 1991;91:429–436.

[44] Palmer, A.G., Cavanagh, J., Wright, P.E., Rance, M. Sensitivity Improvement in Proton-Detected 2-Dimensional Heteronuclear Correlation Nmr-Spectroscopy. *J. Magn. Reson.* 1991;93:151–170.

[45] Wind, R.A., Hu, J.Z., Rommereim, D.N. High-resolution (1)H NMR spectroscopy in organs and tissues using slow magic angle spinning. *Magn. Reson. Med.* 2001;46:213–218.

[46] Hu, J.Z., Rommereim, D.N., Wind, R.A. High-resolution 1H NMR spectroscopy in rat liver using magic angle turning at a 1 Hz spinning rate. *Magn. Reson. Med.* 2002;47:829–836.

[47] Hu, J.Z., Wind, R.A. Sensitivity-enhanced phase-corrected ultra-slow magic angle turning using multiple-echo data acquisition. *J. Magn. Reson.* 2003;163:149–162.

[48] Wind, R.A., Hu, J.Z., Rommereim, D.N. High-resolution 1H NMR spectroscopy in a live mouse subjected to 1.5 Hz magic angle spinning. *Magn. Reson. Med.* 2003;50:1113–1119.

[49] Balaban, R.S., Ceckler, T.L. Magnetization transfer contrast in magnetic resonance imaging. *Magn. Reson. Q* 1992;8:116–137.

[50] Bryant, R.G. The dynamics of water-protein interactions. Annu. Rev. Biophys. Biomol. Struct. 1996;25:29–53.

[51] Chen, J.H., Sambol, E.B., Kennealey, P.T., O'Connor, R.B., DeCarolis, P.L., Cory, D.G., Singer, S. Water suppression without signal loss in HR-MAS 1H NMR of cells and tissues. *J. Magn. Reson.* 2004;171:143–150.

[52] Hoult, D.I. Solvent peak saturation with single phase and quadrature Fourier transformation. *J. Magn. Reson.* 1976;21:337–347.

[53] Hoult, D.I., Richards, R.E. Critical factors in the design of sensitive high resolution nuclear magnetic resonance spectrometers,. *Proc. R. Soc. Lond., Series. A* 1975;344:311–340.

[54] Meiboom, S., Gill, D. Modified spin-echo method for measuring nuclear relaxation times. *Rev. Sci. Instrum.* 1958;29:688–691.

[55] Chen, J.H., Enloe, B.M., Weybright, P., Campbell, N., Dorfman, D., Fletcher, C.D., Cory DG., Singer, S. Biochemical correlates of thiazolidinedione-induced adipocyte differentiation by high-resolution magic angle spinning NMR spectroscopy. *Magn. Reson. Med.* 2002;48:602–610.

[56] Jeener, J. Ampere Summer School, Basko Polje, Yugoslavia. 1971.

[57] Aue, W.P., Bartholdi, E., Ernst, R.R. Two-dimensional spectroscopy. Application to nuclear magnetic resonance. J. Chem. Phys. 1976;64:2229–2246.

[58] Piantini, U., Sorensen, O.W., Ernst, R.R. Multiple quantum filters for elucidating NMR coupling networks. *J. Am. Chem. Soc.* 1982;104:6800–6801.

[59] Rance, M., Sorensen, O.W., Bodenhausen, G., Wagner, G., Ernst, R.R., Wuthrich, K. Improved spectral resolution in cosy 1H NMR spectra of proteins via double quantum filtering. *Biochem. Biophys. Res. Commun.* 1983;117:479–485.

[60] Braunschweiler, L., Ernst, R.R. Coherence transfer by isotropic mixing – application to proton correlation spectrosocpy. J. Magn. Reson. 1983;53:521–528.

[61] Müller, L. Sensitivity enhanced detection of weak nuclei using heternuclear multiple quantum coherence. J. Am. Chem. Soc. 1979;102:4481–4484.

[62] Bodenhausen, G., Ruben, D.J. Natural abundance N-15 NMR by enhanced heternuclear spectroscopy. *Chem. Phys. Lett.* 1980;69:185–189.

[63] Kay, L., Keifer, P., Saarinen, T. Pure absorption gradient enhanced heteronuclear single quantum correlation spectroscopy with improved sensitivity. *J. Am. Chem. Soc.* 1992;114:10663–10665.

[64] Moka, D., Vorreuther, R., Schicha, H., Spraul, M., Humpfer, E., Lipinski, M., Foxall P.J.D., Nicholson, J.K., Lindon, J.C. Magic angle spinning proton nuclear magnetic resonance spectroscopic analysis of intact kidney tissue samples. *Analytical Communications* 1997;34:107–109.

[65] Garrod, S., Humpfer, E., Spraul, M., Connor, S.C., Polley, S., Connelly, J., Lindon J.C., Nicholson, J.K., Holmes, E. High-resolution magic angle spinning H-1 NMR spectroscopic studies on intact rat renal cortex and medulla. *Magn. Reson. Med.* 1999;41:1108–1118.

[66] Griffin, J.L., Walker, L.A., Garrod, S., Holmes, E., Shore, R.F., Nicholson, J.K. NMR spectroscopy based metabonomic studies on the comparative biochemistry of the kidney and urine of the bank vole (Clethrionomys glareolus), wood mouse (Apodemus sylvaticus), white toothed shrew (Crocidura suaveolens) and the laboratory rat. *Comp. Biochem. Physiol. B. Biochem. Mol. Biol.* 2000;127:357–367.

[67] Bollard, M.E., Garrod, S., Holmes, E., Lindon, J.C., Humpfer, E., Spraul, M., Nicholson, J.K. High-resolution (1)H and (1)H-(13)C magic angle spinning NMR spectroscopy of rat liver. *Magn. Reson. Med.* 2000;44:201–207.

[68] Rooney, O.M., Troke, J., Nicholson, J.K., Griffin, J.L. High-resolution diffusion and relaxation-edited magic angle spinning 1H NMR spectroscopy of intact liver tissue. *Magn. Reson. Med.* 2003;50:925–930.

[69] Griffin, J.L., Williams, H.J., Sang, E., Nicholson, J.K. Abnormal lipid profile of dystrophic cardiac tissue as demonstrated by one- and two-dimensional magic-angle spinning (1)H NMR spectroscopy. *Magn. Reson. Med.* 2001;46:249–255.

[70] Griffin, J.L., Troke, J., Walker, L.A., Shore, R.F., Lindon, J.C., Nicholson, J.K. The biochemical profile of rat testicular tissue as measured by magic angle spinning 1H NMR spectroscopy. *FEBS Lett.* 2000;486:225–229.

[71] Wang, Y., Tang, H., Holmes, E., Lindon, J.C., Turini, M.E., Sprenger, N., Bergonzelli G., Fay, L.B., Kochhar, S., Nicholson, J.K. Biochemical Characterization of Rat Intestine Development Using High-Resolution Magic-Angle-Spinning (1)H NMR Spectroscopy and Multivariate Data Analysis. *J. Proteome. Res.* 2005;4:1324–1329.

[72] Tsang, T.M., Griffin, J.L., Haselden, J., Fish, C., Holmes, E. Metabolic characterization of distinct neuroanatomical regions in rats by magic angle spinning 1H nuclear magnetic resonance spectroscopy. *Magn. Reson. Med.* 2005;53:1018–1024.

[73] Lindon, J.C., Holmes, E., Nicholson, J.K. So what's the deal with metabonomics? *Anal. Chem.* 2003;75:384A–391A.

[74] Cheng, L.L., Chang, I.W., Louis, D.N., Gonzalez, R.G. Correlation of high-resolution magic angle spinning proton magnetic resonance spectroscopy with histopathology of intact human brain tumor specimens. *Cancer Res.* 1998;58:1825–1832.

[75] Cheng, L.L., Anthony, D.C., Comite, A.R., Black, P.M., Tzika, A.A., Gonzalez, R.G. Quantification of microheterogeneity in glioblastoma multiforme with ex vivo high-resolution magic-angle spinning (HRMAS) proton magnetic resonance spectroscopy. *Neuro-oncol.* 2000;2:87–95.

[76] Cheng, L.L., Newell, K., Mallory, A.E., Hyman, B.T., Gonzalez, R.G. Quantification of neurons in Alzheimer and control brains with ex vivo high resolution magic angle spinning proton magnetic resonance spectroscopy and stereology. *Magn. Reson. Imaging* 2002;20:527–533.

[77] Tzika, A.A., Cheng, L.L., Goumnerova, L., Madsen, J.R., Zurakowski, D., Astrakas, L.G., Zarifi, M.K., Scott, R.M., Anthony, D.C., Gonzalez, R.G., Black, PM. Biochemical characterization of pediatric brain tumors by using in vivo and ex vivo magnetic resonance spectroscopy. *J. Neurosurg.* 2002;96:1023–1031.

[78] Ratai, E.M., Pilkenton, S., Lentz, M.R., Greco, J.B., Fuller, R.A., Kim, J.P., He, J., Cheng, L.L., Gonzalez, R.G. Comparisons of brain metabolites observed by HRMAS 1H NMR of intact tissue and solution 1H NMR of tissue extracts in SIV-infected macaques. *NMR Biomed.* 2005;18:242–251.

[79] Sitter, B., Sonnewald, U., Spraul, M., Fjosne, H.E., Gribbestad, I.S. High-resolution magic angle spinning MRS of breast cancer tissue. *NMR Biomed.* 2002;15:327–337.

[80] Cheng, L.L., Chang, I.W., Smith, B.L., Gonzalez, R.G. Evaluating human breast ductal carcinomas with high-resolution magic-angle spinning proton magnetic resonance spectroscopy. *J. Magn. Reson.* 1998;135:194–202.

[81] Cheng, L.L., Wu, C., Smith, M.R., Gonzalez, R.G. Non-destructive quantitation of spermine in human prostate tissue samples using HRMAS 1H NMR spectroscopy at 9.4 T. *FEBS Lett.* 2001;494:112–116.

[82] Taylor, J.L., Wu, C.L., Cory, D., Gonzalez, R.G., Bielecki, A., Cheng, L.L. High-resolution magic angle spinning proton NMR analysis of human prostate tissue with slow spinning rates. *Magn. Reson. Med.* 2003;50:627–632.

[83] Burns, M.A., He, W., Wu, C.L., Cheng, L.L. Quantitative pathology in tissue MR spectroscopy based human prostate metabolomics. *Technol. Cancer Res. Treat.* 2004;3:591–598.

[84] Cheng, L.L., Burns, M.A., Taylor, J.L., He, W., Halpern, E.F., McDougal, W.S., Wu, C.L. Metabolic characterization of human prostate cancer with tissue magnetic resonance spectroscopy. *Cancer Res.* 2005;65:3030–3034.

[85] Mahon, M.M., Williams, A.D., Soutter, W.P., Cox, I.J., McIndoe, G.A., Coutts, G.A., Dina R., deSouza, N.M. 1H magnetic resonance spectroscopy of invasive cervical cancer: An in vivo study with ex vivo corroboration. *NMR Biomed.* 2004;17:1–9.

[86] Mahon, M.M., deSouza, N.M., Dina, R., Soutter, W.P., McIndoe, G.A., Williams, A.D., Cox I.J. Preinvasive and invasive cervical cancer: An ex vivo proton magic angle spinning magnetic resonance spectroscopy study. *NMR Biomed.* 2004;17:144–153.

[87] Mahon, M.M., Cox, I.J., Dina, R., Soutter, W.P., McIndoe, G.A., Williams, A.D., deSouza NM. (1)H magnetic resonance spectroscopy of preinvasive and invasive cervical cancer: In vivo-ex vivo profiles and effect of tumor load. *J. Magn. Reson. Imaging* 2004;19:356–364.

[88] Sitter, B., Bathen, T., Hagen, B., Arentz, C., Skjeldestad, F.E., Gribbestad, I.S. Cervical cancer tissue characterized by high-resolution magic angle spinning MR spectroscopy. *Magma* 2004;16:174–181.

[89] Moka, D., Vorreuther, R., Schicha, H., Spraul, M., Humpfer, E., Lipinski, M., Foxall P.J.D., Nicholson, J.K., Lindon, J.C. Biochemical classification of kidney carcinoma biopsy samples using magic-angle-spinning H-1 nuclear magnetic resonance spectroscopy. *J. Pharm. Biomed. Anal.* 1998;17:125–132.

[90] Tate, A.R., Foxall, P.J., Holmes, E., Moka, D., Spraul, M., Nicholson, J.K., Lindon, J.C. Distinction between normal and renal cell carcinoma kidney cortical biopsy samples using pattern recognition of (1)H magic angle spinning (MAS) NMR spectra. *NMR Biomed.* 2000;13: 64–71.

[91] Sitter, B., Autti, T., Tyynela, J., Sonnewald, U., Bathen, T.F., Puranen, J., Santavuori, P., Haltia, M.J., Paetau, A., Polvikoski, T., Gribbestad, I.S., Hakkinen, A.M. High-resolution magic angle spinning and 1H magnetic resonance spectroscopy reveal significantly altered neuronal metabolite profiles in CLN1 but not in CLN3. *J. Neurosci. Res.* 2004;77:762–769.

[92] Millis, K., Weybright, P., Campbell, N., Fletcher, J.A., Fletcher, C.D., Cory, D.G., Singer, S. Classification of human liposarcoma and lipoma using ex vivo proton NMR spectroscopy. *Magn. Reson. Med.* 1999;41:257–267.

[93] Chen, J.H., Enloe, B.M., Fletcher, C.D., Cory, D.G., Singer, S. Biochemical analysis using high-resolution magic angle spinning NMR spectroscopy distinguishes lipoma-like well-differentiated liposarcoma from normal fat. *J. Am. Chem. Soc.* 2001;123:9200–9201.

[94] Garrod, S., Humpher, E., Connor, S.C., Connelly, J.C., Spraul, M., Nicholson, J.K., Holmes, E. High-resolution (1)H NMR and magic angle spinning NMR spectroscopic investigation of the biochemical effects of 2-bromoethanamine in intact renal and hepatic tissue. *Magn. Reson. Med.* 2001;45:781–790.

[95] Waters, N.J., Holmes, E., Waterfield, C.J., Farrant, R.D., Nicholson, J.K. NMR and pattern recognition studies on liver extracts and intact livers from rats treated with alpha-naphthylisothiocyanate. *Biochem. Pharmacol.* 2002;64:67–77.

[96] Bollard, M.E., Xu, J., Purcell, W., Griffin, J.L., Quirk, C., Holmes, E., Nicholson J.K. Metabolic profiling of the effects of D-galactosamine in liver spheroids using (1)H NMR and MAS-NMR spectroscopy. *Chem. Res. Toxicol.* 2002;15:1351–1359.

[97] Griffin, J.L., Walker, L.A., Shore, R.F., Nicholson, J.K. Metabolic profiling of chronic cadmium exposure in the rat. Chem. Res. Toxicol. 2001;14:1428–1434.

[98] Griffin, J.L., Walker, L., Shore, R.F., Nicholson, J.K. High-resolution magic angle spinning 1H-NMR spectroscopy studies on the renal biochemistry in the bank vole (Clethrionomys glareolus) and the effects of arsenic (As3+) toxicity. *Xenobiotica* 2001;31:377–385.

[99] Morvan, D., Demidem, A., Papon, J., Madelmont, J.C. Quantitative HRMAS proton total correlation spectroscopy applied to cultured melanoma cells treated by chloroethyl nitrosourea: demonstration of phospholipid metabolism alterations. *Magn. Reson. Med.* 2003;49: 241–248.

[100] Griffin, J.L., Lehtimaki, K.K., Valonen, P.K., Grohn, O.H., Kettunen, M.I., Yla-Herttuala, S., Pitkanen, A., Nicholson, J.K., Kauppinen, R.A. Assignment of 1H nuclear magnetic resonance visible polyunsaturated fatty acids in BT4C gliomas undergoing ganciclovir-thymidine kinase gene therapy-induced programmed cell death. *Cancer Res.* 2003;63:3195–3201.

[101] Griffin, J.L., Pole, J.C., Nicholson, J.K., Carmichael, P.L. Cellular environment of metabolites and a metabonomic study of tamoxifen in endometrial cells using gradient high resolution magic angle spinning 1H NMR spectroscopy. Biochim. *Biophys. Acta* 2003;1619:151–158.

[102] Laurent, S., Chen, H., Bedu, S., Ziarelli, F., Peng, L., Zhang, C.C. Nonmetabolizable analogue of 2-oxoglutarate elicits heterocyst differentiation under repressive conditions in Anabaena sp. PCC 7120. Proc. Natl. Acad. Sci. U.S.A 2005;102:9907–9912.

The Handbook of Metabonomics and Metabolomics
John C. Lindon, Jeremy K. Nicholson and Elaine Holmes (Editors)

Chapter 5

Chromatographic and Electrophoretic Separations Combined with Mass Spectrometry for Metabonomics

Ian D. Wilson

Department of Drug Metabolism and Pharmacokinetics
AstraZeneca, Mereside, Alderley Park, Macclesfield, Cheshire SK10 4TG, UK

Abstract

The current use of mass spectrometry, in combination with separation techniques such as gas and liquid chromatography and capillary electrophoresis, for metabonomics research is reviewed. Capillary gas chromatography (GC) is a very high resolution separation technique for this type of "global metabolite profiling", especially when GC × GC is performed, but requires extensive sample pre-treatment and derivatisation prior to analysis. High performance liquid chromatography (HPLC) and capillary LC provide high throughput, and generally require minimal sample preparation other than protein precipitation. The recently introduced ultra-performance liquid chromatography (UPLC), which is based on smaller particles and higher pressures than used in conventional HPLC, provides a significant improvement in terms of resolution and sensitivity. Capillary electrophoresis (CE) is another valuable liquid phase technique because of the alternative separation mechanism involved. The attributes of the various combinations are examined and the advantages and limitations of these techniques relative to each other and to NMR spectroscopy are briefly examined.

5.1. Introduction

The use of increasingly powerful and sophisticated instrumental techniques has resulted in enormous changes in the analytical landscape with respect to metabolite profiling. These advanced techniques have enabled the development of the non-targeted methods for "global" metabolite profiling that are essential for metabonomics. Whilst the bulk of the published applications in metabonomics have relied on the use of high field NMR spectroscopy, there is no doubt that other methodologies, particularly those based on mass spectrometric (MS) techniques, especially when linked to separations (often referred to as "hyphenation"), have the ability to contribute to metabonomics studies. Clearly to be useful for metabonomic studies an analytical technique would ideally provide as comprehensive a metabolic profile as possible and would not preferentially select particular classes of metabolites but be equally sensitive to all the compounds in the samples.

Moderate to high throughput is a real advantage for metabonomics as relatively large sample sets are a feature of this type of work and therefore a short analysis time, with minimal sample workup, is to be aimed for. In addition, a method should be sensitive and have a sufficiently broad dynamic range to be able to cope with the wide range of concentrations of metabolites present in the samples. The analytical techniques used should also provide both quantitative results and sufficient structural data to enable rapid and unambiguous biomarker identification.

Such requirements are very demanding and current MS-based methods can achieve some, but not all of them. However, it is arguable that there is no ideal method for obtaining these global metabolite profiles and that MS covers at least some of the bases. Here the current practice of high performance liquid chromatography coupled to mass spectrometry (HPLC-MS) [1] for metabonomics is considered together with gas chromatography-mass spectrometry (GC-MS) [2, 3] and capillary electrophoresis-mass spectrometry (CE-MS) [4, 5].

5.2. Gas Chromatography-Mass Spectrometry

GC-MS represents one of the oldest and most successful hyphenations of separation techniques to a mass spectrometer. Capillary GC provides an efficient and high resolution separation method and there are few practical problems in coupling to the interface of the mass spectrometer. Separations of complex mixtures of the type encountered in metabonomics are usually based on a programmed temperature gradient, with the most volatile components eluting first. In GC-MS there are essentially two modes of ionisation: electron impact ionisation (EI) and chemical ionisation (CI).

The EI technique provides mass spectra where the molecular ion may be weak but the spectra often contain a number of diagnostic fragments that can enable much structural identification to be carried out (especially when combined with searchable databases). CI, being a rather more "gentle" means of ionisation, provides mostly molecular ion information which can be very valuable as it enables the detection/confirmation of the molecular mass of the unknown.

GC provides a highly developed, stable, selective, sensitive and high resolution separation system. This capability is continually being enhanced and, with the introduction of GC-GC separations, combined with ever more powerful MS detectors, including ToF (time of flight) and ToF-ToF instruments, the comprehensive analysis of very complex samples should be possible. The use of ToF enables accurate masses to be obtained with the benefit that atomic compositions can be deduced providing further useful information for structure determination. Capillary GC-MS is in widespread use for metabolomics in areas such as micro-organisms and plants [2] but published applications of GC-MS to mammalian systems are harder to find.

The most obvious disadvantage of GC-based techniques is that the bulk of the components in biofluid or tissue samples are relatively involatile. There is therefore a requirement for a reasonable amount of pre-processing before the sample can actually be analysed, in order to convert the analytes into volatile derivatives before GC analysis is attempted [6]. In the case of a sample such as plasma it would be usual practice to remove interfering plasma proteins by precipitation with three volumes of acetonitrile [3, 7]. An aliquot of the supernatant would then be evaporated to dryness, followed by derivatisation with first methoxylamine hydrochloride (40 mg/mL in pyridine) at 28 °C for 90 min and then *N*-methyl-*N*-(trimethylsilyl)-trifluoracetamide (MSTFA) at 37 °C for 30 min. For aqueous samples such as urine, even though proteins are generally absent, it is still necessary to perform an extraction to enable the analytes to be dissolved in pyridine, as the derivatisation reactions cannot be performed in water. Extraction on to a suitable solid phase extraction (SPE) cartridge is a possible way of obtaining extracts (and can also be used to concentrate samples). However, with such an SPE step there is always the attendant risk of losing important metabolites that are not retained on the phase.

For such a multi-step process it is clearly necessary to institute good quality control (QC) procedures to ensure the validity of the final result. The usual way to do this where single analytes are being analysed is to use internal standards, however, it is difficult to control the extraction, derivatisation, chromatography and detection of hundreds of different analytes from disparate chemical classes, even using a number of internal standards. For this reason, in our studies we have adopted the pragmatic approach of using a pooled sample as a QC. This sample is distributed at random amongst the study samples and taken through the analytical process with them. Following analysis of all of the samples the data are processed via, for example, principal components analysis (PCA) and the closeness of the clustering of the QC samples in the scores plot examined. Providing that all of the QC samples map close together, it can be assumed that the method has performed in a sufficiently reproducible fashion for the data from the study samples to be valid. If, on the other hand, there is a widespread distribution of the QC samples, it is reasonable to assume that there is a problem with the analytical method.

This extensive sample preparation required for GC-MS, together with the relatively long run-times currently associated with GC, means that the technique is relatively low throughput compared to some other technologies. However, this is compensated for to some extent by the very high separation efficiency of the system and the availability of large databases that greatly aid identification of unknowns.

An application of the potential of GC-MS for metabonomic analysis is shown by the example of the analysis of plasma from Zucker (fa/fa) obese and normal Wistar-derived animals given in Figures 5.1 and 5.2. The total-ion-current (TIC) traces shown are for GC-MS with EI (Figure 5.1a and b) and CI (Figure 5.2a and b). The excellent separation of these two classes using PCA on the GC-EI-MS data is shown in Figure 5.3. This figure also includes the data obtained for the QC samples (made by pooling together an aliquot of all of the study samples) which cluster

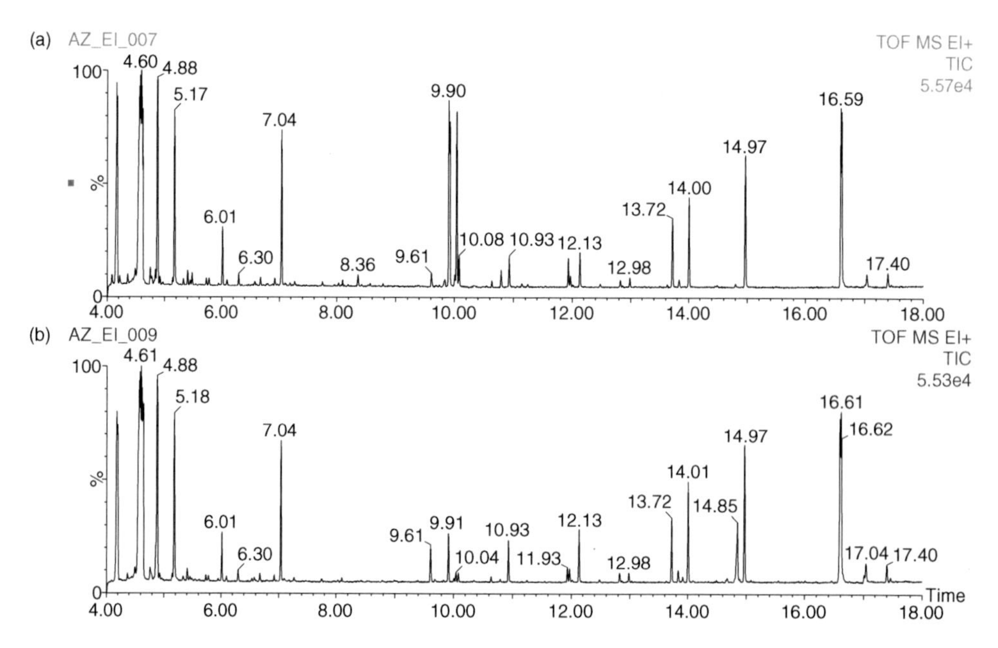

Figure 5.1. Typical TIC traces obtained from GC-EI-MS analysis of plasma obtained from (a) Wistar-derived and (b) Zucker (fa/fa) obese rats.

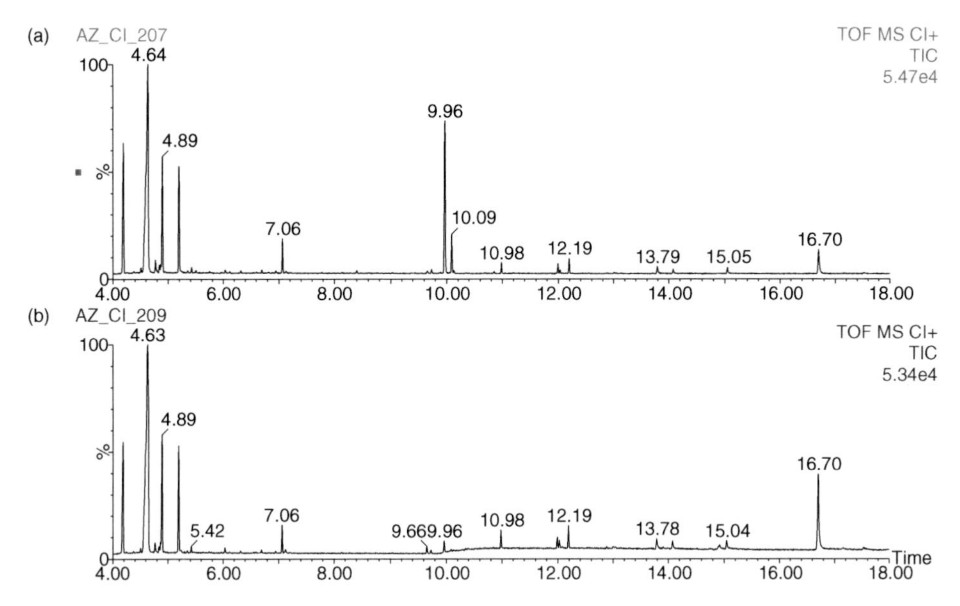

Figure 5.2. Typical TICs obtained from GC-CI-MS analysis of plasma obtained from (a) Wistar-derived and (b) Zucker (fa/fa) obese rats.

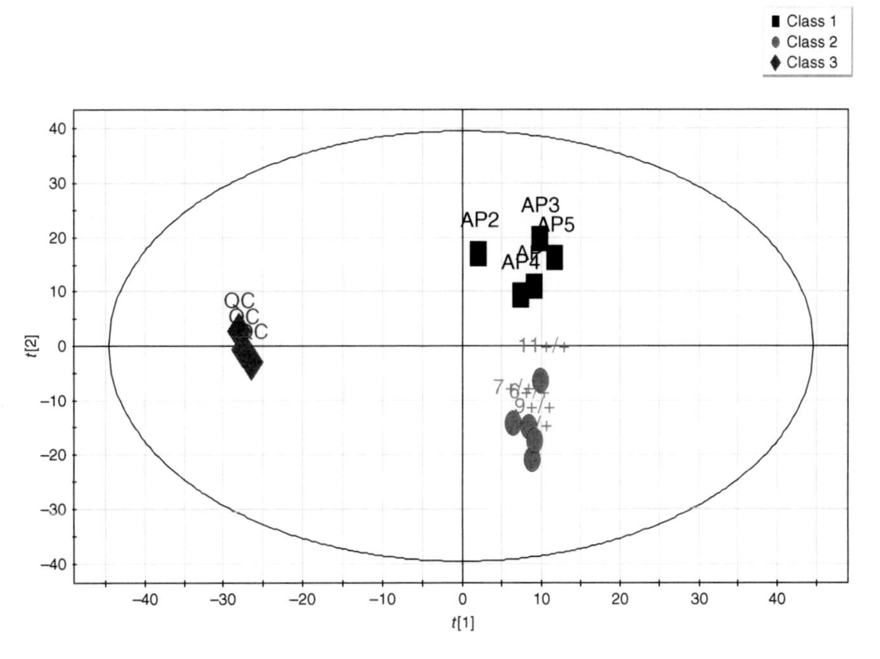

Figure 5.3. Scores plot (component 1 versus component 2) obtained following partial least squares discriminant analysis (PLS-DA) of data derived from GC-EI-MS analysis of plasma samples obtained from Wistar-derived (AP) and Zucker (fa/fa) obese (+/+) rats and QC samples analysed in the same run.

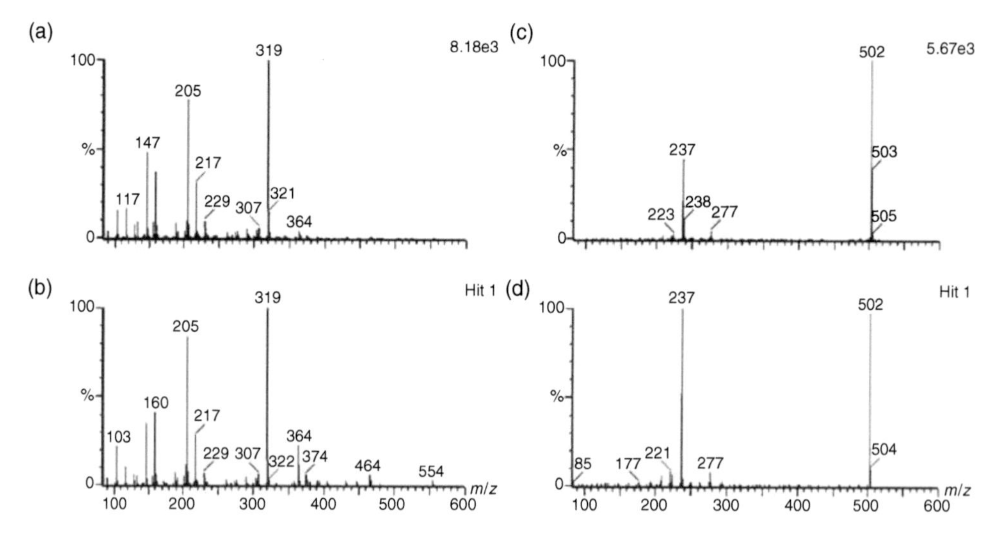

Figure 5.4. Identification of perturbed metabolites from GC-EI-MS analysis: (a) EI-MS for peak eluting at 9.91 min and (b) its library match D-glucose 2,3,4,5,6-pentakis-*O*-(trimethylsilyl-,*O*-methyloxime, (c) EI-MS for peak eluting at 16.58 min and (d) its library match tocopherol (vitamin E), trimethylsilyl derivative.

closely together providing confidence in the overall quality of the data. Figure 5.4 illustrates a typical set of spectra derived from these data for tocopherol.

The capabilities of GC-ToF-MS for obtaining global metabolite profiles of human plasma have recently been demonstrated where, following an automated (closed loop) optimisation, over 950 individual components were detected [3]. This procedure, which was also applied to yeast fermentation broths, involved optimisation of a number of instrumental parameters via a genetic algorithm and provided an almost threefold increase in the number of observable peaks. An example of the type of result possible following such an optimisation is illustrated in Figure 5.5.

A further, major, enhancement in the capabilities of GC-MS to metabolite profiling is the application of multi-dimensional GC × GC (sometimes referred to as "comprehensive GC). In GC × GC, two GC columns are used, the first generally long (typically 30 m) and the second much shorter (e.g. 1.5 m). Compounds eluting from the first column are trapped and focussed cryogenically on the second for a short time followed by a rapid separation on the 2nd column. By using different column chemistries the selectivity of the separation can be modulated, and improved separations can be obtained. Currently applications in metabolite profiling are sparse but in an example GC × GC-ToF-MS was used to examine extracts of spleen from obese NZO and lean C57BL/6 mice [8, 9]. Separations in the first dimension were performed on a 30 m × 250 μm capillary GC column coated with dimethyl polysiloxane (0.25 μm thickness) which separated compounds, using a thermal gradient,

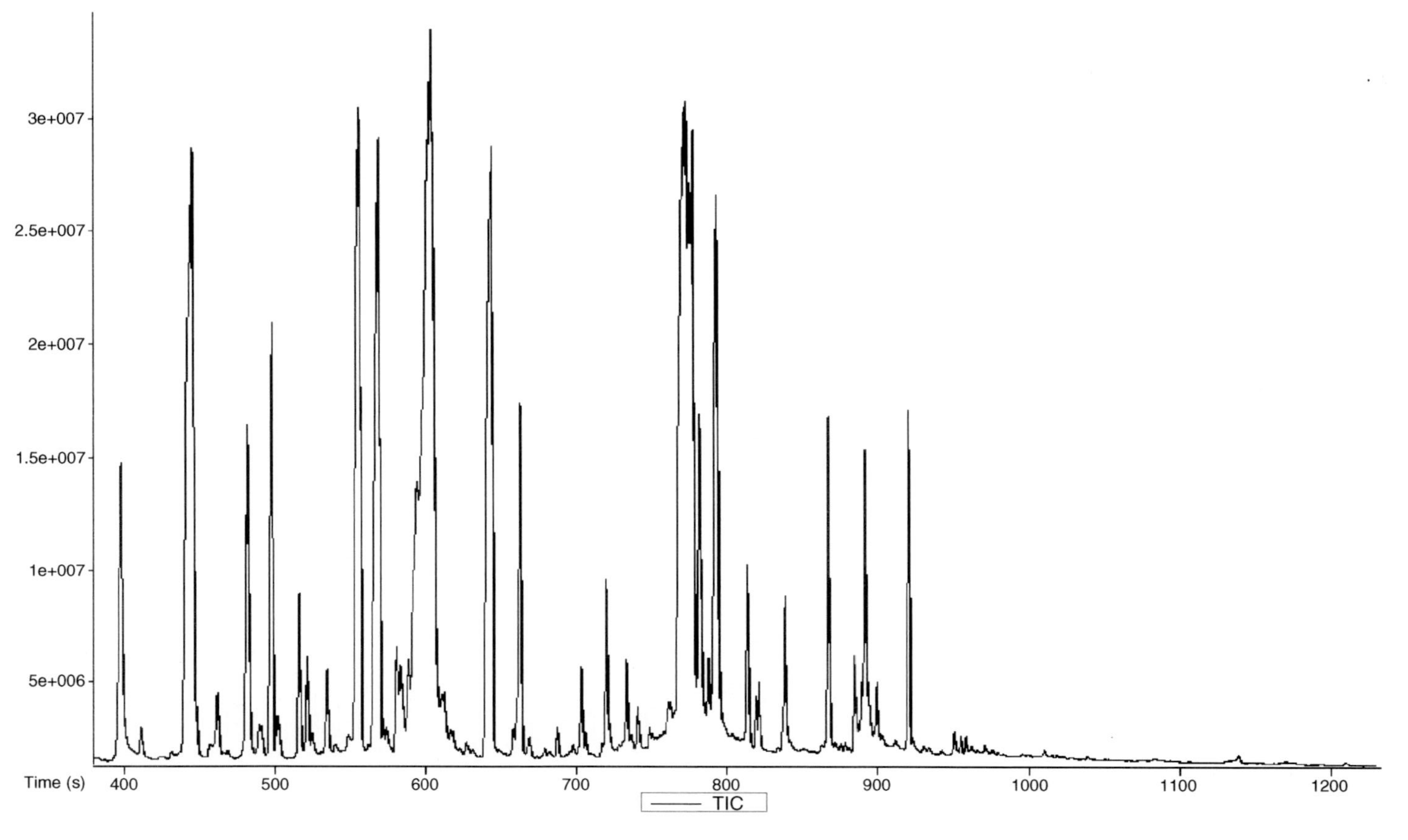

Figure 5.5 A typical total ion chromatogram of human serum using the method optimised as described in Reference 3.

based on volatility. Chromatography in the 2nd dimension was undertaken using a 1.5 m × 100 μm capillary coated with 50% polysilphenylene-siloxane (0.10 μm film thickness), a moderately polar phase, that separated compounds on relative polarity. Compared to conventional GC-ToF-MS (on a 30 m × 250 μm dimethyl-polysiloxane column (0.25 μm film thickness), which enabled the detection of 500 compounds, GC × GC-ToF-MS revealed the presence of some 1200 compounds with a run time of some 65 min.

Despite the current paucity of examples of the use of GC-MS and GC × GC-ToF-MS in metabonomics, there is clearly considerable potential for this powerful analytical technology to make a valuable contribution to the generation of global metabolite profiles and undoubtedly many more examples will be published in the future.

5.3. Liquid Chromatography-Mass Spectrometry

Unlike GC-MS, where the coupling of the separation technique and the spectrometer has proven to be relatively straightforward, the hyphenation of liquid chromatographic separations with mass spectrometers was technically much more demanding. As a result, HPLC-MS as a routine technique is a more recent addition to the bioanalytical tool box and for this reason HPLC-MS has only begun to be used for metabonomics relatively recently [1]. However, HPLC-MS is widely available in bioanalytical laboratories and for this reason, if no other, is destined to become increasingly important in metabonomics studies.

In general, reversed-phase gradient chromatography has been adopted for metabolite profiling work of the sort required for metabonomics studies. This type of separation is compatible with aqueous biological samples such as urine, making analysis possible with minimal sample preparation. More complex samples such as plasma do, however, require more extensive sample preparation. Protein precipitation using about two or three volumes of acetonitrile followed by centrifugation is generally required to prevent irreversible degradation of the HPLC column. Detection to date has usually been performed by electrospray ionisation (ESI), although CI is also an option. Self-evidently, detection in mass spectrometric methods depends upon the ionisation of the molecules in the sample. Because some analytes ionise better in positive ESI and others in negative ESI, it is good practice to analyse the samples using both ionisation modes (usually in separate analytical runs).

Something that has to be borne in mind with all liquid chromatography-mass spectrometry (LC-MS) analyses is that not all molecules ionise equally well leading to differences in sensitivity even when compounds are present in equal molar concentrations. In addition, HPLC-MS studies are complicated by the phenomenon known as "ion suppression" (and also enhancement) whereby the presence of co-eluting substances adversely affects the ionisation of a particular analyte causing its signal to be reduced [10, 11]. In drug analysis in biological fluids, where the identity of the analyte is known, these effects can be studied, minimised through careful sample preparation and chromatographic optimisation and compensated for to some extent by the addition of an internal standard (usually a deuterated version of the analyte itself). However, in a complex mixture of unknown composition, such as a urine or plasma sample, where all of the components may be important, the strategies used for single known analytes are not easily applied. It seems clear therefore that, for a biomarker to be shown to be valid, once the analytes have been identified there is then a need to develop specific and comprehensively validated analytical methods. These methods can then be applied to the samples to confirm that the changes observed, for example, in plasma concentrations of particular substances really do correlate with an observed physiological change.

The very characteristics that make reversed-phase liquid chromatography so well suited to the direct injection of biological fluids with minimal sample preparation also make the technique vulnerable to failure as a result of column degradation and source contamination. Column degradation can reveal itself by changes in column performance (loss of peak shape, changes in retention time) whilst contamination of the ion source usually results in loss of sensitivity. The need for careful monitoring and QC of the methodology in order to ensure the validity of the conclusions is therefore essential if valid data are to be produced. The regular injection of test mixtures through the run can clearly help in determining whether or not system performance is maintained, and internal standards may also have a role to play. However, given the complexity and structural diversity of the molecules present in biological samples then the difficulty of choosing an internal standard(s) is obvious. Once again we have adopted the pragmatic approach of using a pooled sample that is run at regular intervals throughout the analysis in order to monitor the behaviour of the LC-MS analyses undertaken on biofluid samples.

5.4. HPLC-MS for metabonomics

Conventional HPLC separations for metabonomic analysis are undertaken using 4.6 or 3.0 mm i.d. columns, of between 5 and 25 cm in length, packed with 3–5 μm solid materials. These analyses are usually performed using reversed-phase gradients in order to separate the widest range of molecules in a single run. In our own work, the typical chromatographic conditions that we have employed involved separation on a 2.1 mm × 10 cm Symmetry® C18, 3.5 μm column held at 40 °C in a column oven and solvents composed of 0.1% formic acid and acetonitrile. Elution was performed using a two part gradient from 0 to 20% acetonitrile over 0.5–4 min. and then to 95% acetonitrile over the next 4 min. The eluent composition is then held at 95% acetonitrile for 1 min. before returning to 100% aqueous formic acid at 9.1 min. For a column of these dimensions a flow rate of 600 μl/min was used (with 100 μl directed into the ion source of the mass spectrometer), and typically 5–10 μl of rat urine would be injected. This protocol results in an analysis time of ca. 10 min per sample.

The recently published applications of HPLC-MS for metabonomics studies include examples in the investigation of toxicity [12–18], metabotyping, where, for example, strain, gender, aging and diurnal variation in rodents have been studied [19–21] and for disease models [20]. Examples of the results obtained for a typical HPLC-MS analysis of mouse urine (in both +ve and −ve ESI) are shown in Figure 5.6. The TIC traces themselves are not particularly informative, though it is possible to discern differences in the profiles obtained depending upon the ionisation techniques used. However, PCA of the data obtained in this way to compare, for example, different strains of mouse, as shown in Figure 5.7, reveal the power of HPLC-MS as a means of distinguishing animal strains.

Studies on the urine of male and female Zucker (fa/fa) obese rats [20], using both ^{1}H-NMR spectroscopy and HPLC-MS, allowed both diurnal and gender-based differences to be determined. Whilst diurnal effects were marked by increases in taurine, creatinine, allantoin and α-ketoglutarate in the ^{1}H NMR spectra of the Zucker rats for the evening samples, HPLC-MS indicated that (unidentified) ions at m/z 285.0753, 291.0536 and 297.1492 in positive ESI, and 461.1939 in negative ESI were elevated in these samples. Discrimination between male and female Zucker rats was also obtained using both analytical approaches, with HPLC-MS affording the clearest distinction. Thus, in the ^{1}H NMR spectra hippuric acid, succinate, α-ketoglutarate and dimethylglycine were higher in the urine of females compared to males, together with ions at m/z 431.1047, 325.0655, 271.0635 and 447.0946 in positive ESI, and m/z 815.5495 and 459.0985 in negative ESI by HPLC-MS. In addition, a comparison was also performed against samples of urine obtained from a Wistar-derived rat strain with ^{1}H NMR spectroscopic analysis pinpointing higher concentrations of taurine, hippurate and formate and decreased betaine, α-ketoglutarate, succinate and acetate in Zucker versus Wistar-derived rats. The HPLC-MS detected increased hippurate and unidentified ions at m/z 255.0640 and 285.0770 in positive, and 245.0122 and 261.0065 in negative ESI respectively.

More recently, we have used a similar approach of combining both ^{1}H NMR spectroscopy and HPLC-ToF/MS using ESI to study the effect of aging and development [21] in Wistar-derived rats on the profile of endogenous urinary metabolites. Samples were collected from male rats every 2 weeks, from just after weaning at 4 weeks up to 20 weeks of age, and the resulting spectroscopic data were analysed using multivariate data analysis. This enabled age-related metabolite changes to be detected, with urine samples collected at 4 and 6 weeks showing the greatest differences by both analytical techniques. Markers detected by ^{1}H NMR spectroscopy included creatinine, taurine, hippurate and resonances associated with amino acids/fatty acids, which increased with age, whilst citrate and resonances

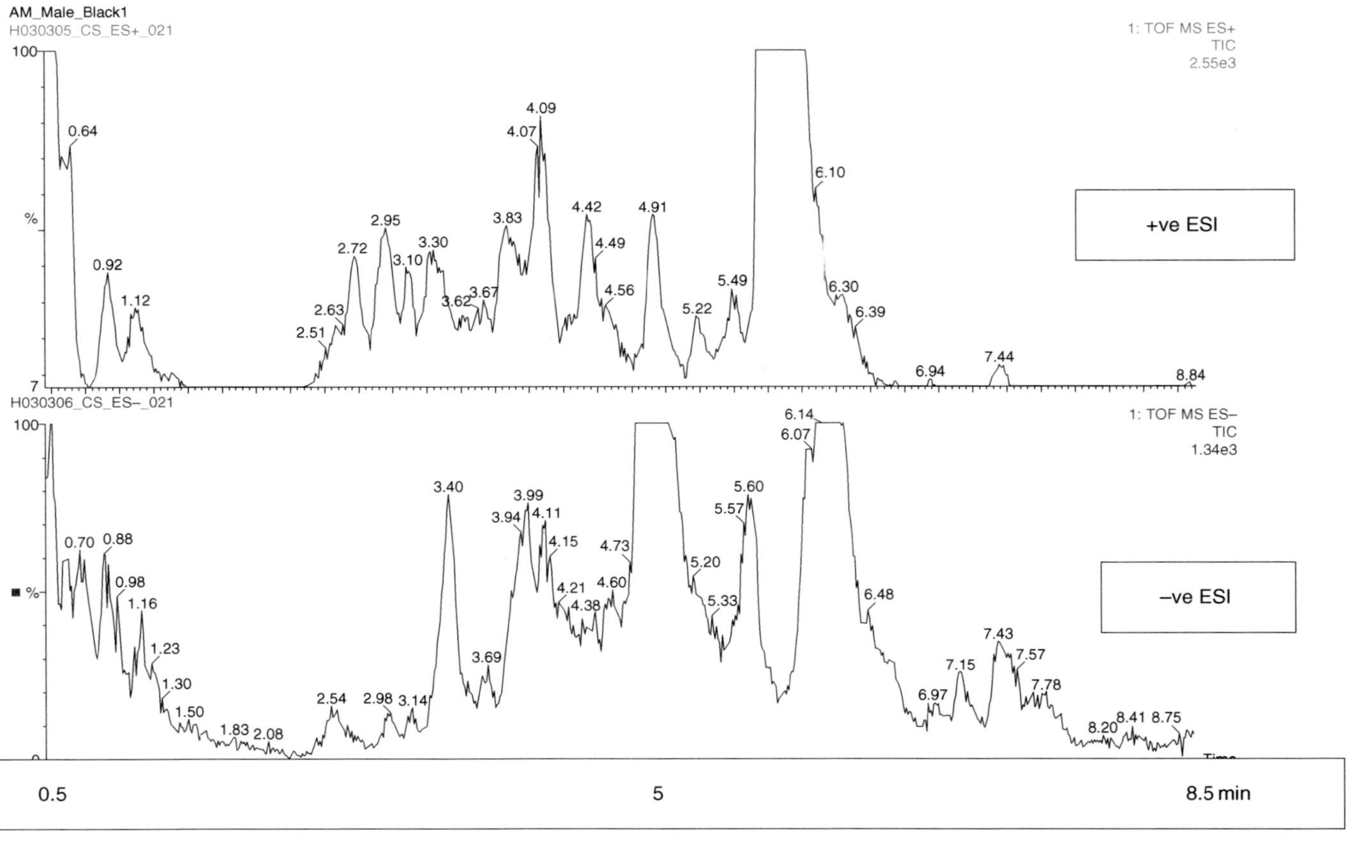

Figure 5.6. Gradient HPLC-oa-ToF-MS of mouse urine. A) total ion current for +ve and B) −ve ion electrospray. The column was a Waters Symmetry C18, 2.1 mm × 10 cm, 3.5 μm packing. Elution was via a linear gradient from 0 to 20% acetonitrile versus 0.1% formic acid in water from 0.5 to 4 min and then 95% acetonitrile at 8 min, followed by a return to 100% aqueous formic acid at 9 min. The column eluent was monitored by ESI-oa-TOF-MS from 100–1400 *m/z*.

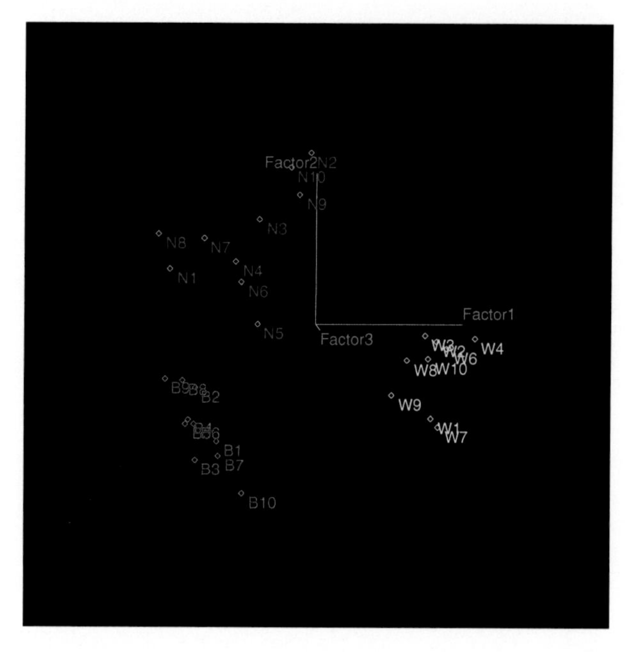

Figure 5.7. Scores plot for the PCA of LC-MS data obtained from urine of male animals from three mouse strains (key: green = nude, yellow = white and red = black). The samples were collected in the morning.

resulting from glucose/myoinositol declined. Interestingly a number of ions were detected with HPLC-MS that were only present in the 4 week urine samples, in both positive and negative ESI. The HPLC-MS analysis also showed age-related increases for a number of compounds, including, for example, carnitine which increased in samples from the older animals. A conclusion from both of these studies in the rat where ^{1}H NMR spectroscopy and HPLC-MS were used was that their use together provided a powerful and complementary approach to sample analysis, and this is also a conclusion of studies in the area of toxicology discussed below.

In addition to the metabotyping experiments described above, there have also been a significant number of applications concerning the investigation of study of toxicity in rats. The first application of HPLC-MS in this area involved an undisclosed candidate drug administered to both male and female rats at two dose levels over a 3-month period [11]. One of the ions detected in the urine of these animals, that appeared to be associated with toxicity in this study, was identified as indican (a metabolite of tryptophan). Identification was made using a combination of accurate mass, MS/MS and comparison with an authentic standard. Shortly afterwards a study of the effects of the drug citalopram was described as part of investigations of phospholipidosis [12]. Rather than use direct injection of the samples, these were subjected to solid phase extraction on a polymeric phase.

The investigation of the nephrotoxicity of heavy metal salts such as uranium nitrate or cadmium chloride using HPLC-MS, with both +ve and −ve ESI, was described [14] using reversed-phase chromatography on an Xterra MS C18 bonded material with a long gradient (ca. 2 h). Markers of toxicity were determined by a manual examination of the data, with compounds such as riboflavin, phenol sulfate and ferrulic acid, amongst others ions identified.

In a series of studies, we have investigated the application of gradient reversed-phase HPLC-ToF-MS, with both + and −ve ESI, for the analysis of urine obtained from rats exposed to a number of nephrotoxins [15–18]. In the first of these studies, an investigation was performed of the effects of the administration of a single 2.0 mg/kg s.c. dose of mercuric chloride to male Wistar-derived rats on the urinary metabolite profiles [15]. In this study, urine was collected for 9 days and then analysed using both HPLC-oa-ToF-MS and ^{1}H NMR spectroscopy. Unsurprisingly both methods of analysis identified marked changes in the pattern of endogenous metabolites with the most pronounced disturbances observed at 3 days post-dose. Thereafter, the metabolite profile gradually returned to a more normal composition. The HPLC-MS-detected markers of nephrotoxicity seen using positive ESI included decreased amounts of kynurenic, xanthurenic and pantothenic acids as well as 7-methylguanine. In addition, an ion at m/z 188, possibly 3-amino-2-naphthoic acid, was observed to increase whilst several unidentified ions at, for example, m/z 297 and 267 decreased after $HgCl_2$ administration. Analysis with negative ESI revealed a number of sulfated compounds such as phenol sulfate and benzene diol sulfate, both of which appeared to decrease in concentration in response to dosing, together with an unidentified glucuronide (m/z 326).

Similar studies on the nephrotoxicity of the immunosuppressant, cyclosporin A[16], involved 9 daily doses of 45 mg/kg/day of the drug. Disturbances of the urinary metabolite profile were only observed after 7-days dosing. The HPLC-MS interpretation of the data was complicated in this instance as a result of the presence of ions derived from cyclosporin, its metabolites and the dosing vehicle which had to be removed prior to PCA. However, as noted with the mercuric chloride example above, an excellent concordance was seen between HPLC-MS and NMR spectroscopy for the time course of toxicity.

A third example of the study of nephrotoxicity using HPLC-MS involved the twice daily administration 60 mg/kg of the antibiotic gentamicin for 7 days [17]. Changes in the pattern of endogenous metabolites were readily detected shortly after the start of dosing using both positive and negative ESI, with the major changes observed beginning on days 5/6 of the study. The MS data showed, as observed for mercuric chloride, reduced xanthurenic acid and kynurenic acid whilst neutral loss experiments also revealed a changed pattern of sulfate conjugation on gentamicin administration. A similar time course was noted when the same samples were analysed using ^{1}H NMR spectroscopy followed by PCA. However, such data require careful interpretation and it is likely that some of the observed changes were due not to the toxicity of the drug on the kidney but to the toxicity of the compound to the gut microflora resulting in changes in the urinary profile of microflora-derived metabolites.

In the same way that differences in patterns of a subset of metabolites were examined in the above study, a recent report described a metabonomic study where constant neutral loss experiments were performed in order to detect only mercapturates present in the urines of human volunteers who were administered 50–500 mg of acetaminophen (paracetamol) [22].

HPLC-ToF-MS was also used to investigate urinary metabolic perturbations associated with D-serine-induced nephrotoxicity [18]. D-Serine is known to cause selective necrosis of the proximal straight tubules in the rat kidney accompanied by aminoaciduria, proteinuria and glucosuria. In these studies, Wistar-derived rats received either D-serine (250 mg/kg i.p.) or vehicle (deionised water), and urine was collected up to 48 h post-dose. Samples were then analysed using reversed-phase gradient HPLC with ToF-MS in both positive or negative ion mode. Changes to the urinary profile were detected at all time points after administration. For example, in negative ESI-MS increased concentrations of serine, an ion at m/z 104.0376 (possibly hydroxypyruvate) and glycerate were detected (the latter being a metabolite of D-serine). Also observed were increased amounts of tryptophan, phenylalanine and lactate, together with decreased methylsuccinic and sebacic acids. In the case of positive ESI, decreased concentrations of xanthurenic acid were noted, together with a general aminoaciduria, including proline, methionine, leucine, tyrosine and valine and an increase in acetylcarnitine.

As well as applications in animal models, HPLC-MS has also been used for applications on human samples for diseases such as type-2 diabetes [23], liver cancer, hepatitis and liver cirrhosis [24]. Although described as metabonomics applications by the authors in fact, in both cases, only a subset of the total metabolic profile was studied (either phospholipids or *cis*-diols) making these studies more akin to conventional class-specific metabolic profiling. However, as the separation methodology employed, including the use of multivariate statistical analysis, is relevant, both applications are briefly described here.

In the application to type-2 diabetes [23] normal phase chromatography was performed on a silica-based "diol" stationary phase using gradient elution. The solvents used for chromatography were mixtures of hexane: propan-1-ol:formic acid:aqueous ammonia (79:20:0.6:0.07 v/v) and propan-1-ol:water:formic acid:aqueous ammonia (88:10:0.6:0.07 v/v) at a flow rate of $0.4\,\text{ml}\cdot\text{min}^{-1}$. With re-equilibration of the column following the gradient the analysis took ca. 60 min per sample. Plasma samples were analysed for phospholipids by ESI-MS/MS, with negative ESI used for profiling and both positive and negative ESI for characterisation and identification.

Multivariate statistical analysis of these data enabled the authors to distinguish between normal and diabetic subjects.

For the analysis of urine from subjects suffering from a variety of liver diseases [24], reversed-phase gradient chromatography was performed using 5 mm ammonium acetate and methanol as solvents on a 25 cm × 4.6 mm i.d. column packed with a 5 μm ODS phase with ESI-MS/MS. Run times, including re-equilibration, were of the order of 70 min. Samples were prepared using extraction onto phenylboronic acid solid-phase extraction cartridges to extract the required *cis*-diols (mainly metabolites of nucleic acids). The data were then analysed by multivariate statistical methods following peak alignment. Identification of the biomarkers found by this approach was using a combination of MS/MS and authentic standards. Some eight such compounds were detected which enabled patients with liver cancer to be distinguished from subjects with other types of liver disease.

5.5. Capillary LC-MS

Alternatives to conventional HPLC such as narrow bore (ca. 2 mm i.d.), micro bore (0.5, 1.0 mm i.d.) and capillary HPLC column formats have also been employed for metabolite profiling [25, 26]. Capillary HPLC offers similar benefits for the analysis of complex mixtures to capillary separations in gas chromatography and, by providing an increased number of theoretical plates via a combination of smaller particles and increased column length, results in greater peak capacity. In addition, sample requirements are modest. However, the use of very long capillaries also requires the use of high operating pressures to force the solvent through the column, and this can prove a limiting factor.

There have been a number of applications of capillary LC to metabonomic and metabolomic assays. In one of these, C18-bonded silica monolithic columns (0.2 mm i.d.), of between 30 and 90 cm in length, were used in combination with MS detection for the analysis of extracts of the plant *Aradopsis thaliana* [25]. In another example capillary LC-MS with 10 cm × 320 μm columns filled with a 3.5 μm C18-bonded packing material was employed for the analysis of urine obtained from Zucker rats [26]. A gradient separation was used in this application (from 0 to 95% acetonitrile with 0.1% aqueous formic acid). In this example approximately twice as many ions were detected for the same samples with the capillary separation compared to conventional HPLC-MS with columns of the same length packed with the same stationary phase whilst consuming a fraction of the sample. Despite the much reduced amount of sample introduced into the system, a much better sensitivity was noted (ca. 100-fold for some metabolites) for the capillary method compared to HPLC-MS on the same samples. The increase in the number of components observed using the capillary separation reflects reduced ion suppression compared to the conventional separation. Interestingly, in this study it was possible to observe diurnal variation in sample composition for both male and female animals using data derived from either HPLC-MS or capillary LC-MS analysis, however, there was little overlap in the ions detected that enabled this classification to be made.

As indicated above, the use of long capillaries requires high operating pressures. Recently the use of such high pressures (20 kpsi), in combination with gradient reversed-phase capillary chromatography, has been reported for the separation of cell lysates of the microorganism *Shewanella oneidnedensis* [27]. The separation was performed at 20 kpsi using a 50 μm i.d. fused silica capillary, 200 cm in length, packed with a stationary phase of 3 μm porous C18-bonded particles. The gradient separation resulted in the detection of more than 5000 metabolites. However, whilst the separation was impressive, the analysis time was long with some 2000 min required for completion.

5.6. Ultra performance Liquid Chromatography-Mass Spectrometry

One of the problems associated with biological samples is their complexity, which places a considerable strain on the ability of the separation system employed to resolve all of the components. Recently chromatography on 1.7 μm stationary phases has been introduced in the form of UPLC (Ultra performance LC)-MS offering a substantial improvement in performance for complex mixture analysis compared to conventional HPLC. The use of such packing materials generally requires the use of much higher operating pressures than are normally encountered in HPLC to achieve high flow rates. However, this combination of high pressure and small particle size results in separations of much greater efficiency than can be obtained using conventional technology. A typical example of the UPLC-MS of mouse urine is shown in the TIC shown in Figure 5.8. The improved resolution and increased number of peaks detected in the UPLC-MS run compared to conventional HPLC-MS is clear when these results are compared to those shown in Figure 5.6.

The first metabonomics application of UPLC-MS was a "functional genomic" investigation involving obtaining metabolic profiling of urines from males and females of two groups of phenotypically normal mouse strains (C57BL19J and Alpk:ApfCD) and a "nude mouse" strain [28]. When compared with conventional HPLC-MS under similar analytical conditions an improved phenotypic classification was seen by the use of UPLC-MS.

As well as the use of UPLC to increase the number of peaks detected in the same run time as a normal HPLC analysis, it can also be used to drastically reduce analysis time for high throughput analyses. An example of this is provided by the use of UPLC with a rapid (1.5 min) reversed-phase gradient applied to the analysis of urine samples from rodents, including normal and Zucker (fa/fa) obese rats and three

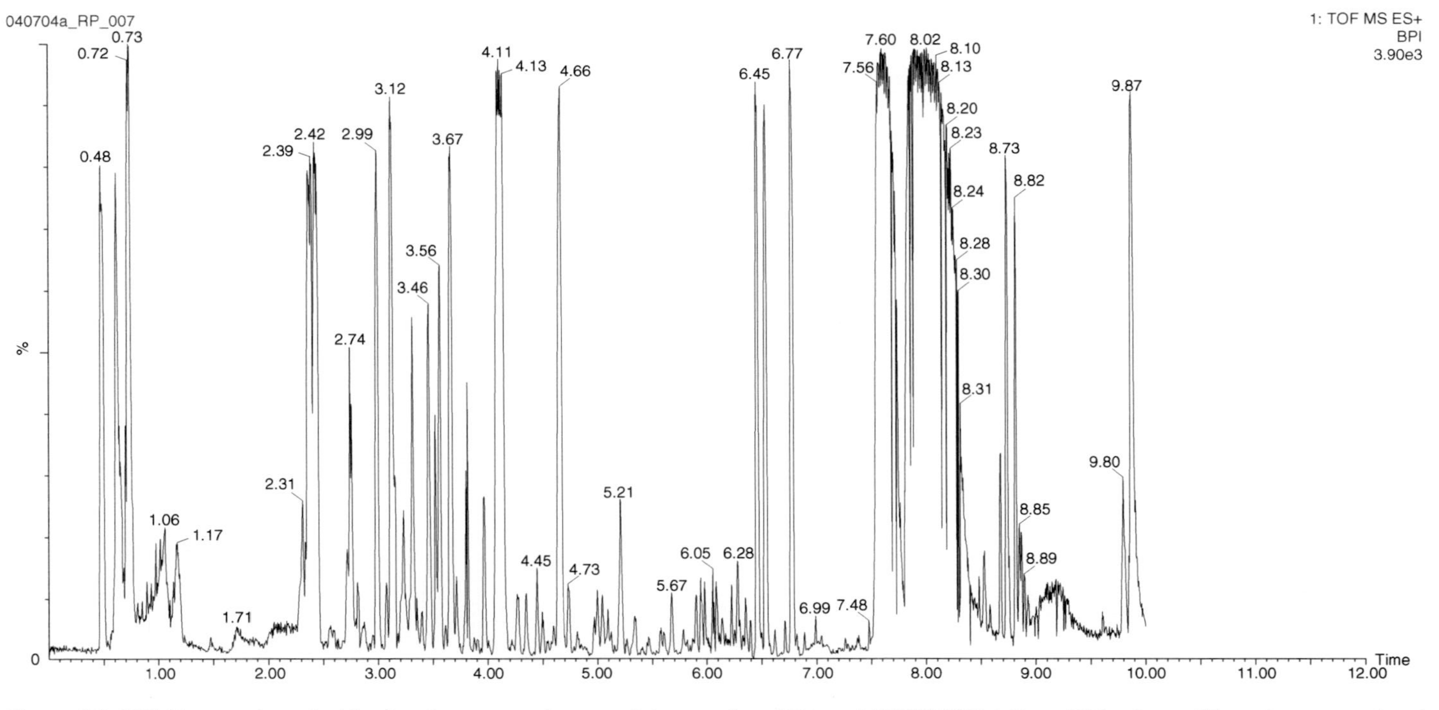

Figure 5.8. UPLC separation of white female mouse urine on a 2.1 mm × 5 cm Waters ACQUITY™ 1.7 μm C18 column. The column was eluted with 0–95% aqueous formic acid: acetonitrile (0.1% formic acid) gradient (10 min) at 500 μL/min. The column eluent was monitored by ESI oa-TOF-MS from 50 to 850 *m/z* in positive ion mode. The sample was collected in the morning.

strains of mice (of both sexes) [29]. This methodology enabled rapid discrimination between age, strain, gender and diurnal variation. The peak capacity and the number of marker ions detected using these fast UPLC separations and oa-TOF MS was found to be similar to that generated by conventional HPLC-MS methods with a 10 min separation.

5.7. Capillary zone electrophoresis-Mass Spectrometry

Another separation technique with considerable potential that can provide metabolite profiles is capillary zone electrophoresis-MS (CZE-MS, CE-MS). To date there have been relatively few published applications of CE-MS, and these have been concentrated in the area of bacterial metabolomics. In these studies, samples have been investigated using "targeted" analyses against a panel of up to ca. 1700 standards [4, 5]. Like capillary HPLC an advantage of CE methods is that they require only very small samples which need little or no preparation for samples such as urine. Because the separation is based on electrophoresis and therefore on charge, a different separation mechanism is utilised compared to HPLC or GC with potential benefits to selectivity and many further applications of the techniques may be anticipated in the future.

5.8. GC-, LC- and CE-MS for metabonomics: A perspective

From the examples provided above it is clear that the combination of MS with any of the three types of separation techniques provides a very powerful addition to the metabolite profiling techniques available for metabonomics. However, all MS-based techniques suffer from the disadvantage that the response of the detector is compound dependent, and can vary widely. Thus, for detection in MS it is necessary to form ions and, if a compound ionises poorly it will not be detected with great sensitivity. Conversely, compounds that ionise well compared to others of similar concentration will dominate the TIC. The widely touted advantage of the great sensitivity of MS compared with many other analytical techniques (in particular, NMR spectroscopy) must therefore be viewed in this context. Because of the compound-dependent response of MS, absolute, rather than relative, quantification is impossible in the absence of an authentic standard, and if the biomarkers are unknown, or not commercially available this can be problematic. In addition, ESI and, to a lesser extent APCI, the most popular ionisation techniques used in liquid-based separations such as HPLC and CE are subject to the phenomenon of ion suppression/ion enhancement. This phenomenon can result in apparent changes in the relative amounts of a component where none have occurred and reinforces the need to identify potential biomarkers and then develop specific and validated bioanalytical methods to confirm their value.

It is also worth noting that GC, LC and CE are not equivalent to each other and each has advantages and disadvantages. Thus the tremendous resolving power and sensitivity of GC and GC × GC-based MS come at the price of a need for extensive sample preparation. The high resolution of GC and GC × GC also requires fairly lengthy analysis times. A useful feature of GC-MS-based metabonomics is that there are already in existence extensive databases of structures to aid in biomarker identification (although these are not yet comprehensive). For liquid chromatography, the advantages include the potential for minimal sample preparation and high throughput. There are currently no equivalents in LC of the extensive databases available for GC making the identification of unknowns more difficult. However, it is fairly easy to scale up HPLC separations to a preparative level and this allows for the isolation and identification of unknowns using a battery of spectroscopic techniques. The widespread availability of HPLC-MS, combined with its relative ease of use, suggests that it will become very widely applied in metabonomics/metabolomic research. However, our own, limited, studies comparing the information derived from the analysis of plasma samples with GC, UPLC and ^{1}H NMR spectroscopy have illustrated the complementary nature of these techniques [7]. We would therefore, like others, advocate a multianalytical-platform approach to global metabolite profiling rather than an over-reliance on a single technology.

Similarly CE-MS, whilst currently less widely available, currently combines the advantages of LC in terms of minimal sample preparation and ease of use with a complementary separation mechanism (based on charge). CE therefore provides a different selectivity which may be advantageous in the analysis of complex mixtures as biofluids.

Clearly, in the case of all of these methods a complex 3D dataset results comprising of retention time, signal intensity and mass. Factors that can complicate the analysis of these data include the potential for drift in retention times and mass accuracy, together with time-dependent changes in detector response. Whilst the discussion of data processing is outside the scope of this article, the use of appropriate quality control measures at all stages of the analytical process is clearly essential if valid and meaningful biomarkers are to be determined.

5.9. Conclusions

The application of hyphenated-MS systems to metabonomics is still at a relatively early stage of development, but there is absolutely no doubt that these techniques will make a substantial contribution to the future development of the field. The ease of use and widespread availability of LC-MS systems will probably ensure that the bulk of the applications will be performed using these instruments. However, the power of GC, especially GC × GC, will ensure an important and continuing role for the technique in this area. Similarly CE-MS, with its high resolution, will also develop into an important tool for metabolite profiling.

References

[1] Wilson, I.D., Plumb, R., Granger, J., Major, H., Williams, R., Lenz, E.M. *J.Chrom. B.* **2005**, *817*, 67–76.

[2] Fiehn, O., Kopka, J., Trethewey, R.N., Willmitzer, L. *Anal. Chem.* **2000**, *72*, 3573–3580.

[3] O'Hagan, S., Dunn, W.B., Brown, M., Knowles, J.D., Kell, D.B. *Anal.Chem.*, **2005**, *77*, 290–303.

[4] Soga, T., Ohashi, Y., Ueno, Y., Naraoka, H., Tomita, M., Nishioka, T. *J. Proteome Res.* **2003**, *2*, 488–494.

[5] Jia, L., Terabe, S. in *Metabolome Analysis*, eds Vaidyanathan, S., Harrigan, G.G., Goodacre, R, Springer, New York, **2005**, pp. 83–101.

[6] Halket, J.M., Waterman, D., Przyborowska, A.M., Patel, R.K.P., Fraser, P.D., Bramley, P.M. *J.Exp. Bot.* **2005**, *56*, 219–243.

[7] Williams, R., Lenz, E.M., Wilson, A.J., Granger, J., Wilson, I.D., Major, H., Stumpf, C., Plumb, R. *Mol. Biosystems,* **2006**, *2*, 174–183.

[8] Welthagen, W., Shellie, R.A., Spranger, J., Ristow, M., Zimmermann, R., Fiehn, O., *Metabolomics*, **2005**, *1*, 65–73.

[9] Shellie, R.A., Welthagen, W., Zrostlikova, J., Spranger, J., Ristow, M. Fiehn, O., Zimmermann, R. *J. Chrom. A.* **2005,** *1086*, 83–90.

[10] Matuszewski, B.K., Constanzer, M.L., Chavez-Eng, C.M *Anal. Chem.*, **1998**, *70*, 882–889.

[11] Law, B., Temesi, D, *J.Chromatogr., B*, **2000**, *748*, 21.

[12] Plumb, R.S., Stumpf, C.L., Gorenstein, M.V., Castro-Perez, J.M., Dear, G.J., Anthony, M., Sweatman, B.C., Connor, S.C., Haselden, J.N. *Rapid Commun. Mass Spectrom*, **2002**, *16*, 1991–1996.

[13] Idborg-Bjorkman, H., Edlund, P.-O., Kvalheim, O.M., Schuppe-Koistinen, I. Jacobsson S.P. *Anal. Chem.* **2003**, *75*, 4784–4792.

[14] Lafaye, A., Junot, C., Ramounet-le Gall, B., Fritsch, P., Tabet, J.-C., Ezan, E., *Rapid Commun. Mass Spectrom.*, **2003**, *17*, 2541–2549.

[15] Lenz, E.M., Bright, J., Knight, R., Wilson, I.D., Major, H. *The Analyst.* **2004**, *129*, 535–541.

[16] Lenz, E.M., Bright, J., Knight, R., Wilson, I.D., Major, H. *J. Pharm. Biomed. Anal.* **2004**, *35*, 599–608.

[17] Lenz, E.M., Bright, J., Knight, R., Westwood, F.R., Davies, D., Major, H. Wilson, I.D. *Biomarkers.* **2005**, *10*, 173–187.

[18] Williams, R.E., Major, H., Lock, E.A., Lenz, E.M: Wilson, I.D. *Toxicology.* **2005**, *207*, 179–190.

[19] Plumb, R.S., Granger, J., Stumpf, C., Wilson, I.D., Evans, J.A., Lenz, E.M. *The Analyst*, **2005**, *128*, 819–823.

[20] Williams, R.E., Lenz, E.M., Evans, J.A, Wilson, I.D., Granger, J.H., Plumb, R., Stumpf, C. *J. Pharm. Biomed. Anal.* **2005**, *38*, 465–471.

[21] Williams, R.E., Lenz, E.M., Lowden, J., Rantalainen, M., Wilson, I.D., *Mol. Biosystems*, **2005**, *1*, 166–180.

[22] Wagner, S., Scholz, K., Donegan M., Burton, L., Wingate, J., Volkel, W., *Anal Chem.* **2005**.

[23] Wang, C., Kong, H., Guan, Y., Yang, J., Gu, J., Yang, S., Xu, G., *Anal. Chem.*, **2005**, *77*, 4108–4116.
[24] Yang, J., Xu, G., Zheng, Y., Kong, Y., Wang, C., Zhao, X., Pang, T., *J. Chromatogr., A*, **2005**, *1084*, 214–221.
[25] Tolstikov, V.T., Lommen, A., Nakanishi, K., Tanaka, A., Fiehn, O. *Anal. Chem.*, **2003**, *75*, 6737–6740.
[26] Granger, J., Plumb, R., Castro-Perez, J., Wilson, I.D., *Chromatographia*, **2005**, *61*, 375–380.
[27] Shen, Y., Zhang, Y., Moore, R.J., Kim, J., Metz, T.O., Hixon, K.K., Zhao, R., Livesay, E.A., Udseth, H.R and Smith, R.D., *Anal. Chem.*, **2005**, *77*, 3090–3100.
[28] Wilson, I.D., Nicholson, J.K., Castro-Perez, J., Granger, J.H., Johnson, K., Smith, B.W., Plumb, R.S. *J. Proteome Res.* **2005**, *4*, 591–598.
[29] Plumb, R.S., Granger, J.H., Stumpf, C.L., Johnson, K., Smith, B.W., Gaulitz, S., Wilson, I.D., Castro-Perez, J. *The Analyst*, **2005**, *130*, 844–849.

The Handbook of Metabonomics and Metabolomics
John C. Lindon, Jeremy K. Nicholson and Elaine Holmes (Editors)

Chapter 6

Chemometrics Techniques for Metabonomics

Johan Trygg[1] and Torbjörn Lundstedt[2,3]

[1]*Research group for Chemometrics, Institute of Chemistry, Umeå University, Sweden*
[2]*Department of Pharmaceutical Chemistry, Uppsala University, Sweden*
[3]*AcurePharma, Uppsala, Sweden*

6.1. Introduction

In biology, as well as in other branches of science and technology, there is a steady trend towards the use of more variables (properties) to characterize observations (e.g. samples, experiments, time points). Often, these measurements can be arranged into a data table, where each row constitutes an observation and the columns represent the variables or factors we have measured (e.g. intensities at a specific wavelength, mass-to-charge ratio, NMR chemical shift). This development generates increasingly complex data tables, which are hard to summarize and overview without appropriate tools. Thus, in this chapter we will try to guide the reader through a chemometrical approach for extracting information out of data.

Chemometrics is an established field in data analysis [1–3] and has proven valuable in the analysis of "omics" data in many applications [4–10]. It includes efficient and robust methods for modelling and analysis of complicated chemical/biological data tables that produce interpretable and reliable models capable of handling incomplete, noisy and collinear data structures. These methods include principal component analysis [11] (PCA) and partial least squares [12–15] (PLS). Chemometrics also provides a means of collecting relevant information through statistical experimental design [16–18]. Therefore, chemometrics can be defined as the information aspect of complex biological and chemical systems.

Chemometrics has grown into a well-established data analysis tool in areas such as multivariate calibration [19, 20] quantitative structure-activity modeling [21, 22], pattern recognition [23–25] and multivariate statistical process monitoring and control [26–28]. Although seemingly diverse disciplines, the common denominator in these application areas is that high complexity data tables are generated and that these can be analysed and interpreted by means of chemometric methods. However, in biology, chemometric methodology has been largely overlooked in favour of traditional statistics. It is not until recently that the overwhelming size and complexity of the "omics" technologies has driven biology towards the adoption of chemometric methods.

There are two main categories of metabonomic studies:

1. Class specific studies, for example disease diagnosis or toxicological classification
2. Dynamic studies, for example the temporal progression of a treatment.

The common theme is that design of experiments (DOE) is used in combination with multivariate analysis (MVA). A brief introduction to the chemometrical approach, DOE and MVA will be given and later illustrated by an example.

6.1.1. Making data contain information – Design of Experiments

The metabonomics approach is more demanding on the quality, accuracy and richness of information in data sets. DOE [16, 17] is recommended to be used through the whole process, from defining the aim of the study to the final extraction of information.

The objective of experimental design is to plan and conduct experiments in order to extract the maximum amount of information in the fewest number of experimental runs. The basic idea is to devise a small set of experiments, in which all pertinent factors are varied systematically. This set usually does not include more than 10 to 20 experiments. By adding additional experiments, one can investigate factors more thoroughly, for example the time dependence from 2 to 5 time points. In addition, the noise level is decreased by means of averaging, the functional space is efficiently mapped and interactions and synergisms are seen.

6.1.2. Extracting information from data – Overview and classification

In metabonomic studies, the observations and samples are often characterized using modern instrumentation such as gas chromatography-mass spectrometry (GC-MS), liquid chromatography-mass spectrometry (LC-MS) and nuclear magnetic resonance (NMR) spectroscopy. The analytical platform is important and largely determined by the biological system and the scientific question. Multivariate analyses based on projection methods represent a number of efficient and useful methods for the analysis and modelling of these complex data. PCA [11] is the workhorse in chemometrics. Using PCA it is possible to extract and display the systematic variation in the data. A PCA model provides a summary or overview of all observations or samples in the data table. In addition, groupings, trends and outliers can also be found.

Hence, such projection-based methods represent a solid basis for metabonomic analysis. Canonical correlation [29], correspondence analysis [30], neural networks [31, 32], Bayesian modeling [33] and hidden Markov models [34] represent additional modelling methods but are outside the scope of this chapter.

6.1.3. Investigating complicated relationships – Discrimination and prediction

Metabonomic studies typically constitute a set of controls and treated samples, including additional knowledge of the samples, for example dose, age, gender and diet. In these situations, it is possible for a more focussed evaluation and analysis of the data. That is, rather than asking the question "what is there?", one can start to ask, "what is its relation to?" or "what is the difference between?". In modelling, this additional knowledge constitutes an extra data table, that is a Y matrix. PLS [14] and Orthogonal-PLS [35–38] (OPLS) represent two modelling methods for relating two data tables. The Y data table can be both quantitative (e.g. age, dose, concentration) and qualitative (e.g. control/treated) data.

6.2. Chemometric approaches to metabonomic studies

The underlying philosophy of chemometrics in combination with the chemometric toolbox can be applied efficiently throughout a metabonomic study. This philosophy is useful from the start of a study (defining the aim) through the whole process to the biological interpretation. This strategy is described step by step below.

6.2.1. Step 1 Definition of aim

It is important to formulate the objectives and goals of the metabonomic study.

- What is previously known?
- What additional information is needed to be known?
- How to reach the objectives, that is what experiments are needed and how to perform them?

6.2.2. Step 2 Study design

6.2.2.1. Class specific studies

The traditional approach to metabonomic disease diagnosis is to identify a group of control observations and another group of observations known to have a specific disease. What is not taken into account is that they may have other, not diagnosed diseases or conditions. Hence, in modelling, disease diagnosis can be regarded as either a two-class or a one-class problem.

Two-class problem: Disease and control observations define two separate classes.
One-class problem: Only disease observations define a class, control samples are too heterogeneous, for example due to other variations caused by diseases, gender, age, diet, lifestyle, genes, unknown factors and so on.

6.2.2.2. Dynamic studies

Metabonomic studies that involve the quantification of the dynamic metabolic response are best evaluated using sequential sampling over an appropriate time course. The evaluation of human biofluid samples is further complicated by a high degree of normal physiological variation caused by genetic and lifestyle differences. Dynamic sampling makes it possible to evaluate and handle the different types of variations such as individual differences in metabolic kinetics, circadian rhythm and fast and slow responders.

6.2.3. Step 2a – Selection of objects

The selection of the objects (e.g. individuals, rats or plants) needs to span the experimental domain in a balanced and systematic manner. To be able to do this, we have to characterize the objects with both measured and observed descriptors. This often includes setting up specific inclusion and exclusion criteria for the study, such as age span (e.g. 18–45 years), body mass index (e.g. 20–30), medicinal chemistry profiles (e.g. lipids, glucose), gender, tobacco habits and use of drugs. In addition to those criteria, additional information regarding each object is collected by questionnaires that include life style factors, food and drinking habits, social situation and so on. This collected information represents a multivariate profile (with K descriptors) for each object that is a fingerprint of its inherent properties.

Geometrically, the multivariate profile represents one point in K-dimensional space, whose position (coordinates) in this space is given by the values in each descriptor. For multiple profiles, it is possible to construct a two-dimensional data table, an X matrix, by stacking each multivariate profile on top of each other. The N rows then produce a swarm of points in K-dimensional space, see Figure 6.1.

6.2.3.1. Projection-based methods

The main, underlying assumption of projection-based methods is that the system or process under consideration is driven by a small number of latent variables (LVs) [39]. Thus projection-based methods can be regarded as a data analysis toolbox, for indirect observation of these LVs. This class of models are conceptually very different from traditional regression models with independent predictor variables.

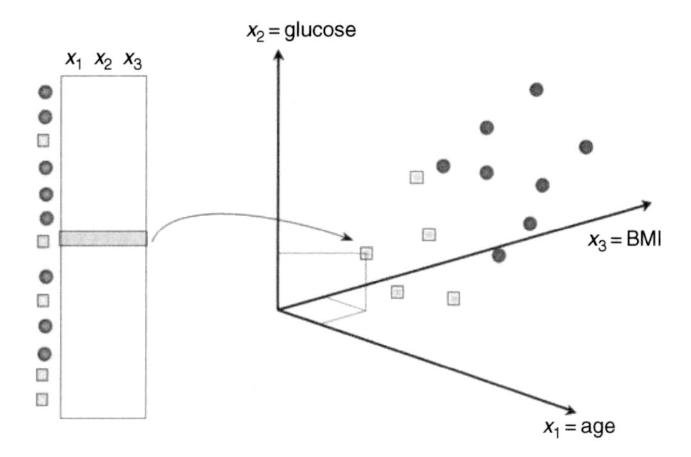

Figure 6.1. Each row (e.g. object or observation) in a K-dimensional data table (here with $K = 3$ variables, designated x_1, x_2, x_3) can be represented as a point in a K-dimensional space (here one point in a three-dimensional space). The coordinates for each object in this multi-dimensional space are given by its three variables, that is a multivariate profile. A data table with N rows then corresponds to a swarm of points. Points that are close to each other have more similar properties than points that lie far apart.

They are able to handle many, incomplete and correlated predictor variables in a simple and straightforward way, hence their wide use.

Projection methods convert the multi-dimensional data table into a low-dimensional model plane that approximates all rows (e.g. objects or observations) in X, that is the swarm of points. The first PCA model component ($t_1 p_1^T$) (see Figure 6.2) describes the largest variation in the swarm of points. The second component models the second largest variation and so on. All PCA components are mutually linearly orthogonal, see Figure 6.2. The scores (T) represent a low-dimensional plane that closely approximates X, that is the swarm of points. A scatter plot of the first two score vectors (t_1 and t_2) provides a summary or overview of all observations or samples in the data table. Groupings, trends and outliers are revealed. The position of each object in the model plane is used to relate objects to each other. Hence, objects that are close to each other have a similar multivariate profile, given the K descriptors. Conversely, objects that lie far from each other have dissimilar properties.

Analogous to the scores, the loading vectors (p_1, p_2) define the relation among the measured variables, that is the columns in the X matrix. A scatter plot, also known as the *loading plot*, shows the influence (weight) of the individual X-variables in

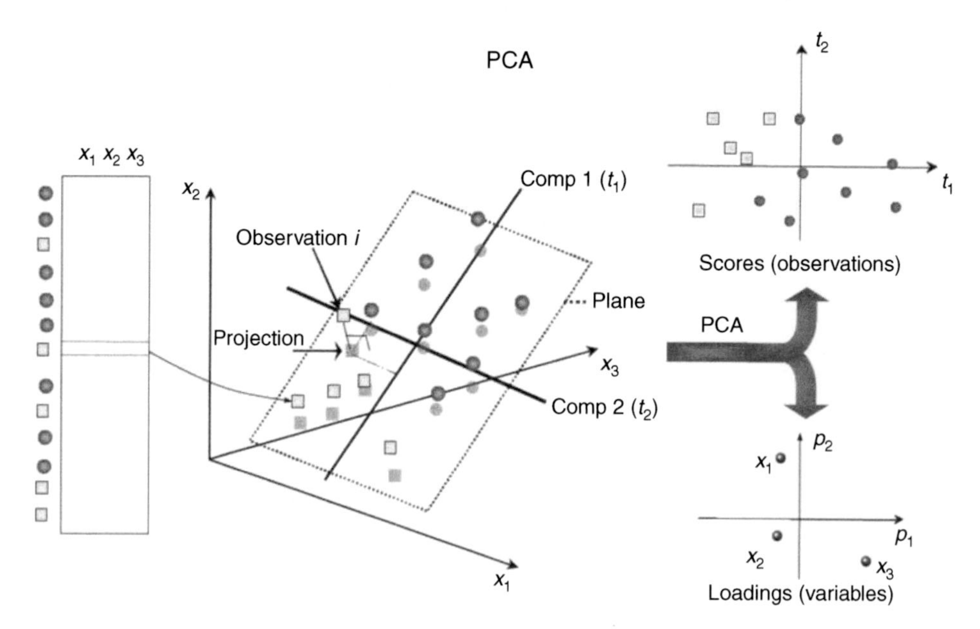

Figure 6.2. A principal component analysis (PCA) model approximates the variation in a data table by a low dimensional model plane. This model plane represents a two-dimensional projection of the multi-dimensional data and provides a score plot, where the relation among the observations or samples in the data table is visualized, for example if there are any groupings, trends or outliers. The loadings plot describes the influence of the variables and the relation among them. An important feature is that directions in the score plot correspond to directions in the loading plot, and vice versa.

the model. An important feature is that directions in the score plot correspond to directions in the loading plot, for example for identifying which variables (loadings) separate different groups of objects (the scores). This is a powerful tool for understanding the underlying patterns in the data. Hence, projection-based methods represent a solid basis for metabonomic analysis.

The part of X that is not explained by the model forms the residuals (E) and represents the distance between each point in K-space and its projection on the plane. The scores, loadings and residuals together describe all of the variation in X.

$$X = \mathrm{TP}^T + E = t_1 p_1{}^T + t_2 p_2{}^T + E$$

6.2.3.2. Multivariate design

The need and usefulness of experimental design in complex systems should be emphasized, because it creates a controlled setting of the environment even though most of the variation between the different objects is uncontrolled. Multivariate design (MVD) [40, 41] is a combination of multivariate characterisation (MVC) [42–44], principal component analysis (PCA) and Design of Experiments (DOE) to

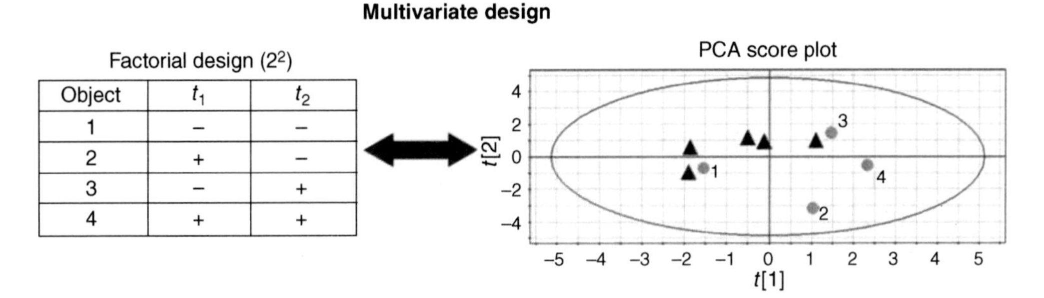

Object	t_1	t_2
1	−	−
2	+	−
3	−	+
4	+	+

Figure 6.3. Four objects (circles) are selected according to a multivariate design that spans the modelled variation.

select a diverse set of objects that represent all objects, that is one that spans the variation. There is a number of different experimental designs that can be applied to span the variation in a systematic way and to obtain well-balanced data. The most commonly used are factorial designs [17] and D-optimal design [45] that fulfil the criteria of balanced data and orthogonality. In MVD, the principal component model scores, for example, t_1 and t_2 are used to select the objects, see Figure 6.3. The selection is based on diversity between the objects.

6.2.4. Step 2b Dynamic sampling

Biological processes are dynamic by nature, that is there is a temporal progression. Some problems are caused by quick and slow responders following intervention or treatment. For this reason, the study design is laid out as sequential samples over an appropriate time course to capture individual trajectories. Sampling period and interval is based on the expected or known time course of the expected effect. In other words, design of experiments is used to maximize the information content and increase the chances of capturing all possible variations of responses. This allows flexibility in the subsequent analysis and an unbiased evaluation of each individual's kinetic profile. This also implies that the often assumed control (or pre-dose) and treated modelling approach is not optimal, as it fails to take into account the individual dynamics, for example slow and fast responders. In addition, for dynamic studies, the traditional control group does not exist. Instead, each individual (object) is its own reference control.

6.2.5. Step 3 Sample preparation and characterisation

In metabonomics, it is important to keep the experimental and biological variation at a minimum. At the same time, the metabolic analysis should be global, quantitative, robust, reproducible, accurate and interpretable. In addition, the physico-chemical diversity of metabolites (amino acids, fatty acids, carbohydrates and organic acids) raises problems for extraction and working up procedures for different analytical techniques. Here, design of experiments represents an important strategy to systematically investigate factors and optimize the experimental protocols. Typical working up procedures for NMR spectroscopy for biofluids and tissue extraction are found in Appendix 4, in the SMRS Policy document [46]. For GC-MS, see References [4, 5].

6.2.6. Step 4 Evaluation of the collected data

In contrast to a ^{1}H-NMR spectrum, data collected from hyphenated instruments such as GC-MS, LC-MS and UPLC-NMR must be processed more extensively before multivariate analysis. The reason is the two-dimensional nature (e.g. chromatogram/mass spectra) of the data for each sample. Curve resolution or deconvolution methods are mainly applied for data processing [47–50] that result in a multivariate profile for each sample. Since a variable in a data table should define the same property over all samples, variability in NMR peak shifts also cause problems for statistical modelling. Because of this, a multitude of different peak alignment methods have been developed [51, 52]. Typically, alignment methods rely upon having a master or reference profile.

Projection-based methods are sensitive to scaling of the variables. Scaling of variables changes the length of each axis in the K-dimensional space. The primary objective of scaling is to reduce the noise in the model, and thereby enhance the information content and quality. Column centring, whereby the mean trajectory is removed from the data, is followed by either no scaling or pareto scaling of the variables. Pareto scaling is recommended for metabonomic data and is done by dividing each variable by the square root of its standard deviation.

Principal component analysis is used to get an overview of the multivariate profiles. Examining the scatter plot of the first two score vectors (t_1 and t_2) reveals the homogeneity of the data, any groupings, outliers and trends. Strong outliers are found as deviating points in the scatter plot. The Hotelling's T_2 region, shown as an ellipse in Figure 6.4 (left), defines the 95% confidence interval of the modelled variation [53]. Outliers may also be detected in the model residuals. The distance to model plot [3] (DModX) can be used and is a statistical test for detecting outliers based on the model residual variance, see Figure 6.4 (right).

Interesting individual observations such as outliers can be examined and interpreted by the contribution plot [54]. It displays the weighted difference between the observation and the model centre. Hence, we can identify what is unique (deviating) for an observation compared to "normality". Similarly, the contribution plot can also be used for comparing different observations.

In the scores plot, Figure 6.5, two groupings are observed (squares and circles). Examining the scatter plot of the first two loading vectors (p_1 and p_2) reveals

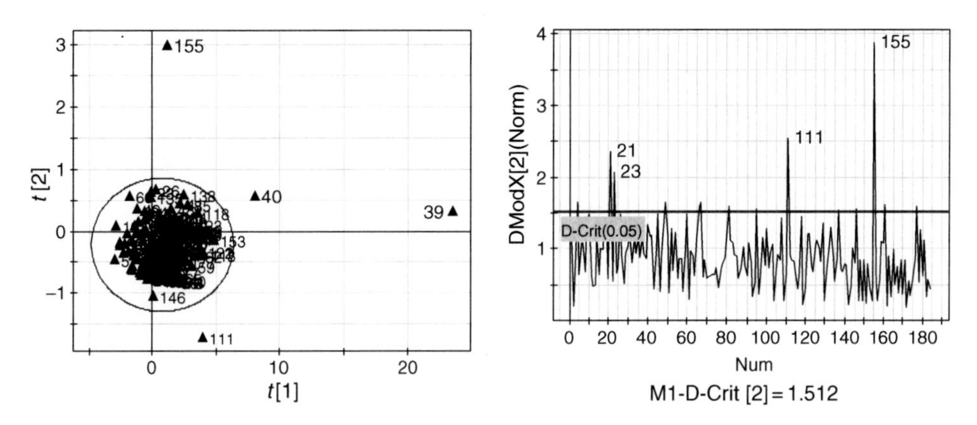

Figure 6.4. In the score plot (left figure), the model is defined by the Hotelling's T^2 ellipse (95% confidence interval) and observations outside the confidence ellipse are considered outliers. Outliers can also be detected by the distance to model parameter, DModX, based on the model residuals (right figure).

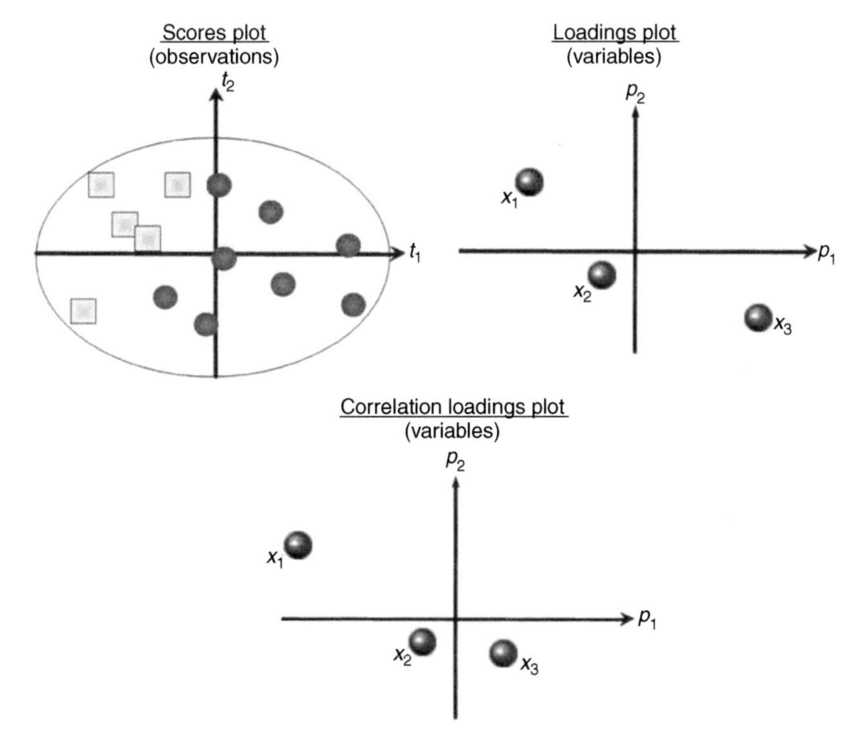

Figure 6.5. The scores plot, the loadings plot and the correlation loadings plot are shown for the first two model components. The scores plot displays an overview of the relationship between the observations (e.g. samples). The loadings plot shows the covariation between each individual variable and the score components. The correlation loadings plot display the correlation between each variable and the score components.

the relationships among the variables. In addition, directions in the scores plot correspond to directions in the loadings plot. This provides the ability to interpret the way the variables are related to a pattern or observations found in the scores plot. This is a powerful tool for understanding the underlying structures in the data.

In Figure 6.5. the loadings plot shows that the variable x_1 is positively covariant to the group marked with squares, and negatively covariant to the group marked with blue circles. Conversely, variable x_3 is positively covariant to the group with circles and negatively covariant to the group marked with squares. A complementary plot to the loadings plot is the correlation loadings plot in Figure 6.5. It reveals the correlation of each variable to the score components in the model. The correlation loadings plot is not dependant upon the scale or size of the variable contrary to the loadings plot that display the covariation.

The loadings plot in Figure 6.5 shows that the x_1 variable has a similar distance from the origin as the x_3 variable. However, in the correlation loadings plot, the x_1 variable has a stronger correlation than the x_3 variable. This means that the x_1 variable has both strong covariation (given by the loadings plot) and strong correlation (given by the correlation loadings plot) with the first two model score components, compared to the x_3 variable.

Compared to the loadings plot, the correlation loadings plot is scale independent.

The prior knowledge gained in Step 2 (Study Design) gives us the ability to separate the observations in at least two different classes. For instance, observations diagnosed with disease vs another group of observations not having the disease. However, knowledge of different types of variations in the collected data can be handled either separately or jointly.

6.2.7. *Soft Independent Modelling of Class Analogy*

The Soft Independent Modelling of Class Analogy (SIMCA) [25] method is a supervised classification method based on PCA. The idea is to construct a separate PCA model for each known class of observations. These PCA models are then used to assign the class belonging to observations of unknown class origin by the prediction of these observations into each PCA class model where the boundaries have been defined by the 95% confidence interval. Observations that are poorly predicted by the PCA class model, hence have large residuals, are classified being outside the PCA model and do not belong to the class.

A SIMCA model, as shown in Figure 6.6 (left), illustrates only one class of observations with strong homogeneity and is well modelled by PCA. This is commonly referred to as the asymmetric case. In Figure 6.6 (right), there are two homogenous classes of observations, each separately modelled by PCA. New observations are predicted into each model, and assigned as belonging to either of the classes, none of the classes or both of the classes.

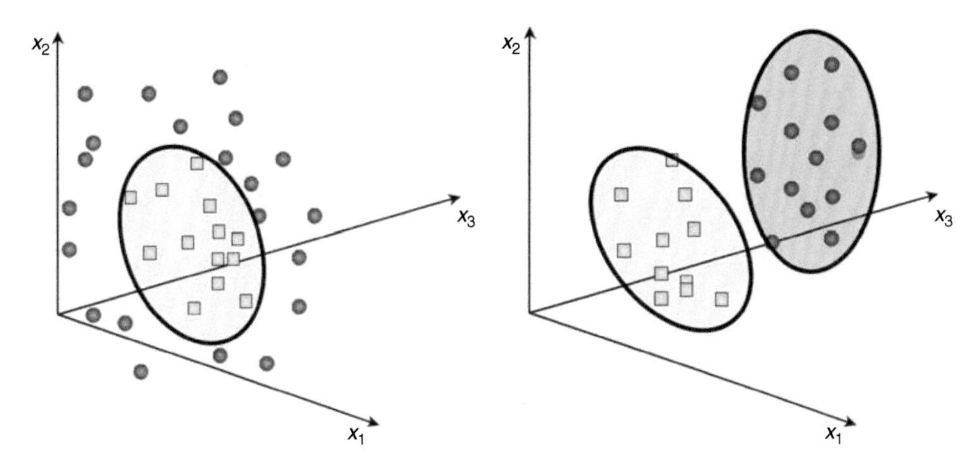

Figure 6.6. Illustration of SIMCA classification. In the left figure, a one class classifier is shown, referred to as the asymmetric case. In the right figure, the SIMCA classification is shown with two classes, separately modelled by PCA.

6.2.8. Partial least squares (PLS) method by projections to latent structures

PLS [12–15] is a method commonly used where a quantitative relationship between two data tables X and Y is sought between a matrix, X, usually comprising spectral or chromatographic data of a set of calibration samples, and another matrix, Y, containing quantitative values, for example concentrations of endogenous metabolites (Figure 6.7). PLS can also be used in discriminant analysis, that is PLS-DA. The Y matrix then contains qualitative values, for example class belonging, gender and treatment of the samples. The PLS model can be expressed by:

$$\text{Model of } X: \quad X = \mathrm{TP}^T + E$$

$$\text{Model of } Y: \quad Y = \mathrm{TC}^T + F$$

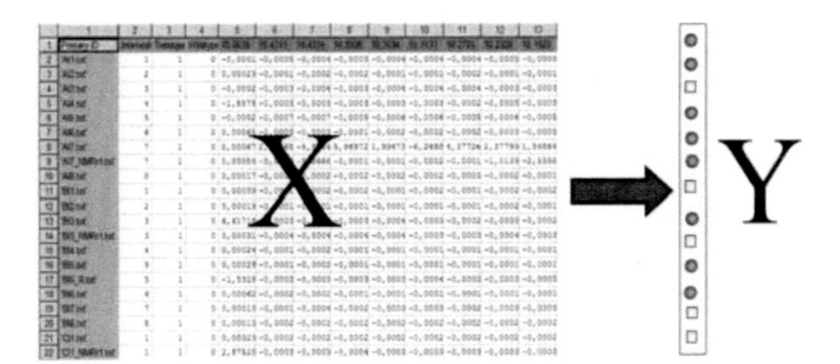

Figure 6.7. Class information can also be used to construct an additional matrix, hereinafter called the Y matrix, consisting of a discrete 'dummy' variable where [1]/[0] indicates the class membership.

The PLS models are negatively affected by systematic variation in the X matrix that is not related to the Y matrix, that is not part of the joint correlation structure between X and Y. This leads to some pitfalls regarding interpretation and has potentially major implications in our selection of metabolite biomarkers, for example positive correlation patterns can be interpreted as negligible or even become negative.

6.2.9. The Orthogonal-PLS method

The OPLS [35] method is a recent modification of the PLS method [14]. The main idea of OPLS is to separate the systematic variation in X into two parts, one that is linearly related to Y and one that is unrelated (orthogonal) to Y. This partitioning of the X-data facilitates model interpretation and model execution on new samples [35]. An OPLS model comprises two modelled variations, the Y-predictive ($T_p P_p^T$) and the Y-orthogonal ($T_o P_o^T$) components. Only the Y-predictive variation is used for the modelling of Y ($T_p C_p^T$).

$$\text{Model of } X: \quad X = T_p P_p^T + T_o P_o^T + E$$

$$\text{Model of } Y: \quad Y = T_p C_p^T + F$$

E and F are the residual matrices of X and Y respectively. OPLS can, analogously to PLS-DA, be used for discrimination (OPLS-DA), see, for instance, Reference [55]. In Figure 6.8, it is shown how additional knowledge, the Y matrix (e.g. gender), is used in the modelling to identify directions in the X model that relate X to Y.

Example study: A food supplement study with dynamic sampling

Step 1 Definition of aim

Investigate the effects on humans of a food supplement by NMR-based metabonomics on plasma samples.

Step 2 Study design

Potential study objects were screened two weeks before the start of the study with a number of inclusion and exclusion criteria (e.g. gender, BMI, age, clinical chemistry) and a questionnaire to provide more in-depth information about lifestyle habits. The information was collected as a multivariate profile of each individual. A PCA was performed on the collected data, followed by an experimental design to select a

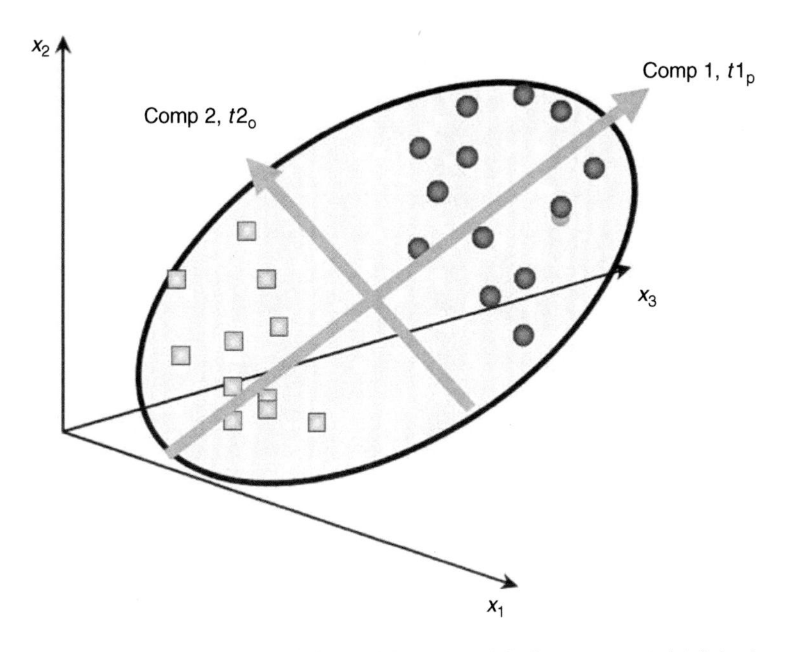

Figure 6.8. A geometrical illustration of the OPLS-DA model. Component 1 ($t1_p$) is the predictive component and displays the between-class ([circles], [squares]) variation of the samples. The corresponding loading profile can be used for identifying variables important for the class separation. Component 2 ($t2_o$) is the Y-orthogonal component and models the within group (intra-class) variation.

diverse set of objects. Four objects were selected, in agreement with a multivariate design; see Figure 6.3, for a deeper analysis of a few specific endogenous metabolites. A dynamic study design was laid out for all of the objects whereby a blood sample was withdrawn at each visit and the plasma prepared, see Figure 6.9.

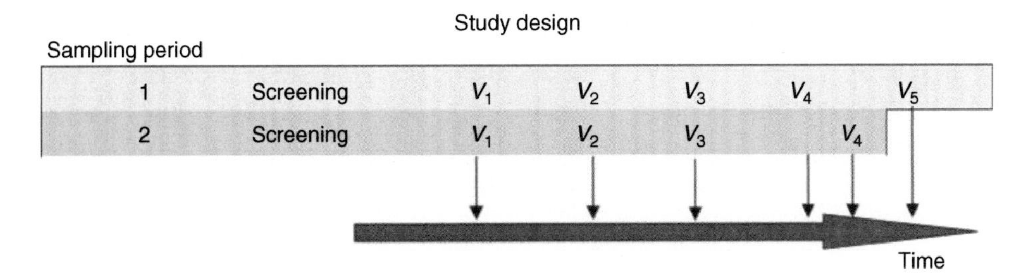

Figure 6.9. Four or five sampling times were set up for the two different sampling periods. The dynamic sampling increases the detection of effect and object differences in metabo-kinetics, for example from slow and fast responders.

Step 3 Sample preparation and characterization
Working up and sample preparation was done according to Standard Operating Procedures (SOP), see References [4, 5, 46].

Step 4 Evaluation of the collected data
Prior to all modelling, column centering was applied to the NMR spectral data. Following this, a PCA model was calculated, to obtain an overview of the data. The scores plot, shows a summary of all samples and this clearly separates the two different sampling t_1 and t_2 periods, see Figure 6.10.

The corresponding loading plot (p_1 and p_2) indicated that there was a problem with peak alignment between the two sampling periods. A line plot of all NMR spectra confirms that this was indeed the case. In addition to the alignment problem, there are also major amplitude differences, see Figure 6.11. Alignment methods can be used to correct for the differential chemical shifts observed. Here, a covariance alignment method was applied.

Following alignment, a new PCA model showed that there still was a separation between each of the two sampling periods, although with minor overlap. Subtracting the screening NMR spectrum from each individual can reduce the amplitude differences between the sampling periods. This is due to the fact that we are interested

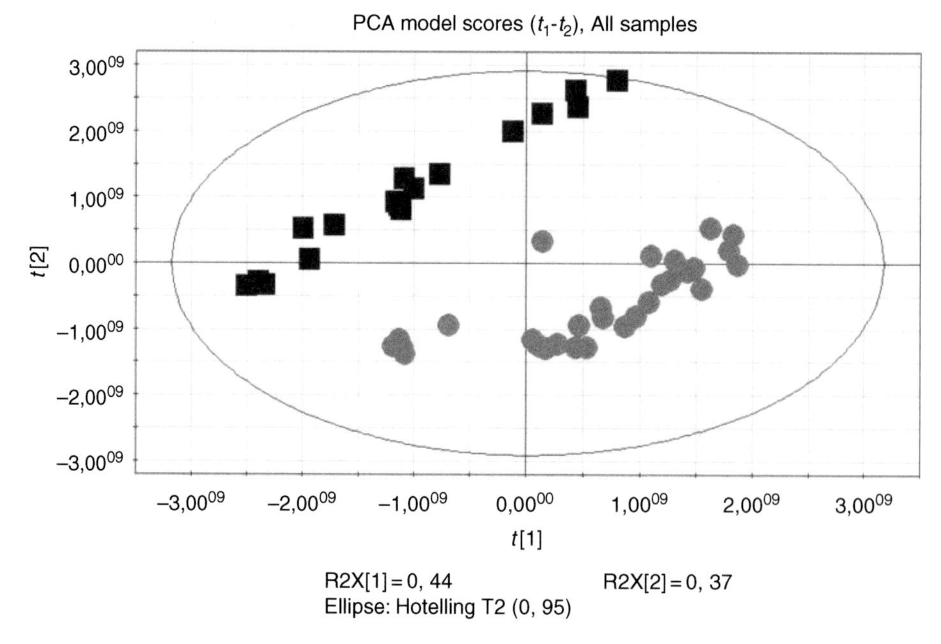

Figure 6.10. PCA scores plot (t_1-t_2) of the NMR spectra shows a clear separation between the sampling periods where squares represent the first sampling period, and circles the second period.

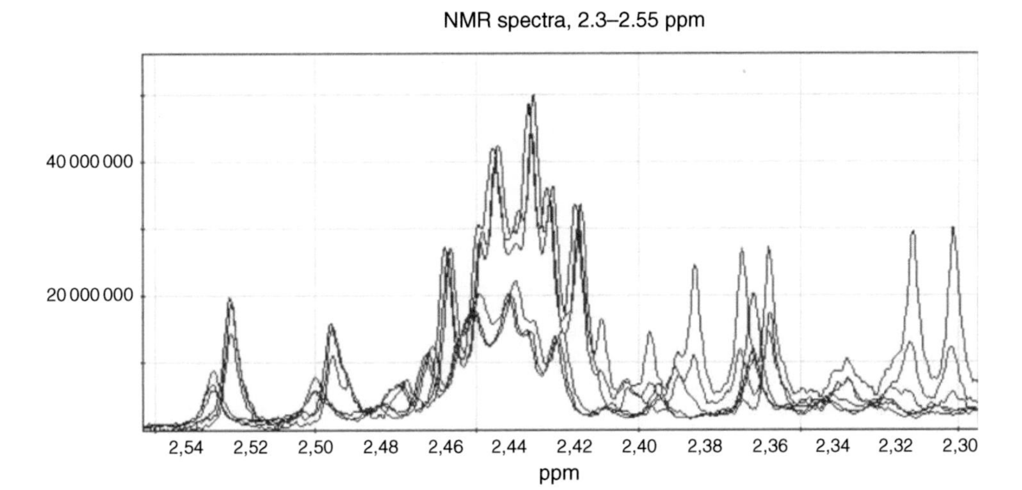

Figure 6.11. NMR spectra from both sampling periods clearly show the reason for the found separation in the score scatter plot.

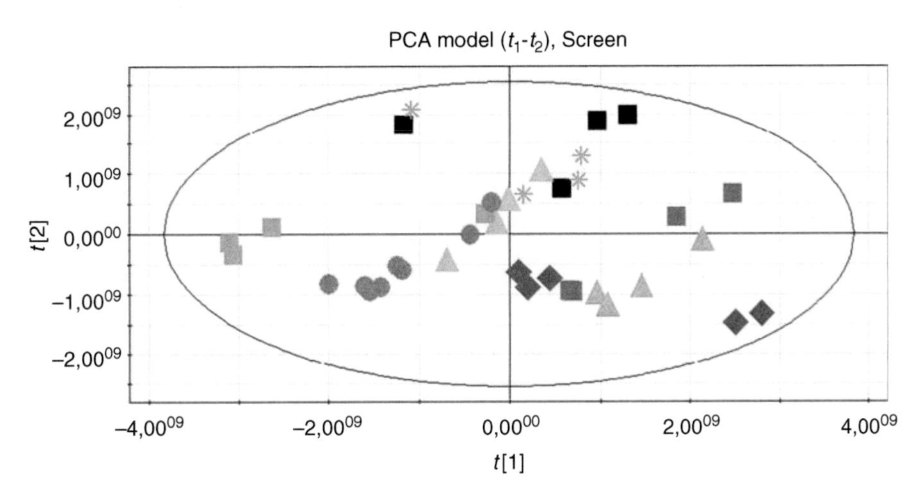

Figure 6.12. PCA score plot (t_1 and t_2) after subtraction of the NMR screening sample from each individual. As shown in the plot this has corrected for the groupings due to different studies.

in modelling the effect of treatment for each individual over time. Again, a PCA model was calculated and its scores plot is found in Figure 6.12.

However, a greater inter-person than intra-person variation is still observed. Each colour corresponds to a different person.

The subtraction of the screening sample helped remove the separation between the sampling periods. However, there is still a problem in that the inter-person

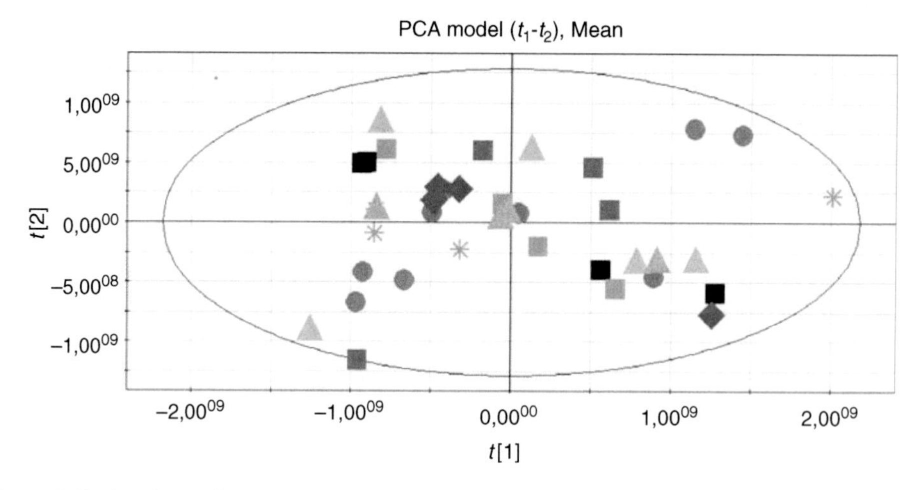

Figure 6.13. As shown in the PCA score plot (t_1-t_2), the systematic differences between objects and sampling periods have been removed by individual mean centring. Each colour corresponds to a different person.

variation is greater than the intra-person variation. This has an adverse influence on evaluating the effect of treatment over all objects. This is also an indication that the screening sample may not be a useful reference sample. One plausible reason may be due to the relatively long time period between the screening sample and the start of the study. It is important that the reference sample used is a biologically equivalent reference point for each object, if not, systematic differences between objects will exist. Hence, the average NMR spectrum for each object was used as their reference point, and the screening sample was excluded from further analysis.

The scores plot of the updated PCA model no longer displays any systematic differences between objects or sampling periods (see Figure 6.13).

However, it becomes clear that all individuals do not have the same behaviour over time following treatment. A number of reasons can exist, for example the absolute effect between individuals can be large, hence those with lower response will be suppressed in the model due to the scale-sensitivity of projection-based models such as PCA. Another reason can be that different individuals have different dynamic responses to treatment, for example quick and slow responders where the main effect for one individual occurs between time point 2 and 3, and for another person between time points 3 and 4.

One way to solve this problem is to create local or separate PCA models for each object, in order to identify the largest effect from start of sampling period (pre-dose). The assumption is that the largest change also reveals the largest effect of treatment. Hence, in the PCA scores plot for each object (individual) (shown in Figure 6.14), a direction of maximum change from the pre-dose sample is identified for one or

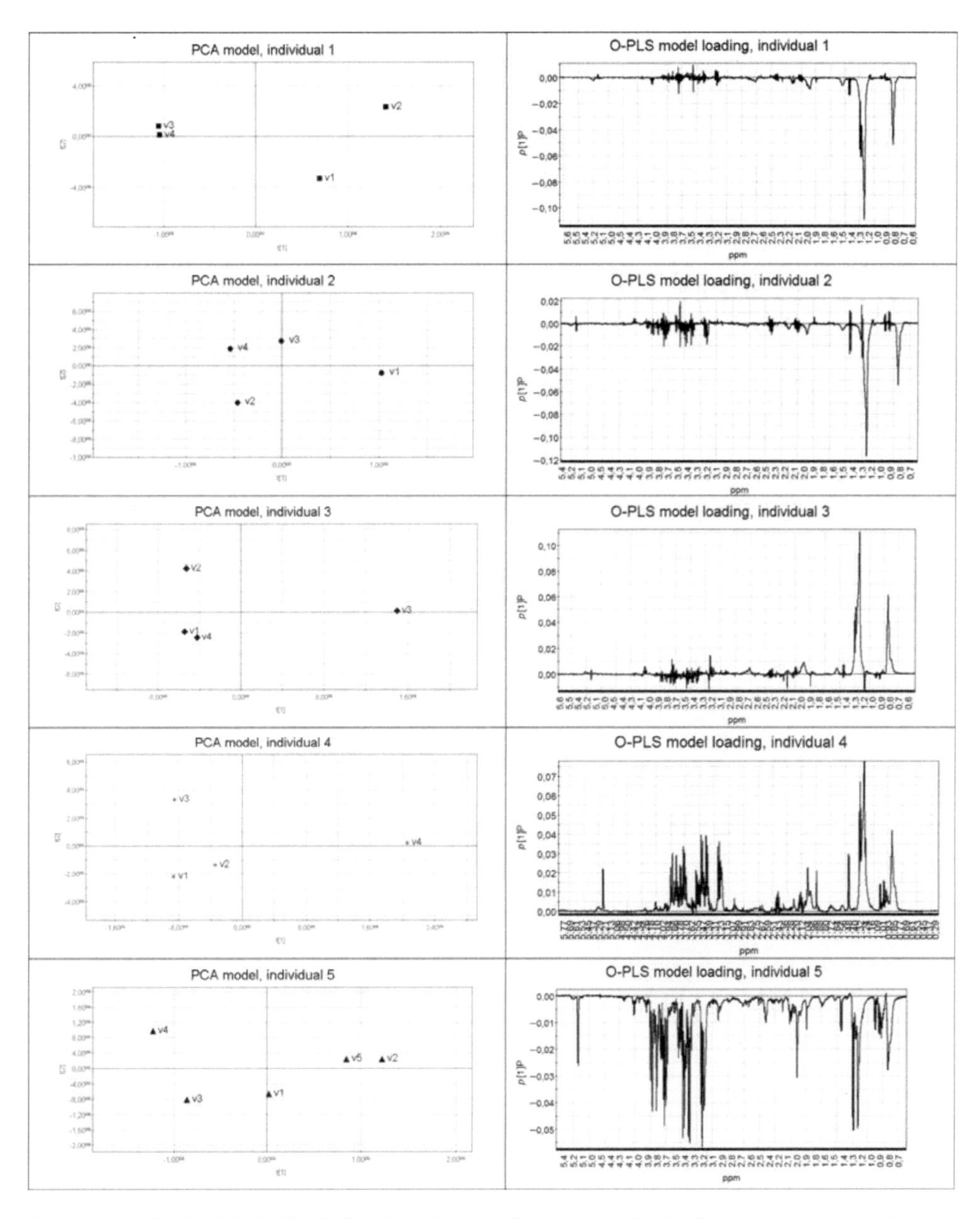

Figure 6.14. On the left, individual trajectories are shown, wherein the largest change over time can be identified. On the right, the OPLS loading profile of the predictive component for each individual is shown. *(Continued)*

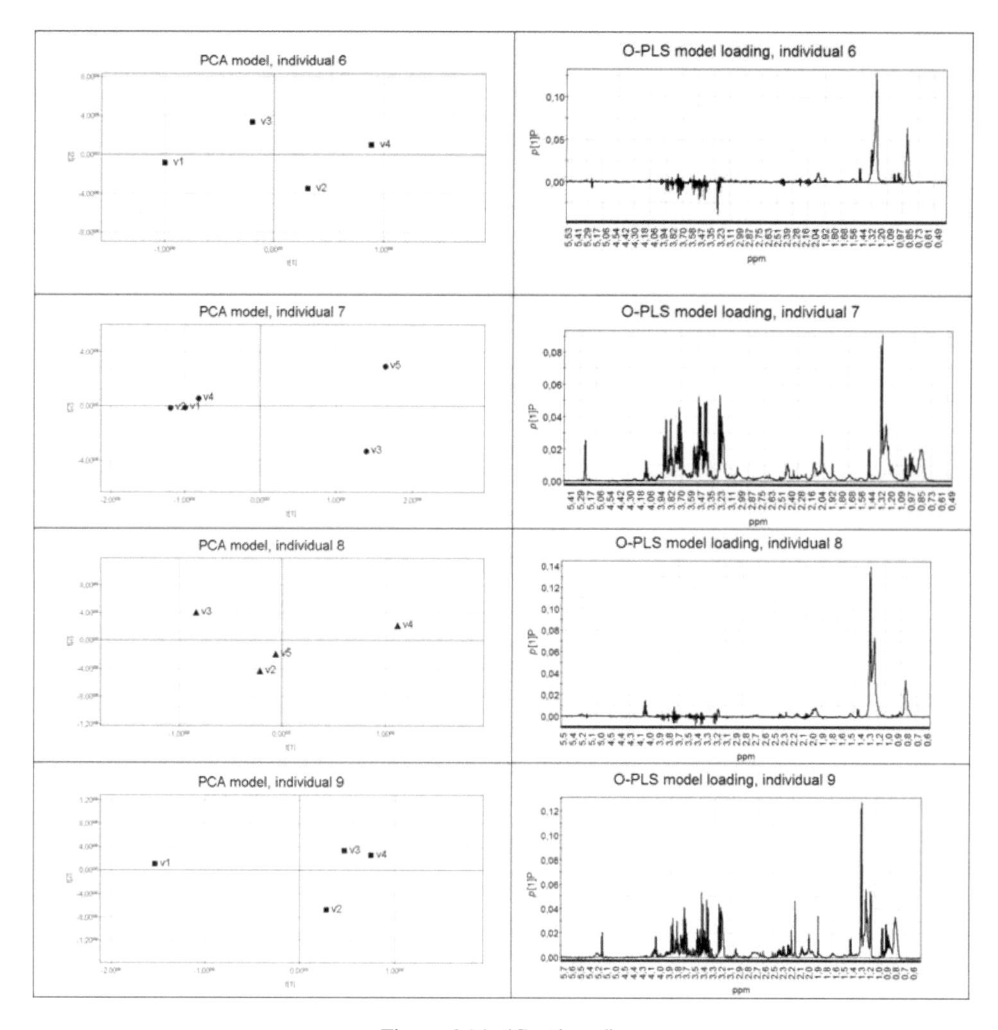

Figure 6.14. (Continued)

several time points by assigning them with a discrete value of one (1), and all others, including the pre-dose sample with the value of zero (0). Following this, an OPLS-DA model was calculated in order to estimate the discriminating loading vector. This was repeated for all objects. The collective set of loading profiles are used to assign similarities between objects. For a summary of the loading vectors, see Figure 6.14.

A visual assessment of the OPLS-DA loading profiles, with lactate as the largest peak, shows two separate groups of profiles with opposite sign of the profile. Individuals 1, 2 and 5 represent one group of loading profiles and the others, individuals 3, 4, 6, 7, 8 and 9 make up the second group. Here it should be strongly emphasized that one should make sure that this observed grouping is not due to the sampling periods, but rather to some underlying phenomenon.

In order to further validate the model, an OPLS-DA model was calculated, based on individuals 3 and 4 only. Those individuals have the most pronounced change in the PCA scores plot, and in addition, they also represent a slow and fast responder (maximum change is seen at different time points, see Figure 6.14). Individuals 6, 7, 8 and 9 were all excluded and used as an external prediction set. The observed vs predicted plot for the model and the prediction set is given in Figures 6.15 and 6.16 respectively. The discriminate line (y-value of 0.25 given by the average of the y-vector used in the OPLS-DA model) means that only one sample is wrongly predicted!

It has to be emphasized that these groupings reflect the maximum change in the PCA model scores, and not necessarily the expected biological effect of treatment.

Early on in this study, four individuals were selected in agreement with a MVD for quantification of a few specific endogenous metabolites by HPLC analysis. Unfortunately, the most prominent metabolite in the loading profile, lactate, was not included as one of those metabolites. As a next step, OPLS modelling was performed with one of those endogenous metabolites as Y and their corresponding NMR measurements as X. Prior to modelling, the average of the endogenous metabolite for each person was removed in the Y-vector.

Individuals 5 and 6 were selected to establish a calibration model between the NMR-spectra and the quantitative measurements. A good model was obtained showing a fair correlation between the quantified concentrations of the metabolite and the calculated concentration (RMSEE = 0.19) see Figure 6.17.

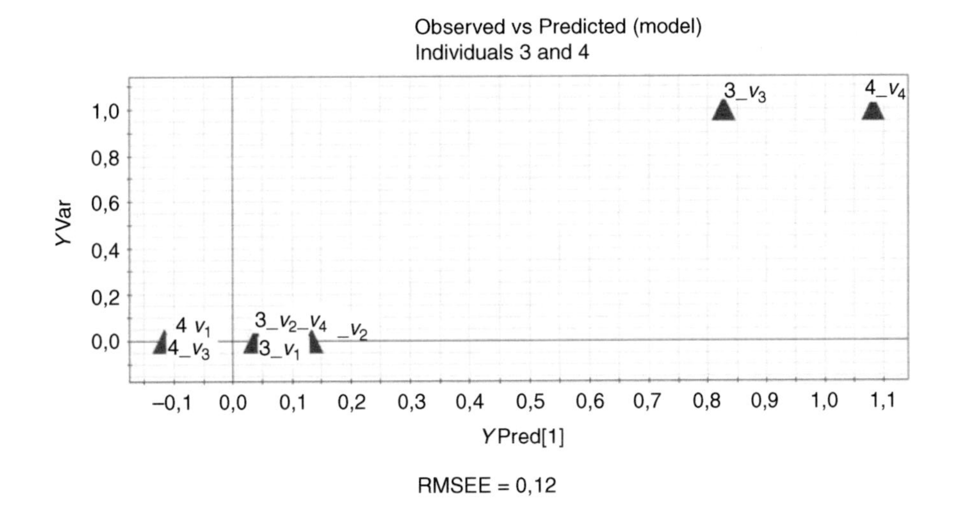

Figure 6.15. The OPLS-DA model shows a clear discrimination between samples having an observed effect to those where no effect was found.

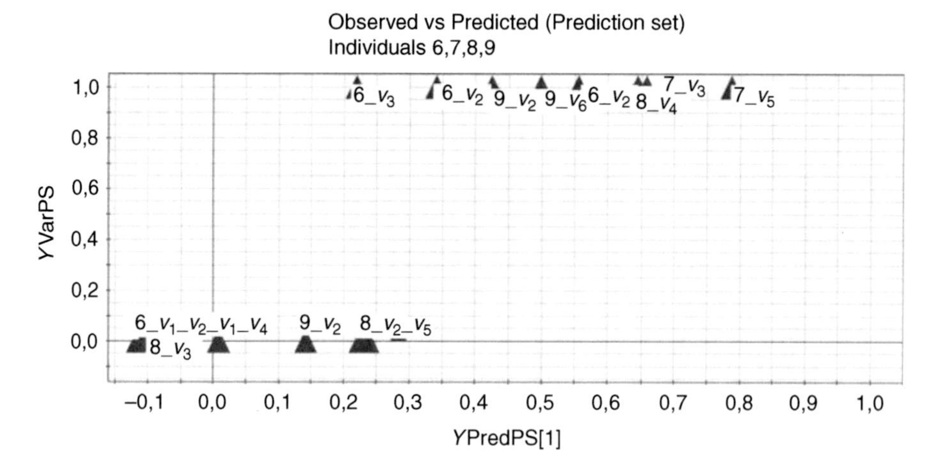

Figure 6.16. The OPLS-DA model predictions of an external test set resulted in only one wrongly predicted sample (6_v_3).

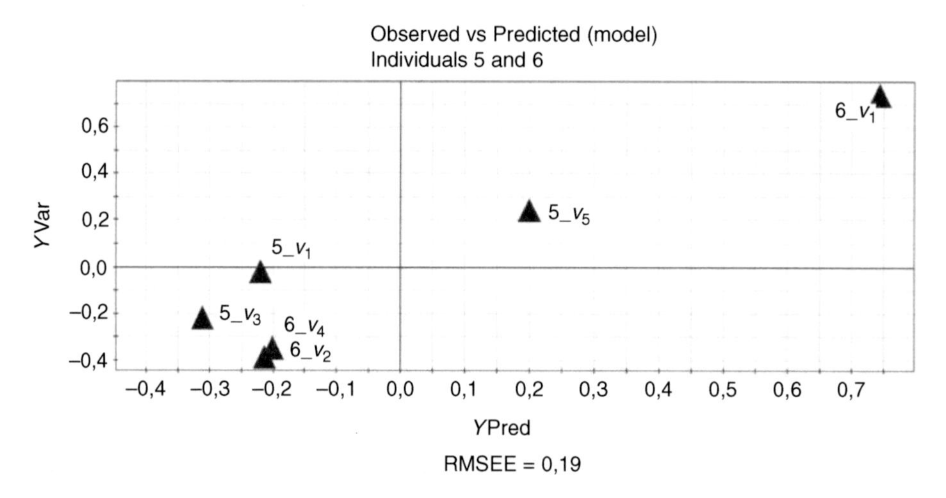

Figure 6.17. The quantified concentrations vs the calculated concentrations of the metabolite by the OPLS calibration model.

The calibration model was used for predictions of the metabolite concentration, for individual 1 and 8 at different time points, from the corresponding NMR-spectra. The predictions obtained vs the quantified metabolite concentration is shown in Figure 6.18. As shown in the figure the calibration model could be used for prediction of new samples with a good result. This indicates that we have identified relevant changes depending on treatment in at least one of the endogenous metabolites by using NMR-spectra.

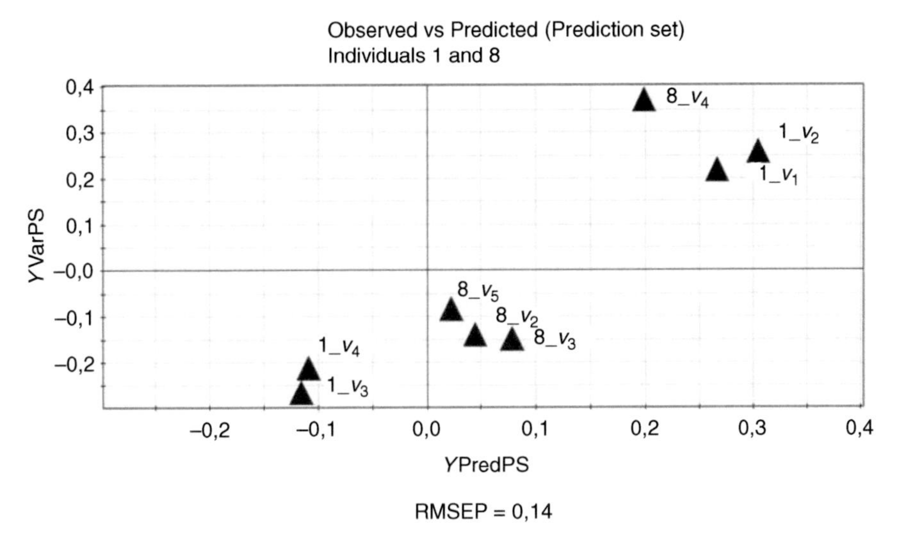

Figure 6.18. The OPLS model predictions of an external test set. The observed quantified concentrations are plotted vs the predicted concentrations of the endogenous metabolite.

6.3. Summary

In this chapter we have tried to guide the reader through a chemometrical approach for extracting information out of complex metabonomic data and this has been illustrated by an example. Wherein the most important findings are the usefulness of dynamic sampling, which provided us with an opportunity to identify slow, medium or fast responders as well as groups of objects showing different response profiles. By using this information all through the evaluation of the data, predictive models could be built on a small number of objects and finally validated by test sets. In the last part of the example we built a calibration model, wherein the NMR-profiles are used as the descriptor and a quantified metabolite was used as response. The endogenous metabolite was quantified by HPLC. This model was validated by an external test set.

The suggested chemometric approach to metabonomic studies is summarized in the following steps;

Step 1: Definition of aim
What is previously known?
What is needed to known?
How to reach those objectives?

Step 2: Study design
 Class specific studies
  Objects or observations (e.g. samples) selected need to span the experimental domain, in a balanced and systematic manner
  Apply multivariate design when selection of objects is possible
 Dynamic studies (investigate temporal progression)
  Sampling over time
Step 3: Sample preparation and characterisation
 Experimental protocol/Analytical technique
Step 4: Evaluation of the collected data
 Data processing, for example GC-MS, LC-MS, LC-NMR
 Overview – PCA
 Classification/Discrimination – SIMCA method & OPLS-DA
  One-class classifier (Control heterogenous)
  Two-class classifier (Control class, Treated class)
 Prediction and biomarker identification

Finally we like to add a short list of questions that always should be asked of a data table independent of approach and/or methods used.

Questions about samples and observations (score plots)
 Are there any outliers?
 Are there groups and/or trends?
 Are there similarities/dissimilarities between samples?
 How do new samples behave?
Questions about variables (loading plots)
 Which variables cause outliers?
 Which variables are responsible for groupings and/or trends?
 Which variables are responsible for class separations?
 How do new variables behave?

6.4. Extensions and future outlook

Systems biology seeks to integrate information from multiple parts of a biological system in a holistic attempt to understand the whole system. A major concern is how to actually integrate multiple blocks of data, for example understand the relation between a data table X (e.g. a set of NMR spectral profiles) and another data table Y (e.g. a set of GC/MS resolved profiles). Current pattern recognition methods based on PLS methods, artificial neural networks, canonical correlation, support vector machines, and so on, all lack the proper model structure to describe these types of data structures, because they focus only on the *X-Y correlation overlap* and not on the *non-overlapping variation* (e.g. Y-orthogonal and X-orthogonal), which, in a biological sense, can be of equal interest. This is a fundamental problem as we certainly can not expect that all variation in NMR and GC/MS profiles co-vary. Here, the OPLS method and extensions thereof represent a good alternative.

6.4.1. O2PLS, an extension of the OPLS model

The OPLS model structure can be extended to include X-orthogonal variation [36, 37], hereafter named O2PLS. The O2PLS model then comprises of three sets of components representing

(i) the joint X–Y variation (given by the T_pP_p and U_pC_p components)
(ii) the Y-orthogonal ($T_oP_o{}^T$) variation
(iii) the X-orthogonal ($U_oC_o{}^T$) variation (Figure 6.19).

$$\text{Model of } X: \quad X = T_pP_p{}^T + T_oP_o{}^T + E$$

$$\text{Model of } Y: \quad Y = U_pC_p{}^T + U_oC_o{}^T + F$$

E and F are the residual matrices of X and Y respectively. This can also be extended to more than two data tables, see for instance Reference [56], and hence it nicely fits into a systems biology framework.

On the left, the Y-orthogonal components are shown, while on the right the X-orthogonal components are illustrated. In the middle section the predictive components are determined and the correlation between the two matrices is calculated.

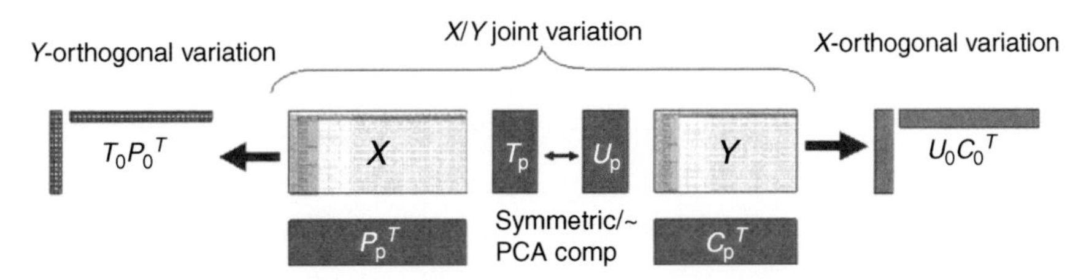

Figure 6.19. A graphical overview of the O2PLS model where also the X-orthogonal variation is modelled. The Y-orthogonal variation ($T_oP_o{}^T$) represents the unique, non-overlapping variation in X, conversely, the X-orthogonal variation ($U_oC_o{}^T$) defines the unique systematic variation in Y. The X/Y joint variation (overlapping between X and Y) is given by the $[T_pP_p{}^T, U_pC_p{}^T]$ components. The O2PLS model structure is bi-directional, meaning that the model can be used for predictions in both directions.

6.4.2. Batch modelling

Batch modelling [26] is routinely being used for analysis of industrial batch process data. A batch process has a finite duration in time, in contrast to a continuous process. By analogy, batch modelling methods are used in metabonomic studies to model the time dependency or dynamics of biological processes, for example the evolution of the effects of a toxic substance in rats. Data collected from such studies produce a three-way data table where each dimensionality represents objects (e.g. rat urine or plant extract samples), variables (e.g. NMR shifts, m/z) and sample time points (see Figure 6.20). Batch modelling is based on modelling two levels, the observation level and the batch level. The observation level shows the dynamics of the biological process of each object over time, see Figure 6.16. For multiple objects (e.g. control rats), it is possible to establish an average trajectory with upper and lower limits based on standard deviations. These indicate the normal development of the object, for example control rats. The established control charts from the

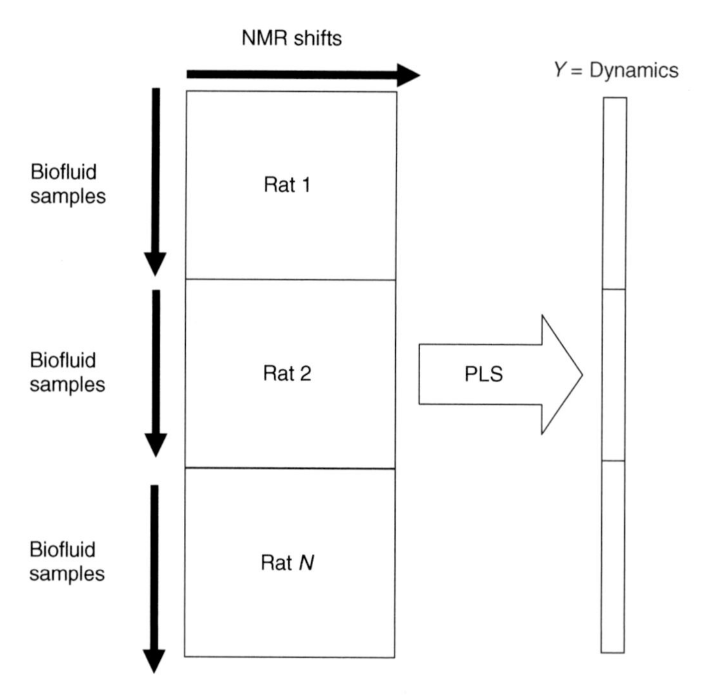

Figure 6.20. In batch modelling, the data is organized as an X-matrix containing blocks of rows where each block represents an object (e.g. a rat). Each row in a block represents the multivariate profile of an observation (e.g. the NMR spectral intensities) at a specific time point. The corresponding row in the Y-matrix contains the dynamics (e.g. the time point). This is followed by an PLS or OPLS model to extract variation from the X matrix related to the dynamics of the system.

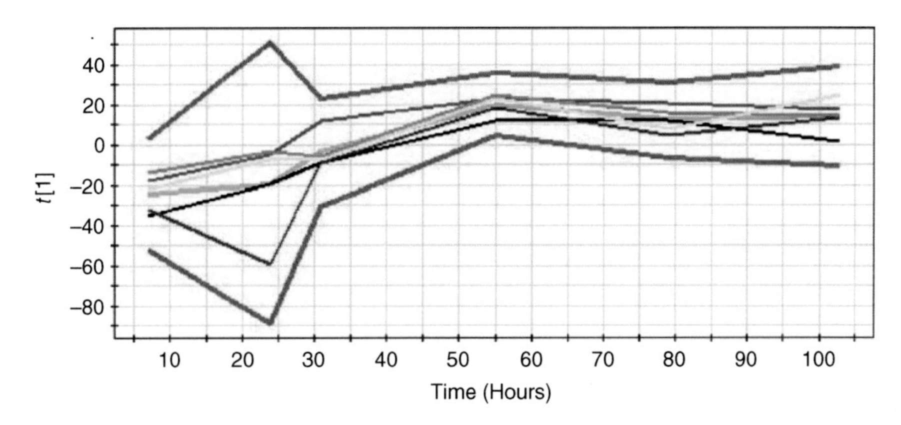

Figure 6.21. Batch control charts can be constructed from a PLS or OPLS batch model score vectors, where each object is shown in a different colour. The black line is the average trajectory and the broader red lines denote upper and lower control limits (plus and minus three times the standard deviation). This control chart is used for detecting deviations from normality.

model can be used to monitor the development of new objects and are used to detect deviations from normality, for example effect of a toxin or drug. Observed deviations from normality can be interpreted by means of contribution plots. Batch modelling is based on the assumption that a control group of objects is followed over the same time period as the treated group. Batch control charts can be constructed from a PLS or OPLS batch model score vectors. The average score trajectory (for each component) with upper and lower control limits (based on standard deviations) indicates the normal dynamic trajectory for a batch. The control chart can be used for detecting deviations from normality (see Figure 6.21).

6.4.3. Hierarchical PCA

The idea behind hierarchical PCA is to block the variables in order to improve transparency and interpretability [57–59]. This method operates on two or more levels, and on each level standard PCA scores and loading plots as well as residuals and their summaries such as DModX are used for interpretation. The procedure can, for two levels, be described as follows (see Figure 6.22). In the first step, in this case is to divide the large matrix into conceptually meaningful blocks and make a separate PCA for each matrix. In the next step the principal components (scores T) from each of these models become the new variables ("super variables") describing the systematic variation from each block. In the final step a PCA model fitted to this data and the hierarchical PCA model is established, see Figure 6.22.

The interpretation of a hierarchical model has to be done in two steps. First, the loading plots of the hierarchical model reveal which of the blocks are most

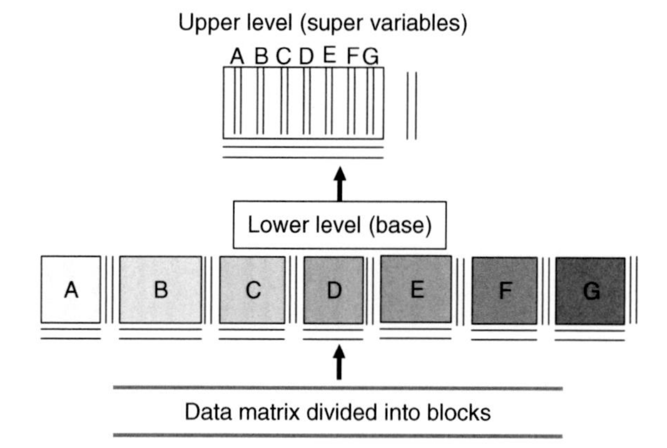

Figure 6.22. H-PCA is shown from the bottom to the top. At the bottom of the figure, the data matrix is divided into blocks. A separate PCA model is calculated for each block and the PCA score components from each model are then combined to form a new matrix, summarising all blocks. This new block of data is then analyzed by a PCA.

important for any groupings that can be seen in the hierarchical score plot. Second, the loading plots for the blocks of interest are studied on the lower level and in the corresponding loading plot the original variables of importance can be identified. Hierarchical PCA is easily extended to hierarchical OPLS or OPLS-DA by adding a Y (response/discriminate) matrix on the upper level.

References

[1] Massart, D.L., Vandeginste, B.G.M., Deming, S.N., Michotte, Y., Kaufman, L., *Chemometrics: A textbook*, Elsevier, Amsterdam, 1988.

[2] Martens, H., Naes, T., *Multivariate Calibration*, Wiley: Chichester, 1989.

[3] Eriksson, L., Johansson, E., Kettaneh Wold, N., Wold, S., *Multi and Megavariate Data Analysis*, Umetrics AB, Umeå 2001.

[4] Jellum, E., Björnson, I., Nesbakken, R., Johannson, E., Wold, S., Classification of human cancer cells by means of capillary gas chromatography and pattern recognition analysis. *J. Chromatography* Nov 6; 217: 231–237, 1981.

[5] Gullberg, J., Jonsson, P., Nordström, A., Sjöström, M., Moritz, T., Design of Experiments: An Efficient Strategy to Identify Factors Influencing Extraction and Derivatization of Arabidopsis Thaliana Samples in Metabolomic Studies with Gas Chromatography/Mass Spectrometry. *Analytical Biochemistry* 331, 283–295, 2004.

[6] Jiye, A., Trygg, J., Gullberg, J., Johansson, A.I., Jonsson, P., Antti, H., Marklund, S.L., Moritz, T., Extraction and GC/MS Analysis of the Human Blood Plasma Metabolome, *Analytical Chemistry* 77: 8086–8094, 2005.

[7] Idborg-Björkman, H., Edlund, P.O., Kvalheim, O.M., Schuppe-Koistinen, I., Jacobsson, S.P., Screening of biomarkers in rat urine using LC/electrospray ionization-MS and two-way data analysis, *Analytical Chemistry* 75: 4784–4792, 2003.

[8] Robertson, D.G., Reily, M.D., Sigler, R.E., Wells, D.F., Paterson, D.A., Braden, T.K., Metabonomics: Evaluation of Nuclear Magnetic Resonance (NMR) and Pattern RecognitionTechnology for Rapid in Vivo Screening of Liver and Kidney Toxicants, *Toxicological Sciences* 57: 326–337, 2000.

[9] Holmes, E., Antti, A., Chemometric Contributions to the Evolution of Metabonomics: Mathematical Solutions to Characterising and Interpreting Complex Biological NMR Spectra. *Analyst* 127: 1549–1557, 2002.

[10] Nicholson, J.K, Connelly, J., Lindon, J.C., Holmes, E., Metabonomics: A Platform for Studying Drug Toxicity and Gene Function. *Nature Reviews Drug Discovery* 1: 153–161, 2002.

[11] Jackson, J.E., *A Users Guide to Principal Components*. Wiley, New York, 1991.

[12] Wold, S., Ruhe, A., Wold, H., Dunn III, W.J. The Collinearity Problem in Linear Regression. The Partial Least Squares Approach to Generalized Inverses. *SIAM J. Sci. Stat. Comput.* 5(3): 735–743, 1984.

[13] Wold, S. Martens, H. Wold, H., Lecture Notes in Mathematics, Proc. Conf. Matrix pencils, Piteå, Sweden, Springer Verlag, Heidelberg, 1983.

[14] Wold, S., Eriksson, L., Sjöström, M., PLS in Chemistry, Schleyer P.V.R (ed.) *Encyclopedia of Computational Chemistry*, John Wiley & Sons, New York, 2006–2016, 1998.

[15] Wold, S., Albano, C., Dunn III, W.J., Edlund, U., Esbensen, K., Geladi, P., Hellberg, S., Johansson, E., Lindberg, W., Sjöström, M., *Multivariate Data Analysis in Chemistry*, NATO ASI Series C 138, D. Reidel Publ. Co., Dordrecht, Holland, 1984.

[16] Lundstedt, T., Seifert, E., Abramo, L., Thelin, B., Nyström, A., Pettersen, J., Bergman, R. Experimental Design and Optimization. *Chem Intel Lab Systems* 42: 3–40, 1998.

[17] Box, G.E.P., Hunter, W.G., Hunter, J.S. *Statistics for Experimenters*, John Wiley & Sons, New York, 1978.

[18] Eriksson, L., Johansson, E., Kettaneh Wold, N., Wikström, C., Wold, S. *Design of Experiments – Principles and Applications*, Umetrics AB, Umeå, 1996.

[19] Gemperline, P.J. Developments in Nonlinear Multivariate Calibration. *Chemometrics and Intelligent Laboratory Systems* 15: 115–126, 1992.

[20] Kowalski, B.R., Seasholtz, M.B., Recent Developments in Multivariate Calibration. *Journal of Chemometrics* 5: 129–145, 1991.

[21] Hellberg, S., Sjöström, M., Skagerberg, B., Wold, S. Peptide Quantitative Structure-Activity-Relationships, a Multivariate Approach. *Journal of Medicinal Chemistry* 30: 1126–1135, 1987.

[22] Lundstedt, T., A QSAR strategy for screening of drugs – and predicting their clinical activity. *Drugs, News & Perspectives* 4(8): 468, 1991.

[23] Wold, S., Albano, C., Dunn III, W.J., Esbensen, K., Hellberg, S., Johansson, E., Sjöström, M. Pattern Recognition: Finding and Using Regularities in Multi-Variate Data, Proc. IUFOST Conf. Food Research and Data Analysis (eds H. Martens, H.M. Russwarm Jr) Applied Science Publishers, London, 1983.

[24] Albano, C., Dunn III, W.J., Edlund, U., Johansson, E., Nordén, B., Sjöström, M., Wold, S. Four Levels of Pattern Recognition. *Analytica Chimica Acta* 103: 429–443, 1978.

[25] Wold, S., Pattern Recognition by Means of Disjoint Principal Components Models. *Pattern Recognition* 8, 127–139, 1976.

[26] Wold, S., Kettaneh, N., Friden, H., Holmberg, A. Modelling and Diagnostics of Batch Processes and Analogous Kinetic Experiments. *Chemometrics and Intelligent Laboratory Systems* 44: 331–340, 1998.

[27] Kourti, T., MacGregor, J.F., Multivariate SPC Methods for Process and Product Monitoring. *Journal of Quality Technology* 28: 409–428, 1996.

[28] Workman, J., Veltkamp, D.J., Burgess, L.W., Process Analytical Chemistry. *Analytical Chemistry* 71: 121R–180R, 1999.

[29] Hotelling, H. The Most Predictable Criterion. *Journal of Educational Psychology* 26: 139–142, 1935.
[30] Greenacre, M.J., *Theory and Applications of Correspondence Analysis*. London: Academic Press, 1984.
[31] Bishop, C.M., *Neural Networks for Pattern Recognition*, Oxford University Press, Oxford, 1996.
[32] Wythoff, B.J., Backpropagation Neural Networks – A Tutorial, *Chemometrics and Intelligent Laboratory Systems* 18(2) 115–155, 1993.
[33] Sivia, D.S. *Data Analysis: A Bayesian Tutorial*. Oxford: Oxford University Press, 1996.
[34] Rabiner, L.R., Juang, B.H., An Introduction to Hidden Markov Models. *IEEE ASSP Magazine*, January, 1986.
[35] Trygg, J., Wold, S., Orthogonal Projections to Latent Structures (O-PLS). *Journal of Chemometrics* 16: 119–128, 2002.
[36] Trygg, J., O2-PLS for Qualitative and Quantitative Analysis in Multivariate Calibration. *Journal of Chemometrics* 16: 283–293, 2002.
[37] Trygg, J., Wold, S., O2-PLS, A Two-block (*X-Y*) Latent Variable Regression (LVR) Method with an Integral OSC Filter, *Journal of Chemometrics* 17: 53–64, 2003.
[38] Cloarec, O., Dumas, M.E., Trygg, J., Craig, A., Barton, R.H., Lindon, J.C., Nicholson, J.K., Holmes, E., Evaluation of the Orthogonal Projection on Latent Structure Model Limitations Caused by Chemical Shift Variability and Improved Visualization of Biomarker Changes in H-1 NMR Spectroscopic Metabonomic Studies, *Analytical Chemistry* 77(2): 517–526, Jan. 15, 2005.
[39] Kvalheim, O.M., The Latent Variable. *Chemometrics Intelligent Laboratory systems* 14: 1–3, 1992.
[40] Wold, S., Sjöström, M., Carlson, R., Lundstedt, T., Hellberg, S., Skageberg, B., Wikström, C., Multivariate Design. *Analytica Chimica Acta* 191: 17, 1986.
[41] Carlson, R., Lundstedt, T., Scope of Organic Synthetic Reactions. Multivariate Methods for Exploring the Reactin Space. An Example of the Willgerodt-Kindler Reaction. *Acta Chemica Scandinavia* B 41: 164, 1987.
[42] Carlson, R., Lundstedt, T., Albano, C., Screening of Suitable Solvents for Organic Synthesis, Strategies for Solvent Selection. *Acta Chemica Scandinavica* B 39: 79, 1984.
[43] Sandberg, M., Sjöström, M., Jonsson, J., A Multivariate Characterization of tRNA Nucleosides. *Journal of Chemometrics* 10: 493–508, 1996.
[44] Oprea, T.I., Gottfries, J., Chemography: The Art of Navigating in Chemical Space. *Journal of Combinatorial Chemistry* 3(2), 157–166, Mar–Apr 2001.
[45] deAguiar, P.F., Bourguignon, B., Khots, M., Massart, D.L., PhanThanLuu, R., D-optimal Designs. *Chemometrics and Intelligent Laboratory Systems* 30(2), 199–210, 1995.
[46] The Standard Metabolic Reporting Structure, Version 2.3, http://www.smrsgroup.org/, January 13 2006.
[47] Jonsson, P., Gullberg, J., Nordström, A., Kowalczyk, M., Sjöström, M., Moritz, T., A Strategy for Extracting Information from Large Series of Non-processed Complex GC/MS Data. *Anal Chem.* 76: 1738–1745, 2004.
[48] Jonsson, P., Bruce, S.J., Moritz, T., Trygg, J., Sjöström, M., Plumb, R., Granger J, Maibaum, E., Nicholson, J.K., Holmes, E., Antti, H., Extraction, Interpretation and Validation of Information for Comparing Samples in Metabolic LC/MS Data Sets. *Analyst* 130: 701–707, 2005.
[49] Halket, J.M., Przyborowska, A., Stein, S.E., Mallard, W.G., Down, S., Chalmers, R.A. Deconvolution Gas Chromatography Mass Spectrometry of Urinary Organic Acids – Potential for Pattern Recognition and Automated Identification of Metabolic Disorders. *Rapid Commun. Mass Spectrom.* 13: 279–284, 1999.
[50] Shen, H.L., Grung, B., Kvalheim, O.M., Eide, I., Automated curve resolution applied to data from Multi-detection Instruments. *Analytica Chimica Acta* 446 (1–2), 313–328, Nov. 19 2001.

[51] Torgrip, R.J.O., Aberg, M., Karlberg, B., Jacobsson, S.P., Peak Alignment Using Reduced Set Mapping, *Journal of Chemometrics* 17(11), 573–582, Nov. 2003.

[52] Vogels, J.T.W.E., Tas, A.C., van den Berg, F., van der Greef, J., A New Method for Classification of Wines Based on Proton and Carbon-13 NMR Spectroscopy in Combination with Pattern Recognition Techniques. *Chemometrics and Intelligent Laboratory Systems* 21, 2–3, 249–258, 1993.

[53] Hotelling, H., The Generalization of Student's Ratio. *Ann. Math. Statist.* 2, 360–378, 1931.

[54] Miller, P., Swanson, R.E., Heckler, C.E., Contribution Plots: A Missing Link in Multivariate Quality Control. *Appl. Math. And Comp. Sci.* 8(4), 775–792, 1998.

[55] Cloarec, O., Dumas, M., Craig, E., Barton, R.H., Trygg, J., Hudson, J., Blancher, C., Gauguier, D., Lindon, J.C., Holmes, E., Nicholson, J., Statistical total correlation spectroscopy: An exploratory approach for latent biomarker identification from metabolic H-1 NMR data sets, *Analytical Chemistry* 77, 1282–1289, 2005.

[56] Eriksson, L., Damborsky, J., Earll, M., Johansson, E., Trygg, J., Wold, S., Three-block Bi-focal PLS (3BIF-PLS) and Its Application in QSAR, SAR QSAR Environmental Research, 15(5–6) 481–499, 2004.

[57] Wold, S., Kettaneh, N., Tjessem, K., Hierarchical Multiblock PLS and PC Models for Easier Model Interpretation and as an Alternative to Variable Selection, *Journal of Chemometrics* 10(5–6), 463–482, Sep.–Dec. 1996.

[58] Eriksson, L., Johansson, E., Lindgren, F., Sjöström, M., Wold, S., Megavariate Analysis of Hierarchical QSAR Data. *Journal of Computer-Aided Molecular Design* 16: 711–726, 2002.

[59] Gunnarsson, I., Andersson, P.M., Wikberg, J., Lundstedt, T., Multivariate Analysis of G Protein-coupled Receptors, *Journal of Chemometrics* 17: 82–92, 2003.

The Handbook of Metabonomics and Metabolomics
John C. Lindon, Jeremy K. Nicholson and Elaine Holmes (Editors)

Chapter 7

Non-linear Methods for the Analysis of Metabolic Profiles

Timothy M.D. Ebbels

Department of Biomolecular Medicine, Faculty of Medicine, Imperial College London, Sir Alexander Fleming Building, South Kensington, London SW7 2AZ UK

7.1. Introduction

Modern metabolic profiling experiments generate large and complex data sets that pose unique data analysis challenges (In this chapter, I use the terms 'metabolic profiling', 'metabonomics' and 'metabolomics' interchangeably). Even for a single biological sample, current chemical analytic methods are capable of generating tens of thousands of descriptive variables reporting on thousands of small molecule metabolites. The most common techniques, such as Nuclear Magnetic Resonance (NMR) spectroscopy and Liquid or Gas Chromatography – Mass Spectrometry (LC/GC-MS), exemplify this problem. Indeed, the ability to profile large numbers of chemically diverse metabolites in an essentially non-selective way is one of their major strengths. Analysis of such data is usually focussed on three main goals: (a) visualisation of overall differences, trends and relationships between samples and variables; (b) determination of whether there is a significant difference between groups related to the effect of interest, and (c) discerning which metabolites (or groups of metabolites) are responsible for these changes. In some cases, two more objectives can be added to these: (d) the construction of models which are predictive of the biological effect for new samples, and (e) the definition of a relationship between the metabolic data and a continuous variable of interest. Multivariate statistical and machine learning tools are the key to achieving these goals in a robust and informative way.

The large number of variables highlighted above are just one of the characteristic features of metabolic profiling data which influence the data analysis procedure.

Typically, variables generated from spectroscopic data are highly correlated. Thus, even when the number of samples exceeds the number of variables, the data matrix will be poorly conditioned, exposing traditional methods such as multiple linear regression to high uncertainties in the calculated model parameters and consequent lack of model robustness. Furthermore, the distributions of metabolic data are often not well approximated by a multivariate normal. This can be due to non-normality in the distribution of individual metabolites (as has been seen in the urine of control rats [1, 2]) or, more commonly, due to the combination of several groups within which the data are normally distributed (e.g. combining data from several time points together). Therefore non-parametric methods which do not constrain the data to normality can be useful. An additional feature peculiar to metabolic profiling data is the difficulty of identifying the data variables with particular chemical species. This is primarily because the samples are usually complex mixtures, many of whose components may not have previously been assigned, though the problem is complicated by issues of sensitivity and peak overlap. Linked to this is the fact that the number of metabolites potentially detectable by these techniques is unknown to the experimenter. This is in contrast with, for example, gene array data, where the number and identity of the measured genes are known. Thus, it is very important that data analysis procedures are equipped with visualisation and interpretation tools that can be used to identify the nature of the metabolites responsible for any biological effects seen.

A number of the characteristics mentioned above argue for the consideration of non-linear methods in modelling metabolic profiling data. For the purposes of this chapter, I define non-linear methods as those which are able to model data on a non-linear manifold or, for classification, those for which the resulting class boundaries are in general non-linear. As an example of the latter, even the simple case of two normally distributed classes with non-identical covariance matrices leads to a non-planar decision surface. Alternatively, when linking metabolic data to the magnitude of some stimulus (e.g. drug dose), a non-linear response is often expected and thus linear models will not capture the full extent of the relationship. Fundamentally, biological processes are non-linear owing to the complex interactions of many entities across multiple levels of biological organisation. Even simple metabolic systems such as linear or branched pathways can be shown to generate highly non-linear responses in theoretical models [3]. Finally, as with all 'omics' technologies, metabolic profiling experiments are sensitive to a myriad of confounding factors. These can be internally generated (such as the competing interactions of multiple biological processes) or due to external influences (e.g. diet, medications etc.). These factors can combine in complex ways to produce metabolic effects which are not linearly related to the stimulus or group difference of interest [4].

There are a number of tradeoffs which must be considered when opting for non-linear over linear methods. Clearly, the main motivation for using a non-linear method will be the existence of non-linear structure in the data which could not be adequately captured by a linear analysis. However, non-linear methods in general employ more tuneable parameters than linear methods, which can lead to difficulties in interpretation and reduced model robustness. In addition, they are usually more susceptible to over fitting and the effects of noise. Thus, they must be applied with caution and the results interpreted with a healthy dose of scepticism.

The methods reviewed here can be broadly divided into unsupervised and supervised approaches. Unsupervised methods attempt to model the data in an exploratory way without using any *a priori* knowledge (e.g. class memberships) to guide the analysis. Tables 7.1 and 7.2 illustrate some of the advantages and disadvantages of such methods for the problems of unsupervised dimension reduction/visualisation and classification respectively. In Table 7.1, the non-linear methods are compared

Table 7.1
Comparison of methods for non-linear dimension reduction and visualisation of metabolic profiling data. Two non-linear methods (MDS/NLM and SOMs) are compared to a standard linear method (PCA)

Property	Linear	Nonlinear	
	Principal components analysis (PCA)[2]	Multidimensional scaling (MDS) / non-linear mapping (NLM)[3]	Self-organising maps (SOMs)[4]
Map non-linear manifold	○	●	●
Straightforward interpretation of map directions	●	○	○
Calculation of further dimensions does not change earlier dimensions	●	○	○
Prediction does not require new model	●	○	●
Deterministic	●	○	○
Can be used for classification	○	○	●
Map has unlimited resolution	●	●	○
Time complexity[1] (training)	$dn_s n_c$	$n_{it} n_s^2 d$	$n_{neigh} n_{node} n_{it} d n_s$
Time complexity[1] (prediction)	dn_c	$n_{it} n_s^2 d$	$n_{node} d$
Space complexity[1]	dn_c	$n_s d$	$n_{node} d$

Notes:
Filled/open circles indicate that the method has/does not have the given property respectively.
1. The complexity expressions show how the time and storage space required to run an algorithm scales with the parameters indicated (usually numbers of samples and variables). For example, a time complexity of n_s^2 means that if the number of samples is doubled, the time required quadruples. The formulae should be regarded as upper bounds on the complexity as in 'big oh' notation (see e.g. [8]). General notation: n_s = no. of samples, d = no. of variables, n_{it} = no. of iterations.
2. PCA: It is assumed that the number of iterations (of, e.g., the NIPALS algorithm) for each of the n_c components is small compared to n_s and d.
3. MDS/NLM: Complexities assume n_{it} random restarts of a gradient descent algorithm.
4. SOMs: n_{node} = no. of nodes in map, n_{neigh} = no. of nodes in neighbourhood.

Table 7.2
Comparison of methods for non-linear unsupervised classification of metabolic profiling data

Property	Hierarchical cluster analysis (HCA)	k-means[1]	Self-organising maps (SOMs)[2]
Representation of hierarchical structure	•	○	○
Not restricted to spherical clusters	•	○	○
Can be used for map-like visualisation	○	○	•
Prediction does not require new model	○	•	•
No. of clusters is chosen automatically	○	○	○
Time complexity (training)	$n_s^2 d$	$n_{it} n_s$	$n_{neigh} n_{node} n_{it} d n_s$
Time complexity (prediction)	$n_s^2 d$	d	$n_{node} d$
Space complexity	$n_s d$	d	$n_{node} d$

Notes:
General notation: n_s = no. of samples, d = no. of variables, n_{it} = no. of iterations.
1. k-means: It is assumed that k is small compared to n_s or d.
2. SOMs: n_{node} = no. of nodes in map, n_{neigh} = no. of nodes in neighbourhood.

to principal components analysis (PCA) as an example of a commonly used linear method. In contrast, supervised approaches use *a priori* knowledge to generate models that are tightly focused on the effects of interest. Table 7.3 compares some of these supervised methods, again including a typical linear method, partial least squares discriminant analysis (PLS-DA) for reference. It should be noted that the tables attempt to provide a broad-brush comparison and cannot fully detail the subtle advantages and disadvantages of each method.

The scope of this chapter is restricted to those methods which have been applied to metabolic profiling data in the literature to date. This necessarily entails the exclusion of a number of interesting methods from the machine learning and multivariate statistical communities (e.g. kernel methods) which have not yet seen such applications. In the case of NMR, I restrict the discussion to high resolution spectra and do not consider *in vivo* magnetic resonance spectroscopy (MRS) applications (see [5, 6] for reviews). In each case, the aim of the chapter is not to present a detailed description, but rather to give the main ideas of the method, followed by some examples of its application in the fields of metabolic profiling. Further, I do not discuss the important issues of data pre-processing, variable selection or statistical validation methods. These are critical components of any pattern recognition approach, whether linear or non-linear and deserve a separate discussion in their own right. Ultimately, the data analyst must remember that the success of any given method depends on its applicability to the problem under consideration. This point of view is founded on the No Free Lunch Theorem [7] which holds that, while there are good specialised approaches for particular problems and good general purpose methods, there is no universally optimal method which will outperform its competitors on all problems.

Table 7.3
Comparison of methods for non-linear supervised classification or regression. Four non-linear methods (ANNs, PNNs/CLOUDS and GPs) are compared to a common linear method (PLS-DA)

Property	Linear	Nonlinear			
	PLS-DA[1]	k Nearest Neighbour (kNN)[2]	Artificial multilayer perceptron neural networks (ANNs)[3]	Probabilistic neural networks (PNNs) / CLOUDS[4]	Genetic Programming (GP)[5]
Non-linear class boundaries	○	●	●	●	●
Easy identification of variables responsible for classification	●	○	○	○	●
Deterministic	●	●	○	●	○
Missing data allowed	●	○	○	●	○
A priori indicators of classification performance	●	○	●	●	○
Time complexity (training)	$n_s d$	1	$n_{it} n_s n_{hid} d$	1	$n_{it} n_{pop} n_{node}$
Time complexity (prediction)	d	$n_s d$	$n_{hid} d$	$n_s d$	n_{node}
Space complexity	d	$n_s d$	$n_{hid} d$	$n_s d$	$n_{pop} n_{node}$

Notes:
General notation: n_s = no. of samples, d = no. of variables, n_{it} = no. of iterations.

1. PLS-DA: it is assumed that the number of components and number of training iterations is small compared to n_s or d.
2. kNN: it is assumed that k is small compared to n_s or d.
3. ANNs: complexities are given for the simple back propagation by gradient descent algorithm, where the network has d input and n_{hid} hidden nodes. It is assumed that the number of output nodes is small compared to n_{hid} or d.
4. PNN/CLOUDS: it is assumed that the smoothing factor, σ is known.
5. GP: the number of iterations, n_{it} can be large and is therefore retained. The number of individuals is denoted by n_{pop} and the number of nodes in each solution by n_{node}. The space complexity is given for the training phase. The space complexity of the prediction phase is n_{node}.

7.2. Unsupervised methods

7.2.1. Hierarchical cluster analysis

One of the simplest and most popular unsupervised methods in post-genomic data analysis is hierarchical cluster analysis (HCA) (see e.g. [8]). This method clusters the data forming a tree diagram, or dendrogram, which shows the relationships between samples. The algorithm begins by computing the distances between all pairs of data points in the multidimensional parameter space. This is usually computed on a Euclidean basis, though other distance metrics such as the Mahalanobis Minkowski definitions are possible. The relative distances between samples are then compared using a similarity index which equals 0 for identity and 1 for the samples separated by the greatest distance. At each iteration the method finds the closest pair of clusters (initially each cluster consists of a single data point) and merges them to form a new cluster. The distances between the new cluster and all other existing clusters are then computed. In the next iteration, the algorithm finds the closest cluster pair from the new set of distances, merges these and so on, until all points are members of a single cluster (see Figure 7.1 left panels). A key detail is that there are many ways of defining the distance or 'linkage' between two clusters. The most popular choices are centroidal, average, single (nearest neighbour) and complete (farthest neighbour) linkage, each of which can lead to a different clustering structure. The most intuitive, centroidal linkage defines the inter-cluster distance as the distance

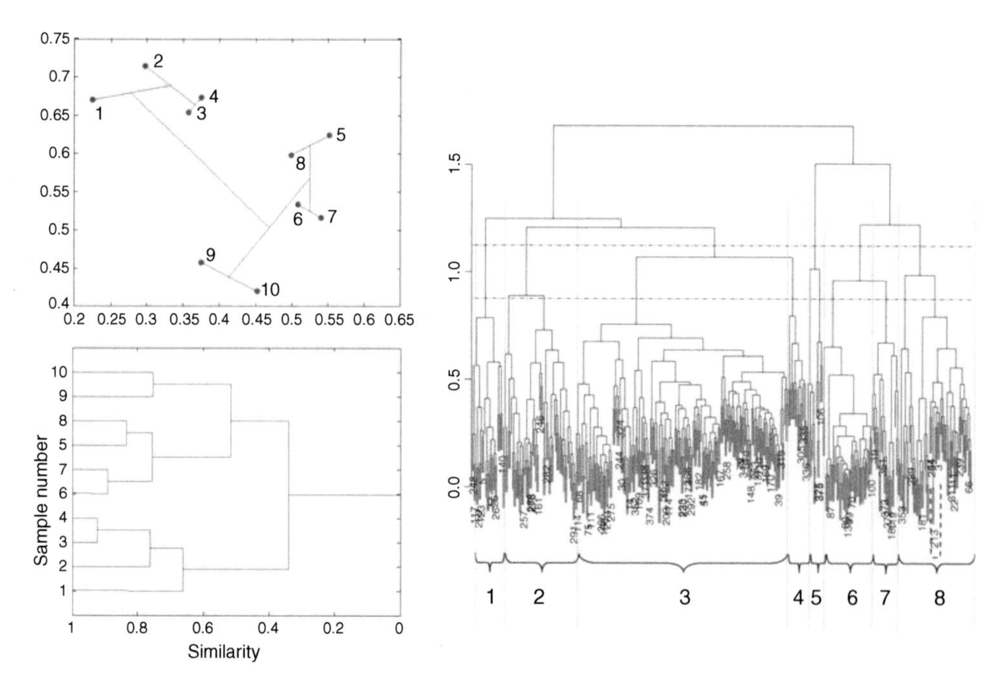

Figure 7.1. Hierarchical cluster analysis (HCA). *Upper left*: An example of 10-point, two-dimensional data set processed with centroidal linkage. The lines connect the centroids of each cluster and indicate the process of cluster joining. *Lower left*: The resulting dendrogram. *Right*: An HCA of spectral regions according to their correlations using NMR spectra of urine from cattle treated with anabolic steroids. (Reprinted with permission from Dumas *et al.* [10]. Copyright (2002) American Chemical Society.)

between the centroids of each cluster (the mean position of all cluster members). Each linkage method skews the algorithm towards particular types of clusters (e.g. chained clusters are favoured by single linkage) and so must be carefully considered when the algorithm is used. In the resulting dendrogram, the similarity value of each node is proportional to the distance between the corresponding clusters. To use HCA for classification, one must decide on a similarity cut-off which breaks the dendrogram into a number of separate clusters. The statistical robustness of the dendrogram structure can be tested by bootstrap resampling procedures in which HCA is performed on many random resamplings of the data and the reproducibility of certain features (e.g. the main branches of the tree) is ascertained. Related to this is the problem that inclusion of new data requires the whole dendrogram to be recomputed and there is no guarantee that the resulting structure will be similar to that generated by the original training set. Interpretation and comparison of dendograms is further complicated by the property that each branch of the tree may be rotated (reversing the order of its child nodes) without changing the displayed clustering structure. Overall, one of the main drawbacks of HCA is that it does not generate diagnostic information on the reasons for the particular clustering structure discovered. For example, HCA alone does not tell us what features are responsible for the main differences between any given pair of subclusters.

HCA is used widely in all areas of science. A recent example of its application in metabonomics [9] used the method to explore a set of 20 toxicology studies. HCA allowed interpretation of the data in terms of the magnitude and site of toxicological effect, and helped to explain misclassifications by other methods. Although clustering of samples is the usual aim of pattern recognition in metabolic profiling studies, one can also apply these methods to clustering the spectral variables. Dumas *et al.* [10] took this approach, using both the signed and absolute value of the Pearson correlation coefficient to analyse NMR spectra from urine from cattle treated with anabolic steroids (see Figure 7.1 right panel). High positive correlations were attributed to structurally related resonances while lower positive and negative correlations were interpreted as deriving from metabolites linked in metabolic pathways or by physiological regulation.

7.2.2. Multidimensional scaling

Also known as non-linear mapping (NLM), multidimensional scaling (MDS) is a dimensionality reduction technique which attempts to find a low dimensional representation of the data in which the inter-point distances are preserved as much as possible (Figure 7.2, left). The primary purpose of MDS is usually visualisation of high dimensional data, although the resulting coordinates can of course be used as input to a classification algorithm. If D_{ij} is the distance between data points i and j in the original, high dimensional space, MDS varies the low dimensional coordinates

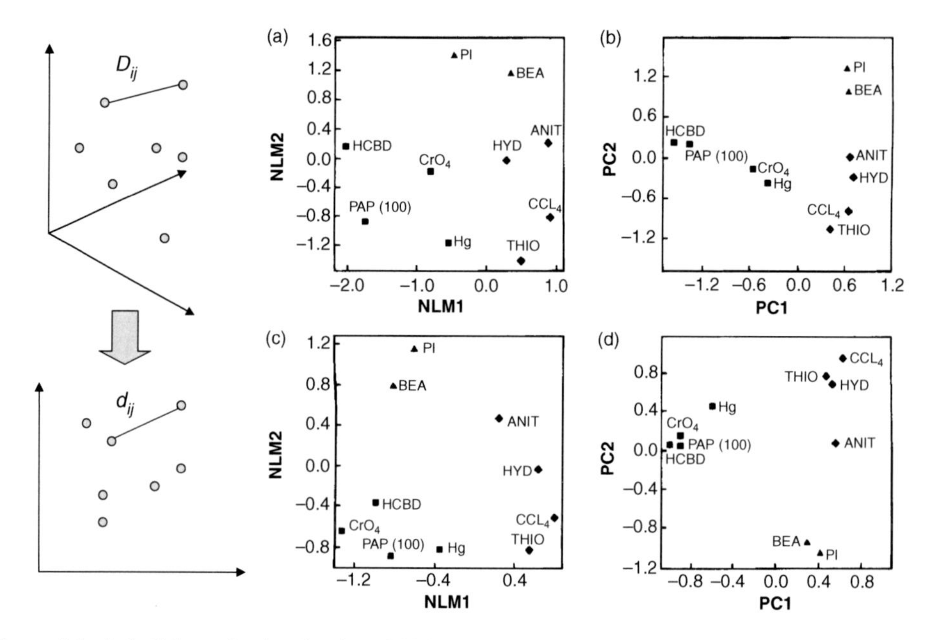

Figure 7.2. *Left*: Schematic showing how MDS attempts to preserve inter-point distances in the high and low dimensional spaces. *Right*: Non-linear maps (a and c) and PCA scores plots (b and d) of the NMR metabolic profiles of rats treated with 10 different toxins. Plots a and b show maps using scored metabolite values, while c and d show maps based on the squared correlation matrix of the original metabolite data. (Reprinted with permission from Gartland *et al.* [13]. Copyright (1991) The American Society for Pharmacology and Experimental Therapeutics.)

of each data point until the new inter-point distances d_{ij} match the D_{ij} as closely as possible as measured by a sum-of-squares error criterion, often termed the 'stress' function. There are several common choices of stress function [8], such as:

$$E_{\text{MDS1}} = \frac{\sum_{i<j} (D_{ij} - d_{ij})^2}{\sum_{i<j} d_{ij}^2}, \tag{7.1}$$

$$E_{\text{MDS2}} = \sum_{i<j} \frac{(D_{ij} - d_{ij})^2}{d_{ij}^2}, \text{ and} \tag{7.2}$$

$$E_{\text{MDS3}} = \frac{1}{\sum_{i<j} d_{ij}} \sum_{i<j} \frac{(D_{ij} - d_{ij})^2}{d_{ij}}. \tag{7.3}$$

The choice of error function determines the emphasis of the algorithm; function (7.1) will be heavily influenced by large discrepancies, while function (7.2) will emphasis large fractional errors. Function (7.3) is a compromise between these two. The latter is similar to the error function used in the closely related Sammon mapping [11]:

$$E_{\mathrm{Sammon}} = \frac{1}{\sum\limits_{i<j} D_{ij}} \sum_{i<j} \frac{(D_{ij} - d_{ij})^2}{D_{ij}}. \tag{7.4}$$

Which of these objective functions is most appropriate for a metabolic profiling analyses will depend on the application. For example, for visualisation of large metabolic changes resulting from a large drug dose in a toxicity study, function (7.1) may be most appropriate, whereas function (7.2) may be more useful for looking at small effects, such as within group variation.

The use of MDS/NLM in metabonomics was exemplified in the first application of pattern recognition approaches to NMR biofluid spectra [12] and thereafter used, along with PCA in several early publications [13–15] for toxicity classification (see Figure 7.2, right). While early work [12, 13, 16] used a small number of metabolites scored by eye according to their increase or decrease with respect to a control group, later publications [14, 15, 17] showed that the NMR spectra themselves could be used if preprocessed using the signal integrated in adjacent spectral regions or bins. The data were known to be distributed on a non-linear manifold in the original space and thus ideal for non-linear techniques such as NLM. However, in metabonomics, the method has been superseded by projection methods such as PCA. The primary reason for this is the difficulty of interpreting the maps in terms of the underlying spectral changes, a crucial point in any exploratory investigation. For example, it is very difficult to define what spectral pattern is associated with an arbitrary position on the map – the transform is one way and the map cannot be inverted. A similar problem occurs with the interpretation of vectors in the map space (perhaps representing changes between control and diseased samples). This, along with the need to recompute the whole map when predicting new data points, and the possibility of suboptimal solutions corresponding to local minima of the stress, outweighed the ability to represent non-linear manifolds in the comparison with linear techniques. In recent years, other non-linear mapping techniques have become available which overcome some of these difficulties. Some examples, briefly described in section 7.4, are 'neuroscale' [18], Isomap [19], local linear embedding [20] and kernel PCA [21], but none of these have yet been applied to metabolic profiling data.

7.2.3. Self-organising maps (SOMs)

The SOM [22, 23] maps a high n-dimensional distribution of data to a low dimensional (usually two-dimensional) array of nodes which enables not only visualisation

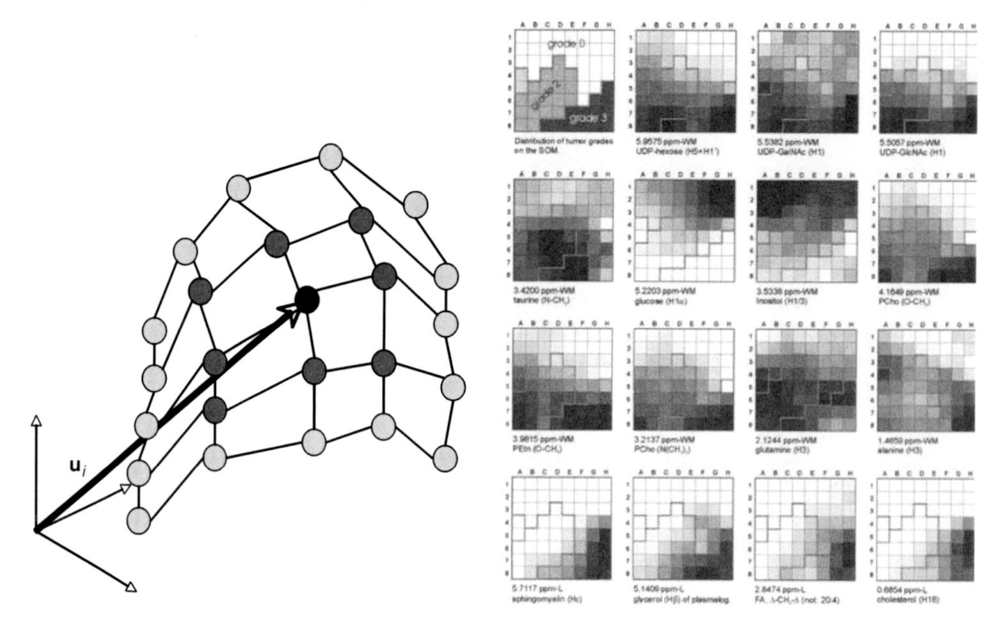

Figure 7.3. *Left*: Schematic diagram of an SOM showing the neighbours of a node (dark shading) and its reference vector $\mathbf{u}_i$. *Right*: SOM analysis of breast tumour data. The top left panel shows the areas on the map assigned to different tumour grades. Each subsequent panel shows the contribution to the SOM reference vectors of a given metabolite resonance, the so-called 'component plane visualisation'. Clear differences can be seen between the metabolites associated with each tumour grade. (Reproduced with permission from Beckonert *et al.* [25]. Copyright 2003 John Wiley and Sons Ltd.)

but also unsupervised classification. Each node i in the array is associated with an n-dimensional reference or 'codebook' vector $\mathbf{u}_i$ (see Figure 7.3, left), the components of which are initially set to random values. During training, each example is presented to the map and the node with the closest reference vector is selected as the 'winning node'. This node and its neighbours are 'activated' – their reference vectors are changed so that they become more closely aligned with the presented example. This is done most strongly for the winning node, and less strongly for the neighbours according to a neighbourhood function (a positive real function which decays as a function of distance). As the rest of the examples are presented, different regions of the two-dimensional array become closely aligned to particular types of training vectors, especially those most common in the training data. The map will tend to have more nodes corresponding to regions of high probability density in n-space and fewer in regions of low probability density. In contrast to methods such as PCA and MDS, a key feature of the SOM is that it is *granular* – the map consists of a discrete array – and mapping to intermediate positions between the nodes is not defined. The extent of this granularity is governed by both the number of nodes and the width of the neighbourhood function. With too few nodes, the map fails to faithfully represent the full extent of the input data distribution, while maps with large numbers of nodes become susceptible to noise and over fitting. Similarly, a broad neighbourhood function may lead to a map which does not discriminate between the different types of input, while a narrow function leads to a map in which the neighbouring nodes are unrelated, losing the desired global ordering. Once the map is trained (the reference vectors are fixed), the mapping position of new data is determined by assigning it to the node with the closest reference vector.

Despite its popularity in other areas of pattern recognition, the SOM has not yet enjoyed widespread use in metabolic profiling applications. One of the earliest applications was in classifying blood plasma lipoprotein profiles [24]. The authors showed how SOMs with 7×8 nodes, trained on 22 variables from 65 profiles, were able to efficiently represent the levels of various plasma lipids, leading to a map indicating relative 'risk' of coronary heart disease. These authors highlighted how profiles that were underrepresented in the training set had to be presented multiple times in order for the map to learn their distribution effectively. In a more recent application [25], the technique was applied to ^{1}H-NMR spectra of breast cancer tissue extracts. An 8×8 SOM was trained on 88 samples using 62 features selected by an algorithm based on k-Nearest Neighbour (kNN) classification. The resulting map showed clear differentiation between tumour grades which was not possible to discern by visual inspection of the NMR spectra alone. These authors also used the 'component plane' visualisation method to show which of their selected metabolites were responsible for the differentiation between tumour grades (see Figure 7.3 *right*). Hirai and co-workers have used SOMs in an integrated analysis of transcriptomic and metabolomic data from *Arabidopsis thaliana* grown under various nutritional stress conditions [26, 27]. The unsupervised clustering ability of the SOM was used to define specific regions of the map associated with correlated transcripts and metabolites with the interpretation that they would be functionally related. Again, component plane visualisations were used to show how the relative levels of transcripts and metabolites in the different map regions differentiated the different treatments. Both these examples serve to show how the versatile clustering and visualisation properties of the SOM could be used to extract useful information from metabolic data.

Despite its flexibility, the SOM does have some disadvantages when compared to other techniques. In the original implementation, the final map solution depended on the order of presentation of the training data. However, the batch-learning version of the algorithm [23], as used by Hirai *et al.*, overcomes this difficulty giving reproducible maps for a given set of training data. Secondly, the size of the map is limited by the number of training data and this can lead to maps which are excessively granular when there are few training points. Finally, and perhaps most importantly, although the algorithm attempts to place similar objects close together on the map, this may not always be possible. Similar clusters may end up in distant parts of the map, especially when the underlying data manifold has a complex nonlinear shape. Therefore, several maps should be viewed, perhaps using resampling techniques to ascertain the most reliable configuration.

7.2.4. k-means clustering

The k-means algorithm has become one of the most widely used clustering approaches finding many applications in post-genomics, especially in the analysis of transcriptomic data. The standard algorithm [8] begins by selecting the desired number of clusters and assigning their centres to randomly chosen training set data points. At each iteration, there are two steps: (1) data points are classified by assigning them to the cluster whose centre is closest and (2) new cluster centres are computed as the centroid of all the points belonging to each cluster. This process continues until both the cluster centres and the class assignments no longer change. The algorithm is simple and fast, requiring just one parameter to be set (the number of clusters). However, the technique inherently looks for compact, spherical clusters, and is therefore not always the method of choice when cluster shapes are expected to be far from multivariate normal.

Despite its popularity in other areas, k-means has not been used widely in metabolic profiling applications, perhaps not only because of the non-normality of some datasets, but also due to the fact that, as a simple clustering method, there are no associated visualisation or diagnostic tools. For the latter reason, k-means is often used in conjunction with other clustering and visualisation methods. For example, one recent application involved the combined use of k-means, tree clustering and MDS in the investigation of plant and marine invertebrate extracts [28].

7.3. Supervised methods

7.3.1. k-Nearest Neighbour classification

The k-Nearest Neighbour (kNN) rule for classification is arguably the simplest of all supervised classification approaches. Test samples are simply classified to the class most frequently occurring amongst the k nearest neighbours in the multidimensional parameter space. Despite its simplicity, the method has a sound theoretical basis in non-parametric density estimation [8] and can often outperform much more sophisticated methods. The method requires only the choice of k, the number of neighbours to be considered when making the classification. Small values of k will select the closest training points which are best able to estimate the correct classification at the test point. However, because of the small numbers, this estimate will be prone to large statistical fluctuations. Conversely, large values of k reduce statistical errors, but allow far-away points to contribute to the classification which may smooth out some of the details of the class distributions. Usually k is chosen as the value which minimises the classification error on some independent validation data or by cross-validation procedures. As with k-means, kNN merely produces a classification for each object without any associated visualisation or interpretation, and as a consequence it is not often used alone. Its competitive classification performance, however, has encouraged its use as a baseline against which other techniques can be compared. These points are illustrated by most metabolic profiling studies using the technique [9, 29, 30].

7.3.2. Artificial neural networks

Neural networks have become one of the most popular methods for pattern recognition and have found myriad applications throughout the biomedical arena. An artificial neutral network (ANN) consists of a layered network of nodes, each of which performs a simple operation on several inputs to produce a single output. In the standard format known as a multilayer perceptron (see Figure 7.4), the input layer feeds the signal to a layer of hidden nodes via weighted connections. In spectral applications, there is usually one input node for each spectral signal or region of the spectrum. Each hidden node receives signals from all the input nodes, which are

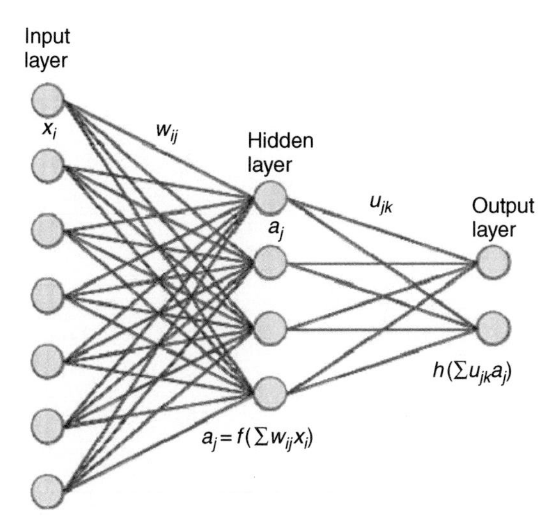

Figure 7.4. Schematic diagram of an ANN with seven input nodes, four hidden nodes and two output nodes. Input values are denoted by x_i, hidden layer weights by w_{ij}, hidden layer outputs by a_j, output layer weights by u_{jk} and the hidden and output layer transfer functions by f and h respectively.

combined as a weighted sum and then subject to a non-linear 'activation function', typically a smoothed step function or sigmoid. This has the effect of applying a soft threshold to the summed inputs, with the weights defining which spectral features each hidden node best responds to. The output nodes combine the signals from the hidden nodes, again as a weighted sum and may or may not apply a non-linear transformation to the result. In classification applications there is typically one output node for each class, plus possibly one corresponding to an 'unknown' class. The class of a new input vector is assigned as that of the output node giving the highest output value. Regression applications normally take the value of a single output node as the dependent variable. A comprehensive account of neural network theory and applications in pattern recognition can be found in the book by Bishop [31].

The use of ANNs is not without its pitfalls. Some experience is required to select the optimal architecture (e.g. number of hidden nodes) and a suitable training algorithm that will avoid local minima in the error function. However, for metabolic profiling applications by far the major disadvantage of ANNs is the difficulty of interpreting the connection weights to understand the 'rule' behind the output behaviour. This is due to the hidden layer introducing two sets of weights between the input and the output nodes. While methods are available to extract the rules encoded by the weights [32, 33], they do not have the simplicity or immediacy of interpretation attainable with other methods.

ANNs have been used extensively in MRS studies (see [6] for a review), mainly in the classification and differentiation of tumours *in vivo*. One of the earliest *ex vivo* applications [34] analysed ^{1}H NMR spectra of perchloric acid extracts from rat tumours. The PCA was used to pre-process the data, taking the top 15 scores as inputs to a four-layer network. Trained with 70 spectra, the network was able to classify all 14 test samples correctly. However, because of the difficulty of interpreting the neural network weights, these authors chose to use factor analysis to discover the biochemical differences producing the classification. A similar approach was later taken by Maxwell *et al.* [35] who classified a sample of 118 brain tumour extracts using rotated PCA vectors as inputs to the neural network. The NMR spectra of tumour cell extracts have also been used to distinguish between drug resistant and drug sensitive cell lines using the neural network approach [36]. The Howells *et al.* data were later reanalysed using both neural and statistical techniques [37]. Variable selection was accomplished using the Bayesian automatic relevance determination method [38] which allows some limited interpretation of the network parameters. In biofluid analysis, one of the earliest applications concerned ^{1}H NMR spectra of urine from rats treated with model toxins [39]. Eighteen metabolites were scored according to their increase or decrease for a group of dosed animals relative to controls and used as inputs to a network with five output nodes corresponding to different sites of toxicity or a non-toxic class. The network successfully classified 6 test samples when trained with 19 training samples. The authors again highlighted the difficulty in interpreting the network in terms of the underlying spectral descriptors. One of the few recent applications of ANNs in metabolic profiling involved the classification of NMR spectra of plant extracts according to the mode-of-action (MOA or biological pathway) affected by 27 different herbicides [40]. In leave-one-out cross validation, the approach was successful in differentiating treated and control samples. The ANN was also able to classify NMR profiles, both in cases where examples of the same MOA were present in the training set and cases where the MOA was entirely novel to the network. The above applications serve to highlight that ANNs may be an excellent choice for non-linear problems when the goal of the analysis is high classification accuracy, but that they have drawbacks when a parsimonious explanation for the classification in terms of the original variables is desired.

7.3.3. Genetic Programming

Approaches to machine learning using evolutionary computing have become very popular because of their flexibility and ability to optimise complex multimodal objective functions. These approaches work analogously to biological evolution by (1) creating a random population of possible solutions, (2) evaluating the fitness of each solution in the population, (3) retaining only the fittest individuals and (4) constructing a new generation of solutions by cloning, mutation and/or recombination of individuals, before returning to step (2). In Genetic Programming (GP) [41] as applied to classification or regression problems, each individual is a mathematical function which combines input variables together into a rule that can be used for categorisation or prediction of continuous quantities. These rules can be visualised as a parse tree (see Figure 7.5, *left*) where each leaf node is an input variable or constant and each internal node is a basic mathematical operation, such as 'add' or 'square root'. Mutation consists of randomly changing the operation performed by an arbitrarily selected internal node or replacing a leaf node by a random new input variable, constant or subtree. Recombination involves selecting a random subtree in each of two parents and swapping them to create two children, a process called crossover. The fitness of an individual is usually measured by the classification success rate or the mean square error of prediction on the training or separate validation set. The whole process is stopped when a suitable criterion has been met, for example maximum number of generations, a given fitness threshold, no further change in the population and so on.

Since the process is not deterministic, the algorithm is usually run many times and an averaging procedure applied to the best rules found in each run. Predictions for new data can easily be found by applying the rules in the same way as for the training set. As the solutions evolve, there is a tendency for the rules to become long and complex ('bloat'). To date this has been avoided by either constraining the maximum size of the parse trees or adding a heuristic penalty to the fitness function

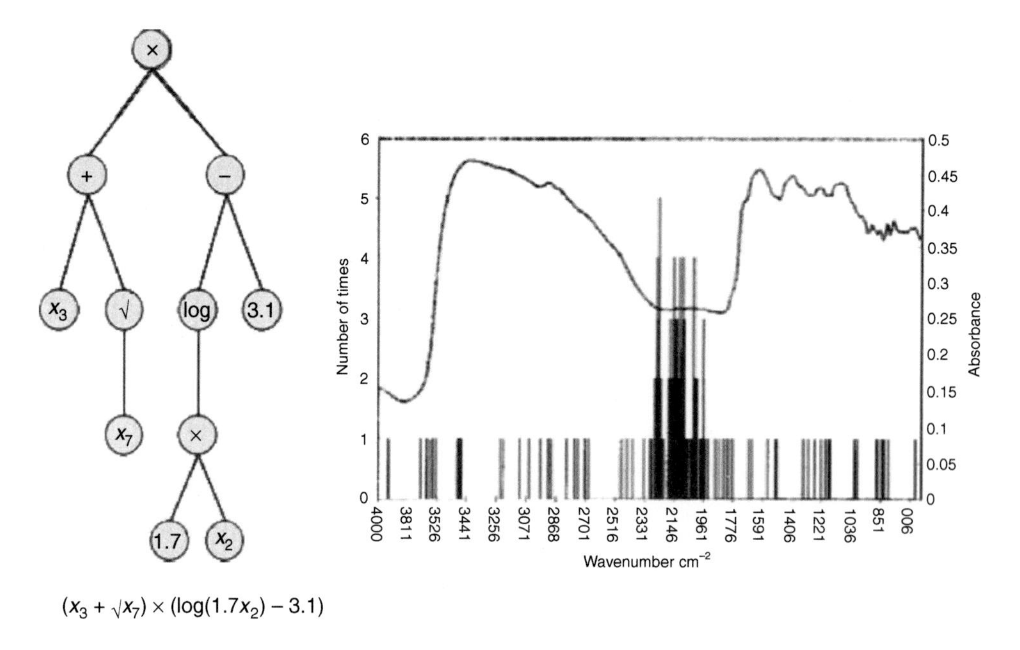

Figure 7.5. *Left* – Example of a GP parse tree and the mathematical expression it represents. *Right* – Variable selection by GP. A histogram of the number of times each variable is selected across 30 GP solutions is overlaid on the average near infra-red spectrum. (From Figure 2 of Johnson *et al.* [44] with kind permission of Springer Sciences and Business Media, Copyright 2000 Kluwer Academic Publishers.)

which favours simple solutions. Compared to other methods, the main advantage of GP is in its interpretability; simple rules using a few metabolites may be derived from spectroscopic data which originally profiled hundreds or thousands of species.

One of the earliest applications of GP to metabolic data [42] involved the analysis of ^{1}H NMR spectra of brain tumour biopsies (a 75 sample subset of data previously analysed by ANNs [35]). The scores of the top 20 principal components, rotated using the varimax procedure, were used as inputs (the terminal nodes) of the GP and the results were compared with a back propagation neural network on a small 10 sample test set. GP was able to obtain comparable classification accuracy to the ANN and the obtained rules used surprisingly few original variables. Interestingly, these authors found that there was little overlap between the variables selected by the neural network and the GP. Following successful use of GPs in the analysis of pyrolysis mass spectral data of fruit juice [43], the method was applied to metabolic profiling and fingerprinting data from several other spectroscopic and chromatographic techniques including FT-IR [44, 45], HPLC [46], non-linear dielectric spectroscopy [47] and various types of MS [48–50]. In most applications the ability of the GP to combine small numbers of explanatory metabolites in simple prediction rules is emphasised. For example, GPs were applied to a HPLC investigation of salicylic acid metabolism in transgenic tobacco plants infected with tobacco mosaic virus [46]. Of the three peaks selected by one of the best GP solutions (from a total of 48), one was indeed salicylate, and the solution allowed a simple low dimensional visualisation of the problem. In another application, GPs were compared with PLS for the prediction of antibiotic concentration in mixtures of *Escherichia coli* and ampicillin using FTIR spectra [45]. Both methods achieve similar degrees of prediction accuracy and both were able to identify the $1767\,cm^{-1}$ absorbance of the beta-lactam ring of ampicillin as an explanatory variable.

Since the GP solution depends on a random population initialisation, the results of many GP runs are usually combined when interpretation is required. The number of times each variable appears in the top ranked rules gives a measure of the variable's importance to the overall problem [44, 45]. This approach, however, ignores the (possibly nonlinear) way in which the variables are combined in each rule. For example, a variable appearing in two rules but with opposite signs will be less easy to interpret than one which always appears with the same sign. A further potential difficulty in the interpretation of the GP-derived rules relates to correlated variables. It is common in spectroscopy for several regions of the spectrum to be highly correlated, perhaps containing signals from the same metabolite. These regions provide redundant information to the classifier and a variable selection method, such as GP, can arbitrarily select any variable from this set. In principle, each GP solution could use a different spectral region with no particular variable being highlighted by frequent use across many rules. This can lead to a situation where the importance of the spectral information could be missed unless individual variables are selected often enough for the frequency histogram to peak (see Figure 7.5, *right*) [44]. Finally, as with all methods using variable selection, many original variables are left unmonitored. Therefore naïve application of the rules can prove dangerous if the new data are not representative of the original training set. For example, new peaks appearing in parts of the spectrum not used by the classification rule may be indicative of an inappropriate sample preparation protocol or sample contamination which may in turn invalidate the use of the rule. Methods not using variable selection often use such 'uninformative' variables as a means of checking that the data do actually conform to the model being used, thus increasing the confidence in the validity of any predictions made.

7.3.4. Probabilistic Neural networks and Classification of Unknowns by Density Superposition

Theoretically, the best performance in classification tasks is obtained with full knowledge of the probability distribution of the data in each class [8]. Using Bayes' rule, classification of novel data is achieved by assigning it to the class whose

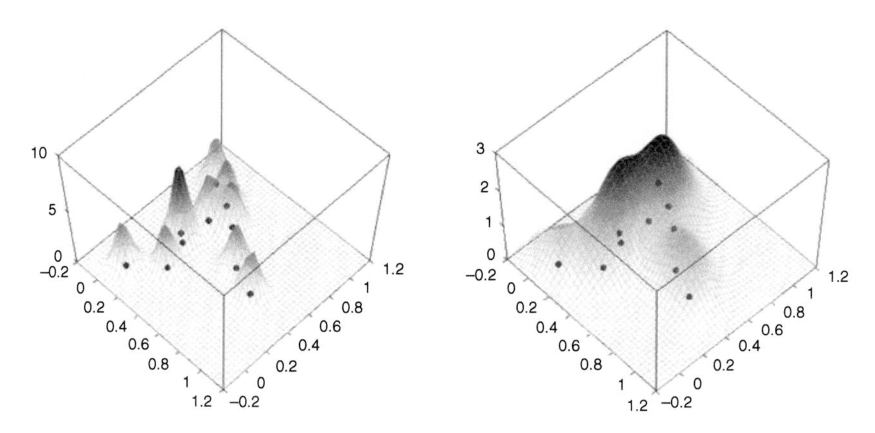

Figure 7.6. Kernel density estimates based on 10 training set data points in two dimensions and using small (*left*) and large (*right*) values of the smoothing parameter σ.

probability density is highest at the location of the new data point. Although techniques are well developed to estimate models employing standard distributions (e.g. Gaussian mixtures), when the data do not fit such standard forms much is gained by attempting a non-parametric estimate. Such estimates can be furnished by kernel density estimators and were the subject of early work in automated classification [51]. A kernel density estimate works by summing the contribution of kernel functions placed at each training set data point (Figure 7.6). Almost any positive real integrable function is admissible as a kernel [52], though by far the most commonly used is the multivariate spherical Gaussian kernel, giving estimates of the form

$$p_A(\mathbf{x}|\mathbf{x}_{i\in A}) = \frac{1}{N_A(2\pi\sigma^2)^{M/2}}\sum_{i\in A}\exp\left(\frac{-|\mathbf{x}-\mathbf{x}_i|^2}{2\sigma^2}\right) \tag{7.5}$$

for the density of class A at point $\mathbf{x}$, where N_A is the number of training set objects $\mathbf{x}_i$ in the class, M is the number of dimensions and σ is the width of the kernel. The choice of the σ parameter, which governs the smoothness of the estimate, is crucial to the success of the technique. Several methods exist to estimate σ, the most successful of which are often those based on cross-validation. The most lucid survey of the subject is still probably that of Silverman [52].

Probabilistic neutral networks (PNNs) achieve a kernel density estimate using a three-layer neural architecture where input nodes (again corresponding to spectral regions or signals) supply their values to every hidden node. Each hidden node corresponds to one training set point and its weights are fixed to the feature vector for that point. The activation function is defined such that presentation of a test vector causes each hidden node to output the value of the kernel, centred at the its training set point, but evaluated at the test point. These values are then passed (unweighted) to the output layer which consists of a single node for each class. These compute the summation of Equation 7.5, outputting the density estimate for each class. Classification of Unknowns by Density Superposition (CLOUDS) [53], originally conceived as a non-neural implementation of the PNN framework, also affords several improvements using the basic kernel density estimate (5). These involve using the density estimate to compute measures of the confidence and uniqueness of the classification and also a new measure of similarity between classes based on the overlap integral of the density estimate [54].

The first application of PNNs to metabolic profiling data was that of Holmes *et al.* [55] who applied the technique to the classification of ^{1}H NMR spectra of urine from rats treated with model toxins. The method outperformed some of the best-known classifiers (SIMCA and ANNs) in classifying the samples into 18 separate site/mechanism categories. PNNs have also been used recently to classify ^{1}H NMR spectra of fish tissue extracts [56]. The first 20 principal components (PC) scores were used as input to the network which was able to classify the samples according to various industrial processing and preservation methods with around 80% accuracy. The PNN approach was extended, as the CLOUDS methodology, by application to the classification of 2844 NMR spectra of urine from rats treated with 19 different model toxins or metabolic stressors in the Consortium for Metabonomic Toxicology (COMET) project [53, 57]. The method was used to classify samples between liver and kidney toxicity, with 77 and 90% success respectively and showed a very low (2%) rate of confusion between these classes. Further, the approach was successful in classifying samples when the data were broken down at the finest level of detail (87 classes, each representing a single treatment at a single time point). In addition it was shown that it is possible to make a pre-selection (i.e. before classes are assigned) of those samples with the most confident and unique classifications and that this significantly improves the success rate.

The CLOUDS approach was recently used as a basis for a toxicology screening ‘expert system’ [54] using a larger set of urinary NMR profiles from the COMET project. Using the kernel density estimate, a novel measure of similarity between classes was developed. This was used to classify a series of 62 treatments as liver or kidney toxins in a leave-one-out procedure where the test data for each treatment comprised those samples which did not fit a model of normal urine. Note that, in contrast to the earlier PNN or CLOUDS approaches mentioned above, no samples from the test treatment were included in the training set at any stage. The CLOUDS-based similarity reflected the overall relationship between the metabolic profiles of two different treatments and could be viewed as a measure of ‘toxin likeness’. The approach was able to correctly classify more than half of the treatments. Most of the remainder gave an ‘unknown’ classification, due to difficulties in assigning a true class from conventional assays, spectral interference due to metabolites of the dosed

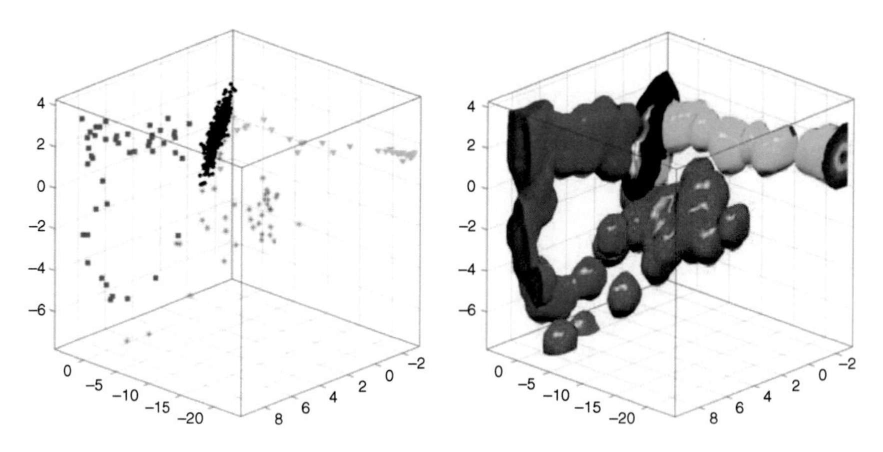

Figure 7.7. Visualisation of a CLOUDS model. *Left*: A plot of the first three principal component scores for samples from controls (black) and from toxins affecting the liver (blue), kidney (red) and the pancreas (green). The spread of the samples reflects the time course of each toxic episode. *Right*: Isosurfaces of constant probability density from a CLOUDS model of each class in the principal component space.

compounds or genuine class overlap. Only three treatments (<5%) were assigned to the wrong organ of effect, a result significant at $p < 10^{-4}$.

The CLOUDS approach can also be used to produce an informative visualisation of the probability densities. Using a dimensionality reduction technique such as PCA, contours of constant probability density (or 'isosurfaces' in three-dimension) can be used to illustrate the density estimates. Figure 7.7 shows an example using data from the COMET project. The distributions of data from toxins affecting three different organs and a control group is shown in the left panel. Isosurfaces of constant probability corresponding to these data are depicted in the right panel, giving a visual impression of the shape of the probability densities. When interpreting these plots, one must remember that the full densities exist in a high dimensional space and we can only visualise a low dimensional projection. Nonetheless, one can clearly see how the densities from different toxins trace a smoothed version of the data point distribution. Overlap between the various classes is visible, as is the ring-like structure exhibited by the time course of the liver toxin. Such an irregular distribution would be poorly modelled by any method relying on assumptions of linear class boundaries or multivariate normality.

Compared to other classification methods, kernel density approaches have several advantages in metabolic profiling applications. For instance, the data are not assumed to be normally distributed and thus complex (possibly disjoint) class distributions can be modelled. Additionally, prior knowledge such as different smoothing parameters for each class or information about the variable correlations can be easily included. Unlike ANNs or GPs, updating the model with new training data is straightforward and fast since no retraining has to be done. The classification probabilities also enhance the interpretation of the results more than what is obtainable from just the assigned classes alone. However, the kernel density approach does suffer some disadvantages, the most important of which is related to the 'curse of dimensionality'. As the number of dimensions increases, the data become very sparsely distributed and the true density cannot be accurately estimated. For example, for n data points rectangularly arrayed on a unit d dimensional hypercube (with $n \geq 2^d$), the nearest neighbour Euclidean distance goes up as $n^{-1/d}$. Taking as an illustration 1024 points, the nearest neighbours in 2 dimensions are separated by 0.03 units, whereas in 10 dimensions, they will be a distance 0.5 apart. It must be noted, however, that an accurate estimate of the true density is not necessarily a prerequisite for good classification. Secondly, in the standard approach, the coordinates of all training set points must be stored. This can result in a substantial computation time and storage requirement if there are large numbers of data points and the dimensionality is high. These problems can be reduced by approaches such as pre-clustering the data or dimension reduction techniques. Finally, interpretation of the density estimates in terms of the spectral signals is difficult due to their high dimensional and irregular nature. Although summaries such as the mean spectrum for each class are informative, new techniques are needed to allow analysis of the biochemical reasons for a classification or class similarity, and this is an area of active research.

7.4. Other methods

Space limitations preclude detailed discussion of other non-linear methods that have or could be applied to metabolic profiling data. In principle almost any method from the diverse fields of machine learning and pattern recognition could be applied, with the possible exception of those designed to work purely with symbol strings such as grammars. Many new methods have been developed specifically to solve some of the aforementioned drawbacks of commonly used techniques. For example, for MDS, the problem of requiring a new run of the algorithm in order to predict new data has been addressed by the 'neuroscale' algorithm [18]. This uses a radial basis function (RBF) neural network to specify the mapping between the high and low dimensional spaces. Once the mapping is known, new data are easily mapped using the RBF framework. However, because the RBF network is not readily invertible, other problems remain, such as the interpretation of the map dimensions. Another development in unsupervised dimension reduction is generative topographic mapping (GTM) which attempts to put the SOM on a sound theoretical basis by specifying a probabilistic model for each node. Other interesting methods to consider in this area include Isomap [19], locally linear embedding [20] and spring embedding [58, 59]. However, to date, no applications of these techniques have been published in the field of metabolic profiling.

Two further sources of non-linear methods are Bayesian approaches and kernel methods. Bayesian methods develop a full probabilistic model of the problem and use statistical inference to predict outcomes. One of the best examples of unsupervised Bayesian classification is the AUTOCLASS algorithm [60] which is able to both assign a class membership probability to each sample and choose the optimum number of classes. The method is, however, constrained to parametric class distributions (such as the multivariate normal) which can limit its scope of application. Bayesian networks are probabilistic graphical models for inferential reasoning and can model systems in many diverse ways. To date, there have been few examples of their application to metabolic profiling data (e.g. [61]) but this seems set to change.

Finally, much interest in the machine learning community has recently centred around kernel methods. These are algorithms in which the data are transformed into a space whose dimensionality is *higher* than the original. With an appropriate transform, the data may be amenable to treatment by linear methods (e.g. classes may become linearly separable). The key feature of kernel methods is that the transformation is done implicitly; the algorithm is written in such a way that the actual high dimensional coordinates are not computed. There are now kernel variants of many traditional linear algorithms, such as PCA [21] and PLS [62]. Although kernel methods have many advantages in terms of modelling non-linear, sparsely distributed and high dimensional data, a serious disadvantage is that there is no principled method for selecting the kernel function (which defines the transformation). Thus, their use depends on expert knowledge of the problem domain and this could be one reason for their lack of application in the metabolic profiling arena.

7.5. Conclusions

In summary, many of the diverse array of non-linear techniques, that have been developed in the machine learning and pattern recognition communities, have been applied to metabolic profiling data. However, there is still scope for many other methods to be examined, in dimensionality reduction, clustering, classification and regression problems. All methods should be applied with caution, with due regard to pitfalls such as over fitting, especially in situations where real world applicability is important. There are already examples in the literature where ‘black box’ application of pattern recognition methods to post-genomic data has led to poorly reproducible results [63]. Therefore, the importance of model validation and interpretation cannot be over emphasised.

The key in metabolic data analysis, as in most fields, is fitting the analysis technique to the problem at hand. No method can be expected to outperform others in all situations. In metabolic profiling we should take advantage of the particular characteristics of metabolic profiles, carefully choosing methods which use these to their advantage. This philosophy applies across both linear and non-linear methods. In general, the two approaches do not compete with each other; rather each will find its own set of niches to which it is well adapted, and the data analyst should be aware of the advantages and disadvantages of the chosen method. Ultimately, one cannot stress enough the requirement that the analyst should understand how the data are generated, including as many experimental nuances and details as possible. Without this knowledge, one cannot expect to be able to make the best marriage between problem and solution, which exemplifies good data analysis.

Acknowledgements

The author would like to thank all members of the Department of Biomolecular Medicine, Faculty of Medicine, Imperial College London. COMET is acknowledged for support and access to the data shown in Figure 7.7. Panels in Figures 7.1–7.3 and 7.5 are reprinted with permission.

References

[1] Ebbels, T. M. D., Holmes, E., Lindon, J. C., and Nicholson, J. K. (2004) Evaluation of metabolic variation in normal rat strains from a statistical analysis of 1H NMR spectra of urine. *J. Pharm. Biomed. Anal. 36*, 823.

[2] Tate, A. R., Damment, S. J. P., and Lindon, J. C. (2001) Investigation of the metabolite variation in control rat urine using 1H NMR Spectroscopy. *Anal. Biochem. 291*, 17.

[3] Hatzimanikatis, V. (1999) Nonlinear metabolic control analysis. *Metab. Eng. 1*, 75–87.

[4] Nicholson, J. K., and Wilson, I. D. (2003) Understanding 'global' systems biology: Metabonomics and the continuum of metabolism. *Nature Reviews Drug Discovery 2*, 668.

[5] Hagberg, G. (1998) From magnetic resonance spectroscopy to classification of tumors. A review of pattern recognition methods. *NMR Biomed. 11*, 148–56.

[6] el-Deredy, W. (1997) Pattern recognition approaches in biomedical and clinical magnetic resonance spectroscopy: A review. *NMR Biomed. 10*, 99–124.

[7] Wolpert, D. H., and Macready, W. G. (1997) No free lunch theorems for optimization. *IEEE Transactions on Evolutionary Computation 1*, 67–82.

[8] Duda, R. O., Hart, P. E., and Stork, D. G. (2000) *Pattern Classification*, 2nd ed., John Wiley & Sons Inc, New York.

[9] Beckonert, O., Bollard, E., Ebbels, T. M. D., Keun, H. C., Antti, H., Holmes, E., Lindon, J. C., and Nicholson, J. K. (2003) NMR-based metabonomic toxicity classification: Hierarchical cluster analysis and k-nearest-neighbour approaches. *Anal. Chim. Acta 490*, 3.

[10] Dumas, M. E., Canlet, C., Andre, F., Vercauteren, J., and Paris, A. (2002) Metabonomic assessment of physiological disruptions using 1H-13C HMBC-NMR spectroscopy combined with pattern recognition procedures performed on filtered variables. *Anal. Chem. 74*, 2261–73.

[11] Sammon, J. W. (1969) A nonlinear mapping for data structure analysis. *IEEE Transactions on Computers C-18*, 401–409.
[12] Gartland, K. P., Sanins, S. M., Nicholson, J. K., Sweatman, B. C., Beddell, C. R., and Lindon, J. C. (1990) Pattern recognition analysis of high resolution 1H NMR spectra of urine. A nonlinear mapping approach to the classification of toxicological data. *NMR Biomed. 3*, 166.
[13] Gartland, K. P., Beddell, C. R., Lindon, J. C., and Nicholson, J. K. (1991) Application of pattern-recognition methods to the analysis and classification of toxicological data derived from proton nuclear-magnetic-resonance spectroscopy of urine. *Mol. Pharmacol. 39*, 629.
[14] Holmes, E., Bonner, F. W., Sweatman, B. C., Lindon, J. C., Beddell, C. R., Rahr, E., and Nicholson, J. K. (1992) Nuclear-magnetic-resonance spectroscopy and pattern-recognition analysis of the biochemical processes associated with the progression of and recovery from nephrotoxic lesions in the rat induced by mercury(Ii) chloride and 2-bromoethanamine. *Mol. Pharmacol. 42*, 922.
[15] Anthony, M. L., Sweatman, B. C., Beddell, C. R., Lindon, J. C., and Nicholson, J. K. (1994) Pattern-recognition classification of the site of nephrotoxicity based on metabolic data derived from proton nuclear-magnetic-resonance spectra of urine. *Mol. Pharmacol. 46*, 199.
[16] Gartland, K. P. R., Beddell, C. R., Lindon, J. C., and Nicholson, J. K. (1990) A pattern-recognition approach to the comparison of pmr and clinical chemical-data for classification of nephrotoxicity. *J. Pharm. and Biomed. Anal. 8*, 963.
[17] Spraul, M., Neidig, P., Klauck, U., Kessler, P., Holmes, E., Nicholson, J. K., Sweatman, B. C., Salman, S. R., Farrant, R. D., Rahr, E., and *et al.* (1994) Automatic reduction of NMR spectroscopic data for statistical and pattern recognition classification of samples. *J. Pharm. Biomed. Anal. 12*, 1215–25.
[18] Lowe, D., and Tipping, M. E. (1997) in *Advances in Neural Information Processing Systems 9* (Mozer, M. C., Jordan, M. I., and Petsche, T., Eds) pp. 543–49, MIT Press.
[19] Tenenbaum, J. B., de Silva, V., and Langford, J. C. (2000) A global geometric framework for nonlinear dimensionality reduction. *Science 290*, 2319–23.
[20] Roweis, S. T., and Saul, L. K. (2000) Nonlinear dimensionality reduction by locally linear embedding. *Science 290*, 2323-+.
[21] Scholkopf, B., Smola, A., and Muller, K. R. (1998) Nonlinear component analysis as a kernel eigenvalue problem. *Neural Computation 10*, 1299–1319.
[22] Kohonen, T. (1990) The self-organizing map. *Proceedings of the IEEE 78*, 1464.
[23] Kohonen, T. (2001) *Self-Organizing Maps*, 3 ed., Springer, New York.
[24] Kaartinen, J., Hiltunen, Y., Kovanen, P. T., and Ala-Korpela, M. (1998) Application of self-organizing maps for the detection and classification of human blood plasma lipoprotein lipid profiles on the basis of 1H NMR spectroscopy data. *NMR Biomed. 11*, 168–76.
[25] Beckonert, O., Monnerjahn, J., Bonk, U., and Leibfritz, D. (2003) Visualizing metabolic changes in breast-cancer tissue using 1H-NMR spectroscopy and self-organizing maps. *NMR Biomed. 16*, 1–11.
[26] Hirai, M. Y., Yano, M., Goodenowe, D. B., Kanaya, S., Kimura, T., Awazuhara, M., Arita, M., Fujiwara, T., and Saito, K. (2004) Integration of transcriptomics and metabolomics for understanding of global responses to nutritional stresses in Arabidopsis thaliana. *Proc. Natl. Acad. Sci. U.S.A. 101*, 10205–10.
[27] Hirai, M. Y., Klein, M., Fujikawa, Y., Yano, M., Goodenowe, D. B., Yamazaki, Y., Kanaya, S., Nakamura, Y., Kitayama, M., Suzuki, H., Sakurai, N., Shibata, D., Tokuhisa, J., Reichelt, M., Gershenzon, J., Papenbrock, J., and Saito, K. (2005) Elucidation of gene-to-gene and metabolite-to-gene networks in arabidopsis by integration of metabolomics and transcriptomics. *J. Biol. Chem. 280*, 25590–5.

[28] Pierens, G. K., Palframan, M. E., Tranter, C. J., Carroll, A. R., and Quinn, R. J. (2005) A robust clustering approach for NMR spectra of natural product extracts. *Magn. Reson. Chem. 43*, 359–65.

[29] Oust, A., Moretro, T., Kirschner, C., Narvhus, J. A., and Kohler, A. (2004) FT-IR spectroscopy for identification of closely related lactobacilli. *J. Microbiol. Methods 59*, 149–62.

[30] Alsberg, B. K., Goodacre, R., Rowland, J. J., and Kell, D. B. (1997) Classification of pyrolysis mass spectra by fuzzy multivariate rule induction-comparison with regression, K-nearest neighbour, neural and decision-tree methods. *Anal. Chim. Acta 348*, 389–407.

[31] Bishop, C. M. (1995) *Neural Networks for Pattern Recognition*, Clarendon Press, Oxford, UK.

[32] Tickle, A. B., Andrews, R., Golea, M., and Diederich, J. (1998) The truth will come to light: Directions and challenges in extracting the knowledge embedded within trained artificial neural networks. *IEEE Transactions On Neural Networks 9*, 1057–68.

[33] Montano, J. J., and Palmer, A. (2003) Numeric sensitivity analysis applied to feedforward neural networks. *Neural Computing & Applications 12*, 119–25.

[34] Howells, S. L., Maxwell, R. J., Peet, A. C., and Griffiths, J. R. (1992) An investigation of tumor 1H nuclear magnetic resonance spectra by the application of chemometric techniques. *Magn. Reson. Med. 28*, 214–36.

[35] Maxwell, R. J., Martinez-Perez, I., Cerdan, S., Cabanas, M. E., Arus, C., Moreno, A., Capdevila, A., Ferrer, E., Bartomeus, F., Aparicio, A., Conesa, G., Roda, J. M., Carceller, F., Pascual, J. M., Howells, S. L., Mazucco, R., and Griffiths, J. R. (1998) Pattern recognition analysis of 1H NMR spectra from perchloric acid extracts of human brain tumor biopsies. *Magn. Reson. Med. 39*, 869–77.

[36] El-Deredy, W., Ashmore, S. M., Branston, N. M., Darling, J. L., Williams, S. R., and Thomas, D. G. (1997) Pretreatment prediction of the chemotherapeutic response of human glioma cell cultures using nuclear magnetic resonance spectroscopy and artificial neural networks. *Cancer Res. 57*, 4196–9.

[37] Lisboa, P. J., Kirby, S. P., Vellido, A., Lee, Y. Y., and El-Deredy, W. (1998) Assessment of statistical and neural networks methods in NMR spectral classification and metabolite selection. *NMR Biomed. 11*, 225–34.

[38] Mackay, D. J. C. (1995) Probable networks and plausible predictions - A review of practical Bayesian methods for supervised neural networks. *Network-Computation In Neural Systems 6*, 469–505.

[39] Anthony, M. L., Rose, V. S., Nicholson, J. K., and Lindon, J. C. (1995) Classification of toxin-induced changes in 1H NMR spectra of urine using an artificial neural network. *J. Pharma. Biomed. Anal. 13*, 205.

[40] Ott, K. H., Aranibar, N., Singh, B., and Stockton, G. W. (2003) Metabonomics classifies pathways affected by bioactive compounds. Artificial neural network classification of NMR spectra of plant extracts. *Phytochemistry 62*, 971–85.

[41] Koza, J. R. (1992) *Genetic Programming: On the Programming of Computers by Means of Natural Selection*, Vol. 1, MIT Press, Cambridge, MA.

[42] Gray, H. F., Maxwell, R. J., Martinez-Perez, I., Arus, C., and Cerdan, S. (1998) Genetic programming for classification and feature selection: Analysis of 1H nuclear magnetic resonance spectra from human brain tumour biopsies. *NMR Biomed. 11*, 217–24.

[43] Gilbert, R. J., Goodacre, R., Woodward, A. M., and Kell, D. B. (1997) Genetic programming: A novel method for the quantitative analysis of pyrolysis mass spectral data. *Anal. Chem. 69*, 4381–89.

[44] Johnson, H. E., Gilbert, R. J., Winson, M. K., Goodacre, R., Smith, A. R., Rowland, J. J., Hall, M. A., and Kell, D. B. (2000) Explanatory Analysis of the Metabolome Using Genetic Programming of Simple, Interpretable Rules. *Genetic Programming and Evolvable Machines 1*, 243.

[45] Goodacre, R. (2005) Making sense of the metabolome using evolutionary computation: Seeing the wood with the trees. *J. Exp. Bot. 56*, 245–54.

[46] Kell, D. B., Darby, R. M., and Draper, J. (2001) Genomic computing. Explanatory analysis of plant expression profiling data using machine learning. *Plant Physiol. 126*, 943–51.

[47] Woodward, A. M., Gilbert, R. J., and Kell, D. B. (1999) Genetic programming as an analytical tool for non-linear dielectric spectroscopy. *Bioelectrochemistry and Bioenergetics 48*, 389.

[48] Taylor, J., Goodacre, R., Wade, W. G., Rowland, J. J., and Kell, D. B. (1998) The deconvolution of pyrolysis mass spectra using genetic programming: Application to the identification of some Eubacterium species. *FEMS Microbiol. Lett. 160*, 237–46.

[49] Goodacre, R., Shann, B., Gilbert, R. J., Timmins, E. M., McGovern, A. C., Alsberg, B. K., Kell, D. B., and Logan, N. A. (2000) Detection of the dipicolinic acid biomarker in Bacillus spores using Curie-point pyrolysis mass spectrometry and Fourier transform infrared spectroscopy. *Anal. Chem. 72*, 119–27.

[50] Goodacre, R., York, E. V., Heald, J. K., and Scott, I. M. (2003) Chemometric discrimination of unfractionated plant extracts analyzed by electrospray mass spectrometry. *Phytochemistry 62*, 859–63.

[51] Parzen, E. (1962) On estimation of a probability density function and mode. *The Annals of Mathematical Statistics 33*, 1065.

[52] Silverman, B. W. (1986) *Density Estimation for Statistics and Data Analysis*, Chapman and Hall Ltd., London.

[53] Ebbels, T., Keun, H., Beckonert, O., Antti, H., Bollard, M., Holmes, E., Lindon, J., and Nicholson, J. (2003) Toxicity classification from metabonomic data using a density superposition approach: 'CLOUDS'. *Analytica Chim. Acta 490*, 109.

[54] Ebbels, T. M. D., Keun, H. C., Beckonert, O., Bollard, E., Lindon, J. C., Holmes, E., and Nicholson, J. K. (submitted).

[55] Holmes, E., Nicholson, J. K., and Tranter, G. (2001) Metabonomic characterization of genetic variations in toxicological and metabolic responses using probabilistic neural networks. *Chem. Res. Tox. 14*, 182.

[56] Martinez, I., Bathen, T., Standal, I. B., Halvorsen, J., Aursand, M., Gribbestad, I. S., and Axelson, D. E. (2005) Bioactive compounds in cod (Gadus morhua) products and suitability of 1H NMR metabolite profiling for classification of the products using multivariate data analyses. *J. Agric. Food. Chem. 53*, 6889–95.

[57] Lindon, J. C., Nicholson, J. K., Holmes, E., Antti, H., Bollard, M. E., Keun, H., Beckonert, O., Ebbels, T. M., Reily, M. D., and Robertson, D. (2003) Contemporary issues in toxicology the role of metabonomics in toxicology and its evaluation by the COMET project. *Toxicol. Appl. Pharmacol. 187*, 137.

[58] Eades, P. (1984) A heuristic for graph drawing. *Congressus Numerantium 42*, 149.

[59] Fruchterman, T. M. J., and Reingold, E. M. (1991) Graph drawing by force-directed placement. *Software – Practice and Experience 21*, 1129.

[60] Cheeseman, P., Kelly, J., Self, M., Stutz, J., Taylor, W., and Freeman, D. (1988) in *Fifth International Conference on Machine Learning,* pp. 54–64, Morgan Kaufmann Publishers, San Francisco.

[61] Li, Z., and Chan, C. (2004) Inferring pathways and networks with a Bayesian framework. *FASEB J. 18*, 746–8.

[62] Lindgren, F., Geladi, P., and Wold, S. (1993) The Kernel Algorithm For PLS. *J. Chemomet. 7*, 45–59.

[63] Baggerly, K. A., Morris, J. S., and Coombes, K. R. (2004) Reproducibility of SELDI-TOF protein patterns in serum: Comparing datasets from different experiments. *Bioinformatics 20*, 777–85.

The Handbook of Metabonomics and Metabolomics
John C. Lindon, Jeremy K. Nicholson and Elaine Holmes (Editors)

Chapter 8

Databases and Standardisation of Reporting Methods for Metabolic Studies

Susanna A. Sansone[1], Michael D. Waters[2] and Mark R. Viant[3]

[1]*EMBL Outstation-Hinxton, European Bioinformatics Institute, Wellcome Trust Genome Campus, Hinxton, Cambridge CB10 1SD UK*
[2]*National Center for Toxicogenomics, National Institute of Environmental Health Sciences, PO Box 12233, MD F1-05, 111 Alexander Drive, Research Triangle Park, North Carolina 27709-2233, USA*
[3]*School of Biosciences, The University of Birmingham, Edgbaston, Birmingham, B15 2TT UK*

8.1. Introduction: The challenges

Functional genomics-based investigations, including metabolic analysis, are complex and large in size, and consequently data management, integration and annotation in this domain present a considerable challenge to biologists and bioinformaticists [1]. In this chapter we describe the major obstacles we need to overcome to fully realize the potential of global metabolic analysis.

8.1.1. Reporting, storing and communicating data

Global metabolic analysis, as in the other omic sciences, are often hard to understand without a lot of annotation that is generally provided as free text. This is commonly known as *metadata*, or data about the data (see Table 8.1 for explanation of common terminologies). Metadata might refer to how the biological material being studied was obtained, grown and/or treated, how the experiment was designed, how the instruments used to process and analyse the biological material were set (parameters and values), or what sorts of analytical procedures were performed on the measurements. Metadata is the key each biologist needs to know about another scientist's data to evaluate the results presented.

Table 8.1
Explanation of common terminologies

Term	Definition
Controlled vocabulary	A predetermined, organized list of words that are used to define or tag content.
eXtensible Mark-up Language (XML)	Mark-up language is in general a language that 'marks-up' the text of a document by putting tags around content to describe what that content is and how to display it. The XML allows designers to create their own customized tags, enabling the definition, transmission, validation, and interpretation of data between applications and between organizations.
Format	A specific pre-established organization of data in a file. The format is established by whatever software created the file (i.e., organized the data) and typically needs to be read by the same or similar program that can interpret the format and present the data to the user.
Interoperability	The ability of software on different machines from different vendors to share data.
Metadata	Data about data; it may include descriptive information about the context, quality and condition, or characteristics of the data.
Object model	A design method in which a system is modelled as a collection of cooperating objects.
Ontology	An explicit formal representation of the knowledge in a subject area; including controlled vocabularies for referring to the concepts and logical statements (that describe what the concepts are and how they can or cannot be related to each other).
Ontology Web Language (OWL)	A semantic mark-up language for publishing and sharing ontologies on the World Wide Web
Reporting structure	An account of the experimental procedures and results, presented usually in detail and in a structured way.
Semantics	The meaning of a string in some language.
Standard	A format, a reporting structure, a controlled vocabulary or an ontology (or others) that has been approved by a recognized standards organization or is accepted as a *de facto* standard by the industry.
Unified Modeling Language (UML)	An open method used to specify, visualize, construct and document especially large, object-oriented methodologies.

A *controlled vocabulary* is a way to insert an interpretive layer of *semantics* between terms used by different biologists to better represent the original intention of the terms used. Therefore, incorporation of controlled vocabularies to describe a particular design of experiment, a sample treatment or an instrument's parameter, would facilitate the interpretation of the procedures. *Ontology* makes that description 'machine-readable', which means that computers would assist the biologist to compare and analyse those descriptions. This assistance is crucial these days, as global metabolic analysis use automated technology, generating very large amounts of data in a short space of time; a new challenge that defies imagination. Very often different software packages output data files in different *formats* and when integrating data from diverse sources it heavily increases the computational costs to interpret experimental procedures and results.

Incorporation of controlled vocabularies or ontology in the annotations of metadata and data can increase the power of computational approaches, providing semantics for features relevant to the interpretation, analysis and integration of global metabolic analysis; whilst the use of common output formats will enable software *interoperability*. Prospective *standards* for global metabolic analysis should be supported and embraced for the same reasons as in other 'omics technologies, making studies more accessible, facilitating comparison, reposition and exchange of data, enabling the extraction of maximum value from data sets, and enabling an assessment of the quality and relevance of a piece of work. From an informatics perspective, in this field, there are no universally accepted standards in terms of data formats, controlled vocabularies or *reporting structures*, although some formats are commonly used, and some common activities such as drug regulatory submission have quite explicit requirements reviewed by [2].

The development of standards is an iterative process and all stakeholders should be included from the onset. It includes research communities in different domains of applications (e.g. toxicology, environment, nutrition), vendors, bioinformaticists, database developers and end users. The latter is a very large category, including other researchers, but also regulatory bodies – when global metabolic data are submitted as part of a new drug application – and journal editors – when metabolic studies are reported as part of a publication. Some attention must be given to the various levels and consequent detail for reporting needs, including journal submissions, public databases and regulatory submissions. One size does not fit all!

It is very important to:

- Clearly define the end users and the goals of the reporting standards. This will help both to avoid unnecessary (and potentially expensive) recording of unnecessary data and metadata and to guarantee that sufficient information is actually captured.
- Obtain a precise statement of the requirements from several groups of end users, which ultimately will increase the chances of community acceptance.

Ultimately these requirements should be translated into a draft *object model* for data storage and an exchange format; controlled vocabularies or ontology should provide semantics for the descriptors. Ideally an object model should reuse existing designs from related standardization projects, both to avoid re-inventing the wheel and to maximize overlap between developing sets of standards. Ideally the model will have a modular flexible design to allow improvements and additions in the future. Furthermore, it is extremely important to validate and verify this model with the deposition and retrieval of data and appropriately associated metadata.

8.1.2. Peak annotation and construction of metabolite libraries

One of the great challenges in global metabolic analysis, whether one is using NMR spectroscopy, mass spectrometry, electrochemical detection or other methods, is poor annotation (metabolite identification) of the observed signals. In a typical NMR study, fewer than 30 metabolites of the estimated few hundred compounds within a spectrum will be unambiguously identified. This limits the metabolic information that can be extracted from the spectral data and restricts the mechanistic insight that could potentially be gained of the biological system under study. Furthermore, peak annotation is a prerequisite for instrument-independent metabolic data, which is both the desired format for the construction of metabolic databases and facilitates the comparison of datasets regardless of the instrument on which they were measured. This problem must be addressed urgently if global metabolic analysis is to fulfil its potential. In fact, the draft recommendations from a Joint SETAC-SOT Pellston Workshop on Emerging Molecular and Computational Approaches for Cross-Species Extrapolations (Portland, USA, July 2004) included to 'establish a public-domain international library of NMR spectra of metabolites and mass spectra of stable isotope-labelled metabolites, using standardised protocols, to aid in the annotation and quantification of metabolomics data'. In principle, such a library could be compared to the data from complex biological samples to annotate the observed metabolite fingerprints. The creation of metabolite libraries is now underway by a limited number of companies, including Bruker (http://www.bruker-biospin.com) and Chenomx Inc (http://www.chenomx.com) and by a number of initiatives in the public sector including the Canadian Human Metabolome Project (http://www.genomeprairie.ca/metabolomics/index.htm) and also a collaboration between Sigma-Aldrich and the University of Birmingham, UK (http://www.cancerstudies.bham.ac.uk/research/nmr). With the considerable time, cost and effort associated with library construction, however, it is imperative that there is community agreement on the standardised analytical parameters and the data formats of these libraries. For the former, parameters to be considered include sample pH, temperature, the identity and concentration of the buffer and so on. In terms of data format, this should conform to the same data and metadata reporting methods as developed for NMR and mass spectra of biological samples. For NMR spectroscopy the construction of a metabolite library is far from trivial, but for LC-MS the challenge is almost overwhelming. However, the benefits that would result from such libraries greatly outweigh the costs.

8.1.3. Data integration in a system biology context

One of the challenges within Systems Biology is to integrate several types of 'omics. This is a new but active area of research in bioinformatics, and necessitates the collaboration with an even broader range of scientific disciplines. Furthermore, to maximize the value of our existing knowledge, these 'omics datasets will need to be linked at the level of metabolic pathways and their regulation by gene products as well as to appropriate metadata that define the experimental context. Studies to date have reported a surprisingly low correlation between the expression of transcripts and the corresponding levels of the translated proteins. This, perhaps, is a reflection of the complexity of the systems that we seek to study. Eventually a database comprising of landscapes of mRNA, protein and metabolite markers associated with established toxicity and pharmacological mode-of-actions in model organisms will be used to determine potentially adverse effects of new products from the chemical or pharmaceutical industries. This would involve identifying exactly which transcriptional, protein and/or metabolic pathways and networks are perturbed, and would provide visualization of the modes of actions based upon, for example, the KEGG database [3]. However, such an endeavour is highly dependent upon community agreement of functional genomics standards. The development of such public databases is a work in progress (see Section 8.3).

8.2. Tackling the challenges

In 2004–2005, a consensus has been emerging for a body of scientists to address the standardization and reporting needs for global metabolic analysis. Many different groups have been working independently and several meetings have been held to address these important topics. In this section we describe some of the key components for building global metabolic standards, providing a brief review of the published, community-vetted developments and a summary of the meetings to date.

8.2.1. Towards a reporting structure

It is important to note from the outset that different reporting scenarios (e.g., journal submissions, public databases and regulatory submissions) will require specific modifications to the basic design given below. Here we arbitrarily decompose the reporting model into three components:

1. *Biological metadata*: a description and unique identifier of the biological sample, including its origin, maintenance or growth conditions, collection protocol and sample preparation. It should maintain a modular design that includes specific details for different fields of research; for example, medical screening or diagnostic tests in a human population would require different information to be captured compared to a controlled laboratory study of drought stress in a plant. This component has the greatest commonality with other reporting structures used in 'omics research (see Section 8.2.3).
2. *Analytical instrument and raw data*: a description of the analytical instrument and parameters used to record the metabolic data, including the raw data files. This component is complicated by the need for several analytical techniques in global metabolic analysis and so must be capable of handling, for example, free induction decays (FIDs) from NMR spectroscopy, mass spectra from direct injection MS, or chromatograms from GC-MS or LC-MS. Processes specific to one particular analytical approach are also included here, which includes technique-specific sample preparation (e.g. derivatisation prior to GC-MS, pH buffering of samples prior to NMR) as well as data pre-processing (e.g. transforming an FID into an NMR spectrum).
3. *Data processing and modelling*: a description of the data processing and analysis of the metabolite data, data visualization methods and validation protocols, including the processed data files. This is the most challenging component to define, in particular, due to overlap with the data pre-processing (which is dependent on the analytical instrument used) and also in terms of whether to include the multivariate or chemometric analyses used in global metabolic analysis. As the technologies used in global metabolic analysis evolve, a minimal goal should be to provide a set of instrument-independent metabolite concentrations.

The first positive step towards a published standard is represented by MIAMET, the Minimum Information About a Metabolomics experiment, a checklist of the information necessary to provide context for global metabolic data [4]. However, it is limited to the plant domain and does not provide the complete formal data description of the specifics required and data items necessary for the development of supportive data handling systems.

To date, the most comprehensive description of what data to capture in an experimental investigation of metabolism (not specifically how to capture it) has been developed and published by the Standard Metabolic Reporting Structure (SMRS, http://www.smrsgroup.org/) group. This group was formed to supply an open, community-driven specification. Its Steering Committee comprises journal editors, industrialists, software developers, government representatives and academics, with clear working targets and practices in plants, microbial systems, the environment, *in vivo* and *in vitro* applications as well as human studies. The group has recognized that capturing study descriptions on a high level is necessary and the myriad of factors that may affect metabolic data should be well documented to allow proper scientific interpretation and evaluation. A public discussion document is available on the public website for comments, while an abridged version has been published [5]. The document comprises all the aspects crucial to any interpretation and therefore subject to consideration for standardization and reporting, such as origin and preparation of the biological sample, the bioanalytical technologies for the measurement of metabolic profiles, and the statistical, chemometric or bioinformatic methods of data analysis. The recommendations also *touch* on the granularity of information required for different reporting needs, including journal submissions, public databases and regulatory submissions. However, time and funding constraints have limited the scope of the SMRS group to initiate other activities such as developing a format for electronic data exchange and ontology.

8.2.2. Object model and exchange format

Some members of the SMRS group are also authors of Architecture for Metabolomics (ArMet; http://www.armet.org). This is a data model for the description of plant global metabolic experiments and their results, which almost meets all the SMRS requirements [6]. ArMet is a component-based structure and it comprises a core set of data items and nine components each of which describes part of the process of a plant global metabolic experiment. This component-based approach enables extensions, so that the complete architecture can be customized. It supports data management and reporting ultimately, but does not deal with data analysis. The ArMet design has been translated into XML with the objective of supporting the exchange of data from different implementations. This data model provides a starting point for the development of data standards for the global metabolic community at large. Two future projects are planning to deploy ArtMet-based systems: MeT-RO (Metabolomics at Rothamsted, www.metabolomics.bbsrc.ac.uk/MeT-RO.htm) and HiMet (users.aber.ac.uk/nwh/Research/Projects/himet.html). However, ArMet does not have an ontology, but some data fields have a complex type, requiring a value and an authority; the authority identifies the ontology or the controlled vocabulary from which the value was taken.

8.2.3. Maximizing synergies

Clearly, with a number of groups already tackling some aspects of the data standardization process, we urgently need to address a global standardization, and, in particular, to develop ontology-based knowledge representations and standard exchange formats. Extensive liaisons with the metabolic communities and stakeholders are key to the success of any such standards-generating project. This process should also get considerable input from standardization activities in the other 'omics domains, such as the Human Proteome Organization (HUPO), Proteomics Standardization Initiative (PSI, psidev.sourceforge.net) and the Microarray Gene Expression Data Society (MGED, www.mged.org) [7, 8]. Both these academic grass-roots communities have joined forces with commercial vendors to address content standards and reporting needs for a single high-throughput technology. A consequence of working within discipline-specific initiatives is the unnecessary duplication of the effort. For example, there are commonalities with proteomics workflows, with reference to techniques such as mass spectrometry, liquid chromatography or capillary electrophoresis. Additionally core biological descriptors relating to the design of the global metabolic investigations, sample generation, growth and treatments are common across the life sciences.

A recent example of outreaching activities across 'omics domains, vendors, pharma, funders and publishers are the MetaboMeetings (http://www.thempf.org/conferences.html), jointly organized by the European Molecular Biology Laboratory (EMBL) – European Bioinformatics Institute (EBI, http://www.ebi.ac.uk), the University of Cambridge (http://www.cam.ac.uk/) and the SMRS group in the Spring and the Summer of 2005. The overall goals of these meetings were twofold: first, to bring together existing groups approaching the standardization of global metabolic investigations from different angles; secondly, to foster 'inter-omics' collaboration, optimizing synergy and avoiding duplications. Discipline-specific initiatives are regarded as important; however, the development of different terminology and data models in isolation will fragment the standards, limiting the potential for data exchange in a functional genomics (or systems biology) context. This latter should not be ignored, as it will significantly affect the structure and content of ontology, file formats and reporting structures, impacting on the development of the software and databases.

The MetaboMeetings have underlined the need for:

- Standards specific to each of the 'omics domains that also function together to support functional genomics investigations.
- Development of data model and exchange formats for the chromatography and the NMR domains.

- A MGED/PSI-like body in this technology domain to bring together representatives of the existing groups.
- A clear agreement to increase the involvement of industry, publishers, funding and regulatory bodies in these standardization activities from the onset.

8.2.3.1. Cross-domain interoperability

The main outcome of the second MetaboMeeting was the need to build upon some specific developments by the HUPO PSI and the MGED Society, seen as relevant to the metabolic community. These include the PSI mzData format, the Functional Genomics Ontology (FuGO) and the Functional Genomics Experiment Object Model (FuGE-OM), here described in detail [9, 10].

- The PSI mzData format captures the use of and data generated by a mass spectrometer (http://psidev.sourceforge.net/ms/) [11]. This *eXtensible Mark-up Language (XML)* based data exchange format is in a mature state with a stable production version 1.05 released in Summer 2005, currently being implemented by all the major databases, mass spectrometry vendors and a number of other companies.
- An all-encompassing FuGO (http://fuge.sf.net/fugo) is now underway to collaboratively develop controlled terminology for transcriptomics, proteomics and global metabolic analysis, as employed in a range of biologically defined domains, to avoid both design errors and redundancies. The decomposition and reuse of the MGED Ontology (MO, http://mged.sourceforge.net/ontologies/index.php) [12] will be part of the process of building FuGO, allowing the MO itself to remain stable, continuing to serve its established user base. The FuGO will be encoded in the *Ontology Web Language* (OWL) format, and developed with the Protégé tool [13]. For an efficient requirement elicitation (know as knowledge elicitation phase), the MGED Reporting Structure for Biological Investigation (RSBI) working group has worked closely with its experimentalists, in the toxicogenomics, nutrigenomics and environmental genomics domains [14]. The group has employed conceptual maps [15] as a method to aid the representation of the conceptual framework.
- The FuGE-OM (http://fuge.sourceforge.net) is designed to store data and annotation arising from high-throughput investigations with complex designs, including those in which more than one type of experimental technique has been used. FuGE is a collaborative effort initiated by the developers of the MGED Microarray Gene Expression (MAGE) and HUPO PSI working groups, set to improve data exchange and data comparison. The use cases and definitions released by the MGED RSBI working group have directly contributed to its development by providing real examples and terminology that bench researchers believe should be reported in a data model. FuGE is a single framework, comprising modules for storing an overview of the investigation structure, protocols and workflows, and

it provides facilities for auditing, giving security settings to objects and specifying references to external databases or ontologies. FuGE-OM is represented in *Unified Modeling Language* (UML), and the object model has been the primary focus for FuGE development, from which the XML schema and toolkit have been automatically generated. FuGE will be the basis for the next version of the MGED Microarray Gene Expression Mark-up Language (MAGE-ML) (Spellman *et al.*, 2002), a standard format for microarray data exchange, and for the PSI proteomics Mark-up Language (PSI-ML).

8.2.3.2. Object model for NMR

Following the MetaboMeetings, there has been considerable progress on developing an NMR data capture object model. A co-ordinated effort for the standardization of NMR data capture has emerged, which has successfully integrated the Cambridge, Aberystwyth and Birmingham efforts (H. Jenkins, D.V. Rubtsov, M.R. Viant, C. Ludwig, U.L. Guenther, J.M. Easton, N.W. Hardy and J.L. Griffin) [16]. At the first MetaboMeeting it was agreed that the University of Cambridge would develop a draft object model for an NMR global metabolic analysis, based somewhat upon SMRS recommendations. During the second MetaboMeeting this model was presented and improved upon by input from multiple academic groups and industry, including members of the SMRS group. During the same period both the University of Aberystwyth and the University of Birmingham each made progress in determining the key data and metadata to capture, the former work as part of an extension to ArMet and the latter including standards for both one- and two-dimensional NMR spectra.

8.2.3.3. The Metabolomics Society MSI

The agreements from the MetaboMeetings to collaborate on reporting guidelines, data formats (including the new NMR object model) and ontology contributions, and the need for political coordination were reported to the Metabolomics Society (http://www.metabolomicssociety.org/) at a meeting jointly organized with the NIH Metabolomics Roadmap Group (http://www.niddk.nih.gov/fund/other/metabolomics2005/) [17]. The Society can clearly draw in large numbers of experimentalists, developers and the large category of end users, and feed in the domain-specific knowledge of a wide range of biological, technical and regulatory experts. As a result, the Metabolomics Society has initiated a process to become an acknowledged authoritative body and prospectively acts as a facilitator for the development of data standards for global metabolic analysis, without restricting or dictating specific practices. The Society has established a Metabolomics Standards Initiative (MSI, http://msi-workgroups.sourceforge.net/) to monitor, co-ordinate and review the efforts of working groups in specialist areas, comprising biological sample context, chemical analysis, data analysis, ontology and data exchange [18]. The standards will be built on the knowledge base generated already from transcriptomics and proteomics efforts and upon the reports and recommendations of all previous global metabolic standardization works.

8.3. Standards in action – a working example

Once community-vetted standards for data communication reach a mature and stable development stage, their implementation in databases and software should be highly encouraged. In this section we provide an example of two public toxicogenomics resources that are leveraging on the MGED data standards to facilitate the establishment an international infrastructure for data exchange.

Scheduled for completion in 2012, the Chemical Effects in Biological Systems (CEBS) knowledgebase is at the end of its third year of development [19]. To promote a systems biology approach to understanding the biological effects of environmental chemicals and stressors, the CEBS knowledgebase is being developed to house datasets from complex data streams in a manner that will allow extensive and complex queries from users. Unified 'omics data representation is achieved through use of a systems biology object model, SysBio-OM [20], employing and extending the MAGE-OM designed for microarray experiments [21]. In the transcriptomics domain, CEBS currently captures MIAME compliant microarray data [22] and permits data export in raw form or in MAGE-ML. ArrayExpress is a public infrastructure for microarray gene expression data, designed to support data in publications [23]. The system is based on the MAGE-OM enabling data import and export in MAGE-ML format. ArrayExpress has been extended to store array-based toxicogenomics data, including conventional toxicology and pathology information [24]. The use of common semantic, consistent annotation and a standard data exchange format, between CEBS and ArrayExpress, will provide the two systems with the ability to support facile, bi-directional data exchange of data, a first step towards the establishment of an infrastructure for toxicogenomics data on an international scale.

8.4. Final remarks

Clearly substantial progress has been made in the past 2 years; however, standards of some depth require a considerable amount of time to develop fully. In the years to come the standardization activities under the Metabolomics Society are likely to lead to international acceptance and we hope to see the establishment of standards-compliant public global metabolic resources.

References

[1] Field, D. and Sansone, S.A. A special issue on data standards. *OMICS: A Journal of Integrative Biology* 10(2):84–93 (2006).

[2] Sansone, S.A., Morrison, N., Rocca-Serra, P., Fostel, J. Standardization initiatives in the (eco)toxicogenomics domain: A review. *Comp. Funct. Genomics*. 8, 633–41 (2004).

[3] Waters, M.D., Fostel, J.M. Toxicogenomics and systems toxicology: Aims and prospects. *Nat. Rev. Genet.* 5:936–48 (2004).

[4] Bino, R.J., Hall, R.D., Fiehn, O., Kopka, J., Saito, K., Draper, J., Nikolau, B.J., Mendes, P., Roessner-Tunali, U., Beale, M.H., Trethewey, R.N., Lange, B.M., Wurtele, E.S., Sumner, L.W. Potential of metabolomics as a functional genomics tool. *Trends Plant Sci.* 9:418–25 (2004).

[5] Lindon, J.C., Nicholson, J.K., Holmes, E., Keun, H.C., Craig, A., Pearce, J.T., Bruce, S.J., Hardy, N., Sansone, S.A., Antti, H., Jonsson, P., Daykin, C., Navarange, M., Beger, R.D., Verheij, E.R., Amberg, A., Baunsgaard, D., Cantor, G.H., Lehman-McKeeman, L., Earll, M., Wold, S., Johansson, E., Haselden, J.N., Kramer, K., Thomas, C., Lindberg, J., Schuppe-Koistinen, I., Wilson, I.D., Reily, M.D., Robertson, D.G., Senn, H., Krotzky, A., Kochhar, S., Powell, J., van der Ouderaa, F., Plumb, R., Schaefer, H., Spraul, M. Standard Metabolic Reporting Structures working group. Summary recommendations for standardization and reporting of metabolic analyses. *Nat. Biotechnol.* 23:833–8 (2005).

[6] Jenkins, H., Hardy, N., Beckmann, M., Draper, J., Smith, A.R., Taylor, J., Fiehn, O., Goodacre, R., Bino, R.J., Hall, R., Kopka, J., Lane, G.A., Lange, B.M., Liu, J.R., Mendes, P., Nikolau, B.J., Oliver, S.G., Paton, N.W., Rhee, S., Roessner-Tunali, U., Saito, K., Smedsgaard, J., Sumner, L.W., Wang, T., Walsh, S., Wurtele, E.S., Kell, D.B. A proposed framework for the description of plant metabolomics experiments and their results. *Nat. Biotechnol.* 22:1601–6 (2004).

[7] Taylor, C.F., Hermjakob, H., Julian, R.K., Garavelli, J.S., Aebersold, R., Apweiler, R. The Work of the Human Proteome Organisation's Proteomics Standards Initiative (HUPO PSI). *OMICS: A Journal of Integrative Biology* 10(2):145–51 (2006).

[8] Ball, C.A. and Brazma A. MGED Standards: Work in Progress. *OMICS: A Journal of Integrative Biology* 10(2):138-44 (2006).

[9] Whetzel, P.L., Brinkman, R.R., Causton, H.C., Fan, L., Field, D., Fostel, J., Fragoso, G., Gray, T., Heiskanen, M., Hernandez-Boussard, T., Morrison, N., Parkinson, H., Rocca-Serra, P., Sansone, S.A., Schober, D., Smith, B., Stevens, R., Stoeckert, C.J. Jr., Taylor, C., White, J., Wood, A. and FuGO Working Group. Development of FuGO: an ontology for functional genomics investigations. *OMICS: A Journal of Integrative Biology* 10(2):199–204 (2006).

[10] Jones, A.R., Pizarro, A., Spellman, P., Miller, M. FuGE: Functional Genomics Experiment Object Model. *OMICS: A Journal of Integrative Biology* 10(2):179–184 (2006).

[11] Pedrioli, P.G., Eng, J.K., Hubley, R., Vogelzang, M., Deutsch, E.W., Raught, B., Pratt, B., Nilsson, E., Angeletti, R.H., Apweiler, R., Cheung, K., Costello, C.E., Hermjakob, H., Huang, S., Julian, R.K., Kapp, E., McComb, M.E., Oliver, S.G., Omenn, G., Paton, N.W., Simpson, R., Smith, R., Taylor, C.F., Zhu, W., Aebersold, R. A common open representation of mass spectrometry data and its application to proteomics research. *Nat. Biotechnol.* 22:1459–66 (2004).

[12] Whetzel, P.L., Parkinson, H., Causton, H.C., Fan, L., Fostel, J., Fragoso, G., Game, L., Heiskanen, M., Morrison, N., Rocca-Serra, P., Sansone, S.A., Taylor, C., White, J., Stoeckert, C.J. Jr. The MGED Ontology: A resource for semantics-based description of microarray experiments. *Bioinformatics* 1;22(7):866–73 (2006).

[13] Noy, N.F., Crubezy, M., Fergerson, R.W., Knublauch, H., Tu, S.W., Vendetti, J., Musen, M.A. Protege-2000: An open-source ontology-development and knowledge-acquisition environment. In *Proceedings of AMIA Annual Symposium* 953 (2003).

[14] Sansone, S.A., Rocca-Serra, P., Tong, W., Fostel, J., Morrison, N., Jones, A.R. and RSBI Members. A strategy capitalizing on synergies: The Reporting Structure for Biological Investigation (RSBI) working group. *OMICS: A Journal of Integrative Biology* 10(2):164–71 (2006).

[15] Castro, A.G., Rocca-Serra, P., Stevens, R., Taylor, C., Nashar, K., Ragan, M.A. and Sansone, S.A. The use of concept maps during knowledge elicitation in ontology development processes—the nutrigenomics use case. *BMC Bioinformatics* 25;7:267 (2006).

[16] Rubtsov, D.V., Jenkins, H., Ludwig, C., Easton, J., Viant, M.R., Guenther, U.L., Griffin, J.L., Hardy N. Proposed Reporting Requirements for the Description of NMR-based Metabolomics Experiments. *Metabolomics* (under review).

[17] Castle, A.L., Fiehn, O., Kaddurah-Daouk, R., Lindon, J.C. Metabolomics Standards Workshop and the development of international standards for reporting metabolomics experimental results. *Brief Bioinform.* 7(2):159–65 (2006).

[18] Fiehn, O., Kristal, B., van Ommen, B., Sumner, L.W., Sansone, S.A., Taylor, C., Hardy, N., Kaddurah-Daouk, R. Establishing reporting standards for metabolomic and metabonomic studies: A call for participation. *OMICS: A Journal of Integrative Biology* 10(2):158–63 (2006).

[19] Waters, M., Boorman, G., Bushel, P., Cunningham, M., Irwin, R., Merrick, A., Olden, K., Paules, R., Selkirk, J., Stasiewicz, S., Weis, B., Van Houten, B., Walker, N., Tennant, R. Systems toxicology and the Chemical Effects in Biological Systems (CEBS) knowledge base. *Environ. Health Perspect.* 111:811–24 (2003).

[20] Xirasagar, S., Gustafson, S., Merrick, B.A., Tomer, K.B., Stasiewicz, S., Chan, D.D., Yost, K.J. 3rd, Yates, J.R. 3rd, Sumner, S., Xiao, N., Waters, M.D. CEBS object model for systems biology data, SysBio-OM. *Bioinformatics.* 20:2004–15 (2004).

[21] Spellman, P.T., Miller, M., Stewart, J., Troup, C., Sarkans, U., Chervitz, S., Bernhart, D., Sherlock, G., Ball, C., Lepage, M., Swiatek, M., Marks, W.L., Goncalves, J., Markel, S., Iordan, D., Shojatalab, M., Pizarro, A., White, J., Hubley, R., Deutsch, E., Senger, M., Aronow, B.J., Robinson, A., Bassett, D., Stoeckert, C.J. Jr., Brazma, A. Design and implementation of microarray gene expression markup language (MAGE-ML). *Genome Biol.* 3:RESEARCH0046 (2002).

[22] Brazma, A., Hingamp, P., Quackenbush, J., Sherlock, G., Spellman, P., Stoeckert, C., Aach, J., Ansorge, W., Ball, C.A., Causton, H.C., Gaasterland, T., Glenisson, P., Holstege, F.C., Kim, I.F., Markowitz, V., Matese, J.C., Parkinson, H., Robinson, A., Sarkans, U., Schulze-Kremer, S., Stewart, J., Taylor, R., Vilo, J., Vingron, M. Minimum information about a microarray experiment (MIAME)-toward standards for microarray data. *Nat. Genet.* 29:365–71 (2001).

[23] Sarkans, U., Parkinson, H., Lara, G.G., Oezcimen, A., Sharma, A., Abeygunawardena, N., Contrino, S., Holloway, E., Rocca-Serra, P., Mukherjee, G., Shojatalab, M., Kapushesky, M., Sansone, S.A., Farne, A., Rayner, T., Brazma, A. The ArrayExpress gene expression database: A software engineering and implementation perspective. *Bioinformatics.* 21:1495–501 (2005).

[24] Mattes, W.B., Pettit, S.D., Sansone, S.A., Bushel, P.R., Waters, M.D. Database development in toxicogenomics: issues and efforts. *Environ Health Perspect.* 112:495–505 (2004).

The Handbook of Metabonomics and Metabolomics
John C. Lindon, Jeremy K. Nicholson and Elaine Holmes (Editors)

Chapter 9

Metabonomics in Preclinical Pharmaceutical Discovery and Development

Donald G. Robertson[1], Michael D. Reily[2] and Glenn H. Cantor[3]

[1]*Metabonomics Evaluation Group, World Wide Safety Sciences, Molecular Profiling, Pfizer Global Research and Development, Ann Arbor MI USA*
[2]*Discovery Biomarkers Group, Pfizer Global Research and Development, Ann Arbor MI USA*
[3]*Department of Discovery Toxicology, Bristol-Myers Squibb, Princeton, NJ USA*

9.1. Introduction

The process of pharmaceutical discovery and development has changed dramatically over the past 10 years. The changes in the industry have profoundly affected the way we assess the efficacy and safety of candidate therapeutic compounds. In the past, discovery and development was largely a linear process with medicinal chemistry at one end and the clinic at the other with pharmacology, absorption, distribution, metabolism, and excretion (ADME), and safety assessment somewhere in the middle. A bottleneck in any one area slowed the whole process down, but made life a little easier for the groups further down the pipeline as the flow of compounds was reduced to the rate allowed by the narrowest bottleneck. That is not the case today. The rate of chemical synthesis changed dramatically with the advent of combinatorial chemistry. High-throughput pharmacology screening (HTS) was adopted to rapidly assess the exponentially expanding libraries of new compounds. Finding a hit is no longer a problem, dealing with too many hits is. The new and improved pipeline now puts tremendous pressure on downstream functions as they try to sort through the array of compounds to find the best candidates to move forward. Avoiding compound attrition is no longer the goal, moving attrition earlier into the process is. Not only has the way we discover and develop drugs changed, but the nature of the therapeutic targets has changed as well. It seems as if the simple targets are rapidly disappearing to be replaced by targets that may be more consequential, but are also significantly more complex. No longer content with simple enzyme inhibitors, we now are going after basic aspects of protein regulation (transcription factors, kinases, etc.) that almost guarantee a host of development issues and safety concerns. If these factors were not bad enough, the industry is now operating in an environment of increased public and regulatory scrutiny, downward pressure on prices, and investor skepticism about performance. The days of running 5-year morbidity and mortality studies for unknown or unproven mechanisms are a thing of the past. Biomarkers of both efficacy and safety are in high demand as a means to shorten and simplify clinical development. The demands are great and the hurdles are high. How are we going to get there?

With this background it is not surprising that one avenue many pharmaceutical companies have been pursuing is the exploration of "omic" technologies to help deliver on the promise of accelerated discovery and development. Transcriptomics and proteomics have received a great deal of attention within the industry as a means to discover new targets, identify biomarkers of efficacy and toxicity, and elucidate mechanisms. Metabonomics is a more recent player within the industry, yet it has received a great deal of attention, being the subject of numerous reviews on its use for pharmaceutically relevant applications [1–12].

Despite the tremendous explosion in metabonomics literature in the past several years, recently a number of papers have raised cautionary warnings with regard to how the technology (and indeed all omic technologies) may be have been oversold or misrepresented as to what it can actually deliver [13–17]. This chapter will attempt to present a balanced view of the technology, emphasizing those areas where metabonomics has already shown promise and those areas where it has yet to deliver with regard to preclinical pharmaceutical applications. It should be understood that the authors are all strong supporters of the technology so some bias will be inevitable, but as heavy users of the technology we are also in a good position to recognize where the weaknesses lie. It is clearly too early in the development of metabonomics for preclinical applications to pass final judgment on its worth but it is fair to say that we have passed the "emerging technology" stage and are now in the critical stage of value determination. The emphasis of this chapter will be on pharmaceutical preclinical applications of metabonomic technology and pharmaceutical clinical applications are covered in more detail in the next chapter.

9.2. Background

Figure 9.1 presents a simplified overview of the drug development process with emphasis on preclinical applications. Different companies have different development strategies and even within a single company development strategies may vary between therapeutic areas. For example, development strategies for oncology and topical products are often quite different from strategies for compounds having

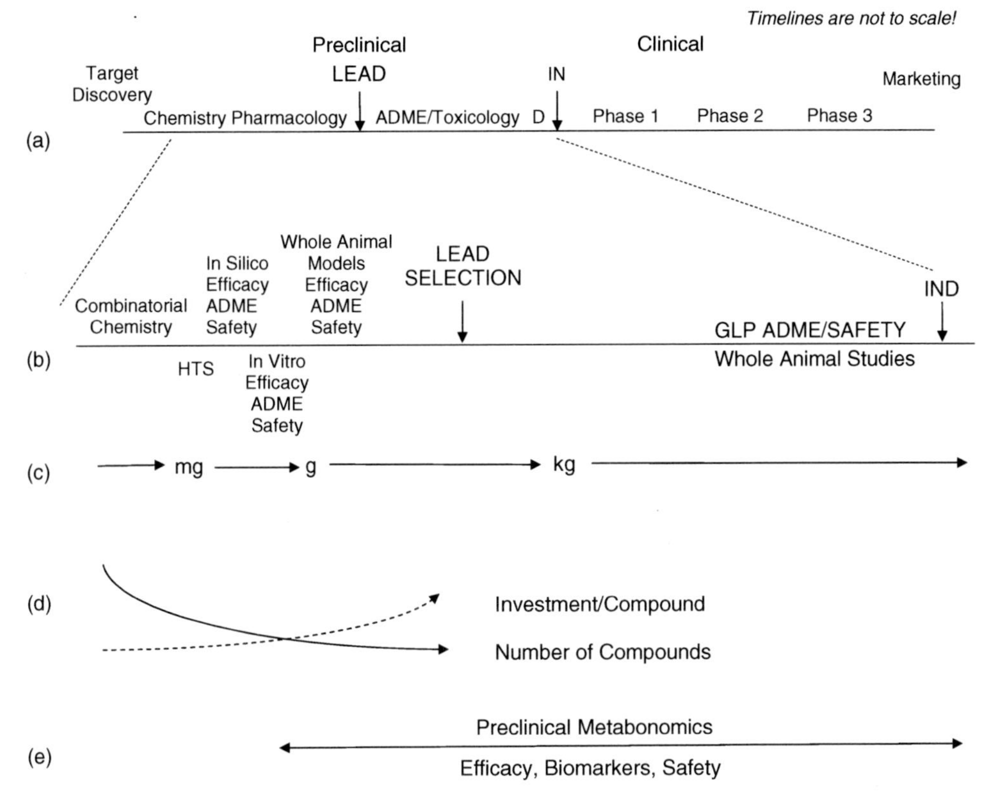

Figure 9.1. Simplified overview of pharmaceutical development. *Line a*: Overview of entire spectrum of pharmaceutical development. *Line b*: Expanded view of the preclinical portion of the pharmaceutical timeline highlighting major preclinical activities. *Line c*: Typical availability of bulk drug quantities. *Line d*: time/numbers/costs dynamic. *Line e*: The realm of preclinical metabonomics.

inflammation or cardiovascular indications. That said, Figure 9.1 provides a fairly generic view of the steps pharmaceutical companies take in developing their products. With regard to metabonomics, bulk drug supply is usually the limiting factor for undertaking *in vivo* studies early on in the process. Of course, in vitro applications could push incorporation of metabonomics even earlier into the development process; however, this has yet to happen to any great extent. Although *in vitro* applications of metabonomic technology have been reported [18–24], they are the exception rather than the rule. Reports of preclinical pharmaceutical *in vitro* applications are even more scarce [25, 26]. This can be expected to change as mass spectrometry (MS) gains increased utilization within the preclinical pharmaceutical community as one of the limiting factors for *in vitro* applications is sample quantity, which is a significant issue for nuclear magnetic resonance (NMR), but much less so for the MS platform. The focus of this chapter will be on whole animal applications, which initiate about the time there is enough bulk available to start *in vivo* experimentation (typically high mg or gram quantities). If studies can be conducted in mice, as little as 50 mg of bulk drug is sufficient for a single dose study at doses up to 300 mg/kg. A more realistic 5-day study (four daily doses) using the same dose range takes only 200 mg or so. While these quantities may make synthetic chemists cringe, the quantities are usually obtainable fairly early in the drug discovery program, typically after HTS and *in silico/in vitro* ADME (solubility, permeability, microsomal, or hepatocyte metabolism, etc.) are complete. If studies need to be conducted in the rat, typically the choice of toxicologists, the drug requirements increase approximately 10-fold. Metabonomics at this stage can potentially generate data for compound differentiation based on efficacy, biomarkers, ADME, or safety. It is important to realize that the technology seldom delivers on all those objectives and sometimes does not deliver on any of them. Despite that, the ability to piggyback the technology onto existing studies (hence avoiding additional studies with attendant increased animal and bulk drug usage) and the potential bonanza of information that can be obtained give the technology a very favorable cost/benefit ratio. Further on in development, the technology adds a significant dimension to problem solving, particularly from a safety perspective. Drug supply is available in quantities to push doses to toxicologically relevant doses (in Good Laboratory Practice (GLP) or ancillary safety studies). Typically at this stage, a safety issue has already been identified and metabonomics may serve as a tool for generating leading biomarkers for safety or for understanding the mechanism of toxicity. Metabonomics studies are generally not limited by bulk drug availability at this point and are carefully designed to generate data to illuminate the toxicity in question. Speed is usually of the essence with panicky discovery/development teams demanding quick answers. Thoughtful study design and additional technologies (magic angle spinning (MAS) or other omic approaches) may be quite useful at this stage, though the larger and more complicated the study, the less likely a speedy resolution. Beyond this "firefighting" stage, a real advantage of metabonomics technology becomes apparent. That is in the realm of translation of preclinical biomarkers to clinical applications. As most (but not all) metabonomics studies are conducted in readily accessible biofluids such as serum (or plasma) and urine, biomarkers identified in these biofluids preclinically can be readily measured clinically as well. In some ways clinical samples are easier to deal with. For example, urine bacterial contamination is much less prevalent in clinical samples compared to preclinical samples. However, the advantage of cleaner samples is more than offset by the lack of ability to rigorously control clinical subjects. Despite these issues, this simple translation of potential biomarkers (as opposed to biopsies frequently required for transcriptomics or proteomic analyses) represents a real advantage of the technology.

9.3. Methods

Metabonomics or metabolomics approaches seek to measure endogenous small molecules or "metabolites" produced by an organism and relate changes in these to state of disease or intoxication. This involves an array of methods ranging from study design right on through to data interpretation. Along the way, methods for sample collection, sample preparation and storage, analytical chemistry, and data analysis all have impact on the quality of the outcome.

9.3.1. Analytical platforms

As with other 'omics technologies, metabonomics is to some degree a fishing expedition – perturb the system and see what surfaces. With a little luck and hard work, the emergent changes will be reproducible for a particular indication; with a bit more luck they will say something about the biology with attendant information about mechanism and perhaps even provide the Holy Grail, a biomarker. Thus, there are two basic outcomes from metabonomics studies. The first of these is basically a spectral or chromatographic output or "pattern" that reflects the overall changes in metabolite concentration without necessarily identifying the actual components that are changing. This type of data, which can be thought of as a metabolic fingerprint [27], can be readily derived from virtually any analytical technique and is nothing more than a numerical representation of an analytical response arising from the milieu of components that comprise the sample. Such data from many samples are readily amenable to multivariate statistical analysis and generation of classification models as described in Chapters 6 and 7 of this book. The second and much more powerful product is a comprehensive and quantitative list of metabolites that are up-regulated or down-regulated, an approach which as been referred to as metabolomics [27] or metabolic profiling [28]. These metabolites can, in turn, be mapped to specific pathways and hence provide biomarkers and/or detailed mechanistic information about a process. This level of detail requires a rigorous analytical approach such as techniques based on NMR or MS, both of which have proven to be powerful tools for metabonomics analysis. It is not realistic to assume that one can ever comprehensively define the entire metabolome, at least with currently available technologies. Nevertheless, it is useful to consider studies primarily aimed at answering the question "Is it possible to differentiate two groups from each other based on changes in their metabolic fingerprint?" as preliminary to and sometimes completely separate from studies aimed at measuring quantitative metabolic changes in specific compound classes or the entire metabolome. In either case, one should cast as broad a net as possible to maximize the potential harvest. For metabonomics, this entails extracting as much detailed information about the metabolome from each sample as possible while maintaining relatively high throughput and to accomplish this, it is imperative that multiple complementary analytical approaches be taken.

As metabonomics transitions from an emerging technology to a useful one, the call for better analytical approaches has been met by adaptation of several well-established analytical tools, especially those based in MS and NMR. This is in no small part due to the power of both NMR and MS to generate reproducible and useful spectral patterns and to directly identify molecular components within complex biological samples. There are numerous reviews that cover technical aspects and metabonomics applications of both NMR [29–31] and MS techniques [27, 28, 32–35]. Early work in the area of endogenous metabolite analysis utilizing NMR spectroscopy [36, 37] and mass spectrometry [38] recognized the extremely rich information that these techniques provide and the relative ease with which it can be acquired. Numerous advancements in both of these technologies [35] have expanded the utility of both NMR and MS to the analysis of increasingly complicated biological systems, including whole biological fluids and tissues.

9.3.2. Analytical equipment considerations

Despite monumental advances in NMR spectroscopy, it is still a very insensitive technique and generally requires at least micrograms of a given metabolite for detection under conditions used in most metabonomics applications. This limitation, along with the need to resolve up to thousands of NMR resonances in a biofluid, dictates that when it comes to NMR-based metabonomics, sensitivity and resolution is the name of the game. Of the many hardware factors that influence both of these parameters [39], magnetic field strength and homogeneity top the list. Basically, the resolution is directly proportional to the field strength and sensitivity is proportional to the 3/2 power of the field strength, so a 600 MHz (14 T) NMR magnet has twice the resolving power and 2.8 times the sensitivity as a 300 MHz (7 T) magnet (all other things being equal). Ultimately, one must choose between affordability and power, as the cost of these magnets increases exponentially with field strength.

From a practical standpoint, the NMR equipment required for preclinical metabonomics applications will have the following characteristics: (1) the highest affordable sensitivity, with 600 MHz NMR being the current *de facto* standard; (2) high homogeneity, with facile automated optimization of line width to below 1 Hz and excellent water suppression specifications; (3) RF channels for observing protons and decoupling ^{13}C; (4) cryogenically cooled probe(s) capable of proton observe and ^{13}C decoupling and equipped with a *Z*-axis gradient; (5) automated sample introduction via a liquids handling robot into a flow cell and automated sample introduction for conventional 3–5 mm diameter NMR tubes, with both systems capable of keeping samples cold while in the queue for analysis; (6) an integrated HPLC-MS system that allows for HPLC-NMR-MS for identification of unknowns (see below);

(7) Although metabonomics *per se* can be done almost exclusively with proton NMR, there are many NMR-active atomic nuclei that are useful for follow-up investigations; thus it is useful to have the ability to observe ^{19}F, ^{13}C, and ^{31}P. A single NMR system with all of these capabilities is pushing the limit, but is possible. Figure 9.2 shows an instrument with all described capabilities except ^{19}F observation.

MS is a much more diverse approach than NMR in that there are many combinations of sample introduction, ionization, and detection approaches that ultimately yield systems with quite distinct capabilities and which provide different types of information. Additionally, the applications of MS to metabonomics analysis is in a greater state of flux than NMR and as such it is difficult to define the "ideal" equipment for MS-based preclinical metabonomics. Unlike NMR, which is primarily used to look at whole biofluids, MS is normally employed as a highly sensitive and selective detector used in conjunction with separation techniques. The most widely used hyphenated MS systems employ either gas chromatography (GC) or high

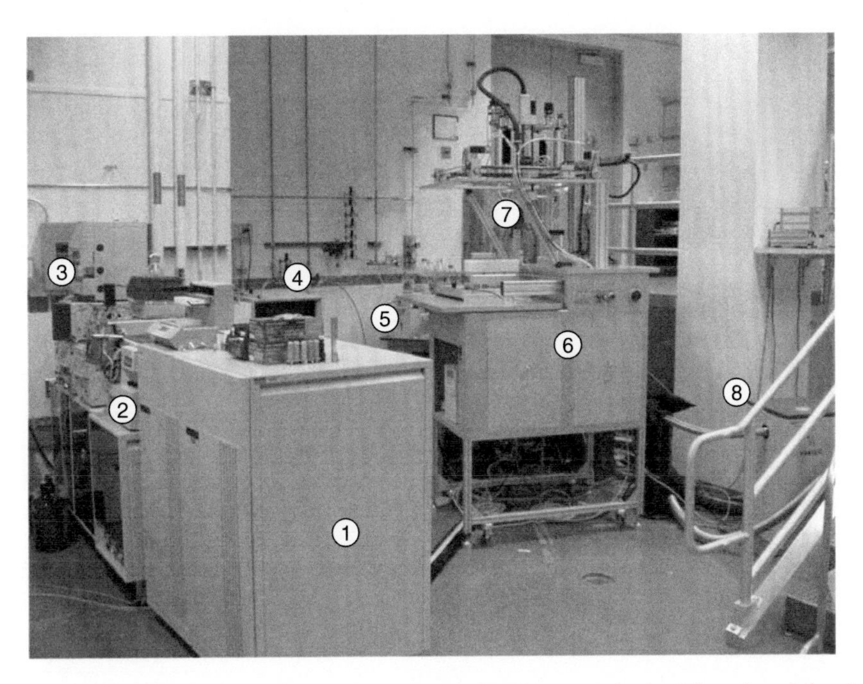

Figure 9.2. Photograph of the metabonomics 600 MHz NMR system in the Pfizer Ann Arbor laboratory. Various components are labeled: (1) NMR console containing all the RF components used for measuring the NMR signal; (2) triple quadrupole mass spectrometer; (3) dual solvent gradient HPLC system with autosampler, stop flow device and UV detector; (4) cryoprobe control console; (5) chilled liquids handler for direct flow injection of samples contained in 96 well plates; (6) 768 position sample changers for conventional NMR tubes with sample cooling provision; (7) NMR cryogenic Dewar containing the superconducting magnet; (8) helium compressor for maintaining the probe coil at 25 K.

performance liquid chromatography (HPLC) as a front-end separation technique. These two chromatographic interfaces are highly complementary and both should be employed when the most thorough metabolic information is desired. In the following paragraphs, we will attempt to highlight some common platforms applicable to preclinical metabonomics and elaborate on their strengths and weaknesses.

The primary benefits of GC-MS include very high chromatographic resolution and well-established libraries of reference compounds (e.g. the NIST Mass Spectral Database has been in existence for nearly 40 years and contains some 175,000 spectra). GC-MS relies on volatilization of target analytes and this generally requires that samples are first extracted then derivatized. This rather tedious sample manipulation can be considered a limitation of the method. In contrast, HPLC utilizes aqueous/organic mobile phases and as a result HPLC-MS requires little or no sample preparation for biofluid analysis. On the downside, the chromatographic resolution is generally much lower than for GC, although recent trends toward HPLC systems operating at very high pressures and utilizing small stationary phase particles have dramatically increased HPLC resolution [40]. HPLC also enjoys many more choices of mobile phase and stationary phase. This is beneficial for optimizing the separation of various analyte classes, but makes it difficult to standardize conditions on which to build useful reference libraries as GC-MS has done. Despite this, efforts are underway to assemble large libraries of ionizable compounds relevant to metabonomics [41]. Accurate mass using time-of-flight (TOF) or Fourier transform-ion cyclotron resonance (FT-ICR) can provide molecular formulas that can then be used to search such databases to rapidly identify unknown components.

An appropriate GC-MS system for metabonomics will include a single quadrupole or TOF design, a coupled flame ionization detector (FID), and an autosampler. The HPLC-MS systems should be capable of soft ionization, either electrospray ionization (ESI) or dual mode design that incorporates both ESI and atmospheric pressure chemical ionization (APCI). To minimize ambiguity in ion assignments, the mass detector should be capable of accurate mass in the 1–3 ppm range and provide a true dynamic range of >3 orders of magnitude. Additionally, the system should be capable of providing MS-MS data; that is, the ability to disintegrate a molecular ion into sub-molecular fragments that can be independently analyzed and linked back to the original molecule. Such MS-MS data can be used in conjunction with accurate mass to reduce isobaric ambiguities.

9.3.3. Fingerprinting methods

Analytical methods that are useful for generating reliable metabolic fingerprints have to be highly reproducible and provide a level of detail that is sufficient to allow differentiation of samples even if only small changes occur. Reproducibility is a key factor and this should be manifested not only within a given sample set, but preferably over long time periods and on different platforms, allowing comparison of datasets acquired in different laboratories and/or across months or years of time. Reproducibility is affected by analytical and sample variability, the latter influenced by both biological variation (which is what metabonomics seeks to capture) and variation introduced by post-collection sample processing (which can confound interpretation of results). Thus, sample processing approaches that require elaborate manipulation of samples on the bench will confound efforts to ensure that metabonomics observations are due to changes in the biological state of the organism and not due to some small variation in reagents, protocol, or laboratory environment.

Proton NMR spectroscopic approaches provide both a highly reproducible fingerprint [42] and a high level of detail, with thousands of spectral lines that arise from hundreds of the most abundant small molecules present in biofluids. Since NMR spectroscopy measures such transitions within the shielded confines of the atomic nucleus, the effect of matrix (i.e. solvent, other analytes, salt, etc.) is less of a factor than in spectroscopies that measure electronic or vibrational transitions. Preparation of biofluid samples for NMR studies typically only requires dilution of the sample with simple buffers containing some deuterium oxide (used as a field lock reference) and a small amount of chemical shift reference [29] and thus avoids the pitfalls of differential sample fractionation that are inherent in chromatographic interfaces or where multiple sample preparation steps are necessary [28]. In the case of urine, a strong solution of phosphate buffer helps to normalize the pH and osmolality variability that is often unrelated to the desired biological endpoint [43], while other more homeostatic fluids such as plasma, serum, or CSF can simply be diluted with isotonic saline or deuterium oxide alone. A major source of variability in an NMR fingerprint is the pH dependence of the chemical shift of protons near ionizable groups. This is predominantly a problem in analysis of urine, which can vary widely in pH, even under normal circumstances. While much of this pH variation is ameliorated by addition of strong phosphate buffer, it is not totally controlled. Recently, an internal standard that is used to measure pH in biological fluids, difluoro-trimethylsilylmethyl phosphonate (DFTMP) has been described that allows in-situ determination of exact pH in urine samples [44]. Another liability of NMR is that it is an inherently insensitive technique. In the context of molecular pattern analysis, 100 micromolar concentrations (tens of micrograms) of analyte must be present in a 0.5 ml sample to meaningfully contribute to a pattern. In summary, with limited sample manipulation, proton NMR spectroscopic approaches provide a matrix-independent, highly reproducible and detailed metabolic fingerprint albeit with low sensitivity.

Unlike NMR, which is typically used to analyze whole biofluids, mass spectrometers are almost always coupled to chromatographic outputs. Thus, the analytical product is an ion chromatogram, which contains mass information as a function of chromatographic retention time. The biggest benefit that MS provides in metabonomic analysis is sensitivity, which can be thousands of times higher than NMR. Whilst this sensitivity is heavily dependent on factors that affect ionization [35], HPLC-MS produces fingerprints that contain thousands of ions from hundreds of components, many of which occur at much lower concentrations than NMR can detect. Dual sources that can interleave ESI and APCI on one analyte stream have been investigated to maximize the number of molecules that ionize [45, 46]. These approaches are now coming onto the market and promise to provide even richer molecular information on each sample. Acquiring data in both positive and negative ionization mode on each sample will also increase the number of MS-observable features. Despite the inherent variability (both long and short term) that chromatography can introduce, useful patterns from which predictive models have been derived have come from HPLC-MS [40, 47, 48] and GC-MS [49]. Recent approaches using FT-MS [28] suggest that it may be possible to infuse biofluids samples directly into a mass spectrometer to obtain molecular fingerprint and/or profile information without the necessity of preliminary chromatography.

A substantial benefit of both NMR- and MS-based fingerprints is the ability to later mine the data for changes in specific molecules. Since the two methods sample different slices of the metabolome, multivariate models created using combined NMR and MS fingerprints will undoubtedly be more powerful than models derived from either approach individually [50, 51]. Ultimately, metabonomic classification models built from animals treated with known toxicants such as those generated in the COMET (Consortium for Metabonomics in Technology) consortium [52] can be used in preclinical toxicology screening paradigms.

9.3.4. Profiling methods

Analytical methods for generating molecular profiles or identification of unknown components must provide quantitative identification of metabolites with good dynamic range. Reproducibility is not nearly as important as with fingerprinting, as long as component assignments are accurate and changes in absolute or relative concentrations can be assessed with a useful degree of precision. Absolute concentrations are preferable and required for assessment of such parameters as total excretion or for evaluation against normal ranges or clinical chemistry results.

The primary parameters that can be readily extracted from NMR data for compound identification are frequency (known as chemical shift in the NMR world), through-bond or scalar coupling, and integrated peak area. The chemical shift of a given molecule is influenced by the intramolecular chemical environment and, therefore, is unique to specific molecules. Scalar coupling between nuclei within a molecule occurs by nuclear-electronic interactions and is manifested over a relatively short range, usually 1–5 bonds. These two parameters make peak assignment relatively straightforward, even in conformational isomers with the same molecular formula. For instance, the para- and meta-isomers of hydroxyphenyl propionic acid have completely unique chemical shift patterns that can be readily distinguished within an NMR spectrum with many hundreds of components. A host of NMR techniques have been developed to correlate peaks at different chemical shifts through scalar interactions that can be used to unambiguously identify and interconnect different atoms within a molecule [29]. The NMR peak area is proportional to concentration and all like nuclei (e.g. protons) essentially have the same extinction coefficient. Thus, relative concentrations of metabolites can be obtained by simple integration and absolute concentrations by comparison with an internal standard. The dynamic range of NMR is only limited by sensitivity and analog-to-digital converter resolution, but can easily achieve six orders of magnitude, with four orders typically obtained in biofluids analysis. In a routine 5 min urine proton NMR spectrum acquired on a cryoprobe-equipped 600 MHz NMR spectrometer, analytes such as 0.5 M urea, 10 mM creatinine, and 50 μM alanine can be quantitatively measured simultaneously. This is, of course, dependent on the ability to at least partially resolve the peak of interest. While longer acquisition times allow for even lower abundance components to be measured, in whole biofluids a practical limit is reached due to the problem of overlap with more abundant species.

Chromatography-hyphenated NMR provides a powerful analytical approach to solving the problem of overlap in complex mixtures and identification of unknown components. Recent technological developments like commercially available flow probes have established HPLC-NMR as a widely recognized tool for metabolite identification [53]. It is a straight-forward matter to use HPLC to fractionate samples while continuously monitoring the NMR spectrum of the eluent. Once the peaks of interest elute, the spectrum obtained is free from interfering peaks and is more analytically useful for identifying the unknown. The analytes need not have UV chromophores to be detected (because the NMR spectrometer is a universal detector) nor does separation need to be complete, since the inherent frequency resolution of the NMR spectrum makes it possible to analyze more than one component in a single spectrum. The biofluid complexity problem is thus reduced chromatographically. Of course, the eluent can be collected in fractions on the outlet side of the NMR flow probe for further work up and analysis. By incorporating an inline mass spectrometer analyzing a small fraction of the HPLC eluent, valuable supplementary information on the eluted analytes can also be obtained. Currently, HPLC systems and mass spectrometers with integrated software to control both the NMR spectrometer and all of the other components are available from major NMR manufacturers.

Mass spectrometry parameters provided by GC-MS and LC-MS for compound identification include mass/charge ratio (m/z) and retention time (RT), where $m =$ mass and $z =$ unit charge (dimensionless). The combination of these two allows for identification of unique "features" expressed as RT-m/z pairs. For a given molecule, there might be many ions that give different m/z values at a single RT that represents "parent" compound, adducts with other constituents in the eluent or fragments generated during the course of ionization. In the case of HPLC-MS analysis of urine, typical data sets include thousands of RT-m/z pairs suitable for semi-quantitative concentration assessment. The so-called "parent" ions are obviously the most desirable and are manifested as the intact protonated or deprotonated molecule, $[M+H]^+$ or $[M-H]^-$, depending upon whether the mass spectrometer is operating in positive or negative mode. When combined with accurate mass detection, the parent ions infer molecular masses accurate enough to generate molecular formulas. With molecular formulas in hand, the task of compound identification is greatly simplified. For GC-MS, analytes are usually derivatized and are ionized using electron impact (EI), which produces an ion pattern that is reproducible for a given molecule, but often does not contain a molecular ion. Even if it did, the extent of derivatization would have to be known to interpret the results. Because of the relatively standard stationary phase and thermal elution protocols, and the reproducible fragmentation that EI provides, the GC-MS data can be used to directly query extensive standard reference libraries to provide reliable molecular assignments. As a result, GC-MS is excellent at quantifying known metabolites (that reside in the reference libraries), but is poor at independent identification of unknown species. Coupling an FID to the GC-MS provides a non-selective readout aligned with the GC-MS ion chromatogram and can improve quantitation precision. As mentioned previously, ionization efficiency depends on many factors, confounding the ability to derive absolute concentrations. Although MS approaches can provide absolute concentrations based on standard curves for individual compounds, this is tedious in practice when matrix and coelution of multiple analytes must be accounted for. Thus, current MS-based metabonomics approaches are largely limited to identifying fold changes in specific metabolites.

9.3.5. Analytical issues in preclinical study design

One can spend a great deal of time and money establishing a metabonomics effort and it all can be undermined by poor study design. Because of the non-selective nature of metabonomics approaches, many factors that go unseen in conventional preclinical studies are revealed with metabonomics. For example, the choice of vehicles, anesthetics, and food can all have influence on the composition of biofluids. Similarly, sample collection, preparation, and storage are also things that must be considered. Finally, a potentially significant factor in many preclinical studies that was discovered using metabonomics approaches is the existence of substantially different phenotypes based on different gut microflora from presumably identically bred laboratory rats [54, 55].

During the in-life portion of any preclinical study it is imperative to keep in mind that whatever goes into the animal can affect what comes out. Drug carrier vehicle choice, often overlooked by newcomers to the field, can produce dramatic and often undesirable results. Not only can certain vehicles show up directly in the analytical data, but they can have their own biochemical consequences and affect the endogenous metabolites that are present in biofluids. Experience has shown that any vehicle containing poly-ethylene glycol (PEG), propylene glycol (PG), or dimethysulfoxide (DMSO) gives rise to strong vehicle-related signals in the urine and should be avoided. Furthermore, PEG and PEG-containing mixtures have been shown to have substantial biochemical consequences of their own [56]. Vehicles with little or no metabonomic consequence include methylcellulose (MC), detergents such as tween, ethanol, saline, phosphate buffer, water, vegetable oils, and carboxymethylcellulose (CMC), or combinations of these. While these vehicles are somewhat benign, they are all not entirely without effect and slight differences between IP corn oil, IP saline, and untreated rats can be noted using principal component analysis (PCA) on urine NMR spectra [39]. Another factor frequently complicating preclinical studies is anesthesia administration, which is normally required for interim blood draws and other invasive procedures. For example, the commonly used ketamine–xylazine combination (but not ketamine alone) causes large transient increases in blood glucose with concomitant changes in other endogenous metabolites (Figure 9.3). Such changes clearly have significant impact on biofluids profiles and potentially on the

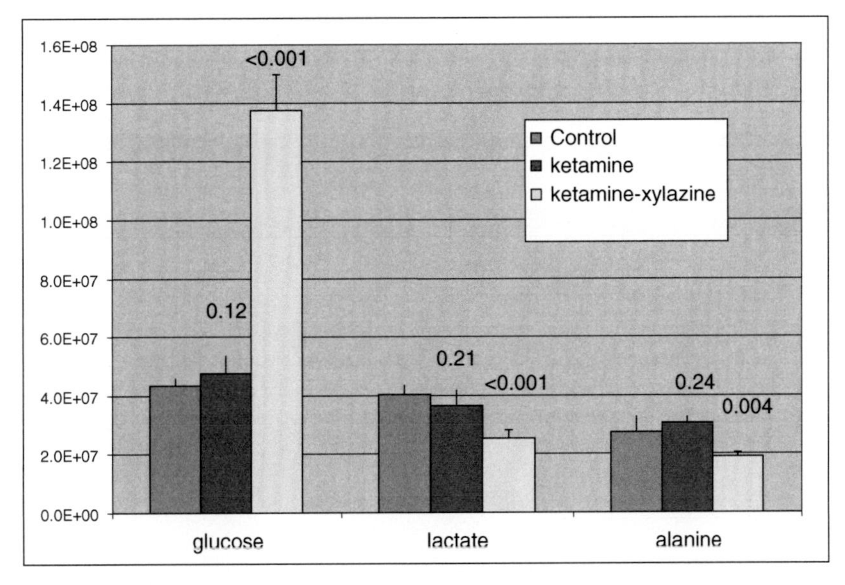

Figure 9.3. Relative concentrations of individual serum components measured from serum ^{1}H CPMG NMR spectra on samples collected 30 min after dosing. Glucose, lactate, and alanine all change significantly when animals are treated with ketamine-xylazine but not ketamine alone. Bar heights correspond to relative concentration, error bars indicate 1 standard deviation and the numbers above the bars indicate the p value between the control and the treatment groups for each metabolite measured.

state of welfare of the treated animals. A recent experience with urine obtained by cystocentesis from animals that were euthanized with pentobarbital resulted in massive peaks in the urine proton NMR spectrum. These were ultimately traced to PG that the barbiturate was formulated with and rendered the urine spectra essentially unusable [11]. The remarkable observation here is that from the moment of injection to death, millimolar concentrations of this excipient made it into the bladder. Experience indicates that isoflurane has less of an effect on biochemical profiles than other commonly used anesthetics. Finally, diet is another variable that has substantial impact on metabonomics measurements [57]. A puzzling if not somewhat amusing example of this was provided in a recent study with guinea pigs. A large unidentified NMR peak was observed during the course of the study and was seen periodically in treatment groups and in control animals. As it turns out, the laboratory Standard Operating Procedures (SOP) for guinea pigs requires that this species be given periodic edible treats outside their normal lab chow. The unknown NMR peak was temporally correlated to a sweet potato treat that was commonly administered in the laboratory [58]. We now avoid the use of sweet potatoes and are on the ready to identify like effects with other dietary diversions.

Sample collection, preparation, and storage represent further opportunities to introduce unwanted variability into metabonomics samples. Since it is difficult to keep feces, hair, food, and dander completely out of urine collection vessels, samples must be kept cold during collection and in the presence of a bacteriostatic agent such as sodium azide. Failure to do so can result in bacterial growth, which consumes the endogenous metabolites and produces lactate, acetate, and other bacterial waste products (Figure 9.4). For plasma, heparin as an anticoagulant is acceptable, whereas commonly used small molecule anticoagulants such as citrate and EDTA must be avoided as they have strong

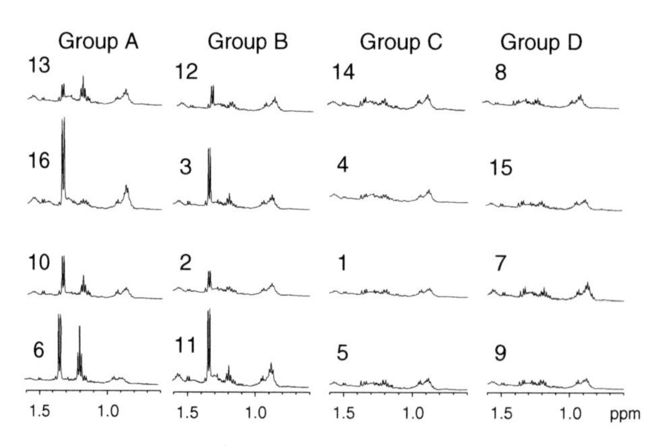

Figure 9.4. Individual urine sample 500 MHz NMR spectra (methyl region) grouped according to treatment. Groups: (A) Room temperature collection, frozen storage; (B) Room temperature collection $+NaN_3$, frozen storage; (C) frozen collection, frozen storage; (D) frozen collection $+NaN_3$, frozen storage. Large lactate signals in groups A and B indicate bacterial contamination.

signals in the proton NMR spectrum. The use of these various agents has not been thoroughly investigated using mass spectrometry. Storage conditions also likely have impact on biofluid metabolite profiles. While there is at least one published study that has investigated the effect of long-term storage parameters on the NMR spectra of blood plasma [59], a systematic examination of different storage conditions on various biofluids has not been carried out from a metabonomics perspective. Until such information is available, it is advisable to use −80 °C freezers for long-term storage and to keep samples cold while in the queue for analysis. Short-term storage should also be considered. Considerable time can be spent with samples on the bench while they are being prepared, or in a queue on the NMR or MS instrument awaiting analysis and provisions to keep these samples cold are prudent.

Clearly, exogenous influences must be carefully considered in study design. All too often, the effect of such agents is dismissed, primarily because traditional endpoints do not directly observe the impact of these as sensitively as does metabonomics. It is interesting to contemplate how much historical variability has been introduced in routine assessments by a lack of understanding of these effects. How much of the expected variability in urinary measurements of organic components is caused by not being aware of the effect of bacterial contamination that is so clearly discriminated using metabonomics? How many costly transcriptomics studies have unresolved outliers that are simply due to differences in gut microflora or post-sweet potato sampling? The answers to these questions may yet be a long time in coming, but metabonomics has, if nothing else, cast light on issues that have long been unknown or overlooked.

9.4. Preclinical efficacy

Metabonomics can enhance understanding of the biochemical defects of animal models of human diseases, leading to discovery of novel drug targets. Although the authors are unaware of published examples of the use of metabonomics to identify specific biochemical defects in animal models that then led to successful discovery of new drugs, it may be premature to expect such publications. Several authors have used metabonomics to characterize endogenous metabolites and metabolic pathways in mouse models, including the *Cln*3 knockout model of juvenile neuronal ceroid lipofuscinosis (Batten's disease) [60] and in several models of cardiac disease [61]. A large-scale multi-institutional project (Biological Atlas of Insulin Resistance) to identify new targets for type-2 diabetes mellitus has been initiated in the United Kingdom. This project will use metabonomic, transcriptomic, and proteomic analyses of diabetic mouse models as well as diabetic human subjects [62, 63].

One potential contribution of metabonomics to preclinical pharmaceutical discovery would be identifying markers of efficacy. Markers could be either single molecules or combinations of multiple molecules with a particular ratio or pattern. Because metabonomics assesses so many small molecules in a non-biased way and then uses chemometrics techniques to identify patterns, it has considerable potential for identifying markers of efficacy. Once identified, these markers could then be identified in a higher throughput way by specifically targeted screening assays.

Another potential contribution would be enhanced understanding of mechanisms of efficacy. As the pharmaceutical industry shifts to discovering targets for complex, multifactorial diseases, it is apparent that the efficacy of new drugs is often due to actions on a variety of pathways and complex interactions among those pathways. Potentially, metabonomics could provide mechanistic insight into why a drug is efficacious, and this understanding could facilitate the discovery of better drugs in the future. Metabonomic analysis of knockout or knockdown mice, which simulate the diverse *in vivo* effects of an inhibitory drug, may be valuable in this regard.

To date, however, metabonomics has played a limited role in understanding the basis of drug efficacy or developing successful efficacy screens, although there are a few publications already. For example, Watkins *et al.* examined the lipid changes induced in (NZO × NON)F1 obese and diabetic mice by the PPAR-gamma agonist rosiglitazone, used for treatment of type-2 diabetes mellitus [64]. Rosiglitazone increased heart cardiolipin, a phospholipid of the inner mitochondrial membrane that increases electron transport efficiency. Interestingly, however, rosiglitazone increased hepatic *de novo* lipogenesis, which was not detected by the standard markers of lipid metabolism, but was identified by the metabonomic analysis. These metabonomic findings could be useful in differentiating efficacious from toxic effects. A PPAR-gamma agonist (or many other compounds in the metabolic therapeutic area) might cause weight loss, and it would be important to distinguish weight loss due to efficacy from weight loss due to toxicity or reduced food consumption.

In an *in vitro* study of the effects of compounds on ionotropic glutamate receptors, Rae *et al.* [65] examined the effects of receptor ligands on the metabonomic response of guinea-pig brain cortical tissue slices. Compounds that altered the ionotropic glutamate receptor by several distinct mechanisms were clearly differentiable, suggesting that this might have utility as an efficacy model.

An alternative approach is to use stable isotope-labeled molecules such as ^{13}C-labeled glucose and then follow their metabolic flux [66]. Using this method, the investigators identified an *in vitro* neoplastic phenotype that included increased and preferential utilization of glucose through the non-oxidative steps of the pentose cycle for nucleic acid synthesis, but limited Krebs cycle glucose oxidation and fatty acid synthesis. They then showed that a variety of anti-cancer drugs altered this neoplastic phenotype and suggested that the method could be used to screen new compounds. Unlike traditional techniques for measuring cell proliferation or apoptosis, this method can also identify specific enzymatic reactions in the metabolic processes and provide mechanistic information on the anti-proliferative action.

9.5. Preclinical toxicity

9.5.1. Principles

Metabonomics can play several roles in preclinical toxicity testing, including helping to design better studies, serving as an endpoint in toxicity screening assays, helping to understand mechanisms of toxicity, and identifying biomarkers that can be useful in clinical studies.

Better study design is important to reduce the numbers of animals used in research, to reduce the expense of preclinical studies, and to avoid misleading results. One confounder of small preclinical studies is the existence of outlying data, often due to inclusion in a study of animals that have pre-existing physiologic differences or chronic, underlying diseases that are different from the other animals in the group. This is a particular problem in studies with non-human primates, which are more genetically and behaviorally diverse than laboratory rodents and sometimes carry underlying diseases or parasites. Additionally, non-human primates are often used repeatedly in a variety of drug trials, with the assumption (not always true) that they return to a naive physiology after several weeks without drug treatment. To make matters worse, group size with non-human primates is often smaller than that with rodents, enhancing the possibility that a single outlier can alter the interpretation of data.

Urinary metabonomics is useful to identify relevant time points for studies. If urine is collected continuously from animals in metabolic cages, then samples are collected during peak toxicity, regardless of when that peak occurs. In contrast, in conventional studies, serum, biopsy, or necropsy samples are collected at discrete time points, often set by tradition or policy rather than relevant biology (which may not be known in early studies). In some cases, conventional toxicology studies have completely missed important toxicologic findings that occurred during periods before the pre-set serum sampling times (see Section 9.5.4). Once the optimal sampling times have been identified, further studies using serum, toxicogenomics, or necropsy endpoints can be better planned. Another source of confusion and wide data variation comes because of the varying time courses of individual animals in reaction to a toxin. Some individuals experience tissue injury early and then undergo reparative processes, while in others, tissue injury occurs later. At a specific time point, the population includes individuals with early injury, late injury, and tissue repair. Accordingly, at a specific time point, analytes such as alanine aminotransferase, used to monitor hepatic damage, can be highly variable and require larger numbers of animals to demonstrate statistically significant effects. One technique to sort out this problem is trajectory analysis, in which the metabonomics of each individual is graphed against time. After correction for different metabonomics starting points, different time-courses, and different magnitudes of effect, the shapes of each individual trajectory curve can be compared [67].

Metabonomics data can be used as an endpoint in toxicology experiments. In some cases, metabonomics can be used as an efficient toxicity screen in preclinical discovery toxicology. This is discussed in more detail in Section 9.5.2. Applications of metabonomics to identify biomarkers for clinical studies are discussed in Section 9.5.3. Perhaps one of the most useful aspects of metabonomics, especially as the ability to identify NMR and MS peaks improves, is to understand mechanisms of toxicity. This is discussed in Section 9.5.4. To date, the number of studies that simply identify changed concentrations of metabolites (or simply changes in spectral patterns without regard for metabolite identification) far exceeds the number of studies that use these changes to understand the mechanisms of toxicity. This will be a major challenge in the future, but, if successful, may be one of the major strengths of metabonomics. Understanding mechanistic toxicity on a biochemical level can lead to revised, more focused testing plans to avoid specific mechanisms of toxicity. Especially valuable to the pharmaceutical industry, mechanistic understanding can explain species-specific adverse effects and lead to knowledgeable, evidence-based risk assessment to predict if these adverse effects will occur in humans.

One of the fundamental challenges of metabonomics is to integrate results, not only across technologies such as NMR and LC-MS but also with the other omics, including transcriptomics, proteomics, and the emerging omics such as glycomics, lipidomics, and others, including interactomics [68]. Changes in endogenous metabolites, either in single molecules or in combinations, are the end-result of a number of changes in levels of mRNA and proteins, together with post-translational changes in proteins. Sometimes, the metabolite changes alone can provide critical insight, but in other cases, especially due to multiple convergent pathways with common endpoints, these changes can be less informative or even misleading.

Technologic improvements will improve our ability to detect the low-concentration metabolites that might play essential roles in biological processes. However, increased sensitivity will bring its own set of problems, including spurious peaks. In a recent study using ultra-high-performance liquid chromatography (UPLC)-MS metabonomics, we found approximately 15,000 peaks, which posed an enormous interpretive problem. To make matters worse, many of the peaks were artifacts caused by adducts formed during the ESI procedure (M. Sanders, personal communication).

One solution to the problem will come from improved chemometrics techniques. Statistical correlative techniques, including statistical total correlation spectroscopy (STOCSY) and statistical heterospectroscopy (SHY), will be especially helpful [63, 69]. STOCSY, in combination with orthogonal projection on latent structure-discriminant analysis (O-PLS-DA), takes a particular peak representing a known metabolite and then identifies other peaks that change in a tightly correlated way with the original peak. Many of the peaks are from the same molecule, of course, and this can be useful in sorting out the multiplicity of peaks. Others are peaks from other molecules in the same pathways. In this way, biologically important peaks that are small and easily confused with background noise can be identified, making pathway analysis much more feasible. SHY, a newly published technique, helps to integrate various techniques. The technique statistically correlates changes in one platform, such as NMR, with changes in another, such as UPLC-MS. This can provide useful confirmation of the robustness of the study as well as help to identify molecules that are only partially characterized by one technique or the other. In principle, SHY can be used to link results from a variety of omics technologies, and thereby enhance overall interpretability of metabonomics data.

Improved sensitivity in analytical techniques, especially MS-based ones, together with the major improvements in chemometrics techniques may also help address one of the major challenges of metabonomics in preclinical toxicology, the gap between analysis of highly toxic model compounds and the more subtly toxic compounds discovered in the pharmaceutical industry. Numerous studies use metabonomics to analyze prototypic toxins, but there are fewer short-term studies that detect and usefully model the subtle long-term effects of pharmaceutical compounds that may look promising at first, but later fail during development.

9.5.2. Screening

9.5.2.1. Background

Toxicity screening was one of the earliest applications envisioned for metabonomics technology [70, 71]. In particular, the application of the technology to rapid throughput *in vivo* toxicity assessment (screening) was an early attraction of the technology [43]. This appeal was particularly timely as the hope for a generic *in vitro* safety screen had diminished markedly. The advantages of *in vitro* screening over *in vivo* screening are significant, including minimal drug requirements and the ability to screen in human cells. Additionally, cellular toxicity studies meant for ferreting out specific biochemical effects in single cell types can be quite powerful. For example, there are a number of approaches (both cellular and non-cellular) for investigating various facets of hepatotoxicity [72]. However, the value of comprehensive *in vitro* screens in routine safety assessment has proved elusive. While human cells would seem to be more reflective of human effects, immortalized cell lines (of any species) frequently respond quite differently from primary cells and the latter are frequently difficult to obtain (particularly human), culture, and maintain (in their differentiated state) for any reasonable period of time. While a few *in vitro* screens for target toxicity endpoints (like genetic toxicity, phototoxicity, dermal irritation, etc.) are routinely employed within the industry [73], a comprehensive *in vitro* safety screen suitable for candidate selection or rejection (prior to *in vivo* studies) has not been forthcoming. All too frequently, endpoints in cultured cells (primary or cell lines) have little translatable meaning to *in vivo* endpoints. Further, the drug concentrations at which measures of cytotoxicity are identified frequently bear little, if any, relationship to *in vivo* exposures. This last point highlights a misunderstanding of the role of safety assessment that is frequently made by non-toxicologists who assume that the role of a toxicologist is simply to identify any and all toxicities associated with a compound. Their view of toxicity is as a "present" or "absent" phenomenon. While this is appealingly simple, it flies in the face of basic biology ignoring the primary tenet of toxicology that Paracelsus identified almost 500 years ago – "the dose makes the poison". Almost every compound developed within the industry is toxic at some dose, what toxicologists need to identify is the therapeutic index, the margin between an efficacious dose and the toxic dose. This is extremely difficult to do *in vitro*. For these reasons, outside the target toxicities like those indicated above, *in vitro* screens have had limited impact within routine pharmaceutical safety assessment and will only be dealt with peripherally.

A significant problem when dealing with safety screening as a subject is that it means different things to different people. The simpler the target(s) being screened against, the simpler the screen. For example, screening a compound for hepatotoxicity (and only hepatotoxicity) is much easier (relatively speaking, of course) than trying to screen for all potential target organs. Additionally, use of screens to predict within a species is much simpler than trying to extrapolate to other species (usually to humans). While one might question the need for screening for rat toxicity, the ability to predict toxicity within rats using an efficient and rapid screen can be extremely useful for reducing the number of expensive chronic GLP studies required. The savings is not just the cost of the studies themselves, but the cost, in terms of both time and money, of all the associated preclinical work occurring concurrently as the drug is made ready for the clinical setting. However, comprehensive safety screens are just one part of the screening picture, indeed they are the least developed part. Focused screens for specific toxicities or target organs are much more amenable to higher throughput formats. Some *in vitro* applications to endpoint specific toxicities have already been mentioned. The use of these focused screens is quite common in pharmaceutical development and employs a variety of *in vitro* and *in vivo* approaches. The typical scenario in which a screen is developed is after a discovery team has identified a dose-limiting toxicity with a lead compound in early pharmacology or safety testing. There then is a strong desire to screen potential back-ups to find compounds that either do not have the problem or at least for which there is an adequate therapeutic index. Obviously, if this can be done *in vitro*, it will be – as that is the quickest and cheapest way to conduct such studies and requires almost negligible bulk drug. However, many toxicities do not lend themselves to *in vitro* assessment, for reasons mentioned above. In these cases, a rapid method for assessing the toxicity with limited animals, and without

Table 9.1
Types of safety assessment screens

Screen	Speed	Cost	*In vitro*?	Metabonomics?	Comment
Traditional Safety Study	Slow	High	No	Yes	The gold standard for safety assessment accepted by all regulatory agencies
Comprehensive[a] Rapid Throughput Screen	Moderate (*in vivo*)	Moderate	No[b]	Yes	Not suitable for regulatory submissions, but useful for internal chemo type assessment or lead prioritization.
Target Organ Assessment	Moderate (*in vivo*)	Moderate	Yes	Yes	*In vitro* is the choice if possible, but many toxicities do not have suitable *in vitro* screens
Endpoint (mutagenicity, phototoxicity, irritation, etc)	Fast	Low	Yes	No[c]	Several *in vitro* screens have been accepted for regulatory submissions, but gaining acceptance for new screens is a challenge.

Notes:
[a] Suitable for multi-target organ assessment and generation of therapeutic indices.
[b] Though claims have been made for *in vitro* screens suitable for comprehensive safety assessment, there has been little acceptance of any to date.
[c] Metabonomics may indeed be applicable to some endpoint screens, but given the speed and low cost of *in vitro* assays, there seems little point to add metabonomics to accepted *in vitro* screens. However, newer screens may make use of metabonomics or other omic technologies.

the need for time-consuming and expensive histopathology represents a real benefit. Table 9.1 presents a brief summary of several types of safety screens as compared to traditional safety studies.

9.5.2.2. Comprehensive safety screening

The ability to take a novel therapeutic, for which little if any *in vivo* data is available, and screen it for safety endpoints across all, or at least multiple, target organs would represent the pinnacle of success for a comprehensive screen. It is also highly improbable. While several *in vitro* approaches employing different cell types (animal or human) under various exposure regimens have been proposed as a way to perform comprehensive screens *in vitro*, they have gained little acceptance within the industry. Additionally, most *in vitro* assays lack the capacity to mimic the pharmacokinetic component of *in vivo* studies. This is particularly important as a primary tool for translating preclinical safety data to human risk assessment is via pharmacokinetic parameters such as Cmax, AUC, $T_{1/2}$, volume of distribution, and so on. Therefore, comprehensive screening has come down to new ways of speeding up – or making more sensitive – traditional safety studies. This goal maybe achieved if the assay can predict longer-term toxicities from shorter-term studies (saving time), higher dose effects from lower dose studies (saving bulk drug synthesis time and cost) or from fewer animals (saving bulk and reducing animal usage). If sensitivity is increased, toxicities that may be completely missed in pharmacology or early safety studies may be identified, driving necessary compound attrition earlier into the discovery process saving significant time and development costs for a compound that was doomed to failure from the start. The most time-consuming and expensive exercise of a traditional toxicity study is histopathology. Therefore, for a comprehensive screen to be faster and cheaper than a traditional safety study, they typically tend to reduce or eliminate histopathology while attempting to retain a similar definitive endpoint. However, histopathology is also the toxicity "gold standard" by which all other endpoints are judged. "Omics" technologies have generated a lot of interest in the toxicology community because of the promise of rapid turnaround of data that would potentially be more sensitive and specific to mechanism. Though this was a bold promise, it has yet to be fully realized for any omic application and metabonomics is no exception. The reasons for this are many, and those issues relevant to metabonomics are dealt with below.

Metabonomics-derived comprehensive safety screens are based on two different approaches. The database approach also known as the classification model approach seeks to establish a spectral database of biofluids from animals treated with known toxins or with defined adverse physiological effects. In theory, the biofluid spectral data from animals treated with a test compound can be compared to the database using one of the several statistical methods and conclusions drawn as to the nature and severity of the observed toxicity. The spectral data comparison can be conducted using chemometric procedures without the need for identification of every peak in the spectrum. While this approach can be very fast, it is based on a prerequisite database, the creation of which is not a trivial undertaking. An alternative approach analyzes the analytical data from control and treated animals, identifying biomarkers associated with the toxicity of interest. The nature of these biomarkers can then be used to generate mechanism-based hypotheses. While, the identification of spectral components is always preferable to not having them, they come at a significant cost of time and human resources involved in generating the identifications.

Database-derived comprehensive toxicity screens: The concept of a metabonomics database screen, also known as a classification model screen, is actually quite straightforward. In simple terms, rats undergoing significant physiological disruption from toxicity, disease, environment, or other factors (age, estrus cycle, etc.) have altered metabolic profiles reflective of their condition and measurable in the various biofluids percolating through the tissues of their bodies. The term "metabolic space" has been used to define the range of variation of the comprehensive profile of all metabolites within an organism. Metabolic space is typically defined using chemometric procedures described elsewhere in this book. Figure 9.5 shows a PCA scores plot of urine NMR spectral data from rats treated with two hepatic and two renal toxins. The separation of the data points of control and treated animals and between rats with different treatments is indicative of metabolic differences reflected in the urine of the animals at the time the samples were collected. Importantly the variation within treatments is much less than the variation between treatments. Also of note is the fact that the two liver toxins separate in metabolic space, as do the two renal toxins, suggestive of differing pathophysiology (which is certainly the case in this instance). In other words, the samples vary in metabolic space according to treatment, which is readily monitored by metabonomics. If one were to examine the loadings driving the principal component (PC) separation one would find that variations in several common metabolites drive these separations – hence the concept

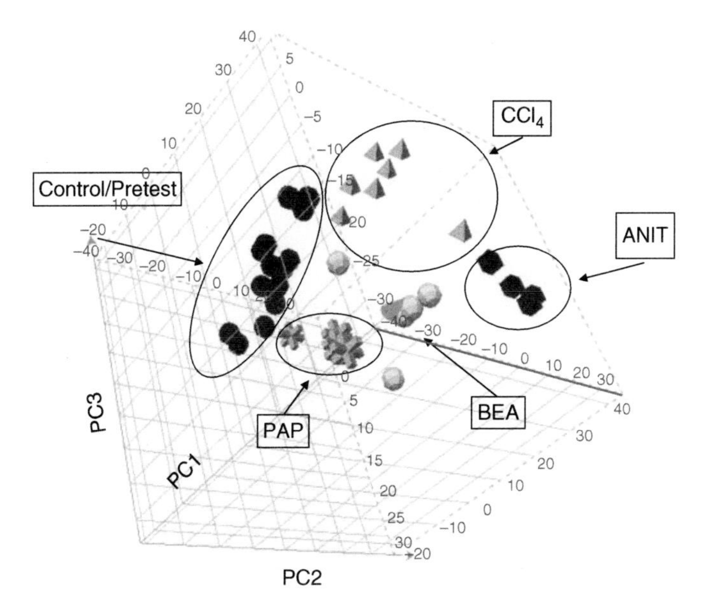

Figure 9.5. A PCA scores plot from urinary NMR data from rats treated with two hepatotoxins (CCl_4 and α-naphthylisothiocyanate (ANIT)) and two different nephrotoxins, bromoethanamine (BEA) and para-aminophenol (PAP).

of the pattern biomarker. This example demonstrates a simplified view of how one might be able to "map out" metabolic space in order to localize samples from treated animals to hepatic toxicity space or renal toxicity space without necessarily requiring metabolite identifications. There are various statistical ways to perform such a task, which are covered in other chapters in this book. Generation of such a database-oriented screen was one of the objectives of the COMET consortium [52, 74]. The effort undertaken to generate the database led to some hard-earned learnings as to the caveats one must consider when developing such a screen. Simply defining the bases for the classifications was a difficult task. Are spectral patterns more related to the associated target organ, the pathophysiology, or the mechanism of toxicity? Making these distinctions was crucial for generating a useful screen. When all was said and done, the product was less than hoped for, but the learnings were invaluable. The bottom line is that metabonomics-based toxicity screens, while probably as useful as any other rapid safety screen, need to be approached and used with caution. These screens are far from automatic with urine in one end and biomarkers and mechanisms popping out the other. There still is a huge and necessary role for the biochemical toxicologist in understanding and interpreting the results.

9.5.2.3. Target organ screens

While comprehensive screening with metabonomics still has a way to go, defined target screening using the technology has shown a great deal of promise. Target organ screening can be roughly divided into two major categories, screens for targets that already have well-established peripheral biomarkers of target toxicity (eg liver and kidney), and those for which peripheral markers have not been as well characterized (e.g. vascular bed, phospholipidosis, adrenal toxicity, etc.). Serum transaminases for hepatic toxicity or serum urea nitrogen or creatinine for renal toxicity are well-established peripheral biomarkers. If developing a screen for either renal or hepatic toxicity, a metabonomics approach would need to demonstrate a clear advantage over these universally accepted biomarkers. These advantages could be in terms of increased sensitivity or selectivity relative to classical serum markers or it may simply be a pragmatic advantage. An example of the latter is the use of urine to monitor hepatic toxicity in the mouse. Limited blood volume makes interim blood draws in the mouse highly problematic; therefore terminal blood samples are usually all that are available for monitoring serum biomarkers of toxicity. Mice produce ample urine for metabonomics studies within a 24-hour period [39]. Therefore metabonomic analyses enable daily monitoring for toxicity in mice without the need for interim necropsies. With regard to increased sensitivity of metabonomics relative to serum biomarkers, sensitivity need not be in terms of classic analytical sensitivity, but in this case may mean biological sensitivity. For example, in work conducted several years ago, Robertson *et al.* demonstrated that the "classic" biliary hepatotoxin α-naphthylisothiocyanate (ANIT) when given as a low single acute dose (10 mg/kg) produced clear metabolic disruption in all four dosed animals within 24 h, while histopathology (on Day 4) revealed changes in only two of four animals and clinical chemistry was uninformative [43]. In this case metabonomics was more "sensitive" in that evidence of altered physiology was evidenced in all four animals while traditional serum chemistry markers (serum bilirubin in this case) suggested no effect and even histopathology suggested only a 50% incidence of effect. This raises the question of whether altered metabolic profiles in and of themselves should be considered a toxic effect or not. This question is controversial, but any potential user should be aware of the ramifications before they embark on utilizing the technology in regulated studies.

With regard to selectivity, early work conducted at Imperial College in London suggested varying urinary metabolic profiles could distinguish localization of nephrotoxicity, papilla vs proximal tubule [75, 76]. In the liver, combinations of various urinary metabolites allowed for differentiation between biliary and parenchymal injury [77]. While this work was never adequately followed up, it does suggest that metabonomics may enable inter-organ localization of injury using a non-invasive peripheral sample, which would represent a clear advancement over traditional markers.

The advantages of metabonomic technology for screening toxicities for which there are no good peripheral biomarkers are more obvious. Vasculopathy is one such toxicity. Although there has been a substantial effort at deriving a peripheral biomarker of vascular toxicity [78], the goal is still illusive. Vascular effects are an all too frequent issue with many drug development programs. Given the focused and segmental nature of the typical drug-induced vascular lesion (non-immune mediated), it was surprising to find that the technology impressively differentiated animals with lesions from animals without lesions even within the same dose group [79]. While the effect was shown to be independent of the concurrent inflammation [80], the lack of specificity of the spectral changes hampers their use in anything but a screening paradigm. However, the metabolic changes must be associated with the lesions for a reason and if that can be elucidated, it may lead to a significant understanding of the complex biochemistry of the lesion. Peroxisome proliferation is another problematic issue in drug development particularly as interest has soared in compounds active at various subtypes of the peroxisome proliferator-activated receptor (PPAR). Connor *et al.* demonstrated metabonomics was a powerful tool for the assessment of peroxisome proliferation via analysis of rat urine samples using a multivariate statistical metabonomic model [81]. Metabonomics-derived screening applications for cardiovascular disease have been well represented in both the preclinical [61, 82] and the clinical literature [83–85]. Similarly oncology applications hold out the promise that the technology may demonstrate clinical utility for screening for this devastating disease [20, 86–88]. Additional areas of promise include applications in diabetes [89, 90], meningitis [91], and urinary tract infections

[92]. While disease screens are not the subject of this chapter, it stands to reason that any metabonomic application that is adequate for disease screening may prove useful for screening for untoward effects of drugs in the same target organs (e.g. drug-induced cancer or cardiovascular toxicity).

Even within target organ applications the use of metabonomics needs to be approached carefully as adequate validation studies are lacking. Gibbs conducted a careful review of several metabonomics and proteomics studies comparing them to traditional markers of safety with regard to specificity and sensitivity [93]. He arrived at the conclusion that while the new methods do a good job at discriminating various mechanisms of toxicity they have a long way to go before they can be considered validated. He also concluded that it was unclear if these new technologies would ever gain acceptance by regulatory agencies. While the former statement is certainly true, the latter observation is simply a reflection of the former. If metabonomics approaches are properly validated and the business case can be made for their use, metabonomics will certainly find its place along side other established technologies and regulatory agencies will be supportive of their development [94].

9.5.2.4. The "normal" rat

One screen that has proven extremely useful in metabonomics applications is the "normal rat". Defining normal is always a tricky exercise, but metabonomics readily defines a multivariate data set from which the biochemical description of normality can be statistically defined. Figure 9.6 shows a representation of a statistically defined normal model derived from data collected by the COMET consortium [52, 74] against which a set of control and pretest urine sample data collected from identical studies conducted in two sister toxicology laboratories (one in Europe and one in the US) was applied. While every effort was made to duplicate the studies identically at the two laboratories, it was clear that one laboratory produced samples that were clearly consistent with the COMET-derived normal model while the other laboratory did not. All traditional measures of toxicity (clinical signs, clinical chemistry, and histopathology) were unremarkable (at least in control animals). Interestingly, the animals from the "abnormal" laboratory were less severely affected by the compound under investigation in the study. Was this because there was a basic biochemical difference in those animals? It was impossible to say from that one data set. However, it raised the intriguing question of inter- and intra-laboratory phenotypic variation in laboratory animal populations and understanding what effect that may have on study results.

The role of gut flora in normal health, metabolism, and (de)toxification of exogenous compounds has been recognized for some time [95–98]. More recently it was recognized that variations in gut flora could arise within animal colonies causing profound effects on urinary metabolic profiles [55, 99]. Robosky *et al.* went on further to demonstrate that altered gut flora could be characteristic of particular colonies from commercially supplied rats leading to profound yet stable differences

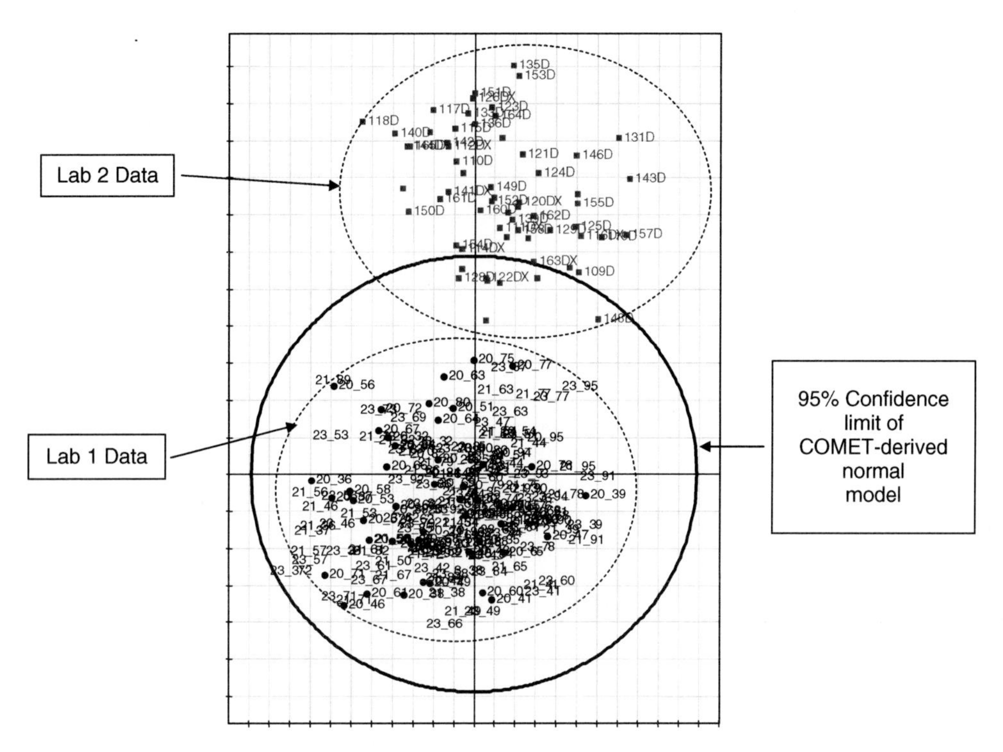

Figure 9.6. Comparison of control/pretest rat urine sample from two laboratories conducting identical experiments using a COMET-derived normal model. All sample from Lab 1 were within the 95% confidence limit of the normal model, while almost all samples from Lab 2 fell outside the normal range. Figure Courtesy of Dr D. Baker.

in urinary metabolic profiles [54]. It is unclear how these variations in gut flora may have confounded animal studies over the years, but the important point is that this variable is now recognized and can be easily monitored for using metabonomics technology and the normal rat model.

9.5.3. Safety biomarkers

There is significant overlap in the literature between work conducted for screening purposes and work conducted to identify biomarkers. After all, in the preclinical setting, one of the main reasons for identifying safety-related biomarkers is for screening purposes. That literature will not be re-examined here. Instead the focus will be higher level in detail.

Like screens, the subject of biomarkers means different things to different people. To some nothing is a biomarker unless it is clearly mechanistically linked and analytically validated. On the other end of the spectrum are those that believe anything that can be statistically associated with an effect, whether it is understood or not, represents a biomarker. With metabonomics the latter type of biomarkers are easy to generate, the former quite a bit more difficult. In practice the validation required for biomarkers depends largely on its end use. If the marker is to be used as a tool to screen preclinical compounds for a particular toxicity, prior to full-scale safety studies, then validation need only consist of those steps necessary to convince the users that there is value added to using the biomarker. On the other hand, a biomarker proposed as a decision-making criterion for use in human clinical trials requires extensive validation, both biologic and analytical. To date, no metabonomics-derived biomarker has met the latter standard. However, that could be said of most other technologies, as validation of a clinical biomarker is a difficult process to complete. Therefore, the role of metabonomics has been largely confined to early stages of safety biomarker discovery and development.

One of the advantages of metabonomics analyses is they are inherently multiplex in nature. The advantage of this is that a pattern of metabolic changes may be more specific than any single molecule would be. For example, physicians and veterinarians have long known that changes in serum sodium can be due to many disorders, but when the ratio of serum sodium to potassium is increased, hyperadrenocorticism (Cushing's disease) is suspected. Conversely, when the ratio of serum sodium to potassium is decreased, adrenal insufficiency (Addison's disease) is suspected. Similarly, observation of a fever in a patient may not be specific enough for a physician to diagnose something in that individual, but combination of that non-specific finding with three or four other similarly non-specific findings may be much more informative. Similarly a pattern of three or four otherwise non-specific molecules may be useful for identifying a specific toxicity.

While this sounds great in theory, it has proved more problematic in practice. While the concept is understood, pattern biomarkers have not been identified to any great extent and there exist plenty of skeptics. The observation that many of the same metabolites change in response to toxins, regardless of the type of toxin or mechanism of action, has jaded many a toxicologist with regard to the technology. These commonly changing metabolites have been come to be known as the "usual suspects" in metabonomics investigations [11]. While the commonality of these changes has been attributed to weight- and diet-related changes [100], diet is not the only reason the "usual suspects" are usual. Instead, a careful understanding of these molecules may prove that they are quite useful as biomarkers though non-specific in nature. This work has only been conducted to a limited extent with things like urinary creatine [101, 102] and urinary taurine [103–105]. Clearly there is much work yet to be conducted in this arena.

One final note is that the use of MS technology for metabonomics (described elsewhere in this book) greatly expands biomarker capability of the technology relative to NMR. In particular, the probability of finding a unique safety-related biomarker (everybody's favorite) increases dramatically. Unfortunately, the use of MS greatly complicates the validation question as the range of metabolites (many unidentified) associated with any toxicity can expand by orders of magnitude.

9.5.4. Mechanisms

Potentially one of the major uses of metabonomics will be enhanced understanding of mechanisms of toxicity. The ability of metabonomics to simultaneously examine changes in many metabolites and the use of chemometrics techniques to summarize or query data make metabonomics particularly suitable for suggesting new hypotheses about mechanisms. While new testable mechanistic hypotheses are very much needed, there have been relatively few studies published to date. The proprietary nature of pharmaceutical industry work suggests that there may be unpublished mechanistic work, but it is clear that this is an underutilized aspect of metabonomics research and a major challenge for the future.

Examples from three different pharmaceutical companies illustrate uses of metabonomics. One example is Mortishire-Smith *et al.*'s analysis of a chemokine receptor 5 antagonist that was found to be hepatotoxic [106]. The metabonomics analysis showed depletion in Krebs cycle intermediates and the appearance of medium chain dicarboxylic acids, which are not normal constituents of urine. The finding of urinary dicarboxylic aciduria suggested the hypothesis of drug-induced impairment of fatty acid metabolism; the hypothesis was subsequently confirmed by other experiments, in which beta-oxidation of [^{3}H]palmitic acid to shorter chain fatty acids was decreased in a dose-dependent manner in *ex vivo* liver slices. This occurred at concentrations of the compound much less than those associated with overt toxicity.

Recently, investigators used metabonomics as part of the risk assessment of muraglitazar, a PPAR-alpha and PPAR-gamma agonist that caused dose-dependent increases in crystalluria and increased urinary bladder papillomas and transitional cell carcinomas that were specifically localized to the ventral portion of the urinary bladder of male rats. The critical question was whether the compound would cause similar neoplasia in humans. Urinary analyses followed by urinary metabonomics found decreases in urinary divalent cations and citrate. This led to the hypothesis that decreased citrate concentration could predispose to the nucleation of urinary calcium phosphate crystals, which could gravitate to the ventral portion of the bladder and cause chronic tissue injury. The metabonomics findings were verified by specific urinary clinical chemical assays. Subsequent studies confirmed that this urinary bladder carcinogenic effect was mediated by urolithiasis, urothelial injury, and compensatory regenerative urothelial hyperplasia, rather than a direct effect on urothelium [107, 108].

Metabonomics can also be useful in understanding mechanisms of adverse effects at later stages of drug research. In a set of studies, metabonomics was used to help understand the mechanistic differences between marketed HIV protease inhibitors (PI) that caused lipodystrophy, insulin resistance, and fat redistribution in human patients and those HIV PI that did not cause these effects [109]. Cultured 3T3-L1 adipocytes were exposed to each compound and cellular metabolites were analyzed by GC- or LC-MS. Of 193 metabolites identified and characterized, the intracellular concentrations of 81 (42%) were significantly reduced by one or more of the PI tested. The metabolites were sugars and their phosphates, amino acids (including all gluconeogenic precursors), glutathione (derived from glutamate), fatty acids and pantothenic acid (essential for fatty acid synthesis), intermediates of the citrate cycle (ATP generation), and a ketone. The drugs associated with clinical lipodystrophy inhibited 24 to 58 metabolites, while the drug not associated with lipodystrophy only altered 3 metabolites. The pattern of metabolite reductions induced by some protease inhibitors is indicative of glucose deprivation with a shift toward gluconeogenesis and glycolysis, together with reductions in intermediate metabolites of lipid synthesis. This may contribute to the lipoatrophy seen as part of the lipodystrophy syndrome. Additional work, including transcriptomics analysis and biochemical studies, suggests that this effect is due to inhibition by some of the HIV protease inhibitors of GLUT4, the transporter responsible for glucose uptake by adipocytes [110, 111].

Metabonomics has been used extensively, together with other omics, in ongoing investigations of mechanisms of vascular injury by phosphodiesterase-4 (PDE-4) inhibitors as described in Section 9.5.2.3. While a clear mechanistic link between urinary metabolite changes and mesenteric vascular changes has not yet been made, the fact that they occur and differentiate animals with lesions from those without them has opened new avenues of mechanistic investigation.

In other studies, a PI-3 kinase inhibitor caused unexplained deaths of rats at high doses. Metabonomics studies found severe and immediate increases in urinary glucose, followed by increases in beta-hydroxybutyrate. This occurred at early time points and was not detected in the conventional toxicity studies that used pre-selected time points for sample collection at later times [112].

Presently, the COMET-2 consortium has formed with the goal of using metabonomics to understand mechanisms of toxicities. The consortium includes investigators from Imperial College, London, Pfizer, Bristol-Myers Squibb, Sanofi-Aventis, and Servier. At present, the group is investigating mechanisms of renal papillary necrosis and of individual variation in hepatotoxicity that cannot be explained by exposure, cytochrome P450, or transporter differences.

9.6. Conclusions

Metabonomics has made significant inroads into the arena of preclinical safety assessment. However, it still has a long way to go before it can be considered standard practice. The role that metabonomics plays within the preclinical setting of the future is not clear and different companies may employ the technology to different ends. Screening was an obvious "quick win" at one point in time but that promise, or at least the speed of adoption, has since diminished. However, the power of the technology to help us understand our animal models was largely unappreciated several years ago and can now be considered a significant deliverable. The role of the technology in understanding mechanisms was seen as a significant advantage of the technology early on and while headway has been made a great deal of promise remains unfulfilled. In short, the technology has moved into the crucial phase of value determination. Some pharmaceutical companies have made and continue to make significant investments into the technology while others are taking a more conservative approach and yet others have adopted a wait-and-see attitude. This should not be considered unusual, as most technologies undergo a similar fate. What is unusual about preclinical metabonomics is that the pharmaceutical industry was an early adopter of the technology and academia is now only starting to catch up. This has lead to a bit of a catch-22 type situation where much of the most significant metabonomics research remains unpublished, because of intellectual property concerns by industrial users, leading to the possible erroneous conclusion by those on the fence with regard to adoption of the technology, that it is of fading interest in the industry. With the NIH road map initiative recognizing metabonomics/metabolomics technology development as a significant consideration for funding [113], we can anticipate another explosion of research, this time academia initiated, that will accelerate the implementation and adoption of the technology on a much broader basis.

References

[1] Lindon, J.C., E. Holmes, and J.K. Nicholson, Metabonomics and its role in drug development and disease diagnosis. Review 32 refs. *Expert Review of Molecular Diagnostics*, 2004. **4**(2): pp. 189–99.

[2] Lindon, J.C., E. Holmes, and J.K. Nicholson, Metabonomics: Systems biology in pharmaceutical research and development. *Current Opinion in Molecular Therapeutics*, 2004. **6**(3): pp. 265–72.

[3] Nicholson, J.K. *et al.*, Metabonomics: A platform for studying drug toxicity and gene function. Review 86 refs. *Nat. Rev. Drug Discov.*, 2000. **1**(2): pp. 153–61.

[4] Nicholson, J.K., J.C. Lindon, and E. Holmes, "Metabonomics": Understanding the metabolic responses of living systems to pathophysiological stimuli via multivariate statistical analysis of biological NMR spectroscopic data. *Xenobiotica*, 1999. **29**(11): pp. 1181–9.

[5] Robosky, L.C. *et al.*, In vivo toxicity screening programs using metabonomics. *Combinatorial Chemistry & High Throughput Screening*, 2002. **5**(8): pp. 651–62.

[6] Shockcor, J.P. and E. Holmes, Metabonomic applications in toxicity screening and disease diagnosis. *Current Topics in Medicinal Chemistry*, 2002. **2**(1): pp. 35–51.

[7] Harrigan, G.G., D.J. Brackett, and L.G. Boros, Medicinal chemistry, metabolic profiling and drug target discovery: A role for metabolic profiling in reverse pharmacology and chemical genetics. *Review 68 refs. Mini-Reviews in Medicinal Chemistry*, 2005. **5**(1): pp. 13–20.

[8] Griffin, J.L. and M.E. Bollard, Metabonomics: Its potential as a tool in toxicology for safety assessment and data integration. *Curr. Drug Metab.*, 2004. **5**(5): pp. 389–98.

[9] Nebert, D.W. and E.S. Vesell, Advances in pharmacogenomics and individualized drug therapy: Exciting challenges that lie ahead. *Eur. J. Pharmacol.*, 2004. **500**(1–3): pp. 267–80.

[10] Valigra, L., Metabolic Profiling: Meet the Latest "Omics". *Drug Discov. Develop.*, 2004. **December**: pp. 39–41.

[11] Robertson, D.G., Metabonomics in toxicology: A review. *Toxicol. Sci.*, 2005. **85**(2): pp. 809–22.

[12] Stoughton, R.B. and S.H. Friend, How molecular profiling could revolutionize drug discovery. *Nat. Rev. Drug Discov.*, 2005. **4**(4): pp. 345–50.

[13] Baker, M., In biomarkers we trust? *Nat. Biotechnol.*, 2005. **23**(3): pp. 297–304.

[14] Pognan, F., Genomics, proteomics and metabonomics in toxicology: Hopefully not "fashionomics". *Pharmacogenomics*, 2004. **5**(7): pp. 879–93.

[15] Bilello, J.A., The agony and ecstasy of "OMIC" technologies in drug development. *Curr. Mol. Med.*, 2005. **5**(1): pp. 39–52.

[16] Witkamp, R.F., Genomics and systems biology–how relevant are the developments to veterinary pharmacology, toxicology and therapeutics? *J. Vet. Pharmacol. Ther.*, 2005. **28**(3): pp. 235–45.

[17] Luhe, A. *et al.*, Toxicogenomics in the pharmaceutical industry: Hollow promises or real benefit? *Mutat. Res.*, 2005. **575**(1–2): pp. 102–15.

[18] Griffin, J.L. *et al.*, Spectral profiles of cultured neuronal and glial cells derived from HRMAS (1)H NMR spectroscopy. *NMR Biomed.*, 2002. **15**(6): pp. 375–84.

[19] Griffin, J.L. *et al.*, Cellular environment of metabolites and a metabonomic study of tamoxifen in endometrial cells using gradient high resolution magic angle spinning 1H NMR spectroscopy. *Biochim Biophys. Acta,* 2003. **1619**(2): pp. 151–8.

[20] Griffin, J.L. and J.P. Shockcor, Metabolic profiles of cancer cells. *Nat. Rev.* Cancer, 2004. **4**(7): pp. 551–61.

[21] Anthony, M.L. *et al.*, 1H NMR spectroscopic studies on the reactions of haloalkylamines with bicarbonate ions: Formation of N-carbamates and 2-oxazolidones in cell culture media and blood plasma. *Chem. Res. Toxicol.*, 1995. **8**(8): pp. 1046–53.

[22] Bailey, N.J. *et al.*, Metabolomic analysis of the consequences of cadmium exposure in Silene cucubalus cell cultures via 1H NMR spectroscopy and chemometrics. *Phytochemistry*, 2003. **62**(6): pp. 851–8.

[23] Hassel, B., U. Sonnewald, G. Unsgard, and F. Fonnum, NMR spectroscopy of cultured astrocytes: Effects of glutamine and the gliotoxin fluorocitrate. *J. Neurochem.*, 1994. **62:** pp. 2187–2194.

[24] Sugiura, Y. *et al.*, Cadmium exposure alters metabolomics of sulfur-containing amino acids in rat testes. *Antioxid Redox Signal*, 2005. **7**(5–6): pp. 781–7.

[25] Verhoeckx, K.C. *et al.*, A combination of proteomics, principal component analysis and transcriptomics is a powerful tool for the identification of biomarkers for macrophage maturation in the U937 cell line. *Proteomics*, 2004. **4**(4): pp. 1014–28.

[26] Boros, L.G., M. Cascante, and W.N. Lee, Metabolic profiling of cell growth and death in cancer: Applications in drug discovery. *Drug Discov. Today*, 2002. **7**(6): pp. 364–72.

[27] Fiehn, O., Combining genomics, metabolome analysis, and biochemical modeling to understand metabolic networks. *Comp. Func. Genom.*, 2001. **2**: pp. 155–168.

[28] Brown, S.C., G. Kruppa, and J.L. Dasseux, Metabolomics applications of FT-ICR mass spectrometry. *Mass Spectrom. Rev.*, 2005. **24**(2): pp. 223–31.

[29] Reily, M.D. and J.C. Lindon, NMR Spectroscopy: Principles and instrumentation, in *Metabonomics in Toxicity Assessment*, J.C. Lindon, D.G Robertson, E. Holmes and J.K. Nicholson, Eds 2005, Taylor & Francis: Boca Raton, FL, USA, pp. 75–104.

[30] Lindon, J.C. *et al.*, Metabonomics technologies and their applications in physiological monitoring, drug safety assessment and disease diagnosis. *Biomarkers,* 2004. **9**(1): pp. 1–31.

[31] Lindon, J.C., J.K. Nicholson, and J.R. Everett, NMR spectroscopy of biofluids. *Ann. Rep. NMR Spect.*, 1999. **38**: pp. 1–88.

[32] Fernie, A.R. *et al.*, Metabolite profiling: From diagnostics to systems biology. *Nat. Rev. Mol. Cell. Biol.*, 2004. **5**(9): pp. 763–9.

[33] Harrigan, G., Metabolic profiling: Pathways in drug discovery. *Drug Discovery Today*, 2002. **7**(6): pp. 351–2.

[34] van der Greef, J., R. van der Heijden, and E.R. Verheij, The role of mass spectrometry in systems biology: Data processing and identification strategies in metabolomics. *Advances in Mass Spectrometry*, 2004. **16**: pp. 145–165.

[35] Robertson, D.G., M.D. Reily, and J.D. Baker, Metabonomics in Preclinical Drug Development. *Expert Opin Drug Metab. Toxicol.*, 2005. **1**(3): pp. 363–376.

[36] Nicholson, J.K., M.J. Buckingham, and P.J. Sadler, High resolution 1H n.m.r. studies of vertebrate blood and plasma. *Biochem. J.*, 1983. **211**(3): pp. 605–15.

[37] Bales, J.R. *et al.*, Use of high-resolution proton nuclear magnetic resonance spectroscopy for rapid multi-component analysis of urine. *Clin. Chem.*, 1984. **30**(3): pp. 426–32.

[38] Fiehn, O., Metabolomics–the link between genotypes and phenotypes. *Plant Mol. Biol.*, 2002. **48**(1–2): pp. 155–71.

[39] Robertson, D.G. *et al.*, Metabonomic Technology as a Tool for Rapid Throughput In Vivo Toxicity Screening, in *Comprehensive Toxicology*, J.P. Vanden Heuvel, G.J. Perdew, W.B. Mattes, and W.F. Greenlee, Eds 2002, Elsevier Science BV: Amsterdam. pp. 583–610.

[40] Wilson, I.D. *et al.*, High resolution "ultra performance" liquid chromatography coupled to oa-TOF mass spectrometry as a tool for differential metabolic pathway profiling in functional genomic studies. *J. Proteome Res.*, 2005. **4**(2): pp. 591–8.

[41] Smith, C.A. *et al.*, METLIN: A metabolite mass spectral database. *Ther. Drug Mon.*, 2005. **27**(6): pp. 747–51.

[42] Keun, H.C. *et al.*, Analytical reproducibility in (1)H NMR-based metabonomic urinalysis. *Chem. Res. Toxicol.*, 2002. **15**(11): pp. 1380–6.

[43] Robertson, D.G. *et al.*, Metabonomics: Evaluation of nuclear magnetic resonance (NMR) and pattern recognition technology for rapid in vivo screening of liver and kidney toxicants. *Toxicol. Sci.*, 2000. **57**(2): pp. 326–37.

[44] Reily, M.D. *et al.*, DFTMP, an NMR reagent for assessing the near-neutral pH of biological samples. *J. Am. Chem. Soc.* 2006, in press.

[45] Siegel, M. *et al.*., Evaluation of a dual electrospray ionization/atmospheric pressure chemical ionization source at low flow rates (approx. 50uL/min) for the analysis of both highly and weakly polar compounds. *J. Am. Soc. Mass Spec.*, 1998. **9**: pp. 1196–203.

[46] Syage, J.A. *et al.*, Atmospheric pressure photoionization. II. Dual source ionization. *J Chromatogr A*, 2004. **1050**(2): pp. 137–49.

[47] Jonsson, P. *et al.*, Extraction, interpretation and validation of information for comparing samples in metabolic LC/MS data sets. *Analyst*, 2005. **130**(5): pp. 701–7.

[48] Williams, R.E. *et al.*, D-Serine-induced nephrotoxicity: A HPLC-TOF/MS-based metabonomics approach. *Toxicology*, 2005. **207**(2): pp. 179–90.

[49] Jonsson, P. *et al.*, High-throughput data analysis for detecting and identifying differences between samples in GC/MS-based metabolomic analyses. *Anal. Chem.*, 2005. **77**(17): pp. 5635–42.

[50] Williams, R.E. *et al.*, A combined (1)H NMR and HPLC-MS-based metabonomic study of urine from obese (fa/fa) Zucker and normal Wistar-derived rats. *J. Pharm. Biomed. Anal.*, 2005. **38**(3): pp. 465–71.

[51] Williams, R. *et al.*, The metabonomics of aging and development in the rat: An investigation into the effect of age on the profile of endogenous metabolites in the urine of male rats using 1H NMR and HPLC-TOF MS. *Molec. Biosyt.*, 2005. **1:** pp. 166–175.

[52] Lindon, J.C. *et al.*, The Consortium for Metabonomic Toxicology (COMET): Aims, activities and achievements. *Pharmacogenomics*, 2005. **6**(7): pp. 691–9.

[53] Lindon, J.C., J.K. Nicholson, and I.D. Wilson, Directly coupled HPLC-NMR and HPLC-NMR-MS in pharmaceutical research and development. *J. Chromatogr. B Biomed. Sci. Appl.*, 2000. **748**(1): pp. 233–58.

[54] Robosky, L.C. *et al.*, Metabonomic identification of two distinct phenotypes in Sprague-Dawley (Crl:CD(SD)) rats. *Toxicol. Sci.*, 2005. **87**(1): pp. 277–84.

[55] Gavaghan, C.L. *et al.*, Directly coupled high-performance liquid chromatography and nuclear magnetic resonance spectroscopic with chemometric studies on metabolic variation in Sprague–Dawley rats. *Anal. Biochem.*, 2001. **291**(2): pp. 245–52.

[56] Beckwith-Hall, B.M. *et al.*, NMR-based metabonomic studies on the biochemical effects of commonly used drug carrier vehicles in the rat. *Chem. Res. Toxicol.*, 2002. **15**(9): pp. 1136–41.

[57] Vigneau-Callahan, K.E. *et al.*, Characterization of diet-dependent metabolic serotypes: Analytical and biological variability issues in rats. *J. Nutr.*, 2001. **131**(3): pp. 924S–932S.

[58] Pennie, W., S. Beushausen, and D. Robertson, Applications of genomics, proteomics and metabonomics in predictive and mechanistic toxicology, in *Principles and methods in Toxicology: Fifth Edition*, A.W. Hayes, Editor. In Press, Taylor & Francis: Philadelphia.

[59] Deprez, S. *et al.*, Optimisation of collection, storage and preparation of rat plasma for 1H NMR spectroscopic analysis in toxicology studies to determine inherent variation in biochemical profiles. *J. Pharm. Biomed. Anal.*, 2002. **30**(4): pp. 1297–310.

[60] Pears, M.R. *et al.*, High resolution 1H NMR-based metabolomics indicates a neurotransmitter cycling deficit in cerebral tissue from a mouse model of batten disease. *J. Biol. Chem.*, 2005. **280**(52): pp. 42508–14.

[61] Jones, G.L. *et al.*, A functional analysis of mouse models of cardiac disease through metabolic profiling. *J. Biol. Chem.*, 2005. **280**(9): pp. 7530–9.

[62] Anonymous. Biological Atlas of Insulin Resistance *http://www.bair.org.uk/*. 2006.

[63] Cloarec, O. *et al.*, Statistical total correlation spectroscopy: An exploratory approach for latent biomarker identification from metabolic 1H NMR data sets. *Anal. Chem.*, 2005. **77**(5): pp. 1282–9.

[64] Watkins, S.M. *et al.*, Lipid metabolome-wide effects of the PPAR gamma agonist rosiglitazone. *J. Lipid. Res.*, 2002. **43**(11): pp. 1809–17.

[65] Rae, C. *et al.*, A metabolomic approach to ionotropic glutamate receptor subtype function: A nuclear magnetic resonance in vitro investigation. *J. Cereb. Blood Flow Metab.*, 2006. **26**(8): pp. 1005–17.

[66] Boros, L.G., D.J. Brackett, and G.G. Harrigan, Metabolic biomarker and kinase drug target discovery in cancer using stable isotope-based dynamic metabolic profiling (SIDMAP). *Curr. Cancer Drug Targets*, 2003. **3**(6): pp. 445–53.

[67] Bollard, M.E. *et al.*, Comparative metabonomics of differential hydrazine toxicity in the rat and mouse. *Toxicol. Appl. Pharmacol.*, 2005. **204**(2): pp. 135–51.

[68] Vidal, M., Interactome modeling. *FEBS Lett.*, 2005. **579**(8): pp. 1834–8.

[69] Crockford, D.J. *et al.*, Statistical heterospectroscopy, an approach to the integrated analysis of NMR and UPLC-MS data sets: Application in metabonomic toxicology studies. *Anal. Chem.*, 2006. **78**(2): pp. 363–71.

[70] Anthony, M.L. *et al.*, Classification of toxin-induced changes in 1H NMR spectra of urine using an artificial neural network. *J. Pharm. Biomed. Anal.*, 1995. **13**(3): pp. 205–11.

[71] Holmes, E. *et al.*, Development of a model for classification of toxin-induced lesions using 1H NMR spectroscopy of urine combined with pattern recognition. *NMR Biomed.*, 1998. **11**(4–5): pp. 235–44.

[72] Farkas, D. and S.R. Tannenbaum, In vitro methods to study chemically-induced hepatotoxicity: A literature review. *Curr. Drug Metab.*, 2005. **6**(2): pp. 111–25.

[73] Spielmann, H. and M. Liebsch, Lessons learned from validation of in vitro toxicity test: From failure to acceptance into regulatory practice. *Toxicol. In Vitro*, 2001. **15**(4–5): pp. 585–90.

[74] Lindon, J.C. *et al.*, Contemporary issues in toxicology the role of metabonomics in toxicology and its evaluation by the COMET project. *Toxicol. Appl. Pharmacol.*, 2003. **187**(3): pp. 137–46.

[75] Anthony, M.L. *et al.*, Pattern recognition classification of the site of nephrotoxicity based on metabolic data derived from proton nuclear magnetic resonance spectra of urine. *Mol. Pharmacol.*, 1994. **46**(1): pp. 199–211.

[76] Gartland, K.P., F.W. Bonner, and J.K. Nicholson, Investigations into the biochemical effects of region-specific nephrotoxins. *Mol. Pharmacol.*, 1989. **35**(2): pp. 242–50.

[77] Beckwith-Hall, B.M. *et al.*, Nuclear magnetic resonance spectroscopic and principal components analysis investigations into biochemical effects of three model hepatotoxins. *Chem. Res. Toxicol.*, 1998. **11**(4): pp. 260–72.

[78] Kerns, W. *et al.*, Drug-induced vascular injury – a quest for biomarkers. *Toxicol. Appl. Pharmacol.*, 2005. **203**(1): pp. 62–87.

[79] Robertson, D.G. *et al.*, Metabonomic assessment of vasculitis in rats. *Cardiovasc. Toxicol.*, 2001. **1**(1): pp. 7–19.

[80] Slim, R.M. *et al.*, Effect of dexamethasone on the metabonomics profile associated with phosphodiesterase inhibitor-induced vascular lesions in rats. *Toxicol. Appl. Pharmacol.*, 2002. **183**(2): pp. 108–9.

[81] Connor, S.C. *et al.*, Development of a multivariate statistical model to predict peroxisome proliferation in the rat, based on urinary 1H-NMR spectral patterns. *Biomarkers*, 2004. **9**(4–5): pp. 364–85.

[82] Akira, K., M. Imachi, and T. Hashimoto, Investigations into biochemical changes of genetic hypertensive rats using 1H nuclear magnetic resonance-based metabonomics. *Hypertens Res.*, 2005. **28**(5): pp. 425–30.

[83] Brindle, J.T. *et al.*, Rapid and noninvasive diagnosis of the presence and severity of coronary heart disease using 1H-NMR-based metabonomics. *Nat. Med.*, 2002. **8**(12): pp. 1439–44.

[84] Brindle, J.T. *et al.*, Application of chemometrics to 1H NMR spectroscopic data to investigate a relationship between human serum metabolic profiles and hypertension. *Analyst*, 2003. **128**(1): pp. 32–6.

[85] Sabatine, M.S. *et al.*, Metabolomic identification of novel biomarkers of myocardial ischemia. *Circulation*, 2005. **112**(25): pp. 3868–75.

[86] Fan, T.W., A.N. Lane, and R.M. Higashi, The promise of metabolomics in cancer molecular therapeutics. *Curr. Opin. Mol. Ther.*, 2004. **6**(6): pp. 584–92.

[87] Odunsi, K. *et al.*, Detection of epithelial ovarian cancer using 1H-NMR-based metabonomics. *International Journal of Cancer*, 2005. **113**(5): pp. 782–8.

[88] Griffiths, J.R. and M. Stubbs, Opportunities for studying cancer by metabolomics: Preliminary observations on tumors deficient in hypoxia-inducible factor 1. *Advances in Enzyme Regulation*, 2003. **43**: pp. 67–76.

[89] Wang, C. *et al.*, Plasma phospholipid metabolic profiling and biomarkers of type 2 diabetes mellitus based on high-performance liquid chromatography/electrospray mass spectrometry and multivariate statistical analysis. *Anal. Chem.*, 2005. **77**(13): pp. 4108–16.

[90] Yang, J. *et al.*, Discrimination of Type 2 diabetic patients from healthy controls by using metabonomics method based on their serum fatty acid profiles. *J. Chromatogr. B Analyt Technol. Biomed. Life Sci.*, 2004. **813**(1–2): pp. 53–8.

[91] Coen, M. *et al.*, Proton nuclear magnetic resonance-based metabonomics for rapid diagnosis of meningitis and ventriculitis. *Clin. Infect. Dis.*, 2005. **41**(11): pp. 1582–90.

[92] Van, Q.N. *et al.*, The use of urine proteomic and metabonomic patterns for the diagnosis of interstitial cystitis and bacterial cystitis. *Dis. Markers*, 2003. **19**(4–5): pp. 169–83.

[93] Gibbs, A., Comparison of the specificity and sensitivity of traditional methods for assessment of nephrotoxicity in the rat with metabonomic and proteomic methodologies. *Journal of Applied Toxicology*, 2005. **25**(4): pp. 277–95.

[94] MacGregor, J.T., The future of regulatory toxicology: Impact of the biotechnology revolution. *Toxicol. Sci.*, 2003. **75**(2): pp. 236–48.

[95] Mikov, M., The metabolism of drugs by the gut flora. *Eur. J. Drug Metab. Pharmacokinet*, 1994. **19**(3): pp. 201–7.

[96] Rowland, I.R., Interactions of the gut microflora and the host in toxicology. *Toxicol. Pathol.*, 1988. **16**(2): pp. 147–53.

[97] Rowland, I.R., Factors affecting metabolic activity of the intestinal microflora. *Drug Metab. Rev.*, 1988. **19**(3–4): pp. 243–61.

[98] Boxenbaum, H.G. *et al.*, Influence of gut microflora on bioavailability. *Drug Metab. Rev.*, 1979. **9**(2): pp. 259–79.

[99] Phipps, A.N. *et al.*, Effect of diet on the urinary excretion of hippuric acid and other dietary-derived aromatics in rat. A complex interaction between diet, gut microflora and substrate specificity. *Xenobiotica*, 1998. **28**(5): pp. 527–37.

[100] Connor, S.C. *et al.*, Effects of feeding and body weight loss on the 1H-NMR-based urine metabolic profiles of male Wistar Han rats: Implications for biomarker discovery. *Biomarkers*, 2004. **9**(2): pp. 156–79.

[101] Clayton, T.A. *et al.*, An hypothesis for a mechanism underlying hepatotoxin-induced hypercreatinuria. *Arch. Toxicol.*, 2003. **77**(4): pp. 208–17.

[102] Clayton, T.A. *et al.*, Hepatotoxin-induced hypercreatinaemia and hypercreatinuria: Their relationship to one another, to liver damage and to weakened nutritional status. *Arch. Toxicol.*, 2004. **78**(2): pp. 86–96.

[103] Waterfield, C.J. *et al.*, The correlation between urinary and liver taurine levels and between pre-dose urinary taurine and liver damage. *Toxicology*, 1993. **77**(1–2): pp. 1–5.

[104] Waterfield, C.J. *et al.*, Taurine, a possible urinary marker of liver damage: A study of taurine excretion in carbon tetrachloride-treated rats. *Arch. Toxicol.*, 1991. **65**(7): pp. 548–55.

[105] Timbrell, J.A. and C.J. Waterfield, Changes in taurine as an indicator of hepatic dysfunction and biochemical perturbations. Studies *in vivo* and *in vitro*. *Adv. Exp. Med. Biol.*, 1996. **403**: pp. 125–34.

[106] Mortishire-Smith, R.J. *et al.*, Use of metabonomics to identify impaired fatty acid metabolism as the mechanism of a drug-induced toxicity. *Chem. Res. Toxicol.*, 2004. **17**(2): pp. 165–73.

[107] Tannehill-Gregg, S. *et al.*, Preliminary evidence of urolithiasis with muraglitazar-related urinary bladder carcinogenesis in male rats. *Toxicological Sciences*, 2006. **90**(Supplement): p. 428.

[108] Sanderson, T. *et al.*, Carcinogneicity studies of muraglitazar, a dual PPAR alpha and gamma agonist, in mice and rats. *Toxicological Sciences*, 2006. **90**(Supplement): pp. 429.

[109] Flint, O.P. *et al.* Characterization of HIV protease inhibitors using large-scale metabolomics of an adipocyte cell model *in vitro*. Presented at the *3rd International AIDS Society Conference on HIV Pathogenesis*. 2005. Rio de Janeiro.

[110] Parker, R.A. *et al.*, Endoplasmic reticulum stress links dyslipidemia to inhibition of proteasome activity and glucose transport by HIV protease inhibitors. *Mol. Pharmacol.*, 2005. **67**(6): pp. 1909–19.

[111] Hertel, J. *et al.*, A structural basis for the acute effects of HIV protease inhibitors on GLUT4 intrinsic activity. *J. Biol. Chem.*, 2004. **279**(53): pp. 55147–52.

[112] Robertson, D.G., Metabonomics in Preclinical Safety Evaluation: Where are We. *Toxicological Sciences*, 2006. **90**(Supplement): p. 456.

[113] Anonymous. The NIH RoadMap Initiative *http://nihroadmap.nih.gov/initiatives.asp*. 2006.

The Handbook of Metabonomics and Metabolomics
John C. Lindon, Jeremy K. Nicholson and Elaine Holmes (Editors)

Chapter 10

Applications of Metabonomics in Clinical Pharmaceutical R&D

Jeremy R. Everett

Pfizer Global R&D, Sandwich CT Kent UK

10.1. Introduction

The pharmaceutical industry is facing a number of key challenges at the present time, and these can be summarised as –

- pharmaceutical R&D costs have risen sharply over the past 10 years
- output has been level across the industry at around four new drug mechanisms launched per year but only one new blockbuster drug mechanism (i.e. greater than $1 billion per year sales) launched per year
- patent expiries are biting into the sales of many companies and generic launches are taking most of the sales of an off-patent compound within 18 months of patent expiry
- the viability of drug discovery and development as a business proposition is being questioned
- price-to-earnings ratios are significantly lower than 10 years ago.

These challenges and problems have a number of causes including:

- high attrition rates (low project and/ or compound survival)
  - insufficient attention to target "druggability" assessment/validation
  - insufficient attention to compound early safety assessment
  - insufficient attention to translational medicine approaches

- increasing regulatory hurdles to drug approvals
- increasing commercial hurdles
  - good generic drugs are now available in many areas
  - more stringent cost-benefit analyses by Regulators and Payers
- public expectations of full efficacy and the fact that the precautionary principle is applied with respect to safety for all patients, which is difficult to achieve in diverse populations like human beings.

The pharmaceutical industry is responding in a number of ways including:

- mergers and acquisitions
- growth of biological agents at the expense of small drug molecules
- in-licensing of pharmaceutical agents from other pharmaceutical companies and biotechnology companies
- investments in various technologies to improve efficiency and effectiveness including:
  - genomics, proteomics, metabonomics, imaging, biomarker discovery
  - file enrichment, parallel chemistry, high-throughput screening, structural biology
- outsourcing to Western countries to increase flexibility
- outsourcing to lower cost countries to decrease costs.

Metabonomics is potentially of key importance to the future of the pharmaceutical industry as it can attack one of the main causes of low productivity: compound or product attrition. Attrition is the failure of a compound to progress to the next stage of the drug discovery and development process because of safety or efficacy concerns, problems with delivery to the site of action, pharmacokinetic, cost or intellectual property issues. Metabonomics has shown itself capable of providing key information on compounds which can help decide if a particular compound has a liability. The major areas of impact to date include:

- assessment of the toxic potential of compounds, particularly, but not exclusively, in pre-clinical studies (see Chapter 9)
- disease diagnosis across many clinical areas: inborn errors of metabolism has been an area of intense focus in the past and indeed metabonomics methods were used in the discovery of a new inborn error of metabolism recently, but metabonomics is applied to a broad swathe of disease areas today (see Chapters 12–15)
- assessment of drug efficacy.

The metabonomics approach is typically pursued by the search for biomarkers in a biofluid chosen for study. Biomarkers are physical, functional or biochemical characteristics that can be used to measure the progress of disease, aging, toxicity or the effects of treatment. In a metabonomics context, these biomarkers may be represented by changes in the levels of either individual metabolites observed in a particular biofluid, or changes in the patterns of the levels observed for two or more metabolites. In pharmaceutical R&D, biomarkers are important in the following contexts:

- establishing Confidence in Safety (CIS) for a particular target (gene product), with which a drug interacts, or CIS for a particular compound which interacts with that target;
- establishing Confidence in the Rationale (CIR) that modulating a particular target in the body in a particular fashion (agonist, inverse agonist, antagonist, inhibitor) will provide the desired effect for the clinical indication in mind; this is sometimes known as target discovery and target validation;
- establishing Confidence in Differentiation (CID) between the effects of a new, hopefully superior drug and those of an older pharmaceutical agent;
- establishing Confidence in the Pharmacology (CIP) in the compounds being tested, that is confidence that a particular compound which is active *in vitro* against a target, will actually modulate the target in the desired fashion in an *in vivo* context (animal or human) and further that effects seen pre-clinically in animals will translate into equivalent effects in human patients; this is the important area of translational pharmacology.

Establishing confidence in the various areas mentioned above is critical to reducing the severe attrition or loss of compounds that is experienced in every pharmaceutical company due to lack of safety, lack of efficacy or lack of differentiation from competitor products. Thus the availability of predictive biomarkers for these key features from metabonomics and other technologies such as genomics, proteomics and bio-imaging is very valuable to pharmaceutical R&D workers. This has never been truer than at present (2006) since the industry is facing very significant challenges to its long-term future; drug discovery and development costs are of the order of $1 billion per drug launch, less than 1 in 10 compounds entering development from drug discovery make it to the marketplace, and of those compounds that make it all the way, a significant number fail to generate a sufficient return on investment. A recent publication by Stoughton and Friend [1] predicted that molecular profiling by metabonomic, proteomic and transcriptomic methods could revolutionise the effectiveness of drug discovery and development in the future, but might require different thinking in terms of organisation than the traditional pharmaceutical R&D model.

10.2. Clinical applications of metabonomics in pharmaceutical R&D to establish Confidence in Rationale and Confidence in Safety

10.2.1. Recent reviews

A number of reviews of clinical applications of metabonomics have appeared recently, for example one of the earliest was the review by Lindon *et al.* which defined metabonomics and gave a number of clinical and pre-clinical applications [2]. In particular, the recent review from Reily *et al.* [3] on applications of metabonomics in pharmaceutical R&D is recommended, along with those from Lindon, Nicholson *et al.* [4, 5, 6], van der Greef *et al.* [7] and Breau and Cantor [8]. This chapter will focus on the more recent literature, from 2000 to 2005.

10.2.2. Background work

Lenz and co-workers studied the proton NMR (^{1}H NMR) spectra of plasma and urine samples from a group of 12 healthy male subjects on two occasions, 14 days apart [9]. The subjects were fed a standard diet. As expected, little inter-subject or inter-sampling date variation was seen in the plasma data. However, in the case of the urine spectra, there was considerable inter-subject variation but less intra-subject variation. The authors concluded that it was possible to collect consistent metabonomic data in clinical studies. Subsequently, the same authors conducted a study to compare urine samples from two studies: the first involving 120 healthy, male and female British subjects and the second involving 30 healthy British and Swedish subjects. Both cultural and dietary influences exhibited themselves in terms of degree of fish eating and adherence to the Atkins diet. The authors concluded that great care would need to be taken in the interpretation of biomarkers of disease and response to drug therapy in human populations [10].

10.2.3. Disease diagnosis and treatment

Roy and co-workers have demonstrated that an LC-MS approach to metabonomic and proteomic analysis of human sera showed that the technology could differentiate between 19 controls and age- and sex-matched sera from 19 patients with rheumatoid arthritis, using both metabonomic and proteomic approaches. This study is part of a longitudinal rheumatoid arthritis study by SurroMed that will enrol over 250 patients over 3 years and measure disease progression by routine clinical laboratory tests, cellular phenotyping as well as metabolic and proteomic profiling by this LC-MS method [11].

Yang and co-workers [12] have described how pattern recognition (PR) methods applied to the output from gas chromatographic analysis of serum fatty acid profiles were able to discriminate between healthy controls and type-2 diabetes patients with greater specificity and sensitivity than in the absence of the PR. A similar approach was taken by Fiehn and Spranger [13, 14] as part of a review assessing the use of metabonomics to discover metabolic profiles associated with human disease. Wang and co-workers have also shown [15] that LC/MS followed by multivariate statistical analysis can be successfully applied to the metabolic profiling of plasma phospholipids in type-2 diabetes mellitus. The application of an orthogonal signal correction filtered model highly improved the class distinction and predictive power of PLS-DA models. With this methodology, it was possible not only to differentiate type-2 diabetes mellitus from controls, but also to discover and identify the potential biomarkers with LC/MS/MS. The work shows that LC/MS combined with multivariate statistical analysis is a complement or an alternative to NMR for metabonomics applications.

The Yang group has also published a strategy for applying high-performance liquid chromatography (HPLC) and LC-MS-MS to metabonomics research [16]. One of the key problems to be solved in this strategy is to match the peaks between the chromatograms. A peak alignment algorithm was developed to match the chromatograms before pattern recognition. The strategy has been applied to liver diseases, and the false-positive diagnosis of liver cancer instead of hepatocirrhosis or hepatitis was reduced to 7%. The structures of eight potential biomarkers were given for distinguishing liver cancer from hepatocirrhosis and hepatitis.

Murphy *et al.* [17] showed that high-resolution magic angle spinning (HR-MAS) ^{1}H NMR spectroscopy is a useful tool for assessing intact liver tissue during the whole process of human liver transplantation. This work is the first reported application of HR-MAS ^{1}H NMR spectroscopy to human liver biopsy samples, and enabled a determination of liver metabolic profiles before removal from donors, during cold perfusion and after implantation into recipients. Several interesting results were obtained: glycerophosphocholine (GPC, a phospholipid degradation product) was seen to decrease for all livers but one, suggesting increased cell turnover. Interestingly, in the graft that developed primary graft dysfunction, GPC did not vary, probably reflecting a lower degree of cellular activity, and GPC was proposed as a new biomarker for liver function. Second, the clearance of University of Wisconsin (UW) preservation solution was seen to be efficient only for non-fatty livers.

Prabakaran *et al.* [18] applied parallel transcriptomics, proteomics and metabonomics technologies to generate molecular profiles of post-mortem human brain tissue from 10 schizophrenics and 10 healthy controls. Principal component analysis (PCA) showed that white and grey matter proteomics profiles from schizophrenia brains differed significantly from control brains and could be used to classify the disease status. Almost half the altered proteins identified by proteomics were associated with mitochondrial function and oxidative stress responses. This was mirrored by transcriptional and metabolite perturbations. The same samples used for the proteomics investigation were analysed by high resolution magic angle spinning (HRMAS) ^{1}H NMR spectroscopy and PCA to characterize metabolic profiles in intact brain tissue. Out of ca. 60 identifiable metabolites, 10 were significantly altered ($P < 0.05$) between schizophrenia and control tissues in white matter. In grey matter, a similar pattern was observed using PCA, but did not reach significance using univariate statistical analysis. The authors propose that oxidative stress and the ensuing cellular adaptations are linked to the schizophrenia disease process and hope that this new understanding may advance approaches to treat, diagnose and prevent schizophrenia and related syndromes. This study gives an important insight into the power of molecular profiling methods to not only diagnose but understand disease processes.

Amyotrophic lateral sclerosis (ALS) is a thickening of tissue in the motor tracts of the lateral columns and anterior horns of the spinal cord and results in progressive muscle atrophy that starts in the limbs. Kaddurah-Daouk and co-workers reviewed [19] the power of electrochemical detection when combined with LC or LC-MS analysis of human serum samples to readily distinguish the serum from control and ALS patient serum and also to monitor a biomarker corresponding to the uptake of Riluzole in ALS patients and the effect of drug treatment.

The use of NMR spectroscopy of biofluids for diagnosing inborn errors of metabolism has been reviewed [20] and is also included later in the volume, and as an example a new type of inborn error was identified [21].

Van *et al.* [22] have used MS- and NMR-based metabonomics methods combined with a novel pattern recognition system based on self-organising cluster maps, to diagnose interstitial cystitis and bacterial cystitis relative to unaffected patients with a success rate of ca. 84%.

^{1}H NMR-based metabonomics [23] was applied to lumbar cerebrospinal fluid (CSF) samples collected prospectively from a cohort of patients with bacterial, fungal or viral meningitis and from control subjects without neurological disease and from ventricular CSF samples from patients with ventriculitis associated with an external ventricular drain, and from control subjects. The PCA clearly distinguished patients with bacterial or fungal meningitis (11 patients) from patients with viral meningitis (12) and control subjects (27) and clearly distinguished patients with post-surgical ventriculitis (5) from post-surgical control subjects (10). Metabolites of microbial and host origin that were responsible for class separation were determined. Metabonomic data also correlated with the onset and course of infection in a patient with two episodes of bacterial ventriculitis and with response to therapy in another patient with cryptococcal meningitis. It was concluded that metabonomic analysis is rapid, requires minimal sample processing, and is not targeted to specific microbial pathogens, making the platform potentially suitable for use in the diagnostic laboratory. This pilot study indicates that metabonomic analysis of CSF is potentially a more powerful diagnostic tool than conventional laboratory indicators for distinguishing bacterial from viral meningitis and for monitoring therapy. This should have important implications for reduced empirical use of antibiotics, reduced treatment duration and enhanced treatment efficacy.

Sharma *et al.* [24] used NMR spectroscopy to quantify metabolites in seminal plasma from subjects injected with a new male contraceptive called RISUG (a copolymer of styrene maleic anhydride dissolved in dimethyl sulphoxide), and in seminal plasma from normal ejaculates. No significant difference in the concentration of citrate was observed between the groups, indicating that the prostate was not affected by RISUG. The concentrations of glucose, lactate, glycerophosphorylcholine and choline were significantly lower ($P < 0.01$) in subjects injected with RISUG compared with controls. Citrate:lactate and glycerophosphorylcholine:choline ratios were significantly lower in RISUG-injected subjects than in controls ($P < 0.01$), which was interpreted as indicating the occurrence of partial obstructive azoospermia. The study reported that intervention with RISUG in the vas deferens for as long as 8 years was safe and did not lead to prostatic diseases.

10.3. Summary and future outlook

Metabonomics is starting to have an impact upon disease diagnosis, and upon monitoring the efficacy and safety of drug treatment for humans. At present, there are relatively few examples of clinical applications of metabonomics in pharmaceutical R&D compared with the numerous pre-clinical examples, but this will undoubtedly change rapidly in the future. One factor that will positively impact on this is the emergence of large-scale biological sample banks or Biobanks with the capabilities of storing millions of frozen human biological samples including biofluids. More ready access to this treasure trove of samples will serve to stimulate more large-scale studies of human metabonomics and indeed the trend has already begun [25].

References

[1] Stoughton, R.B., Friend, S.H. Innovation: How molecular profiling could revolutionize drug discovery. *Nature Reviews Drug Discovery* (2005), 4(4), 345–350.

[2] Lindon, J.C., Nicholson, J.K., Holmes, E., Everett, J.R. Metabonomics: Metabolic Process Studies by NMR Spectroscopy of Biofluids. *Concepts in Magnetic Resonance* (2000), 12(5), 289–320.

[3] Reily, D., Robertson, D.G., Delnomdedieu, M., Baker, J.D. High resolution NMR of biological fluids: Metabonomics applications in pharmaceutical research and development. *American Pharmaceutical Review* (2003), 6(3), 105–109.

[4] Lindon, J.C., Holmes, E., Bollard, M.E., Stanley, E.G., Nicholson, J.K. Metabonomics technologies and their applications in physiological monitoring, drug safety assessment and disease diagnosis. *Biomarkers* (2004), 9(1), 1–31.

[5] Lindon, J.C., Holmes, E., Nicholson, J.K. Metabonomics: Systems biology in pharmaceutical research and development. *Current Opinion in Molecular Therapeutics* (2004), 6(3), 265–272.

[6] Lindon, J.C., Holmes, E., Nicholson, J.K. Metabonomics and its role in drug development and disease diagnosis. *Expert Review of Molecular Diagnostics* (2004), 4(2), 189–199.

[7] van der Greef, J., Stroobant, P., van der Heijden R. The role of analytical sciences in medical systems biology. *Current Opinion in Chemical Biology* (2004), 8(5), 559–565.

[8] Breau, A.P., Cantor, G.H. Application of metabonomics in the pharmaceutical industry. *Metabolic Profiling* (2003), 69–82.

[9] Lenz, E.M., Bright, J., Wilson, I.D., Morgan, S.R., Nash, A.F.P. A 1H NMR-based metabonomic study of urine and plasma samples obtained from healthy human subjects. *Journal of Pharmaceutical and Biomedical Analysis* (2003), 33(5), 1103–1115.

[10] Lenz, E.M., Bright, J., Wilson, I.D., Hughes, A., Morrisson, J., Lindberg, H., Lockton, A. Metabonomics, dietary influences and cultural differences: A 1H NMR-based study of urine samples obtained from healthy British and Swedish subjects. *Journal of Pharmaceutical and Biomedical Analysis* (2004), 36(4), 841–849.

[11] Roy, S.M., Anderle, M., Lin, H., Becker, C.H. Differential expression profiling of serum proteins and metabolites for biomarker discovery. *International Journal of Mass Spectrometry* (2004), 238(2), 163–171.

[12] Yang, J., Xu, G., Hong, Q., Liebich, H.M., Lutz, K., Schmuelling, R.-M., Wahl, H.G. Discrimination of Type 2 diabetic patients from healthy controls by using metabonomics method based on their serum fatty acid profiles. *J. Chromatography, B: Analytical Technologies in the Biomedical and Life Sciences* (2004), 813(1–2), 53–58.

[13] Fiehn, O., Spranger, J. Use of metabolomics to discover metabolic patterns associated with human diseases. *Metabolic Profiling* (2003), 199–215.

[14] Fiehn, O., Spranger, J. Can metabolomics be used for assessing nutritive-dependent human diseases? *Phytochemistry Reviews* (2003), Volume Date 2002, 1(2), 223–230.

[15] Wang, C., Kong, H., Guan, Y., Yang, J., Gu, J., Yang, S., Xu, G. Plasma Phospholipid Metabolic Profiling and Biomarkers of Type 2 Diabetes Mellitus Based on High-Performance Liquid Chromatography/Electrospray Mass Spectrometry and Multivariate Statistical Analysis. *Analytical Chemistry* (2005), 77(13), 4108–4116.

[16] Yang, J., Xu, G., Zheng, Y., Kong, H., Wang, C., Zhao, X., Pang, T. Strategy for metabonomics research based on high-performance liquid chromatography and liquid chromatography coupled with tandem mass spectrometry. *Journal of Chromatography, A* (2005), 1084(1–2), 214–221.

[17] Duarte, I.F., Stanley, E.G., Holmes, E., Lindon, J.C., Gil, A.M., Tang, H., Ferdinand, R., Gavaghan McKee, C., Nicholson, J.K., Vilca-Melendez, H., Heaton, N., and Murphy, G.M. Metabolic Assessment of Human Liver Transplants from Biopsy Samples at the Donor and Recipient Stages Using High-Resolution Magic Angle Spinning 1H NMR Spectroscopy, *Analytical Chemistry* (2005), 77, 5570–5578.

[18] Prabakaran, S., Swatton, J.E., Ryan, M.M., Huffaker, S. J., Huang, J. T.-J., Griffin, J.L., Wayland, M., Freeman, T., Dudbridge, F., Lilley, K.S., Karp, N.A., Hester, S., Tkachev, D., Mimmack, M. L., Yolken, R. H., Webster, M.J., Torrey, E.F., Bahn, S. Mitochondrial dysfunction in schizophrenia: Evidence for compromised brain metabolism and oxidative stress. *Molecular Psychiatry* (2004), 9(7), 684–697.

[19] Kaddurah-D. R., Beecher, C., Kristal, B.S., Matson, W. R., Bogdanov, M., Asa, D.J. Bioanalytical advances for metabolomics and metabolic profiling. *PharmaGenomics* (2004), 4(1), 46–48, 50, 52.

[20] Moolenaar, S.H., Engelke, U.F.H., Wevers, R.A. Proton nuclear magnetic resonance spectroscopy of body fluids in the field of inborn errors of metabolism. *Annals of Clinical Biochemistry* (2003), 40(1), 16–24.

[21] Moolenaar, S.H., Gohlich-Ratmann, G., Engelke, U.F., Spraul, M., Humpfer, E., Dvortsak, P., Voit, T., Hoffmann, G.F., Brautigam, C., van Kuilenburg, A.B., van Gennip, A., Vreken, P., Wevers, R.A. Beta-Ureidopropionase deficiency: A novel inborn error of metabolism discovered using NMR spectroscopy on urine. *Magnetic Resonance in Medicine* (2001), 46(5), 1014–7.

[22] Van, Q.N., Klose, J.R., Lucas, D.A., Prieto, D.A., Luke, B., Collins, J., Burt, S. K., Chmurny, G.N., Issaq, H. J., Conrads, T.P., Veenstra, T.D., Keay, S.K. The use of urine proteomic and metabonomic patterns for the diagnosis of interstitial cystitis and bacterial cystitis. *Disease Markers* (2003), Volume Date 2003–2004, 19(4,5), 169–183.

[23] Coen, M., O'Sullivan, M., Bubb, W.A., Kuchel, P.W., Sorrell, T. Proton nuclear magnetic resonance-based metabonomics for rapid diagnosis of meningitis and ventriculitis. *Clinical Infectious Diseases* (2005), 41(11), 1582–1590.

[24] Sharma, U., Chaudhury, K., Jagannathan, N.R., Guha, S.K. A proton NMR study of the effect of a new intravasal injectable male contraceptive RISUG on seminal plasma metabolites. *Reproduction.* 122, 431–6 (2001).

[25] Dumas, M.-E., Maibaum, E.C., Teague, C., Ueshima, H., Zhou, B., Lindon, J.C., Nicholson, J.K., Stamler, J., Elliott, P., Chan, Q., and Holmes, E. Assessment of Analytical Reproducibility of 1H NMR Spectroscopy Based Metabonomics for Large-Scale Epidemiological Research: the INTERMAP Study, *Analytical Chemistry* (2006), 78, 2199–2208.

The Handbook of Metabonomics and Metabolomics
John C. Lindon, Jeremy K. Nicholson and Elaine Holmes (Editors)

Chapter 11

Exploiting the Potential of Metabonomics in Large Population Studies: Three Venues

Burton H. Singer[1], Jürg Utzinger[2], Carol D. Ryff[3], Yulan Wang[4] and Elaine Holmes[4]

[1]*Office of Population Research, Princeton University, Wallace Hall, Princeton, NJ 08544, USA*
[2]*Department of Public Health and Epidemiology, Swiss Tropical Institute, P.O. Box, CH-4002 Basel, Switzerland*
[3]*Department of Psychology and Institute on Aging, University of Wisconsin, 2245 MSC, 1300 University Ave, Madison, WI 53706, USA*
[4]*Department of Biomolecular Medicine, Faculty of Medicine, Imperial College London, Sir Alexander Fleming Building, South Kensington, London SW7 2AZ, UK*

11.1. Introduction

The purpose of this chapter is to illustrate the potential of ^{1}H nuclear magnetic resonance (NMR)-based metabonomics [1] for advancing multiple avenues of scientific research. We focus on three broad areas that, while related to each other, represent quite different applications of this new technology. In each, we describe both human and animal studies, giving emphasis to the role of biomarker transferability between the two and the capacities of NMR-based metabonomics to focus analyses within highly heterogeneous populations. Each of our three areas also includes illustrations of research where metabolic profiling has already been employed as well as examples where such methods have not yet been used. The common denominator of the latter is that they could all substantially benefit from adopting a metabonomics approach. A further important feature of the examples where we envisage metabonomics to have uniquely valuable potential is the delineation of close linkages between physical and social environmental characteristics and their metabolic signatures.

The first venue of metabonomic application we describe pertains to *disease risk and pathogenesis*. Here we focus on the potential of NMR-based metabonomics to improve diagnostic capabilities, including repertoires of early warning indicators of disease risk and pathogenesis as well as to enhance capabilities for monitoring and surveillance of disease control programs. Our first illustration pertains to tropical regions of the world where infection with multiple parasites (i.e. polyparasitism) is the norm, rather than the exception [2, 3, 4, 5, 6, 7]. Although such parasitic load is pervasive in developing countries, several major obstacles render it a challenging issue, including the need for distinct diagnostic methods (often based on various kinds of biofluids) for different infections, the lack of sensitivity and specificity for many diagnostic assays, and the fact that several diagnostic approaches are invasive, labor intensive, and costly. Thus, we examine whether metabolic profiles, derived from urine, stool, and/or blood samples, can be used for simultaneous diagnosis of multiple parasitic infections and with a common level of resolution. We illustrate the scope of the problem with polyparasitism in rural communities in the western part of Côte d'Ivoire. A missing link for developing parallel diagnoses from metabolic profiles is the fact that metabolic signatures of even single parasitic infections are unknown. Thus, as an important first step, we summarize findings from animal studies of infections with *Schistosoma mansoni* (in mice) and *Schistosoma japonicum* (in hamsters). We then indicate a promising strategy for moving from animals to humans with single parasite infections (i.e. travelers and tourists returning from the tropics) as well as polyparasitism in endemic settings.

In the context of human aging, and analogous to polyparasitism, we note that co-morbidity involving multiple chronic conditions is the norm rather than the exception in elderly populations in North America and Europe [8], and probably also in other parts of the industrialized world. The concept of allostatic load was introduced [9] and operationalized [10, 11] as an indicator of wear and tear on multiple biological systems as they adapt and respond, within the individual, to life's demands (represented as complex phenotypes). Operationalizations of allostatic load, using limited sets of biomarkers, have been used as an early warning system for later-life morbidity, disability, and even mortality. Here we review the literature focused on biomarker levels across multiple physiological systems and how they can be incorporated into prediction rules for later life incident disease, disability, and mortality [12, 13, 14, 15]. A missing ingredient in these studies, however, is measurement of the functional status of such fundamental systems as energy metabolism, fat metabolism, and especially gut microflora metabolism [16, 17] all of which are essential for life, but until recently, have been difficult to assess on a large scale in human populations. Thus, we emphasize how more nuanced early warning systems could be developed based on metabolic profiling. Illustrating this promise, we note that the relationship between standard panels of biomarkers and the added value, for predictive purposes, of metabolic profiles has already been established in a clinical population with coronary heart disease [18]. At the animal level, we further note studies that have identified the metabolic signature of insulin resistance phenotypes [19]. In addition, a variety of animal studies have identified metabolic biomarkers of disease progression in a range of neurodegenerative and psychological disorders. This has been dramatically augmented by recent human studies [20, 21] characterizing metabolic signatures of schizophrenia. Extension of these kinds of investigations to heterogeneous general populations, where diverse endpoints are of interest, is feasible and represents an important challenge for metabonomic technology.

The second broad venue we consider pertains to the *metabolic signatures of physical and psychosocial experience.* Our emphasis is on life challenges and psychosocial experience, where a substantial literature has already documented associations between cumulative adversity and downstream morbidity and mortality. Our first illustration in this section involves metabolic profiles of animals exposed to restraint stress and physical shaking paradigms in both acute and chronic modes [22]. Then we identify metabolic signatures of infant rats exposed to moderate periods of maternal separation, and older (3–4 months of age) rats exposed to a water avoidance test. At issue here was whether or not animals exposed to maternal separation in infancy were better able to cope with the later life challenge [23]. Considerably more complex and naturalistic behavioral phenotypes, such as those observed in visible burrow systems in rats and mice [24], are also considered as a useful extension of this line of inquiry. These investigations characterize naturally occurring defensive and aggressive behaviors from which metabolic profiles could be richly informative. At the human level, we examine a growing literature focused on specification of psychosocial phenotypes in longitudinal surveys and assessment of their biological signatures [25, 26, 27]. A key objective therein is measurement of cumulative adversity – for example in the work place, or in social relationships, or in socioeconomic hardship – and identification of related biological correlates, which in combination may be predictive of subsequent disease outcomes [28].

However, human studies also reveal growing interest in the phenomenon of resilience – that is, instances in which those exposed to cumulative adverse challenges are, nevertheless, able to cope successfully and remain in positive states of health. Their biomarker signatures can be anticipated to differ from those lacking the resilience resources and who are overwhelmed with adverse experiences [29]. We illustrate these ideas with a look at cumulative social relational and economic profiles in a 40-year longitudinal study and their links to allostatic load [25]. A related example examines the biological correlates of psychological well-being and ill-being, showing evidence of largely distinctive biological signatures between the two [30]. Metabolic signatures of these diverse experiential and well-being profiles would offer an important new window on human functioning that is considerably more nuanced, and focused on different systems, than standard panels of biomarkers. Our last example in this section brings genetics into the inquiry by considering the role of protective social environments in contexts of known genetic risk [31, 32, 33]. Here we are interested in the underlying processes that prevent individuals with genetic predispositions for a diversity of diseases from actually progressing to disease states [34, 35]. A metabonomics approach could richly advance these important agendas.

The third broad venue we consider focuses on coarse level *metabolic distinctions in population studies*. Comparisons are made at the country level between the United States, China, and Japan. Analogous comparisons are carried out for subpopulations within countries – for example Shanxi in northern China and a predominantly Japanese population in Hawaii. Further comparisons involve the Yanomamo in Venezuela, Xinguano in Brazil, and Chicago in the United States. Even with metabolic representations that do not discriminate between genetic, dietary, and a diversity of cultural and lifestyle features, there is clear evidence that metabonomic studies on heterogeneous human populations in natural settings could have substantial payoff. Alternatively, population studies incorporating dietary factors and currently benefiting from the application of metabonomic analysis are the International Multicenter Epidemiological Study (INTERSALT) [36] and the International Study of Macro/micronutrients and Blood Pressure (INTERMAP) [37] studies on the relationship of population blood pressure levels with diet and lifestyle [38, 39].

We also indicate the kinds of complex psychosocial phenotypes that can be utilized in such metabonomic studies in the future. As such, these investigations will provide useful input to the growing catalogue of susceptibility genes for various chronic diseases, and accompanying evidence that many with these genotypes do not ultimately move into disease states [33]. Returning to animal studies, we conclude this section with a look at well-characterized, free-ranging baboon troops in Tanzania and Kenya [40, 41, 42]. Much is known about the physical environmental challenges confronting these baboons as well as about their social relationships. How social relationships among adult females influence the survival of offspring has been a primary focus. The life histories of these baboons (detailed behavioral phenotypes in naturalistic settings) offer a useful analog of long-term social relational profiles tracked in human population studies, and thereby a rich forum within which to examine cross-species metabolic profiling.

11.2. Disease risk and pathogenesis

11.2.1. Parasitic load

As noted in our introduction, a number of challenges surround research on polyparasitism. First, different kinds of parasitic infections require distinct diagnostic methods, often based on various kinds of biofluids. For example, fecal samples are used for diagnosis of intestinal parasites, whereas Giemsa-stained thick-and-thin blood films are employed for diagnosis of *Plasmodium*, which is the causative agent of malaria [43, 44]. Second, most diagnostic assays lack sensitivity and specificity, particularly for diagnosis of low infection intensities, which is a typical feature for a diversity of epidemiological settings. For example, the population prevalence and intensity of infection with *Plasmodium* in highly urbanized areas of the tropics is low, and hence microscopy of Giemsa-stained blood films misses those infections that are below a critical detection threshold [43]. In China, where large-scale control programs against schistosomiasis have been implemented and sustained over the past several decades, fecal examinations using the Kato-Katz method fail to diagnose all infections [45]. Third, several diagnostic approaches are invasive (e.g. examination of cerebrospinal fluid plays a crucial role for diagnosis of human African trypanosomiasis [46]), somewhat labor intensive, costly, and require well-trained personnel, as interpretation demands considerable expertise, particularly at low levels of infection intensities.

In light of the above challenges, we describe in-depth studies carried out over the last decade in the region of Man, western Côte d'Ivoire, with the objectives: (i) to enhance our understanding of the epidemiology of single and multiple species parasitic infections, and (ii) to develop an integrated approach for risk profiling and spatial targeting of disease control interventions. Our study area is bound by 07°07′ and 07°36′ N latitude and stretches from 07°24′ and 07°50′ E longitude, and hence covers an area of $\sim$2,500 km^2 [47, 48]. The town of Man is located in the heart of the study area at an altitude of 320 m above sea level, some 600 km north-west of Abidjan, the economic capital of Côte d'Ivoire (Figure 11.1). With an estimated 120,000 inhabitants, the town of Man is a typical mid-sized city of Côte d'Ivoire. One-third of the households are engaged in urban agriculture, primarily along the banks of the Kô River, which crosses the town from north to south. Dominant crop systems consist of small-holder rice plots, larger irrigated rice paddies, and vegetable gardens [49]. In surrounding villages, situated at altitudes between 200 and 1300 m above sea level, subsistence farming is the primary day-to-day activity with farmers cultivating banana, cassava, maize, manioc, vegetables and yams, and, on the mountain slopes, rain-fed rice. Coffee and cocoa serve as the most important cash crops in this area. The climate is tropical with two distinct seasons; the rainy season between April and October and the dry season stretches from November to March, with annual precipitation usually above 1000 mm. Given the diverse ecology and the remoteness and difficult accessibility of many villages in this study area, the huge diversity of ethnic groups and the wide array of different local languages are not surprising. The most important ethnic groups comprise the Guerre, Toura, Wobe, and Yacouba, whereas immigrants from neighboring countries (i.e. Burkina

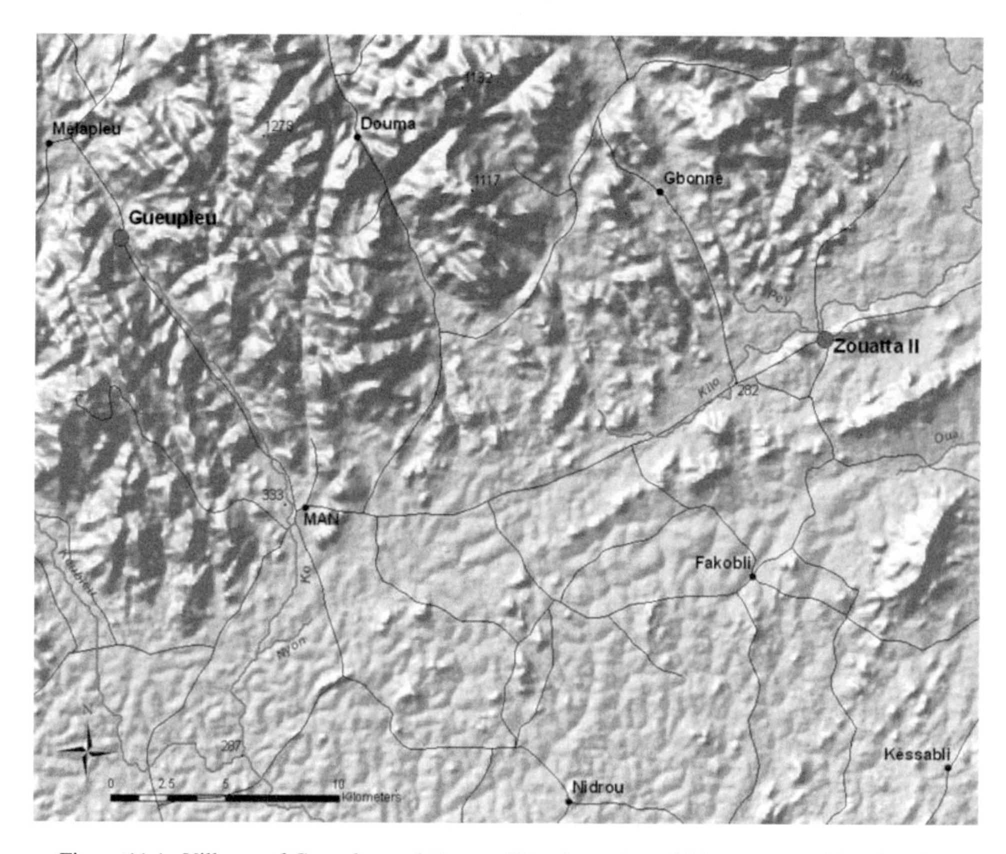

Figure 11.1. Villages of Gueupleu and Zouatta II in the region of Man, western Côte d'Ivoire.

Faso and Mali) represent significant numbers in both rural and urban population censuses.

As documented already in the mid-1980s, the region of Man represents the most important focus of intestinal schistosomiasis due to *S. mansoni* in Côte d'Ivoire [50]. Schistosomiasis is a chronic and debilitating disease, which contributes to anemia, delays physical and cognitive development in childhood and eventually leads to death if left untreated [51, 52]. Our first cross-sectional study carried out in the region of Man, in 1997, focused on 209 schoolchildren aged between 8 and 16 years from 3 neighboring villages including Gueupleu (see Figure 11.1), and confirmed the high endemicity of *S. mansoni*. Employing a rigorous diagnostic approach (i.e. microscopic examination of four Kato-Katz thick smears derived from four consecutive stool specimens [53]), we documented a *S. mansoni* prevalence of 92.3% [54]. The Kato-Katz technique allows the simultaneous diagnosis of soil-transmitted helminths, that is *Ascaris lumbricoides*, *Trichuris trichiura*, and hookworms, although care is needed in the examination of the latter, as parasite eggs tend to disintegrate rapidly once the thick smears have been prepared [55]. Up-to-date reviews on the global distribution and huge public health significance of soil-transmitted helminthiasis are available [51, 56]. In our study population, we noted high prevalence of both hookworm (60.8%) and *A. lumbricoides* (38.3%), whereas infections with *T. trichiura* were found in only four children (1.9%). Polyparasitism was very common [54].

Importantly, we found a significant association between *S. mansoni* and reported blood in stool; children with a moderate or heavy infection (more than 100 eggs per gram of stool) were 2.9 times more likely to report the presence of blood in their stool over the past four weeks when compared to none or only lightly infected counterparts. Encouraged by this finding, a large study was carried out in 1998 in all 134 primary schools in the study area. In the first step, a pre-tested questionnaire was administered through the education system and teachers systematically interviewed all schoolchildren attending grades 4–6 about common symptoms and diseases. Within 5 weeks, filled-in questionnaires were returned by 121 schools (90.3%), with over 12,000 pupils interviewed. In the second step, a cross-sectional epidemiological survey was carried out in a random sample of 60 schools. Two consecutive stool samples were collected from previously interviewed children, processed by the Kato-Katz method and examined for *S. mansoni* and soil-transmitted helminths. Complete questionnaire and parasitological survey data were obtained from 5047 children. The overall prevalence of *S. mansoni* was found to be 54.4% with a strong geographical heterogeneity, ranging from 4.0 to 94.0% at the unit of the school [47]. These data allowed the drawing of a spatially explicit risk map for *S. mansoni*, which is important for guiding disease control interventions in a cost-effective manner. This line of investigation, with a strong emphasis on risk mapping and prediction, using a combination of geographic information system, remote sensing techniques, and Bayesian spatial statistics, has been pursued further. Predictive risk maps for *S. mansoni* [57], hookworm [48] and *S. mansoni*–hookworm co-infections [58] have been generated and can guide spatial targeting of control interventions, and can be utilized for monitoring changes longitudinally.

Over time, we further expanded our research emphasis on polyparasitism, with epidemiological investigations increasingly focusing on entire communities, rather than schoolchildren only. Our initial study was carried out in Gueupleu (see Figure 11.1). The epidemiological design was a cross-sectional survey with all inhabitants invited to provide a single fecal specimen. Two different diagnostic approaches were employed, namely (i) the Kato-Katz method for diagnosis of *S. mansoni* and soil-transmitted helminths as described above, and (ii) an ether-concentration method applied to sodium acetate–acetic acid–formalin (SAF)-fixed stool samples [59, 60]. The latter samples were examined under a light microscope by experience laboratory technicians for *S. mansoni*, soil-transmitted helminths, and intestinal protozoa (i.e. *Blastocystis hominis*, *Chilomastix mesnili*, *Endolimax nana*, *Entamoeba coli*,

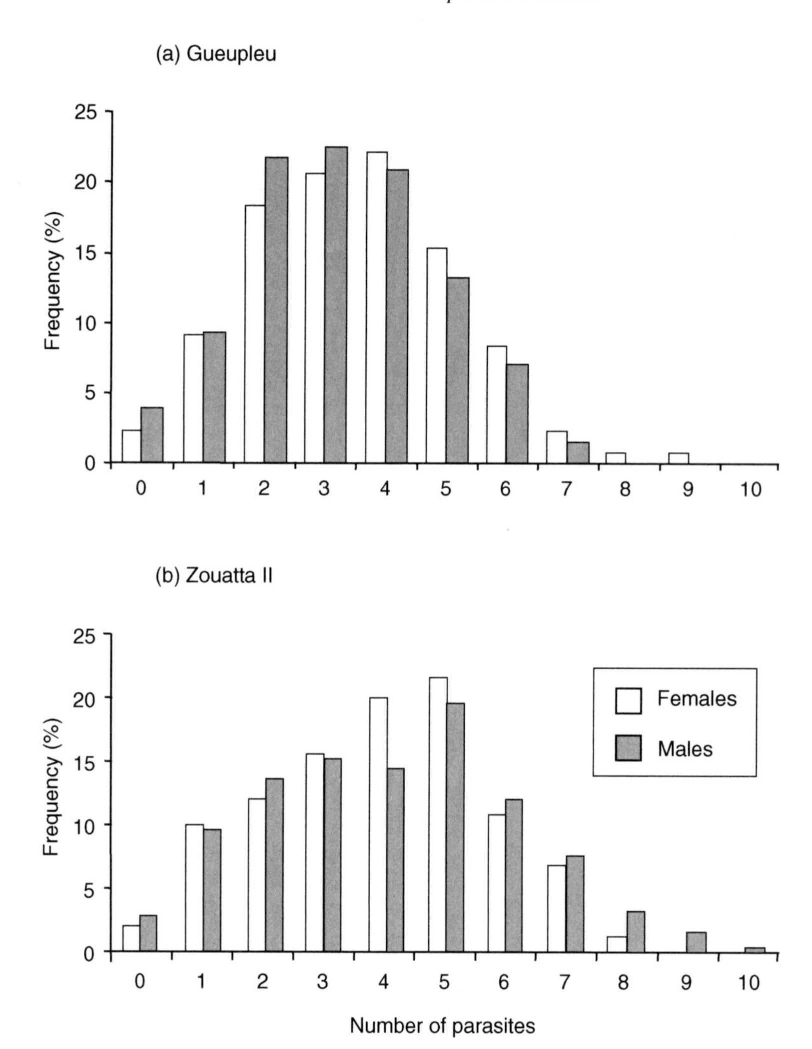

Figure 11.2. Parasite load exemplified by the extent of polyparasitism among males (■) and females (□) in two villages in the region of Man, western Côte d'Ivoire (see map in Figure 11.1).

Entamoeba hartmanni, *Entamoeaba histolytica/E. dispar*, *Giardia duodenalis*, and *Jodamoeba bütschlii*). Overall, 260 people had complete data records and we found that two-thirds of the population sample harbored at least three parasites concurrently [61]. Figure 11.2a shows the frequency distribution of polyparasitism in this population sample, stratified by sex.

In 2002, a cross-sectional parasitological survey was carried out in Zouatta II, located some 30 km east of the town of Man (see Figure 11.1). We were able to obtain complete data records from 500 participants, that is multiple fecal specimens diagnosed by the Kato-Katz technique and an ether-concentration method, and a single thick-and-thin blood film examination [44]. As explained above, the stool samples were used for diagnosis of *S. mansoni*, soil-transmitted helminths and intestinal protozoa, whereas the blood films were diagnosed for *Plasmodium* [43]. The frequency distribution of polyparasitism in this population sample, stratified by sex, is depicted in Figure 11.2b. Three or more parasites were found in three-quarters of the population.

Study participants also provided urine specimens and small amounts (3–4 ml per person) were frozen, and transferred on dry ice to Imperial College London for subsequent ^{1}H NMR analysis. Our goal was to identify biomarkers from the acquired NMR spectra that might give rise to a specific parasitic infection. This proved an extremely challenging exercise, probably explained by the sheer complexity of the dataset (i.e. 500 participants, aged 5 days to 91 years, both sexes, different nutritional status, high and diverse parasitic load, etc.). Regarding age, it is interesting to note that different age groups seem to cluster tightly when considering a three-dimensional metabolic space, as shown in Figure 11.3.

Thus, what is needed now is assessment of distinct metabolite profiles across different age groups, socioeconomic status and environmental elevation. Carrying out such profiling requires that we can clearly identify the metabolic signatures of a single species infection in humans and, in a sequential step, the signature of dual and multiple infections. As discussed in more detail in a subsequent section below, ^{1}H NMR signatures of *S. mansoni* have only been characterized in mice, thus far.

11.2.2. Metabolic signatures of schistosome infection

As a first step in identifying the metabolic signature of a schistosome infection, we used a *S. mansoni*-mouse model [62]. Ten mice were each infected with 80

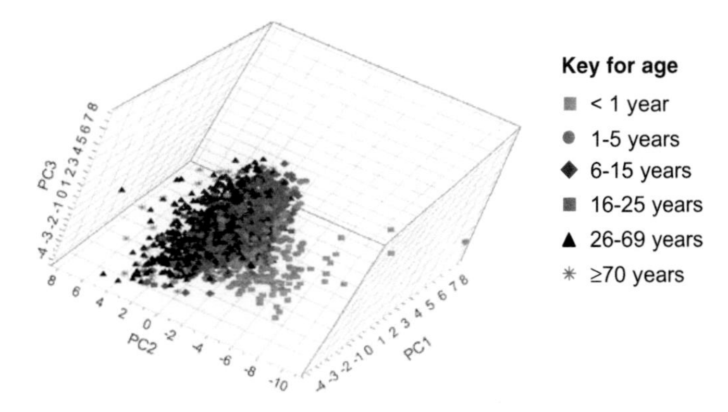

Figure 11.3. Three-dimensional PCA projections of ^{1}H NMR spectra from urine samples in Zouatta II, Côte d'Ivoire: age distribution.

Markers: Liver damage, gut microbes, energymetabolism, fat metabolism

Metabolite	Chemical shift (ppm)	Changes direction	Importance of contribution (rank)
Trimethylamine	δ 2.88	↑	7.547 (1)
Phenylacetylglycine	δ 7.36, δ 3.76, δ 3.68	↑	3.196; 2.957; 2.767 (2)
Taurine	δ 3.24, δ 3.28, δ 3.40, δ 3.44	↓	1.664; 1.608; 1.159; 6.504 (3)
Citrate	δ 2.54, δ 2.70	↓	3.553; 2.782 (4)
Acetate	δ 1.92	↓	2.663 (5)
Creatine	δ 3.04, δ 3.92	↑	5.671; 4.234 (1)
β-alanine	δ 2.32, δ 3.32	↑	3.308; 1.934; (2)
Pyruvate	δ 2.36	↑	2.946 (3)
2-oxoisocaproate	δ 0.92, δ 2.64	↓	2.746; 1.064 (4)
Tryptophan	δ 7.24, δ 7.28, δ 7.52, δ 7.71	↑	2.391; 1.945; 0.914; 0.999 (5)
Hippurate	δ 7.84, δ 7.64, δ 7.56, δ 3.96	↓	2.137; 1.005; 0.415; 0.997 (6)
Butyrate	δ 0.88, δ 1.56	↓	1.487; 1.484 (7)
p-cresol glucuronide	δ 3.88, δ 3.84, δ 7.08, δ 7.04	↑	1.487, 1.438, 1.303, 1.217 (8)
Succinate	δ 2.44	↓	1.350 (9)
Malonate	δ 3.16	↓	1.204 (10)
Propionate	δ 1.04, δ 2.16	↓	1.093; 1.717 (11)
2-oxoglutarate	δ 2.46,δ 3.00	↓	0.890; 0.704 (12)
Alanine	δ 1.48	↓	0.889 (13)
2-oxoisovalerate	δ 1.12	↓	0.619 (14)
3-D-hydroxybutyrate	δ 1.20	↓	0.554 (15)

Figure 11.4. Metabolic signature of a *S. mansoni* infection in mice, color coded according to the biological significance of the discriminating metabolite. Variable importance parameters were generated from partial least squares – discriminant analysis as detailed in [62].

S. mansoni cercariae, and urine samples were collected 49 and 56 days post-infection. A corresponding set of urine samples were collected from each of 10 uninfected control mice. The metabolic signature of a *S. mansoni* infection in mice is summarized in Figure 11.4.

Changes in gut microbes, energy metabolism, and fat metabolism are clearly in evidence. Liver damage is an expected consequence of such infection, and it is reflected here by decreases in taurine and acetate and increases in creatine, *β*-alanine, and tryptophan. With this characterization of a mouse infection, the next step is to characterize *S. mansoni* infection in humans with no other parasitic infection (e.g. among tourists and travelers returning from the tropics, who only were exposed for a short time, but have a laboratory-confirmed infection as revealed by the presence of parasite eggs in their feces). The analogous studies should then be carried out for hookworms (separately for *Necator americanus* and *Ancylostoma duodenale*), other soil-transmitted helminths (e.g., *A. lumbricoides* and *T. trichiura*), as well as a host of other intestinal parasites. Characterization of *S. mansoni*–hookworm co-infection needs to be studied in an animal model, and thereby provide clues for what to anticipate in humans infected with those two parasites concurrently. This sequence of analyses can carry us a considerable distance toward the longer term objective of parallel diagnoses of multiple parasitic infections in endemic areas in the tropics. Some of the complexity associated with this line of inquiry is already manifest in a study of *S. japonicum* infection in the Syrian hamster, to which we now turn.

Wang *et al.* [63] infected two separate batches of male Syrian hamsters with 100 *S. japonicum* cercariae. At days 34–36 post-infection, urine was obtained from the infected hamsters and from corresponding sets of non-infected control hamsters. One set of hamsters was infected during the active period of the year, and the second set was infected during what would normally be a time of hibernation. As indicated in Figure 11.5, we already see differences in metabolic profiles between the control hamsters as a function of active *versus* hibernation period. The interesting feature of these differences, in the context of polyparasitism, is that when pooling data from both experiments one is seeing two superimposed profiles: the hibernation effect and the *S. japonicum* effect.

In addition to the coarse discrimination via principal component analysis (PCA, see Chapter 6), we find that the main biochemical effects of a *S. japonicum* infection in the Syrian hamster are reduced levels of urinary tricarboxylic acid cycle intermediates, including citrate and succinate, and increased levels of pyruvate, a range of microbial-related metabolites such as hippurate, *p*-cresol glucuronide, phenylacetylglycine, and trimethylamine. This is quite similar to what was observed in *S. mansoni* infections in the mouse, except that in the hamster there was inhibition of manufacture of short-chain fatty acids. Given the extensive, and now resurging, range of *S. japonicum* infection in China [45], also accompanied by polyparasitism with other intestinal parasites, it would be important to identify the metabolic profile of *S. japonicum* in humans. Then animal studies, followed by human studies, of metabolic profiles of co-infection are in need of attention. We

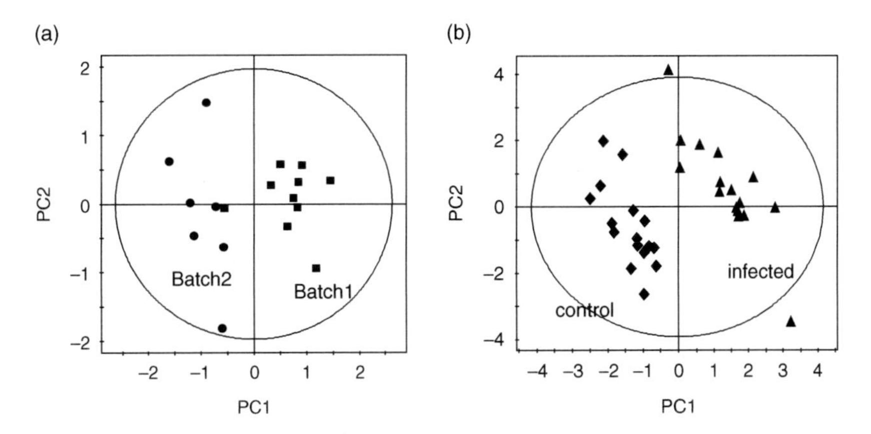

Figure 11.5. Two-dimensional PCA projections of ^{1}H NMR spectra from urine in (a) control hamsters in the hibernation (■) and active period (•) and (b) after removing interbatch differences. A clear differentiation between the control (▲) and infected (♦) hamsters can be observed (Source: [63]).

believe that this iterative process employing suitable host-parasite models kept under well-controlled laboratory environments and "real-life" investigations in humans hold promise for biomarker discovery and metabolic profiling of parasitic infections and disease states.

11.2.3. Allostatic load and aging: Early warning biomarkers

A wide range of biomarkers, reflecting activity in multiple biological systems (e.g. neuroendocrine, immune, cardiovascular, and metabolic), have been found to prospectively predict disability, morbidity, and mortality outcomes in elderly populations. Levels of these biomarkers, singly or in combination, may serve as an early warning system of risk for future adverse health outcomes. Using data from the MacArthur Study of Successful Aging, 13 biomarkers were examined as predictors of mortality occurrence over a 12-year period in a sample of men and women ($N =$ 1189) aged 70–79 years at enrollment into the study [15]. Biomarkers examined in analyses included markers of neuroendocrine functioning (urinary epinephrine, norepinephrine, cortisol, dehydroepiandrosterone), immune activity (serum C-reactive protein (CRP), fibrinogen, interleukin-6 (IL-6), albumin), cardiovascular functioning (systolic and diastolic blood pressure), and metabolic activity (high-density lipoprotein (HDL) cholesterol, total to HDL cholesterol ratio, glycosylated hemoglobin). Recursive partitioning techniques [64, 15] were used to identify sets of combinations of different biomarkers that were associated with a high risk of mortality over the 12-year period. Of the 13 biomarkers examined, almost all entered into one or more combinations of high-risk conditions. Prediction rules for individuals were based on their combinations of high risk conditions. The recursive partitioning-based rules can also be interpreted as an operationalization of the notion of allostatic load, as described in a prior study [11].

There are interesting gender differences associated with combinations of biomarkers at elevated but, in some instances still, sub-clinical levels, which predict the earliest deaths after the baseline assessments. For the men at highest risk of mortality, there is a cluster of five biomarkers that occur together at elevated levels, namely (i) CRP, (ii) IL-6, (iii) fibrinogen, (iv) norepinephrine, and (v) epinephrine. In this subgroup of men, 71.7% have all five of these biomarkers at elevated risk levels; 97.2% have at least four of the five biomarkers; and 100% have three or more of the five biomarkers at elevated risk levels.

For women at highest mortality risk, there is a cluster of four biomarkers that occurred together at elevated levels: (i) systolic blood pressure, (ii) CRP, (iii) IL-6, and (iv) glycosylated hemoglobin. All four of these biomarkers were present for 17.2% of the highest risk women, and two smaller biomarker clusters of systolic blood pressure, glycosylated hemoglobin, and CRP or IL-6 occurred in 34.5 or

37.9% of the women, respectively. Systolic blood pressure was present at elevated risk levels in all (100%) of the women with the shortest times to death after baseline data collection.

Direct gender comparisons reveal that elevated systolic blood pressure occurs in 100% of the highest risk women's combinations of conditions and in only 17% of the men's combinations. Fibrinogen, norepinephrine, and epinephrine, individually and in combination, dominate the men's combinations, but do not even occur on the women's sets of highest risk conditions. CRP and IL6 occurred frequently in the highest mortality risk combinations of both men and women.

Analogous to the focused coronary heart disease study on a clinical population [18], it would be highly desirable to do metabolic profiling from serum samples on the much more diverse population of the MacArthur Study of Successful Aging. Formulating and testing mortality prediction rules using ^{1}H NMR spectra alone, and comparing their performance with the prediction rules based on the above-mentioned 13 biomarkers [15] would be very instructive. Ultimately, we would like to formulate mortality prediction rules using both metabolic profiles and the 13 biomarkers. We conjecture that such rules should encapsulate much more subtle gradations of mortality risk than any of the extant techniques.

11.2.4. Metabolic profiling of insulin resistance

Many of the large human population studies of biomarkers and subsequent disease risk (as described above) include measurements of levels of glycosylated hemoglobin as an indicator of possible insulin resistance (IR) [65]. Before considering metabolic profiles in such studies that also suggest the presence of IR, it is necessary to at least be able to identify the metabolic signature of IR phenotypes in controlled animal studies. To this end, [19] tested the effects of dietary changes (5–40% fat content) on plasma and urine ^{1}H NMR profiles in inbred mouse strains selected for their resistance (BALB/c) and susceptibility (129S6) to IR and nonalcoholic fatty liver disease (NAFLD). The authors showed that gut microbial metabolism alters the metabolism of the mammalian host. Indeed, a specific metabotype with low plasma phosphocholine and high gut microbiota–mediated urinary excretion of methylamines is associated with predisposition to IR and NAFLD. Dumas *et al.* [19] have described the first diet-induced toxicity mechanism triggered by symbiotic microbiota. This is another example of the "thin line between gut commensal and pathogen", a phenomenon noted by several prior investigators [66, 67].

It would be useful, as a next step, to superimpose the dietary interventions with chronic psychological stress challenges in mice. This might provide useful suggestions of what to look for in the metabolic signature of psychosocial–dietary phenotypes in humans and their linkage to IR.

11.2.5. Metabolic investigations of neuropathological disease

There is a paucity of suitable biomarkers for assessing disease progression in neurodegenerative and psychological disorders. The identification of reliable biomarkers for disease progression and response to therapeutic intervention, particularly those that can indicate detection of neurodegenerative diseases in an early or even pre-symptomatic stage, are essential. Both toxicological experiments and studies on genetically modified animal models have provided invaluable insight into the etiology of a wide spectrum of neurodegenerative disorders [68]. For example, 6-hydroxydopamine (6-OHDA) in rats [69] and 1-methyl-4-phenyl-1,2,3,6-tetrahydropyridine (MPTP) in non-human primates [70] have been commonly used to produce animal models of Parkinson's disease (PD) via specific degeneration of the nigrostriatal cells. The R6/2 mouse model for Huntington's disease (HD), containing 150 CAG repeats, is consistent with the juvenile onset form of the disease [71]. The NMR-based metabonomic studies in the R6/2 mouse characterized the metabolic signature of HD in several tissues and body fluids (urine, plasma, skeletal muscle, striatum, cerebral cortex, cerebellum, and brain stem) at 4, 8, and 12 weeks of age [72]. Choline levels were observed to decrease in most of the neuroanatomical regions analyzed in the R6/2 mouse, whereas glycerophosphocholine increased, suggesting a pro-catabolic phenotype in the R6/2 mouse model. This has also been shown in a small number of HD patients where glycerophosphocholine levels correlated with disease progression [73]. Currently the European Huntington's Disease Network is focused on collaborating across European cohorts to establish biomarkers of HD and is actively employing a systems biology approach combining transcriptomic, proteomic, and metabonomic data from HD patients and age- and sex-matched controls.

The characterization of biomarkers for schizophrenia in accessible body fluids such as plasma or urine offers exciting possibilities of developing new early or pre-symptomatic therapeutics in addition to identifying high risk individuals or subgroups. The ^{1}H NMR spectroscopic analysis of CSF samples from 135 drug-naive, first-onset, minimally treated paranoid schizophrenia patients and healthy control individuals showed differences in the levels of glucose and a consistent pH dependent shift of the glutamate resonances [20]. This study also showed normalization of the CSF profiles toward the control group, but only in those patients where antipsychotic therapy intervention was made during the first psychotic episode. A second metabonomic study investigating the differences in blood plasma between female monozygotic twins, discordant for schizophrenia, showed a correlation between low density lipoprotein (LDL) and very low density lipoprotein (VLDL) levels and global functioning scores [21]. For both of these studies the subtlety of the metabolic signature of schizophrenia was extracted from the complexity of human biofluid profiles using partial least squares discriminant analysis (PLS-DA) with an in-built orthogonal filter for removing confounding systematic variation. This and closely related mathematical modeling methods are undergoing continuous development and show a great deal of promise for epidemiological applications.

11.3. Metabolic signatures of physical and psychosocial experience

11.3.1. Response to stress

In an attempt to identify the changes in metabolic profiles over time as a result of both acute and chronic psychological challenges, Teague *et al.* [22] employed restraint stress and physical shaking paradigms in both acute and chronic modes to Sprague-Dawley (SD) rats. Metabolic profiles were obtained over time on two scales: hours and weeks. At the coarse level of PCA with scores plots depicted in two dimensions, we already see evidence of changes in metabolites released in response to an acute stress at 1 h, 3 h, and 6 h post-challenge. For chronically stressed animals there are distinct shifts observed at days 9, 21, 35, and 44 after initiation of daily restraint stress – and, in a sub-group of animals, augmented by some physical shaking – challenges intermixed with quiescent periods. The interesting conclusion regarding the metabolic changes was that the rats were adapting to the stressful challenges. The study of Teague *et al.* [22] did not follow the rats long enough to observe incident disease conditions, nor did it include challenges to which the rats were clearly not adapting. These amendments would be necessary to distinguish what we would regard as metabolic profiles of allostatic load from the dynamics of metabolite release that should more properly be associated with allostasis, a process of maintaining homeostasis [74, 75]. The important feature of this study, from our perspective, is that it provides clear evidence of the sensitivity of metabolic systems to psychological stressors and indicates the potential for associating more complex psychosocial phenotypes – in animals and humans – with distinctive metabolic profiles.

Metabolic effects of stress in the form of moderate periods of maternal separation of newborn rats have been assessed using metabonomic techniques. Rat pups were removed from their cages and their mothers, and placed in an isolated cage in an adjacent room at a temperature of 32.0 °C (±0.5 °C) at post-natal days 2–14 for 180 min daily. At the end of each separation period, pups were returned to their home cage and rolled in the soiled home cage bedding before being reunited with their mother. Blood plasma samples were obtained from the maternal separated group and control group at 3–4 months later. It was discovered that rats in maternal separated group were metabolically different from rats in the corresponding control group. Exposure to the stress of isolation from the mother caused a decrease in levels of total lipoproteins and increased levels of amino acids, glucose, lactate, creatine and citrate. Changes in these metabolites are likely to result from an increase in the secretion of hormones such as glucocorticoids and catecholamines, which are well documented to respond to stressful challenges [76, 77].

In the same study, animals were exposed to a stress at a later stage of development (3–4 months of age), using a water avoidance test, with or without having undergone the maternal separation stress. Rats exposed to water avoidance alone showed a clearer metabolic differentiation from a matched group of control rats than those exposed to both maternal separation and water avoidance. This observation could imply that rats that had been previously stressed may have become desensitized to stress and would, thus, accommodate secondary stress better than those that had not been exposed to this first early life stressor [23].

Dietary intervention was also assessed to address if a long chain polyunsaturated fatty acid (LC-PUFA)–enriched dietary intervention can reverse the effects of stress-induced metabolic changes. The results showed that dietary intervention had a measurable metabolic effect on rats exposed to stresses. Rats fed with LC-PUFA diets followed by exposure to the water avoidance stress presented relative higher levels of VLDL and lower levels of HDL than those exposed to water avoidance alone and without the LC-PUFA diet. The LC-PUFA dietary intervention was unable to reverse the metabolic changes induced by stress, but did alter the response to some extent [23].

11.3.2. Complex behavioral phenotypes: Visible burrow systems

Given the substantial array of animal studies that identify the metabolic signature of single challenges/interventions, the pressing need is to expand the available catalogue of profiles to include considerably more complex phenotypes. In the context of psychosocial phenotypes, it would be useful to carry out a series of studies in rats exposed to more naturalistic challenges than the restraint stress and physical shaking paradigms in the study of Teague *et al.* [22]. The visible burrow system, extensively utilized to study behaviors of rats in more natural habitats, would seem like an appropriate setting in which to do metabolic profiling over time. The extant visible burrow studies characterizing normal defensive and aggressive behaviors in rats [24] lack any simultaneous metabolic profiling. In more general terms, the flexibility to introduce considerable variation in environments in visible burrow systems, linked to metabolic profiling over time, could be the basis for building an important catalogue of responses in animals that could be very suggestive of what to look for in human studies on large heterogeneous populations.

11.3.3. Cumulative economic and social experience and allostatic load

We illustrate the linkage between psychosocial phenotypes and their biological signature in a sub-sample of participants in the Wisconsin Longitudinal Study

(WLS) (UW-Madison). The WLS is a longitudinal survey, comprising an initial randomly selected population sample of 10,317 men and women, who graduated from Wisconsin high schools in 1957. Baseline survey data were obtained from the original respondents in 1957, and repeated cross-sectional surveys were carried out in 1975, 1992–93, and 2004–05. An additional round of sociodemographic survey information, biomarkers and supplementary questionnaires pertaining to social relationships, were collected from a representative sub-sample of 106 individuals (1%) in 1997–98.

Using the rich life history data available in WLS, we created pathways using information from two life domains, cumulative economic standing and cumulative social relationship profiles. This allowed us to investigate whether the degree of adversity relative to advantage in these two domains was linked to the probability of exhibiting multiple physiological indicators (i.e. allostatic load) that predict later-life incident cardiovascular disease and decline in physical and cognitive functioning. Social relationships were a central feature of our pathway specifications because of the substantial literature linking the interpersonal realm to mortality outcomes and, when positive, to protective factors [78, 79].

A parental bonding scale was employed to assess the extent to which children had caring as well as supportive and affectionate relationships with their parents [80]. Separate "Father Caring" and "Mother Caring" scales were used. The underlying working hypothesis is that the experience of uncaring and even abusive interactions with one or both parents governs negative social relationship pathway/phenotype. Moreover, it is hypothesized to contribute to wear and tear on multiple biological systems, with the effects becoming manifest in later life [81].

With regard to appraisal of midlife relationships, four aspects of connection to a spouse or significant other were considered, namely (i) emotional, (ii) sexual, (iii) intellectual, and (iv) recreational modes of intimacy. The rationale for including these aspects of connections is that they are likely to contribute to cumulative relationship profiles, which in turn might be associated with later-life biological indicators of wear and tear on the human body. Assessment was by means of the Personal Assessment of Intimacy in Relationships (PAIR) inventory, according to Schaefer and Olson [82]. While the emotional and sexual subscales represent the most intimate forms of connection between two people, the intellectual and recreational subscales accentuate mutually enjoyed experience, companionship, and the scope of shared communication. The PAIR inventory seeks to identify the degree to which each partner feels intimate in each of the four relational areas.

Formal specification of relationship pathways were done according to the following three steps. In the first step, individuals were cross-classified as either positive (+) or negative (−) on the "Father Caring" and "Mother Caring" components of the parental bonding scale according to whether their score on the respective scale is above or below the median. Each person will, thus, have a ("Father Caring, Mother

Caring") pair of valences which can be either (−,−), (−,+), (+,−), or (+,+). In the next step, scores on the emotional (E) and sexual (S) subscales of the PAIR inventory were combined into an E + S score, representing probes of the most personal and intimate aspects of spousal ties. Analogously, the intellectual (I) and recreational (R) scores were combined into an I + R score. Persons were scored as positive (+) or negative (−) on each of the combined scales according to whether they were above or below the median on the respective combination. Each individual had a pair of (E + S, I + R) valences which could also be any one of the four options (−,−), (−,+), (+,−), or (+,+). In the third and final step, a person was defined to be on the negative *relational pathway* – equivalently, *negative relational phenotype* – if he/she scored (−,−) on the ("Father Caring", "Mother Caring") pair and/or on the E/S and I/R pair in adulthood. An individual was defined to be on a *positive relational pathway* – equivalently, *positive relational phenotype* – if he/she has at least one + on the "Father Caring", "Mother Caring" pair and at least one + on E + S or I + R. Thus, the positive path requires some positive relational experience with one or both parents in childhood and at least one of the combined forms of intimacy in adulthood. This pathway again underscores the cumulative nature of positive emotional experience with significant others in childhood and adulthood.

With regard to economic indicators, we used household income assessments at two points in the life course as the basis for generating economic pathways. Household income in 1957 for WLS respondents indicated the economic circumstances of their parents. These figures were assembled at the first wave of WLS data collection and are derived from Wisconsin state tax records. We also used respondents' household income, assessed in 1992–93, when they were 52–53 years of age. Individuals were defined to have been in a positive childhood economic environment (+) [or respectively, negative (−)] if the household income of parents in 1957 was at or above [or respectively, below] the median household income for the state in that year. Analogously, at age 52–53, respondents were defined to be in positive (+) [or respectively, negative (−)] economic circumstances if their household income was at or above [or respectively, below] the median household income for the state in 1992–93.

Composite pathways were then constructed based on the combination of both social relational and economic circumstances in both childhood and adulthood. Thus, those with the greatest cumulative adversity were those who had persistent economic disadvantage from childhood to adulthood (−,−) *and* who were on a persistently negative social relationship pathways from childhood to adulthood. Those with the greatest cumulative advantage has positive profiles on economic circumstances both in childhood and adulthood (+,+) *and* were on a persistently positive social relationship pathway. The full combination of pathways included those with mixed adversity and advantage as well.

Table 11.1
Risk-zone and system designation for individual biomarkers

Risk zone	System
Highest quartile	
Systolic blood pressure (SBP) (>= 148 mm Hg)	Metabolic
Diastolic blood pressure (DBP) (>= 83 mm Hg)	Metabolic
Waist-hip Ratio (WHR) (>= 0.94)	Metabolic
Ratio total cholesterol/HDL (>= 5.9)	Metabolic
Glycosylated hemoglobin (>= 7.1%)	Metabolic
Urinary cortisol (>= 25.7 ug/g creatinine)	HPA-axis
Urinary norepinephrine (>= 48 ug/g creatinine)	Sympathetic nervous system
Urinary epinephrine (>= 5 ug/g creatinine}	Sympathetic nervous system
Lowest quartile	
HDL cholesterol (<= 37 mg/dl)	Metabolic
DHEA-S (<= 350 ng/ml)	HPA-axis

Abbreviations: HDL = High-density lipoprotein cholesterol; HPA = hypothalamus–pituitary–adrenal axis; DHEA = dehydroepiandrosterone.

With these cumulative phenotypes at hand, we utilized a system of 10 biomarkers and elevated risk zones/levels for them [13], as indicated in Table 11.1.

We then calculated an allostatic load score [10, 13] designed to be an early warning indicator of possible dysregulation in one or more of the biological systems designated. We defined allostatic load score for an individual to be: [Number of biomarkers in the above list for which the individual satisfies the stated inequality]. Contributions from different biological systems are all weighted equally in this scoring of AL. Higher AL scores are interpreted as reflecting more extensive wear and tear on the systems being directly measured and possibly on other systems (e.g. the immune system) that communicate with those we are measuring but for which there was no assessment in the WLS. The ALs at or above three have been shown to predict elevated incidence of cardiovascular disease, decline in physical and cognitive functioning, and mortality at older ages [13].

Figure 11.6 shows the associations between elevated risk zone AL scores at ages 59–60 and the various economic and social relational pathways/phenotypes. As predicted, those showing the greatest likelihood of high allostatic load (69%) were individuals with persistent economic adversity and who also were on the negative social relationship pathway. Comparatively, only 36% of individuals on the counterpoint pathway (persistent economic advantage, positive relationships in childhood and adulthood) had high allostatic load. Illustrating the possible protective influence of good social relationships vis-à-vis persistent economic adversity, we note than only 22% of these individuals had high allostatic load.

Such findings are limited by the small sample size and relatively crude behavioral phenotypes. More pertinent to the present chapter is the coarse set of resting

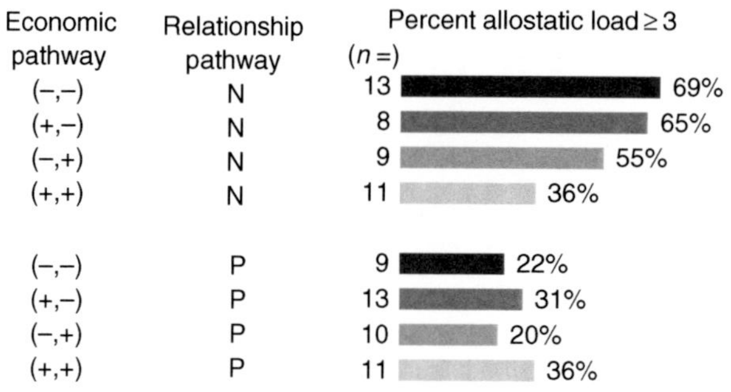

Economic pathway	Relationship pathway	(n =)	Percent allostatic load ≥ 3
(–,–)	N	13	69%
(+,–)	N	8	65%
(–,+)	N	9	55%
(+,+)	N	11	36%
(–,–)	P	9	22%
(+,–)	P	13	31%
(–,+)	P	10	20%
(+,+)	P	11	36%

Figure 11.6. Economic and social relationship phenotypes linked to allostatic load. [P = positive relational phenotype, N = negative relational phenotype, + = positive economic phenotype, – = negative economic phenotype] (Source: [25]).

biomarker levels that comprise allostatic load. A metabonomics approach would thus offer significant refinements in understanding possible costs, measured in metabolic profiles, associated with these various life histories that illustrate not only cumulative adversity, but its possible modulation by protective social relational phenotypes.

11.3.4. Psychological profiles and biomarkers

In the context of mental health, we raise the question of whether psychological well-being and ill-being comprise opposite ends of a bipolar continuum, or whether they are best construed as separate, independent dimensions. Bipolarity predicts "mirrored" biological correlates (i.e. well-being and ill-being correlate similarly with biomarkers, but show opposite directional signs), whereas independence predicts "distinct" biological correlates (i.e. well-being and ill-being have different biological signatures). Multiple aspects of well-being and ill-being were assessed in a sample of aging women in the Wisconsin Community Relocation Study [83] ($N = 135$, mean age = 74 years) on whom neuroendocrine and cardiovascular factors were also measured. Well-being and ill-being measures were significantly linked with diverse biomarkers.

The positive phenotypes for this study were responses on scales of six dimensions of eudaimonic well-being, namely (i) autonomy, (ii) environmental mastery, (iii) personal growth, (iv) positive relations with others, (v) purpose in life, and (vi) self-acceptance. These were operationalized using Ryff's theoretical integration of numerous formulations of positive functioning [84]. Positive forms of hedonic well-being were assessed with the positive affect scale of the Positive and Negative Affect Schedule (PANAS) inventory [85] and the positive affect scale of the short-form Mood and Anxiety Symptom Questionnaire (MASQ) [86]. The PANAS scale consists of 10 items that gather information about the respondent's affective state (scale range = 10–50). Items for positive affect included feeling interested, excited, strong, enthusiastic, proud, alert, inspired, determined, attentive, and active. The MASQ scale included 14 items (scale range = 14–70) that capture joy-in-living aspects of positive affect (e.g. felt happy, cheerful, optimistic, up, looked forward to things with enjoyment, having a lot of fun). Respondents were interviewed about how much they had felt this way with a recall period of one week. High scores indicate high positive affect.

Finally, psychological ill-being, the negative phenotype, was measured in terms of four different assessments, namely (i) negative affect, (ii) depressive symptoms, (iii) trait anxiety, and (iv) trait anger. These are described in detail elsewhere [30, 83].

Linking these phenotypes to the same set of biomarkers used in Table 11.1, augmented by weight, we find that the overall pattern of effects was more strongly supportive of the distinct, as opposed to the mirrored, hypothesis. For seven biomarkers, three neuroendocrine (cortisol, DHEA-S, and norepinephrine) and four cardiovascular (HDL cholesterol, total/HDL cholesterol, systolic blood pressure, and waist–hip ratio), significant correlations with well-being (or ill-being) were not accompanied by significant associations with ill-being (or well-being, respectively) and the *same* biomarker. Moreover, of the correlations that reflected the distinct pattern, the majority (69%) were instances in which measures of well-being were significantly correlated with biomarkers, but no parallel effects were evident for measures of ill-being. Stated another way, psychological well-being showed a more pervasive and distinct biological signature than was evident for psychological ill-being. However, it is important to note that these effects were evident for only a subset of the indicators of eudaimonic well-being (e.g. positive relations with others, purpose in life, personal growth). Surprisingly few significant associations were found for hedonic well-being.

Alternatively, for two biomarkers, namely weight and glycosylated hemoglobin, there was strong evidence of mirrored biological patterns. For both, it was the same measure of well-being (i.e. positive relationships with others) that showed significant differences with the opposite directional pattern obtained for measures of ill-being. Specifically, those with better social relationships had lower weight, while those with higher depressive symptoms had higher weight. Similarly, those with better social relationships had lower levels of glycosylated hemoglobin, while those with higher levels of negative affect, trait anxiety, and trait anger had higher levels of the same biomarker. In addition, epinephrine was the single biomarker showing positive associational patterns both with well-being (positive relations with others) and ill-being (trait anger).

Although nearly all of the above findings conformed to the anticipated pattern that higher well-being would be associated with lower biological risk, and conversely that higher ill-being would be associated with high biological risk, there were several anomalies, the most notable of which was that higher levels of negative affect, anxiety, and anger were significantly correlated with lower levels of systolic blood pressure. Whether this is a replicable finding remains to be seen, but we note that a community survey of persons aged between 77 and 99 years found that systolic hypotensive individuals had higher levels of negative affect than systolic normotensives [87]. The central challenge for the future is to assess whether there are distinct metabolic profiles associated with the above phenotypes. An affirmative answer would provide the first characterization of the metabolic signatures of positive and negative well-being.

11.3.5. Protective social environments in contexts of genetic vulnerability

The vast and growing literature on susceptibility genes for diverse diseases is linked to complementary studies focused on identifying the pathways from pre-disposing genes to disease states. While this is a natural research progression, it ignores the interesting feature that most people with susceptibility genes do not progress to disease endpoints. As one illustration of this point, two recent papers focusing on anti-social behavior [31] and depression [32] emphasized gene *X* environment interactions in which childhood maltreatment and possession of susceptibility genes (having a functional polymorphism associated with deficiency in monoamine oxidase A (MAOA) activity, or being a short allele (s/s) homozygote in the promoter region of the serotonin transporter (5-HTT) gene) were associated with these outcomes. What both papers, and many others like them, neglect to point out is that the majority of people with the adverse environment and genetic predisposition for, respectively, conduct disorders or depression do not ultimately manifest these conditions. This suggests that there are likely positive environmental influences that negate the negative predispositions and childhood social environments. Identification of such nurturing environments and assessment of gene expression patterns associated with them is a topic still waiting development.

One instance of positive social environments negating the negative effects of being a type s/s 5-HTT homozygote and experiencing severe maltreatment as a young child is demonstrated in the recent study of Kaufman and colleagues [88]. What we lack at the present time is an understanding of the underlying protective biology (gene expression patterns, metabolic signatures) associated with the effects of nurturing and social support for the *a priori* disadvantaged individuals. Obtaining and interpreting the metabolic signatures of adverse early childhood gene *X* environment disadvantage in the Dunedin (New Zealand) population studied by Caspi *et al.* [31, 32] might shed light on these issues.

11.4. Metabolic distinctions in populations studies

11.4.1. Spectra by country and region: Principal component analyses

At the coarsest level, we ask whether it is possible to discriminate between distinct populations on the basis of broad metabolic characteristics. Operationally, this question can be reduced to asking whether, on the basis of a straightforward PCA of ^{1}H NMR spectra, projection of the data onto a plane specified by the first two principal components – or into a three-dimensional space specified by the first three principal components – produces distinct population-specific clusters of points. An affirmative answer has been given in samples comparing United States (Chicago metropolitan area), Chinese (Guangxi province), and Japanese (Aito town) populations [39], and samples from a northern Chinese population (Shanxi) and a Hawaiian (Honolulu) population comprised of participants of predominantly Japanese origin based on the LC-MS profiling method. Although the specific combinations of metabolites responsible for discrimination at the level of a coarse PCA can be associated with complex combinations of genetic, physical, and social, environmental, dietary, and microbial influences, this coarse-grained separation suggests that much finer distinctions should be possible. Figure 11.7 shows the two-dimensional PCA plots (for more explanation of PCA, please see Chapter 6) indicating separation between the Chicago, Guanxi, and Aito town populations. Figure 11.8 indicates the distinctiveness between the Shanxi (north Chinese) and the Hawaiian (Honolulu) populations.

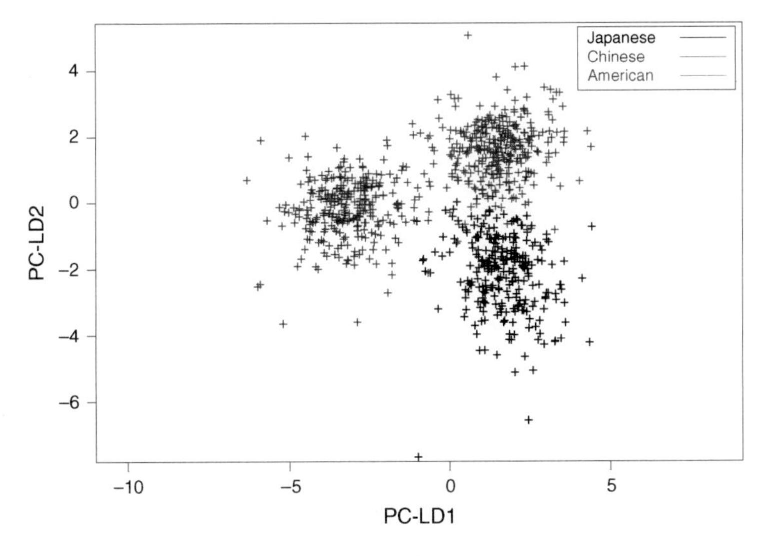

Figure 11.7. Two-dimensional PCA projections of ^{1}H NMR spectra from urine samples in Japanese, Chinese, and American populations in the INTERMAP study (Source: [39]).

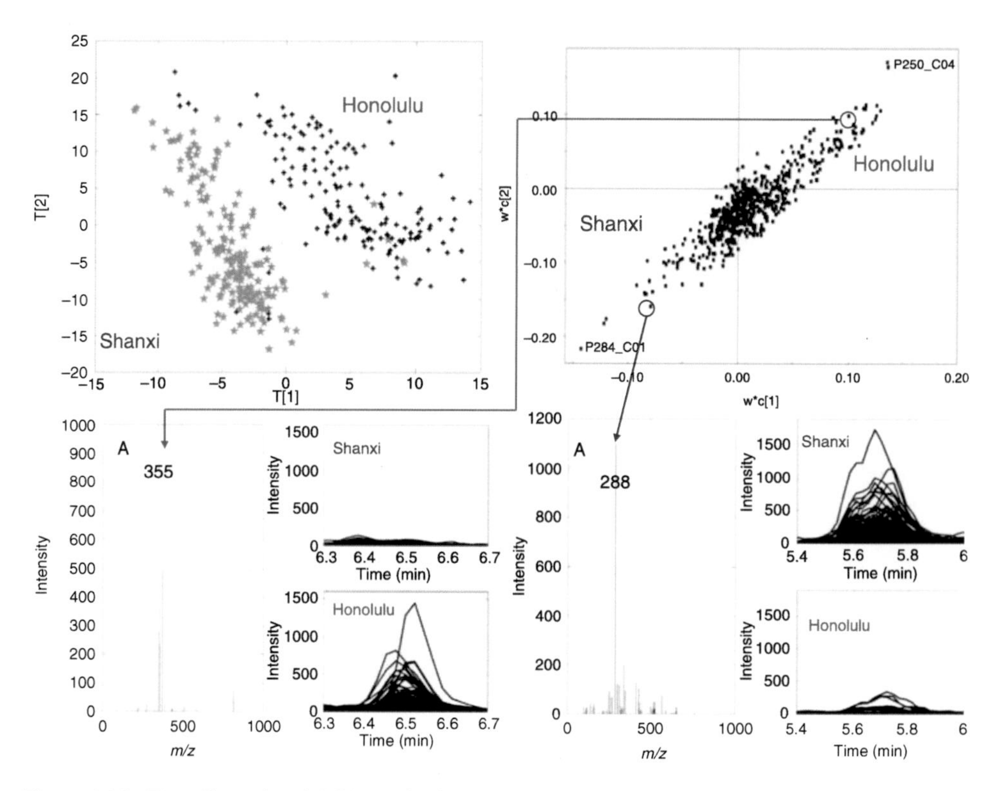

Figure 11.8. Two-dimensional PCA projections and loadings plot of LC-MS spectra distinguishing Chinese and Hawaiian populations (Source: Data from INTERMAP study [89]).

The corresponding loadings plots in this figure show differential distributions for two of the metabolites that dominate the metabolic separation.

The data for these analyses were based on ^{1}H NMR spectra from 24-hour urine samples collected as part of the INTERMAP study [39]. This study was originally launched in 1996 to investigate the relationship of multiple dietary variables to blood pressure. The 4680 adult men and women, aged 40–59 years, were selected from 17 population samples in China, Japan, the United Kingdom, and the United States [37].

On a broader scale, the INTERSALT study [36] was initiated in 1984 to investigate the linkage between sodium intake and hypertension. In industrialized localities, there is a strong relationship in which higher sodium excretion (resulting from high salt intake) is associated with elevated systolic and diastolic blood pressure. Further, blood pressure tends to increase with increasing age. The INTERSALT study also included data collection on two of the world's most unacculturated (with modern customs) populations: the Yanomamo Indians of Venezuela and the Xinguano Indians of Brazil [36]. The Yanomamo have very low salt intake (<0.5 g per day), do not show rising blood pressure with increasing age, and do not exhibit vascular disease. They also have very high aldosterone excretion, which has been documented to play an important role in maintaining blood pressure at lower levels. The Xinguano show very low sodium excretion, but higher than the Yanomamo. They also maintain lower blood pressure throughout their lives and have an average body mass index (BMI) level lower than the average BMI level in any of the western populations in the INTERSALT sample.

With this background at hand, it is interesting to compare the metabolic profiles of the Yanomamo and Xinguano with each other and to juxtapose with a western population that we would anticipate to have a very different signature.

Although the metabolic profile of the Chicago population is clearly differentiated from those of the South American Indians, subtle metabolic differences arise even between populations such as the Yanomamo and Xinguano, based on a combination of genetic, dietary, and lifestyle-related factors. This suggests that phenotypes that include subtle lifestyle variations will have different metabolic signatures. The challenge ahead is to specify these phenotypes, and build up a catalogue of corresponding metabolic profiles with accompanying interpretations.

In a somewhat more detailed decomposition of human populations, we consider PCA plots for urine spectra from sub-groups in Zouatta II, Côte d'Ivoire, a population discussed previously in Section 11.2.1. Figure 11.9 reveals discrimination between men with a heavy *S. mansoni* infection with men who had no *S. mansoni* infection at the cross-sectional survey carried out in May 2002.

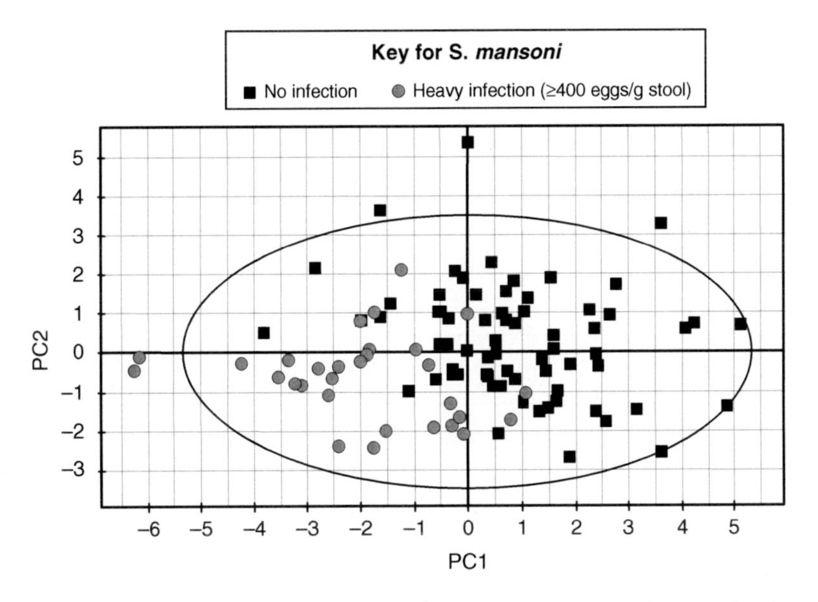

Figure 11.9. Two-dimensional PCA projections of ^{1}H NMR spectra on urine samples from men in Zouatta II, Côte d'Ivoire: uninfected vs heavy infection with *S. mansoni*.

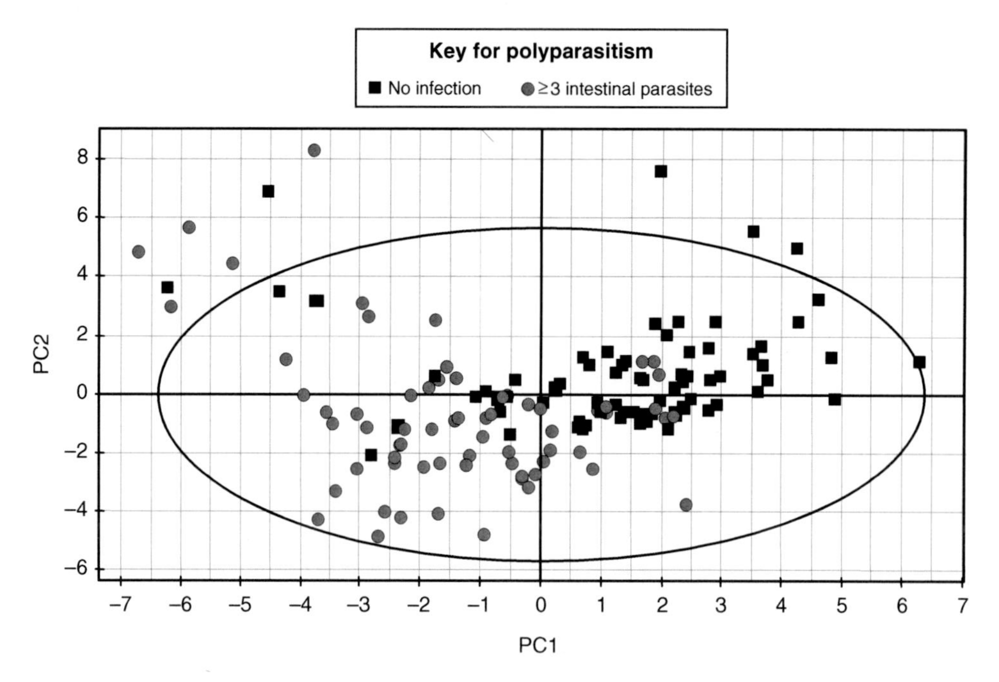

Figure 11.10. Two-dimensional projection of ^{1}H NMR spectra on urine samples from men in Zouatta II, Côte d'Ivoire: Uninfected vs diagnosed with three or more intestinal parasites.

Further, given the pervasiveness of polyparasitism in tropical regions of the world, it is also useful to know that men infected with any combination of at least three intestinal parasites out of a list of 11 such parasites are clearly distinguishable from those having no intestinal parasites (Figure 11.10). The challenge ahead is the identification of the metabolic signature of each kind of parasitic infection by itself, and sequentially the signature of combinations of parasites present in co-infections. The vital role of animal studies that can guide inquiry into the human subjects has been stressed before.

11.4.2. Complex phenotypes in large populations

In the context of polyparasitism, it will be important to extend the mouse and hamster studies with an experimental schistosome infection (discussed in Section 11.2), to include infection with a range of other intestinal parasites. These studies should establish the extent to which multiple parasitic infections can be identified in parallel, thereby allowing for much simplified diagnostics in populations such as Zouatta II, Côte d'Ivoire (Figure 11.1). Moving from animals to humans, we need to obtain metabolic profiles of single parasitic infections, analogous to what has been – and should be – done with animals. Knowledge gained from our initial *S. mansoni*-mouse and *S. japonicum*-hamster studies should be suggestive of what to expect in the subsequent human studies. Finally, we require characterization of spectra in studies of humans with pairs and triples of parasite co-infection. This is an essential step toward interpretation of the ^{1}H NMR spectra that is currently in hand from a population sample obtained from a typical village in the western part of Côte d'Ivoire, where polyparasitism is pervasive as signified by the frequency distributions in Figure 11.2b.

There is an extensive literature relating psychosocial phenotypes, assessed in large human populations, to levels of biomarkers, such as those in the panel in Table 11.1, augmented by fibrinogen and the inflammatory markers, CRP and IL-6 [90]. This is analogous to the part of the metabonomic investigation of coronary heart disease [18], where the use of some of these same biomarkers is used for diagnostic purposes. We need the analogue of the metabolic profiles in the above-mentioned coronary heart study to ascertain the extent to which much more nuanced biological signatures of psychosocial phenotypes can be identified. The Midlife in the United States (MIDUS) study is a particularly advantageous site for such an investigation, as it contains the most extensive array of psychosocial assessments of any of the major population studies, and covers an age range from 25 to 85 years. The MIDUS study also contains the first United States national assessment of diverse forms of cognitive functioning. Later-life cognition is being associated with psychosocial phenotypes in MIDUS, and there is an opportunity here for a unique study that relates metabolic profiles to psychosocial *X* cognitive performance combinations. We also anticipate that the metabolic profiles, linked to the psychosocial and cognitive phenotypes, can be used to develop better prediction rules than heretofore for later-life morbidity and mortality, as well as for prolonged periods of high functioning.

Finally, we observe that large human population studies such as the WLS (UW-Madison), MIDUS (MIDUS), Whitehall II (UCL Dept of Epidemiology & Public Health), the British Birth Cohort Studies [91, 92, 93], the Health and Retirement Survey (HRS) (National Institute on Aging), to name only a few examples, contain very heterogeneous sets of individuals. Thus, in order to obtain sharp discrimination in metabolic profiles between different phenotypes, it will be advisable to compare profiles across extreme groups. Despite the fact that the coarse-level PCA comparisons exhibited in Section 11.3 indicate differences in what are clearly quite heterogeneous populations, identifying the signatures of specific psychosocial profiles and distinguishing them from the influences of dietary variation and a diversity of behavioral patterns could be difficult. It would appear that our best strategy for ultimately distinguishing the metabolic profiles of sub-populations with subtle differences in phenotype is to work from sharp distinctions with extreme groups toward the more nuanced smaller differences in phenotypic features.

11.4.3. Behavioral phenotypes in naturalistic settings: Social relations in well-characterized baboons

Analogous to the longitudinal human studies described in the previous sections, Silk *et al.* [41] used 16 years of behavioral data from a well-characterized population of wild baboons in the Amboseli basin of Kenya to ascertain whether or not positive social relationships among adult females in a troop enhanced survival of their offspring. Female–female relationships include frequent grooming, close spatial proximity, and occasional acts of coalitionary support. Positive social support has been hypothesized to be valuable to adult females because they enhance prospects of obtaining support from others in within-group contests and to increase tolerance from more powerful group members. The study of Silk *et al.* [41] demonstrates that supportive social bonding is positively associated with infant survival, independent of the effects of dominance rank in the troop, group membership, and physical environmental conditions. An understanding of the biological substrates of such behavior could be obtained from metabolic profiling and comparison of female baboons who have positive, as opposed to negative, behavioral phenotypes. In a closely related context, there is already a start on assessing the biological substrates of successful vs unsuccessful pregnancies, which could also be related to social bonding, in that longitudinal assays have been collected on three sets of steroid hormones: estrogens, progestins, and glucocorticoids. It has recently been shown [94] that fecal estrogens in wild baboons predict impending fetal loss starting two months before the externally observed loss.

Although we previously emphasized the linkage between social relationships and health outcomes in humans, it is not yet clear whether the mechanisms that underlie the character of social bonds and infant survival in baboons are the same. Nevertheless, the findings of Silk *et al.* [41] are consistent with data on the beneficial health effects of positive social support in humans [95, 26]. It would be useful to compare metabolic profiles over time for positive and negative social relationship phenotypes in both humans and baboons. This would add a detailed biological underpinning to the behavioral evidence that positive social relationships have adaptive value for primate females.

11.5. Summary and discussion

We have covered wide territories in this chapter, all in the interest of exploiting the potential of NMR-based metabonomics. In this section, we summarize what was learned from the three broad venues, giving emphasis to the rich interplay between human and animal studies in moving these areas forward. At the human level, we began our section on disease risk and pathogenesis with a demonstration of the extent of parasitic load carried by individuals in selected communities living in rural parts of western Côte d'Ivoire. This prototypic example of polyparasitism highlighted the need for parallel diagnostic assays of multiple parasitic infections with comparable levels of resolution. Metabolic profiling has the potential to respond to this challenge, but since our current knowledge about the signatures of parasitic infections is in its infancy, this theme requires focused attention if significant advances are to be made. Encouraging results have been obtained via the characterizations of the metabolic signatures of *S. mansoni* and *S. japonicum* infections in mice and Syrian hamsters, respectively. Further, metabolic characterizations of *Trypanosoma brucei brucei* and *Plasmodium berghei* infections in rodents seem to be eminently feasible, and could represent a significant advance in our understanding of parasitic infections more generally. A significant and relatively novel aspect of such studies is the identification of three-way interactions between parasites, host animals, and gut microflora. We have already seen the beginnings of these cross-talks in the animal studies with an experimental schistosome infection described in Section 11.2. It would seem that this is only the beginning of a vast uncharted frontier.

A second example pertained to neuropathological disease where the need for suitable biomarkers to assess disease progression and response to therapeutic intervention is essential. Responsive to this need, we described animal models of PD and HD in which metabolic signatures of tissues and body fluids are providing fine-grained analyses of disease progression. At the human level, we noted related plans to adopt a systems biology approach combining transcriptomic, proteomic, and metabonomic data in the European Huntington's Disease Network. Additional examples pertained to recent studies of the metabolic signature of schizophrenia, an area in which metabonomic applications show great promise for epidemiological applications.

Our discussion of allostatic load and aging summarized extant evidence that has linked multiple biomarkers (neuroendocrine, immune, and cardiovascular) to subsequent disability, morbidity, and mortality in elderly populations. However, at present, these studies assess large molecule biomarkers, and thus neglect the functional status of the fundamental systems of energy metabolism, fat metabolism, and gut microflora metabolism. Bringing small molecule metabolic profiles to these investigations has the potential to provide more nuanced early warning systems. The gains in diagnostic precision for coronary heart disease exhibited by metabolic profiles [18], for example, are indicative of the added insight that will accrue from adopting metabonomic approaches. In these areas as well there is promising interplay with animal studies, such as those that have identified the metabolic signature of IR phenotypes.

Following the above examples, we then moved to examine the potential of NMR-based metabonomics to advance knowledge of how life challenges and psychosocial experience affects the organism. Such work constitutes a window on what are likely experiential precursors to the above-described processes of disease risk and pathogenesis. We began this section with studies of metabolic profiles in animals exposed to both acute and chronic stress (restraint stress, physical shaking). Related investigations illustrated the metabolic effects of maternal separation of newborn rats, showing changes in lipoproteins, amino acids, glucose, lactate, creatine, and citrate that likely result from increased secretion of hormones, such as glucocorticoids and catecholamines. These investigations also showed that dietary intervention had a measurable effect on exposure to stress. At the level of more complex behavioral phenotypes, we called for investigations of visible burrow systems in mice, rats, or hamsters, with accompanied collection of biofluids, which would allow for longitudinal metabolic profiling, punctuated by profiles, for example, of aggressive or defensive behaviors.

In human studies, we looked at the biological signature of cumulative economic and social relational adversity (and advantage) using the allostatic load battery of biomarkers. Preliminary findings showed greater evidence of physiological wear and tear among those exposed to greater cumulative adversity, but also pointed to the possible protective influences of nurturing and supportive social relationships in the face of persistent economic hardship. This work, however, is based on a coarse set of resting biomarkers and thus would be greatly enhanced by more refined profiles of metabolites. Because psychiatric disorders (e.g. schizophrenia) are now being studied with regard to their metabolic signature, we also summarized research showing that diverse indicators of psychological well-being and psychological distress show distinctive biological correlates. But again, this work employs large molecule indicators of neuroendocrine regulation and cardiovascular risk. As such, it would be substantially advanced by shifting the focus to metabolic signatures of psychological strengths and vulnerabilities. Similar advances could accrue to those currently mapping the neurochemical substrates of psychological resilience and vulnerability [29].

We extended the theme of protective psychosocial environments by drawing attention to the fact that many with genetic predisposition for particular diseases and psychological disorders never progress to disease status, sometimes even in the face of environmental stressors [32]. Metabolic profiling offers great potential for clarifying the protective health mechanisms that underlie psychosocial supports and resources, and thereby offers a central role in the formulation of more effective health promotion and disease prevention strategies.

Our third primary section pertained to large population studies where we demonstrated that it is possible to discriminate between distinct populations on the basis of broad metabolic characteristics. These undoubtedly reflect complex combinations of genetic, physical, and social, environmental, dietary, and microbial influences. They point toward the potential of a new field of metabonomic epidemiology, which could offer interpretable metabolic profiles linked to psychosocial phenotypes and parasitic co-infections in large human populations. Associating psychosocial phenotypes – for example, based on the kinds of social relationship histories and measures of psychological functioning discussed in Section 11.3 – with distinct metabolic signatures is imminently feasible. It could proceed, for example, with stored urine and blood samples from several large-scale surveys carried out in the United States and the United Kingdom [91, 92, 96, 97, 98, 99, 93]. Although it would be desirable to have longitudinal psychosocial and ^{1}H NMR spectroscopic data collected simultaneously on large populations, the present state of available data is very unbalanced. Surveys such as WLS and the British Birth Cohort Studies are rich in psychosocial information, in some instances running over periods of 50–60 years. In contrast, collection of biofluids such as urine and blood is usually available for at most two rounds of surveys and, even then, on relatively small sub-samples of much larger study populations. Nevertheless, it is imperative that the profiling process be initiated soon on these well-characterized populations if these opportunities are to be realized.

And again, we emphasize the importance of animal studies, which of necessity are a prelude to developing the capacity to interpret the spectra associated with complex phenotypes. An important step forward in this direction would be studies involving multiple naturalistic challenges over a substantial portion of mouse and rat lives (visible burrow systems, as noted earlier in this chapter). Extending this work to non-human primates, such as the baboon populations described in Section 11.4.3, would be highly informative.

Overall, we have described a challenging research agenda that needs to be driven by extensive cross-talk between human and animal studies. Of paramount importance will be the specification of complex phenotypes that represent the critical psychosocial and physical environmental features that map onto distinctive metabolic signatures. In large population studies, the extensive heterogeneity in life course experiences and environmental exposures makes the construction of taxonomies of these complex phenotypes possible. Herein lies an important opportunity with the potential to vastly expand the domain of metabonomics as the basis for biomarker identification of both protective and damaging experiences and their subsequent physical and mental health outcomes.

Acknowledgments

We thank Prof. John C. Lindon for inviting us to write this chapter. Thanks are addressed to Mr Christian Beck-Woerner for generating Figure 11.1. Burton H. Singer and Carol Ryff acknowledge financial support from the National Institute on Aging Grant PO1 – AG020166 and National Institutes of Health Grant

MO1 – RR03186 (to General Clinical Research Center, University of Wisconsin). Jürg Utzinger acknowledges financial support from the Swiss National Science Foundation (project no. PPOOB-102883), and Yulan Wang acknowledges financial support from Nestle Ltd.

References

[1] Nicholson, J. K., Connelly, J., Lindon, J. C., and Holmes, E. (2002) Metabonomics: A platform for studying drug toxicity and gene function. *Nat. Rev. Drug Discov.*, *1*, 153–161.

[2] Buck, A. A., Anderson, R. I., and MacRae, A. A. (1978) Epidemiology of poly-parasitism. I. Occurrence, frequency and distribution of multiple infections in rural communities in Chad, Peru, Afghanistan, and Zaire. *Tropenmed. Parasitol.*, *29*, 61–70.

[3] Ashford, R. W., Craig, P. S., and Oppenheimer, S. J. (1992) Polyparasitism on the Kenya coast. 1. Prevalence, and association between parasitic infections. *Ann. Trop. Med. Parasitol.*, *86*, 671–679.

[4] Churge, R. N., Karumba, N., Ouma, J. H., Thiongo, F. W., Sturrock, R. F., and Butterworth, A. E. (1995). Polyparasitism in two rural communities with endemic *Schistosoma mansoni* infection in Machakos District, Kenya. *J. Trop. Med. Hyg.*, *98*, 440–444.

[5] McKenzie, F. E. (2005) Polyparasitism. *Int. J. Epidemiol.*, *34*, 221–222.

[6] Hotez, P. J., Molyneux, D. H., Fenwick, A., Ottesen, E., Ehrlich Sachs, S., and Sachs, J. D. (2006) Incorporating a rapid impact package for neglected tropical diseases with programs for HIV/AIDS, tuberculosis, and malaria. *PLoS Med.*, *3*, e102.

[7] Utzinger, J., and de Savigny, D. (2006) Control of neglected tropical diseases: Integrated chemotherapy and beyond. *PLoS Med.*, *3*, e112.

[8] Jaur, L., and Stoddard, S. (1999) *Chartbook on Women and Disability in the U.S.* Washington, DC: US National Institute on Disability and Rehabilitation Research.

[9] McEwen, B. S., and Stellar, E. (1993) Stress and the individual: Mechanisms leading to disease. *Arch. Intern. Med.*, *153*, 2093–2101.

[10] Seeman, T. E., Singer, B. H., Rowe, J. W., Horwitz, R. I., and McEwen, B. S. (1997) Price of adaptation – allostatic load and its health consequences. MacArthur studies of successful aging. *Arch. Intern. Med.*, *157*, 2259–2268.

[11] Singer, B., Ryff, C. D., and Seeman, T. E. (2004) Operationalizing allostatic load. In J. Schulkin (Ed.), *Allostasis, Homeostasis, and the Costs of Physiological Adaptation* (pp. 113–149). New York, NY: Cambridge University Press.

[12] D'Agostino, R. B., Grundy, S., Sullivan, L. M., Wilson, P., for the CHR Risk Prediction Group (2001) Validation of the Framingham coronary heart disease prediction scores: Results of a multiple ethnic groups investigation. *JAMA*, *286*, 180–187.

[13] Seeman, T. E., McEwen, B. S., Rowe, J. W., and Singer, B. H. (2001) Allostatic load as a marker of cumulative biological risk: MacArthur studies of successful aging. *Proc. Natl. Acad. Sci. U.S.A.*, *98*, 4770–4775.

[14] Malik, S., Wong, N. D., Franklin, S. S., Kamath, T. V., L'Italien, G. J., Pio, J. R., *et al.* (2004) Impact of the metabolic syndrome on mortality from coronary heart disease, cardiovascular disease, and all causes in United States adults. *Circulation*, *110*, 1245–1250.

[15] Gruenewald, T. L., Seeman, T. E., Ryff, C. D., and Singer, B. H. (2006) Combinations of biomarkers predictive of later life mortality. *Proc. Natl. Acad. Sci. (in press).*

[16] Backhed, F., Levy, R. E., Sonnenburg, J. L., Peterson, D. A., and Gordon, J. L. (2005) Host-bacterial mutualism in the human intestine. *Science, 307*, 1915–1920.

[17] Nicholson, J. K., Holmes, E., and Wilson, I. D. (2005) Gut microorganisms, mammalian metabolism and personalized health care. *Nat. Rev. Microbiol., 3*, 431–438.

[18] Brindle, J. T., Antti, H., Holmes, E., Tranter, G., Nicholson, J. K., Bethell, H. W. L., *et al.* (2002) Rapid and noninvasive diagnosis of the presence and severity of coronary heart disease using ^{1}H-NMR-based metabonomics. *Nat. Med., 8*, 1439–1444.

[19] Dumas, M.-E., Barton, R. H., Toye, A., Cloarec, O., Blancher, C., Rothwell, A., *et al.* (2006a) Metabolic profiling reveals a contribution of gut microbiota to fatty liver phenotype in insulin-resistant mice. *Proc. Natl. Acad. Sci. U.S.A, 103*, 12511–12516.

[20] Holmes, E. Tsang, T. M., Huang, J. T. J., Leweke, M., Koethe, D., Gerth, C. W. *et al.* (2006) Evidence that early intervention may impact on disease progression and outcome in schizoprenia. *PLoS Med.*, 3 *327*: 1–9.

[21] Tsang, T. M., Huang, J. T. J., Holmes, E., and Bahn, S. (2006b) Metabolic profiling of plasma from discordant schizophrenia twins: correlation between lipid signals and global functioning in female schizophrenia patients. *J. Proteome Res.* 5, 756–760.

[22] Teague, C. R., Dhabhar, F. S., Beckwith-Hall, B., Holmes, E., Powell, J., Cobain, M., *et al.* (2006). Metabonomic studies on the physiological effects of acute and chronic psychological stress in Sprague-Dawley rats. *[in review].*

[23] Wang, Y., Holmes, E., Tang, H., Lindon, J. C., Sprenger, N., Turini, M. E., *et al.* (2006b) Experimental metabonomic model of dietary variation and stress interactions. *J. Proteome Res.*, 5, 1535–1542.

[24] Blanchard, D. C., Spencer, R. L., Weiss, S. M., Blanchard, R. J., McEwen, B., and Sakai, R. R. (1995) Visible burrow system as a model of chronic social stress: Behavioral and neuroendocrine correlates. *Psychoneuroendocrinology, 20*, 117–134.

[25] Singer, B., and Ryff, C. D. (1999) Hierarchies of life histories and associated health risks. *Ann. N. Y. Acad. Sci., 896*, 96–115.

[26] Seeman, T. E., Singer, B. H., Ryff, C. D., Dienberg Love, G., and Levy-Storms, L. (2002) Social relationships, gender, and allostatic load across two age coherts. *Psychosom. Med., 64*, 395–406.

[27] Steptoe, A., and Marmot, M. (2003) Burden of psychosocial adversity and vulnerability in middle age: Association with biobehavioral risk factors and quality of life. *Psychosom. Med., 65*, 1029–1037.

[28] Steptoe, A. Brydon, L. and Kunz-Ebrecht, S. (2005) Changes in financial strain over three years, ambulatory blood pressure, and cortisol response to awakening. *Psychosom. Med., 67*, 281–287.

[29] Charney, D. S. (2004) Psychobiological mechanisms of resilience and vulnerability: Implications for successful adaptation to extreme stress. *Am. J. Psychiatry, 161*, 195–216.

[30] Ryff, C. D., Singer, B. H., and Dienberg Love, G. (2004) Positive health: Connecting well-being with biology. *Philos. Trans. R. Soc. Lond. B, Biol. Sci., 359*, 1383–1394.

[31] Caspi, A., McClay, J., Moffitt, T. E., Mill, J., Martin, J., Craig, I. W., *et al.* (2002) Role of genotype in the cycle of violence in maltreated children. *Science, 297*, 851–854.

[32] Caspi, A., Sugden, K., Moffitt, T. E., Taylor, A., Craig, I. W., Harrington, H., *et al.* (2003) Influence of life stress on depression: Moderation by polymorphism in the 5-HTT gene. *Science, 301*, 386–389.

[33] Ryff, C. D., and Singer, B. H. (2005) Social environments and the genetics of aging: Advancing knowledge of protective health mechanisms. *J. Gerontol. B Psychol. Sci. Soc. Sci., 60* (Spec. No. 1), 12–23.

[34] Nicholson, J. K., and Wilson, I. D. (2003) Understanding 'global' systems biology: Metabonomics and the continuum of metabolism. *Nat. Rev. Drug Discov.*, *2*, 668–676.

[35] Clayton, T. A., Lindon, J. C., Cloarec, O., Antti, H., Charuel, C., Hanton, G., *et al.* (2006) Pharmaco-metabonomic phenotyping and personalized drug treatment. *Nature*, *440*, 1073–1077.

[36] Stamler, R. (1991) Implications of the INTERSALT study. *Hypertension 17* (1 Suppl.): 16–20.

[37] Stamler, J., Elliott, P., Dennis, B., Dyer, A. R., Kesteloot, H., Liu, K., *et al.* (2003) INTERMAP: background, aims, design, methods, and descriptive statistics (nondietary). *J. Hum. Hypertens.* 17: 591–608.

[38] Teague, C., Holmes, E., Maibaum, E., Nicholson, J., Tang, H., Chan, Q., *et al.* (2004) Ethyl glucoside in human urine following dietary exposure: Detection by ^{1}H NMR spectroscopy as a result of metabonomic screening of humans. *Analyst*, *129*, 259–264.

[39] Dumas, M.-E., Malbaum, E. C., Teague, C., Ueshima, H., Zhou, B. F., Lindon, J. C., *et al.* (2006b) Assessment of analytical reproducibility of ^{1}H NMR spectroscopy based metabono-.mics for large-scale epidemiological research: The INTERMAP study. *Anal. Chem.*, *78*, 2199–2208.

[40] Altmann, J., and Alberts, S. C. (2003) Variability in reproductive success viewed from a life history perspective in baboons. *Am. J. Hum. Biol.*, *15*, 401–409.

[41] Silk, J. B., Alberts, S. C., and Altmann, J. (2003) Social bonds of female baboons enhance infant survival. *Science*, *302*, 1231–1234.

[42] Sapolsky, R. M. (2005) The influence of social hierarchy on primate health. *Science*, *308*, 648–652.

[43] Moody, A. (2002) Rapid diagnostic tests for malaria parasites. *Clin. Microbiol. Rev.*, *15*, 66–78.

[44] Raso, G., Luginbühl, A., Adjoua, C. A., Tian-Bi, N. T., Silué, K. D., Matthys, B., *et al.* (2004) Multiple parasite infections and their relationship to self-reported morbidity in a community of rural Côte d'Ivoire. *Int. J. Epidemiol.*, *33*, 1092–1102.

[45] Utzinger, J., Zhou, X. N., Chen, M. G., and Bergquist, R. (2005) Conquering schistosomiasis in China: The long march. *Acta Trop*, *96*, 69–96.

[46] Lejon, V., and Büscher, P. (2005) Cerebrospinal fluid in human African trypanosomiasis: A key to diagnosis, therapeutic decision and post-treatment follow-up. *Trop. Med. Int. Health*, *10*, 395–403.

[47] Utzinger, J., N'Goran, E. K., Ossey, Y. A., Booth, M., Traoré, M., Lohourignon, K. L., *et al.* (2000). Rapid screening for *Schistosoma mansoni* in western Côte d'Ivoire using a simple school questionnaire. *Bull World Health Organ*, *78*, 389–398.

[48] Raso, G., Vounatsou, P., Gosoniu, L., Tanner, M., N'Goran, E. K., and Utzinger, J. (2006a) Risk factors and spatial patterns of hookworm infection among schoolchildren in a rural area of western Côte d'Ivoire. *Int. J. Parasitol.*, *36*, 201–210.

[49] Matthys, B., N'Goran, E. K., Koné, M., Koudou, B. G., Vounatsou, P., Cissé, G., *et al.* (2006) Urban agricultural land use and characterization of mosquito larval habitats in a medium-sized town of Côte d'Ivoire. *J. Vector Ecol.*, (in press).

[50] Doumenge, J. P., Mott, K. E., Cheung, C., Villenave, D., Chapuis, O., Perrin, M. F., *et al.* (1987) *Atlas of the Global Distribution of Schistosomiasis*. Talence, Geneva: CEGET-CNRS, OMS/WHO.

[51] Utzinger, J., and Keiser, J. (2004) Schistosomiasis and soil-transmitted helminthiasis: Common drugs for treatment and control. *Expert Opin. Pharmacother.*, *5*, 263–285.

[52] King, C. H., Dickman, K., and Tisch, D. J. (2005) Reassessment of the cost of chronic helmintic infection: A meta-analysis of disability-related outcomes in endemic schistosomiasis. *Lancet*, *365*(9470), 1561–1569.

[53] Katz, N., Chaves, A., and Pellegrino, J. (1972) A simple device for quantitative stool thick-smear technique in schistosomiasis mansoni. *Rev. Inst. Med. Trop. Sao Paulo*, *14*, 397–400.

[54] Utzinger, J., N'Goran, E. K., Esse Aya, C. M., Acka Adjoua, C., Lohourignon, K. L., Tanner, M., *et al.* (1998) *Schistosoma mansoni*, intestinal parasites and perceived morbidity indicators in schoolchildren in a rural endemic area of western Côte d'Ivoire. *Trop. Med. Int. Health, 3*, 711–720.

[55] Geerts, S., and Gryseels, B. (2000). Drug resistance in human helminths: Current situation and lessons from livestock. *Clin. Microbiol. Rev., 13*, 207–222.

[56] Bethony, J., Brooker, S., Albonico, M., Geiger, S. M., Loukas, A., Diemert, D., *et al.* (2006) Soil-transmitted helminth infections: Ascariasis, trichuriasis, and hookworm. *Lancet, 367*, 1521–1532.

[57] Raso, G., Matthys, B., N'Goran, E. K., Tanner, M., Vounatsou, P., and Utzinger, J. (2005) Spatial risk prediction and mapping of *Schistosoma mansoni* infections among schoolchildren living in western Côte d'Ivoire. *Parasitology, 131*, 97–108.

[58] Raso, G., Vounatsou, P., Singer, B. H., N'Goran E, K., Tanner, M., and Utzinger, J. (2006b) An integrated approach for risk profiling and spatial prediction of *Schistosoma mansoni*-hookworm coinfection. *Proc. Natl. Acad. Sci. U.S.A, 103*, 6934–6939.

[59] Allen, A. V. H., and Ridley, D. S. (1970) Further observations on the formol-ether concentration technique for faecal parasites. *J. Clin. Pathol., 23*, 545–546.

[60] Marti, H., and Escher, E. (1990) SAF–an alternative fixation solution for parasitological stool specimens. *Schweiz. Med. Wochenschr., 120*, 1473–1476.

[61] Keiser, J., N'Goran, E. K., Traoré, M., Lohourignon, K. L., Singer, B. H., Lengeler, C., Tanner, M., Utzinger, J. (2002) Polyparasitism with *Schistosoma mansoni*, geohelminths, and intestinal protozoa in rural Côte d'Ivoire. *J. Parasitol.*, 88, 461–466.

[62] Wang, Y., Holmes, E., Nicholson, J. K., Cloarec, O., Chollet, J., Tanner, M., *et al.* (2004) Metabonomic investigations in mice infected with *Schistosoma mansoni*: An approach for biomarker identification. *Proc. Nat. Acad. Sci. U.S.A., 101*, 12676–12681.

[63] Wang, Y., Utzinger, J., Xiao, S. H., Xue, J., Nicholson, J. K., Tanner, M., *et al.* (2006a) System level metabolic effects of a *Schistosoma japonicum* infection in the Syrian hamster. *Mol. Biochem. Parasitol., 146*, 1–9.

[64] Hastie, T., Friedman, J. H., and Tibshirani, R. (2001) *Elements of Statistical Learning: Data Mining, Inference, and Prediction.* New York, NY: Springer-Verlag.

[65] Khaw, K.-T., Wareham, N., Luben, R., Bingham, S., Oakes, S., Welch, A., *et al.* (2001) Glycated haemoglobin, diabetes, and mortality in men in Norfolk cohort of European Prospective Investigation of Cancer and Nutrition (EPIC-Norfolk). *British Medical Journal, 322*, 1–6.

[66] Gilmore, M. S., and Ferrette, J. J. (2003). The thin line between gut commensal and pathogens. *Science, 299*, 1999–2002.

[67] Falkow, S. (2006) Is persistent bacterial infection good for your health? *Cell, 124*, 699–702.

[68] Turmaine, M., Raza, A., Mahal, A., Mangiarini, L., Bates, G. P. and Davies, S. W. (2000) Nonapoptotic neurodegeneration in a transgenic mouse model of Huntington's disease. *Proc. Natl. Acad. Sci. U.S.A.* 97, 8093–8097.

[69] Ungerstedt, U. (1968) 6-Hydroxy-dopamine induced degeneration of central monoamine neurons. *Eur. J. Pharmacol.* 5:107–110.

[70] Langston, J. W., Forno, L. S., Rebert, C. S., and Irwin, I. (1984) Selective nigral toxicity after systemic administration of 1-methyl-4-phenyl-1,2,5,6-tetrahydropyrine (MPTP) in the squirrel monkey. *Brain Res.* 292:390–394.

[71] Bates, G. (2000) Huntington's disease. In reverse gear. *Nature, 404*, 944–955.

[72] Tsang, T. M., Woodman, B., McLoughlin, G. A., Griffin, J. L., Tabrizi, S. J., Bates, G. P. *et al.* (2006a) Metabolic characterization of the R6/2 transgenic mouse model of Huntington's disease by high-resolution MAS ^{1}H NMR spectroscopy. *J. Proteome Res.* 5, 483–492.

[73] Underwood, B. R., Broadhurst, D., Dunn, W. B., Ellis, D. I., Michell, A. W., Vacher, C., *et al.* (2006) Huntington disease patients and transgenic mice have similar pro-catabolic serum metabolite profiles. *Brain* 129:877–886.

[74] McEwen, B. (2002) Sex, stress, and the hippocampus: Allostasis, allostatic load and the aging process. *Neurobiol. Aging*, *23*, 921–939.

[75] Korte, S. M., Koolhaas, J. M., Wingfield, J. C., and McEwen, B. S. (2005) The Darwinian concept of stress: Benefits of allostasis and costs of allostatic load and the trade-offs in health and disease. *Neurosci. Biobehav. Rev.*, *29*, 3–38.

[76] Cassar-Malek, I., Listrat, A. and Picard, B. (1998) Hormonal regulation of muscle fibres characteristics: A review. *Prod. Anim.*, *11*, 365–377.

[77] Welberg, L. A. M., and Seckl, J. R. (2001) Prenatal stress, glucocorticoids and the programming of the brain. *J. Neuroendocrinol.*, 13, 113–128.

[78] Taylor, S. E., Repetti, R. L. and Seeman, T. (1997) Health psychology: What is an unhealthy environment and how does it get under the skin? *Annu. Rev. Psychol.*, *48*, 411–447.

[79] Ryff, C. D. and Singer, B (2000) Interpersonal flourishing: A positive health agenda for the new millennium. *Pers. Soc. Psychol. Rev.*, *4*, 30–44

[80] Parker, G., Tupling, H., and Brown, L. B. (1979) Parenteral bonding instrument. *Br. J. Med. Psych.*, *52*, 1–10.

[81] Singer, B., Friedman, E., Seeman, T., Fava, G. A., and Ryff, C. D. (2005) Protective environments and health status: Cross-talk between human and animal studies. *Neurobiol. Aging*, *26* (Suppl. 1), 113–118.

[82] Schaefer, M. T., and Olson, D. H. (1981). Assessing intimacy – the PAIR inventory. *J. Marital Fam. Ther.*, *7*, 47–60.

[83] Ryff, C. D., Dienberg Love, G., Urry, H. L., Muller, D., Rosenkranz, M. A., Friedman, E. M., *et al.* (2006) Psychological well-being and ill-being: Do they have distinct or mirrored biological correlates? *Psychother. Psychosom.*, *75*, 85–95.

[84] Ryff, C. D. (1989) Happiness is everything, or is it? Explorations on the meaning of psychological well-being. *J. Pers. Soc. Psychol.*, *57*, 1069–1081.

[85] Watson, D., Clark, L. A., and Tellegen, A. (1988). Development and validation of brief measures of positive and negative affect: The PANAS scales. *J. Pers. Soc. Psychol.*, *54*, 1063–1070.

[86] Ryff, C. D., and Keyes, C. L. M. (1995). The structure of psychological well-being revisited. *J. Pers. Soc. Psychol.*, *69*, 719–727.

[87] Jorm, A. (2001) Association of hypotension with positive and negative affect and depressive symptoms. *Br. J. Psychiatry*, *178*, 553–555.

[88] Kaufman, J., Yang, B. Z., Douglas-Palumberi, H., Houshyar, S., Lipschitz, D., Krystal, J. H., *et al.* (2005) Social supports and serotonin transporter gene moderate depression in maltreated children. *Proc. Nat. Acad. Sci. U.S.A.*, *101*(49), 17316–17321.

[89] Jonsson, P., Bruce, S. J., Moritz, T., Trygg, J., Sjöström, M., Plumb, R., *et al.* (2005) Extraction, interpretation and validation of information for comparing samples in metabolic LC/MS data sets. *Analyst*, *130*, 701–707.

[90] Seeman, T. E., Crimmins, E., Huang, M. H., Singer, B., Bucur, A., Gruenewald, T., *et al.* (2004) Cumulative biological risk and socio-economic differences in mortality: MacArthur studies of successful aging. *Soc. Sci. Med.*, *58*, 1985–1997.

[91] Center for Longitudinal Studies. *The British Cohort Study*, from http://www.cls.ioe.ac.uk/studies.asp Section=000100020002 (accessed: 30 May 2006).

[92] Medical Research Council. *The British 1946 Birth Cohort Study*, from http://www.nshd.mrc.ac.uk/ (accessed: 30 May 2006).

[93] Power, C., and Elliott, J. (2006) Cohort profile: 1958 British birth cohort (National Child Development Study). *Int. J. Epidemiol.*, *35*, 34–41.

[94] Beehner, J. C., Nguyen, N., Wango, E. O., Alberts, S. C. and Altmann, J. (2006) The endocrinology of pregnancy and fetal loss in wild baboons. *Horm. Behav.*, *49*, 688–699.

[95] Taylor, S. E., Klein, L. C., Lewis, B. P., Gruenewald, T. L., Gurung, R. A. R., and Updegraff, J. A. (2000) Biobehavioral responses to stress in females: Tend-and-befriend, not fight-or-flight. *Psychol. Rev.*, *107*, 411–429.

[96] MIDUS. *Midlife in the United States*, from http://midus.wisc.edu/scopeofstudy.php (accessed: 30 May 2006).

[97] National Institute on Aging. *The Health and Retirement Study*, from http:// hrsonline.isr.umich.edu/ (accessed: 30 May 2006).

[98] UCL Dept of Epidemiology and Public Health. *Whitehall II Study*, from http://www.ucl.ac.uk/whitehallII/ (accessed: 30 May 2006).

[99] UW-Madison.*Wisconsin Longitudinal Study*, from http://www.ssc.wisc.edu/wlsresearch (accessed: 30 May 2006).

The Handbook of Metabonomics and Metabolomics
John C. Lindon, Jeremy K. Nicholson and Elaine Holmes (Editors)

Chapter 12

Metabolite Profiling and Cardiovascular Disease

David J. Grainger

Department of Medicine, Cambridge University, Box 157, Addenbrooke's Hospital, Cambridge CB2 2QQ, UK

12.1. Introduction

Obtaining an accurate diagnosis for a wide range of diseases remains as challenging a problem for the delivery of healthcare as finding treatments that work. Indeed, improving diagnostic and therapeutic capabilities are becoming increasingly entwined, reflecting the trend towards personalised medicine. Decades ago, broad classical diagnoses such as "diabetes" led to vast numbers of individuals with similar symptoms receiving essentially similar treatments, even if the molecular causes differed dramatically between individuals.

Today, we recognise that not everyone presenting with the same cluster of symptoms will respond equally well to the same treatment. There are at least two distinct reasons for this: most obviously, two different diseases with different causes might result in indistinguishable symptoms, yet require very different interventions to cure. Equally importantly, every individual is unique, and each may respond differently to the same drug, even when sharing the same disease entity.

As a result, an array of molecular diagnostic tests has been adopted to refine broad diagnostic definitions. Insulin-dependent and non-insulin-dependent diabetes mellitus can be readily distinguished, for example by measuring plasma levels of insulin. For many diseases, the diagnostic criteria are defined in terms of normal ranges for a plethora of plasma components. Consequently, smaller and more homogeneous groups of individuals are picked out, and therapeutic strategies optimised for their treatment.

Looking to the future, a wide range of studies is already underway examining the impact of the genetic make-up of the individual on their response to particular drugs. Although few such "pharmacogenomic" tests have reached the clinic, it seems certain that before long the physician will be able to tailor the choice of therapeutic agent, and even the dose, to the particular individual. For example, well-studied polymorphisms in enzymes responsible for drug metabolism in the liver, or transport proteins in the kidney, can dramatically affect the response to certain drugs in particular individuals.

Personalised medicine may be good for the healthcare consumer, but it presents a headache for traditional pharmaceutical development strategies. As the trend continues, market sizes for individual drugs will almost certainly shrink, as the physician gains tools to rationally determine who will respond best to a particular intervention. Current, highly resource intensive, paradigms for drug development and registration cannot be sustained below a certain threshold for the average market size for a new therapeutic agent.

In contrast, personalised medicine offers new hope to diagnostics developers. An ever-greater share of healthcare spending will be directed towards properly identifying the disease status of the individual and the appropriate therapeutic intervention to prescribe. Better still, the need to measure hundreds or even thousands of parameters simultaneously to obtain a personalised diagnosis opens up a whole new approach to diagnostics: profiling. New technologies for measuring large numbers of analytes simultaneously, and thereby making available to the physician an accurate description of the particular individual, raise the exciting possibility of new, high-margin profiling diagnostic (or "pronostic") tests.

Profiling diagnostics come in essentially two flavours: the first represents a bottom-up approach in which a series of analytes known to be associated with a particular disease are combined to repeatedly refine the diagnosis [1]. In some sense, this just reflects the continuation of the present trend of identifying more and more disease "biomarkers", although the ability to assemble a biomarker profile still requires technology to allow multiple analyses to be performed in parallel at acceptable cost.

The second type, however, breaks the mould of existing molecular diagnostic tests entirely. Here, a profile is created by measuring thousands of analytes, but without any reference to the particular disease state [2]. Indeed, in an ideal world, the analytes composing the profile are selected entirely randomly. Pattern recognition algorithms are then used to identify signatures within the profile which are always present in individuals with a particular disease, or who respond to a particular treatment, but are always absent in healthy people or non-responders. Crucially, the success of such approaches, which have been called "molecular fingerprinting" [3] depends only on the data density of the profile and not at all on knowing the specific molecular identity of any of the analytes.

12.2. Current approaches to molecular fingerprinting

Until recently, the most advanced approaches to molecular fingerprinting were based on genomics (the study of complete collections of genes), at the level of either DNA polymorphisms or gene expression into mRNA. This most likely reflects the relative ease with which parallel assays could be developed for genomic applications. Today, it is possible to determine haplotypes at tens of thousands of polymorphisms or to estimate the level of expression of thousands of genes simultaneously, for example by using gene arrays. In principle, the exquisite specificity of complementary base pairing of nucleic acids facilitates the rapid and specific measurement of many different determinations in parallel.

Unfortunately, even expression profiling (which is more dynamic than examining the almost static genomic fingerprint) provides only a partial description of the current status of a biological organism. Changes in gene expression partly reflect the genome sequence of the individual, and partly reflect the environmental influences on the organism but environmental stimuli are often reflected through a number of distorting "mirrors": for example, dietary changes may be sensed by proteins to create intracellular signals which are integrated with other environmental and genetic cues before being transduced to a change in gene expression through co-ordinate regulation of transcription factors. As a result, expression profiling provides only a very indirect (and often unresponsive) description of the current physiological status of the organism and its response to environmental stimuli [4].

In contrast, measuring metabolites (such as sugars, fats, amino acids, vitamins and so forth) provides a more direct, and hence more dynamic, snapshot of the current physiological status of the organism. In some sense, the metabolite profile reflects the genome through a number of distorting "mirrors". Thought of in this way, gene expression profiling and metabolic profiling are complementary and any comprehensive biological fingerprint must ultimately encapsulate both kinds of information. Proteomics lies somewhere between these two extremes (Figure 12.1). Gene expression patterns are a major (but certainly not the only) determinant of protein expression, while proteins function as important regulators of metabolite levels (for example, by acting as enzymes in intermediary metabolism) and also of gene expression (by acting as transcription factors). This complex interdependency of the major profiling techniques is illustrated in Figure 12.1.

In many senses, the type of analyte (whether DNA, RNA, protein or metabolite) is not germane to the concept of profiling diagnostics. The separation of profiles into

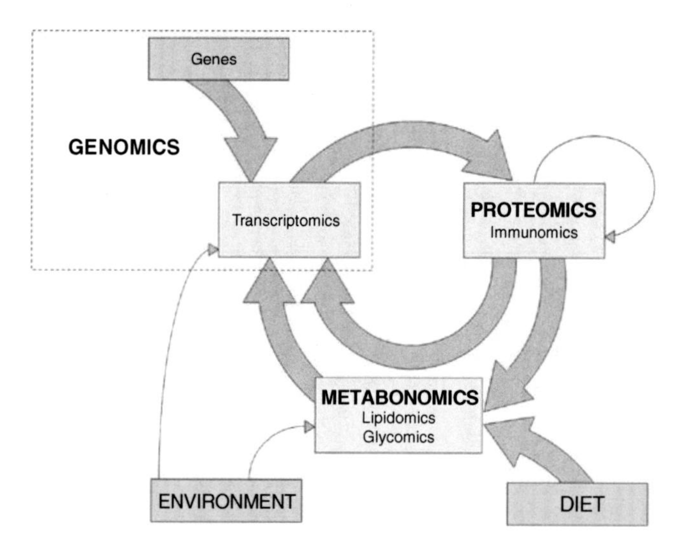

Figure 12.1. The relationship between genomics, proteomics and metabonomics, reflecting the genetic and environmental determinants of physiology.

gene expression, protein expression or metabolite arrays instead reflects the need to use different technological approaches to gather information in each of these domains rather than any scientific justification for the subdivisions. The interconnections (albeit complex and non-linear) between the "omics" disciplines (Figure 12.1) make it likely that if a diagnostic signature can be found in one domain, then it will be reflected into the others.

From this perspective, selection of the most appropriate tools to develop a profiling diagnostic for a particular disease depends on the relative cost per unit of data obtained by each of the different methods and the reproducibility of the data obtained relative to the expected perturbations in the molecular signature associated with the disease. Of course, with the whole concept of profiling diagnostics still in its infancy, anything but a subjective assessment of these criteria remains out of reach, but the amount of relevant datasets on which to base such a decision is increasing all the time.

Set against these criteria, current implementations of metabolite profiling, whether based on chromatography and mass spectrometry or on NMR spectroscopy, stack up very favourable compared to other profiling techniques such as proteomics or gene expression analysis. NMR spectroscopy in particular is highly reproducible: a typical NMR spectrum of a human serum sample varies by only 1% or so across the information-dense region of the spectrum, even on repeated measures 14 days apart [5]. Set against such methodological reproducibility, coefficients of variation in metabolite levels between individuals are typically 5% or more. In contrast, the expression levels of very few genes differ between individuals to a greater extent than the replicate reproducibility of even the best gene array analyses.

12.3. Profiling diagnostics for coronary heart disease

Coronary heart disease (CHD) is the biggest cause of morbidity and mortality in the Western world, responsible for more than 100,000 deaths in the UK alone in 2003, some 40% of which were defined as "premature" (occurring in individuals under 70 years of age) [6].

As early as the first decade of life the architecture of the blood vessel wall in the arteries that supply the heart begins to change. The smooth muscle cells, which are normally restricted to the muscular tunica media layer of the artery, migrate into the expanded intima and narrow the vessel lumen. Over many years, leukocytes (and, in particular, macrophages and T cells) are recruited to the intima, establishing a persistent local inflammatory response. Cholesterol and various other fats also begin to accumulate in the expanded intima, forming an early atherosclerotic plaque (Figure 12.2) [7]. Throughout this period, the disease can progress insidiously without any overt clinical symptoms. These changes in vessel wall architecture are

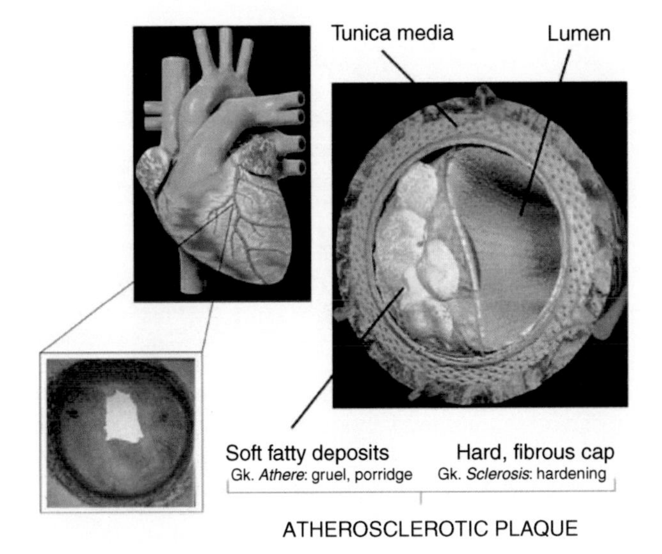

Figure 12.2. The anatomy of coronary heart disease. The commonest form of heart disease results from atherosclerotic deposits in the coronary arteries, which supply blood to the myocardium (top left). A section through a diseased coronary artery (lower left) shows the reduction in lumen diameter (stenosis), reducing blood flow, as a consequence of the deposition of yellowish lipid-rich material into the vessel intima. These components are illustrated diagrammatically (right panel).

not restricted to the coronary arteries (although certain blood vessels are more prone to disease than others for reasons that remain incompletely understood), and the prevalence of such atherosclerotic changes is very high: as many as 80% of cadavers exhibit lipid-rich plaques in their arteries [8].

As the plaque grows, and so the lumen narrows reducing blood flow, the heart muscle can become locally ischaemic resulting in chest pains and breathlessness that typifies the condition of angina pectoris. In many individuals, the plaque remains stable and while the symptoms of angina may be debilitating the condition does not become life-threatening [9]. For these individuals, various medical and surgical options are available: lipid lowering drugs, such as the statin class, can reduce further plaque expansion and may in some cases even cause regression of existing plaques, while surgical interventions include angioplasty (where a balloon catheter is used to physically enlarge the vessel lumen) and bypass grafting (in which vein grafts are used to provide a collateral circulation around the blockage).

Unfortunately, in some individuals, for reasons that are still poorly understood, the atherosclerotic plaque can become unstable. Plaques with large lipid cores or with a high density of macrophages (secreting proteases which destroy the extracellular matrix) appear to be particularly vulnerable [10]. As a result, the plaque ruptures, bringing the thrombogenic core composed of lipid and necrotic cell debris into contact with the blood, resulting in rapid formation of a large, occluding thrombus. Within seconds, the entire blood flow through the artery can be cut off. If the blood vessel is not rapidly recanalised, the nearby myocardium is starved of oxygen and begins to die (a process termed "myocardial infarction"). If too large a part of the heart muscle is lost, the individual dies as a result.

Although effective medical and surgical interventions exist, one of the biggest problems associated with the prevention and treatment of CHD is obtaining a definitive diagnosis. To date, invasive imaging techniques (such as angiography) remain the most prevalent method used to diagnose the presence of atherosclerotic plaques within the coronary circulation (and some plaques, whose bulk is eccentric to the lumen and therefore not stenotic, remain invisible even on angiography). Worse still, the diagnostic approaches routinely used in the clinic poorly distinguish between the chronic stable plaques which may cause angina but are unlikely to cause myocardial infarction from the vulnerable, unstable plaques which may rupture and kill the sufferer unless aggressive intervention is made.

As a result, clinical management of cardiovascular disease depends on assessing risk rather than on definitive diagnosis. Clinicians measure a range of different parameters (usually beginning with low cost, non-invasive measures such as serum cholesterol, and progressing to more expensive or more invasive tests such as treadmill ECGs, perfusion imaging or angiography), then apply algorithms to calculate risk scores on which the decision to intervene is based (Figure 12.3).

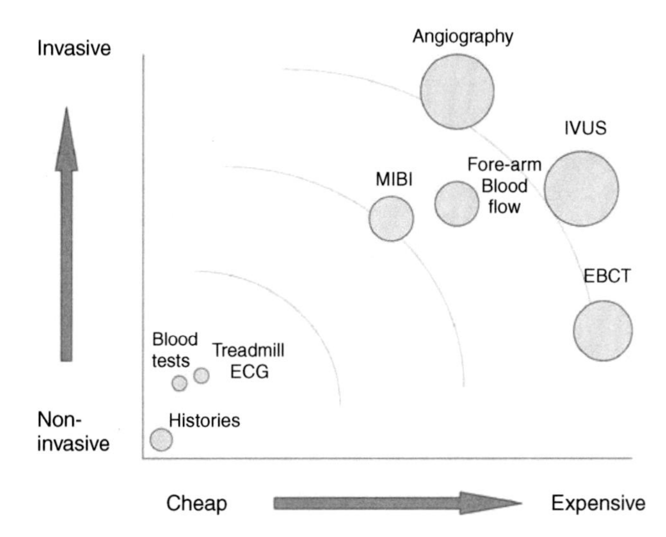

Figure 12.3. Comparison of existing modalities for the diagnosis of heart disease, or evaluation of risk of cardiovascular events. Existing diagnostic tools are compared on the basis of resource implications and degree of invasiveness. The size of the circles represents the relative diagnostic capability of each approach. Ideally, any new profiling diagnostic would be relatively low cost, non-invasive and powerful.
Abbreviations: ECG, electrocardiogram; MIBI, myocardial perfusion scan using 2-methoxy isobutyl isonitrile; IVUS, intravascular ultrasound; EBCT, electron-beam computed tomography.

The problem with this approach is that any combination of risk factors that achieves a significant enrichment of myocardial infarction sufferers into the "high risk group" results in the exclusion of the majority of those who will suffer a myocardial infarction. Worse still, the greater the enrichment sought, the smaller the percentage of events that are captured. Clearly, adopting this approach uses available resources for the prevention of CHD more efficiently, but leaves behind a very significant public health burden in the untreated cases.

The only solution is to improve diagnosis of the presence of coronary atherosclerosis (using tests which are less invasive and less expensive, so they can be applied more widely; Figure 12.3) and ultimately to provide tests which distinguish those with unstable atherosclerotic disease who will go on to have a myocardial infarction event, from those with chronic stable disease. The potential clinical and financial benefits from the development of such a test are enormous.

As a result, the diagnosis of CHD is an appealing target for the development of new profiling diagnostics, or pronostics. Current approaches can only reliably identify the presence of atherosclerotic lesions (e.g. using angiography to visualise stenotic lesions; Figure 12.4), and then only using modalities which are very resource intensive, highly invasive or more likely both (Figure 12.3). In marked contrast, a profiling diagnostic, at least in principle, offers the opportunity to overcome the

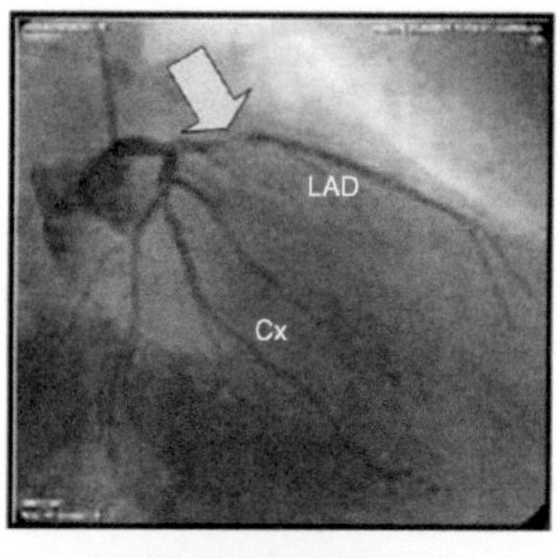

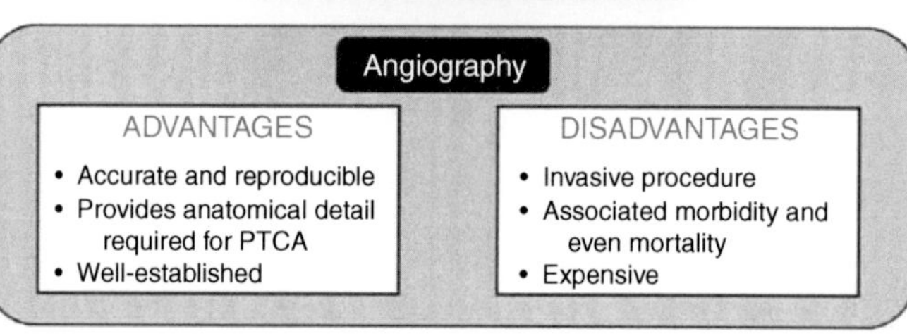

Figure 12.4. An example of coronary angiography. A typical coronary angiogram is shown (upper panel), where two of the three major coronary arteries are visible as dark lines on this X-ray image due to the injected contrast medium. A single large stenosis is visible (arrow) in the left anterior descending (LAD) artery. The circumflex artery (Cx) appears normal. The advantages and disadvantages of angiography for the diagnosis of CHD are tabulated below.
Abbreviations: PTCA, percutaneous transluminal coronary angioplasty.

limitations of angiography by providing information about plaque stability as well as size, in a non-invasive format at an acceptable cost.

What sort of profiling tool would be best suited to this task? Over and above the general considerations outlined in the previous section, there is ample reason to believe that metabolic profiling will provide useful information about CHD status. Many of the best characterised univariate correlates of the disease are metabolites (such as cholesterol and various triglycerides, for example). Furthermore, there is already evidence that the complex metabolic disturbances characterised as the Metabolic Syndrome are a risk factor for heart disease [11]. On this basis, it seems logical to select metabolic profiling for the diagnosis of CHD.

12.4. Metabonomics of coronary heart disease

Since the primary aim of applying profiling to CHD was to improve diagnosis (rather than elucidate new molecular mechanisms involved in the pathophysiology), fingerprinting approaches (which do not emphasise the importance of identifying all the metabolites responsible for classification of cases and controls) seem most appropriate.

The advantages and disadvantages of the most common methods of generating a metabolic profile have been extensively discussed elsewhere [4]. In general, NMR spectroscopy is likely to be superior in most applications to chromatographic separation followed by mass spectrometry (liquid chromatography-mass spectrometry (LC-MS), gas chromatography-mass spectrometry (GC-MS) or similar approaches) when the primary aim is fingerprinting. This reflects the extraordinary replicate reproducibility of the spectrum, even using complex biological fluids such as serum, as well as the zero bias of the technique (essentially all metabolites are detectable in proton NMR spectroscopy, provided only that there is sufficient material present).

On this basis, we have performed a pilot study to investigate the diagnostic utility of a metabolic profile obtained using simple one-dimensional proton NMR spectroscopy. Blood samples were collected from consecutive patients presenting at Papworth Hospital (Cambridgeshire, UK) for a diagnostic angiogram, who had either normal coronary arteries (NCA) with no evidence of stenosis in any of the major coronary branches (designated NCA subjects) or severe atherosclerotic disease defined as at least 50% stenosis of all three major coronary arteries (designated tricuspid valve dysplasia (TVD) subjects). Serum was prepared using standardised protocols and stored at −80 °C until analysed [12]. For our pilot analysis, approximately 40 samples were collected for each of the two groups. A metabolic profile was then obtained by 600 MHz proton NMR spectroscopy, using selective irradiation to suppress the water signal, followed by phasing and baseline correction with reference to the lactate peak at 1.33 ppm. The resulting spectra were data reduced to 256 integral segments ("bins") of 0.04 ppm width, with the integral of bins from 4.5 to 6.5 ppm set to zero to avoid any impact from variable water suppression or proton exchange to the abundant urea signal. Finally, the signal in the remaining 208 bins was normalised to total spectral area [13].

The resulting metabolite profiles from all 80 individuals were remarkably similar on superficial examination (Figure 12.5), reflecting the highly conserved nature of mammalian intermediary metabolism. However, closer inspection revealed subtle differences between the individuals. We then applied chemometric methodology to the full dataset to identify any reproducible multivariate signatures that distinguished the NCA and TVD subjects. Supervised Projection to Latent Structures Discriminant Analysis (PLS-DA) following application of a single round of Orthogonal Signal Correction (OSC) revealed a signature capable of almost completely separating the NCA and TVD groups [13]. Importantly, external validation of this model (using small external hold-out sets) suggested that the model was not over-fitted and likely represented a general biochemical signature of heart disease. Further investigations strongly suggested that the signature was not primarily the result of the gender differences between the two groups nor any differences in their medication history.

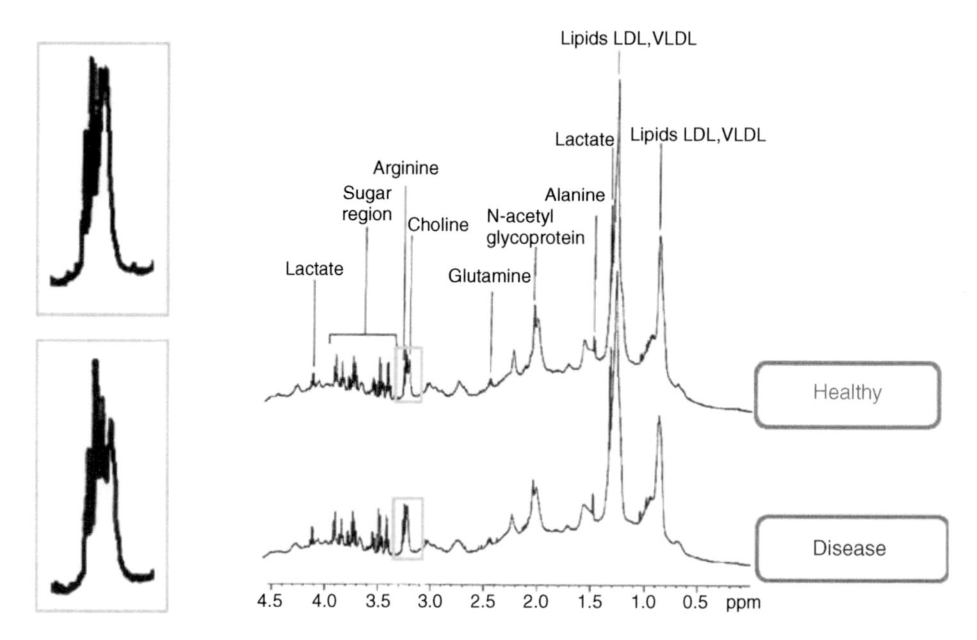

Figure 12.5. Serum metabolic profiles. Two metabolic profiles, derived by one-dimensional 600 MHz proton NMR spectroscopy of serum samples as previously described [13], are shown. The lower profile represents serum from an individual with no angiographic evidence of heart disease, while the upper profile represents serum from an individual with severe (TVD) heart disease. Grossly the profiles are similar, but careful examination reveals potentially important differences (insets).

Although the primary purpose of this study was to evaluate the possibility that a metabolic fingerprint could be used clinically to provide a non-invasive diagnosis of the presence of CHD, we were able to draw some conclusions about the likely molecular nature of the metabolites responsible for distinguishing the groups. The region of the spectrum around 1.3 ppm seems to be particularly important, reflecting subtle differences in the lipoprotein compositions between TVD and NCA individuals [5, 13]. It is suggestive of alterations in the fatty acid composition of the triglyceride component of the lipoproteins, but one-dimensional NMR spectroscopy is largely unable to distinguish individual fatty acids (and in particular to resolve isomers, such as double bond positions in mono- and polyunsaturated fats). Such findings are hardly surprising: even crude lipoprotein fractionations based on density (such as very low density lipoprotein (VLDL), low density lipoprotein (LDL) and high density lipoprotein (HDL) cholesterol measurements) provide some useful diagnostic information, although clearly the NMR-derived metabolite profile has better diagnostic potential than these crude fractionations. Our findings are also consistent with those of Otvos and colleagues, who have used NMR to characterise lipoprotein profiles in subjects with and without CHD [14, 15].

The region around 3.22 ppm also seems to contribute to the separation of the two groups, with a peak presenting choline or a choline derivative significantly larger among the NCA subjects [13]. Since choline-containing phosopholipids are relatively enriched in HDL fractions, and since HDL-cholesterol is known to be depressed in subjects with CHD, this finding is also consistent with our understanding of lipid metabolism defects in heart disease [5].

It is important to emphasise, however, the contribution of other metabolites that have not been identified (at least some of which are unlikely to be lipid components). The multivariate nature of the predictive model is likely to underlie the excellent (and potentially clinically useful) separation of the two groups that we obtained, demonstrating the major strength of fingerprinting when diagnosis is the primary aim of the investigation. The extent to which the diagnostic potential of this multivariate fingerprint can be replicated through identification of the contributing metabolites, followed by validation of separate assays for each of them, remains to be determined. If very many individual biomarkers are contributing to the fingerprint, it remains possible that a pronostic test could be clinically and commercially viable as the principal diagnostic approach for CHD.

Encouraging though the results of this initial study were [13], the design (comparing individuals with no evidence of disease by angiography, with the most severe sufferers) maximises the likelihood of obtaining separation. In practice, however, the extent of atherosclerotic disease is a continuum, and separating those with clinically significant disease from healthy individuals is a much more challenging task. In a separate study, attempting to use the same 600 MHz NMR-derived serum metabolite profile to distinguish individuals with CHD of different severity (based on a crude scoring system) suggested that some distinction could be made, but there remained considerable overlap between the groups [13].

These studies have focussed on providing a non-invasive test which replicates the outcome of invasive angiography. However, it is clear that angiography itself is an imperfect diagnostic for CHD. As noted above, the extent of vessel stenosis may be an unreliable indicator of the likelihood of plaque rupture and hence acute myocardial infarction. Plaque composition, rather than size, may be the most important determinant of clinical outcome.

A recent study by Sabatine and colleagues used a metabolite profiling approach to identify novel biomarkers associated with myocardial ischaemia [16], as opposed to coronary vessel stenosis. In a small study of 36 subjects, 18 of whom exhibited inducible ischemia on treadmill exercise, they generated a metabolite profile of plasma using an LC-MS approach, which provides a tabulated list of metabolites and their relative concentrations. Comfortingly, they identified an increase in lactic acid and changes in metabolites of adenosine monophosphate (AMP) in both cases and controls, but more interestingly an impressive functional pathway trend analysis identified a selective alteration in citric acid metabolism associated with the presence of inducible ischemia [16]. Such a study illustrates the impressive power of analysing relatively small groups of subjects using cutting-edge multivariate statistical analysis tools.

In a similar way, we have used NMR-derived metabolic profiles to search for signatures associated with high blood pressure. This analysis clearly demonstrated that metabolic covariates of blood pressure exist, although the fingerprint nature of the model curtailed our ability to identify specific molecular biomarkers that may have revealed new insight into the pathophysiology of hypertension [17].

Continuing this trend to examine metabolic signatures associated with physiological (rather than anatomical) cardiovascular end-points will likely be a productive direction in which to proceed. Ultimately, it will be interesting to determine whether any pre-existing metabolic phenotype (or "metabotype") can be identified as being at greater risk of acute myocardial infarction. Such studies are inherently more difficult to design and implement: it is difficult to identify population sub-groups at high absolute risk of acute myocardial infarction from whom to collect blood samples. As a result, either large study cohorts or long durations from sampling to cardiovascular event (or both) are required to power the study adequately. One viable approach may be to construct pseudoprospective cohorts using stored blood samples from large epidemiological studies such as EPIC, in which many hundreds of incident cardiovascular events are recorded, and from which several matched control samples can be identified for each case [18].

Another useful application of metabolic profiling is to investigate the similarities between widely used animal models of disease and the human pathology being modelled. We have recently reported the first such direct comparison, in the neurodegenerative disease arena, comparing Huntington's Disease with a transgenic mouse model [19]. Interestingly, similar metabolic perturbations suggestive of a systemic pro-catabolic tendency were identified in both mice and humans providing much needed validation of the rodent model. A combined proteomic and metabolomic analysis of apoE-deficient mice, a widely used model of vascular lipid lesion development [20, 21], from the laboratory of Xu, has recently been published [22], but the focus on local changes in the blood vessel wall (as opposed to the systemic changes in blood studied in humans) precludes any direct comparison. They reported a decline in alanine (although difficulties associated with pH changes and carbamate formation make interpretation of changes in amino acid levels on the basis of NMR-derived metabolic profiles more challenging) and depletion of adenosine nucleotides in blood vessels from apoE-deficient mice prior to the development of major fatty deposits. A direct comparison of the changes observed during atherogenesis in mice and man will likely be informative, and provide much needed information on the likely validity (or otherwise) of the large number of animal models of vascular lipid lesion formation.

12.5. The next steps

Early studies published to date have done little more than illustrate the potential for applying metabolite profiling to CHD. In particular, the populations studied have been rather small and limited to the comparison of relatively extreme phenotypes. Despite these limitations is already clear that metabolic signatures exist which have the potential to improve our capability to diagnose the presence of CHD prior to invasive imaging. Even more recent studies [23], which suggested that the degree of diagnostic capability present in an NMR-derived serum metabolic profile was less than previously demonstrated [13], nevertheless demostrate a statistically significant association. The reduced power in this latter study likely results from their inappropriate combination of disparate clinical cohorts, as opposed to the suggestion that the original study [13] over-estimated its power due to gender bias, since the original study had incorporated gender into the modelling strategy. More interestingly, Kirschenlohr and colleagues suggest that statin use may decrease the diagnostic power of serum metabolite profiling (since <5% of the subjects in the Brindle report [13] were using statins). Since statins are intended to normalise the lipoprotein profile, and since subtle differences in lipoprotein composition dominated the original separation reported by Brindle *et al.* [13], the findings of Kirschenlohr tentatively support the concept that the differences in lipoprotein composition reported by Brindle *et al.* were causal rather than merely a consequence of disease status. Cleary, more extensive studies of better design than the Kirschenlohr study [23] are urgently required to definitively answer these questions.

On the basis of these findings, we implemented a much larger study called Metabonomics and Genomics in Coronary Artery Disease (MaGiCAD), in which blood samples and extensive background clinical data were obtained on 1234 subjects arriving at Papworth Hospital for a diagnostic coronary angiogram [24, 25]. Considerable care was taken to ensure that these subjects recruited to the study were representative of the entire population undergoing angiography, so that the eventual results of the study could be generalised. While it is common practice in laboratory scale clinical studies to use highly selected patient populations, this can lead to an over-estimate of the likely utility of any biomarker identified, resulting in disappointing diagnostic performance when studies are translated into real-world clinical populations.

In addition to the blood sample (from which DNA, serum and plasma are prepared) and urine sample (which can also be subjected to metabolite profiling), we have collected an extensive lifestyle questionnaire [25], as well as the results of conventional biochemistry tests applied to these patients. This will allow a direct comparison of the diagnostic capability of any metabolic signature we identify with more conventional markers of disease risk, such as lipoprotein density fractions, clotting parameters or inflammatory markers such as C-reactive protein (CRP).

Most importantly, the heart disease phenotype of the study subjects has been comprehensively determined. First, the angiograms have been independently read by two experienced clinicians independent from the clinical management of the patient (to eliminate any possible bias in the interpretation of the angiography data based on other clinical factors known to the reader). From these independent angiogram reads, we have constructed a detailed "map" of the distribution and extent of atherosclerotic changes evident from the angiogram [25], as well as a number of simple classifications based on published criteria [26, 27]. Secondly, we have recorded the history of previous cardiovascular events (including myocardial infarction and stroke) for the patients, and monitor all of them for future incident events (both fatal and nonfatal). In this way, we hope to identify not only metabolic signatures associated with anatomical disease, but also any patterns which distinguish stable disease from those at high risk of plaque rupture and acute myocardial infarction.

Within the subjects recruited into MaGiCAD will be individuals who have no coronary atherosclerotic disease visible by angiography, and thus serve as controls for comparison with the diseased individuals. However, in many cases these subjects with apparently normal coronary arteries had suffered symptoms consistent with heart disease (such as angina, exercise-induced electrocardiographic changes or breathlessness). As a result, we have also recruited into the study 100 individuals not undergoing diagnostic angiography as an important additional control group [25].

Recruitment to MaGiCAD began in October 2001 and closed in December 2005. Metabolite profiles for all of these subjects are being generated by NMR spectroscopy, and for a smaller subset of the samples by GC-MS, and early results from the trial are expected to be available before the end of 2006. One of the most obvious outcomes from the study will be a replication, on a much larger patient population, of the original observation that metabolite profiling can be used to provide a clinically useful non-invasive diagnosis of CHD. However, the broad design of MaGiCAD allows very many more conclusions to be drawn. For example, we will be able to compare the power of diagnostic tests based on metabolite profiling with the current crop of conventional biochemical biomarkers of heart disease. We will also be able to compare the application of metabolite profiling of urine and serum (since a recently completed study in Parkinson's Disease suggests that the metabolite profile of urine in these patients contains more diagnostic information than the serum profile from the same individuals). Similarly, we can compare (for the first time on such a large population) the diagnostic information obtained from a fingerprinting method such as NMR spectroscopy compared with a more targeted separation-based approach.

Another interesting angle will be the generation of a broader "multi-omics" profile for at least a subset of the subjects within MaGiCAD. By constructing a genomic profile (using chip-based SNP analysis), a proteomic profile (by differential two-dimensional-gel electrophoresis applied to serum samples) and a metabolic profile, we will be able to determine whether, and to what extent, the three domains contribute additional diagnostic information. Only this kind of comparison can identify the lowest cost approach to obtaining a "sufficiently data rich" profile to reliably diagnose the presence of CHD [4]. Indeed, as newer, potentially lower cost, profiling platform technologies arise (such as Differential Megaplex Immunoassay for profiling the immune system, a process we term "immunomics"), then MaGiCAD will provide an ideal resource for comparing their capabilities with the more conventional profiling domains, as well as with panels of simple biochemical markers such as those in routine clinical use today.

Ultimately, if metabolic profiling (or any other pronostic test) can improve the non-invasive diagnosis of CHD this will be of great value to the clinical management of the disease, and pay a large public health dividend. Even more enticing, however, is the possibility that hitherto unsuspected mechanisms in the pathophysiology of CHD may be uncovered. The pilot-scale studies published to date suggest that perturbations in lipid metabolism alone (already well known to play a central role in the development of atherosclerotic disease) may be sufficient to provide a useful diagnostic test equivalent in power to angiography. However, it will be interesting to see whether other important metabolic changes can be identified in the much larger studies which will report over the coming years. Even if the eventual conclusion is that the existing triumvirate of lipid metabolism, thrombosis and inflammation are indeed the key (or only) pathogenic mechanisms in atherosclerotic vascular disease, a more detailed understanding of just how they are perturbed may still aid the development of new classes of therapeutic agents.

References

[1] Assmann, G., Schulte, H. Identification of individuals at high risk for myocardial infarction. *Atherosclerosis*. 1994, 110 Suppl: S11–21.

[2] Archacki, S., Wang, Q. Expression profiling of cardiovascular disease. *Hum. Genomics*. 2004, 1:355–370.

[3] Yanagisawa, K., Xu, B.J., Carbone, D.P., Caprioli, R.M. Molecular fingerprinting in human lung cancer. *Clin. Lung Cancer*. 2003, 5:113–118.

[4] Grainger, D., Nicholson, J. Metabonomics and Metabolomics. In *Encyclopedia of Molecular Cell Biology and Molecular Medicine* (Ed. R.A. Meyers). 2005. Wiley-VCH, Weinheim.

[5] Grainger, D., Mosedale, D., Holmes, E., Nicholson, J. Metabonomics as a tool for understanding lipid metabolism. In Unravelling Lipid Metabolism with Microarrays (Eds A. Berger, M.A. Roberts). 2005. Marcel Dekker, New York, pp 405–422.

[6] Peterson, S., Peto, V., Scarborough, P., Rayner, M. *Coronary heart disease statistics*. British Heart Foundation, London. 2005 (http://www.heartstats.org)

[7] Ross, R. The pathogenesis of atherosclerosis – an update. *N. Engl. J. Med.* 1986, 314:488–500.

[8] Shirani, J., Yousefi, J., Roberts, W.C. Major cardiac findings at necropsy in 366 American octogenarians. *Am. J. Cardiol.* 1995, 75:151–156.

[9] Weissberg, P., Rudd, J. Atherosclerotic biology and epidemiology of disease. In *Textbook of Cardiovascular Medicine* (Ed. Topol, E.). 2002 (2nd Edition). Lippincott, Williams & Wilkins, Philidelphia.

[10] Falk, E. Why do plaques rupture? *Circulation*. 1992, 86: III30–42.

[11] Sundstrom, J., Riserus, U., Byberg, L., Zethelius, B., Lithell, H., Lind, L. Clinical value of the metabolic syndrome for long term prediction of total and cardiovascular mortality: Prospective, population based cohort study. *BMJ*. 2006, 332: 878–882.

[12] Grainger, D.J., Kemp, P.R., Metcalfe, J.C., Liu, A.C., Lawn, R.M., Williams, N.R., Grace A.A., Schofield, P.M., Chauhan, A. The serum concentration of active transforming growth factor-beta is severely depressed in advanced atherosclerosis. *Nature Med.* 1995, 1: 74–79.

[13] Brindle, J.T., Antti, H., Holmes, E., Tranter, G., Nicholson, J.K., Bethell, H.W., Clarke, S., Schofield, P.M., McKilligin, E., Mosedale, D.E., Grainger, D.J. Rapid and noninvasive diagnosis of the presence and severity of coronary heart disease using 1H-NMR-based metabonomics. *Nature Med.* 2002, 8: 1439–1444.

[14] Otvos, J.D., Jeyarajah, E.J., Bennett, D.W. Quantification of plasma lipoproteins by proton nuclear magnetic resonance spectroscopy. *Clin. Chem.* 1991, 37: 377–386.

[15] Kuller, L., Arnold, A., Tracy, R., Otvos, J., Burke, G., Psaty, B., Siscovick, D., Freedman, D.S., Kronmal, R. Nuclear magnetic resonance spectroscopy of lipoproteins and risk of coronary heart disease in the cardiovascular health study. *Arterioscler Thromb. Vasc. Biol.* 2002, 22: 1175–1180.

[16] Sabatine, M.S., Liu, E., Morrow, D.A., Heller, E., McCarroll, R., Wiegand, R., Berriz G.F., Roth, F.P., Gerszten, R.E. Metabolomic identification of novel biomarkers of myocardial ischemia. *Circulation*. 2005, 112: 3868–3875.

[17] Brindle, J.T., Nicholson, J.K., Schofield, P.M., Grainger, D.J., Holmes, E. Application of chemometrics to 1H NMR spectroscopic data to investigate a relationship between human serum metabolic profiles and hypertension. *Analyst*. 2003, 128: 32–36.

[18] Yuyun, M.F., Khaw, K.T., Luben, R., Welch, A., Bingham, S., Day, N.E., Wareham, N.J. A prospective study of microalbuminuria and incident coronary heart disease and its pronostic significance in a British population: The EPIC-Norfolk study. *Am. J. Epidemiol.* 2004, 159: 284–293.

[19] Underwood, B.R., Broadhurst, D., Dunn, W.B., Ellis, D.I., Michell, A.W., Vacher, C., Mosedale, D.E., Kell, D.B., Barker, R.A., Grainger, D.J., Rubinsztein, D.C. Huntington disease patients and transgenic mice have similar pro-catabolic serum metabolite profiles. *Brain*. 2006, 129: 877–886.

[20] Plump, A.S., Smith, J.D., Hayek, T., Aalto-Setala, K., Walsh, A., Verstuyft, J.G., Rubin, E.M., Breslow, J.L. Severe hypercholesterolemia and atherosclerosis in apolipoprotein E-deficient mice created by homologous recombination in ES cells. *Cell*. 1992, 71: 343–353.

[21] Zhang, S.H., Reddick, R.L., Piedrahita, J.A., Maeda, N. Spontaneous hypercholesterolemia and arterial lesions in mice lacking apolipoprotein E. *Science*. 1992, 258: 468–471.

[22] Mayr, M., Chung, Y.L., Mayr, U., Yin, X., Ly, L., Troy, H., Fredericks, S., Hu, Y., Griffiths, J.R., Xu, Q. Proteomic and metabolomic analyses of atherosclerotic vessels from apolipoprotein E-deficient mice reveal alterations in inflammation, oxidative stress, and energy metabolism. *Arterioscler Thromb. Vasc. Biol.* 2005, 25: 2135–2142.

[23] Kirschenlohr, H.L., Griffin, J.L., Clarke, S.C., Rhydwen, R., Grace, A.A., Schofield, P.M., Brindler, K.M., Metcalfe, J.C. Proton NMR analysis of plasma is a weak predictor of coronary artery disease. *Nature Med.* 2006, 12: 705–710.

[24] Mosedale, D.E., Smith, D.J., Aitken, S., Schofield, P.M., Clarke, S.C., McNab, D., Goddard, H., Gale, C.R., Martyn, C.N., Bethell, H.W., Barnard, C., Hayns, S., Nugent, C., Panicker, A.,

Grainger, D.J. Circulating levels of MCP-1 and eotaxin are not associated with presence of atherosclerosis or previous myocardial infarction. *Atherosclerosis.* 2005, 183: 268–274.

[25] http://www.magicad.org.uk.

[26] Ringqvist, I., Fisher, L.D., Mock, M., Davis, K.B., Wedel, H., Chaitman, B.R., Passamani, E., Russell, R.O. Jr., Alderman, E.L., Kouchoukas, N.T., Kaiser, G.C., Ryan, T.J., Killip, T., Fray, D. Pronostic value of angiographic indices of coronary artery disease from the Coronary Artery Surgery Study (CASS). *J. Clin. Invest.* 1983, 71:1854–1866.

[27] Califf, R.M., Phillips, H.R. 3rd, Hindman, M.C., Mark, D.B., Lee, K.L., Behar, V.S., Johnson, R.A., Pryor, D.B., Rosati, R.A., Wagner, G.S. Pronostic value of a coronary artery jeopardy score. *J. Am. Coll. Cardiol.* 1985, 5: 1055–1063.

The Handbook of Metabonomics and Metabolomics
John C. Lindon, Jeremy K. Nicholson and Elaine Holmes (Editors)

Chapter 13

The Role of NMR-based Metabolomics in Cancer

Leo L. Cheng[1] and Ute Pohl[2]

[1]*Departments of Radiology and Pathology, Massachusetts General Hospital, Harvard Medical School, Boston, Massachusetts, U.S.A.*
[2]*Department of Histopathology, Addenbrooke's Hospital, Cambridge, U.K.*

13.1. In the era of "-omics"

Biological science in the 21st century has been marked by the amending of well-recognized scientific branches into newly created disciplines with the suffix of "-omics." As you may have already noticed, in less then half a decade such popularized disciplines as genomics and proteomics have been embraced by the scientific community as well as the mass media, although the exact scopes and dimensions of these disciplines are still in the process of development. Not surprisingly, the topic of this book, metabolomics – a junior member of the "-omics" family – and the connection of metabolomics with the aspect of this chapter, human oncology, are even less concretely defined. To facilitate our discussions in this chapter we wish to define current cancer metabolomics as: a study of the global variations of metabolites, and a measurement of global profiles of metabolites from various known metabolic pathways under the influence of oncological developments and progressions. We wish to emphasize the keyword in this definition is "global," which is clearly different from other adjectives, such as "individual."

This seemingly simple definition may not be obvious when you try translating it into clinical research. In fact, if you proposed to search for cancer metabolite profiles before the turn of the 21st century, you would probably be branded as conducting

Note: We thank Kate Jordan for editorial assistance. L.L.C. is partially supported by grants: PHS/NIH CA095624 and DOD W81XWH-04-1-0190.

"fishing expeditions." Fortunately, the expedition phrase has gradually faded away from today's scientific colloquy and the disengagement between perceptions of common sense and the rules in scientific pursuits have been somewhat lessened in regard to metabolomics. As evidence, if you browse the website of the National Institute of Health (NIH) of the United States, you will find the word "metabolomics" in the title of Requests for Applications (RFAs) published by the agency ("Metabolomics Technology Development" http://grants.nih.gov/grants/guide/rfa-files/RFA-RM-04-002.html and "Genomic, Proteomic, and Metabolomic Fingerprints as Alcohol Biomarkers" http://grants.nih.gov/grants/guide/rfa-files/RFA-AA-06-001.html). In addition, during the preparation of this chapter, a workshop titled "Frontiers in Metabolomics for Cancer Research" was organized in October 2005 by the National Cancer Institute at NIH.

This change of heart regarding "-omics" came as a direct result of the birth of genomics. Technological inventions, such as those accompanying the human genome project, have created this new scientific branch that captures the front pages of mass media news as well the covers of scientific journals. After the completion of the human genome project, the innovative use of the suffix "-omics" was created to reflect the processes of searching for correlations between the large arrays of genome data measured by these new technologies, with physiological and pathological conditions. Genomics, rather than genetics, has granted "licenses" to many pursuits of clinical importance that would otherwise be rejected by scientific communities because of their exploratory nature and the lack of hypotheses that could be clearly delineated. In other words, the introduction of these new "-omics" branches has ratified formerly inconceivable concepts, allowing scientists a greater breadth of exploration in areas into which they have only vague insight; these explorations do not come at the expense of good science, however, and the "-omics" studies still rely on sound data collection and valid, measurable parameters.

There are many genomics examples of gene expression studies for oncology that can be found in the literatures in the past 5 years. For instance, in a seminal paper published in Nature in 2000, C.M. Perou *et al.* reported results of DNA microarrays of 8102 human genes analyzed using breast tissue samples from 42 individuals [1]. Among the measured genes, they analyzed 1753 (22% of 8102) of them in an effort to establish "molecular portraits of human breast tumours," as shown in Figure 13.1. In this cluster analysis map each row represents a single gene, while each column came from a measured sample. Without getting into details it may be appreciated that, with the appearance of clustered red and green dots, the map seems to indicate the existence of certain patterns that are not completely random. In hindsight, considering research development experiences since the publication of the paper, one can ponder a number of issues regarding the published study that may have direct implications to the topic of this chapter. First, cognizant of the biological variations among individuals, how representative would the "portraits"

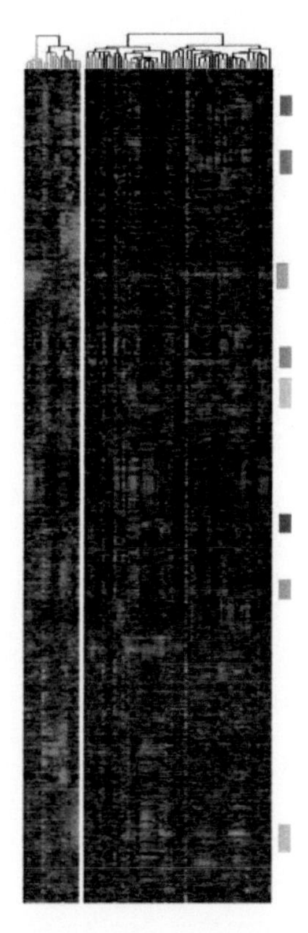

Figure 13.1. The 1753-gene cluster diagram revealing gene expression in 84 experimental samples of breast cancer reported by C.M. Perou *et al.* [1]. Each row represents a single gene, and each column an experimental sample. The colored bars on the right identified the gene clusters corresponding to (from the top): endothelial, stromal/fibroblast, basal epithelial, B-cells, adipose-enriched/normal breast, macrophage, T-cells, and luminal epithelial cells.

be if a number of tumors were only collected from a single case? Secondly, how valuable would the gene profiles be without knowing the pathological details from which the profiles were derived? These issues arise because of the existence of heterogeneity, a major characteristic of human malignancy, that is the pathological features and the amount of these features vary among different regions in the same tumor, and the study was conducted on homogenized tissue blocks [2]. Later studies have indicated that these problems could be resolved to satisfactory degrees. The first issue is administrative and somewhat easy to solve as long as multiple clinical cases are available to the study. To solve the second issue, a colleague of ours, D.C. Sgroi, has demonstrated the need for and the feasibility of sample preparation via a special protocol. With the assistance of laser capture microdissection techniques, Sgroi and his colleagues removed pathologically identified specific cell types from

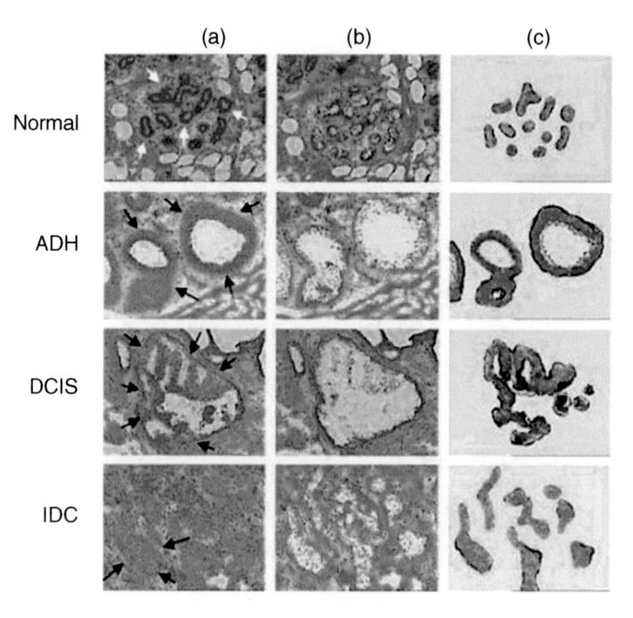

Figure 13.2. Laser capture microdissection identified normal breast epithelium (white arrows) and abnormal epithelium (black arrows) from atypical ductal hyperplasia (ADH), ductal carcinoma in situ (DCIS), and invasive ductal carcinoma (IDC) from a single breast specimen. Images of precapture (lane (a)), postcapture (lane (b)), and the captured epithelial compartments (lane (c)) are shown.

36 breast cancer patients (Figure 13.2), and subjected these removed cells to 12,000-gene cDNA microarray analyses [3]. By analyzing the most varied 1940 genes (16% of 12,000), they were able to propose genomic profiles for tumors of different grades, as shown in Figure 13.3. Although there are many other studies that may be discussed, these two reports contain some common features that may be appreciated for our subsequent discussion. First, more than 80% of measured parameters did not reach the final analyses, and second, at least, more than half of the genes that composed the final cluster maps were of unknown functions, according to the most conservative estimation.

This seemingly "new" concept for research designs (deviating from the dominated and accepted structure of hypothesis-driven research approaches) has in fact always been employed in medical science and medicinal practice during the progression of modern medicine. For instance, considering histopathology and patient prognostication, and even equipped with the most up-to-date knowledge of anatomic pathology, it is almost impossible for us to imagine how one would propose any intelligent hypothesis that could relate cellular morphological changes with disease processes at the dawn of the discovery of optical microscopes, or even today. Nevertheless, that did not prevent observation-based anatomic pathology from becoming a discipline that dictates almost every aspect of oncological practice to this day (as will

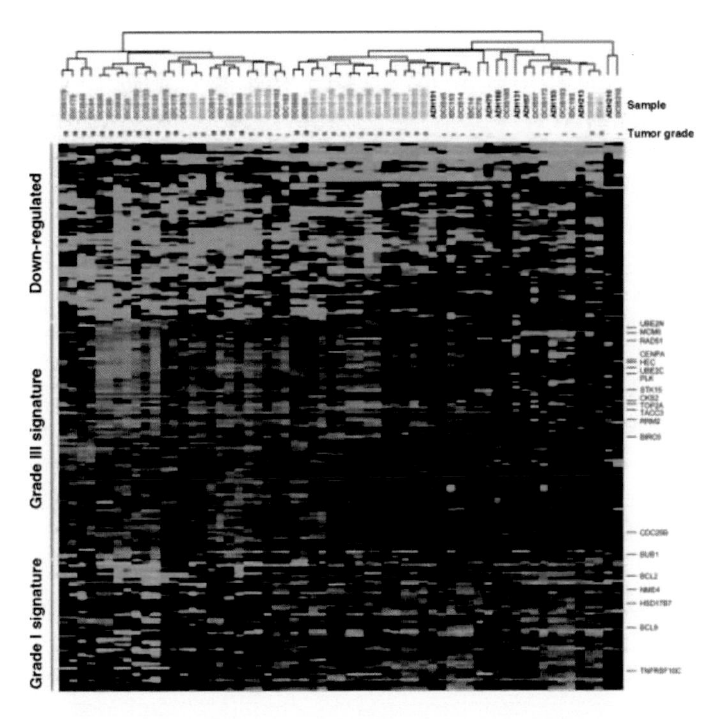

Figure 13.3. Two-dimensional clustering of 61 samples and the top 200 genes correlating with tumor grade.

be discussed later in this chapter). Thus, we can appreciate the emphasis on the criteria of hypothesis-driven research as the philosophical practices accompanying the blooming of sciences in the 20th century. In addition, we should also realize the vital value of scientific explorations and derivations from these observations both in the history and at present. And finally, as hypothesis-driven research was a valuable tool of the past, so can we appreciate the transition in the 21st century to this new era of "-omics" and the immense scientific potential this era holds.

13.2. Current metabolomics

Examining the active field of current metabolomics from the same perspective that you may use to scrutinize the status of genomics, you may realize that the state of current metabolomic activities is fundamentally different from contemporary genomics or proteomics. To fully appreciate the status and utilize the capability of current metabolomics, we need to dwell a bit on the still young history of genomics.

The development of genomics relied almost entirely on innovations in the technologies of molecular biology. One such innovation which is now widely used was the creation of gene microarrays, as exemplified in the previously discussed human breast cancer studies. The fundamental characteristic of these new array paradigms is their ability to measure a very large number of parameters at the same time. These parameters can either be the investigation of thousands of genes from one sample or the simultaneous comparison of the expression of one gene for hundreds of tissue samples. These investigations produce massive amounts of data in a magnitude that was never before witnessed in the human pursuits of biological science. In the context of human pathology, this situation may resemble the overwhelming sensation that our predecessors might have experienced when optical microscopes were first introduced into gross anatomy. The rules of engagement were changed. Thus, the transition to morphological examination more than a century ago required digestion and evaluation efforts of many generations, in many decades, to present us with the current prominent status of anatomic pathology. While optical microscopes allowed pathological evaluations to intrude into smaller physical spaces, genomics expanded our ability to evaluate a vast amount of genetic parameters, for many of which our current genetic knowledge cannot provide insight. Hence, we may expect that the efforts of the coming generations may be necessary to fully comprehend the functional importance of these genes; although hopefully the ever-increasing pace of developments in science and technology may facilitate these digestion and evaluation efforts. For example, we notice a recent publication on *interactome* and *transcriptome* in Nature (11 August 2005) [4]. In this work, M. Vidal and colleagues expanded the previous concept of the interactome for studies of protein–protein interactions, also known as *interactomics*, into evaluations of relationships of gene expression profiles, now known as *integrative interactomics*, to identify the functions of genes in *Caenorhabditis elegans*.

Fortunately, the current status of metabolomics is less complicated than those of genomics and proteomics, which mostly present either dots of individual genes or proteins with *unknown* linkages among them, or genes/proteins with indiscernible functions that require interactomics to reveal them. As already stated in the beginning of this chapter, our interpretation of and emphasis for current metabolomics is focused on evaluations of global variations and global profiles of metabolites from various *known* metabolic pathways in relation to physiological and pathological processes. This "known" vs "unknown" difference, in our opinion, is the unique characterization of metabolomics that philosophically differentiates it from the current concepts of genomics and proteomics.

This "known" status is due solely to the fact that, up till now, there have not been fundamental technological breakthroughs in the innovation of metabolite analysis. This situation is very unlike the fields of genomics and proteomics where breakthrough technical innovations have been witnessed in the past years. Methodologies used in current metabolite analyses were well-established in past decades. Because of this the metabolic knowledge currently studied by the methodologies of analytic chemistry is known, at least individually, for each metabolite based on characterization through many tests and proven metabolic pathways. These methodologies currently include nuclear magnetic spectroscopy (NMR), and mass spectroscopy, the latter usually with a gas or liquid chromatography separation stage.

You may agree with the concept that the words "genomics" and "proteomics" first serve to capture the images of scientific frontiers whose constitution and significance are still unclear. Second these terms emphasize the need to consider all the measurable parameters simultaneously. If you concur with these statements, then from our above discussion you may realize that the same concept cannot be applied to the parallel expression – metabolomics – without alterations. At the present, conducted with the above-mentioned well-tested methodologies for metabolite analyses, the chance for the discovery of new metabolites in vast quantities as witnessed in the other "-omics" fields is relatively slim. However, this should not discourage the development of metabolomics from the philosophical aspect of the "-omics" concept on global interpretations of the organization of the current knowledge on the overall metabolite pathways. The task of primary importance in current metabolomics is to interpret the global metabolite alterations collectively in the context of their overall changes in relation to human physiological and pathological conditions. For the purpose of this chapter, it is to better understand the global inter-connectivity of metabolites from various known metabolic pathways during the development and progression of human malignancies. Such a concept reflects and agrees well with the somewhat ancient idea that the human body is a united organic entity within which all the biological processes are inter-connected and balanced to produce measurable metabolite profiles. It is this overall profile, likely not a single metabolite, that may alter cell chemistry and be informative as indicative of disease. We also wish to emphasize that our statement regarding the slim chance for discovering new metabolites with the current analytic methods merely reflects our assessment of the limitations of the existing analytic methodologies. It by no means indicates we should stop expanding our knowledge on human metabolism. On the contrary, given the numbers of new genes and proteins being discovered, we believe the numbers of metabolites unknown to us may be too large to estimate.

A metabolite, according to the Oxford English Dictionary, is "a substance that is a substrate or product of a metabolic reaction, or that is necessary to a metabolic reaction; *esp.* an intermediate or end product of a metabolic pathway (Draft Revision Dec. 2001)." The pictorial interpretations of this simple definition can be very complicated for many of us to grasp. Demonstrations of the current knowledge on human metabolism can be found in various maps and charts of metabolic pathways; charts such as the very well known one comprehensively generated by Dr Donald E. Nicholson of Leeds, England, in collaboration with the International Union of Biochemistry & Molecular Biology and with Sigma-Aldrich. Since any substance registered in such a chart, and possibly millions of the others that are still unknown to us, can potentially be the fair targets and subjects of metabolomics, the complex future of metabolomics seems impossible to comprehend at this moment. On the other hand, if we again consider that under normal conditions all of these pathways in a human body are initialized by dietary intakes, then such a puzzling chart may be simplified greatly to carbohydrates, lipids, proteins, amino acids, and so on. Thus, it seems that any measurement of these substances, using any modality taught in college classes of analytical chemistry, can be potential subjects of and tools for metabolomics. In reality, we need to point out that, presently in 2006, the bulk of the so-called metabolomics data associated with human oncology is supplied by NMR analyses. We attribute this observation to the fact that there has been an established NMR community actively pursuing identification of cancer-related metabolites far prior to the era of "-omics." We hope that by now you have noticed that whenever we mention metabolomics, we define it as "the current metabolomics" or "metabolomics of today." This is due particularly to the above-mentioned fact of the prior existence of analytical technologies in metabolite chemistry. Many parts of our discussion should be viewed as applicable only to the present status of metabolomics. Although it is hard to predict the era to come, we are certain that technical developments will change the face as well as the contents of metabolomics in the near future. The situation of current and future metabolomics is concisely summarized by the previously mentioned NIH RFA:

> Technologies currently in use for metabolomic analysis include NMR, chromatography and mass spectrometry, each of which has significant limitations in quantification, scope, and/or throughput. No one technology can effectively measure, identify and quantify, with sufficient sensitivity and precision, the diverse range of metabolites and their dynamic fluctuations in cells. An integrated set of technologies is needed to address the entire spectrum of challenges for metabolomics. Ideally, new technologies should yield quantitative, comprehensive data and be applicable to achieving anatomical resolution at the cellular and subcellular level ("Metabolomics Technology Development" http://grants.nih.gov/grants/guide/rfa-files/RFA-RM-04-002.html).

13.3. Current oncology

13.3.1. Introduction

Any clinical procedure in modern medicine, including oncology, can be characterized into one of the following categories for disease management: screening, diagnosis, or therapy. Developments in modern medicine, therefore, have been spinning around advancements in these areas. Surveying the current status of oncology, the impacts of new scientific knowledge and technology innovations are evident. Unfortunately the consequence of new knowledge and its impact on the life or the quality of life for patients may not always be as pleasant as expected. For instance, with the realization of the relationship between certain genetic mutations and breast cancer development, prophylactic mastectomy (preventive removal of breasts) has been offered to women who are considered to be at high risk. However, harboring certain genetic mutations does not mean that cancer is present, or will always occur. Hence, it may be argued that prophylactic mastectomy might not be necessary for all women in whom these mutations are detected. Therefore, it can be best argued that in such a case new and more accurate disease markers are urgently needed; specifically markers that are particularly sensitive in early detection of cancer and preferably pre-cancerous lesions. The pursuit of how best to incorporate scientific developments into oncological practices to serve patients and not merely to treat diseases presents as not only a medical or technological topic, but a medical philosophical and ethical issue as well.

Examining the effects of emerging disciplines on disease management, we find that histopathology (the current "gold standard" in cancer diagnosis) is no longer viewed as an optimal diagnostic tool. Histopathology provides accurate, literal pictures of the existing state of specimens under evaluation. However, its ability is limited when it comes to providing information that may delineate the preceding condition of the individual patient prior to the particular histopathological presentation or to suggesting the possible or probable disease course for an individual condition, partly due to the limited nature of morphological appearances that may mask the actual character of the disease. Histopathology works statistically well in providing disease interpretations for a class of conditions or a group of patients. Particularly, the histopathological approach to cancer diagnosis has served oncology sufficiently in the past, when the oncological needs for decision-making on therapies were a direct result of symptoms that had displayed clinical significance. However, these past successes have been challenged in the current era of cancer screening, when patient presentations are asymptomatic and the diseases are early enough to defy the statistical efficiency of the morphological pathology. With the creation of new testing protocols there is now a societal demand for early diagnosis and individual disease management. For this reason, if you immerse yourself in oncological discussion these days, often what you will hear is not how to treat this group of patients, but rather what is the best treatment plan for this *individual*. This situation can best be illustrated by some examples.

13.3.2. Prostate cancer

When discussing the impact of the new screening era on oncology, no other type of cancer is more divisive than prostate cancer. The status of prostate cancer as a common, controversial malignancy has not changed since the second half of the 1980s when the blood test of prostate specific antigen (PSA) for cancer screening was introduced. In the United States alone, in 2005, it was estimated that more than 635 men were diagnosed with the disease daily, and another 83 lost their lives to the disease [5]. Meanwhile, considerable concerns have been expressed regarding the adverse effects of surgical intervention for tumors that may never likely become life threatening, as the >150 radical prostatectomies performed each day in the United States result in impotence for >90 men, and/or incontinence of urine for >45 men.

Facing this complicated situation involving a common disease, instead of posting clinical questions that intrigue us, we encourage you to ponder simultaneously a number of factors related to prostate cancer: the increased incidence rate achieved with PSA screening led to the increase of "indolent" cancers in PSA screening populations; the seemingly steady death rate over the history (Figure 13.4); the adverse effects experienced by considerable numbers of patients as results of therapies; and the large body of autopsy evidence that shows about 20–30% of men before the PSA testing era harbored indolent prostate cancer in their gland in non-prostate-cancer-related deaths. How should all these facts play into decision-making in the prostate cancer clinic, both in terms of providing the most appropriate care for an individual patient and for the interests of healthcare costs of the society?

These issues were realized by the National Institute of Cancer (NCI) in the United States in the 1990s. At the request of the then NCI director, Dr Richard Klausner, the NCI Prostate Cancer Progress Review Group (PRG) was formed in 1997 with more than 20 prominent scientists and patient advocates on prostate cancer. After more than one year of work, the PRG published a report titled "Defeating Prostate Cancer: Crucial Directions for Research"

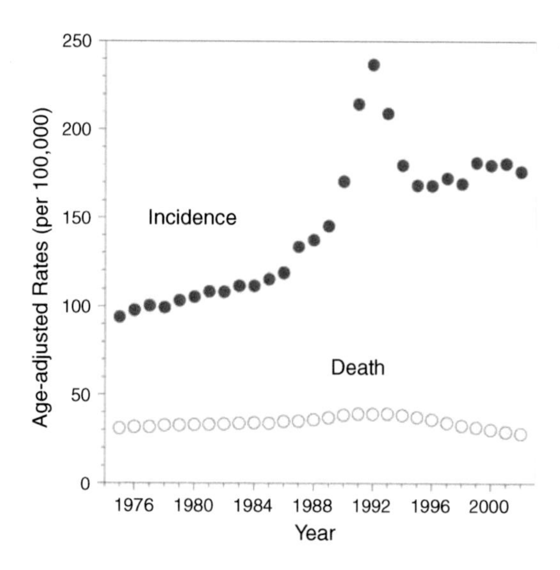

Figure 13.4. Statistical data on human prostate cancer in the United States (Based on the data kindly provided by A. Jemal, also see [6]).

(http://prg.nci.nih.gov/pdfprgreports/1998prostate.pdf). In the PRG Report the panel notes that "[b]ased on autopsy studies it is estimated that as many as 1 in 4 men 30 years of age may harbor a small focus of PCa in their glands." Furthermore, "[t]he basic problem . . . is that while about 11 percent of all men will contract clinically significant PCa in their lifetime, only 3.6 percent will die of it. The challenge is to determine which of the 7 to 8 percent can safely go untreated." The Report cites "our inability to distinguish 'indolent' from 'aggressive' carcinomas," concluding that this can "lead to the adverse consequences of over-treatment" if "[the] tumor is one that will remain 'quiescent,' or clinically insignificant throughout [the patient's] natural life span."

The utility of PSA testing in detecting clinically significant prostate tumors has been clinically proven. However, in the PSA screening era, most newly diagnosed prostate cancer (more than 70% at conservative estimations) belong to a similar group according to Gleason histological grading, which current histopathology cannot further sub-categorize without additional surgical intervention. Furthermore, tumors of clinical significance may be neither lethal to their hosts, nor indicative for "definitive therapies" that may increase "morbidity, particularly incontinence and/or impotence," as being noted by the PRG report. In fact, the PRG report estimates that currently in the United States >30% of prostate cancer patients are over-treated. The urgent task set forth by the PRG report is to discover "new and better markers than PSA for the early diagnosis of patients who harbor fast-progressing and virulent forms of prostate cancer." These markers are expected to "refine" early detection with PSA testing by identifying cancers able to kill their hosts if left untreated, double-checking for overlooked virulent cancers, and in doing so furnishing "prognostic markers that can guide the therapy of patients in an individualized fashion". To realize this goal, the PRG advised the development and validation of molecular assays with the comments that "[e]valuation of limited disease in the prostate is prone to extensive sampling error caused by heterogeneity. The availability of micro-dissection techniques and/or aspiration biopsy, coupled with molecular analyses (i.e. RT-PCR) or array analysis), could provide new approaches to conventional tissue analysis." If these words from the 1990s are translated into today's vernacular, they likely indicate the future advances in prostate cancer clinic will be discovered through "-omics".

13.3.3. Breast cancer

The numbers of NCI Progress Review Groups (PRGs) have grown to 12 that cover almost all types of human malignancies (http://planning.cancer.gov/disease/plans.shtml#prg). However, before the turn of the 21st century, there were only two issued in 1998. One of them was the previously discussed prostate cancer PRG, and the other, commissioned in the same period and published at the same time, was titled "Charting the Course: Priorities for Breast Cancer Research" (http://prg.nci.nih.gov/pdfprgreports/1998breastcancer.pdf).

The status of breast cancer in women is very similar to that of prostate cancer in men. In the United States, breast cancer has always been the most frequently diagnosed cancer in women, and was only reduced to the second leading cause of cancer death in 1980s after the increase of lung cancer incidence [6].

In this era of public awareness and improved screening technologies, breast cancer, similar to prostate cancer, presents its own controversial issues. Controversies in the care of breast cancer patients can be visualized and understood, step-by-step, by following the practices in a comprehensive breast health clinic.

Very often one to several suspicious lesions are identified by self-examination, annual physical examination, and/or a mammography. Since no non-invasive diagnostic modality is now available, everyone agrees that a biopsy, whether a fine-needle aspiration (FNA) or more often a core needle biopsy (both possible in a doctors' office), is necessary.

Unfortunately, this may be the last universal agreement among caregivers, breast cancer patients, and their physicians regarding treatment options. Disagreements arise immediately at the interpretation of the observation of many FNA and core needle biopsy specimens. Discrepancies in these observations often result in open biopsies (operating room procedures) for these FNA and/or core needle inconclusive patients, who represent a very large population. Focused on the histopathology issues of breast cancer, the breast cancer PRG report recognized that light-microscope-based diagnostic and prognostic "criteria are suboptimal," and recommended that "markers should be sought that will signal the presence and identity of specific types of lesions, indicate their prognosis if left untreated, and predict the likelihood that they will respond to particular types of therapy. Clinically useful markers will most likely be identified and characterized first in tissue samples obtained during biopsy or surgical procedures."

Another controversy in today's breast cancer clinic deals with the uncertainty of how to deal with the greatly increased number of ductal carcinoma in situ (DCIS) patients in the mammographic screening era. Twenty years ago DCIS represented only <1–5% of detected breast malignancies. The introduction and widespread availability of mammographic screening, which has greatly improved our ability to detect non-palpable and asymptomatic breast cancer, has increased the reported incidence of DCIS detection to $>50\%$ of all malignancies discovered by mammogram-directed biopsies. This increase has generated a spectrum of challenges to current surgical pathology, ranging from evaluation of breast biopsy specimens to direction of post-surgery therapy. For instance, it is known that frozen section intra-operative evaluation, while generally reliable in diagnosing palpable breast masses, is of limited use in the diagnosis of non-palpable DCIS. The vast number of cases itself has also indicated that DCIS, rather than a simple breast cancer subtype, represents a collection of lesions with varied malignant potential and morphological heterogeneity. Thus, the inter-observer reproducibility of histopathological diagnosis for DCIS is poor. Mastectomy is considered a curative treatment for DCIS, but its radical nature has recently been criticized as unnecessary overtreatment for many women with mammogram-detected, early stage DCIS. Ideally, all women diagnosed with non- or less-aggressive DCIS could successfully be treated with breast-conserving treatment (BCT), but this course may not be successful for highly aggressive lesions. Commonly histopathology cannot classify variations in DCIS morphology, assess tumor aggressiveness, nor direct adjuvant systematic therapies (ASTs) for individual patients, particularly when the objectives for intervention emphasize consideration of a patient's comfort and quality of life. Regarding DCIS, the PRG summarized that "[t]hese women will have near normal survival but may experience short and long-term morbidity from treatment. DCIS is seriously understudied from a disease- and patient-focused outcomes perspective." The PRG further recommends research to integrate patient-focused data "with biological prognostic information to make the best treatment decisions" for these DCIS patients.

The increased number of tumors detected at early stages have also generated another controversy. Currently, treatments for tumors at early stages (such as DCIS, and low grades) occupy an entirely different arena from what they did in the early 1980s predominated by total mastectomy. Breast-conserving treatment is considered to be a major aim of breast cancer treatment in the 21st century. The new BCT approach is clinically and scientifically sound, since it has been shown that patients with a grade I tumor have an 85% chance of surviving for 10 years after diagnosis; however, a grade III tumor reduces the chance of 10-year survival to 45%. An 85% likelihood for survival at 10 years following diagnosis can be considered a great achievement in our battle against breast cancer in light of cancer epidemiology. However, in reality, clinicians and patients are more interested to know, not cancer statistics, but if THIS patient will survive for 10 years. Each patient represents 100% of her own statistics, and unfortunately, due to the extreme heterogeneity of the disease, the current pathology cannot differentiate the 85% survival from the 15% mortality group. Uncertainty as to the nature and aggressiveness of an individual woman's cancer, and the need to rely on historical statistical data, can result in undertreatment or overtreatment of the *individual* woman. For example, chemotherapy is routinely given to women with breast tumors over 1 cm and negative lymph nodes even though only 3–5% of those treated are expected to benefit. Conversely, chemotherapy is routinely withheld from women with tumors under 1 cm and negative lymph nodes, even though some will ultimately develop metastatic disease.

New approaches to characterize tumors that better reflect their biological behavior are sorely needed. These needs were concisely summarized by the PRG as the opportunities in breast cancer research to: "Develop new methods to diagnose clinically significant breast disease and predict clinical outcome better than conventional histologic examination and the few available biomarker assays;" and to discover "biomarkers that predict the clinical outcome of precancerous and cancerous breast lesions if left untreated (i.e. prognostic factors) with a high degree of certainty."

This list of controversies related to the breast cancer clinic goes on, but we will stop after the next one concerning post-surgery AST. Current pathology, developed from observations over the last 100 years, has identified many prognostic factors for invasive breast cancer, such as tumor size, differentiation status, lymph node stage, vascular invasion, and so on. However, these factors have proven to be too generalized in that they fail to consider the biochemical characteristic of an individual tumor. Since almost all the AST currently used in breast cancer clinic – chemo- and radiotherapy – are known to have association with morbidity, personalized therapeutic protocols based on tumor biomolecular signatures are clearly the hope of future breast cancer clinics, especially in the age of BCT. In addition, although there are a number of oncological drugs available, the current AST chemotherapies are still generally conducted on a trial-and-error basis for an individual patient. If one agent fails, valuable time in the window of treatment is lost. Clinicians thus face an urgent need to measure and understand the biological nature of a specific tumor in order to predict the potential efficacy of a particular agent for an individual patient. Again, the breast cancer PRG urges researchers to "learn more about the biology of breast cancer for the purpose of predicting clinical course and predicting response to therapy;" and to discover "biomarkers that predict the response of precancerous and cancerous breast lesions to specific types of therapy (i.e. predictive factors) with a high degree of certainty."

The arguments for and discussions of the need for new and better disease markers for every step of oncological clinics can be made for every type of cancer. Biomarkers that are urgently needed include those that are particularly sensitive in early detection of cancer and precancerous lesions, as well as biomarkers that may test the current oncological hypothesis that cancers may be controlled as chronic disorders, rather than being intervened as acute diseases.

13.4. From histological to molecular pathology

The discovery and invention of cancer screening protocols were the fruits of intensive cancer research efforts in the 1970s and 1980s that resulted in advancements of knowledge in cancer biology and developments in biomedical technologies. The contributions of these screening practices in the identification of malignancies at their early stages, and in the improvement of disease control and patients' survival, were evident. For instance, at present, the percentage of women diagnosed with breast cancer is 1.5 times higher than it was in 1980, while the percentage of women who die of the disease has been reduced by about 15% from its value during the same period. Similarly, as previously discussed, with the assistance of blood PSA tests, the detection of incidences of prostate cancer at relatively early stages has increased greatly. Thus, if we look narrowly at the ratio between the greatly increased incidences over the seemingly almost unaltered death numbers due to the disease, it may be argued that the invention of PSA testing has guided many asymptomatic prostate cancer patients to seek early treatments, and the "cure rate" of the disease has improved. However, as already discussed, there are some troubling facts related to these early diagnoses that should lead us to pause and evaluate the current status of the clinic and the clinical impacts of early diagnoses. This data illustrates the fact that morphology-based histological pathology served oncological practice sufficiently to the end of the 1980s, or even the beginning of the 1990s, prior to the establishment of the current concepts and practices of cancer screening. In those days, oncological clinics received patients either with palpable masses or internal masses which had grown to become functionally destructive. In other words, before the start of the cancer-screening era, oncological clinics dealt with relatively later stage diseases. From these cases, pathological evidences and experiences were also accumulated from the corresponding stages of the diseases. Therefore, the criteria for cancer diagnoses that matured in those days are based heavily on later stage histological pathologies and need to be updated to include new observations of vast amounts of tumors diagnosed at early stages.

Fortunately, crises and opportunities often come hand-in-hand. Developments in molecular biology, particularly in cancer biology, coincide with the progression of cancer screening evolution. These developments have greatly enlarged our breadth of knowledge, or, more conservatively, have begun to unveil mysteries surrounding the biological mechanism behind these morphological alterations. Applications for these developments in clinical practices have created new pathological branches that form current molecular pathology. A number of these new biological markers have been utilized in clinical evaluations. For instance, many common names, such as estrogen receptor (ER), P53, and so on, have been integrated into comprehensive pathological evaluations for their empirical values in the patient prognostications of certain malignancies.

To systematically analyze the potential utility of cancer biological information – bioinformatics – obtained from genomics, proteomics, and/or metabolomics in assisting histopathology in oncological clinics, it may be helpful to analyze the relationship among these scientific branches. To begin these analyses, we need to assume that the metabolic activities in malignant cells are different from those in normal cells. We consider this hypothesis to be scientifically sound, reasonable, and self-evident.

Based on this hypothesis, we wish to consider the sequence of events in cancer development and the roles metabolites play. Changes in metabolic profiles are the results of certain active metabolic pathways. For these pathways to be active, active enzymes are needed. To have enzyme activities, enzymes need to be synthesized. Furthermore, to synthesize enzyme proteins requires the existence of messenger RNAs (mRNAs), and thus the correct DNAs. On the other hand, with mutated DNA the required mRNA may not be available for the synthesis of the enzymatic proteins; thus as a result of the DNA mutation, certain metabolic pathways are blocked. We can only present this basic molecular biology concept as a chain of events; however, we wish to emphasize that by presenting in this way we do not suggest that there are measurable temporal sequences and delays in association with genes, proteins, and metabolites. Viewing the human body as a united entity, without the exact knowledge regarding the dynamic rates of these molecular (genomic, proteomic, and metabolomic) transformations, we can speculate that the entire process may occur simultaneously and be coupled with feedback loops and parallel processes. This means that a logically clean chain of events is not always discernable, although we suspect that there must be a time window during which the presentation of malignancy transformations may be detectable by molecular biology means, while the revelation in cellular morphology is still uncertain. Thus, quantitative evaluations (proteomics, metabolomics) of the biological activities of an individual lesion may provide more sensitive and predictive parameters that may be able to subcategorize histomorphology results.

From this above reasoning, we can likely conclude that in order to observe certain metabolic profiles during or after malignant transformations, their corresponding gene players need to be in the right time and place, that is these genomic factors may be the necessary condition for the formation of the related metabolomic results. Hence, the processes of malignancy formations and progressions will require every player involved in all these "-omics," and every interfering factor in the entire duration, to be in harmonic association. Factors from all these disciplines reflect the same biological process from different angles, and at different points in the malignancy formation. In the larger picture of genomics, proteomics, and metabolomics, genomics has a stronger predictive capability than proteomics or metabolomics, while the latter two are viewed more as studies of the *current state* of the malignancy (although this may change as our understanding of complex gene relationships develops). A complete assessment of the strengths and weaknesses of parameters measured from each discipline, and their correlations with histopathological observations and oncological realities, presents great potential for the improvement of cancer characterization to meet the needs of the current screening era.

Having discussed the general connections between all these disciplines and their relationship with human oncology, we would like to clearly state the importance of histopathology and its predominant role in the present practice of oncology. We are compelled to do so because one of us, as a clinical pathologist, reads pathology slides under microscopes everyday; more importantly, we believe new disciplines should complement, not directly contradict, histopathology. Current histopathology, as previously explained, is the gold standard for the diagnosis of malignancy and the prognostication of a patient. The knowledge of this field is based on the collective and continuous contributions of human efforts over the past centuries. The very basic "do no harm" principle of modern medicine requires us to practice oncology exclusively based on the recommendations of histopathological conclusions until there is convincing clinical proof that other modalities can bring more benefit to the patients. Acknowledging that histopathology is a very subjective tool that leads to a lot of gray area, we conclude that histopathology is an excellent point from which to launch research in the areas of the "-omics" and, with time, new quantitative forms of measurements will be developed.

13.5. Chemical detection of cancer

To be trained as a board-certified pathologist requires one's earnest efforts for more than a decade. However, after a few minutes with an effective pathologist anyone with genuine interest may be able to appreciate the unique cellular pattern presentation of cancer observable under a microscope. A nice depiction of the disease can be found in the image of a crab on the cover of the first issue of Nature Reviews Cancer, where leg-like arrays of diseased cells spreading into the surrounding of the lesion pictorially exemplified why "cancer" gets its name from the Greek word for "crab." Seeing cancer cells and the "crab" formation under a microscope irrefutably displays the presence of the condition. In contrast to visual presentations of the disease, the idea of diagnosis based on molecular biology or chemistry requires more imagination. Nevertheless, evidence of associations between the presence of certain proteins/metabolites and clinical conditions of cancers has been broadly studied and reported. For instance, in 1993 C.E. Mountford and colleagues reported an interesting study involving magnetic resonance spectroscopy (MRS) analysis of lymph node metastasis in a rat model. From their study they found MRS to be sensitive enough to suggest the existence of malignant cells in lymph nodes, what they termed "micrometastases," that were not apparent even when the entire node was serially sectioned and examined by histology. However, the magnetic resonance (MR) conclusions were confirmed with the development of malignancy by xenografting nodal tissue into nude mice [7]. Studies such as this are extremely clinically relevant. In 1990, a retrospective study in Lancet revealed that, "Serial sectioning of lymph nodes judged to be disease-free after routine examination revealed micro-metastases in an additional 83 (9%) of 921 breast cancer subjects [8]."

The utility of biological markers in human cancer classification for the purpose of predicting patient outcomes have also been demonstrated in human clinical studies. An example of such evaluations in human brain tumors was published by one of us with D.N. Louis and colleagues in 2003. The study investigated the use of gene expression profiling of a 12,000 gene microarray in classifying a set of 50 high-grade gliomas (28 glioblastomas and 22 anaplastic oligodendrogliomas). By developing a prediction model, results of the study indicated that the inclusion of genomic profiles could produce brain tumor classification criteria that were more objective, explicit, and consistent with patient clinical outcomes than assessments made from standard pathology alone [9].

Chemical and molecular biological detections of cancer have also evolved in more clearly "seeing" tumours with the development of various molecular imaging methodologies, including infrared, mass spectroscopy, and magnetic resonance. The advantage of these imaging approaches is that they preserve the concept of recognizing the anatomic structures central to pathology evaluations by expanding the visible wavelength used in histopathology into the invisible domains of human vision. However, at present, there are still many limiting trade-off aspects regarding these attempts. Commonly, a better spatial resolution (or anatomic details) comes at the expense of a limited spectral resolution (differentiating different chemicals) for any imaging modality because of the intrinsic competition between signals and noise.

Discussions on applications for genomics and proteomics in human malignancy, as rapidly emerging fields, are beyond the scope of this monograph on metabolomics. However, we wish to point out that the differences between proteomics and metabolomics, in our opinion, are not as fundamental as the differences between them and genomics. To simplify the difference between proteomics and metabolomics, if metabolomics concentrates on the understanding of molecules that are considered to be monomers (for instance, amino acids), then proteomics focuses on the research of some kind of polymers of these monomers (i.e. proteins). Therefore, any method developed in analytical chemistry, or found in an instrumental analysis textbook, can have its potential applications in both proteomics and metabolomics studies. However, there are exceptions.

One exception, a method that may only be used for metabolomics but not for proteomics analyses, happens to be the major topic that we will discuss in the rest of this chapter, that is cancer pathology measured with intact tissue high resolution magnetic resonance spectroscopy (HR MRS). In general, this approach is capable of analyzing cellular metabolites and tissue histopathology from the same sample. This methodology eliminates the confounding factor of tumor heterogeneity seeing a correspondence between molecular profiles and histopathological patterns created with separate samples from the same clinical case. We select this chemical methodology to illustrate applications for metabolomics on human cancer for several reasons. First, since the methodology was proposed by one of us in the mid-1990s, it has been rapidly utilized by many research laboratories around the world on a variety of human diseases including many types of malignancies. Second, we have not yet seen any other metabolic approach (mass spectroscopy, liquid chromatography, or infrared, etc.) that can demonstrate the same degree of relevance to clinical oncology in terms of the potential to establish one-to-one correlations between metabolite profiles and tissue pathologies, for almost all of the other approaches require procedures of chemical extractions of tissue samples that have not yet been characterized by histopathology.

13.6. Magnetic resonance spectroscopy and cancer

Before we engage in the discussion of intact tissue HR MRS for metabolite identification and utilization in cancer metabolomics, we wish to overview briefly the field of MRS in cancer prior to the discovery of this methodology.

Magnetic Resonance Spectroscopy has also been known as NMR to chemists and physicists decades before the medical use of magnetic resonance imaging (MRI). The NMR signals are extremely sensitive to changes in chemical environments and have been widely applied in physical and chemical analyses before the invention of MRI. For the topic of this chapter (see Chapter 3 for more details of the technology) we will confine ourselves to discussion of *ex vivo* tissue MRS and will not traverse the enormous field of medical imaging.

Prior to the "-omics" era, MRS had shown the ability to quantify metabolites of different cell types and detect relatively small populations of abnormal cells. Diagnostic *in vivo* proton MRS studies of human malignancies have aimed at characterization of lesions and assessment of tumor grades and stages. However, these attempts have not yet reached the critical status where clinical decisions can be based solely on these measurements.

It has long been considered that fundamental *in vivo* improvements rely on the more accurate characterization and quantification of tumor metabolites measurable *ex vivo* at high magnetic field strength with high spectroscopic resolution. *Ex vivo* studies started even before the mid-1980s, aiming to correlate cancer pathologies with spectroscopically measurable metabolite alterations, to elucidate the details of cancer metabolites so as to better understand tumor biology and to improve designs of *in vivo* methodologies. The *ex vivo* studies conducted prior to the introduction of intact tissue HR MRS utilized conventional NMR methodology designed to analyze aqueous homogeneous solutions. Unfortunately, such an approach rests on the false assumption that fundamental physical differences between aqueous solutions and non-liquid tissues are relatively insignificant. In reality, substantial differences exist between aqueous solutions and intact tissues. As an unavoidable result, low spectral resolution data observed with non-liquid tissues generally preclude measurement of individual metabolites.

The invalidity of the above assumption (i.e. that intact biological tissues and aqueous solutions may be treated similarly) did not go unrecognized before the discovery of intact tissue high resolution methodology. To overcome the issue of low resolution observed with intact tissue samples, and to achieve measurement of individual metabolites at high spectral resolution, tissue metabolites were analyzed in solutions of chemical extractions. Unfortunately, there were other difficulties or uncertainties associated with extraction approaches. First, understandably, the measured spectral results depend on the applied extraction procedures and their completeness. Since the assurance of completeness may not be tested by any experimental design that involves the procedure itself, it may be accurate to state that extractions may alter measurable metabolites to an unknown degree. Second, and more importantly, as previously discussed, heterogeneity of human malignancies limits the usefulness of extraction approaches because the procedure prevents histopathological evaluation of the same specimen. Hence, even equipped with perfectly resolved spectra and precisely quantified metabolite concentrations, no one could rationalize the pathology compositions that generated this data. Regrettably, an extremely large body of extract studies on human malignancies, which could be misleading to various degrees, exists in the literature. In short, we will never know the exact pathological details corresponding with various chemical and biological profiles obtained for many types of human malignancies in hundreds, maybe thousands, of publications. Applications of conventional NMR on studies of animal models may not be assessed with equal alarm, as tumor heterogeneity often does not play a critical role in those models, sometimes referred to as "living Petri dishes."

13.7. Development of intact tissue MRS

During his post-doctoral studies at the Massachusetts General Hospital and Harvard Medical School, one of us discovered that spectral resolution of proton (^1H) NMR of intact tissue could be greatly enhanced by subjecting the unaltered intact tissue sample to an NMR technique known as magic angle spinning (MAS), which was previously developed for chemical studies of solids. This line-narrowing technique by mechanical sample rotation can generate spectral resolution sufficient for the identification and quantification of individual metabolites in *intact* biological tissue without needing to produce solutions of tissue extractions. In a later study, with a critical observation made by D.C. Anthony, a neuropathologist and the then acting Chief of Pathology at Boston's Children Hospital, it was found that under a moderate rotation (for instance, below 3 kHz for brain tumor tissues) tissue pathological structures did not undergo severe alterations [10]. This meant routine histopathology could be conducted with samples after spectroscopy analyses, if the pathologist is aware of certain possible artifacts introduced by the mechanical spinning. Only after this additional discovery was the HRMAS methodology established as a capable choice for recording sample metabolite concentrations and quantitative pathology from the same specimens in order to correlate spectroscopic data with quantitative pathology. This methodology has since been termed high resolution magic angle spinning proton magnetic resonance spectroscopy (HRMAS ^{1}H MRS). For readers who wish to know more about the methodology, its physical description and experimental design, please see Chapter 4 of this book.

HRMAS ^{1}H MRS has multifold advantages. For instance, histopathology evaluations are the impressions of pattern recognitions by a trained observer, whereas spectroscopy examinations can express a medical condition by numerical values once the links between data of spectroscopy and known pathologies are established. Therefore, a computer can readily assume further analyses of spectroscopy results, while computerization of histopathological evaluations is still difficult for us to envision. Furthermore, the above-mentioned numerical nature of spectroscopic evaluation of tissue status may be more objective and repeatable than the "subjective" nature of histopathological evaluations. Even more importantly, HRMAS ^{1}H MRS will likely constitute biological parameters for patient prognostication that will be complementary to or more sensitive/accurate than those in current clinical practice based on cellular morphology.

It also needs to be realized that, although drastic improvements in tissue metabolic analyses have been fostered by the introduction of this method, the discussed tissue MRS is not the "silver bullet" that can solve all the remaining issues with cancer diagnosis. There are still many intrinsic limitations associated with intact tissue spectroscopy analyses. For example, utilizing this analytical approach alone will not solve the widely recognized "sampling error" problem associated with histopathology, that is histopathology results can only be as good as a measure of features that present themselves on the evaluated tissue slides. Due to the heterogeneous nature of human malignancies, false negative conclusions may be reached if the sampled tissue specimens contain no cancer cells, particularly during biopsies. This clinical issue clearly cannot be overcome by utilizing a different analytical protocol. In fact, we suspect that similar to histopathology, any molecular pathology evaluation will also encounter sampling errors of its own kind. Similarly, as we previously reported that different regions possess different metabolite profiles that correlate with differences in tissue pathologies observed in a case of human brain tumor [10], we suspect that heterogeneities in genomic and proteomic profiles may also exist, hence these are also susceptible to sampling errors. Interesting questions to consider are how much overlap exists among these heterogeneities, which one is less heterogeneous throughout the lesion, and which one may have the largest field effect that extending the malignant information beyond the perimeters of the optical or human vision visible lesions. We will discuss and illustrate this concept of field effects with examples in the following sections.

13.8. From tissue MR spectra to cancer metabolomics

The status of cancer metabolite studies can best be presented by reviewing current publications of research using intact tissue MRS. First of all, these studies have covered many types of human cancers including: brain, prostate, cervix, breast, kidney, and sarcomas. Next, we notice the gradual shift in the nature of studies away from "proof-of-concept" examinations, with small numbers of subjects, toward evaluations of clinical populations with cases number ranging from 50+. We also observe a gradual change in the overall perception of the MRS methodology. This MRS methodology is now recognized for its unique advantage in establishing correlations between cellular metabolites and tissue pathologies measured from the same sample. Finally, a trend in the research design of extreme metabolomics importance is also emerging: the traditional NMR approach of attempting to relate changes in single metabolites with the disease conditions has been substituted by the concept of analyzing the *global* metabolite profiles and seeking to reveal the changes in these *overall* profiles that are associated with the physiological and pathological conditions of interest. It is precisely this last trend, which agrees with our definition of metabolomics, that has convinced us that metabolomics has reached the status where it can claim membership in the "-omics" family. These methodological evolutions in the studies of human malignancies illustrate the maturation process through which a technical discovery may ultimately deliver important health advances.

There is a twofold motivation for investigations of NMR-visible metabolites in biological tissues. The original inspiration responded to the conviction that these metabolites might be of pathological importance and useful to improving the accuracy of disease diagnoses. Research endeavors toward this direction started in the late 1970s and early 1980s. Subsequently, in the 1990s, rapid technical innovations in MR imaging motivated researchers to try and understand the spectroscopic features observed in the newly developed localized *in vivo* MRS, and attempt to design new *in vivo* strategies for spectroscopy measurement of targeted specific metabolites. Both motivating factors consider human malignant conditions the most suitable systems to test hypotheses and develop techniques because samples from the identifiable lesions associated with the diseases can be either removed for *ex vivo* analysis or localized from MR images for *in vivo* evaluations. Furthermore, imaging visible sites of lesions also allows the identification and selection of "normals," or disease-free sites from the same subject, such as the contralateral site in the cases of brain tumors, to be used as paired controls to reveal disease-related metabolite alterations. For these reasons, and also due to the relative cardiac/respiratory motion stability and tissue homogeneity of the brain compared to other organs, human brains and brain tumors have been the main spectroscopy research targets, although brain tumors represent only about 1% of all malignancies.

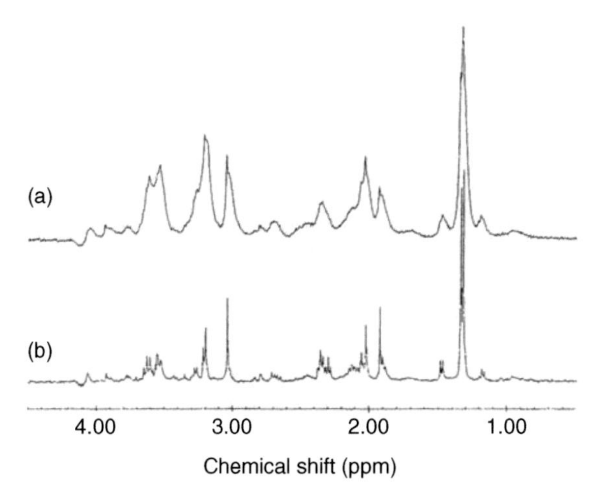

Figure 13.5. A comparison of brain tissue proton MR spectra obtained with a sample (a) static; and (b) with 2.5 kHz rotation at the magic angle (54°44′ away from the direction of the magnet field).

Since studies of the brain have dominated the development of *in vivo* MRS in both concept and technology, it was logical that a number of initial intact tissue HRMAS MRS studies of human oncology were devoted to analyses of human brain tumors. Another advantage of studies with brain tissue that makes measurement of cellular metabolites less challenging is the lack of adipose tissue, which can mask spectral regions of interest in tissue such as breast tissue. Therefore, without extensive masking of cellular metabolite signals by the presence of fatty acid peaks, the net effect of signal peak narrowing produced by the application of the HRMAS technique was evident and easy to visualize, which contributed to the general acceptance of the approach as a valid innovation for analyses of intact human pathological tissues (Figure 13.5). Later, with the acceptance of the concept, it was proven that using HRMAS even in breast tissues the peak widths from fat signals could be reduced to result in spectral windows where metabolite signals were visible and quantifiable.

The benefit and rational of studying brain tumors encountered one critical logistic issue that was somewhat unique to this particular malignancy: tissue availability. Unlike treatment of any other malignant conditions, neurosurgeons, or more precisely patients, cannot afford to remove additional brain tissue just to create a clean surgical margin of malignancy. Hence, instead of using scalpels, the neurosurgeons often resort to the assistance of suction tips to remove minimal but absolutely necessary amounts of cancerous tissue. Hence, excess tumor material after pathological diagnosis is often very limited and not readily available for research. This perhaps also explains our observation of the literature on HRMAS tissue spectroscopy studies of human brain tumors. Most studies evaluate different aspects of the feasibility of methodology with small patient numbers (e.g. for each type and grade), such as the validity of measuring spectral and pathological data from the same specimens and the correlations between *in vivo* and *ex vivo* spectroscopic observations. We have not yet seen a report of an *ex vivo* tissue study with significantly large patient numbers that can address the issue of biological variations. Technical developments and feasibility studies addressing issues encountered during studies of a particular malignancy have also been seen in reports on other more common types of cancers, for instance the effects of tissue degradation on the measured cellular metabolite profiles, and the developments of slow rate HRMAS methodology during human prostate cancer studies.

The effect and function of HRMAS can be visualized as "squeezing" a broad resonance peak into a center band surrounded by a number of side bands distanced by the sample rotation rate, as illustrated in Figure 13.6. These side bands can severely hinder the interpretation of observations if they fall into regions where metabolite signals of interest reside. Therefore, in common practice, sample rotation rates are chosen such that they are fast enough to "push" the first pair of side bands at each side of the center band outside the spectral regions of interest, which means the utilization of rotation rates between 3 and ~5 kHz for proton spectra depending on

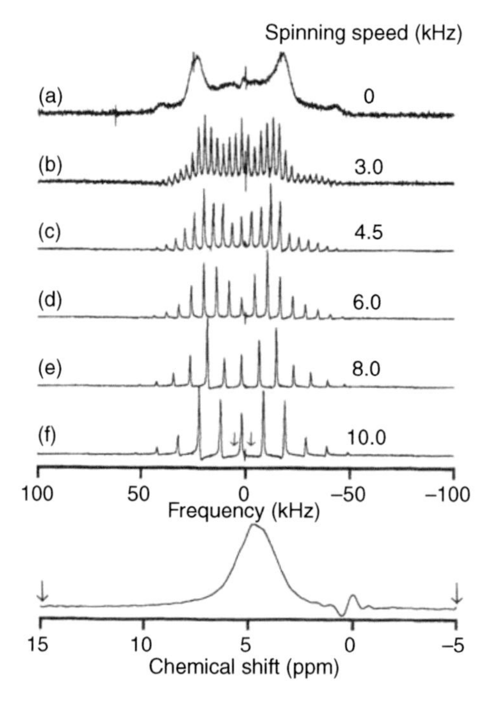

Figure 13.6. An illustration of the effects of MAS on the appearance of spectral patterns measured with crystalline powder samples of barium chlorate monohydrate at room temperature with sample spinning rate from 0 to 10 kHz.

the used magnetic field strength. At these rotation rates pathological identification of cancer cells is possible, but some tissue types display evidence of the centrifugal stress-related artifactual alterations on the samples. For instance, in prostate tissue the glandular structures, particularly the glandular spaces, were significantly altered. These pathological structures are extremely important for the establishment of correlations with tissue metabolic profiles. Therefore, maximal preservation and accurate histopathological quantification is sought. To achieve this, following the basic principles of physics, one needs to consider reducing the structural damage from the centrifugal stress of sample rotation by using reduced rotation rates. Without involving too much detail on the physics of HRMAS, experimental results have shown that for many types of human tissue a reduction in rotation rates to less than 1 kHz does not decrease the ability of HRMAS to produce spectra of narrow peaks, while the details of tissue pathological structures are largely preserved (see Chapter 4). Several studies on the development of slow HRMAS methodologies have been reported, and these methodologies have been utilized in evaluations of large patient populations.

It should be realized that results of these technical developments and feasibility evaluations are likely to be tissue type specific, hence the applicability to different types of tissue samples needs to be tested individually. For example, results published by one of us on techniques of slow HRMAS that had been developed for human prostate cancer analyses also worked well with brain tissues since both tissue types are characterized by lack of adipose components [11, 12]. The proposed scheme is rendered ineffective, however, with the reported slow rotation rates, on samples rich in adipose tissue, such as those of breast or skin. On the other hand, samples of mostly connective tissue nature may be more durable to mechanical stresses, hence their pathological structures are less vulnerable to higher rates of HRMAS sample rotation.

The above discussion directly addresses an important issue related to evaluation of tissue metabolite profiles and pathological quantities from the same sample. We have rationalized the tremendous significance of such a sequential analytic protocol in the evaluations of human cancers due to the heterogeneous characteristics of malignant diseases. While we have emphasized the capability of the HRMAS MRS approach in addressing this fundamental need, we are also aware that among the several dozens of research articles of HRMAS MRS on human cancers, only a handful of them were conducted according to this protocol. The rest of them followed the "classical" approach where results of metabolites and pathologies are obtained from adjacent samples, as in studies involving tissue chemical extractions. Although our previous arguments on utilizing the same samples are still vital and valid, many reports did show correlations between the measured tissue metabolite concentrations and the patient cancer status so clearly and convincingly with adjacent samples that we would be hard pressed to discount them. To reconcile the seeming conflict between the concept and the reality one must consider, these observations might suggest a "field effect," that is the malignancy-related tissue metabolite profiles, or cancer metabolomics, are delocalized from cancer cells and extend to their surrounding histopathologically benign tissues. Although the suggestion of the existence of "field effects" in cancer metabolomics can only be considered a hypothesis, it is certainly not a foreign concept, particularly regarding the widely accepted and studied cancer-stroma effects in today's molecular oncology.

An additional troubling issue in current literature is a deviation from our "global" concept of metabolomics. In many studies, a number of individual metabolites were measured in an attempt to associate them with the malignant condition. Although these studies may not represent the perfect approach for metabolomics, they contain valid observations. These results likely reflect the fact that many metabolites previously linked to malignancy could indeed be the major contributors to the characteristic metabolite profiles of a particular condition. This is verified by the following example, which we hope illustrates our suggested approach for studies of human cancer metabolomics.

Recently, one of us published a report on the study of human prostate cancer with intact tissue HRMAS MRS [13]. In short, this study included 199 prostate tissue samples obtained from 82 prostate cancer patients after prostatectomy. The study was conducted by first analyzing samples with HRMAS MRS, and then with quantitative pathology. Metabolite concentrations were calculated from the spectroscopy data and their profiles were generated by statistics with principal component analysis (PCA). Quantitative pathological evaluations of these tissue samples indicated that 20 samples of the total 199 samples from prostate cancer patients contained cancerous glands, while the rest ($n = 179$) represented histologically benign tissue obtained from cancerous prostates. This was not surprising as it agrees with the frequency seen in biopsy in prostate cancer clinic and reflects the infiltrative, heterogeneous nature of prostate cancer. Following spectroscopy and pathology analyses, tissue metabolite profiles were correlated with quantitative pathology findings using linear regression analysis, and evaluated against patient pathological statuses by using analysis of variance (ANOVA). Paired-t-tests were then used to show the ability of tissue metabolite profiles to differentiate malignant from benign samples obtained from the same patient (Figure 13.7), and correlated with patient serum PSA levels. Finally, metabolite profiles obtained from histologically benign tissue samples were used to delineate a subset of less aggressive tumors and predict tumor perineural invasion within the subset.

Reflecting on the discussions in this section, we wish to emphasize a number of points within this example that directly address our concept of cancer metabolomics.

First, it is feasible and clinically important to conduct intact tissue studies of a large patient population, from which results may be revealed that potentially revolutionize oncological treatment.

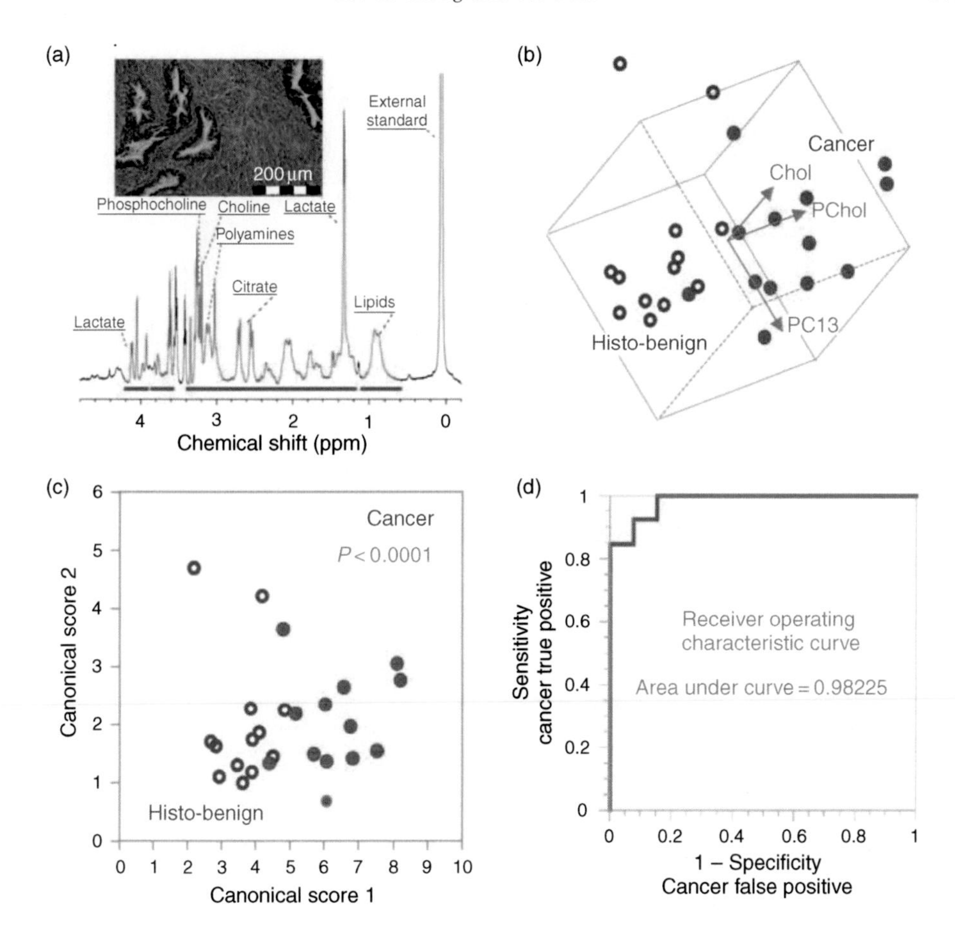

Figure 13.7. (a) The HR MAS ^{1}H MR spectrum of intact tissue obtained from the removed prostate of a 61-year-old patient with Gleason score 6, T2b tumors. Histopathology analysis of the tissue sample (insert) after its spectroscopy measurement revealed that the sample contained 40% histopathologically defined benign epithelium and 60% stromal structures, with no identifiable cancerous glands. Cellular metabolites mentioned in the text are labeled on the spectrum. The 36 most intense resonance peaks or metabolite groups above the horizontal bars were selected for analyses, while the other regions were excluded from the calculation, partly due to surgery-related alcohol contamination. (b) Three-dimensional plot of Principal Component 13 (PC13 correlates linearly with vol% of cancer cells in tissue samples) vs phosphocholine (Pchol) vs choline (Chol). Cancerous and histologically benign (histo-benign) tissue samples from 13 patients can be visually separated in the observation plane. The paired Student's t-test (cancer vs. histo-benign from the same patients) results for PC13, PChol and Chol are: 0.012, 0.004, and 0.001. Only results from these 13 patients could be evaluated with paired tests, for other cancer positive samples were collected from patients with whom no histo-benign samples were analyzed. (c) The canonical plot resulting from discriminant analysis of the three variables in Figure 13.7b presents the maximum separation between the two groups. (d) The resulting receiver operating characteristic (ROC) curves indicates the accuracy of using the three variables in Figure 13.7b. to positively identify cancer samples.

Second, histomorphological evaluations of the same samples are critical for the correct interpretation of spectroscopic data and to define cancer metabolomic profiles.

Third, correct data statistical analysis is crucial to generate tissue metabolomic profiles. There are many bio-statistical approaches that one can use to achieve this purpose, and in the example, PCA was used. The selection of PCA for the analyses of spectroscopy data in the example was based on the aim of the work: to correlate spectral metabolite profiles with tissue pathologies and patient clinical statuses. Based on our definition of cancer metabolomics, malignant pathological processes manifest simultaneous changes in multiple measurable metabolites, and a change in a single metabolite may not represent the underlying process. To test this hypothesis, PCA attempts to identify combinations (principal components or PCs) of the measured concentrations that may reflect distinct pathological processes if they exist in the set of the samples. Positive contributions of certain metabolites indicate the elevations of these metabolites within the component (process), and negative contributions suggest suppressions. For instance, the study has shown that the cancer-related PCs 13 and 14 both had the metabolites phosphocholine (PChol) and choline (Chol) as their major positive contributors in agreement with current *in vivo* and *ex vivo* MRS literature descriptions of the relationship of these metabolites with malignancy.

Finally, the reported capability of certain metabolite profiles (e.g. represented as principal components) to differentiate various pathological tumor stages, measured with histopathologically benign samples from cancerous prostates, seems to support the hypothesis of the existence of "field effects." However, the scale and the extent of these effects are still largely undefined.

13.9. Future directions and implications

In summary, we have organized this chapter in light of the potential metabolomics applications to malignancy diagnosis with tissue samples. We consider intact tissue HRMAS MRS is still the best, and the only known, method that can provide both metabolite and pathology data from the same samples. Because our aims are orientated around this specific aspect of extreme oncological importance, we did not include discussions on chemical measurements of body fluids or other liquid samples, and their potential usages in clinic.

Within this specific scope of the chapter, with the apparently sufficient developments of the spectroscopy methodology, the emphasis of cancer metabolomics in the near future will be focused on answering four key issues.

In connection with evaluations of current histopathology and within the parameters thus defined, the first task of cancer metabolomics is to test the sensitivities of metabolite profiles in subcategorizing tumor conditions for patient outcomes in the same histopathological groups. Secondly, research in this field will attempt to identify metabolomic profiles reflecting the status of tumor biology that can be used to indicate the tumor's sensitivity to a particular treatment protocol. The third task will likely concentrate on the identification of metabolomic profiles that may be more sensitive than histopathology evaluations in indicating pre-cancerous conditions. Finally, we envision that in order to gain more comprehensive understanding on cancer biology in terms of disease mechanisms and treatment strategies, increasing research efforts will focus on aspects that connect metabolomics with genomics and proteomics, particularly in clinical and pre-clinical areas where the concepts of non-invasive diagnoses of human malignancies with molecular imaging have been forcefully pursued. With the future of metabolomics now at hand it is reasonable to assume that cancer metabonomics will have major contributions to the history of medical science.

References

[1] Perou, C.M., Sorlie, T., Eisen, M.B., van de Rijn, M., Jeffrey, S.S., Rees, C.A., Pollack, J.R., Ross, D.T., Johnsen, H., Akslen, L.A., Fluge, O., Pergamenschikov, A., Williams, C., Zhu, S.X., Lonning, P.E., Borresen-Dale, A.L., Brown, P.O., Botstein, D. Molecular portraits of human breast tumours. *Nature* 2000;406(6797):747–752.

[2] Perou, C.M., Jeffrey, S.S., van de Rijn, M., Rees, C.A., Eisen, M.B., Ross, D.T., Pergamenschikov, A., Williams, C.F., Zhu, S.X., Lee, J.C., Lashkari, D., Shalon, D., Brown, P.O., Botstein, D. Distinctive gene expression patterns in human mammary epithelial cells and breast cancers. *Proc. Natl. Acad. Sci. U.S.A* 1999;96(16):9212–9217.

[3] Ma, X.J., Salunga, R., Tuggle, J.T., Gaudet, J., Enright, E., McQuary, P., Payette, T., Pistone, M., Stecker, K., Zhang, B.M., Zhou, Y.X., Varnholt, H., Smith, B., Gadd, M., Chatfield, E., Kessler, J., Baer, T.M., Erlander, M.G., Sgroi, D.C. Gene expression profiles of human breast cancer progression. *Proc. Natl. Acad. Sci. U.S.A* 2003;100(10):5974–5979.

[4] Gunsalus, K.C., Ge, H., Schetter, A.J., Goldberg, D.S., Han, J.D., Hao, T., Berriz, G.F., Bertin, N., Huang, J., Chuang, L.S., Li, N., Mani, R., Hyman, A.A., Sonnichsen, B., Echeverri, C.J., Roth, F.P., Vidal, M., Piano, F. Predictive models of molecular machines involved in Caenorhabditis elegans early embryogenesis. *Nature* 2005;436(7052):861–865.

[5] Jemal, A., Tiwari, R.C., Murray, T., Ghafoor, A., Samuels, A., Ward, E., Feuer, E.J., Thun, M.J. Cancer statistics, 2004. *CA Cancer J. Clin.* 2004;54(1):8–29.

[6] Jemal, A., Murray, T., Ward, E., Samuels, A., Tiwari, R.C., Ghafoor, A., Feuer, E.J., Thun, M.J. Cancer statistics, 2005. *CA Cancer J. Clin.* 2005;55(1):10–30.

[7] Mountford, C., Lean, C., Hancock, R., Dowd, S., Mackinnon, W., Tattersall, M., Russell, P. Magnetic resonance spectroscopy detects cancer in draining lymph nodes. *Invasion & Metastasis* 1993;13:57–71.

[8] Bettelheim, R., Price, K., Gelber, R., Davis, B., Castiglione, M., Goldhirsch, A., Neville, A. The International (Ludwig) Breast Cancer Study Group, Prognostic importance of occult axillary lymph node micrometastases from breast cancer. *Lancet* 1990;335:1565–1568.

[9] Nutt, C.L., Mani, D.R., Betensky, R.A., Tamayo, P., Cairncross, J.G., Ladd, C., Pohl, U., Hartmann, C., McLaughlin, M.E., Batchelor, T.T., Black, P.M., von Deimling, A., Pomeroy, S.L., Golub, T.R., Louis, D.N. Gene expression-based classification of malignant gliomas correlates better with survival than histological classification. *Cancer Res.* 2003;63(7):1602–1607.

[10] Cheng, L.L., Anthony, D.C., Comite, A.R., Black, P.M., Tzika, A.A., Gonzalez, R.G. Quantification of microheterogeneity in glioblastoma multiforme with ex vivo high-resolution magic-angle spinning (HRMAS) proton magnetic resonance spectroscopy. *Neuro-oncol.* 2000;2(2):87–95.

[11] Taylor, J.L., Wu, C.L., Cory, D., Gonzalez, R.G., Bielecki, A., Cheng, L.L. High-resolution magic angle spinning proton NMR analysis of human prostate tissue with slow spinning rates. *Magn. Reson. Med.* 2003;50(3):627–632.

[12] Burns, M.A., Taylor, J.L., Wu, C.L., Zepeda, A.G., Bielecki, A., Cory, D.G., Cheng, L.L. Reduction of Spinning Sidebands in Proton NMR of Human Prostate Tissue With Slow High-Resolution Magic Angle Spinning. *Magnetic Resonance in Medicine* 2005;54(1):34–42.

[13] Cheng, L.L., Burns, M.A., Taylor, J.L., He, W., Halpern, E.F., McDougal, W.S., Wu, C.L. Metabolic characterization of human prostate cancer with tissue magnetic resonance spectroscopy. *Cancer Res.* 2005;65(8):3030–3034.

[14] Cheng, L.L., Ma, M.J., Becerra, L., Ptak, T., Tracey, I., Lackner, A., Gonzalez, R.G. Quantitative neuropathology by high resolution magic angle spinning proton magnetic resonance spectroscopy. *Proc. Natl. Acad. Sci. U.S.A* 1997;94(12):6408–6413.

[15] Zheng, L. *Solid state NMR studies of proton diffusion* [PhD]. Waltham: Brandeis University; 1993.

The Handbook of Metabonomics and Metabolomics
John C. Lindon, Jeremy K. Nicholson and Elaine Holmes (Editors)

Chapter 14

NMR Spectroscopy of Body Fluids as a Metabolomics Approach to Inborn Errors of Metabolism

Udo F.H. Engelke, Marlies Oostendorp and Ron A. Wevers*

Radboud University Nijmegen Medical Center, Laboratory of Pediatrics and Neurology, NL-6500 HB Nijmegen, The Netherlands

14.1. Introduction

Nuclear magnetic resonance(NMR) spectroscopy has been used to analyze metabolites, drugs and toxic agents in body fluids and reviews on these topics have been published [1–3]. Proton NMR (^{1}H-NMR) spectroscopy has also been successfully applied to the field of inborn errors of metabolism [4]. The technique is essentially non-selective and provides an overview of proton-containing metabolites. Moreover, simultaneous quantification of many metabolites over large concentration ranges can be done easily in a short time-frame. Because of these features proton NMR spectroscopy easily picks up a metabolic derangement or an abnormality in the metabolome. The technique either as a standalone approach or in combination with liquid chromatography and/or mass spectroscopy is at the frontiers of metabolomics and systems biology approaches.

14.1.1. Diagnostic approach of inborn errors of metabolism

Inborn errors of metabolism (IEM) form a considerable group of genetic diseases. The majority is due to defects of single genes coding for specific enzymes. Figure 14.1a shows a schematic representation of a metabolic pathway. In most of

*Address correspondence to this author at: Radboud University Nijmegen Medical Centre, Laboratory of Pediatrics and Neurology, Reinier Postlaan 4, 6525 GC Nijmegen, The Netherlands.

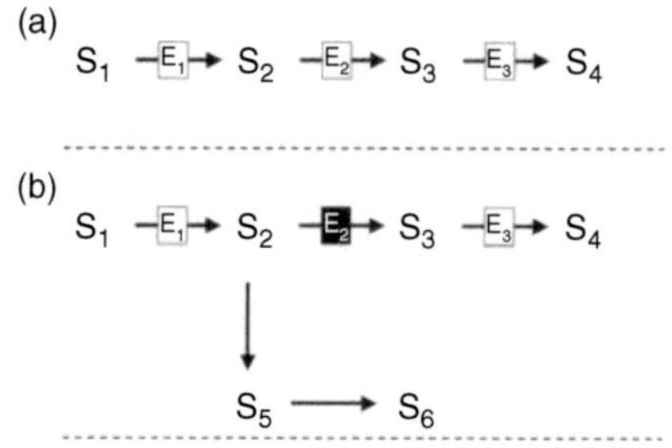

(a)
$S_1 \xrightarrow{E_1} S_2 \xrightarrow{E_2} S_3 \xrightarrow{E_3} S_4$

(b)
$S_1 \xrightarrow{E_1} S_2 \xrightarrow{E_2} S_3 \xrightarrow{E_3} S_4$

$S_5 \longrightarrow S_6$

Figure 14.1. (a) A schematic representation of a metabolic pathway. Enzymes E_1, E_2 and E_3 catalyze the reactions converting the metabolite S_1 into S_2, S_3 and S_4. (b) A defect in enzyme E_2 (black square) will give rise to the accumulation of metabolites S_2 and to a lesser extent S_1 and to a decreased availability of metabolites S_3 and S_4. Also unusual alternative metabolites like S_5 and S_6 may occur.

the disorders, problems arise due to accumulation of substances, which are toxic or interfere with normal function (Figure 14.1b, metabolites S_1 and S_2). Clinical signs may also relate to the decreased availability of metabolites behind the metabolic block (Figure 14.1b, metabolites S_3 and S_4). Also alternative pathways may metabolize an accumulating metabolite. Such pathways may play a minor role or no role at all in normal metabolism. In a patient with a metabolic block, alternative pathways may result in unusual metabolites (Figure 14.1b, metabolites S_5 and S_6). These often are characteristic for the disease and therefore diagnostically important. Also they may play a role in the pathophysiology of the disease.

One of the methods to diagnose IEM is the detection of abnormal metabolite concentrations in body fluids such as urine, blood plasma, blood serum or cerebrospinal fluid (CSF). From these concepts it is easy to understand that the diagnosis of an IEM may rely on (1) an abnormal accumulation of a specific metabolite, (2) the presence of metabolites that are normally not present in body fluids or (3) a decreased concentration of a metabolite that is always present in the body fluid. Laboratories specialized in diagnosing IEM use a variety of analytical methods to measure various metabolic groups. These are, for instance, GC-MS for the detection of organic acids, ion exchange chromatography for the detection of amino acids and high performance liquid chromatography (HPLC) for the detection of purines and pyrimidines. These techniques are selective and sample pretreatment such as derivatization and extraction procedures are often necessary. The ^{1}H-NMR spectroscopy is not selective and, because most metabolites contain protons, this technique may be considered as an alternative analytical approach for diagnosing known, but also as yet unknown, IEM.

The detection limit for body fluid NMR spectroscopy depends on a number of factors such as field strength, number of protons contributing to a resonance and the region of the spectrum where the resonance is observed. In general, the detection limit is in the low micromolar range in the less crowded regions of the spectrum.

For example, the detection limit at S/N = 3 for the 6 equivalent protons in dimethyl sulfone (observed as a singlet resonance at 3.14 ppm in blood plasma and CSF) was determined to be 0.5 μmol/L (500 MHz, 128 transients; using a conventional 5 mm TXI probehead operating at ambient temperature) [5]. Another study showed detection limits between 2 and 40 μmol/L for various metabolites in unconcentrated plasma samples measured at 600 MHz and acquiring 132 transients [6].

Metabolite quantification in plasma and CSF can be performed by adding a known concentration of 3-(trimethylsilyl)-2,2,3,3-tetradeuteropropionic acid (TSP) to the sample and concentrations can be expressed as μmol/liter. For urine, the singlet resonance of creatinine at 3.13 ppm (pH = 2.50) can be used as a concentration reference. Metabolite concentrations in urine are often expressed per mmol creatinine. For several compounds good correlations have been obtained with other analytical techniques. In CSF and blood plasma, quantitative data for alanine, threonine and lactic acid correlated well with data obtained with conventional techniques [6, 7]. Excellent linearity, recovery and reproducibility have been found for *N*-acetyltyrosine, *N*-acetyltryptophan and *N*-acetylaspartic acid in urine, normally not analyzed in conventional screening methods [8].

In this chapter, we provide a literature overview of the current applications of NMR spectroscopy to metabolic diseases.

14.2. Methods

14.2.1. Sample collection, preparation and choice

NMR spectroscopy has been used to investigate a wide range of body fluids [2]. However, when used as a diagnostic tool for inborn errors of metabolism, NMR spectra are mainly recorded on blood plasma, blood serum, urine or CSF samples.

Blood is collected by venipuncture into standard vials containing either lithium heparin or ethylene diamine tetra acetate (EDTA) as anti-coagulant. Subsequently, plasma can be separated from the cells by centrifugation. When EDTA is used to prevent clotting, extra resonances can be observed in the NMR spectrum due to complexes formed between EDTA and Ca^{2+} and Mg^{2+} ions present in plasma [9]. The CSF can be obtained by lumbar puncture and has to be centrifuged to remove cells. If NMR analysis cannot be performed immediately after collection of blood, urine or CSF, the samples have to be kept frozen at −20 °C or lower until further usage.

Whereas urine is measured directly, plasma and serum are usually deproteinized. The plasma protein concentration (approximately 70 g/l) results in broad overlapping signals that obscure resonances from low-molecular mass metabolites of interest [6]. Binding of the commonly used internal standard trimethylsilyl-2,2,3,3-tetradeuteropropionic acid (TSP-d_4) on proteins has been shown to have a negative influence on quantification [10]. Additionally, resonances for small, protein-bound metabolites are broadened due to T_2-relaxation processes, making accurate quantification difficult [11]. The effect of deproteinization on the ^{1}H NMR spectrum of a control plasma sample is illustrated in Figure 14.2. Normal CSF has a rather low protein concentration (<500 mg/l) and NMR measurements on both native [12–17] and deproteinized samples have been reported [7].

Various techniques are available for plasma or serum deproteinization, including extraction with acetone or acetonitrile, perchloric acid precipitation, solid phase extraction chromatography and ultrafiltration. Of these methods, ultrafiltration using a 10 kDa cut-off filter (e.g. Sartorius) is best suited for metabolic studies [18]. Before use, the filter has to be washed by centrifugation of water to remove glycerol that is present in several commercially available filters. The main disadvantage of deproteinization is possible loss of protein-bound metabolites. However, excellent correlation between ^{1}H-NMR and enzymatically determined lactate concentrations was found [6], even though lactate is assumed to interact with plasma proteins and so become partly NMR invisible [19].

The pH of the sample has an important influence on the observed chemical shifts and also determines whether acidic or basic protons are detectable. Unfortunately, a large range of pH values has been used in the literature. The physiological pH and pH of 2.5 have been used most frequently. The pH of urine varies considerably between samples. The pH of CSF samples increases significantly upon standing. To improve intersample reproducibility and to allow comparison between different samples, standardization of pH was initially proposed by Lehnert and Hunkler in 1986 [20]. They chose pH 2.5 because the slope of the titration curve is expected to be minimal for all organic acids and because the proton concentration is not yet so high as to cause hydrolysis of certain compounds. Chemical shifts in this review are given for pH 2.5 unless mentioned otherwise.

Finally, an aliquot of TSP-d_4 in ^2H_2O has to be added to the sample to provide a chemical shift ($\delta = 0.00$) and concentration reference as well as a deuterium lock signal. Alternatively, an internal reference or the residual solvent signal can be used as chemical shift reference. In this case, addition of pure ^2H_2O is sufficient.

The ionic strength of body fluids varies considerably and may be high enough to adversely affect the tuning and matching of the RF circuits of a spectrometer probe, particularly at high field strengths. Therefore, it may be necessary to retune the spectrometer probe for each sample. Automatic tune and match (ATM) probes are available.

Traditionally the approach in diagnosing IEM is to make use of urine as the body fluid of first choice. The kidneys excrete most water-soluble small molecules but some, such as glucose, are reabsorbed. The urine NMR spectrum is by far the most complex body fluid spectrum allowing the diagnosis of most inborn errors of metabolism. However, there may be a case for separate CSF NMR studies in

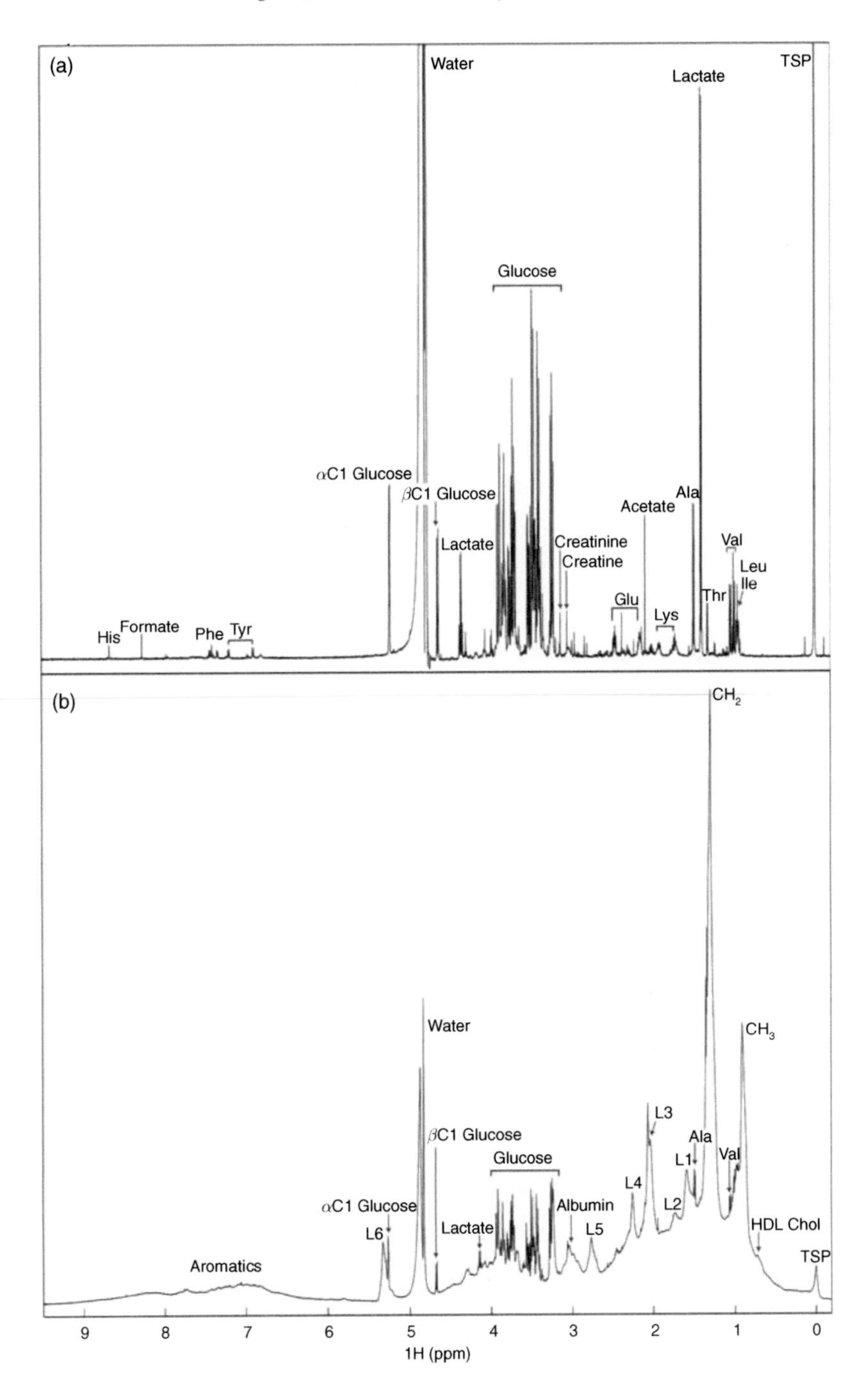

Figure 14.2. The 500 MHz ^{1}H-NMR spectrum of a plasma sample (a) after, and (b) before deproteinization through a 10 kDa filter.

patients suspected to suffer from neurometabolic disease. A paper by Wolf *et al.* [21] demonstrated the presence of the dipeptide *N*-acetylaspartylglutamic acid (NAAG) in CSF of two patients with a novel neurometabolic disease while NAAG could not be demonstrated in the urine NMR spectra of the patients. Only a few studies have addressed the NMR spectra of body fluid cells. Studies are available on leucocytes and erythrocytes [22]. The composition of the intracellular fluid differs considerably from the body fluid composition. Several metabolites are unable to leave the cell, as they cannot pass the outer cell membrane. Phosphorylated compounds may serve as an example. Inborn errors where such compounds accumulate may be easily diagnosed in homogenates of leucocytes. However, such examples have not yet been presented in the literature.

14.2.2. One- and two-dimensional NMR spectroscopy techniques

The number of observed metabolites in body fluids is largely dependent on the magnetic field strength of the NMR spectrometer. At higher fields, the spectral dispersion and sensitivity increase, allowing assignment of many metabolites that cannot be detected at lower fields. Therefore, working at the highest field available will provide the most complete metabolic information. Currently, most studies are performed on 400, 500 or 600 MHz spectrometers, but higher field strengths up to 800 MHz have been used [23–25]. Additionally, low frequency spinning of the sample can be used to improve spectral resolution. Reported measurement temperatures are found between 20 and 37 °C.

A wide range of one-dimensional (1D), two-dimensional (2D) and even higher dimensional NMR experiments are available to the modern NMR operator. Especially in the field of structural biology, multidimensional experiments are important for successful determination of protein or nucleic acid structures [26, 27].

For studying inborn errors of metabolism, 1D ^{1}H NMR spectra are most frequently employed. Additionally, several homo- and heteronuclear 2D experiments can be recorded, which are particularly helpful in assisting the assignment of resonances. The most commonly used experiments for diagnosing metabolic diseases will now be discussed. The information content of the spectra will be demonstrated using spectra recorded on a sample containing pure methionine (Figure 14.3). A schematic overview of the experiments is in Table 14.1.

14.2.2.1. One-dimensional NMR spectroscopy

Since proton NMR is relatively sensitive and because protons are present in almost every metabolite, a 1D, single pulse ^{1}H NMR experiment with solvent presaturation is a good starting point when diagnosing a possible inborn error of metabolism. When overlap is not too severe and metabolite concentrations are high enough

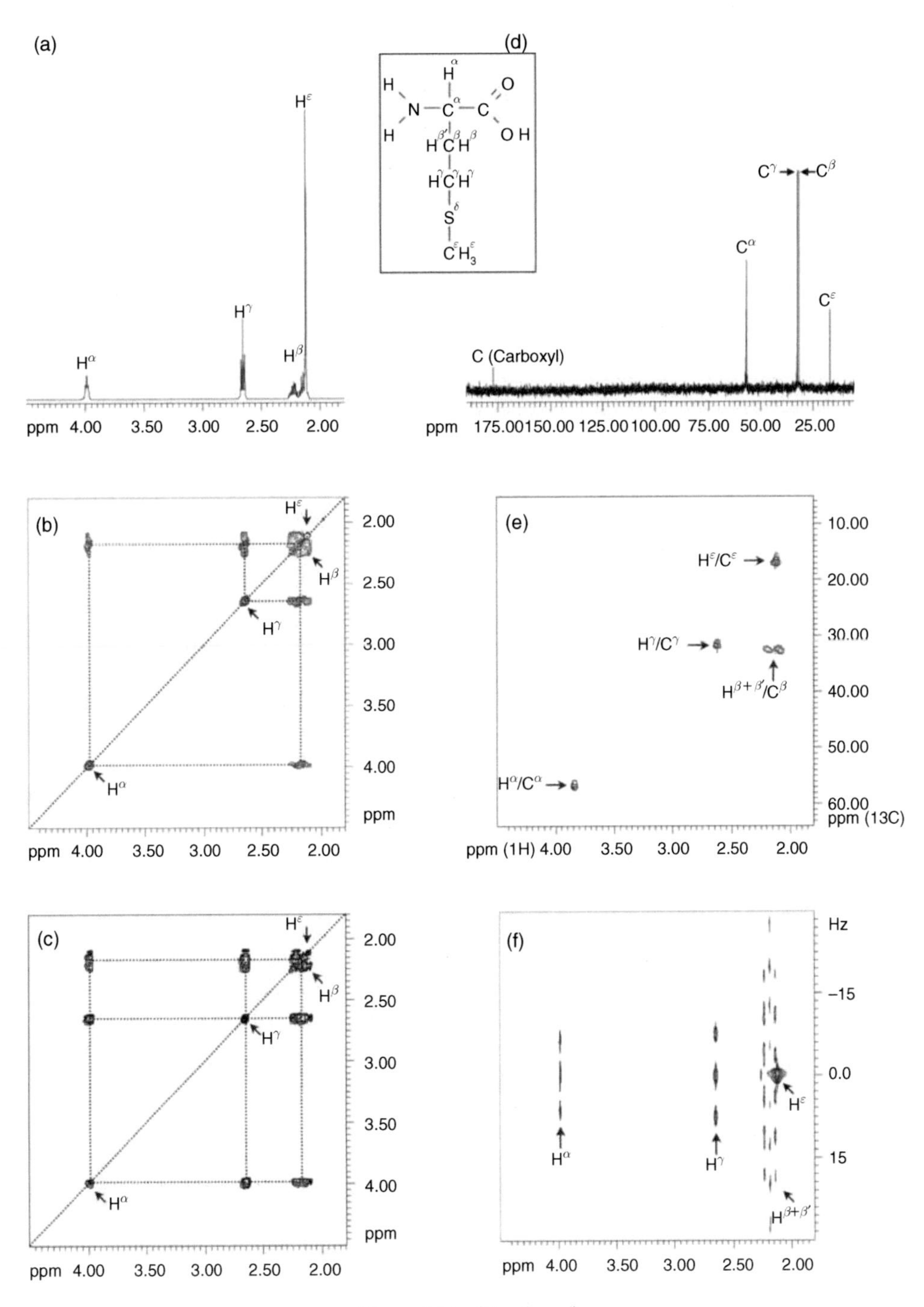

Figure 14.3. (Continued)

(>2–40 μmol/L), the chemical shifts and J-coupling constants can be determined directly and subsequent quantification of these metabolites is possible. The singlet resonance at 2.13 ppm from the ε methyl protons of methionine is most suited to quantify this metabolite in body fluids (Figure 14.3a). However, assignment can be

---

Figure 14.3. (Continued) Six different 500 MHz NMR spectra of methionine.

A: 1D ^{1}H-NMR spectrum

S-$C^{\varepsilon}H_3$: The methyl protons on C^{ε} give a singlet resonance at 2.13 ppm. This singlet corresponds to the three equivalent protons of the methyl group. The fact that it is a singlet illustrates that the neighboring atom (the S atom) does not carry any proton(s). In body fluids the 2.13 ppm resonance is often used to quantify methionine. $C^{\gamma}H_2$: Due to the J-coupling with the two neighboring protons on C^{β}, the protons on C^{γ} give a triplet resonance at 2.63 ppm ($^{\gamma}J_{\gamma,\beta} = 7.3$ Hz). $C^{\beta}H_2$: The protons on C^{β} give two multiplets at 2.10 and 2.20 ppm. This is explained by the fact that C^{α} is an asymmetrical carbon atom, and the two methylene protons on C^{β} are therefore diastereotopic; consequently, they give separate resonances. Due to the J-coupling with the proton on C^{α} and the protons on C^{γ}, we find two multiplets. $C^{\alpha}H$: Due to the J-coupling with the two diastereotopic methylene protons on C^{β}, the proton on C^{α} gives a disturbed triplet (doublet/doublet) resonance at 3.95 ppm. The peak area for the various resonances in methionine is proportional to the number of protons attached to the various C atoms.

B: 2D ^{1}H-^{1}H COSY spectrum

Chemical shift and scalar coupling can be obtained simultaneously for methionine by detecting the off-diagonal cross peaks, which are symmetric with respect to the diagonal. The diagonal and cross peaks connected by dashed lines indicate which protons of methionine have a scalar coupling. The diagonal peak of the two protons on C^{β} forms a corner of two squares, as these protons are coupled both to the proton on C^{α} and to those on C^{γ}.

C: 2D ^{1}H-^{1}H TOCSY spectrum

Compared with the COSY spectrum (Figure 14.3b) two new signals have now appeared. These provide evidence of a correlation between the protons on C^{α} and C^{γ}.

D: 1D ^{1}H-decoupled ^{13}C spectrum

In the 1D ^{1}H-decoupled ^{13}C NMR spectrum, $C^{\alpha}, C^{\beta}, C^{\gamma}$ and C^{ε} from methionine give resonances at 56.7, 32.5, 31.7 and 16.7 ppm, respectively. Note: This 1D ^{1}H-decoupled ^{13}C spectrum was recorded at pH 7.0.

E: 2D ^{1}H-^{13}C HSQC spectrum

This type of NMR spectroscopy reveals which carbons and protons in methionine are chemically bonded. For methionine, the ^{1}H/^{13}C shift values for the correlation peaks are 3.86/56.7 (α), 2.20; 2.10/32.5 (β), 2.65/34.7 (γ) and 2.13/16.7 (α). Note: This HSQC spectrum of methionine was recorded at pH 7.0.

F: 2D ^{1}H J-resolved spectrum

The J-multiplets are displayed along the F_1 dimension, while the chemical shifts are shown in the F_2 dimension. This minimizes overlap and allows better determination of coupling constants, which also are detectable in 1D ^{1}H-NMR. The protons on C^{γ} give a triplet resonance at 2.63 ppm with a J-coupling of 7.3 Hz ($^{\gamma}J_{\gamma,\beta} = 7.3$ Hz)

Table 14.1
Overview of different experiments used in the diagnosis of inborn errors of metabolism

Experiment	Description
1D ^{1}H (Figure 14.3a)	Single pulse with solvent presaturation. Routinely used to determine chemical shifts, J-values and metabolite quantities (only for well-resolved peaks).
^{1}H-^{1}H COSY (Figure 14.3b)	Used to establish which protons are spin-coupled. Helpful in resolving overlap and to assign crowded regions.
^{1}H-^{1}H DQF-COSY	COSY variant with greater resolution. Cross peaks near the diagonal can be resolved. Sensitivity is 50% less compared to COSY.
^{1}H-^{1}H TOCSY (Figure 14.3c)	COSY variant displaying relayed connectivities. Used as an aid in assignment.
^{1}H J-resolved (Figure 14.3f)	Separates J-splitting and chemical shifts on to two orthogonal axes. Reduced overlap and more accurate determination of J-values and chemical shifts.
^{1}H 1D Spin-echo	Spectral simplification based on difference in relaxation times. Not suitable for quantification.
1D ^{1}H-decoupled ^{13}C (Figure 14.3d)	More resonances resolved due to larger ^{13}C spectral width than ^{1}H. Suffers from inherent low sensitivity and low natural abundance.
^{1}H-^{13}C HSQC (Figure 14.3e)	Displays connectivities between chemically bonded ^{1}H and ^{13}C nuclei. Helpful when assigning complicated spectra.

rather difficult when high order spin coupling patterns are present or when spectral regions are crowded. Furthermore, presaturation of the water signal can result in loss of nearby peaks.

14.2.2.2. Two-dimensional NMR spectroscopy

Two-dimensional NMR experiments can provide additional information to solve overlap problems and allow identification of metabolites that otherwise remain undetected. They are based on the couplings between magnetic nuclei. These can be dipolar (through space) or scalar (through bond) couplings. Here, only experiments from the latter category will be discussed.

14.2.2.3. ^{1}H-^{1}H correlation spectroscopy

The first homonuclear 2D experiment is correlation spectroscopy (COSY). The underlying concept is coherence transfer from one spin to another via J-coupling [28]. It reveals the network of spin–spin couplings in each molecule and can therefore be used to aid in spectral assignment. The peaks under the diagonal of a COSY spectrum correspond to the peaks normally observed in a 1D spectrum. The off-diagonal peaks (cross peaks) provide the information on which protons are spin coupled. From the COSY spectrum of methionine, for instance, it is clear that the α -proton is J-coupled to the β-protons (Figure 14.3b). Unfortunately, there is a 90° phase difference between the diagonal and the cross peaks, which makes it impossible to correct the phase of a COSY spectrum in such a way that all peaks have absorption phase. When the phase of the cross peaks is corrected to pure absorption, the diagonal peaks display dispersive line shapes in both dimensions, and their long dispersion tails can conceal nearby cross peaks. A solution is to record a Double Quantum Filtered COSY (DQF-COSY) experiment, in which both diagonal and cross peaks have absorption line shapes. Consequently, the spectral resolution is improved and cross peaks close to the diagonal can be readily detected. Also, chemical shifts and coupling constants can be determined more easily. However, a disadvantage is that the sensitivity is reduced compared to a regular COSY, because only two out of four scans lead to cross peaks and the other two lead to diagonal peaks [28]. Another solution to the phase problem is displaying the conventional COSY spectrum in magnitude mode. All peaks appear absorptive, although cross peaks near the diagonal are still likely to be obscured. In the literature, COSY spectra of body fluids have been widely used in patients with IEM, for example in fucosidosis [8], 3-hydroxy-3-methylglutaryl CoA lyase deficiency [29], ribose 5-phosphate isomerase deficiency [30] and Salla disease [8, 31].

14.2.2.4. Total Correlation Spectroscopy

The COSY spectra of complex mixtures like body fluids can still show a substantial amount of overlap. Resulting ambiguities in the spectral assignment may be resolved by detecting multiple relayed connectivities using a Total Correlation Spectroscopy (TOCSY) experiment. For instance, in an AMX spin system cross peaks are observed between A and X, where both spins are coupled to spin M, but not to each other. This allows detection of long range interactions (>3 bonds), that are usually too weak in normal COSY spectra. Other advantages are a higher sensitivity for larger molecules and absorption line shapes for both diagonal and cross peaks. The number of observed cross peaks depends on the length of the applied TOCSY mixing sequence. A spectrum recorded with a short mixing period mainly contains cross peaks of directly coupled spins, whereas a long mixing time gives rise to relayed cross peaks. In the case of methionine, the TOCSY spectrum (Figure 14.3c), for instance, shows cross peaks between the α-proton and the γ-protons, which were not present in the regular COSY (cf. Figure 14.3b). A 1D version of TOCSY spectra at 750 MHz has been used to assign peaks in spectra from seminal fluid by Spraul *et al.* [32]. Furthermore, 1D TOCSY has been used to identify 3-ureidoisobutyric acid as an accumulating compound in ureidopropionase deficiency [33]. To our knowledge there are no examples of successful application of 2D TOCSY to the field of inborn errors of metabolism.

14.2.2.5. J-resolved spectroscopy (JRES)

Another useful homonuclear 2D experiment is the J-resolved (J-res) experiment, which separates the J fine structure from the chemical shifts. In the resulting spectrum, the J-multiplets are displayed along the F_1 dimension, while the chemical shifts are shown in the F_2 dimension. This minimizes overlap and allows better determination of coupling constants. Furthermore, a proton-decoupled 1D spectrum can be obtained by projection of the 2D J-resolved spectrum onto the F_2 axis. The JRES spectrum of methionine is shown in Figure 14.3f [30]. The 600 MHz 2D J-res ^{1}H-NMR spectra from urine and blood plasma samples was published by Foxall *et al.* [34]. In the literature, JRES ^{1}H-NMR spectra of body fluids have been used in patients with IEM, for example ribose 5-phosphate isomerase deficiency [30], maple syrup urine disease [24] and 2-methylacetoacetyl CoA thiolase deficiency [35].

14.2.2.6. Spin-echo spectra

Proton spectra can be also simplified by making use of the differences in relaxation behavior between different molecules. This principle forms the basis of spin-echo experiments. In spin-echo spectra, signals arising from relatively large molecules (i.e. protein signals in intact plasma) are attenuated due to their shorter relaxation times. Furthermore, even-numbered multiplets are displayed as negative peaks, whereas singlets and odd-numbered multiplets appear positive [35, 36]. Spin-echo experiments can therefore be helpful when assigning complex spectra. However, the method is less suited for quantification due to signal loss during long echo times. Since plasma samples are routinely deproteinized, spin-echo experiments are not recorded often nowadays when trying to diagnose an inborn error of metabolism.

14.2.2.7. ^{13}C NMR Spectroscopy

With the advances made in NMR spectrometer technology, ^{13}C NMR becomes more and more feasible. Despite its low intrinsic sensitivity ($\sim$ 60 times less than ^{1}H) and low natural abundance (1.1%), ^{13}C NMR can provide a wealth of additional information. This is mainly because of the large range of observed chemical shifts, resulting in many resolved signals, and because precursors of the metabolite of interest can be isotopically enriched. This last option enhances sensitivity and can simplify the study of metabolic pathways [37]. Another advantage of ^{13}C NMR spectroscopy is the absence of the water resonance, making solvent suppression no longer necessary.

The simplest ^{13}C method used in the field of inborn errors of metabolism is the 1D ^{1}H-decoupled ^{13}C experiment. Proton decoupling results in enhanced sensitivity due to collapsing of multiplets into singlets and due to the Nuclear Overhauser Effect (NOE; up to 4 times for ^{13}C). By comparing the decoupled spectrum with the corresponding ^{1}H-coupled spectrum, structural information can be obtained about the number of protons bound to each carbon [37]. As an example, the 1D proton-decoupled ^{13}C spectrum of methionine is shown in Figure 14.3d. Although 1D ^{13}C spectra of biofluids are not commonly recorded yet, their effective use in studying several inborn errors of metabolism, including galactosemia [38], dimethylglycine dehydrogenase deficiency [39], 2-hydroxyglutaric aciduria [40], Canavan disease [41, 42] and argininosuccinic aciduria [43], has been shown.

14.2.2.8. Heteronuclear Single Quantum Coherence spectroscopy

An example of an experiment that combines ^{1}H and ^{13}C spectroscopy is the Heteronuclear Single Quantum Coherence (HSQC). Since proton detection is employed, it is much more sensitive than a carbon-detected experiment. The resulting spectrum reveals which carbons and protons are chemically bonded (see Figure 14.3e for methionine). Hence, it can be used to facilitate spectral assignment. For instance, the assignment of 3-methylglutaconic acid, 3-hydroxyisovaleric acid and 3-methylglutaric acid in urine from a patient with 3-methylglutaconic aciduria type I (OMIM: 250950) [44]. Furthermore its effectiveness in diagnosing Canavan disease has been suggested [42].

14.2.3. Data processing and spectral interpretation

The NMR data are typically processed by Fourier transformation following apodization and zero filling of the free induction decay (FID). The phase is then corrected to obtain absorption line shapes and if necessary, a subsequent base-line correction can be performed. Metabolites can be quantified by fitting their resonances and the internal concentration reference signal with Lorentzian line shapes. Subsequently the concentrations are calculated from the relative integrals of the fitted resonances and the known concentration of the internal standard. For blood and CSF, metabolite concentrations are expressed in μmol/l. For urine, the creatinine singlet at 3.13 ppm (3 equivalent protons) is commonly used as internal reference and metabolite concentrations are usually reported in μmol/mmol creatinine.

The NMR spectrum of a body fluid taken from a patient with an inborn error of metabolism can be different from a control spectrum in three ways. First, unusual metabolites can be observed. Secondly, certain metabolites that are always present can be totally or partially absent. Thirdly, a novel constituent of the body fluid occurs in increased concentration compared to an age-related reference range. For relatively sparse spectral regions, this information can be directly obtained from a

^{1}H 1D spectrum and consequently lead to the diagnosis. However, body fluid spectra are usually very complex (especially for urine), showing hundreds to thousands of resonances in which small biochemical changes may be easily lost. Automatic data reduction and pattern recognition methods reduce this wealth of information and can therefore provide an alternative way for the diagnosis of inherited metabolic diseases [24, 45]. The first step can be to divide the 1D NMR spectrum into small sections, which provide a series of descriptors for each sample. These descriptors are then used to map the spectrum using methods such as non-linear mapping (NLM) or principal component analysis (PCA). The resulting 2D map displays clusters of similar samples. A nice separation from control urine samples was obtained for patients suffering from cystinuria, oxalic aciduria, porphyria, Fanconi syndrome and 5-oxoprolinuria [45]. It should be noted that different classification parameters can result in different clustering. Therefore, several maps may be required when screening for distinct metabolic diseases.

14.3. NMR spectra of body fluids

14.3.1. Healthy individuals and databases of model compounds

14.3.1.1. Urine

Figure 14.4 shows a part of the 500 MHz ^{1}H-NMR spectra (4.4–0.8 ppm) of urine from a healthy child and a healthy adult. In both spectra, the major metabolites and the peaks with chemical shift at pH 2.5 are creatinine (singlet: 3.13 ppm; singlet: 4.29), creatine (singlet: 3.05; singlet: 4.11), citric acid ($[AB]_2$ spin system: 2.99; 2.83 ppm), dimethylamine (triplet: 2.71), alanine (doublet: 1.51) and lactic acid (doublet: 1.41). Several sources with peak assignments, chemical shift and multiplicity for a wide range of metabolites are available in the literature [1, 20, 46, 47]. For the interpretation of urine NMR spectra, the age dependency in the concentration of many metabolites is of relevance. For instance, high concentrations of betaine (singlet: 3.27; singlet: 3.94) and *N, N*-dimethylglycine (DMG) (singlet: 2.93) can be observed in the urine 1D ^{1}H spectrum from the child (Figure 14.4B). Especially in the first month of life values up to 1500 μmol/mmol and 550 μmol/mmol creatinine may be found for betaine [48] and DMG respectively [39]. In adult urine, the concentration of betaine and DMG generally is lower with a maximal normal value of 80 μmol/mmol creatinine and 26 μmol/mmol creatinine, respectively [39]. Betaine and DMG are metabolites that will not be detected using conventional screening techniques, and NMR studies have reported age-related reference values for these metabolites [39, 48]. Besides age, other factors such as gender, medication and diet may cause severe variations in body fluid NMR spectra. These factors will be discussed in Section 14.3.5.

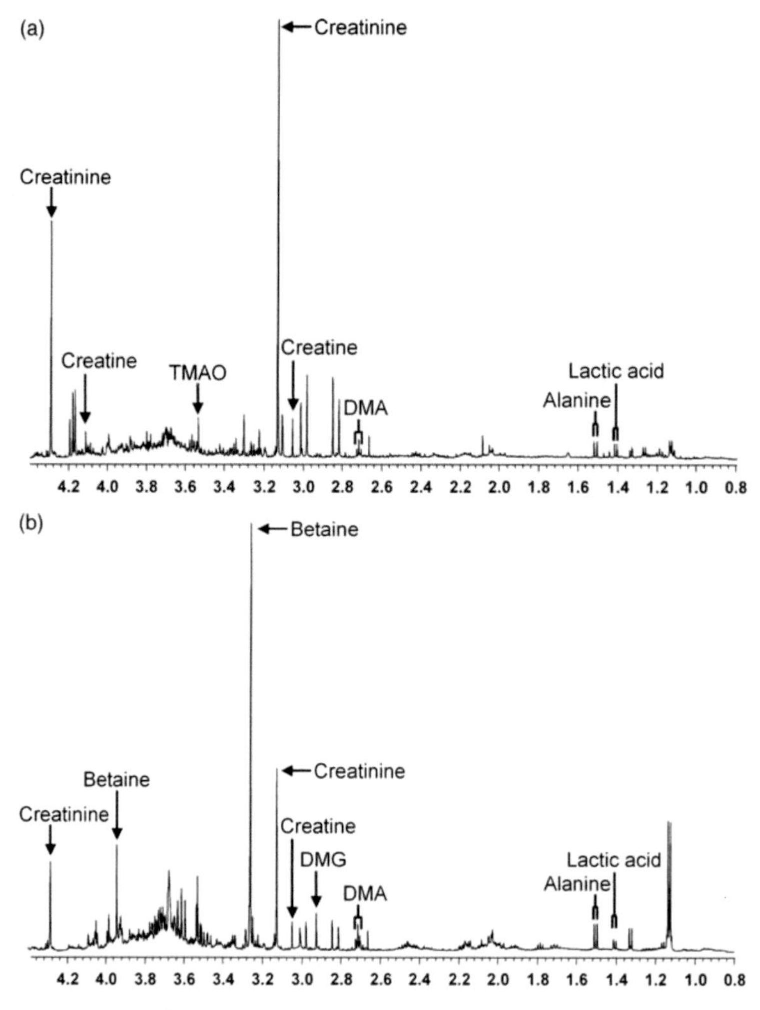

Figure 14.4. The 500 MHz ^{1}H-NMR spectrum of urine from a healthy adult (a) and a healthy child (b). High concentrations of betaine and *N, N*-dimethylglycine can be observed in the urine from the child (b). These concentrations are normal for the ages of the subjects.

14.3.1.2. Plasma/serum

In deproteinised blood plasma, the major metabolites and the peaks with chemical shift at pH 2.5 are lactic acid (doublet: 1.41 ppm; quartet: 4.36), citric acid ($[AB]_2$ spin system: 2.99; 2.83 ppm), glucose (doublet of doublets: 3.23; various: 3.30–3.95; doublet: 4.63; doublet: 5.22), acetic acid (singlet: 2.08) (Figure 14.2a). Due to the high concentration of glucose (5 mmol/l), identification of metabolites in the region between 3.30 and 3.95 ppm is difficult with 1D ^{1}H NMR spectroscopy. For instance, the singlet resonance of glycine at 3.72 ppm, caused by the two protons of the methylene group, is hard to identify at physiological concentration (concentration range: 100–384 μmol/l). The NMR data of 38 identified metabolites and 14 unidentified peaks measured in deproteinized blood plasma at pH 2.5 have been published [6]. Recently, the singlet resonance at 3.14 ppm that was described as unknown in this blood plasma study has been identified as dimethyl sulfone ($DMSO_2$) [5]. Several other reports about assignments in human blood plasma have been published [23, 49].

14.3.1.3. Cerebrospinal fluid

In the literature, there are several ^{1}H-NMR studies of CSF [7, 14, 16, 17, 50]. Sweatman *et al.* [14] reported assignments of 138 resonances in ^{1}H-NMR spectra of human CSF. These derived from 46 metabolites, and the assignments were achieved by the measurement of a combination of 2D experiments (JRES and COSY spectra). A standardized method for recording 1D ^{1}H-NMR spectra from CSF was published by Wevers *et al.* [7]. Quantitative analysis by NMR vs conventional techniques showed good correlation for alanine, threonine, valine and lactic acid. Resonances from 50 metabolites in 40 CSF were found in this study. Furthermore, 16 unidentified resonances were reported. *Myo*-inositol (triplet: 3.27; triplet: 4.05) in CSF is strongly age-dependent. The concentration of *myo*-inositol may be as high as 700 μmol/L in the CSF of young children (0–3 months), while in adults the upper reference range limit is estimated to be <30 μmol/L.

14.3.2. NMR spectra of known inborn errors of metabolism

Figure 14.5 shows an example of a 1D ^{1}H-NMR spectrum of a urine sample from a patient with an IEM. The patient has alkaptonuria (OMIM: 203,500). The disease is caused by a deficiency of the enzyme homogentisic acid oxidase in tyrosine metabolism (Figure 14.6, enzyme H). The characteristic metabolite for this disease is homogentisic acid. Normally this compound cannot be detected in body fluids. The concentration in this patient's urine amounted to 2500 μmol/mmol creatinine.

Table 14.2 provides a list of available papers on metabolic diseases studied with NMR spectroscopy. In many studies, peak assignments were achieved by the measurement of 2D experiments, measurement of the pure model compound, adding the pure model compound to the body fluid or specific resonance shifts upon pH changes. In the next pages, we provide an overview of some key publications on several groups of inherited metabolic disorders studied with NMR spectroscopy. The available literature indicates the power of NMR spectroscopy for the field of inborn errors of metabolism. This review shows that at least 100 different IEM can be diagnosed by NMR spectroscopy of urine. For most of these diseases the handbook

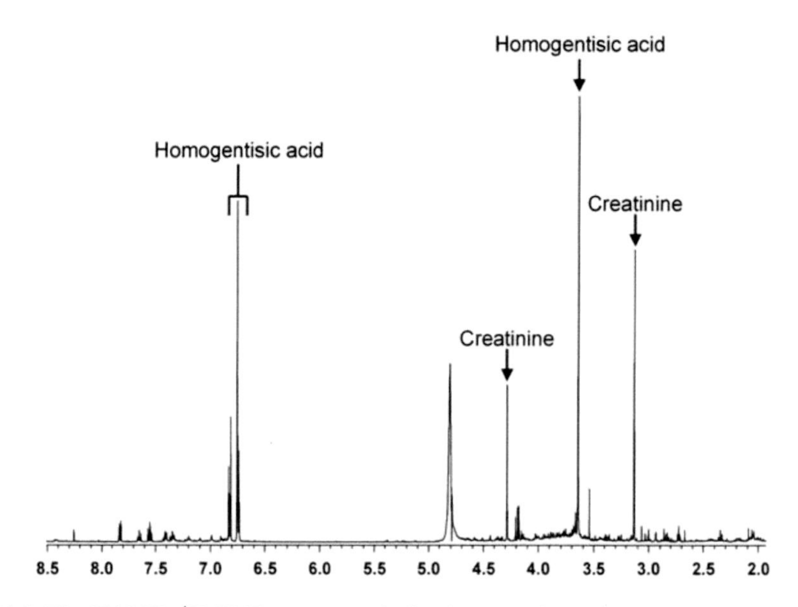

Figure 14.5. The 500 MHz ^{1}H-NMR spectrum of urine from a patient with alcaptonuria. The spectrum shows a high concentration of homogentisic acid (2500 μmol/mmol creatinine) normally not detected in urine.

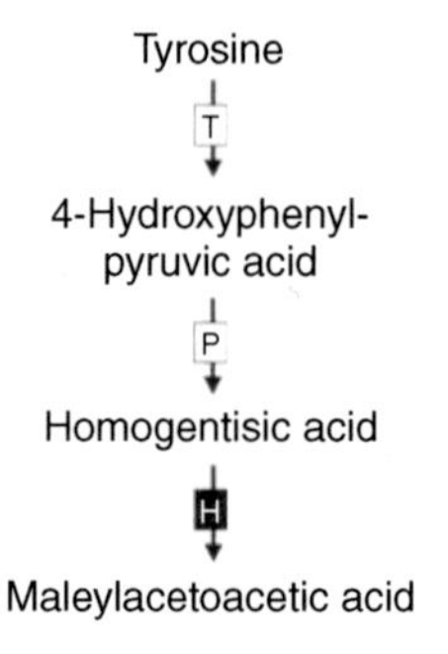

Figure 14.6. Metabolic pathway of tyrosine. Homogentisic acid oxidase deficiency (black square) results in accumulation of homogentisic acid in patients with alkaptonuria. T: tyrosine transaminase (EC 2.6.1.5); P: 4-hydroxyphenylpyruvate dioxygenase (EC 1.13.11.27); H: homogentisic acid oxidase (EC 1.13.11.5)

of NMR spectroscopy in inborn errors of metabolism shows the characteristic parts of the NMR spectrum [47].

14.3.2.1. Amino acid disorders

Amino acids are metabolites containing an amino group (NH_2) and a carboxylic acid group (COOH). They are extensively metabolized and several inborn errors of metabolism are located in amino acid metabolism. One of such pathways is

Table 14.2
Inborn errors of metabolism studied using body fluid NMR spectroscopy

Inborn errors of metabolism	References
Amino acid disorders	
Aminoacylase deficiency, Type I	[97, 98]
Argininosuccinic acid lyase deficiency	[43, 47, 51, 99]
Citrullinaemia	[8, 47, 51]
Cystathionine beta synthase deficiency	[47]
Cystinuria	[45, 95, 100]
Hawkinsinuria	[47, 101]
Histidinaemia	[7, 47]
Homocystinuria	[102–104]
Hyperlysinaemia II	[47]
Lysinuric protein intolerance	[47]
Methionine adenosyl transferase deficiency	[5, 47]
Non-ketotic hyperglycinaemia	[20, 47]
Ornithine carbamoyl transferase deficiency	[47, 51]
Phenylketonuria	[20, 47, 105]
Prolidase deficiency	[47, 106]
Prolinaemia II	[47, 107]
Sarcosinaemia	[47]
Tyrosinaemia I	[8, 47]
Carbohydrate-related disorders	
Congenital Defect of Glycosylation type IIb	[47, 52]
Galactosaemia	[30, 38, 47]
L-Arabinosuria	[47, 108]
Ribose 5-phosphate isomerase deficiency	[30, 47, 109]
Lysosomal diseases	
β-Mannosidosis	[8, 47]
α-Mannosidosis	[8, 47]
Aspartylglucosaminuria	[8, 47]
Fucosidosis	[8, 47]
G_{M1}-gangliosidosis	[8, 47]
G_{M2}-gangliosidosis (Sandhoff Disease)	[8, 47]
Salla disease	[8, 31, 47, 110]
Sialidosis	[8, 47]
Organic acidurias	
β-Ketothiolase deficiency	[20, 47]
2-Aminoadipic aciduria	[47]
2-Hydroxyglutaric aciduria	[20, 24, 40, 47, 111]
2-Methyl-3-hydroxybutyryl CoA dehydrogenase deficiency	[47]
2-Methylacetoacetyl CoA thiolase deficiency	[35]
3-Hydroxy-3-methylglutaryl CoA lyase deficiency	[20, 29, 112]

(Continued)

Table 14.2 (Continued)

Inborn errors of metabolism	References
3-Methylcrotonyl CoA carboxylase deficiency, isolated	[7, 20, 47, 113]
3-Methylglutaconic aciduria, Type I	[44, 114]
4-Hydroxybutyric aciduria	[47]
5-Oxoprolinuria	[6, 45, 47, 111, 112, 115]
Alkaptonuria	[47, 95, 116–118]
Canavan disease	[7, 8, 41, 42, 47, 119]
Cytochrome c oxidase deficiency	[120]
Glutaric aciduria type I	[36, 47]
Glutaric aciduria type II	[20, 47, 121]
Isovaleric acidaemia	[20, 36, 47, 122, 123]
Lactic acidosis	[20, 47]
Malonic aciduria	[47]
Maple syrup urine disease (=branched-chain ketoaciduria)	[20, 24, 36, 47, 112]
Methylmalonic aciduria	[20, 36, 47, 122, 124–126]
Mevalonic aciduria	[47]
Oxalic aciduria	[45, 112]
Propionic acidaemia	[20, 36, 36, 47, 124, 125, 127]
Pyruvate dehydrogenase E_1 deficiency	[120]
Pyruvate dehydrogenase E_3 deficiency	[120]
Lipid metabolism	
Cerebrotendinous xanthomatosis	[47]
Glycerol kinase deficiency	[47, 95]
Medium-chain acyl CoA dehydrogenase (MCAD) deficiency	[128]
Short-chain acyl CoA dehydrogenase (SCAD) deficiency	[47]
Purines/Pyrimidines Disorders	
Adenosine deaminase deficiency	[47, 55]
Adenylosuccinate lyase deficiency	[47, 55]
Beta-ureidopropionase deficiency	[33, 47, 62]
Dihydropyrimidinase deficiency	[47, 55, 129, 130]
Dihydropyrimidine dehydrogenase deficiency	[47, 55, 129, 130]
Molybdenum cofactor deficiency	[47, 55]
Purine nucleoside phosphorylase deficiency	[47, 55]
UMP synthase deficiency	[47]
Xanthine dehydrogenase deficiency	[47, 55]
Vitamin-related disorders	
Biotinidase deficiency	[20, 47]
Cobalamin Disorder Cbl-C	[131]
Vitamin B_{12} deficiency	[17]
Others	
Cytochrome b deficiency	[120]
Dimethylglycine dehydrogenase deficiency	[39, 47, 61, 132]

Table 14.2 (Continued)

Inborn errors of metabolism	References
Fanconi syndrome	[45]
French type sialuria	[8, 47]
Guanidinoacetate methyltransferase deficiency	[47, 59]
Huntington's disease	[133]
Hypoacetylaspartia	[134]
N-Acetylaspartylglutamate accumulation in CSF	[21]
Porphyria	[45]
Trimethylaminuria (fish-odour syndrome)	[47, 58, 135–140]
Wilson disease	[141]

the biosynthesis of urea (urea cycle). Investigation of urea cycle disorders by ^{1}H-NMR spectroscopy was published for the urine of patients with citrullinaemia (OMIM: 215,700), argininosuccinic aciduria (OMIM: 207,900) and ornithine carbamoyl transferase deficiency (OMIM: 311,250) [51]. The diagnostic metabolites citrulline and *N*-acetylcitrulline could be observed in the urine ^{1}H-NMR spectra of all patients with citrullinaemia. These observations agree with those of Engelke *et al.* [8], who reported an urinary citrulline and *N*-acetylcitrulline concentration of 825 and 518 μmol/mmol creatinine respectively in a patient with argininosuccinate synthetase deficiency (citrullinemia). The urine 1D spectrum of the patients with argininosuccinic aciduria shows the characteristic metabolite argininosuccinate. Orotic acid was detected in urine samples from three out of four patients with ornithine carbamoyl transferase deficiency.

For the diagnosis of amino acid disorders CSF is seldom used. The metabolic profile of CSF may confirm a diagnosis. Table 14.2 gives an overview of the inborn errors involving amino acid metabolism that have been studied with body fluid NMR spectroscopy.

14.3.2.2. Carbohydrate-related disorders

Carbohydrates may be found in body fluids in a soluble form as monosaccharides, disaccharides, oligosaccharides and mucopolysaccharides and also as glycolipids. The NMR spectrum of these molecules can be very complex. A paper by Moolenaar [30] gives a table with resonances typical for a variety of monosaccharides and disaccharides. Oligosaccharides accumulate in the urine in several lysosomal diseases. Specific oligosaccharides have been isolated from thin layer chromatography plates or otherwise purified and could subsequently be characterized using NMR spectroscopy. In some cases this has contributed to finding the primary defect in such disease. The most recent example is the identification of a tetrasaccharide Glc(alpha1-2)Glc(alpha1-3)Glc(alpha1-3)Man structure that contributed to the identification of the primary defect in the subtype IIb of the CDG syndrome [52] (Congenital Disorder of Glycosylation IIb) (OMIM: 606,056).

Engelke *et al.* have claimed the diagnosis of at least 5 different lysosomal diseases in native urine with NMR spectroscopy [8]. The ganglioside concentration in the urine of patients with G_{M1} and G_{M2} gangliosidosis escaped detection by NMR in the native urine sample probably because of the low concentration. The group of the mucopolysaccharidoses form a subgroup of the lysosomal diseases. The accumulating mucopolysaccharides or glycosaminoglycans are long chains of repeating saccharide units. Hochuli *et al.* have studied these compounds with 2D NMR spectroscopy but direct diagnosis in the native urine is not feasible [53].

14.3.2.3. Organic acids

Organic acids are indicators of organic acidurias associated with various inborn errors of metabolism. In 1985, Iles *et al.* [36] studied the organic acid metabolites in urine from patients with propionic acidemia, methylmalonic aciduria, branched chain ketoaciduria (maple syrup urine disease), isovaleric acidemia and glutaric aciduria type I. The study provides a list with ^{1}H-NMR chemical shifts (at pH 6.0) of 29 metabolites observed in the urine of the patients with the five selected organic acidurias. Another list containing chemical shifts of many organic acid metabolites was published by Lehnert and Hunkler [20]. They measured urine samples and model compounds with NMR spectroscopy under standard pH conditions (pH 2.50). They investigated urine samples of 11 different inborn errors of metabolism involving organic acids including three diseases from leucine metabolism. A recent study gives the body fluid NMR characteristics for all known inborn errors of leucine metabolism [44]. The ^{1}H-NMR spectroscopy of urine can easily discriminate between these inborn errors of leucine metabolism and provide a correct diagnosis. Table 14.2 gives an overview of the available studies on organic acidurias.

14.3.2.4. Purines and pyrimidines

More than 12 enzyme defects in the metabolic pathways of purines and pyrimidines are known to occur [54]. Wevers *et al.* [55] have studied body fluids from 25 patients with 9 different inborn errors in these pathways. Characteristic abnormalities could be demonstrated in the 1D ^{1}H-NMR spectra of urine samples of all patients. The only exception was case of adenine phosphoribosyl transferase deficiency in which the accumulating metabolite, 2,8-dihydroxyadenine, could not be seen under the NMR conditions used. In 2,8-dihydroxyadenine all protons exchange rapidly with water and therefore this metabolite is NMR invisible. The authors have provided a list of the most important ^{1}H-NMR resonances from relevant purine and pyrimidine metabolites measured at pH 2.5.

14.3.2.5. Others

NMR spectroscopy can be used to find diseases characterized by the presence of metabolites that cannot be detected with conventional screenings methods. The available literature shows that NMR spectroscopy is the method of choice to diagnose patients with trimethylaminuria or "fish odor syndrome" (OMIM: 602,079). This is a malodor syndrome caused by the accumulation of trimethylamine (TMA), due to a deficiency of the liver enzyme flavin-containing monooxygenase 3 (FMO3). In urine of affected patients the TMA concentration is increased, while the trimethylamine *N*-oxide (TMAO) concentration is decreased. NMR spectroscopy enables the simultaneous determination of these metabolites in urine (TMA: doublet at 2.98 ppm and TMAO: singlet at 3.54 ppm).

Patients with milder mutations can only be diagnosed with a loading test as the TMA concentration in a random urine sample may be fully normal. Loading tests have been described with TMA [56], with choline [57] and with fish meal [58]).

Guanidinoacetate methyltransferase (GAMT) deficiency (OMIM: 601,240) is a disorder in the biosynthesis of creatine. ^{1}H-NMR spectroscopy of CSF can be used for diagnosing this disorder [47, 59]. In CSF, creatine (singlet resonances at 3.05 and 4.11 ppm) and creatinine (singlet resonances at 3.13 and 4.29 ppm) are completely absent. Normally, they are always present in CSF (reference value: creatine 25–70 μmol/L and creatinine 20–100 μmol/L [59]). Guanidinoacetic acid (GAA) is the characteristic accumulating metabolite in GAMT deficiency, the oldest of the three creatine biosynthesis disorders known to date. To our knowledge GAA has not yet been described as a helpful diagnostic metabolite in *in vitro* NMR spectra of GAMT patients (δ GAA: 3.98 ppm; singlet). Accumulation of GAA in GAMT patients (CSF: $\pm$1 μmol/L; plasma: $\pm$10 μmol/L [60]; urine: $\pm$400 μmol/mmol creatinine [59]) is not visible in NMR spectra due to overlap of other resonances and the detection limit of the technique.

14.3.3. Novel diseases identified with body fluid NMR

Three novel diseases were discovered by body fluid NMR spectroscopy, for example dimethylglycine dehydrogenase deficiency (OMIM: 605,850), β-ureidopropionase deficiency (OMIM: 606,673) and a severe hypomyelination associated with increased levels of *N*-acetylaspartylglutamate in CSF. In two other cases, the technique was strongly involved in the elucidation of the molecular defect, for example ribose 5-phosphate isomerase deficiency (OMIM: 608,611) and congenital disorder of glycosylation type IIb (OMIM: 606,056). The NMR findings in these novel diseases are briefly described below.

14.3.3.1. Dimethylglycine dehydrogenase deficiency (OMIM: 605,850)

In urine and plasma NMR spectra of an adult patient with a fish odor and muscle weakness, high singlet resonances were observed at 2.93 and 3.80 ppm. The ^{1}H-NMR spectroscopy of model compounds revealed that *N, N*-dimethylglycine (DMG) caused these resonances. Addition of pure DMG to the urine sample and ^{13}C-NMR spectroscopy of the patient's urine and an authentic reference solution of DMG confirmed that the two singlets were caused by DMG. The concentration of DMG amounted to 457 μmol/mmol creatinine in urine (reference range: 1–26) and 221 μmol/L in serum (reference range: 1–5). The high concentration of DMG was caused by a deficiency of the enzyme dimethylglycine dehydrogenase (DMGDH) converting DMG to sarcosine. A homozygous missense mutation was found in the DMGDH gene of the patient [61].

14.3.3.2. β-Ureidopropionase deficiency (OMIM: 606,673)

A new defect in pyrimidine metabolism found with *in vitro* ^{1}H-NMR spectroscopy was reported by Assmann *et al.* [62]. Characteristic resonances of 3-ureidopropionic acid (triplet: 2.57 ppm; triplet: 3.36 ppm) and 3-ureidoisobutyric acid (doublet: 1.13 ppm; multiplet: 2.69 ppm; AB protons: 3.26 ppm nd 3.29 ppm) were observed in high concentrations in the urine spectrum (861 and 718 μmol/mmol creatinine, respectively) [33]. The 1D and 2D NMR techniques were used to identify these metabolites. The accumulation of these metabolites was caused by a deficiency of the enzyme β-ureidopropionase (UP) converting 3-ureidoisobutyric acid and 3-ureidopropionic acid to, respectively, β-aminoisobutyric acid and β-alanine. In this patient, Vreken *et al.* [63] found compound heterozygosity for two splice acceptor site mutations in the UPB1 gene, IVS1-2A-G and IVS8-1G-A. Van Kuilenburg *et al.* described a group of patients with this defect [64].

14.3.3.3. Severe hypomyelination associated with increased levels of N-acetylaspartylglutamate in CSF

Wolf *et al.* [21] described two unrelated girls with a severe hypomyelination of the brain. Defects in the proteolipid protein (PLP) gene causing *X*-linked Pelizaeus-Merzbacher disease (PMD) could be ruled out. NMR spectroscopy of the CSF revealed an increased concentration of NAAG, a putative neurotransmitter that is thought to play a role in intercellular signaling in the brain. The NAAG concentration amounted to 95 and 197 μmol/L in the two patients (reference range: $<$4.4 μmol/L). The molecular and genetic basis of this novel disease remains as yet unsolved. This is the first inborn error of metabolism where a first key for understanding the molecular defect was found with NMR spectroscopy of CSF.

14.3.3.4. Ribose 5-phosphate isomerase deficiency (OMIM: 608,611)
The combination of *in vivo* NMR spectroscopy of the brain with body fluid *in vitro* NMR spectroscopy led to the recognition and characterization of the first ever inborn error of polyol metabolism. *In vivo* NMR spectroscopy of the brain with a severe white matter disease showed unknown resonances between 3.5 and 4.0 ppm. Body fluid NMR spectroscopy identified increased amounts of arabitol and ribitol in all body fluids of the patient. Arabitol and ribitol in urine amounted to 1800 and 190 μmol/mmol creatinine (reference range respectively: <98 and <11). In CSF, arabitol and ribitol concentrations were 5100 and 1200 μmol/L (reference range respectively: <39 and <5 [30]). It could be shown that polyols caused the abnormal resonances in the brain MR spectrum. This led to the discovery of the defective enzyme ribose 5-phosphate isomerase, an enzyme in the pentose phosphate pathway. Mutations in the corresponding gene could be demonstrated in the DNA [65].

14.3.3.5. Congenital disorder of Glycosylation type IIb (OMIM: 606,056)
The group of van Coster found a patient who constantly excreted an unknown tetrasaccharide in the urine. This compound could be shown with thin-layer chromatography. After isolation NMR spectroscopy helped to identify this compound as Glc(α 1–2)Glc(α 1–3)Glc(α 1–3)Man. This led the way to the characterization of a novel subtype of the so-called CDG syndrome (Congenital Disorder of Glycosylation type IIb). The defect is in the enzyme glucosidase I (GCS1) involved in biosynthesis of *N*-glycans. The defect could be confirmed at the enzyme level and the molecular genetic level [52]. Abnormal resonances from the accumulating tetrasaccharide can be directly observed in a random sample of the patient [47].

14.3.4. Novel pathway: Methanethiol metabolism

Using 1D and heteronuclear 2D NMR techniques, Engelke *et al.* [5] have shown that $DMSO_2$ is a "regular" constituent of CSF (Figure 14.7) and plasma. The $DMSO_2$ derives from dietary sources, intestinal bacterial metabolism, and from human endogenous metabolism. In man, methanethiol is converted into $DMSO_2$. Increased plasma $DMSO_2$ concentrations are found in patients with methionine adenosyltransferase I/III (MAT I/III) deficiency (OMIM: 250,850). These findings led to the elucidation of a novel metabolic pathway that was not known to occur in man (Figure 14.8).

14.3.5. Drugs and food components in body fluids

Besides age, other factors such as medication and diet may cause severe variations in body fluid NMR spectra and hinder proper spectrum interpretation and quantification. Some examples are described below.

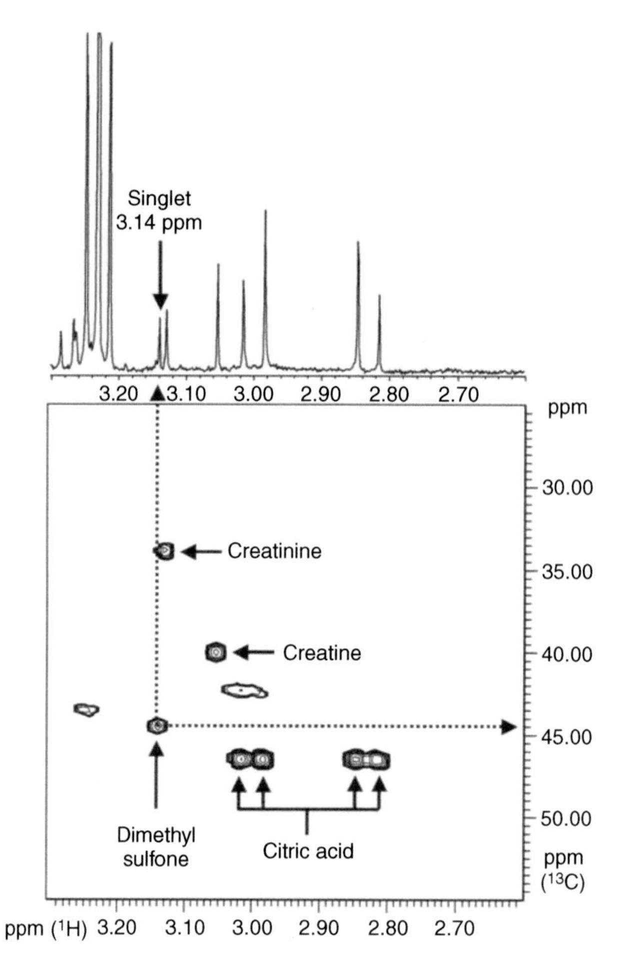

Figure 14.7. The 1D ^{1}H (upper) and ^{1}H-^{13}C HSQC (lower) spectrum of a representative CSF sample (pH = 2.5). Dimethyl sulfone showed a peak with the coordinates 3.14 (^{1}H)/44.3 (^{13}C). The concentration of $DMSO_2$ (singlet 3.14 ppm) amounted to 18 μmol/L.

14.3.5.1. Drugs

Valproic acid (Depakene®, Valproate, Valrelease®) Several NMR studies [66–68] reported urine samples from patients who took the anticonvulsant drug valproic acid. In man, valproic acid is extensively metabolized into several products, and some are excreted into the urine. The urine ^{1}H-NMR spectra of these patients show many high resonances from 3-hydroxyvalproic acid and valproic acid glucuronide. Azaroual *et al.* [67] reported a valproic acid intoxification identified by NMR spectroscopy.

Vigabatrin (Sabril®) The anti-epileptic drug vigabatrin is often used in patients suspected to have an inborn error of metabolism. The ^{1}H-NMR urine spectra

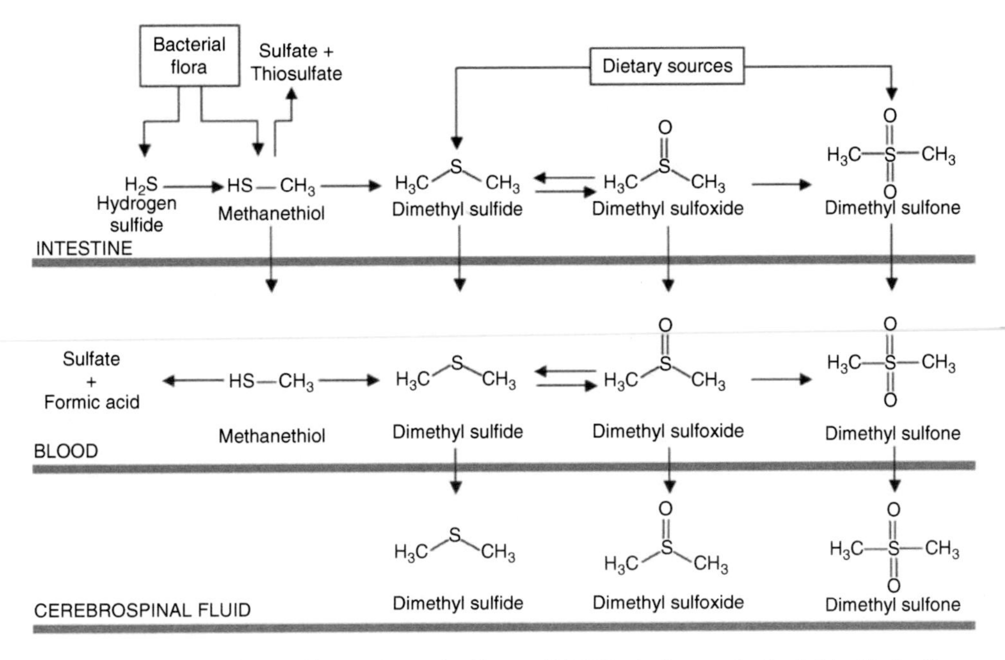

Figure 14.8. The partial putative pathway leading to $DMSO_2$ in human body fluids. Both, dietary sources and putative endogenous metabolic pathways can potentially contribute the presence of $DMSO_2$ in human body fluids.

of vigabatrin-treated patients show high vigabatrin resonances between 2.0 and 2.1 ppm caused by CH_2 protons (multiplet). This vigabatrin multiplet interferes with the *N*-acetyl resonances. Therefore correct interpretation and quantification of the *N*-acetyl region of the ^{1}H-NMR spectrum is impossible.

Ethosuximide (Zaronthin®) Ethosuximide is a drug used to control seizures. A patient with guanidinoacetate methyltransferase deficiency was treated with this drug and resonances of ethosuximide (triplet: 0.86 ppm; singlet: 1.28 ppm; multiplet: 1.65 ppm and AB: 2.68 ppm) could be observed in the CSF ^{1}H-NMR spectrum of this patients [59]. Due to the enzyme deficiency, creatine and creatinine resonances were absent in this spectrum as expected.

Dimethyl sulfone (Methylsulfonylmethane, MSM) NMR spectroscopy was performed on CSF from a 3-year-old boy with mental retardation and loss of acquired skills. A high concentration of $DMSO_2$ (singlet at 3.14 ppm) was observed in his 1D ^{1}H-NMR CSF spectrum. At the time of the lumbar puncture the patient had used a dietary supplement containing $DMSO_2$ [5].

Other drugs that are observed in body fluid NMR spectra are acetaminophen (Paracetamol®) [11, 68–70], ampicillin [71], carbamazepine (Epitol®, Tegretol®)

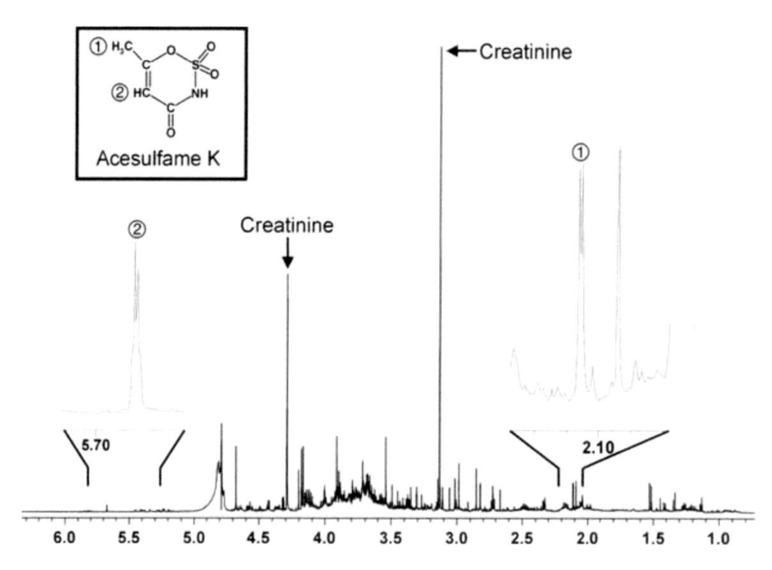

Figure 14.9. The 500 MHz 1D ^{1}H-NMR spectrum of urine from a healthy control after drinking Coca Cola light. Acesulfame K showed a doublet at 2.11 ppm and a quartet at 5.67 ppm.

[68], diethylcarbamazine (Hetrazan®) [72] and penicillin [11] and the drug vehicle, propylene glycol [50]. In some cases also metabolites of the drugs can be observed.

14.3.5.2. Food components

Acesulfame K Acesulfame K is a sweetener about 200 times sweeter than sucrose. Acesulfame K is not metabolized and it can be found in the urine ^{1}H-NMR spectra of persons who drink light products containing Acesulfame K, for example cola light (Figure 14.9). Other food ingredients or their metabolites that are observed in body fluid NMR spectra are ethanol in alcoholic drinks (triplet: 1.17 ppm; quartet: 3.64 ppm), TMAO in fresh fish (singlet: 3.54 ppm), trigonelline in coffee (singlet: 4.44 ppm; triplet: 8.10 ppm; singlet: 9.19 ppm) and Vitamin C (multiplet: 3.75 ppm; multiplet: 4.07 ppm; doublet: 4.97 ppm).

14.4. Future perspectives

Although many inborn errors of metabolism can be readily diagnosed with NMR, the technique is not yet routinely available in the majority of hospital laboratories. This is likely due to the high cost of the equipment and the fact that spectrum interpretation is usually not straightforward, requiring highly trained personnel. Several ongoing developments, which may make the application of body fluid NMR in diagnosing inborn errors of metabolism more successful and hopefully also more accessible, will be discussed in this chapter. Sections 14.4.1–14.4.3 will deal with advances made in NMR hardware technology, while Sections 14.4.4–14.4.5 will cover some of the progress in spectral analysis software and pulse sequence design.

14.4.1. Detection of other nuclei

Protons are highly abundant in many metabolites and have a high sensitivity, making them the most widely used nuclei in the study of inborn errors of metabolism. However, several other NMR active nuclei, which can provide important additional information, are present as well. Although ^{19}F and ^{31}P have a favorable natural abundance and relative sensitivity, there are few phosphorous- or fluorine-containing metabolites present in body fluids. On the contrary, ^{13}C and ^{15}N are present in many metabolites, but suffer from their low inherent sensitivity and natural abundance. These disadvantages may be overcome with the use of cryogenic probe technology, which was first presented by Styles *et al.* in 1984 [73]. By cooling the receiver coil and preamplifier to approximately 20 K with liquid helium, the resistance of the coil is significantly reduced. Consequently, the thermal noise is decreased by approximately a factor 4. This leads to a corresponding increase in the signal-to-noise ratio (S/N) per scan or, for the same S/N, a 16-fold reduction in acquisition time when compared to a normal probe. The improved sensitivity is such that ^{13}C spectra with good S/N can be obtained for body fluid samples with acceptable acquisition times [74]. The higher sensitivity obtained by using cryoprobe technology can be advantageous for proton NMR as well. With the resulting decrease in detection limit it is likely that NMR can diagnose many additional inborn errors of metabolism, where relatively low amounts of metabolites accumulate. It furthermore may facilitate the discovery of new diseases.

The low natural abundance of several nuclei can also be used as an advantage in metabolic investigations, since their weak background signal allows the study of the metabolism of labeled compounds. An example of this technique was already given in 1975 by Tanaka *et al.* [75], who elucidated the metabolic pathway of ^{13}C-labelled valine in a boy suffering from methylmalonic acidemia by administering the patient orally with DL-[α-^{13}C]-valine and DL-[α, β-^{13}C]-valine.

14.4.2. Linking NMR with other analytical techniques

On one hand, NMR has the advantage that many metabolites can be detected simultaneously and that no preselection is required. On the other hand, this is also a disadvantage, since it can cause highly complex spectra with a considerable amount of overlap. By linking NMR to HPLC it is possible to separate and directly identify the compounds present in complex mixtures. The first example employing coupled 500 MHz HPLC-NMR was presented in 1992 by Spraul *et al.* [76], who investigated the metabolism of the drug ibuprofen in urine samples. Both stopped-flow and continuous-flow experiments were performed. In 1995, the first set up combining NMR with both HPLC and mass spectrometry (MS) was presented [77] and the stopped-flow method with parallel NMR and MS detection has now become standard [78]. In body fluid research, HPLC-NMR and HPLC-NMR-MS have mainly been used to investigate the metabolic fate of different drugs and potential drug candidates [25, 76, 79–82]. Both methods have proven to be valuable tools to obtain structural and pharmaceutical data of metabolites present in complex matrices with a single experiment. In the field of inborn errors of metabolism, it is expected that the combination of NMR with liquid chromatography and/or MS can also be helpful in elucidating (partially) unknown pathways. Furthermore, the detection of new metabolites and their associated diseases may become less complicated.

14.4.3. Microprobes

Over the years, NMR has undergone some major developments, but the conventional 5 mm sample tube used today is still essentially the same as in the 1960s. Historically, relatively large amounts of sample were needed to compensate for the low inherent NMR sensitivity. Nowadays, ultra-high-field spectrometers equipped with cryoprobe technology are becoming available, making measurement of mass-limited samples theoretically possible. The sensitivity of these samples can be further enhanced by using microprobes. With decreasing diameter, d, of the detector coil, the NMR mass sensitivity (S/N per mole) increases to a first approximation with 1/d [83, 84]. A 5-fold increase in mass sensitivity, which corresponds to a 25-fold reduction in measurement time compared to a conventional 5 mm probe, was reported for a 1 mm high-resolution probe with 2.5 μl active sample volume [85, 86]. Other advantages of using low-volume probes are a reduction of approximately a factor 100 in the use of expensive deuterated solvents and a decreased residual solvent signal [85]. Furthermore, microprobes are very well suited for hyphenation with several separation techniques, including HPLC and capillary electrophoresis [87].

In body fluid NMR, the application of microprobes is very valuable when only a minor amount of fluid is available. Its effective use in studying a few microliters of rodent CSF and blood plasma has been demonstrated [85, 86]. An additional advantage when studying metabolic disorders in small animal models (e.g. in "knock-out" animals) is that enough body fluid material for microprobe NMR investigations can be obtained without sacrificing the animal. Although microprobes have not yet been applied to study inborn errors of metabolism in man, their enhanced sensitivity and small sample volumes can be promising in this field as well.

14.4.4. Diffusion-edited NMR spectroscopy

The presence of macromolecules in blood plasma, serum and CSF can hamper NMR analysis and they are therefore often removed prior to NMR measurement, as described previously. Unfortunately, this can result in loss of metabolites and furthermore inhibit the study of many possible interactions. Diffusion-edited spectroscopy applied to intact samples provides an alternative for spectral simplification [88, 89]. Under normal conditions, molecular diffusion is not proportional to observed spectral intensities. However, when pulsed field gradients are applied during the experiment, diffusion can cause changes in signal intensities. When compared to a conventional spectrum, fast diffusing molecules such as low molecular weight metabolites will show a much larger intensity difference than slowly diffusing molecules such as lipoproteins. Diffusion-edited spectroscopy can therefore effectively suppress small molecule resonances by a factor 100 or more, making it possible to directly study macromolecules [89, 90]. A spectrum displaying only low molecular weight metabolites is obtained by subtracting the macromolecular spectrum from a total spectrum recorded without gradients. For a plasma sample, the resulting spectrum is highly similar to a spectrum obtained after plasma deproteinization by ultrafiltration [89]. Since diffusion-edited NMR does not require any sample preparation, the experimental time required can be significantly reduced (up to 30 mins in the case of ultrafiltration), making the use of NMR as a diagnostic tool in clinical chemistry more attractive. The application of diffusion-edited spectra to aid resonance assignment in studies of diseases has also been suggested [88, 91]. Additionally, a combination of diffusion-dited NMR spectroscopy with chemometric methods in toxicological investigations has been reported [92].

In contrast to spin-echo sequences, which can be applied to filter out macromolecule resonances as well, metabolites can be readily quantified with diffusion-edited NMR spectroscopy. However, TSP-d_4 cannot be used as a concentration reference in this case, since it is partially invisible due to its interaction with proteins. Formate can, for instance, be used as an alternative.

14.4.5. Automation

In complex mixtures like body fluids, each component generates its own NMR spectrum, which possibly overlaps with other compounds in the mixture and makes analysis often difficult. Besides using liquid chromatography coupled to NMR as discussed in Section 14.4.2, this problem can also be solved with automated signal acquisition and special analysis software combined with spectral databases, which can greatly facilitate spectral interpretation and diagnosis of inborn errors of metabolism [93]. Automated NMR measurements require that the apparatus be equipped with a sample changer and that it is capable of self-locking and shimming.

Nowadays, probes that also perform automatic tuning and matching are becoming available as well. Next, the computer has to start acquisition of the appropriate experiments. In one suggested automation scheme, first a 1D ^{1}H spectrum is recorded and processed, which is then used to detect deviations from normal spectra by principal component analysis incorporating a database with control spectra [93]. Subsequently, a 2D COSY is recorded to identify metabolites that are indicative of the possible inborn errors of metabolism. The success of this approach was demonstrated by correct diagnosis of a few inherited metabolic diseases [93].A somewhat different set-up used only 1D ^{1}H spectra to directly identify compounds by chemical shift and database comparison [94]. Specialized deconvolution software performs the automatic peak assignment and subsequent metabolite quantification based on peak areas. In total, 1000 urine samples, including 69 samples from patients with 13 different inborn errors of metabolism, were analyzed in less than 2 min per sample. All healthy controls were classified as normal, whereas 66 of the pathological samples were correctly diagnosed [94]. A third chemometric method capable of automated analysis is a neural network trained to distinguish 1D ^{1}H spectra of patients with a specific disorder from healthy individuals [95, 96]. A neural network is created by supplying data for which the outcome is already known and can include many different diseases. A diagnostic accuracy of almost 100% has been accomplished in the case of the metabolic disorder cystinuria [95].

Automated spectral acquisition and analysis can greatly accelerate spectrum interpretation and facilitate clinical diagnosis by providing clear "yes or no" answers. A second advantage is that it permits NMR measurements to be performed by non-specialists. Therefore, it is likely that further development of these methods can make NMR spectroscopy more attractive as a routine diagnostic tool in hospital laboratories.

Although all new developments discussed above show promising results, it is expected that especially automated data acquisition and analysis will provide a large step forward in making NMR spectroscopy a suitable method for standard clinical analysis. It is also likely that a combination of two or more of the described techniques will be employed within several years, incorporating each of their separate advantages.

References

[1] Lindon, J.C., Nicholson, J.K., Everett, J.R., NMR spectroscopy of biofluids. Webb, G.A., editor Annual reports on NMR spectroscopy London: Academic Press 1999;38:1–8.

[2] Nicholson, J.K., Wilson, I.D. High Resolution Proton Magnetic Resonance Spectroscopy of Biological Fluids. *Progress in Nuclear Magnetic Resonance Spectroscopy* 1989;21:449–501.

[3] Lindon, J.C., Nicholson, J.K., Holmes, E., Everett, J.R. Metabonomics: Metabolic processes studied by NMR spectroscopy of biofluids. *Concepts in Magnetic Resonance* 2000;12:289–320.

[4] Moolenaar, S.H., Engelke, U.F., Wevers, R.A. Proton nuclear magnetic resonance spectroscopy of body fluids in the field of inborn errors of metabolism. *Ann. Clin. Biochem.* 2003;40:16–24.

[5] Engelke, U.F., Tangerman, A., Willemsen, M.A., Moskau, D., Loss, S., Mudd, S.H., Wevers, R.A. Dimethyl sulfone in human cerebrospinal fluid and blood plasma confirmed by one-dimensional (1)H and two-dimensional (1)H-(13)C NMR. *NMR Biomed.* 2005;18:331–6.

[6] Wevers, R.A., Engelke, U., Heerschap, A. High-resolution 1H-NMR spectroscopy of blood plasma for metabolic studies. *Clin. Chem.* 1994;40:1245–50.

[7] Wevers, R.A., Engelke, U., Wendel, U., de Jong, J.G., Gabreels, F.J., Heerschap, A. Standardized method for high-resolution 1H-NMR of cerebrospinal fluid. *Clin. Chem.* 1995;41:744–51.

[8] Engelke, U.F., Liebrand, van Sambeek, M.L., de Jong, J.G., Leroy, J.G., Morava, E., Smeitink, J.A., Wevers, R.A. N-acetylated metabolites in urine: Proton nuclear magnetic resonance spectroscopic study on patients with inborn errors of metabolism. *Clin. Chem.* 2004;50:58–66.

[9] Nicholson, J.K., Buckingham, M.J., Sadler, P.J. High resolution ^{1}H n.m.r. studies of vertebrate blood and plasma. *Biochem. J.* 1983;211:605–15.

[10] Kriat, M., Confort Gouny, S., Vion Dury, J., Sciaky, M., Viout, P., Cozzone, P.J. Quantitation of metabolites in human blood serum by proton magnetic resonance spectroscopy. A comparative study of the use of formate and TSP as concentration standards. *NMR Biomed.* 1992;5:179–84.

[11] Connor, S., Everett, J., Nicholson, J.K. Spin-echo proton NMR spectroscopy of urine samples. Water suppression via a urea-dependent T2 relaxation process. *Magn. Reson. Med.* 1987;4:461–70.

[12] Maillet, S., Vion-Dury, J., Confort-Gouny, S., Nicoli, F., Lutz, N.W., Viout, P., Cozzone, P. Experimental protocol for clinical analysis of cerebrospinal fluid by high resolution proton magnetic resonance spectroscopy. *Brain Res. Protocols* 1998;3:123–34.

[13] Lutz, N.W., Maillet, S., Nicoli, F., Viout, P., Cozzone, P.J. Further assignment of resonances in ^{1}H NMR spectra of cerebrospinal fluid (CSF). *FEBS Lett.* 1998;425:345–51.

[14] Sweatman, B.C., Farrant, R.D., Holmes, E., Ghauri, F.Y., Nicholson, J.K., Lindon, J.C. 600 MHz 1H-NMR spectroscopy of human cerebrospinal fluid: Effects of sample manipulation and assignment of resonances. *J. Pharm. Biomed. Anal.* 1993;11:651–64.

[15] Subramanian, A., Gupta, A., Saxena, S., Gupta, A., Kumar, R., Nigam, A. *et al.* Proton MR CSF analysis and a new software as predictors for the differentiation of meningitis in children. *NMR Biomed.* 2005;18:213–25.

[16] Bell, J.D., Brown, J.C., Sadler, P.J., Macleod, A.F., Sonksen, P.H., Hughes, R.D., Williams, R. High resolution proton nuclear magnetic resonance studies of human cerebrospinal fluid. *Clin. Sci.* (London) 1987;72:563–70.

[17] Commodari, F., Arnold, D.L., Sanctuary, B.C., Shoubridge, E.A. 1H NMR characterization of normal human cerebrospinal fluid and the detection of methylmalonic acid in a vitamin B12 deficient patient. *NMR Biomed.* 1991;4:192–200.

[18] Daykin, C.A., Foxall, P.J., Connor, S.C., Lindon, J.C., Nicholson, J.K. The comparison of plasma deproteinization methods for the detection of low-molecular-weight metabolites by (1)H nuclear magnetic resonance spectroscopy. *Anal. Biochem.* 2002;304:220–30.

[19] Bell, J.D., Brown, J.C., Kubal, G., Sadler, P.J. NMR-invisible lactate in blood plasma. *FEBS Lett.* 1988;235:81–6.

[20] Lehnert, W., Hunkler, D. Possibilities of selective screening for inborn errors of metabolism using high-resolution 1H-FT-NMR spectrometry. *Eur. J. Pediatr.* 1986;145:260–6.

[21] Wolf, N.I., Willemsen, M.A., Engelke, U.F., van der Knaap, M.S., Pouwels, P.J., Harting, I. *et al.* Severe hypomyelination associated with increased levels of N-acetylaspartylglutamate in CSF. Neurology 2004;62:1503–8.

[22] Sze, D.Y., Jardetzky, O. Determination of metabolite and nucleotide concentrations in proliferating lymphocytes by 1H-NMR of acid extracts. *Biochim. Biophys Acta* 1990;1054:181–97.

[23] Nicholson, J.K., Foxall, P.J., Spraul, M., Farrant, R.D., Lindon, J.C. 750 MHz ^{1}H and ^{1}H-^{13}C NMR spectroscopy of human blood plasma. *Anal. Chem.* 1995;67:793–811.

[24] Holmes, E., Foxall, P.J., Spraul, M., Farrant, R.D., Nicholson, J.K., Lindon, J.C. 750 MHz 1H NMR spectroscopy characterisation of the complex metabolic pattern of urine from patients with inborn errors of metabolism: 2-hydroxyglutaric aciduria and maple syrup urine disease. *J. Pharm. Biomed. Anal.* 1997;15:1647–59.

[25] Sidelmann, U.G., Braumann, U., Hofmann, M., Spraul, M., Lindon, J.C., Nicholson, J.K., Hansen, S.H. Directly coupled 800 MHz HPLC-NMR spectroscopy of urine and its applications to the identification of the major phase II metabolites of tolfenamic acid. *Anal. Chem.* 1997;69:607–12.

[26] Cavanagh, J., Fairbrother, W.J., Palmer III, A.G., Skelton, N.J. *Protein NMR Spectroscopy: Principles and Practice* (Academic Press Inc., New York, 1996).

[27] Wijmenga, S.S., Van Buuren, B.N.M. The use of NMR methods for conformational studies of nucleic acids. *Prog. Nucl. Magn. Reson. Spec.* 1998;32:287–387.

[28] Van de Ven, F.J.M. *Multidimensional NMR in liquids. Basic principles and experimental methods*. Wiley-VCH Verlag GmbH, 1995.

[29] Iles, R.A., Jago, J.R., Williams, S.R., Chalmers, R.A. 3-Hydroxy-3-methylglutaryl-CoA lyase deficiency studied using 2-dimensional proton nuclear magnetic resonance spectroscopy. *FEBS Lett.* 1986;203:49–53.

[30] Moolenaar, S.H., van der Knaap, M.S., Engelke, U.F., Pouwels, P.J., Janssen Zijlstra, F.S., Verhoeven, N.M. *et al.* In vivo and in vitro NMR spectroscopy reveal a putative novel inborn error involving polyol metabolism. *NMR Biomed.* 2001;14:167–76.

[31] Sewell, A.C., Murphy, H.C., Iles, R.A. Proton nuclear magnetic resonance spectroscopic detection of sialic acid storage disease. *Clin. Chem.* 2002;48:357–9.

[32] Spraul, M., Nicholson, J.K., Lynch, M.J., Lindon, J.C. Application of the one-dimensional TOCSY pulse sequence in 750 MHz 1H-NMR spectroscopy for assignment of endogenous metabolite resonances in biofluids. *J. Pharm. Biomed. Anal.* 1994;12:613–8.

[33] Moolenaar, S.H., Gohlich Ratmann, G., Engelke, U.F., Spraul, M., Humpfer, E., Dvortsak, P. *et al.* beta-Ureidopropionase deficiency: A novel inborn error of metabolism discovered using NMR spectroscopy on urine. *Magn. Reson. Med.* 2001;46:1014–7.

[34] Foxall, P.J., Parkinson, J.A., Sadler, I.H., Lindon, J.C., Nicholson, J.K. Analysis of biological fluids using 600 MHz proton NMR spectroscopy: Application of homonuclear two-dimensional J-resolved spectroscopy to urine and blood plasma for spectral simplification and assignment. *J. Pharm. Biomed. Anal.* 1993;11:21–31.

[35] Williams, S.R., Iles, R.A., Chalmers, R.A. Spin-echo and 2-dimensional 1H nuclear magnetic resonance studies on urinary metabolites from patients with 2-methylacetoacetyl CoA thiolase deficiency. *Clin. Chim. Acta* 1986;159:153–61.

[36] Iles, R.A., Hind, A.J., Chalmers, R.A. Use of proton nuclear magnetic resonance spectroscopy in detection and study of organic acidurias. *Clin. Chem.* 1985;31:1795–801.

[37] Fan, T.W.M. Metabolite profiling by one- and two-dimensional NMR analysis of complex mixtures. *Prog. Nucl. Magn. Reson. Spec.* 1996;28:161–219.

[38] Wehrli, S.L., Berry, G.T., Palmieri, M., Mazur, A., Elsas, L., Segal, S. Urinary galactonate in patients with galactosemia: Quantitation by nuclear magnetic resonance spectroscopy. *Pediatr. Res.* 1997;42:855–61.

[39] Moolenaar, S.H., Poggi Bach, J., Engelke, U.F., Corstiaensen, J.M., Heerschap, A., de Jong, J.G. *et al.* Defect in dimethylglycine dehydrogenase, a new inborn error of metabolism: NMR spectroscopy study. *Clin. Chem.* 1999;45:459–64.

[40] Bal, D., Gryff Keller, A. 1H and 13C NMR study of 2-hydroxyglutaric acid and its lactone. *Magn. Reson. Chem.* 2002;40:533–6.

[41] Bal, D., Gryff Keller, A., Gradowska, W. Absolute configuration of N-acetylaspartate in urine from patients with Canavan disease. *J. Inherit. Metab. Dis.* 2005;28:607–9.

[42] Krawczyk, H., Gradowska, W. Characterisation of the 1H and 13C NMR spectra of N-acetylaspartylglutamate and its detection in urine from patients with Canavan disease. *J. Pharm. Biomed. Anal.* 2003;31:455–63.

[43] Krawczyk, H., Gryff Keller, A., Gradowska, W., Duran, M., Pronicka, E. 13C NMR spectroscopy: A convenient tool for detection of argininosuccinic aciduria. *J. Pharm. Biomed. Anal.* 2001;26:401–8.

[44] Engelke, U.F., Kremer, B., Kluijtmans, L., van der Graaf, M., Morava, E., Loupatty, F.J., Wanders, R.J., Moskau, D., Loss, S., van den Bergh, E., Wevers, R.A. NMR Spectroscopic studies on the late onset form of 3-methylglutaconic aciduria type 1 and other defects in leucine metabolism. *NMR Biomed* 2006;19:271–8.

[45] Holmes, E., Foxall, P.J., Nicholson, J.K., Neild, G.H., Brown, S.M., Beddell, C.R. *et al.* Automatic data reduction and pattern recognition methods for analysis of 1H nuclear magnetic resonance spectra of human urine from normal and pathological states. *Anal. Biochem.* 1994;220:284–96.

[46] Zuppi, C., Messana, I., Forni, F., Rossi. C., Pennacchietti, L., Ferrari, F., Giardina B. 1H NMR spectra of normal urines: Reference ranges of the major metabolites. *Clin. Chim. Acta* 1997;265:85–97.

[47] Moolenaar, S.H., Engelke, U.F., Hoenderop, S.M.G.C., Sewell, A.C., Wagner, L., Wevers, R.A. Handbook of ^{1}H-NMR spectroscopy in inborn errors of metabolism, 1 edn, Heilbronn: SPS Verlagsgesellschaft, 2002, ISBN 3-936145-49-0.

[48] Davies, S.E.C., Woolf, D.A., Chalmers, R.A., Rafter, J.E.M., Iles, R.A. Proton nmr studies of betaine excretion in the human neonate: Consequences for choline and methyl group supply. *J. Nutr. Biochem.* 1992;3:523–30.

[49] Tang, H., Wang, Y., Nicholson, J.K., Lindon, J.C. Use of relaxation-edited one-dimensional and two dimensional nuclear magnetic resonance spectroscopy to improve detection of small metabolites in blood plasma. *Anal. Biochem.* 2004;325:260–72.

[50] Petroff, O.A., Yu, R.K., Ogino, T. High-resolution proton magnetic resonance analysis of human cerebrospinal fluid. J. Neurochem. 1986;47:1270–6.

[51] Burns, S.P., Woolf, D.A., Leonard, J.V., Iles, R.A. Investigation of urea cycle enzyme disorders by 1H-NMR spectroscopy. *Clin. Chim. Acta* 1992;209:47–60.

[52] De Praeter, C.M., Gerwig, G.J., Bause, E., Nuytinck, L.K., Vliegenthart, J.F., Breuer, W. *et al.* A novel disorder caused by defective biosynthesis of N-linked oligosaccharides due to glucosidase I deficiency. *Am. J. Hum. Genet.* 2000;66:1744–56.

[53] Hochuli, M., Wuthrich, K., Steinmann, B. Two-dimensional NMR spectroscopy of urinary glycosaminoglycans from patients with different mucopolysaccharidoses. NMR Biomed. 2003;16:224–36.

[54] Blau, N., Duran, M., Blaskovics, M.E., Gibson, K.M. Physician's guide to the laboratory diagnostics of metabolic diseases, 2 ed. Berlin Heidelberg: Springer-Verlag, 2003.

[55] Wevers, R.A., Engelke, U.F., Moolenaar, S.H., Brautigam, C., de Jong, J.G., Duran, R. *et al.* 1H-NMR spectroscopy of body fluids: Inborn errors of purine and pyrimidine metabolism. *Clin. Chem.* 1999;45:539–48.

[56] al Waiz, M., Ayesh, R., Mitchell, S.C., Idle, J.R., Smith, R.L. Trimethylaminuria: the detection of carriers using a trimethylamine load test. *J. Inherit. Metab. Dis.* 1989;12:80–5.

[57] Marks, R., Greaves, M.W., Prottey, C., Hartop, P.J. Trimethylaminuria: the use of choline as an aid to diagnosis. *Br. J. Dermatol.* 1977;96:399–402.

[58] Podadera, P., Areas, J.A., Lanfer Marquez, U.M. Diagnosis of suspected trimethylaminuria by NMR spectroscopy. *Clin. Chim. Acta* 2005;351:149–54.

[59] Schulze, A., Hess, T., Wevers, R., Mayatepek, E., Bachert, P., Marescau, B. *et al.* Creatine deficiency syndrome caused by guanidinoacetate methyltransferase deficiency: diagnostic tools for a new inborn error of metabolism. *J. Pediatr.* 1997;131:626–31.

[60] Schulze, A., Ebinger, F., Rating, D., Mayatepek, E. Improving treatment of guanidinoacetate methyltransferase deficiency: Reduction of guanidinoacetic acid in body fluids by arginine restriction and ornithine supplementation. *Mol. Genet. Metab.* 2001;74:413–9.

[61] Binzak, B.A., Wevers, R.A., Moolenaar, S.H., Lee, Y.M., Hwu, W.L., Poggi Bach, J. *et al.* Cloning of dimethylglycine dehydrogenase and a new human inborn error of metabolism, dimethylglycine dehydrogenase deficiency. *Am. J. Hum. Genet.* 2001;68:839–47.

[62] Assmann, B., Gohlich-Ratmann, G., Brautigam, C., Wagner, L., Moolenaar, S., Engelke, U. *et al.* Presumptive ureidopropionase deficiency as a new defect in pyrimidine catabolism found with in vitro H-NMR spectroscopy. *J. Inherit. Metab. Dis.* 1998;21 (Suppl. 2).

[63] Vreken, P., van Kuilenburg, A.B., Hamajima, N., Meinsma, R., van Lenthe, H., Gohlich Ratmann, G. *et al.* cDNA cloning, genomic structure and chromosomal localization of the human BUP-1 gene encoding beta-ureidopropionase. *Biochim. Biophys Acta* 1999;1447:251–7.

[64] van, Kuilenburg, A.B., Meinsma, R., Beke, E., Assmann, B., Ribes, A., Lorente, I. *et al.* beta-Ureidopropionase deficiency: an inborn error of pyrimidine degradation associated with neurological abnormalities. *Hum. Mol. Genet.* 2004;13:2793–801 .

[65] Huck, J.H., Verhoeven, N.M., Struys, E.A., Salomons, G.S., Jakobs, C., van der Knaap, M.S. Ribose-5-phosphate isomerase deficiency: New inborn error in the pentose phosphate pathway associated with a slowly progressive leukoencephalopathy. *Am. J. Hum. Genet.* 2004;74:745–51.

[66] Meshitsuka, S., Koeda, T., Muro, H. Direct observation of 3-keto-valproate in urine by 2D-NMR spectroscopy. *Clin. Chim. Acta* 2003;334:145–51.

[67] Azaroual, N., Imbenotte, M., Cartigny, B., Leclerc, F., Vallee, L., Lhermitte, M., Vermeersch, G. Valproic acid intoxication identified by 1H and 1H-(13)C correlated NMR spectroscopy of urine samples. *MAGMA* 2000;10:177–82.

[68] Komoroski, E.M., Komoroski, R.A., Valentine, J.L., Pearce, J.M., Kearns, G.L. The use of nuclear magnetic resonance spectroscopy in the detection of drug intoxication. *J. Anal. Toxicol.* 2000;24:180–7.

[69] Spraul, M., Hofmann, M., Lindon, J.C., Farrant, R.D., Seddon, M.J., Nicholson, J.K., Wilson, I.D. Evaluation of liquid chromatography coupled with high-field 1H NMR spectroscopy for drug metabolite detection and characterization: The identification of paracetamol metabolites in urine and bile. *NMR Biomed.* 1994;7:295–303.

[70] Godejohann, M., Tseng, L.H., Braumann, U., Fuchser, J., Spraul, M. Characterization of a paracetamol metabolite using on-line LC-SPE-NMR-MS and a cryogenic NMR probe. *J. Chromatogr. A* 2004;1058:191–6.

[71] Connor, S.C., Everett, J.R., Jennings, K.R., Nicholson, J.K., Woodnutt, G. High resolution 1H NMR spectroscopic studies of the metabolism and excretion of ampicillin in rats and amoxycillin in rats and man. *J. Pharm. Pharmacol.* 1994;46:128–34.

[72] Jaroszewski, J.W., Berenstein, D., Slok, F.A., Simonsen, P.E., Agger, M.K. Determination of diethylcarbamazine, an antifilarial drug, in human urine by 1H-NMR spectroscopy. *J. Pharm. Biomed. Anal.* 1996;14:543–9.

[73] Styles, P., Soffe. N.F., Scott, C.A., Cragg, D.A., Row, F., White, D.J., White, P.C.J. A high-resolution NMR probe in which the coil and preamplifier are cooled with liquid helium. *J. Magn. Reson.* 1984;60:397–404.

[74] Keun, H.C., Beckonert, O., Griffin, J.L., Richter, C., Moskau, D., Lindon, J.C., Nicholson, J.K. Cryogenic probe 13C NMR spectroscopy of urine for metabonomic studies. *Anal. Chem.* 2002;74:4588–93.

[75] Tanaka, K., Armitage, I.M., Ramsdell, H.S., Hsia, Y.E., Lipsky, S.R., Rosenberg, L.E. [13C]Valine metabolism in methylmalonicacidemia using nuclear magnetic resonance: propinonate as an obligate intermediate. *Proc. Natl. Acad. Sci. U.S.A.* 1975;72:3692–6.

[76] Spraul, M., Hofmann, M., Dvortsak, P., Nicholson, J.K., Wilson, I.D. Liquid chromatography coupled with high-field proton NMR for profiling human urine for endogenous compounds and drug metabolites. *J. Pharm. Biomed. Anal.* 1992;10:601–5.

[77] Pullen, F.S., Swanson, A.G., Newman, M.J., Richards, D.S. Online liquid chromatography-nuclear magnetic resonance-mass spectrometry – a powerful spectroscopic tool for the analysis of mixtures pf pharmaceutical interest. *Rapid Comm. Mass Spectrom* 1995;9:1003–6.

[78] Lindon, J.C. HPLC-NMR-MS: Past, present and future. *Drug Discov. Today* 2003;8:1021–2.

[79] Lindon, J.C., Nicholson, J.K., Wilson, I.D. Direct coupling of chromatographic separations to NMR spectroscopy. *Prog. Nucl. Magn. Reson. Spec.* 1996;29:1–49.

[80] Burton, K.I., Everett, J.R., Newman, M.J., Pullen, F.S., Richards, D.S., Swanson, A.G. On-line liquid chromatography coupled with high field NMR and mass spectrometry (LC-NMR-MS): A new technique for drug metabolite structure elucidation. *J. Pharm. Biomed. Anal.* 1997;15:1903–12.

[81] Corcoran, O., Spraul, M. LC-NMR-MS in drug discovery. *Drug Discov. Today* 2003;8:624–31.

[82] Spraul, M., Freund, A.S., Nast, R.E., Withers, R.S., Maas, W.E., Corcoran, O. Advancing NMR sensitivity for LC-NMR-MS using a cryoflow probe: application to the analysis of acetaminophen metabolites in urine. *Anal. Chem.* 2003;75:1536–41.

[83] Peck, T.L., Magin, R.L., Lauterbur, P.C. Design and analysis of microcoils for NMR microscopy. *J. Magn. Reson.* B 1995;108:114–24.

[84] Olson, D.L., Peck, T.L., Webb, A.G., Magin, R.L., Sweedler, J.V. High-resolution microcoil 1H-NMR for mass-limited, nanoliter-volume samples. *Science* 1995;270:1967–70.

[85] Schlotterbeck, G., Ross, A., Hochstrasser, R., Senn, H., Kuhn, T., Marek, D., Schett, O. High-resolution capillary tube NMR. A miniaturized 5-microL high-sensitivity TXI probe for mass-limited samples, off-line LC NMR, and HT NMR. *Anal. Chem.* 2002;74:4464–71.

[86] Griffin, J.L., Nicholls, A.W., Keun, H.C., Mortishire Smith, R.J., Nicholson, J.K., Kuehn, T. Metabolic profiling of rodent biological fluids via 1H NMR spectroscopy using a 1 mm microlitre probe. *Analyst* 2002;127:582–4.

[87] Wolters, A.M., Jayawickrama, D.A., Sweedler, J.V. Microscale, N.M.R. *Curr. Opin. Chem. Biol.* 2002;6:711–6.

[88] Liu, M., Nicholson, J.K., Lindon, J.C. High-resolution diffusion and relaxation edited one- and two-dimensional 1H NMR spectroscopy of biological fluids. *Anal. Chem.* 1996;68:3370–6.

[89] de Graaf, R.A., Behar, K.L. Quantitative 1H NMR spectroscopy of blood plasma metabolites. *Anal. Chem.* 2003;75:2100–4.

[90] Liu, M., Tang, H., Nicholson, J.K., Lindon, J.C. Use of ^{1}H NMR-determined diffusion coefficients to characterize lipoprotein fractions in human blood plasma. Magn. Reson Chem. 2002;40:S83–S88.

[91] Liu, M., Nicholson, J.K., Parkinson, J.A., Lindon, J.C. Measurement of biomolecular diffusion coefficients in blood plasma using two-dimensional 1H-1H diffusion-edited total-correlation NMR spectroscopy. *Anal. Chem.* 1997;69:1504–9.

[92] Beckwith Hall, B.M., Thompson, N.A., Nicholson, J.K., Lindon, J.C., Holmes, E. A metabonomic investigation of hepatotoxicity using diffusion-edited 1H NMR spectroscopy of blood serum. *Analyst* 2003;128:814–8.

[93] Spraul, M., Eichhoff, U. NMR analysis of mixtures using hyphenation techniques and software. Russian *Chem. Bull. Int. Ed.* 2003;52:2529–38.

[94] Wishart, D.S., Querengesser, L.M.M., Lefebvre, B.A., Epstein, N.A., Greiner, R., Newton, J.B. Magnetic Resonance Diagnostics: A new technology for high-throughput clinical diagnostics. *Clin. Chem.* 2001;47:1918–21.

[95] Bamforth, F.J., Dorian, V., Vallance, H., Wishart, D.S. Diagnosis of inborn errors of metabolism using 1H NMR spectroscopic analysis of urine. *J. Inherit. Metab. Dis.* 1999;22:297–301.

[96] Dayhoff, J.E., DeLeo, J.M. Artificial neural networks: Opening the black box. *Cancer* 2001;91:1615–35.

[97] Sass, J.O., Schweitzer-Krantz, S., Moebus, R., Engelke, U.F., Wevers, R.A., Omran, H. Aminoacylase I deficiency: A "new" inborn error of metabolism? *J. Inherit. Metab. Dis.* 2005;28:50.

[98] Van Coster, R.N., Lissens, W., Giardina, T., Engelke, U.F., Smet, J., Seneca, S. *et al.* An Infant with large amounts of several N-acetylated amino acids in the urine has aminoacylase 1 deficiency. *J. Inherit. Metab. Dis.* 2005;28:50.

[99] Burns, S.P., Iles, R.A. An investigation of argininosuccinic acid anhydrides in argininosuccinic acid lyase deficiency by 1H-NMR spectroscopy. *Clin. Chim. Acta* 1993;221:1–13.

[100] Pontoni, G., Rotondo, F., Spagnuolo, G., Aurino, M.T., Carteni Farina, M., Zappia, V., Lama, G. Diagnosis and follow-up of cystinuria: use of proton magnetic resonance spectroscopy. *Amino Acids* 2000;19:469–76.

[101] Lehnert, W., Stogmann, W., Engelke, U., Wevers, R.A., van den Berg, G.B. Long-term follow up of a new case of hawkinsinuria. *Eur. J. Pediatr.* 1999;158:578–82.

[102] Brown, J.C., Mills, G.A., Sadler, P.J., Walker, V. 1H NMR studies of urine from premature and sick babies. *Magn. Reson. Med.* 1989;11:193–201.

[103] Burns, S.P., Iles, R.A., Ryalls, M., Leonard, J.V. Methylgenesis from betaine in cystathionine-beta-synthase deficiency. *Biochem. Soc. Trans.* 1993;21:455S.

[104] Lundberg, P., Dudman, N.P., Kuchel, P.W., Wilcken, D.E. 1H NMR determination of urinary betaine in patients with premature vascular disease and mild homocysteinemia. *Clin. Chem.* 1995;41:275–83.

[105] Yamaguchi, S., Koda, N., Eto, Y., Aoki, K. Rapid screening of metabolic disease by proton NMR urinalysis. *Lancet* 1984;2:284.

[106] Moolenaar, S.H., Engelke U.F., Abeling N.G., Mandel H., Duran M., Wevers R.A. Prolidase deficiency diagnosed by 1H NMR spectroscopy of urine. *J. Inherit. Metab. Dis.* 2001;24:843–50.

[107] Farrant, R.D., Walker, V., Mills, G.A., Mellor, J.M., Langley, G.J. Pyridoxal phosphate deactivation by pyrroline-5-carboxylic acid. Increased risk of vitamin B6 deficiency and seizures in hyperprolinemia type II. *J. Biol. Chem.* 2001;276:15107–16.

[108] Onkenhout, W., Groener, J.E., Verhoeven, N.M., Yin, C., Laan, L.A. L-Arabinosuria: A new defect in human pentose metabolism. *Mol. Genet. Metab.* 2002;77:80–5.

[109] van der Knaap, M.S., Wevers, R.A., Struys, E.A., Verhoeven, N.M., Pouwels, P.J., Engelke U.F. *et al.* Leukoencephalopathy associated with a disturbance in the metabolism of polyols. *Ann. Neurol.* 1999;46:925–8.

[110] Baumkotter, J., Cantz, M., Mendla, K., Baumann, W., Friebolin, H., Gehler, J., Spranger, J. N-Acetylneuraminic acid storage disease. *Hum. Genet.* 1985;71:155–9.

[111] Bal, D., Gradowska, W., Gryff Keller, A. Determination of the absolute configuration of 2-hydroxyglutaric acid and 5-oxoproline in urine samples by high-resolution NMR spectroscopy in the presence of chiral lanthanide complexes. *J. Pharm. Biomed. Anal.* 2002;28:1061–71.

[112] Neild, G.H., Foxall, P.J., Lindon, J.C., Holmes, E.C., Nicholson, J.K. Uroscopy in the 21st century: High-field NMR spectroscopy. *Nephrol. Dial. Transplant* 1997;12:404–17.

[113] Moürmans, J., Bakkeren, J., de Jong, J., Wevers, R., van Diggelen, O.P., Suormala, T. *et al.* Isolated (biotin-resistant) 3-methylcrotonyl-CoA carboxylase deficiency: Four sibs devoid of pathology. *J. Inherit. Metab. Dis.* 1995;18:643–5.

[114] Engelke, U.F., Kremer, B., de Jong, J., Schuuring J., van den Berg, A., Loupatty, F. *et al.* Leukoencephalopathy in late onset 3-methylglutaconic aciduria. *J. Inherit. Metab. Dis.* 2004;27:80.

[115] Ghauri, F.Y., Parkes, H.G., Nicholson, J.K., Wilson, I.D., Brenton, D.P. Asymptomatic 5-oxoprolinuria detected by proton magnetic resonance spectroscopy. *Clin. Chem.* 1993;39:1341.

[116] Yamaguchi, S., Koda, N., Yamamoto, H. Analysis for homogentisic acid by NMR spectrometry, to aid diagnosis of alkaptonuria. *Clin. Chem.* 1989;35:1806–7.

[117] Yamaguchi, S., Koda, N., Ohashi, T. Diagnosis of alkaptonuria by NMR urinalysis: rapid qualitative and quantitative analysis of homogentisic acid. *Tohoku J. Exp. Med.* 1986;150:227–8.

[118] Morava, E., Kosztolanyi, G., Engelke, U.F., Wevers, R.A. Reversal of clinical symptoms and radiographic abnormalities with protein restriction and ascorbic acid in alkaptonuria. *Ann. Clin. Biochem.* 2003;40:108–11.

[119] Burlina, A.P., Kunnecke, B., Seelig, J., Burlina, A.B. 1H-NMR spectroscopy of cerebrospinal fluid and urine in Canavan's disease. Proceedings of the 11th Annual Meeting os Society of Magnetic Resonance in Medicine 1992;2:3416.

[120] Rafter, J.E., Chalmers, R.A., Johnson, A., Iles, R.A. Diagnosis of respiratory chain defects using 1H n.m.r. spectroscopy in vitro. *Biochem. Soc. Trans.* 1990;18:551–2.

[121] Burns, S.P., Holmes, H.C., Chalmers, R.A., Johnson, A., Iles, R.A. Proton NMR spectroscopic analysis of multiple acyl-CoA dehydrogenase deficiency – capacity of the choline oxidation pathway for methylation in vivo. *Biochim Biophys Acta* 1998;1406:274–82.

[122] Yamaguchi, S., Koda, N., Eto, Y., Aoki, K. Quick screening and diagnosis of organic acidemia by NMR urinalysis. *J. Pediatr.* 1985;106:620–2.

[123] Lorek, A.K., Penrice, J.M., Cady, E.B., Leonard, J.V., Wyatt, J.S., Iles, R.A. *et al.* Cerebral energy metabolism in isovaleric acidaemia. *Arch. Dis. Child Fetal Neonatal Ed.* 1996;74:F211–F213.

[124] Kodama, S., Sugiura, M., Nakao, H., Kobayashi, K., Miyoshi, M., Yoshii, K. *et al.* 1H-NMR studies of urine in propionic acidemia and methylmalonic acidemia. *Acta Paediatr. Jpn* 1991;33:139–45.

[125] Iles, R.A., Chalmers, R.A., Hind, A.J. Methylmalonic aciduria and propionic acidaemia studied by proton nuclear magnetic resonance spectroscopy. *Clin Chim. Acta* 1986;161:173–89.

[126] Iles, R.A., Chalmers, R.A. Nuclear magnetic resonance spectroscopy in the study of inborn errors of metabolism. *Clin. Sci.* (London) 1988;74:1–10.

[127] Davies, S.E., Iles, R.A., Stacey, T.E., de Sousa, C., Chalmers, R.A. Carnitine therapy and metabolism in the disorders of propionyl-CoA metabolism studied using 1H-NMR spectroscopy. *Clin. Chim. Acta* 1991;204:263–77.

[128] Rafter, J.E., Chalmers, R.A., Iles, R.A. Medium-chain acyl-CoA dehydrogenase deficiency: A 1H-n.m.r. spectroscopic study. *Biochem. Soc. Trans.* 1990;18:912–3.

[129] Wevers, R.A., Engelke, U., Rotteveel, J.J., Heerschap, A., de Jong, J.G., Abeling, N.G. *et al.* 1H NMR spectroscopy of body fluids in patients with inborn errors of purine and pyrimidine metabolism. *J. Inherit. Metab. Dis.* 1997;20:345–50.

[130] Schmidt, C., Engelke, U.F., Wevers, R.A., Anninos, A., Feyh, P., Hofmann, U. *et al.* Proficiency testing results of different methods of pyrimidine analysis. *J. Inherit. Metab. Dis.* 2004;27:235.

[131] Van Hove, J.L., Damme Lombaerts, R., Grunewald, S., Peters, H., Van Damme, B., Fryns, J.P. *et al.* Cobalamin disorder Cbl-C presenting with late-onset thrombotic microangiopathy. *Am. J. Med. Genet.* 2002;111:195–201.

[132] Tassani, M., Valensin, G., Hayek, J., Engelke, U.F., Wevers, R.A. Dimethylglycine dehydrogenase deficiency. *J. Inherit. Metab. Dis.* 2005;28:257.

[133] Nicoli, F., Vion Dury, J., Maloteaux, J.M., Delwaide, C., Confort Gouny, S., Sciaky, M., Cozzone, P.J. CSF and serum metabolic profile of patients with Huntington's chorea: A study by high resolution proton NMR spectroscopy and HPLC. *Neurosci. Lett.* 1993;154:47–51.

[134] Boltshauser, E., Schmitt, B., Wevers, R.A., Engelke, U., Burlina, A.B., Burlina, A.P. Follow-up of a child with hypoacetylaspartia. Neuropediatrics 2004;35:255–8.

[135] Abeling, N.G., van Gennip, A.H., Bakker, H.D., Heerschap, A., Engelke, U., Wevers, R.A. Diagnosis of a new case of trimethylaminuria using direct proton NMR spectroscopy of urine. *J. Inherit. Metab. Dis.* 1995;18:182–4.

[136] Maschke, S., Wahl, A., Azaroual, N., Boulet, O., Crunelle, V., Imbenotte, M. *et al.* 1H-NMR analysis of trimethylamine in urine for the diagnosis of fish-odour syndrome. *Clin. Chim. Acta* 1997;263:139–46.

[137] Mazon, R.A., Gil Setas, A., Berrade, Z.S., Bandres E.T., Wevers, R., Engelke, U., Zschocke, J. Trimetilaminuria o sindrome del olor a pescado. Nueva mutacion genica y primer caso documentado en Espana. [Primary trimethylaminuria or fish odor syndrome. A novel mutation in the first documented case in Spain]. *Med. Clin.* (Barc) 2003;120:219–21.

[138] Murphy, H.C., Dolphin, C.T., Janmohamed, A., Holmes, H.C., Michelakakis, H., Shephard, E.A. *et al.* A novel mutation in the flavin-containing monooxygenase 3 gene, FM03, that causes fish-odour syndrome: Activity of the mutant enzyme assessed by proton NMR spectroscopy. *Pharmacogenetics* 2000;10:439–51.

[139] Matsushita, K., Kato, K., Ohsaka, A., Kanazawa, M., Aizawa, K. A simple and rapid method for detecting trimethylamine in human urine by proton NMR. Physiol Chem Phys Med NMR 1989;21:3–4.

[140] Eugene, M., Diagnostic du "fish odour syndrome" par spectrometrie RMN proton des urines. [Diagnosis of "fish odor syndrome" by urine nuclear magnetic resonance proton spectrometry]. *Ann. Dermatol. Venereol.* 1998;125:210–2.

[141] Nicholson, J.K., Gartland, K.P. 1H NMR studies on protein binding of histidine, tyrosine and phenylalanine in blood plasma. *NMR Biomed.* 1989;2:77–82.

The Handbook of Metabonomics and Metabolomics
John C. Lindon, Jeremy K. Nicholson and Elaine Holmes (Editors)

Chapter 15

A Survey of Metabonomics Approaches for Disease Characterisation

John C. Lindon and Elaine Holmes

Department of Biomolecular Medicine, Faculty of Medicine, Imperial College London, Sir Alexander Fleming Building, South Kensington, London SW7 2AZ UK

15.1. Introduction

In Chapters 12 and 13, the application of metabonomics techniques for the diagnosis of cardiovascular disease and for investigations in cancer have been reviewed and the now comprehensive literature on using NMR spectroscopic techniques for diagnosing inborn errors of metabolism caused by single gene defects has been described in Chapter 14. In this chapter, a survey is given of a wide miscellany of other disease investigations, mainly but not exclusively, studied by ^{1}H NMR spectroscopy of biofluids.

Genomic approaches have been used extensively to attempt disease diagnosis, but mechanistic connectivities between the observed gene expression variation and the disease processes often still need to be established. However, most diseases are multi-factorial in origin, with a range of both genetic and environmental causes and in fact there are very few diseases that are wholly caused by genetic factors even though the initial pathology could be derived from a single gene defect (e.g. Huntington's disease). This con cept has been discussed by Bompressi *et al.* [1], and Figure 15.1 encapsulates this range of disease causes.

Similarly, a number of studies have highlighted differences in patterns of protein expression in disease compared to controls, but again often these biomarker combinations are of proteins of unknown function or even identity.

Humans have hundreds of functionally specialised cell types that can interact in different ways with environmental factors, and thus influence disease development, and incidentally also modulate the effects of drugs, and it is therefore plausible the

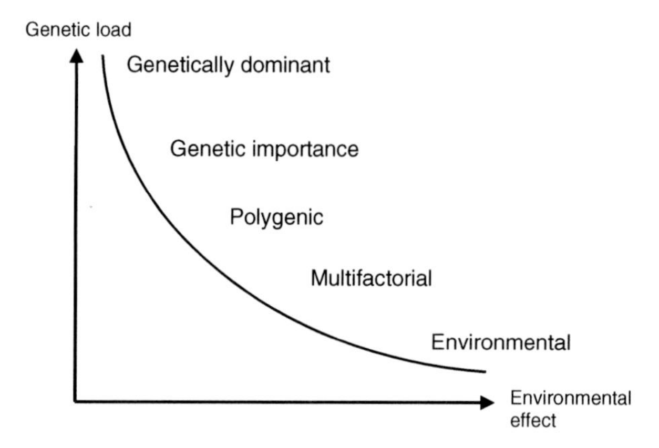

Figure 15.1. The relationship between disease and genetic and environmental factors. Adapted from Bomprezzi *et al. J. Med. Genet.* 40, 553 (2003), who quoted that "in theory there are no diseases completely free from the influence of both genetic and environmental factors" [1].

metabonomics might provide a novel means of modelling these interactions and understanding their significance. However, the relationships between "endogenous" metabolic processes (coded in the genome and intrinsic to cellular function) and foreign compound and food metabolism are poorly understood, especially with respect to environmental factors such as age, gender, tobacco smoking and so on. This problem has been recognised and means for addressing it have been proposed [2]. The concept of "global systems biology" has been proposed and this attempts to integrate multivariate biological information to better understand the interaction of genes with the environment. However, the measurement is difficult at the analytical level and the modelling of such diverse information sets is complex at the bioinformatic level, where in both cases a wide variety of techniques is possible.

Species such as humans have an extra level of complexity, and can be considered "superorganisms". This is because mammals coexist with an internal ecosystem of diverse symbiotic microbiota and parasites that have interactive metabolic processes with the host. It has thus been concluded that novel mathematical, statistical and analytical approaches will be needed to measure and fully model metabolic compartments in interacting cell types and genomes that are connected by such co-metabolic processes in symbiotic mammalian systems [3].

15.2. Healthy individuals

In order to understand the significance of altered biochemistry in diseases, it is necessary to evaluate normal human biochemical variation. This topic is covered in Chapter 11 in more detail, but some selected studies are highlighted here. One study has tried to assess the feasibility of metabonomics in clinical studies [4]. A ^{1}H NMR-based metabonomic analysis was performed on plasma and urine samples obtained from a group of 12 healthy male subjects on two occasions 14 days apart. The subjects were fed a standard diet, and plasma and urine samples were obtained on both days. The ^{1}H NMR spectra were obtained from all the urine and plasma samples, and were analysed using principal components analysis (PCA). In plasma there was relatively little variability between subjects and between study days. In the case of endogenous urinary metabolite profiles there was considerable inter-subject variability, but less intra-subject variation. In addition, in all subjects, diurnal variation was seen in the urine sample profiles. From this, the authors suggested that it would be possible to collect consistent metabonomics data in clinical studies.

A further study assessed the feasibility and comparability of metabonomic data in clinical studies conducted in different countries without dietary restriction [5]. A ^{1}H NMR-based metabonomic analysis was performed on urine samples obtained from two separate studies, both including male and female subjects. The first was on a group of healthy British subjects ($n = 120$), whilst the second was on healthy subjects from two European countries (Britain and Sweden, $n = 30$). The subjects were asked to provide single, early morning urine samples collected on a single occasion. The ^{1}H NMR spectra obtained for urine samples were inspected visually and interpreted using principal components analysis (PCA). This revealed outlier samples within the urine collections and displayed interesting differences, revealing characteristic dietary and cultural features between the subjects of both countries, such as high trimethylamine-*N*-oxide (TMAO) excretion in the Swedish population from eating fish, and high taurine excretion, due to the Atkins diet. This study confirmed that the endogenous urinary profile is subject to distinct cultural and severe dietary influences and that great care needs to be taken in the interpretation of any biomarkers of disease for diagnostic purposes.

One aspect of human metabolism variation that needs to be considered in detail is the presence of gut microflora [6]. These interact extensively with the host through metabolic exchange and co-metabolism of substrates. Such metabolome–metabolome interactions are poorly understood, but might be implicated in the aetiology of many human diseases. The importance of the gut microbiota in influencing the disposition, fate and toxicity of drugs in the host has also been assessed, and it was clear that appropriate consideration of individual human gut microbial activities will be a necessary part of any future personalised health-care efforts, including disease diagnosis and the assessment of therapeutic regimes.

15.3. Effects of nutrition supplements and natural remedies

In many situations where a clinical diagnosis is required, the subject will already be taking a variety of therapeutic agents for a range of conditions, both acute and chronic, and this is more likely as the age of the subject increases. Whilst the effect of drug–drug interactions and the masking of symptoms by drug therapy is well known, the problem of self-administered natural and herbal remedies and the widespread use of vitamins and supplements should also be considered.

Metabonomics has been used to evaluate a range of herbal products and natural food additives. For example, one study described the first metabonomic approach to determining biochemical modifications following dietary intervention in humans using soy isoflavones [7]. An approach based on chemometric analysis of ^{1}H NMR spectra of blood plasma was used to investigate metabolic changes following dietary intervention with soy isoflavones in healthy pre-menopausal women under controlled environmental conditions. Clear differences in the plasma lipoprotein, amino acid and carbohydrate profiles were observed following soy intervention, suggesting a soy-induced alteration in energy metabolism.

A second study also utilising high-resolution ^{1}H NMR spectroscopy in conjunction with chemometric methods (discriminant analysis with orthogonal signal correction) was applied to the study of human biological responses to chamomile tea ingestion [8]. Daily urine samples were collected from volunteers during a 6-week period incorporating a 2-week baseline period, 2 weeks of daily chamomile tea ingestion, and a 2-week post-treatment phase. Although strong inter-subject variation in metabolite profiles was observed, clear differentiation between the samples obtained before and after chamomile ingestion was achieved on the basis of increased urinary excretion of hippurate and glycine with a depleted creatinine concentration. Samples obtained up to 2 weeks after daily chamomile intake formed an isolated cluster in the discriminant analysis map, from which it was inferred that the metabolic effects of chamomile ingestion were prolonged during the 2-week post-dosing period, as shown by the partial least squares (PLS) scores plot given in Figure 15.2. This study highlighted the potential for metabonomic technology in the assessment of nutritional interventions, despite the high degree of variation from genetic and environmental sources.

Indeed, the natural extension to this approach is that adopted by the practice of Chinese traditional medicine where a complex mixture of natural materials is used to treat a disease rather than the use of a single powerful drug. The systems biology approach promises to be a useful tool for increasing the understanding of the mechanism and effectiveness of the Chinese traditional approach [9].

15.4. Genetically modified animal models of disease and genetic disorders

Many diseases have been modelled using animal species and one of the recent promising approaches has been to genetically modify animals, usually mice, by "knocking-in" or "knocking-out" specific genes or mutating a specific gene.

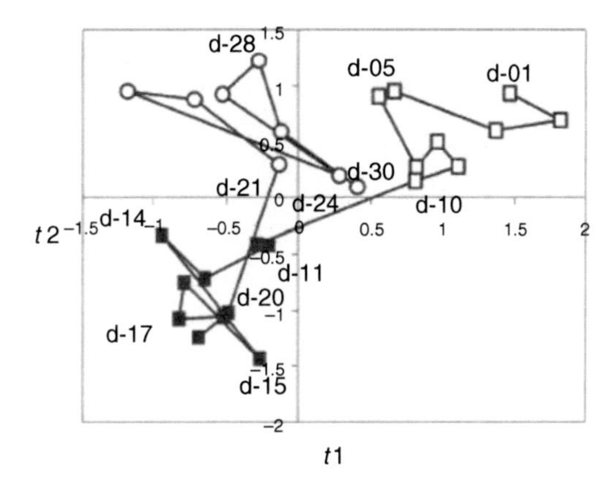

Figure 15.2. Trajectory of PLS scores plot of t[1] versus t[2] for ^1H NMR data after orthogonal signal correction showing the dynamic progression of the urinary metabolite composition and highlighting the separation between, before (open squares), during (filled squares) and after (open circles) chamomile tea ingestion. d indicates the day of sampling, where days d01–d10 represent the pre-chamomile phase, d11–d20 the period of chamomile ingestion, and d21–d30 the recovery or post-chamomile phase [8]. Reproduced with permission.

Examples include the well-known insulin-resistance strains of mice with two main knock-out forms IRS-1 and IRS-2, or mice with a cystic fibrosis gene inserted. Metabonomics can be used to assess how effective the genetic alteration has been and to evaluate the altered phenotype as a function of the gene change.

The functional genomics of a given pathology induced in a genetically modified animal is complex given the biochemical effects that occur when the expression of a given gene or protein is altered. A method of providing metabolite profiles for a disease using pattern recognition coupled with ^1H NMR spectroscopy and exemplified using the mdx mouse, a model of Duchenne muscular dystrophy (DMD) has been suggested [10]. Using ^1H MAS NMR spectroscopy of tissues, dystrophic tissue had distinct metabolic profiles not only for cardiac and other muscle tissues, but also in the cerebral cortex and cerebellum, where the role of dystrophin is still controversial. These metabolite levels were expressed as ratios to demonstrate the effectiveness of the approach at separating dystrophic from control tissue. The cardiac taurine/creatine ratio was 2.08 ± 0.04 in the mdx animals and 1.55 ± 0.04 in the controls with $p < 0.005$. Similarly, the cortex discriminating ratio comprised phosphocholine/taurine, 1.28 ± 0.12 (mdx), 0.83 ± 0.05 (control), $p < 0.01$, and the distinguishing cerebellum ratio was glutamate/creatine 0.49 ± 0.03 (mdx), 0.34 ± 0.03 (controls) with $p < 0.01$. This technique produced new metabolic biomarkers for following disease progression and also demonstrated that many metabolic pathways are perturbed in dystrophic tissue.

One recent study has investigated mouse models of various cardiac diseases [11] in which metabolic profiling of cardiac tissue through high resolution MAS ^{1}H NMR spectroscopy (see Chapter 4) combined with multivariate statistics. The data sets included metabolic profiles from mouse models of DMD, two models of cardiac arrhythmia, and one of cardiac hypertrophy. The metabolic profiles demonstrated that the strain background is an important component of the global metabolic phenotype of a mouse, providing insight into how a given gene deletion may result in very different responses in diverse populations. Despite these differences associated with strain, multivariate statistics were capable of separating each mouse model from its control strain, demonstrating that metabolic profiles could be generated for each disease and it was thought that this approach could be an effective and rapid method of phenotyping mouse models of disease.

An alternative approach is to use mass spectrometry (MS) as an analytical approach and this has been widely employed in studies of inborn errors of metabolism. One recent example is that of Lesch-Nyhan syndrome (LNS) which is caused by a severe deficiency of hypoxanthine-guanine phosphoribosyltransferase (HPRT). This is clinically characterised by self-injurious behaviour and nephrolithiasis, and the latter is treatable with allopurinol, an inhibitor of xanthine oxidase which converts xanthine and hypoxanthine into uric acid. In the HPRT gene, more than 200 different mutations are known and *de novo* mutation occurs at a high rate. Thus, there is a great need to develop a highly specific method to detect patients with HPRT dysfunction by quantifying the metabolites related to this enzyme. A simplified urease pretreatment of urine, gas chromatography-mass spectrometry (GC-MS), and a stable isotope dilution method were applied to quantify hypoxanthine, xanthine, urate, guanine and adenine in 100 μl or less of urine or eluate from filter-paper-urine strips by additional use of stable isotope-labelled guanine and adenine as the internal standards [12]. In this procedure, the recoveries were above 93% and linearities of the analytes were satisfactory. Compared to control means, in four patients with proven disease, hypoxanthine was elevated 8.4–9.0-fold, xanthine 4–6-fold, guanine 1.9–3.7-fold, whilst adenine was decreased. Because of the allopurinol treatment for all the four patients, their level of urate was not elevated, but orotate was increased, although uracil was unchanged as compared with the control value. It was concluded that even in the presence of treatment with allopurinol, patients with this disease can be chemically diagnosed by this procedure.

15.5. Metabonomics in disease studies

15.5.1. Diabetes and obesity

In human beings, insulin deficiency is a relatively common condition which can have serious, complex and far-reaching effects if not treated, and this condition is called *diabetes mellitus*. Diabetes is characterised by polyuria, weight loss in spite of increased appetite, high plasma and urinary levels of glucose, metabolic acidosis, ketosis and coma. The condition is very complex but the fundamental effects of insulin deficiency are: (i) reduced transport of glucose into muscle and other peripheral tissues, whilst at the same time the liver increases excretion of glucose into the blood and (ii) accelerated lipid catabolism (to acetyl-CoA) and decreased lipid synthesis. The muscles and other tissues thus become starved of glucose, whilst highly elevated levels of glucose are found in the urine and plasma. This deficiency in glucose leads to ketosis and acidosis in the same manner as fasting. In many cases, diabetes can be controlled by the administration of insulin. However, patients who develop diabetes in later life are often obese and these individuals do not respond so well to insulin therapy because of an apparent decrease in the number of insulin receptors on their adipose cells (they become insulin resistant).

Some very early metabonomic studies used ^{1}H NMR spectroscopy to analyse the composition of the urine and plasma of a number of diabetic patients [13]. Marked elevations in the plasma levels of the ketone bodies and glucose were seen after insulin withdrawal. The levels of these metabolites, as well as lactate, valine and alanine, were also measured by standard clinical chemistry methods, in order to test the accuracy of the NMR method. In general, the NMR results were in good agreement with the conventional assay results except for acetoacetate. Later, another study reported further results including NMR assays of plasma that showed that the observed levels of certain organic acids such as lactate could be low due to binding to plasma proteins [14]. The CH_3 and CH_2 resonances of the very low density lipoprotein (VLDL) and chylomicron fatty acyl groups in plasma ^{1}H NMR spectra decreased significantly in intensity relative to the CH_3 signal of high density lipoprotein (HDL) and low density lipoprotein (LDL), indicating the rapid metabolism of the mobile pool of triglycerides in VLDL and chylomicrons. Similar marked increases in the levels of ketone bodies and glucose were observed in the urine of the same diabetic patient. A related study showed that human cerebrospinal fluid (CSF) ^{1}H NMR spectra were different between three normal subjects and two mildly diabetic people, with the spectra of the latter group showing elevations in glucose and glycine [15].

The NMR spectrum of the plasma of a non-insulin-dependent obese diabetic was shown to be dominated by signals from glucose and mobile triglycerides in lipoproteins with no signals visible from the ketone bodies. In uncontrolled diabetes, the plasma levels of "free fatty acids", triglycerides and lipoproteins rise markedly due to decreased removal of triglycerides from the plasma into fat depots. This effect was also observed in the case of a diabetic with hypertriglyceridaemia and Type IV hyperlipoproteinaemia [13], showing the very elevated triglyceride levels decreasing as the patient responded to a better diet and insulin regime.

On the basis of these studies, it was concluded that NMR spectroscopy was a useful tool in the diagnosis of diabetes, the monitoring of the development of the disease and the monitoring of patient therapy. No other technique could provide a simultaneous assay on all three ketone bodies, glucose, the lipoproteins and other important biochemicals in one fast and non-destructive test.

Much more recently, capillary gas chromatography combined with pattern recognition analysis of the data has been used for the diagnosis of type-2 diabetes mellitus based on serum lipid metabolites. The results suggested that serum fatty acid profiles, using pattern recognition methods to improve the sensitivity and specificity, might provide an effective approach to the discrimination of type-2 diabetic patients from healthy controls [16].

Liquid chromatography/mass spectrometry (LC/MS) followed by multivariate statistical analysis has also been successfully applied to plasma phospholipid metabolic profiling in type-2 diabetes mellitus [17]. The PCA and partial least squares discriminant analysis (PLS-DA) models were tested and compared for the ability to separate the disease and control groups. The application of an orthogonal signal correction filtered model improved the class distinction and predictive power of PLS-DA models. It was also possible to identify the potential biomarkers with LC-MS-MS. The results are summarised in Figure 15.3 which shows PCA scored plots based on LC-MS data of phospholipids from diabetic and control subjects. Selecting spectral data based on Fisher weights to choose the most discriminating parameters improved the class separation.

Turning to animal models of the disease, several studies have used the Zucker obese rat model. One study has used capillary high performance liquid chromatography (HPLC) and orthogonal acceleration TOF-MS in a metabonomic investigation of urine obtained from male and female Zucker rats [18] as models for diabetes. The results obtained using capillary HPLC for performing the chromatographic separation were also compared with the analysis of the same samples by conventional HPLC. Capillary HPLC-MS gave increased sensitivity (ca. 100-fold) despite using a fraction of the sample volume consumed by HPLC (0.5 μl vs 10 μl). Capillary HPLC-MS also provided a greater peak count (ca. 3000 ions compared to 1500 for HPLC) for the same analysis time compared to conventional HPLC-MS. These data were processed using PLS-DA enabling markers of diurnal variation to be detected for urine samples from both male and female rats.

High visceral fat deposition and intramyocellular lipid levels (IMCL) are both associated with the development of type-2 diabetes. The relationship between these factors has been explored in diet- and glucocorticoid-induced models of insulin resistance [19]. In the diet-induced model, lean and fa/fa Zucker rats were fed either normal or high-fat chow over a 4-week period. Fat distribution, IMCL content in the tibialis anterior muscle and whole body insulin resistance were measured before and after the 4-week period and metabonomics data sets were collected using

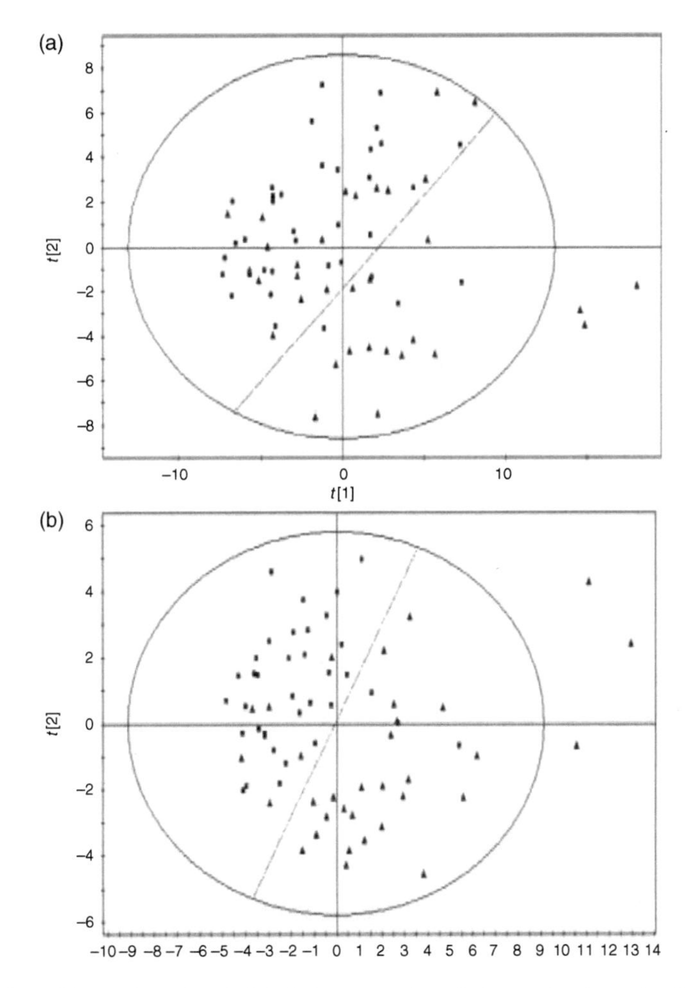

Figure 15.3. The PCA plot based on LC-MS determination of phospholipid species from 34 diabetes mellitus type-2 patients and 35 controls (a) without screening by Fisher weight, (b) with screening by Fisher weight. The solid ellipse denotes the 95% significance limit. The square points are from controls and the triangle points are from diabetes mellitus type-2 patients [17]. Reproduced with permission.

both *in vivo* and plasma ^{1}H NMR spectroscopy. The high fat diet–induced increase in IMCL in the muscle was strongly correlated with visceral fat accumulation and greater glucose intolerance in both groups. The increase in muscle IMCL to visceral fat accumulation was threefold greater for fa/fa rats. In the glucocorticoid-induced model, insulin sensitivity was impaired with dexamethasone. *In vivo* fat accumulation and muscle IMCL content measurements were combined with *ex vivo* analysis of plasma and muscle tissue. The dexamethasone treatment had minimal effects on visceral fat accumulation while increasing muscle IMCL levels compared with controls. Dexamethasone increased plasma glucose by twofold and increased the saturated fatty acid content of plasma lipids. The lipid composition of the muscle was unchanged by dexamethasone treatment, indicating that the relative increase in IMCL observed *in vivo* resulted from a decrease in lipid oxidation. Although, visceral fat might influence IMCL accumulation in the context of dietary manipulations, a causal relationship still remained to be determined. It was considered that dexamethasone-induced insulin resistance likely results from a different mechanism, independent of visceral adiposity.

15.5.2. Arthritis

This family of diseases was first studied many years ago using ^{1}H NMR spectroscopy to measure the levels of a variety of endogenous components in synovial fluid aspirated from the knees of patients with osteoarthritis (OA, $n = 10$), rheumatoid arthritis (RA, $n = 18$) and traumatic effusions (TE, $n = 11$) patients [20].

The spin-echo 400 MHz ^{1}H NMR spectra from synovial fluid showed signals of a large number of endogenous components although many potential markers of inflammation could not be monitored because of their low concentrations or because of the slow molecular dynamics of macromolecular species. The low molecular weight endogenous components showed a wide patient-to-patient variability and showed no statistically significant correlation with disease state. However, correlations were reported between the disease states and the intensities of the *N*-acetyl signals from acute phase glycoproteins such as α_1-acid glycoprotein and haptoglobin. These molecules have a high molecular weight and their signals would not be expected to be observed in spin-echo ^{1}H NMR spectra. However, most glycoproteins contain *N*-acetylneuraminic acid or *N*-acetylglucosaminc units and the *N*-acetyl signals from these units are NMR detectable, if the branched carbohydrate side chains are sufficiently mobile, as appears to be the case. Correlations between the disease state and the levels and type of triglyceride in the synovial fluid were also reported since triglyceride CH_3, CH_2 and vinylic CH group signals were observed in the spectra. In OA, the CH_3 and CH_2 levels were very low compared with RA or TE, and in addition, it was reported that the ratio of intensities CH_2:CH_3 was lower in OA than in RA or TE, thus implying a shorter chain length for the fatty acyl chains. The levels of triglycerides in RA were slightly lower than in TE and thus, on the basis of literature data for triglyceride levels in TE, the RA levels are greater than those expected in controls.

In addition, the ^{1}H NMR spectra of the synovial fluid of a female patient with seronegative erosive RA and of another female patient with sarcoidosis and independent inflammatory OA were followed over the course of several months and standard clinical tests were performed on paired blood serum samples taken at the same time [20]. It was found that the levels of triglyceride CH_3, CH_2 and =CH

NMR signals, and those from glycoprotein *N*-acetyl groups and creatinine all correlated well with one another, and with standard clinical measures of inflammation. The correlation of disease state with creatinine level was of particular interest, and the altered triglyceride composition and concentration in OA was suggested as a potential marker for the disease.

Parkes *et al.* have also conducted ^{1}H NMR studies on synovial fluid from patients with RA [21, 22]. In the first study, the nature of the non-transferrin-bound iron ions in the synovial fluid of patients with inflammatory joint disease was investigated since this form of iron has the adverse effect of stimulating free-radical lipid peroxidation involving oxygen. The 500 MHz spin-echo ^{1}H NMR spectra showed that the incubation of patient synovial fluid with the ferric iron chelator, desferrioxamine, resulted in small increases in intensity of the resonances from citrate, indicating that at least some of the non-transferrin-bound iron is present as ferric citrate. In a second study, 500 MHz spin-echo ^{1}H NMR spectroscopy was used to detect the production of formate and a low molecular weight, *N*-acetyl-containing oligosaccharide, derived from the oxygen radical–mediated depolymerisation of hyaluronate, in the synovial fluid of patients with RA, during exercise of the inflamed joint. Gamma radiolysis of rheumatoid synovial fluid and of aqueous hyaluronate solutions was also shown to produce formate and the oligosaccharide species. It was proposed that the hyaluronate-derived oligosaccharide and formate could be novel markers of reactive oxygen radical injury during hypoxic reperfusion injury in the inflamed rheumatoid joint.

^{13}C NMR spectroscopy has also been employed to monitor synovial fluid from patients with arthritis [23]. In contrast to the ^{1}H NMR studies described above, signals could now be seen from hyaluronic acid, the main determinant of the viscoelasticity of the synovial fluid, even though the molecular weight is in the region 500–1600 kD. The ^{13}C NMR spectra of synovial fluids from patients with RA $(n = 9)$, OA $(n = 7)$, joint trauma and controls $(n = 9$ total, with controls being samples taken from cadavers immediately after death) were compared with one another and with spectra of authentic hyaluronic acid, both before and after the incubation of the latter with hyaluronidase, an enzyme which depolymerises the hyaluronic acid. The synovial fluid NMR spectra from the patients with RA had sharper signals for the $C-1$ and $C-1'$ carbons of hyaluronic acid than those from the OA patients, which in turn exhibited sharper signals than those from the cadavers or the joint trauma patients. Since depolymerisation of the hyaluronic acid was observed to be accompanied by a decrease in the half-band widths of its ^{13}C resonances, the degree of polymerisation of hyaluronic acid was deduced to decrease in the order: controls/joint trauma patients > OA patients > RA patients. Since it is known that the consequence of hyaluronate depolymerisation may be articular cartilage damage, it was concluded that ^{13}C NMR spectroscopy may be a useful method for studying these clinical relevant biophysical changes in synovial fluid.

More recently, an animal model of OA has been investigated by a combination of ^{1}H NMR spectroscopy and multivariate data analysis. Urine samples were obtained from osteoarthritic male Hartley guinea pigs ($n = 44$) at 10 and 12 months of age, treated from 4 months onwards with variable vitamin C doses and from healthy male Strain 13 guinea pigs ($n = 8$) at 12 months of age, treated with vitamin C [24]. An NMR fingerprint was identified that reflected the osteoarthritic changes in the guinea pigs and the component metabolites found indicated that energy and purine metabolism are of major importance in OA. Metabolic fingerprinting also allowed detection of differences in OA-specific metabolites induced by different dietary vitamin C intakes.

15.5.3. Neurological diseases

The chemical composition of the CSF gives a good indication of the health or otherwise of the central nervous system and therefore biofluid NMR studies of such diseases initially focused on CSF.

A ^{1}H NMR spectroscopic study on the CSF of 19 controls, 42 patients with disk herniations and 23 with cerebral tumours has been reported [25]. A semi-quantitative analysis of metabolites was conducted but "relative concentrations" were derived by dividing a given metabolite signal intensity by that of lactate. Since the concentration of the latter metabolite can vary quite widely in normal CSF, let alone in pathological conditions, this could be a misleading procedure. No significant differences were found between the CSFs of controls and disk herniation patients, but the CSFs of the tumour patients were found to have significantly lower glucose/lactate ratios and higher valine/lactate ratios than in the controls. Population overlap prohibited diagnosis based on any one ratio, and so, in a subsequent study [26] with an enlarged patient population ($n = 143$), discriminant analysis using 16 metabolite concentration ratios was performed to see if diagnoses could be made by NMR spectroscopy. The results demonstrated good predictive capability, except for tumour diagnosis, and the poor discrimination of tumours was ascribed to their inherent variability.

The ^{1}H NMR spectra of CSF from three infants with bacterial meningitis have been measured and showed greatly increased lactate, but lowered glucose signals, relative to normal adults [15], with a complete absence of signals from citrate.

From 23 Alzheimer's disease (AD) patients and controls, 500 and 600 MHz ^{1}H NMR data on post-mortem CSF have been obtained [27]. The main differences between the spectra of the two groups were found to be in the region where the resonances of aspartate, *N*-acetyl aspartate, citrate, glutamate and methionine occur. The PCA showed that separation of the two groups was possible and that citrate was the principal marker, with citrate levels in the AD patients much reduced compared with the controls. Patient age and the time interval between death and autopsy were examined to see whether these factors might account for the differences between the AD patients and the control groups. Allowing for these factors, the inter-group differences were reduced but still significant ($p < 0.05$). It was hypothesised that the reduction in CSF citrate found in the AD patients may be due to the reductions in pyruvate dehydrogenase (PDH) reported in the parietal cortex and temporal cortex of AD patients. A decrease in PDH activity could result in a drop in mitochondrial citrate and a corresponding drop in the secretion of citrate to the CSF from the CNS.

In a more recent study, the global high-concentration metabolite composition of CSF has been correlated with patient outcome after sub-arachnoid haemorrhage using multivariate statistics of ^{1}H NMR spectroscopic data [28]. Following sub-arachnoid haemorrhage the most significant complication is sustained cerebral vascular contraction (vasospasm), which may result in terminal brain damage from cerebral infarction, but the biochemical cause of vasospasm remains poorly understood. In total, 16 patients with aneurysmal sub-arachnoid haemorrhage were compared with 16 control patients who required a procedure where CSF was obtained. Multivariate statistics readily distinguished the disease from the heterogeneous control group, even when only those controls with blood contamination in the CSF were used. Using PCA and orthogonal signal correction, vasospasm was correlated to the concentrations of lactate, glucose and glutamine. These pattern recognition models of the NMR data also predicted the Glasgow Coma Score (54% within $+/-1$ of the actual score on a scale of 1–15 for the whole patient group), Hunt and Hess SAH severity score (88% within ±1 of the actual score on a scale of 1–5 for the aSAH group) and cognitive outcome scores (78% within ±3 of the actual score on a 100% scale for the whole patient group).

A parallel transcriptomics, proteomics and metabonomics approach has been employed on human brain tissue to explore molecular disease signatures of schizophrenia [29]. Almost half the altered proteins identified by proteomics were associated with mitochondrial function and oxidative stress responses. This was mirrored by transcriptional and metabolite perturbations. Cluster analysis of transcriptional alterations showed that genes related to energy metabolism and oxidative stress differentiated almost 90% of schizophrenia patients from controls, while confounding drug effects could be ruled out. It was proposed that oxidative stress and the ensuing cellular adaptations are linked to the schizophrenia disease process and that this new disease concept might advance approaches to treatment, diagnosis and disease.

A number of studies have employed NMR-based metabonomics to investigate animal models of neurological disorders. These include a study of focal inflammatory central nervous system lesions induced by micro-injection of a replication-deficient recombinant adenovirus expressing TNF-α or IL-1β cDNA into the brains of Wistar rats as a model of multiple sclerosis [30]. These animals were compared with a group of naive rats and a group of animals injected with an equivalent null adenovirus. Urine samples were collected 7 days after adenovirus injection, when the inflammatory lesion was maximally active. The PCA and PLS discriminant analysis of the urine ^{1}H NMR spectra revealed significant differences between each of the cytokine adenovirus groups and the control groups. For the TNF-α group, the main differences lay in citrate and succinate, while for the IL-1β group, the predominant changes occurred in leucine, isoleucine, valine and *myo*-inositol. Thus, it was possible not only to identify urinary metabolic vectors that separated rats with inflammatory lesions in the brain from control animals, but also to distinguish between different types of brain inflammatory lesions.

Many spinocerebellar ataxias (SCAs) are caused by expansions of cytosine–adenosine–guanine (CAG)-trinucleotide repeats that encode abnormal stretches of polyglutamine. SCA3, or Machado-Joseph disease (MJD), is the commonest dominant inherited ataxia, with pathological phenotypes showing a CAG-triplet repeat length of 61–84. A mouse model of SCA3 containing the MJD gene with a CAG triplet expansion of 84 repeats has been produced, these mice having been shown previously to possess a mild progressive cerebellar deficit. NMR-based metabonomics identified a number of metabolic perturbations in these mice, including a consistent increase in glutamine concentration in both tissue extracts of the cerebellum and cerebrum (see Figure 15.4) and in spectra obtained from intact tissue using magic angle spinning (MAS) ^{1}H NMR spectroscopy. As well as an increase in

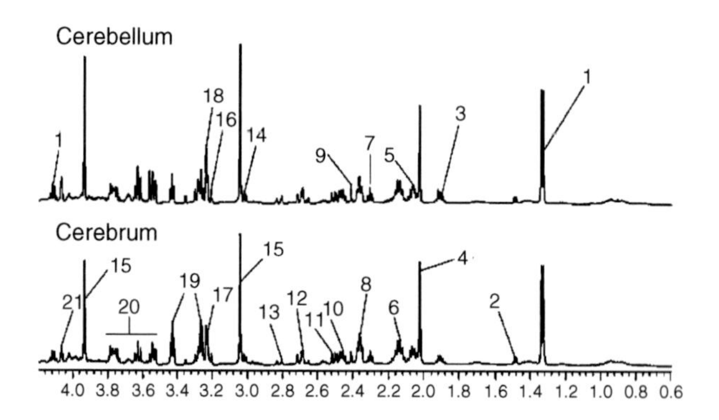

Figure 15.4. High-resolution 600 MHz ^{1}H NMR spectra of aqueous extracts of cerebellum and cerebrum tissue taken from an SCA3 mouse. Each set of resonances represent a different chemical moiety within a given metabolite, with the resonance intensity being proportional to the total concentration of the metabolite. Key: (1) CH_3 of lactate; (2) CH_3 of alanine; (3) $-CH_2$ GABA; (4) *N*-acetylaspartate (NAA); (5) glutamate $-CH_2$ (predominantly); (6) glutamine $-CH_2$; (7) GABA $-CH_2$; (8) glutamate $-CH_2$; (9) succinate CH; (10) glutamine $-CH_2$; (11) glutamate $-CH_2$; (12) aspartate and NAA; (13) aspartate and NAA; (14) GABA $-CH_2$; (15) creatine; (16) choline; (17) phosphocholine; (18) glycerophosphocholine; (19) taurine; (20) *myo*-inositol (with minor contributions from glutamate and glutamine); and (21) *myo*-inositol. [31]. Reproduced with permission.

glutamine, both brain regions demonstrated decreases in GABA, choline, phosphocholine and lactate (representing the summation of lactate *in vivo*, and *post mortem* glycolysis of glucose and glycogen). The metabolic changes were discussed in terms of the formation of neuronal intra-nuclear inclusions associated with SCA3 [31].

A recent study has investigated a mouse model of Batten disease [32], one of a range of neuronal ceroid lipofuscinoses primarily affecting children, where the underlying pathogenesis of the disease is unknown despite some of the causative genes being identified. A knock-out model (Cln3 null mutant) has been used to study this disease through the application of MAS ^{1}H NMR spectroscopy of different brain regions. The null mice were characterised by increased glutamate and decreased GABA in aqueous extracts from three brain regions, and the changes were consistent with the reported altered expression of genes involved in neurotransmitter cycling between glutamate/glutamine and the production of GABA. In addition, variations in the levels of myo-inositiol, creatine and *N*-acetylaspartate were also seen and all of the changes were distinct from those seen during a normal ageing process.

Finally, NMR methods have been used to study whether there are differences in the urinary metabolic profiles between behaviourally distinct groups of rats, namely, dominant and submissive. Dominant-submissive relationships were established in rat pairs competing for access to a feeder filled with sweetened milk. Dominant rats spend significantly longer amounts of time at the feeder than do their submissive partners. During a 2-week period, rats were tested for the dominant-submissive relationship. At the end of the second week, behavioural groups of rats were selected and urine was collected during a 3.5 h time period. PCA of the NMR data revealed galactose as a discriminating factor between rats classified as dominant and those classified as submissive. Independent measurements of galactose showed that the amount present in the urine correlated with the time spent in the feeder zone, thereby supporting the validity of the model, and since the galactose probably arose from the milk feed, proving that the dominant rats ingested more milk [33].

15.5.4. Renal diseases

The first study of renal failure patients analysed the ^{1}H NMR spectra of the blood plasma from nine patients with chronic renal failure during dialysis, two patients in the early stages of renal failure and six normal controls [34]. For patients on acetate dialysis, the method clearly showed how the acetate was accumulated and metabolised during the course of the dialysis, as well as allowing changes in the relative concentrations of endogenous plasma components to be monitored. In 9 of the 11 cases, it was reported that plasma betaine levels were elevated but this was probably a misassignment for TMAO since no resonance was observed for the CH_2 group of betaine. It was later shown [35] that the *N*-methyl resonances of betaine and TMAO are very close at normal plasma pH and that it is the TMAO level that is elevated in renal failure patients. A subsequent ^{1}H, ^{13}C and ^{14}N NMR study of the plasma and urine from 16 chronic renal failure patients [36] showed that the plasma levels of TMAO correlated with those of urea and creatinine, suggesting that the presence of TMAO is closely related to the degree of renal failure and another study also showed increased levels of lactate in the plasma, ascribed to metabolic disturbances (mainly acidoses) associated with decreased renal function [35]. The acidosis either occurs because of decreased renal tubular bicarbonate reabsorption or because of retention of acids. Elevated creatinine levels were found in the plasma of renal failure patients (levels of $>1000\,mmol \cdot l^{-1}$ are possible, compared with $45–120\,mmol \cdot l^{-1}$ for normal subjects). Patients undergoing haemodialysis were differentiated by the presence of elevated dimethylamine, whilst glycine was predominantly raised in the plasma of the peritoneal dialysis patients.

The "uraemic syndrome" is associated with a complex set of biochemical and pathophysiological changes that still today remain poorly understood. The first application of 750 MHz ^{1}H NMR spectroscopy for studying the biochemical composition of plasma from patients on haemodialysis found increased plasma levels of metabolites including methylhistidine, glycerol, choline, TMAO, dimethylamine and formic acid [37]. The concentrations of these metabolites, and ratios to others, varied with the type of dialysis therapy. For example, the biochemical composition of plasma from patients on peritoneal dialysis was remarkably consistent whereas pre- and post-haemodialysis significant fluctuations in the levels of TMAO, glucose, lactate, glycerol, formic acid and lipoproteins (VLDL, LDL and HDL) were observed. Patients taking paracetamol-containing medication long term for pain control had abnormally high plasma levels of paracetamol glucuronide, which might correlate with the presence of anorexia and wasting in this group. Elevated concentrations of glycoprotein fragments were also observed and might relate to the presence of high levels of *N*-acetyl-glucosaminidase (NAG), and other glycoprotein cleaving enzymes in the plasma, originating from damaged kidney cells. These observations led to the hypotheses that the biochemical changes associated with uraemic symptoms and dialysis therapy include the elevation and incomplete removal by dialysis, of a group of toxic methylamine species, the dialysis-dependent alterations in the metabolism of groups of physiologically important compounds such as choline and the lipoproteins and the generation of formic acid.

A statistical analysis of the ^{1}H NMR spectral data from the urines of 52 patients with glomerularnephritis (GN) and 8 healthy volunteers on a pseudo-nephrological diet has been reported [38]. The aim of this detailed study was to determine both the diagnostic and the prognostic value of NMR urinalysis in this condition. Peak intensities in the spectra were normalised relative to the creatinine signal and were given as so-called "excretion indices". This approach was selected because creatinine is not secreted or re-absorbed in the renal tubules and is only filtered in the renal glomeruli. Thus daily creatinine excretion is stable for each person and is largely independent of external factors such as the protein content of meals. Using the excretion indices eliminated the dependence of signal intensities on the glomerular filtration rate and specifically revealed changes in metabolite concentration due to renal tubular dysfunction. The urine of the patients was characterised by increased levels of amino acids, ketone bodies, TMAO, dimethylamine and lactate but decreased levels of citrate and α-ketoglutarate. The elevated amino acids, as well as TMAO and dimethylamine, were interpreted as being due to renal tubular interstitial changes. Progression of the disease was shown to be accompanied by a change in the amino aciduria from non-essential amino acids to essential amino acids. The NMR urinalysis was found to be more sensitive than standard biopsy techniques in certain cases and was capable of establishing tubular and papillary damage in patients with clinically preserved kidney function, especially those with the nephrotic variant of the disease.

Foxall and co-workers have reported a ^{1}H NMR study of the fluid from the cysts of six patients with autosomal dominant polycystic kidney disease (ADPKD) [39]. ADPKD in adults is characterised by the slow progressive growth of cysts in the kidney which are lined by a single layer of renal tubular epithelium. When these cysts reach 1–5 cm in size they can significantly distort the kidney, and disrupt both the blood supply and the renal function, and ADPKD is one of the common causes for renal transplantation in adults. Little was known about the exact biochemical composition of cyst fluids prior to this study, or about the relationship between cyst fluid composition and the pathogenesis of the disease. The NMR spectra of the cyst fluids were assigned by standard methods developed earlier for other biofluids and by the use of spin-echo and 2D J-resolved experiments. The spectra revealed a number of unusual features and showed the cyst fluids to be distinct from both blood plasma and urine. Isoleucine, lysine, threonine and valine were present at millimolar concentrations. High concentrations of acetate, lactate, succinate, creatinine and dimethylamine were also found in the cyst fluids, and in ratios different from those of blood plasma or urine. Glucose concentrations varied from 3.4 to 9.6 mM, and the majority of the fluids contained NMR signals from the *N*-acetyl groups of mobile glycoprotein carbohydrate side chains. Unusually, the fluids from all six patients contained high levels of ethanol, which was not related to consumption of alcoholic beverages or drug preparations. In general there was little variation in the composition of the cyst fluids as revealed by NMR spectroscopy, although the protein signal intensity did vary somewhat. It was hypothesised that this constancy of composition reflected the chronic nature of the accumulation of the cyst fluid and a long turnover time of the cyst components, which thus has the effect of averaging the compositions. The unique biochemical composition of the cyst fluids was ascribed to abnormal transport processes occurring across the cyst epithelial wall, reflecting polarity reversal of the cystic epithelium.

Intact kidney tissue samples of normal and spontaneously hypertensive rats have been analysed by ^{1}H MAS NMR spectroscopy and PCA [40]. Radial components (cortex, outer stripe of the outer medulla, inner stripe of the outer medulla, and papilla) were sampled from various regions across the kidney from multiple animals in order to establish inter- and intra-animal variability. The effects of temperature on the NMR spectra were also measured using studies at 2 and 30 °C. Papilla was differentiated from the other tissue types, and this variation by tissue type was greater than the effect of temperature on the samples. This study also revealed long-term stability issues of tissue storage at −80 °C. The greatest differentiation between normal and hypertensive rats was found in the cortex and the regions in the NMR spectra that were correlated with this variation were identified.

15.5.5. Liver disease

One early study on liver disease was the use of ^{31}P NMR spectroscopy of bile from 17 patients with primary biliary cirrhosis of the liver and from 14 clinically healthy men [41]. Multistage fractional duodenal catheterisation was used to obtain the bile, and the cystic and hepatic portions were examined separately.

In a recent study, a strategy has been developed for applying HPLC and LC-MS-MS to a metabonomics study of liver diseases. One of the key problems to be solved was the matching of the peaks between the chromatograms and so a peak alignment algorithm was developed to match the chromatograms before the multivariate statistical data processing. As an application example, this strategy was applied to metabonomics on liver diseases, and the false-positive result diagnosis of liver cancer from hepatocirrhosis and other hepatitis diseases was effectively reduced to 7.4% [42]. Based on the pattern recognition approach, several potential biomarkers were found and further identified by LC-MS-MS experiments. The structures of eight potential biomarkers were given for distinguishing the liver cancer from the hepatocirrhosis and hepatitis diseases. Because of the selective absorption on the phenylboronic acid affinity column used, the potential biomarkers had to be metabolites with a *cis*-diol group, and the nucleosides were the urinary major compounds in this class.

15.5.6. Organ transplantation

The clinical course following renal transplantation is rarely simple and early dysfunction of the allograft presents a diagnostic dilemma, frequently requiring histopathological examination of a tissue biopsy specimen to determine its aetiology. Because a renal or liver biopsy is a highly invasive procedure, it cannot be repeated frequently, and considerable effort has been directed towards developing non-invasive methods of monitoring allograft function.

^{1}H NMR spectroscopy at 500 and 600 MHz has been used to study the urine samples from 33 kidney transplant patients over a period of 14 days post-operation [43]. For each of the patients, the NMR data were correlated with clinical observations, graft biopsy pathology and conventional renal function tests. All of the patients received prednisolone and cyclosporine-A as immunosuppressive treatment, although none of the patients showed clinical or histopathological signs of cyclosporine-A nephrotoxicity. The aim of the work was to determine whether or not NMR spectroscopy would be a useful method of monitoring renal transplant function. A statistically significant increase in urinary TMAO was found for patients with graft dysfunction when compared with either normal controls or patients with good graft function. However, there was an overlap between the TMAO concentrations of each of the groups, indicating that TMAO measurements alone would not be a reliable marker of graft dysfunction. Urinary levels of dimethylamine were also elevated in the patients with graft dysfunction, but this also was not statistically significant. In patients with clinical evidence of renal ischaemia and acute tubular necrosis, the NMR spectra of their urines showed high levels of lactate, acetoacetate, hippurate and acetoacetate but minimal aminoaciduria. It was concluded that NMR spectroscopy was a useful method for probing biochemical abnormalities following renal transplantation, and that the urinary metabolite profiles monitored by NMR appeared to parallel the graft function.

A similar study reported an analysis of the ^{1}H NMR spectra of urine and plasma from 39 patients following renal transplantation [44]. Peak heights of several urinary metabolite signals were compared with the peak height of the creatinine signal for two groups of patients, divided according to their renal function. Many other NMR parameters were also measured, but it was suggested that the use of three parameters could provide a means of discriminating between patients with rejection, cyclosporin-A nephrotoxicity or cyclosporin-A overdose. The 3 parameters were: (i) the ratio TMAO/creatinine in the urine, (ii) the presence of TMAO in plasma and (iii) the ratio of a peak at $\delta_H \sim 3.7$ to that of creatinine in the urinary ^{1}H NMR spectrum. The signal at $\delta_H \sim 3.7$ was assigned to a polyethylene glycol (PEG) with a molecular weight of around 1000 Da, considered to be released from the glyceride of oleic acid and PEG used in the formulation of the cyclosporin-A, and hence was a marker of drug administration.

From the above two studies, it is apparent that urinary TMAO concentrations are an important marker of renal transplant dysfuntion, although the sensitivity and specificity of the measurement are currently insufficient for clinical diagnosis.

The success of heart transplantation has improved with the chronic use of the immunosuppressant agent cyclosporin-A to suppress rejection processes, and the better treatment of acute rejection episodes when they do occur. However, the early detection of acute cardiac graft rejection can rely on invasive and iterative right ventricular endomyocardial biopsies. In early studies, the line widths of lipoprotein resonances in plasma NMR spectra were measured in an attempt to correlate these with heart graft rejection after transplantation [45, 46]. The NMR measurements were made on plasma from transplant patients who had received a donor organ between 10 and 1700 days prior to this investigation. The line width values were found to vary according to the endomyocardial-biopsy-based Billingham grading of rejection status. Thus it was observed that post-transplantation, the line width values of all patients were lower than those of controls. It was also apparent that there was a direct relationship between the severity of rejection grading and the NMR peak line widths. A third report from the same group contained the results of 410 measurements on a total of 46 patients [47] that gave similar but improved discrimination, and it was concluded that ^{1}H NMR spectroscopy of blood plasma could be a useful non-invasive test to assist in the diagnosis of rejection.

There is a report that the areas of two glycoprotein signals in the spin-echo ^{1}H NMR spectra of blood plasma from heart transplant patients correlated with a standard echocardiography parameter used to monitor rejection [48]. Using the sum of the areas of the *N*-acetyl signals of *N*-acetyl glucosamine and *N*-acetyl neuraminic acid moieties of plasma glycoproteins (NAG + NANA) divided by that of the methyl signal of alanine, and plotting this against isovolumetric left ventricular relaxation times (IVRT, measured by Doppler echocardiography), gave a good correlation in five patients and an acceptable correlation in three, but only a poor correlation in five further patients. It was noted that infections and inflammatory states unrelated to rejection interfered with the correlation. This initial work was followed up with a second report which followed a group of 18 heart transplant recipients over 2.5 years [49]. A similar approach was taken but this time the ratio of the total signal from "glycosylated residues" to that of the "lipoprotein" CH_3 signal was used as well as the (NAG + NANA)/alanine ratio. Again, it was concluded that this NMR approach could be useful in the non-invasive diagnosis of mild to moderate acute heart graft rejection.

de Certaines and co-workers have reported some preliminary results on kidney and liver transplant patients [50]. Plasma line width measurements were reported on 28 samples from 10 kidney transplant patients at up to 15 days post-transplantation, with average significantly lower than those of the controls. They also studied plasma line widths in 21 patients who had undergone bone marrow transplants for medullar aplasia and leukaemia. In this case, the lowest line widths were seen in patients who did not have a graft reaction.

The first application of high resolution MAS ^{1}H NMR spectroscopy to human liver biopsy samples, allowing a determination of their metabolic profiles before removal from donors, during cold perfusion and after implantation into recipients has recently been published [51]. The assignment of peaks observed in the spectra was aided by the use of 2D *J*-resolved, TOCSY and ^{1}H-^{13}C HMQC spectra. The spectra were dominated by resonances from triglycerides, phospholipids and glycogen and from a variety of small molecules including glycerophosphocholine (GPC), glucose, lactate, creatine, acetate, amino acids and nucleoside-related compounds such as uridine and adenosine. In agreement with histological data obtained on the same biopsies, two of the six livers were found to contain high amounts of triglycerides by NMR spectroscopy (see Figure 15.5), which also indicated that these tissues contained

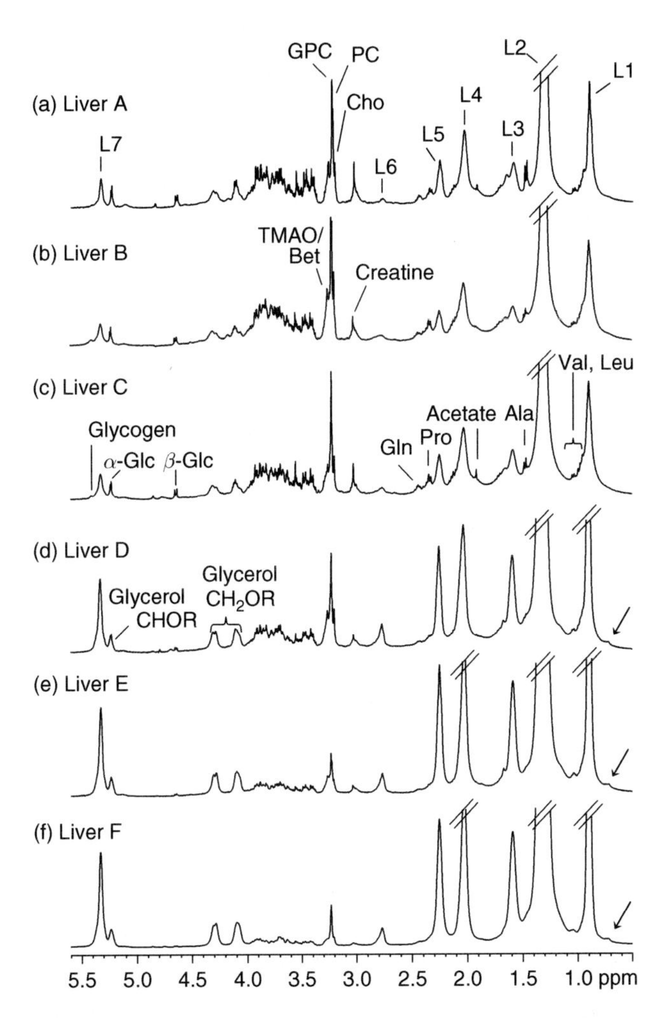

Figure 15.5. The 400 MHz ^1H HR-MAS NMR spectra of human liver biopsies from different donors (rotation rate 4 kHz). Arrows in d–f indicate a signal possibly assigned to cholesterol. Assignment: L1, lipid CH_3; L2, lipid $(CH_2)_n$; L3, lipid CH_2CH_2CO; L4, lipid CH_2–CH=CH; L5, lipid CH_2CO; L6, lipid CH=CH–CH_2–CH=CH; L7, lipid CH=CH; Cho, choline; PC, phosphocholine; GPC, glycerophosphocholine; TMAO, trimethylamine-*N*-oxide; Bet, betaine; Glc, glucose; Val, valine; Leu, leucine; Ala, alanine; Gln, glutamine [51].

a higher degree of unsaturated lipids and a lower proportion of phospholipids and low molecular weight compounds. Additionally, proton T_2 relaxation times indicated two populations of lipids, a higher mobility triglyceride fraction and a lower mobility phospholipid fraction, the proportions of which changed according to the degree of fat content. Glycerophosphocholine was found to decrease from the pre-transplant to the post-transplant biopsy of all livers except for one of the livers with a histologically confirmed high lipid content and this might represent a biomarker of liver function post-transplantation. The NMR signals produced by the liver preservation solution were clearly detected in the cold perfusion–stage biopsies of all livers but remained in the post-transplant spectra of only the two livers with a high lipid content and were prominent mainly in the graft that later developed primary graft dysfunction. This study showed biochemical differences between livers used for transplants that can be related to the degree and type of lipid composition. This technology might therefore provide a novel screening approach for donor organ quality and a means to assess function in the recipient after transplantation.

15.5.7. Infectious diseases

Malaria is caused by several species of *Plasmodium* parasites which enter the human bloodstream after penetration of the skin by Anopheline mosquitos. An early study reported 270 MHz ^{1}H NMR spectroscopic and lactate concentration measurements on the blood sera from 20 Nigerians seropositive to *Plasmodium,* 13 seronegative Nigerians and 6 healthy Japanese controls [52]. Significantly lower lipid line widths and higher lactate concentrations were reported for the sera of the malaria positive group than for the other two groups. No lipoprotein concentration measurements were reported.

Schistosomiasis, a chronic and debilitating parasitic disease, affects approximately 200 million people in the developing world and imposes a substantial public health and economic burden. Accurately diagnosing at the individual level, monitoring disease progression and assessing the impact of pharmacological interventions at the population level are of prime importance for controlling the disease. Using a *Schistosoma mansoni* mouse model, the infection has been investigated using metabolic profiling, employing ^{1}H NMR spectroscopy and multivariate pattern recognition techniques [53]. Ten mice were infected with *S. mansoni* and urine samples were collected 49 and 56 days post-infection. Urine samples were also obtained from 10 uninfected control mice at the same time. The metabolic signature of an *S. mansoni* infection was shown to consist of reduced levels of the tricarboxylic acid cycle intermediates, including citrate, succinate and 2-oxoglutarate, and increased levels of pyruvate, suggesting stimulated glycolysis. A disturbance of amino acid metabolism was also associated with an *S. mansoni* infection, as indicated by depletion of taurine, 2-oxoisocaproate and 2-oxoisovalerate and elevation of tryptophan in the urine.

A range of microbial-related metabolites, including trimethylamine, phenylacetylglycine, acetate, p-cresol glucuronide, butyrate, propionate and hippurate, were also perturbed in the *S. mansoni* infection, indicating disturbances in the gut microbiota. This study highlighted the potential of metabolic profiling to enhance understanding of biological responses to parasitic infections. It also holds promise as a basis for novel diagnostic tests with high sensitivity and specificity and for improved disease surveillance and control.

^1H NMR spectroscopy has been applied to both lumbar CSF collected retrospectively from a cohort of patients with bacterial, fungal or viral meningitis and from control subjects and also to ventricular CSF from patients with ventriculitis associated with an external ventricular drain and from appropriate control subjects [54]. The data were interpreted using PCA, and this was clearly able to distinguish patients with fungal or bacterial from those with viral meningitis and from controls. It also allowed a distinction between post-surgical ventriculitis patients and controls. The metabolites that were responsible for these class differences were identified and shown to be of both host and microbial origin. The metabonomic data also correlated with the onset and time-course of infection in a single patient with two episodes of bacterial ventriculitis and with the response to therapy in another patient with cryptococcal meningitis. The technique was shown to be rapid, to require minimal sample pre-treatment and to be independent of having to target specific microbial pathogens.

A new approach for the identification of pathological changes in scrapie-infected Syrian hamster brains using Fourier transform infrared microspectroscopy has been reported. Using computer-based pattern recognition techniques and imaging, infrared maps with high structural contrast were obtained. This strategy permitted comparison of spectroscopic data from identical anatomical structures in scrapie-infected and control brains. Consistent alterations in membrane state-of-order, protein composition, carbohydrate and nucleic acid constituents were detected in scrapie-infected tissues. Cluster analysis performed on spectra of homogenised medulla oblongata and pons samples also reliably separated uninfected from infected specimens. This method was said to provide a useful tool not only for the exploration of the disease process but also for the development of rapid diagnostic and screening techniques of transmissible spongiform encephalopathies [55].

15.5.8. Infertility

Extensive assignment data for the ^1H NMR spectra of seminal fluid and its component secretions, prostatic and seminal vesicle fluids, have been published [56], and spectra from normal controls have been compared with those from patients with vasal aplasia (obstruction of the *vas deferens* leading to blockage of the seminal vesicles) and those with non-obstructive infertility. The ^1H NMR spectra of the seminal fluid from patients with non-obstructive infertility (all had sperm antibodies) were similar to those of normal subjects. However, the ^{1}H NMR spectra of the seminal fluid from patients with vasal aplasia were grossly different from those of normal subjects and corresponded closely to those of prostatic secretions from normals, due to the lack of seminal vesicle secretion into the fluid.

In the ^{1}H NMR spectra of the vasal aplasia patients, signals from amino acids were either absent or present at very low levels. This is due to the fact that in normal seminal fluid, the amino acids are derived from prostatic peptidase activity on seminal vesicle peptides. Similarly, choline is at a low level or absent in the seminal fluid from vasal aplasia patients, as it derives (indirectly) from the seminal vesicle component. Significant differences were observed between the normal and the vasal aplasia patient groups for the molar ratios citrate/choline and spermine/choline. It was concluded that ^{1}H NMR spectroscopy of seminal fluid could prove to be a useful clinical procedure for the detection and differential diagnosis of vasal aplasia, which otherwise requires minor surgery and/or ultrasound examination.

15.5.9. Cancer

Although Chapter 13 of this book has addressed many of the studies of cancer that have involved metabonomics, that chapter was principally about high resolution MAS NMR studies of tissue samples. Recently a paper has appeared that has used ^{1}H NMR spectroscopy of serum coupled with multivariate statistical approaches to distinguish women with epithelial ovarian cancer (EOC) from healthy controls [57]. In this study, ^{1}H NMR spectra were measured on blood serum from 38 patients with EOC prior to surgery, 12 patients with benign ovarian cysts and 53 healthy women (both pre-and post-menopausal). The spectra were interpreted using PCA and Soft Independent Modelling of Class Analogy (SIMCA). The PCA gave 100% separation of all 38 serum samples from the EOC patients, from all of the pre-menopausal normal samples and from all the spectra derived from patients with benign ovarian disease. Also it was possible to separate 37 out of 38 cancer samples from 31 out of 32 of the post-menopausal control sera. The SIMCA analysis, as visualised using a Cooman's plot, showed that sera from EOC patients, from patients with benign cysts and from the post-menopausal controls did not share a common multivariate metabolic parameter space.

From analysis of significant spectral regions, it was shown that patients with and without disease could be separated based on intensity differences in the ^{1}H NMR spectra at 2.77 and 2.04 ppm. In addition, comparing post-menopausal subject sera to those from patients with EOC, the most discriminating region of the spectra was around 3.7 ppm, but also the regions at 2.25 and 1.18 ppm were important. The authors concluded that the approach deserved further evaluation.

15.5.10. Drug overdose

One of the first studies in this area was by Petroff and co-workers in 1986 [58], who reported a careful and quantitative analysis of the ^{1}H NMR spectra of six CSF samples from three patients, including a 34-year-old man presenting with seizures several hours after injecting heroin and cocaine while intoxicated with alcohol, and a 7-month-old girl who presented as a febrile, cyanotic hypotensive in a coma. This latter patient rapidly developed high intracranial pressure and status epilepticus, and a liver biopsy indicated Reye's syndrome. The ^{1}H NMR spectrum of the CSF of the adult showed elevated signals for citrate, inositol, creatinine/creatine and lactate. The interpretation of the abnormal levels of these and other metabolites in both patients was based on known CSF composition/clinical correlations. It was concluded that NMR analysis of CSF was potentially of great importance in pathophysiological investigations, not least because of its speed and the ability to monitor a wide range of diagnostically important metabolites.

15.6. Concluding remarks

From this chapter and the preceding chapters, it can be seen that metabolic profiling in general and metabonomics in particular have a significant role to play in the

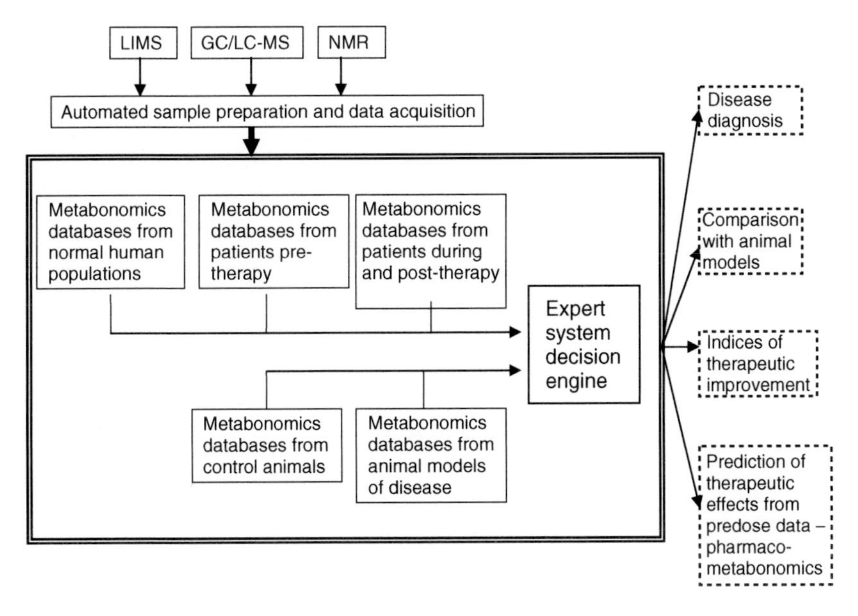

Figure 15.6. Incorporation of metabonomics-based expert systems into clinical diagnosis and for evaluation of disease models and the effectiveness of therapeutic procedures [59].

differential diagnosis of diseases and also in an understanding of the underlying biochemical changes seen in many diseases. The complex interplay between the host metabolic processes under genetic control, the effects of food and nutrition and the effects of the gut microflora all have to be considered as they have complex co-modifying roles. One way in which metabonomics might play a role in disease diagnosis is encapsulated in Figure 15.6 [59].

References

[1] Bomprezzi, R., Kovanen, P.E. and Martin, R. New approaches to investigating heterogeneity in complex traits. *Journal of MedicalGenetics*, **40**, 553–558 (2003).

[2] Nicholson, J.K. and Wilson, I.D. Understanding 'global' systems biology: Metabonomics and the continuum of metabolism. *Nature Reviews Drug Discovery*, **2**, 668–676 (2003).

[3] Nicholson, J.K., Holmes, E., Lindon, J.C. and Wilson, I.D. The challenges of modeling mammalian biocomplexity. *Nature Biotechnology*, **22**, 1268–1274 (2004).

[4] Lenz, E.M., Bright, J., Wilson, I.D., Morgan, S.R. and Nash, A.F.P. A H-1 NMR-based metabonomic study of urine and plasma samples obtained from healthy human subjects. *Journal of Pharmaceutical and Biomedical Analysis*, **33**, 1103–1115 (2003).

[5] Lenz, E.M., Bright, J., Wilson, I.D., Hughes, A., Morrisson, J., Lindberg, H. and Lockton, A. Metabonomics, dietary influences and cultural differences: A H-1 NMR-based study of urine samples obtained from healthy British and Swedish subjects. *Journal of Pharmaceutical and Biomedical Analysis*, **36**, 841–849 (2004).

[6] Nicholson, J.K., Holmes, E. and Wilson, I.D. Gut microorganisms, mammalian metabolism and personalized health care. *Nature Reviews Microbiology*, **3**, 431–438 (2005).

[7] Solanky, K.S., Bailey, N.J.C., Beckwith-Hall, B.M., Davis, A., Bingham, S., Holmes, E., Nicholson, J.K. and Cassidy, A. Application of biofluid H-1 nuclear magnetic resonance-based metabonomic techniques for the analysis of the biochemical effects of dietary isoflavones on human plasma profile. *Analytical Biochemistry*, **323**, 197–204 (2003).

[8] Wang, Y.L., Tang, H.R., Nicholson, J.K., Hylands, P.J., Sampson, J. and Holmes, E. A metabonomic strategy for the detection of the metabolic effects of chamomile (Matricaria recutita L.) ingestion. *Journal of Agricultural and Food Chemistry*, **53**, 191–196 (2005).

[9] Wang, M., Lamers, R.J.A.N., Korthout, H.A.A.J., van Nesselrooij, J.H.J., Witkamp R.F., van der Heijden R., Voshol, P.J., Havekes, L.M., Verpoorte, R. and van der Greef J. Metabolomics in the context of systems biology: Bridging traditional Chinese medicine and molecular pharmacology. *Phytotherapy Research*, **19**, 173–182 (2005).

[10] Griffin, J.L., Williams, H.J., Sang, E., Clarke, K., Rae, C. and Nicholson, J.K. Metabolic profiling of genetic disorders: A multi-tissue H-1 nuclear magnetic resonance spectroscopic and pattern recognition study into dystrophic tissue. *Analytical Biochemistry*, **293**, 16–21 (2001).

[11] Jones, G.L.A.H., Sang, E., Goddard, C., Mortishire-Smith, R.J., Sweatman, B.C., Haselden, J.N., Davies, K., Grace, A.A., Clarke, K. and Griffin, J.L. A functional analysis of mouse models of cardiac disease through metabolic profiling. *Journal of Biological Chemistry*, **280**, 7530–7539 (2005).

[12] Ohdoi, C., Nyhan, W.L. and Kuhara, T. Chemical diagnosis of Lesch-Nyhan syndrome using gas chromatography-mass spectrometry detection. *Journal of Chromatography B*, **792**, 123–130 (2003).

[13] Nicholson, J.K., O'Flynn, M.P., Sadler, P.J., Macleod, A.F., Juul, S.M. and Sönksen P.H. Proton nuclear magnetic resonance studies of serum, plasma and urine from fasting normal and diabetic subjects. *Biochemical Journal*, **217**, 365–375 (1984).

[14] Bell, J.D., Brown, J.C.C. and Sadler, P.J. NMR studies of body fluids. *NMR in Biomedicine*, **2**, 246–256 (1989).

[15] Bell, J.D., Brown, J.C.C., Sadler, P.J., Macleod, A.F., Sönksen, P.H., Hughes, R.D. and Williams, R. High resolution proton nuclear magnetic resonance studies of human cerebrospinal fluid. *Clinical Science*, **72**, 563–570 (1987).

[16] Yang, J., Xu, G.W., Hong, Q.F., Liebich, H.M., Lutz, K., Schmulling, R.M. and Wahl, H.G. Discrimination of Type 2 diabetic patients from healthy controls by using metabonomics method based on their serum fatty acid profiles. *Journal of Chromatography B*, **813**, 53–58 (2004).

[17] Wang, C., Kong, H.W., Guan, Y.F., Yang, J., Gu, J.R., Yang, S.L. and Xu, G.W. Plasma phospholipid metabolic profiling and biomarkers of type 2 diabetes mellitus based on high-performance liquid chromatography/electrospray mass spectrometry and multivariate statistical analysis. *Analytical Chemistry*, **77**, 4108–4116 (2005).

[18] Granger, J., Plumb, R., Castro-Perez, J. and Wilson, I.D. Metabonomic studies comparing capillary and conventional HPLC-oa-TOF MS for the analysis of urine from Zucker obese rats. *Chromatographia*, **61**, 375–380 (2005).

[19] Korach-Andre, M., Gao, J., Gounarides, J.S., Deacon, R., Islam, A. and Laurent, D. Relationship between visceral adiposity and intramyocellular lipid content in two rat models of insulin resistance. *American Journal of Physiology-Endocrinology and Metabolism*, **288**, E106–E116 (2005).

[20] Williamson, M.P., Humm, G. and Crisp, A.J. ^{1}H nuclear magnetic resonance investigation of synovial fluid components in osteoarthritis, rheumatoid arthritis and traumatic effusions. *British Journal of Rheumatology*, **28**, 23–27 (1989).

[21] Parkes, H.G., Allen, R.E., Furst, A., Blake, D.R. and Grootveld, M. Speciation of non-transferrin bound iron ions in synovial fluid from patients with rheumatoid arthritis by proton nuclear magnetic resonance spectroscopy, *Journal of Pharmaceutical and Biomedical Analysis*, **9**, 29–32 (1991).

[22] Parkes, H.G., Grootveld, M., Henderson, E.B., Farrell, A. and Blake, D.R. Oxidative damage to synovial fluid from the inflamed rheumatoid joint detected by ^{1}H NMR spectroscopy. *Journal of Pharmaceutical and Biomedical Analysis*, **9**, 75–82 (1991).

[23] Albert, K., Michele, S., Günther, U., Fial M., Gall, H. and Saal, J. ^{13}C NMR investigation of synovial fluids, *Magnetic Resonance in Medicine*, **30**, 236–240 (1993).

[24] Lamers, R.J.A.N., De Groot, J., Spies-Faber, E.J., Jellema, R.H., Kraus, V.B., Verzijl, N., Te Koppele, J.M., Spijksma, G.K., Vogels, J.T.W.E., van der Greef, J. and van Nesselrooij, J.H.J. Identification of disease- and nutrient-related metabolic fingerprints in osteoarthritic guinea pigs. *Journal of Nutrition*, **133**, 1776–1780 (2003).

[25] Koschorek, F., Gremmel, H., Stelten, J., Offermann, W. and Liebfritz, D. Cerebrospinal fluid: Detection of tumours and disk herniations with MR spectroscopy. *Neuroradiology*, **167**, 813–816 (1988).

[26] Leibfritz, D., Koschorek, F., Stelten, J. and Offermann, W. CSF: Detection of tumours and disk herniations with MRS, in *Magnetic Resonance Spectroscopy of Biofluids, a new tool in clinical biology*, Ed. J.D. de Certaines, World Scientific, Singapore. pp. 167–178 (1989).

[27] Ghauri, F.Y.K., Nicholson, J.K., Sweatman, B.C., Wood, J., Beddell, C.R., Lindon, J.C. and Cairns, N.J. NMR Spectroscopy of human post mortem cerebrospinal fluid: Distinction of Alzheimer's disease from control using pattern recognition and statistics. *NMR in Biomedicine*, **6**, 163–167 (1993).

[28] Dunne, V.G., Bhattachayya, S., Besser, M., Rae, C. and Griffin, J.L. Metabolites from cerebrospinal fluid in aneurysmal subarachnoid haemorrhage correlate with vasospasm and clinical outcome: A pattern-recognition H-1 NMR study. *NMR in Biomedicine*, **18**, 24–33 (2005).

[29] Prabakaran, S., Swatton, J.E., Ryan, M.M., Huffaker, S.J., Huang, J.T.J., Griffin, J.L., Wayland, M., Freeman, T., Dudbridge, F., Lilley, K.S., Karp, N.A., Hester, S., Tkachev, D., Mimmack, M.L., Yolken, R.H., Webster, M.J., Torrey, E.F. and Bahn, S. Mitochondrial dysfunction in schizophrenia: evidence for compromised brain metabolism and oxidative stress. *Molecular Psychiatry*, **9**, 684–697 (2004).

[30] Griffin, J.L., Anthony, D.C., Campbell, S.J., Gauldie, J., Pitossi, F., Styles, P. and Sibson, N.R. Study of cytokine induced neuropathology by high resolution proton NMR spectroscopy of rat urine. *FEBS Letters*, **568**, 49–54 (2004).

[31] Griffin, J.L., Cemal, C.K. and Pook, MA. Defining a metabolic phenotype in the brain of a transgenic mouse model of spinocerebellar ataxia 3. *Physiological Genomics*, **16**, 334–340 (2004).

[32] Pears, M.R., Cooper, J.D., Mitchison, H.M., Mortishire-Smith, R.J., Pearce, D.A. and Griffin, J.L. High resolution ^{1}H NMR-based metabolomics indicates a neurotransmitter cycling defect in cerebral tissue from a mouse model of Batten disease. *Journal of Biological Chemistry*, **280**, 42508–42514 (2005).

[33] Leo, G.C., Caldwell, G.W., Crooke, J., Malatynska, E., Cotto, C., Hastings, B., Scowcroft., Hall, J., Browne, K. and Hageman, W. The application of nuclear magnetic resonance-based metabonomics to the dominant-submissive rat behavioral model. *Analytical Biochemistry*, **339**, 174–178 (2005).

[34] Grasdalen, H., Belton, P.S., Pryor, J.S. and Rich, G.T. Quantitative proton magnetic resonance of plasma from uraemic patients during dialysis, *Magnetic Resonance in Chemistry*, **25**, 811–816 (1987).

[35] Holmes, E., Foxall, P.J.D. and Nicholson, J.K. Proton NMR analysis of plasma from renal failure patients: Evaluation of sample preparation and spectral-editing methods. *Journal of Pharmaceutical and Biomedical Analysis*, **8**, 955–958 (1990).

[36] Bell, J.D., Lee, J.A., Sadler, P.J., Wilkie, D.R. and Woodham, R.H. Nuclear magnetic resonance studies of blood plasma and urine from subjects with chronic renal failure: Identification of trimethylamine-N-oxide. *Biochimica Biophysica Acta*, **109**, 101–107 (1991).

[37] Lindon, J.C., Nicholson, J.K. and Everett, J.R. NMR spectroscopy of biofluids. *Annual Reports on NMR Spectroscopy* (Ed. Webb G.A.), **38**, 1–88 (1999).

[38] Knubovets, T.L., Lundina, T.A., Sibeldina, L.A. and Sedov, K.R. ^{1}H NMR urinalysis in glomerulonephritis: A new prognostic criterion, *Magnetic Resonance Imaging*, **10**, 127–134 (1992).

[39] Foxall, P.J.D., Price, R.G., Jones, J.K., Neild, G.H., Thompson, F.D. and Nicholson, J.K. High resolution proton magnetic resonance spectroscopy of cyst fluid from patients with polycystic kidney disease. *Biochimica Biophysica Acta*, **1138**, 305–314 (1992).

[40] Huhn, S.D., Szabo, C.M., Gass, J.H. and Manzi, AE. Metabolic profiling of normal and hypertensive rat kidney tissues by hr-MAS-NMR spectroscopy. *Analytical and Bioanalytical Chemistry*, **378**, 1511–1519 (2004).

[41] Khristianovitch, D.S., Reshetnyak, V.I., Loginov, A.S., Yushmanov, V.E., Tumanyan, M.A. and Sibeldina, L.A. ^{31}P-NMR spectroscopy of human liver and bile. *Byulleten' Éksperimental'noi Biologii i Meditsiny*, **106**, 678–681 (1988).

[42] Yang, J., Xu, G.W., Zheng, W.F., Kong, H.W., Wang, C., Zhao, X.J. and Pang, T. Strategy for metabonomics research based on high-performance liquid chromatography and liquid chromatography coupled with tandem mass spectrometry. *Journal of Chromatography A*, **1084**, 214–221 (2005).

[43] Foxall, P.J.D., Mellotte, G.J., Bending, M.R., Lindon, J.C. and Nicholson J.K. NMR spectroscopy as a novel approach to the monitoring of renal transplant function. *Kidney International*, **43**, 234–245 (1993).

[44] Le Moyec, L., Pruna, A., Eugene, M., Bedrossian, J., Idatte, J.M., Huneau, J.F. and Tomé, D. Proton nuclear magnetic resonance spectroscopy of urine and plasma in renal transplantation follow-up. *Nephron*, **65**, 433–439 (1993).

[45] Eugene, M., de Certaines, J., Le Moyec, L., Le Rumeur, E., Desruennes, M., Lechat, P. and Cabrol, C. Spectroscopie du plasma par resonance magnetique nucléaire du proton au cours de la transplantation cardiaque. *Comptes Rendu Academie de Science Paris*, **307 (III)**, 41–45 (1988).

[46] Eugene, M., Le Moyec, L. and de Certaines, J.D. Proton NMR spectroscopy of plasma in organ transplantation, in *Magnetic Resonance Spectroscopy of Biofluids, a new tool in clinical biology*, Ed. de Certaines J.D., World Scientific, Singapore, pp. 179–197, (1989).

[47] Eugene, M., Le Moyec, L., de Certaines, J.D., Desruennes, M., Le Rumeur, E., Fraysse, J.B. and Cabrol, C. Lipoproteins in heart transplantation: proton magnetic resonance spectroscopy of plasma, *Magnetic Resonance in Medicine*, **18**, 93–101 (1991).

[48] Pont, H., Vion-Dury, J., Kriat, M., Mouly-Bandini, A., Sciaky, M., Viout, P., Confort-Gouny, S., Messana, T., Goudart, M., Monties, J.R. and Cozzone, P.J. NMR spectroscopy of plasma during acute rejection of transplanted hearts. *The Lancet*, **337**, 792–793 (1991).

[49] Vion-Dury, J., Mouly-Bandini, A., Viout, P., Sciaky, M., Confort-Gouny, S., Monties, J. and Cozzone, P. Early detection of heart transplant rejection using cardiac echography combined with the assay of glycosylated residues in plasma by proton NMR spectroscopy. *Comptes Rendus de l'Academie des Science Serie III*, **315**, 479–484 (1992).

[50] de Certaines, J.D. High resolution magnetic resonance spectroscopy in clinical biology, *Leukemia*, **4**, 667–670 (1990).

[51] Duarte, I.F., Stanley, E.G., Holmes, E., Lindon, J.C., Gil, A.M., Tang, H., Ferdinand, R., Gavaghan-McKee, C., Nicholson, J.K., Vilca-Melendez, H., Heaton, N. and Murphy, G.M. Metabolic assessment of human liver transplants from biopsy samples at the donor and recipient stages using high resolution magic angle spinning ^{1}H NMR spectroscopy. *Analytical Chemistry*, **77**, 5570–5578 (2005).

[52] Nishina, M., Hori, E., Matsushita, K., Takahashi, M. and Ohsaka, A. ^{1}H-NMR spectroscopic study of serums from patients with malaria, *Physiological Chemistry and Physics and Medical NMR*, **20**, 269–271 (1988).

[53] Wang, Y.L., Holmes, E., Nicholson, J.K., Cloarec, O., Chollet, J., Tanner, M., Singer, B.H. and Utzinger, J. Metabonomic investigations in mice infected with Schistosoma mansoni: An approach for biomarker identification. *Proceedings of the National Academy of Sciences of the United States of America*, **101**, 12676–12681 (2004).

[54] Coen, M., O'Sullivan, M., Bubb, W.A., Kuchel, P.W. and Sorrell, T. Proton nuclear magnetic resonance-based metabonomics for rapid diagnosis of meningitis and ventriculitis. *Clinical and Infectious Diseases*, **41**, 1582–1590 (2005).

[55] Kneipp, J., Lasch, P., Baldauf, E., Beekes, M. and Naumann, D. Detection of pathological molecular alterations in scrapie-infected hamster brain by Fourier transform infrared (FT-IR) spectroscopy. *Biochimica Biophysica Acta*, **1501**, 189–199 (2000).

[56] Lynch, M.J., Masters, J., Pryor, J.P., Lindon, J.C., Spraul, M., Foxall, P.J.D. and Nicholson, J.K. Ultra high field NMR spectroscopic studies on human seminal fluid, seminal vesicle and prostatic secretions. *Journal of Pharmaceutical and Biomedical Analysis*, **12**, 5–19 (1994).

[57] Odunsi, K., Wollman, R.M., Ambrosone, C.B., Hutson, A., McCann, S.E., Tammela, J., Geisler, J.P., Miller, G., Sellers, T., Cliby, W., Qian, F., Keitz, B., Intengan, M., Lele, S

and Alderfer, J.L. Detection of epithelial ovarian cancer using 1H-NMR-based metabonomics. *International Journal of Cancer*, **113**, 782–788 (2005).

[58] Petroff, O.A.C., Yu, R.K. and Ogino, T. High-resolution proton magnetic resonance analysis of human cerebrospinal fluid, *Journal of Neurochemistry*, **47**, 1270–1276 (1986).

[59] Lindon, J.C., Holmes, E. and Nicholson, J.K. Metabonomics and its role in drug development and disease diagnosis. *Expert Review of Molecular Diagnostics*, **4**, 189–199 (2004).

The Handbook of Metabonomics and Metabolomics
John C. Lindon, Jeremy K. Nicholson and Elaine Holmes (Editors)

Chapter 16

Metabolic Profiling: Applications in Plant Science

Richard N. Trethewey and Arno J. Krotzky

metanomics GmbH and metanomics Health GmbH Tegeler Weg 33, 10589 Berlin, Germany

16.1. Introduction

Metabolite profiling has a long history in the plant sciences, no doubt because the investigation of metabolism has remained at the forefront of this discipline since the groundbreaking work of Calvin. The first plant profilers were focused on applying the technology to commercial objectives: the development of herbicides with novel modes of action. However, as the technologies of profiling developed, and the entry-level cost reduced, the 1990s saw a new wave of metabolite profiling studies from plant scientists. In addition, the rapid progress in plant genomics coupled to the ability to generate a wide range of transgenic plants has provided the possibility for directed experimentation in plant metabolism in a way that is not possible in the medical, pharmaceutical or environmental sciences. Today we have reached a "golden age" of metabolite profiling where there are around 25 groups working across the world on applying metabolite profiling technologies to an ever increasing range of biological questions in the plant sciences. Indeed, in many cases metabolite profiling technologies themselves have been pioneered and sharpened on plant applications. In this chapter, we shall briefly introduce the technologies of plant metabolite profiling and focus on reviewing the major contemporary applications.

16.2. The plant metabolome

16.2.1. How big is the plant metabolome?

Plants are static, anchored in the ground: they cannot move to seek food or avoid environmental or biological stresses. This requires that plants be self-sufficient and manifests itself in the complexity of plant metabolism. In principle, metabolites can be divided into three functional groups. There are the molecules that participate in metabolite pathways; necessary chemical structures that are created by enzymatic processes. The second functional group are the small molecules which serve as building blocks for more complex macromolecules such as DNA, protein or polysaccharides. Finally, there are a range of small molecules that serve a signalling or defense function and transmit information within or across cells. It is a matter of definition about which small molecules constitute metabolites, and the biological community will need to sharpen these definitions in the future. However, following the nomenclatures established recently for genomics, the term "metabolome" has recently acquired acceptance for the total metabolite complement of a cell, tissue or organism [1]. Given the fact that plant metabolism is more extensive and complex than bacterial or animal metabolism it is very likely that plant metabolomes are particularly large. Further, it is becoming increasingly clear that the specificity of enzymes involved in secondary metabolism, present only in plants, is often low [2]. Consequently for any given enzyme involved in secondary metabolism, for example glucosyltransferases or P450 monooxygenases, there can be multiple metabolite substrates and products. Estimates of metabolome size for an individual species vary in the 5,000–25,000 range, whilst in total more than 100,000 metabolites have been identified in the plant kingdom [3, 4]. This latter figure may well be less than 10% of the total as recent discoveries in natural product chemistry indicate that we only have a superficial view of the total diversity [5].

16.2.2. Regulation of the plant metabolome

The regulation of metabolism, or more specifically the regulation of the relative abundance of metabolites and small molecules, is a highly complex process which is today only poorly understood. Part of the problem is that the study of metabolic flux in plants is both technically very demanding and has been generally neglected [6, 7]. Further, steady state measurements of metabolite levels, if limited in the number of time points or the range of the metabolite network covered, do not ultimately enable conclusions about changes in flux to be drawn. Even in humans, where the golden age of biochemistry has long passed, we have only a rudimentary understanding of metabolic regulation.

Whilst most metabolic study focuses on the analysis of individual pathways, it is already clear that metabolism is best studied as a network [8]. In particular, the importance of signalling has been historically underestimated and there is increasing evidence of the extent to which changes in metabolite levels can cause changes in gene or protein function. These changes may in turn influence the performance of the metabolic network and the levels of individual metabolites. Further, the importance of small molecules as signals between the cell and the environment and between cells of different organs is becoming increasingly recognized [9, 10]. Metabolite profiling is one of the best contemporary approaches to generate the data required in order to study metabolism as a network [11]. As the acceptance of the need to study metabolism on a broad network basis increases then so will the range and number of applications of metabolite profiling technology.

16.3. Metabolite profiling technologies

The terms "metabolite profiling" or "metabolomics" are used with respect to a wide range of methods and technologies in the plant sciences. Essentially, two different classes of metabolite profiling can be distinguished: targeted and broad. What are today termed targeted metabolite profiling methods often have very old origins, and are focused on the study of a small number of chemically similar metabolites. Often targeted profiling methods can be fully calibrated and deliver an absolute quantification of the abundance of the metabolite under study and in many cases have a long history of application (e.g. in the study of plant secondary metabolism). Increasingly there are papers appearing that claim to be applying modern metabolomics techniques but behind the façade is simply an old targeted analysis method. The major innovation in the last years has been the development of broad metabolite profiling methods. These differ from the targeted methods in several key aspects: (i) a wider range of chemistries are covered, often many hundreds of analytes are studied in a single analysis; (ii) detection and measurement occurs for chemistries which differ in abundance levels by many orders of magnitude; (iii) typically more than half of the analytes detected are currently not known in their chemical structure; and (iv) absolute calibration is not possible or can be done only on a limited scale, and thus semi-quantitative data is provided on each metabolite (i.e. abundance with respect to control measurements). Whilst the jump to following hundreds of metabolites in broad metabolite profiling methodologies has been a big one, and it seriously challenges our capabilities for data processing and interpretation, it should be remembered that there are no methods that are currently capable of delivering data on more than a small fraction of the metabolome in a single analysis. In contrast to transcriptomics and proteomics, where only one chemical class of molecule is being analyzed, metabolomics has to analyze molecules with a very wide range of chemical and physical properties. This fact makes it unlikely that we will see in the future a single methodology that is capable of providing a read out on the complete metabolome.

The fact that in contemporary plant metabolomics, broad metabolite profiling methods give rise to profiles where a significant number of metabolites are unknown in chemical structure represents a considerable challenge [1, 12]. There will have to be concerted action from the plant scientific community to address this problem through the generation of libraries of standard substances and analytical information [1]. Nevertheless it is still productive to follow the "known unknowns" in metabolomic analysis as if they deliver reliable, statistically sound, data then they can be used in multivariate analysis to classify particular groups. In addition, the known unknowns have the potential to be biomarkers for particular biological processes. Once identified, such biomarkers can then be subject to targeted investigation in order to elucidate the underlying chemical structure. In the case of industrial applications of metabolite profiling, it is important to precisely define which metabolites need to be measured in a particular project and adapt methods to give an optimal detection and quantification of these key compounds. In commercially orientated projects it is normally the case that key metabolites will be available as standards, and calibration to enable quantification can therefore be performed. In the case of plant biotechnology projects, primary metabolism is often at the forefront of interest, for example fatty acids, amino acids, organic acids, sugars and vitamins. Certain specific projects may target secondary metabolites (e.g. wine constituents), or particular signalling or regulatory molecules (e.g. phytohormones) may play a key role. Therefore it is sufficient for most commercial applications to work with metabolite profiles that contain a few hundreds of metabolites, only a fraction of the complete metabolome. Nevertheless the interest in industry in new innovative metabolite profiling methodologies that enhance metabolite coverage is high.

Whilst there has been a long historical tradition in the use of nuclear magnetic resonance (NMR) spectroscopy for plant metabolite profiling [13], mass spectrometry (MS) has now overtaken NMR as the state of the art in plant metabolomics, primarily because of the generally higher sensitivity of the latter technique. There are, however, a confusing range of hardware and methods available to the metabolite profiler [14] and we will only summarize the key technologies here.

16.3.1. Gas chromatography-Mass spectrometry-based metabolite profiling

The workhorse of contemporary plant metabolite profiling is no doubt gas chromatography coupled to mass spectrometry (GC-MS). GC-MS-based profiling has a long and extensive history in the medical sciences [15, 16]. Sauter, working at BASF AG, was the first to pick up on this methodology and apply it to a problem in the plant sciences (Section 16.4; [17]). However, initially the field was limited by the separation capabilities of GC columns, the expense of the MS systems, and ultimately the affordability of computational power. This restricted the early adoption of the technology to a few groups that had the financial resources and necessary technical expertise. In the mid-1990s these barriers slowly disappeared due to the massive drop in the cost of computational power, the consequent improvement in analytical software and significant developments in the engineering of GC-MS systems. In the 1990s GC-MS became a standard technology across a wide range of laboratories and crucially no longer required highly trained and dedicated technical staff to operate. GC-MS is now the ideal starting point for any research group looking for a relatively cheap entry into the field of metabolite profiling. Indeed, most of the applications of broad metabolite profiling discussed in this chapter have utilized variations on the GC-MS methodologies developed at the Max Planck Institute for Molecular Plant Physiology (Section 16.5; [18]).

Whilst GC-MS is now a robust benchtop technology, it is important to note that there continue to be substantial new innovations in this technology [19]. One particularly important development which was first commercialized in the late 1990s is the coupling of GC to a time-of-flight (TOF) mass spectrometer. TOF has the key advantage that it scans faster than conventional quadrupole technology allowing either improved deconvolution of peaks in complex mixtures or shorter run times. In addition, TOF technology can determine mass-to-charge ratios with a higher accuracy than is possible with quadrupole systems. The group of Fiehn has been the pioneer of the use of GC-MS(TOF) technology in plant metabolomics (e.g. [20, 21]) and it will no doubt be adopted by more groups as the cost of the technology falls.

Attracting even more interest is the recent development of two-dimensional GC-MS [22]. With this technology it is possible to perform two separate GC separations in a single analysis, thus allowing isomers to be separated which would otherwise be submerged into a single peak. Both quadrupole [23] and TOF technology [24] can be used for mass detection, although the faster scanning times of TOF are certainly advantageous. Applications of this new technology to the analysis of sandalwood oil [25], tobacco essential oils [26], coffee bean volatiles [27] and ginseng [28] have been reported. The first example where the potential of GC $\times$ GC TOF for broad metabolite profiling has been demonstrated has recently been published [29]. The authors reported that they could resolve 1200 compounds in mouse spleen tissue with this technology, whereas conventional one-dimensional GC-MS TOF could only resolve around 500 compounds at a similar quality threshold. The amount of raw data that is generated in two-dimensional GC analysis significantly exceeds that from one-dimensional analysis, and Shellie *et al.* [30] have extended the initial work on mice spleens to include an investigation of different statistical approaches to evaluate the profiles of obese mice in comparison to control samples.

In addition to improvements on the hardware side there have recently been important new initiatives with respect to GC-MS databases that will foster the wider adoption of this technology. Schauer *et al.* [31] have published a collection of mass spectra and retention times from identified and non-identified plant, mammalian and bacterial metabolites. In total some 2000 "tags" for individual metabolites are currently included, derived from both quadrupole and TOF mass spectrometers. There are around 360 known compounds, and the authors hope that the availability of this resource will speed up the identification of novel compounds through effort across different laboratories. The same group have also released a database GMD,

The Golm Metabolome Database, where the results of GC/MS metabolite profiling experiments are freely disseminated [12]. The intention here is that free access to experimental data across the metabolomics community will foster the improvement of bioinformatics methods for data analysis.

16.3.2. Liquid chromatography-Mass spectrometry-based metabolite profiling

GC-MS is now widely established as a key metabolite profiling technology but it has one profound limitation: only metabolites that are volatile, or can be made volatile through derivitization, can be analyzed. Whilst a range of derivitization methods have been developed and the range of compounds that can be analyzed by GC-MS has been extended, chemical derivitization can be a lengthy process that introduces sources of error. Therefore, many groups have been working on establishing methods for liquid chromatography mass spectrometry (LC-MS) and the potential for this technology has attracted much attention and discussion in the literature [32]. High performance liquid chromatography (HPLC) has long been used in a wide range of targeted metabolite profiling approaches, particularly for the investigation of plant secondary metabolism. As MS detectors and couplings to LC became available in the 1990s, a large number of targeted methods were adapted to the use of LC-MS due to the higher sensitivity and specificity (e.g. saponins [33]; isoprenoids [34]). Nevertheless it is only recently that improvements in both LC separation technologies and MS interfaces has brought us to the point where broad LC-MS metabolite profiling is now feasible [35]. Indeed, developments in the ionization techniques available for LC-MS such as atmospheric pressure ionization (API), atmospheric pressure chemical ionization (APCI), atmospheric pressure photoionization (APPI) and electrospray ionization (ESI) have been particularly impressive [36].

LC-MS application development has been driven primarily by requirements in the pharmaceutical industry, attracted by the sensitivity and precision of the technology for studying drugs or xenobiotics and their metabolic products. To date, the only published example of broad LC-MS metabolite profiling in the plant sciences has been that of Roepenack-Lahaye *et al.* [37]. This group was able to follow some 2000 different signals in *Arabidopsis* leaves and roots when using capillary LC coupled to TOF-MS. Some of these signals could be assigned to particular metabolites or classes. Experience at the company metanomics, represented by the authors of this chapter, has shown that it is possible to develop and implement wide LC-MS-based metabolite profiling in high throughput (Figure 16.1). Therefore, it is to be anticipated that there will soon be a sharp increase in the number of publications on broad LC-MS methods. In addition, very recent improvements in liquid chromatography, the so-called ultra performance LC (UPLC) might prove to be particularly effective in metabolomics studies where complex mixtures are analyzed [38, 39].

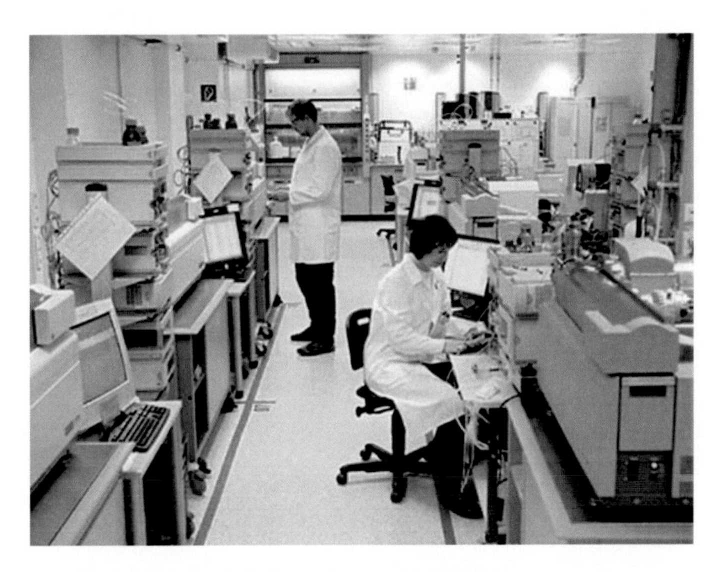

Figure 16.1. High-throughput LC-MS/MS at metanomics.

16.3.3. NMR-based metabolite profiling

NMR is the oldest metabolite profiling technique that has been used in the plant sciences, with the first applications being published in the 1970s. However, the number of groups that were able to work with this technology remained relatively low and therefore the potential of this approach was never truly realized. Although NMR spectroscopy is typically less sensitive than MS its strength lies in the specific identification and quantitative analysis of plant metabolites either *in vivo* or in simple plant extracts. The concept of metabolite fingerprinting, that is the rapid classification of samples without accurate quantification at the metabolite level, was first established with NMR spectroscopy and this has proven to be a fast, rapid and effective tool to discriminate groups of related samples [40, 41, 42, 43]. In the case of metabolite profiling with ^{1}H NMR spectroscopy typically 20–40 known metabolites can be detected and quantified in an unfractionated plant extract. When information is available on specific metabolites such data can be used to generate biological hypothesis about the changes observed in certain groups (e.g. study of cadmium toxicity in *Silene cucubalus* cell cultures, [44]. One of the key advantages of NMR spectroscopy is the relative ease with which it is possible to discriminate isotopes, and much experimentation has focused on the use of stable isotope labelling with ^{2}H-, ^{13}C- and ^{15}N-labelled precursors to investigate metabolic flux.

The limitation in sensitivity has led to a relative neglect of NMR-based metabolite profiling in the plant sciences in comparison to the emerging MS-based technologies. Although, we do not see this changing, NMR will continue to play a major role in structural elucidation, possibly in direct combination with LC/MS techniques (e.g. [45, 46]) and in the analysis of metabolic flux [6]. The history and status of NMR spectroscopic profiling in plants has been the subject of several recent expert review articles [47, 48, 13, 49, 50].

16.3.4. Other technologies

Whilst it is to be expected that GC-MS, LC-MS and NMR will remain the mainstay of plant metabolite profiling for the forseeable future, there are other technologies that may find application in the future. Particularly promising is the use of capillary electrophoresis couple to mass spectrometry (CE-MS) which is suitable for the analysis of polar water-soluble metabolites [51]. To date, exploration of the application of CE-MS to metabolomics has been essentially within one group, so a wider validation of this technology still has to occur. However, this group has recently published a study that indicates the suitability of the technology to plant metabolomics [52]. In this paper the authors reported that they could successfully measure in rice leaves the levels of 88 metabolites involved in glycolysis, the tricarboxylic acid cycle, the pentose phosphate pathway, photorespiration and amino acid biosynthesis. However, the authors noted that the method is currently not suited for the measurement of sugars as the highly akaline buffer solution used to ionize sugars interferes with MS detection. Nevertheless the small sample volumes and the amenability to miniaturization and automation may mean that in due time CE-MS based systems come to play a considerable role in plant metabolomics in the future.

One technology that has attracted much hype is also one of the most expensive: Fourier-transform ion cyclotron resonance mass spectrometry (FT-ICR-MS, also FTMS; [53, 54]). The potential advantage of this technology is the mass accuracy, and the proponents of the technology argue that this is sufficient to allow the chromatographic step to be excluded. There are now several publications where data from FTMS has been used in plant metabolomics studies [55, 56, 57, 58, 59]. However, to date there has not been a single publication that validates this methodology, not least due to the fact that extensive ion suppression effects during direct ionization of complex biological mixtures are well described [60, 61]. Therefore the results of these studies need to be considered with some caution, at least with respect to quantitative aspects. A further limitation of direct injection FTMS is that differences in individual isomers, which are widespread amongst metabolites, cannot be detected. Recently, the first studies where FTMS has been coupled to LC have been published and such systems can be expected to be extremely powerful [62]. Nevertheless it remains to be seen how widely and rapidly LC-FTMS will be adopted due to the need to develop dedicated software, the technical expertise required to run the systems and the overall expense.

Finally, several groups have reported on other technologies for metabolic fingerprinting. Such methods do not bring the same expectation of good quantification of individual metabolites, but offer advantages in terms of throughput. For example Dunn *et al.* [63] have reported on the use of automated electrospray TOF MS for the characterization of tomato fruit extracts. Further, Johnson *et al.* have demonstrated that Fourier transform infrared spectroscopy can be used to fingerprint silage fermentations [64] and salt-stressed tomatoes [65], whilst Gidman *et al.* [66] have used this technology to investigate plant–plant interactions. It remains to be seen whether such "quick and dirty" fingerprinting approaches find widespread application, or whether more robust metabolite profiling approaches will be followed from the outset of projects.

16.3.5. Technologies for industrial applications

For industrial applications of metabolite profiling, there are three key requirements which significantly influence the choice of technology: robust quantification, economic reliability and high-throughput operation. To date, these requirements have been fully met only by GC-MS, LC-MS and NMR, and most commercial organizations are focused on these technologies. Many industrial applications require that samples are processed within very defined and strict timelines and thus the importance of robustness has to be carefully considered. There is often a trade-off between the innovative potential of new technologies and their ability to offer sustained performance. At metanomics we have made excellent experience operating a laboratory of 50 benchtop GC-MS quadrupole systems (Figure 16.2), with sustained

Figure 16.2. High-throughput GC-MS at metanomics: 50 systems are in operation.

"up time" availability of each system of over 95%, measured over at least 5 years of operation. Recent improvements in LC-MS technologies also mean that similar stability can now be achieved with this technology too.

The key challenge for high-throughput operation is that there are currently no "shrink-wrapped" software solutions for metabolomics. Thus all commercial companies involved in plant metabolomics have invested heavily in the development of software suites to support all steps in the process from experimental design, through laboratory operation, quality control and data validation, and on to data analysis and interpretation. However, despite this significant barrier to entry, metabolite profiling is actually more suited to high-throughput operation than proteomics or transcriptomics. Indeed, at metanomics around 400,000 peaks are identified, integrated and validated in chromatograms each day, and in total more than 1.1 million profiles have now been processed.

16.3.6. Standardization in metabolite profiling experiments

The speed of adoption and the value generated by all new technologies is enhanced when widely supported standards can be developed and implemented. A range of different groups have spotted this necessity and there are currently several different initiatives underway to generate a framework for recording and reporting metabolite profiling experiments. Probably, the most comprehensive analysis has been provided by the Standard Metabolomics Reporting Structure (SMRS) Group (www.smrsgroup.org) which has recently published a position paper along with a synopsis of the conclusions in Nature Biotechnology [67]. The publication was co-authored by representatives of 25 different commercial and academic organizations and provides a vision for an open, community-driven specification for the standardization and reporting of metabolic studies. In particular, detailed policy directions are provided for three main areas: the origins of biological samples, the methodologies used to obtain data and the analytical techniques to which the data are subjected. Specifics include recommendations on the naming of compounds, the documentation of the treatment of living biological samples, appropriate data processing steps and guidance on statistical models. Following on from the first National Institute of Health (NIH) Metabolomics Standards Workshop in August 2005, a decision was made to merge the actitivities of the SMRS group with that of the newly formed Metabolomics Society (www.metabolomicssociety.org) in order to make more progress towards a generally accepted set of standards.

Aligned with the fact that the plant community played a key role in pioneering metabolomics, it was also early in addressing the needs for standardization [1]. In 2004, Jenkins *et al.* [68] published a framework for the description of plant metabolomics experiments and their results based on their experiences with a data model that they developed called ArMet (architecture for metabolomics). This data model included the entire course of an experiment from initial planning to final analysis of results. The authors noted that the adoption of such formal conventions allows the proper integration of experimental results, the repetition of experiments between laboratories and provides the foundation for systems which allow data storage and transmission [69].

16.4. Agrochemical development

The first application of metabolite profiling to plant research, in the area of crop protection, was pioneered in the early 1980s at the German chemical company, BASF AG by Sauter. This work was summarized in 1991 in a landmark publication in the American Chemical Society Symposium Series [17]. The paper entitled "Metabolic profiling of plants – A new diagnostic technique" described how GC-MS analysis had been applied to assist the classification of mode of action in herbicide research. Barley plants were treated with four chemically unrelated herbicides and the so-called "response profiles" generated. These were defined as the ratio of changes of all metabolites in comparison to untreated control material. At this time, around 200 peaks were included into the profile and Sauter was the first to work with both known and unknown substances for diagnostic purposes. The response profiles of the four herbicides were found to be completely different, a fact that was obvious on visual inspection of the results. By investigating the known components in more detail, Sauter and his colleagues were able to show that the response profiles were characteristic and could be linked directly to knowledge about the individual mode of action. The authors concluded that the technique was suitable for rapid fingerprinting in order to classify compounds, and in particular, that it was able to identify novel chemistries which demonstrated a new mode of action. It was predicted that the classification would become more powerful in proportion to the number of different response patterns that were available. The publication of the paper reflected the fact that this approach had become a standard tool in the herbicide research at BASF (Sauter, personal communication).

Grossmann, also working at BASF, has generalized the concept of a response profile to include a range of data on the phenotypical end points following disruption of molecular, biochemical or physiological processes by herbicide application. In a recent publication he describes this concept, which he names "Physionomics" [70]. Specifically, he has extended the response profile to include bioassays that differentiate between the responses of multiple organization units (plant, tissue, meristematic cells, organelle), developmental stages, types of metabolism (phototrophic, heterotrophic) and physiological processes in the plant. Experience has shown that the joint interpretation of physionomic and metabolite profiling data increases the success rate in classifying new biologically active compounds and can go a long way towards pinpointing novel mode of actions.

BASF has not been alone in exploring the value of metabolite profiling in crop protection research. Scientists at American Cyanamid were able to develop a system for the classification of herbicide mode of action using ^{1}H NMR spectroscopy and artificial neural networks [71]. This procedure had the advantage of simplicity in that signals could be taken from crude plant extracts without any further sample processing, as is required for GC-MS analysis. However, it suffers from the limitation that the number of metabolites covered is relatively low due to sensitivity problems and this therefore decreases the chance that a novel mode of action can be pinpointed. Nevertheless Ott *et al.* [71] were able to demonstrate that they could build an expert system that was able to classify 19 of the most relevant mode of actions for corn.

It is to be anticipated that the application of metabolite profiling in agrochemical development will increase. There have recently been publications which describe the application of expression profiling to the characterization of herbicide mode of action (e.g. 2,4-dichlorophenoxyacetic acid (2,4-D) [72], flufenacet [73]) and such work would be greatly enhanced if metabolite profiling data could be integrated. Further, Lange *et al.* [34] have demonstrated that LC-MS-based profiling of polar metabolites can be used to make a detailed assessment of potential herbicide targets in the isoprenoid biosynthesis pathway.

16.5. Metabolite engineering in plants

Perhaps the most developed application of metabolite profiling in the plant sciences is the study of metabolic pathways and their manipulation through metabolic engineering. The field of metabolic engineering became possible through the development of capabilities in the late 1980s to genetically transform a wide range of important plant species. As the 20th anniversary of plant metabolite engineering approaches, it must be said that there have been very many failures and only a small number of spectacular successes [74, 75].

The primary reason for the lack of success lies in our limited understanding of metabolic regulation (Section 16.2.2). In addition, there is still today a strong tendency for metabolite engineers to assume that their work is by definition limited to a particular pathway. Many studies focus only on the end point of the metabolic pathway in question, and perhaps the intermediates directly leading up to the final product. However, the fact that metabolic pathways do not operate in isolation and are part of metabolic, genetic and physiological networks must never be underestimated. If the flux through one pathway is modulated there are very often further consequences as a result of alterations in the levels of individual metabolites. There are now many examples of this from plant metabolite engineering: following antisense inhibition of adenylate kinase in potato tubers there was a surprising accumulation in starch [76]; inhibition of the major enzyme in starch biosynthesis in potato tubers, ADPGlucose pyrophosphorylase gave rise to altered amino acid levels [77]; lines of *Arabidopsis* which accumulated polyhydroxybutyric acid were found to have multiple pleiotropic changes in metabolism [78]. It is surely no coincidence that some of the most successful examples of plant metabolite engineering have been where completely new pathways have been introduced. In these cases the metabolic engineer needs only to ensure that adequate substrate is available, achieve sufficient expression of the introduced enzymes and avoid that the enzyme products be degraded by endogenous mechanisms. Further, it is relatively unlikely that intermediates in the new pathway serve a signalling role in the host system or that they have a toxic effect. Positive examples where novel metabolic pathways have been introduced into plants include the production of Vitamin A in rice [79] or the engineering of bioplastic production in Brassica [80].

The conclusion from the early experience with metabolic engineering must be that the potential for multiple pleiotropic changes needs to be evaluated early in the development of a project using a broad technique such as metabolite profiling. The investigation of such pleiotropic changes offers the chance to identify novel regulatory mechanisms in the metabolite network, and if initial approaches do not reach their target, it may also provide the opportunity to generate new hypothesis on how to tackle the problem. In addition, in commercial projects, early information on the scope of changes that occur in response to engineering contributes to the ongoing risk assessment of such projects. In particular, there is increasing public demand for more information on the extent of metabolic changes in modified varieties [81].

The first group to explore the possibilities for applying metabolite profiling to plant metabolite engineering was that of Willmitzer. In 1996, Trethewey set about adapting the methods of Sauter *et al.* [17] to the study of the metabolic engineering of primary metabolism in potato tubers. Initial results with the method quickly showed that exceedingly complex chromatograms with many hundreds of distinct peaks could be obtained from potato tuber tissue. However, the choice of potato tubers also highlighted one of the limitations of the silylation derivatization method of Sauter: it generates multiple isomers of the abundant sugars. Therefore, the method was modified in two respects. First, a phase separation was introduced to separate the polar and the non-polar metabolites even though this doubled the number of chromatographic analyzes that needed to be performed. Second, a two-step derivitization procedure was introduced that reduced the isomer complexity. The resulting potato tuber method, along with an extensive validation, was published by Roessner *et al.* [18], who showed that it enabled the qualitative and the quantitative determination of more than 150 tuber metabolites including sugars, sugar alcohols, dimeric and trimeric saccharides, amines, amino acids and organic acids. The technical variability of the method was assessed by determining the behaviour of 25 defined metabolites drawn from different chemical classes. Fortunately, the technical variability was found to be an order of magnitude lower than the biological variability observed between different potato tubers. The standard deviations that were reported were in the 10–25% range for the majority of the metabolites which were comparable with those obtained using other technologies for metabolite analysis. Since this publication, the methods have been adopted by a wide range of groups and adapted to the study of a range of different species and tissues providing another layer of validation for this innovation.

The first application of GC-MS-based metabolite profiling methods to metabolic engineering was the analysis of potato tubers which had been manipulated in starch and sucrose metabolism. Roessner *et al.* [82, 18] profiled lines that overexpressed a yeast invertase, specifically in the tuber [83, 84] and thus had enhanced capacity to cleave sucrose to glucose and fructose. Rather than enhancing starch metabolism, as had been hoped when the project was initiated, the elevated sugar levels triggered an induction of glycolysis through a signalling mechanism which at the time was not understood. These conclusions had taken many months of painstaking biochemical assaying to piece together. However, when the GC-MS protocol was applied to the material it was found that all the conclusions could have been obtained in a single analysis procedure taking less than one week. In addition, as the GC-MS profiles were open and covered metabolites that would not normally have been assayed in a sucrose/starch project, a range of further unexpected pleiotropic changes were found. For example, there was an accumulation of shikimate and maltose and a large reduction in inositol in these tubers. This led to new hypotheses about the interconnection of primary metabolism in potato tubers.

This pioneering work was then extended to cover the analysis of seven further transgenic potato tuber genotypes subject to metabolite engineering in the pathway of starch biosynthesis [82, 77, 85] Given the amount of data that was generated and the complexity of the changes in the profiles, the authors adopted data mining approaches such as hierarchical cluster analysis and principal component analysis (PCA) in order to gain an overview of the results. Although all genotypes shared a common end phenotype, a reduction in the final starch content, the cluster analysis revealed that each genotype separated into distinct clusters indicating that the underlying profiles were divergent. The scientific development of this field and the impact of metabolite profiling have been reviewed in more detail in a dedicated article [86]. This early application of metabolite profiling to metabolic engineering demonstrated the potential to speed up the scientific cycle from hypothesis through experimentation to follow-on hypothesis and illustrated how the range and scope of potential new discoveries can be increased.

Further examples of the application of metabolite profiling to metabolic engineering have now been published. Morino *et al.* [87] have generated rice calli that overexpress a feedback insensitive alpha subunit of anthranilate synthase and thus are capable of overproducing tryptophan. They used HPLC metabolite profiling to compare the aromatic metabolites between these overproducing strains and the wild type background. The methods used generated a total of 71 chromatographic peaks and the authors were, in this case, able to show that the approximately 80-fold increase in tryptophan was quite specific with only a limited effect on the other peaks. One of the minor metabolites found to be elevated by the profiling of the transgenic calli was purified and found to be a previously unknown indole alkaloid glucoside. In a further investigation of indole 3-acetic acid (IAA) levels, an increase in both free IAA and a range of conjugates was found indicating that there is some interdependency between engineering tryptophan levels and the abundance of this important plant hormone. Encouraged by these results, transgenic potato plants were created and analyzed for changes in their metabolite profiles [88]. As in the rice calli, the authors concluded that the effects of the transformation were limited to tryptophan and IAA, which were increased in upper parts of shoots by 2–20-fold and 8.3–39-fold respectively. Surprisingly, despite these large changes in IAA and tryptophan, no effect on plant growth or development was found.

Kristensen *et al.* [89] have used the metabolic engineering of dhurrin in transgenic *Arabidopsis* plants to address the question of whether such approaches can be specific. The cyanogenic glucoside dhurrin is produced from tyrosine and the entire biosynthetic pathway was created in *Arabidopsis* by transferring three genes from *Sorghum bicolor*. The engineered *Arabidopsis* plants accumulated an impressive 4% of their dry weight in dhurrin. Phenotypical analysis along with metabolite profiling and transcript profiling revealed only marginal unexpected alterations. The metabolite profiling was performed by a targeted LC-MS analysis of mature rosette leaves which provided data on 17 metabolites. When only the first gene in the biosynthetic pathway, responsible for producing p-hydroxybenzylglucosinolate from tyrosine, was introduced the same low impact on other parameters was observed, with accumulation of the glucosinolate at 3% of the dry weight. However, when the first two genes of the pathway were introduced the plants became stunted and there were numerous alterations in transcript and metabolite profiles. This is probably because the first two genes generate cyanohydrin which is toxic and labile, and the plant responds by activating de-toxification mechanisms which lead to the accumulation of numerous glucosides. The authors concluded that it is possible to perform targeted metabolic engineering with only marginal effects on the metabolome and transcriptome although it clearly makes a difference which metabolites are being synthesized and at what level. Nevertheless these results are based upon very narrow profiles with only a rudimentary quantification and therefore there is a need to investigate such material with more sophisticated profiling technologies.

The broadest investigation of a metabolome following metabolic engineering has recently been published by Catchpole *et al.* [90]. In this study the authors have used GC-MS TOF to study a wide range of metabolites found in the tubers of field grown conventional and genetically modified potatoes. They first performed a “fingerprinting” of the material and then undertook more detailed profiling of metabolites where the fingerprinting results indicated that significant quantitative differences may have occurred. Overall the authors concluded that the transgenic potatoes were, apart from some targeted changes, substantially equivalent in metabolite contents to non-engineered varieties. The transgenic material studied were potatoes which had been engineered to synthesize the polymer fructan and it could be that this pathway is quite removed from the regulatory signals and networks associated with primary metabolism.

In conclusion, there are now many examples where metabolite profiling has demonstrated a utility in the area of plant metabolic engineering. There is considerable debate as to how precise metabolic engineering strategies can and will be, and it will certainly be the task of broad metabolite profiling to investigate this question further [91]. Nevertheless, first examples indicate that precision metabolic engineering is possible, but that metabolite profiling technology should be utilized early in such projects to determine whether multiple pleiotropic changes are also generated.

16.6. Plant secondary metabolism

In the long term, plant secondary metabolism studies are likely to develop into one of the most important application areas for metabolite profiling. Research into plant secondary metabolites interfaces with a range of different biological and medical disciplines not least because secondary metabolites possess numerous properties that find application in pharmaceuticals, crop protection, dyes, flavours and fragrances. Thus in addition to the scientific interest in this field there is likely to be a growing commercial interest that will foster the application of profiling technologies.

Whilst still at an early stage there have now been several studies that demonstrate the potential of metabolite profiling in the study of plant secondary metabolism. As already introduced, our understanding of secondary metabolism is generally much weaker than that of primary metabolism [92] and most studies have focused on defining pathways and intermediates [94].

To date, one of the best examples of the impact of metabolite profiling has been on our understanding of lignin biosynthesis and the phenylpropanoid pathway [93]. The exact pathway of lignin biosynthesis remains unclear despite the best efforts of several large groups [94, 95]. Targeted metabolite profiling, in combination with the production of transgenic plants altered in candidate enzymes, has enabled hypotheses over the precise role of different enzymes to be tested [96, 97]. For example, it has recently been proposed that caffeoyl CoA is the *in vivo* substrate of caffeoyl CoA *O*-methyltransferase, but that caffeic acid is not a substrate *in vivo* for caffeic acid *O*-methyltransferase. Such very specific results have been extended with a broader metabolite profiling study on phenolic metabolites in transgenic alfalfa lines altered in lignin biosynthesis [98]. Soluble phenolics, wall-bound phenolics and soluble and core lignin were analyzed in transgenic alfalfa with genetically down-regulated *O*-methyltransferase genes. Given the chemical nature of these substances, HPLC profiling was performed and the resulting metabolic phenotypes were differentiated using PCA. The authors found that differences in the leaf metabolite profiles between wild type and transgenic plants could not be resolved. However, the stem profiles of phenolic compounds could be clearly distinguished. The authors concluded that such broader profiling is a useful tool for monitoring the consequences of changes in the lignin biosynthetic pathway.

A further area where metabolite profiling has great potential is in the study of flavonoid and isoflavonoids, not least due to the much debated impact on human health of such phytoestrogens [99, 100]. Kachlicki *et al.* [101] prepared extracts from the roots of three lupine species (*Lupinus albus, L. angustifolius, L. luteus*) and analyzed them using both LC/UV and LC/ESI/MS. They were able to observe up to 20 isomeric isoflavone conjugates: these were di- and monoglycosides of genistein and 2′-hydroxygenistein with different patterns of glycosylation, both at oxygen and carbon atoms whilst some glycosides were acylated with malonic acid. Wu *et al.* [102] have used the same technologies to study the isoflavone contents of Edamame and Tofu soy varieties, from which the immature fresh soybeans or the mature soybean seeds are consumed, respectively. In total, 16 isoflavones were identified, including three aglycones, three glycosides, two glycoside acetates and eight glycoside malonates. Whilst these studies were primarily technical in nature, focusing on the development of methodologies for isoflavone profiling they indicate how these methodologies will be used in the future in conjunction with biological or medical experimentation. Examples of such studies have recently been published using less powerful methodologies where the contents of isoflavones have been investigated during the course of the germination of soybean seeds [103] or where the distribution of soy isoflavones have been followed in human tissues [104]. It is to be anticipated that the resolution of such work will be greatly enhanced once MS-based profiling methodologies are utilized.

A first glimpse of the potential for metabolite profiling to aid the metabolic engineering of secondary pathways has now been provided by work on the biosynthesis of isoflavone phytoestrogens. The key enzyme in this pathway is believed to be isoflavone synthase which converts naringenin to genistein glycoconjugates in legumes [105, 106]. Transgenic *Arabidopsis* plants have been created where both isoflavone synthase and a second enzyme chalcone isomerase, which catalyses the step prior to naringenin formation, have been introduced. Targeted metabolite profiling of flavonoid compounds with LC-MS revealed that overexpression of the chalcone isomerase alone resulted in a three-fold increase in the content of the major glycoconjugates of the flavonols kaempferol and quercitin. However, as is often the case in metabolic engineering, the desired result of elevated genistein levels was not obtained when both genes were introduced. However, metabolite profiling revealed that the levels of flavonols were reduced indicating that novel metabolic reactions did indeed occur in the double transgenic lines. The application of wider metabolite profiling technologies might reveal more information about the changes in metabolic flux following heterologous overexpression of isoflavone synthase and chalcone isomerase.

Further examples of the application of targeted metabolite profiling to the study of secondary metabolism include isoprenoids [34], oxylipins [107, 108], sesquiterpene lactones [109], taxane diterpenoids [110], taxoids [111], alkaloids [112, 113], saponins [114] and volatile compounds [115, 116, 117]. Recently, metabolite profiling approaches led to the surprising discovery of four phenolic amides in potato tubers [118]. The fact that these metabolites had not previously been identified in such a well-studied food crop illustrates the power of non-targeted metabolite profiling technologies. New approaches for the metabolite profiling of phytohormones, a class of metabolites often present in very low abundance, have also been reported [119].

Finally, metabolite profiling of secondary metabolites is likely to play in the future a significant role in the analysis of commercial botanical products to ensure productivity, efficacy and quality. For example, Wang *et al.* [120], working at the AndroScience Corporation, have developed an LC-MS method for the profiling of black cohosh (*Cimicifuga racemosa*) herbal products. They selected triterpene glycosides for the targeted profiling as these appeared to be one of the major compound groups in *Cimicifuga* species. The "fingerprint profiles" that they generated were quite different according to which species or product they analyzed. In some cases the authors reported that the total amount of triterpene glycosides found in commercial products were either in excess of or lower than that claimed by the manufacturers. In conclusion, Wang *et al.* proposed that their methodology was suitable for routine application in the manufacture and quality control of black cohosh products. Similarly, Demirci *et al.* [121] have developed HPLC methods for the quality control of saponins in the leaves of medicinal Ivy (*Hedera helix*), and Bailey *et al.* [122] have used high-field ^1H-NMR spectroscopy and chemometrics techniques for data analysis to discriminate commercial feverfew preparations. Woo *et al.* [123] working at Unigen Inc have investigated ginsenoside biosynthesis using LC-MS to profile the 27 putative ginsenosides found in ginseng (*Panax ginseng* C.A. Meyer). The authors generated 993 different ginseng hairy root lines, profiled them and were able to select lines which produced 4–5 times the amount of ginsenoside found in wild type populations. They were also able to identify lines which varied in the ratio of individual ginsenosides. It is likely that cell cultures will play an increased role in the production of specific secondary metabolites in the future and that metabolite profiling will become the analysis method of choice [124].

16.7. Complex problems in plant science

As the technologies of metabolite profiling have matured, applications in plant science beyond the direct study of metabolism have started to develop. Indeed there has been a rapid acceleration in the application of metabolite profiling to complex problems in plant science in the last 3 years. Under complex problems are included the study of plant development, yield and nutrition along with the response and adaptation to biotic and abiotic stresses. In all of these areas, metabolism and metabolites play a key role. However, the interpretation of metabolite data requires the integration of other physiological, phenotypical or environmental data and we are just at the beginning of using metabolite profiling data in a holistic context in plant biology.

16.7.1. Development and yield

It is well known that the metabolic needs of a plant vary according to tissue and developmental stage. Historically, broad metabolite profiling was developed for leaf and tuber tissue and it is only recently that other tissues and systems have been investigated. For example, plant roots are known to continuously produce and secrete compounds into their immediate rhizosphere although the diversity of compounds involved and the regulation of the process are not fully understood. Walker *et al.* [125] developed an *in vitro* system that allowed them to monitor root metabolism and secretion following application of a range of elicitors (salicylic acid, jasmonic acid, chitosan and two fungal cell wall elicitors). They then used HPLC to profile the resulting exudates and could measure and quantify some 289 possible secondary metabolites. The chemical structures of 10 of these (butanoic acid, *trans*-cinnamic acid, *o*-coumaric acid, *p*-coumaric acid, ferulic acid, *p*-hydroxybenzamide, methyl *p*-hydroxybenzoate, 3-indolepropanoic acid, syringic acid and vanillic acid) were determined using NMR spectroscopy and were found to be present at concentrations where antimicrobial or anti-fungal activity might be expected. Meanwhile, working with *Arabidopsis*, Narasimhan *et al.* [126] have estimated using "rhizosphere metabolomics" (in this case targeted LC-MS methods that cover 125 secondary metabolites) that phenylpropanoids account for 84% of the secondary metabolites exuded from roots.

Of particular value to our understanding of development and physiology will be methods that increase the overall resolution of metabolite profiling at the tissue or cellular level. For example, Fiehn [20] has reported on metabolomic analysis of phloem exudates of *Cucurbita maxima*. Using GC-MS TOF, over 400 components could be resolved using deconvolution algorithms and 90 metabolites could be tentatively identified. The impact of assimilation rates on metabolite levels was investigated over a period of 4.5 days. The author was surprised that the method did not allow a diurnal rhythm to be distinguished in any of the phloem metabolites followed, although distinct metabolite profiles were observed for each of the individual plants involved in the study. Further investigation will be required to distinguish whether these effects were an artefact of the experimental methods or genuinely reflect large individual to individual variation.

Arguably, the application of metabolite profiling to the study of metabolic regulation during development is most advanced in leaf tissue. Using a mature GC-MS protocol the diurnal changes in primary metabolites in potato leaves (*Solanum tuberosum* cv. Desiree) were analyzed and the data extended with the results from conventional spectrophotometric assays [127]. The authors found that there were 56 metabolites which showed significant changes in the diurnal period, including several metabolites which were not previously known to cycle. Using PCA analysis of the metabolite profiles at the different time points the authors concluded that there is a very tight control of central metabolic pathways and that metabolite patterns change only progressively through a diurnal period. Jeong *et al.* [128] have applied GC-MS profiling to the study of the sink/source transition in leaves of the tree quaking aspen (*Populus tremuloides* Michx.). These researchers found about 250–300 peaks that could be resolved and were able to identify around a quarter of them as participants in primary metabolism. Interestingly, two-thirds of the identified metabolites were found to exhibit greater than fourfold variations during the course of the sink/source transition. Many of the observations fitted with contemporary physiology: for example, sucrose concentrations increased during leaf expansion whereas hexoses showed a peak at mid-expansion, whilst amino acids generally decreased during leaf expansion. The authors concluded that the ability to profile metabolic shifts that reflect changes in photoassimilation, transport and storage processes is a key step towards ultimately understanding the genetic basis for carbon partitioning and growth. Further, the fact that they could identify metabolite markers for cell wall and membrane biogenesis and that these could be interpreted in the context of carbon and nitrogen utilization opens up the possibility to study carbon sinks in more detail and to use such markers widely in the breeding of *Populus*. Tarpley *et al.* [129] have defined a set of biomarker metabolites that characterize the development of a rice plant, by examining metabolite composition during initial tiller development of rice (*Oryza sativa* L.). Some 21 metabolite markers were found to characterize this developmental period and were responsible for 83% of the overall metabolic variation. The authors conclude that such a relatively small set of metabolite markers has the potential to be applied in rapid comparative screening of rice plants to distinguish changes in developmental status or to monitor the impact of different environmental or genotypical factors.

The concept of using metabolite profiling to determine metabolite biomarkers for physiological processes has been developed further by Diaz *et al.* [130]. Leaf senescence is an important developmental stage in annual plants and it facilitates the mobilization and recycling of key nutrients to the developing seeds. As such it is ideally suited for investigation through metabolite profiling. The authors did not apply modern profiling methodologies but created their metabolite profiles of around 15 components based on the results from an amino acid analyzer and several dedicated assays. Therefore, this work is not state of the art in terms of technology but, as it was possible for the authors to identify biomarkers for senescence in *Arabidopsis*, it does represent the cutting edge of this field of plant biology. The authors worked with recombinant inbred lines of *Arabidopsis* and they showed that the proportions of gamma-aminobutyric acid, leucine, isoleucine, aspartate and glutamate correlated with both the senescence phenotype of the lines and their age. In particular, they found that the glycine/serine ratio was a sensitive indicator that could detect the onset of senescence before any phenotypic changes could be observed.

Metabolite profiling will not be restricted to the analysis of plant tissues in isolation. Desbrosses *et al.* [131] have developed GC-MS methods for studying plant–microbe interactions in the model legume, *Lotus japonica.* In the study, these authors report on the application of their method to the analysis of nodules, roots, leaves and flowers of symbiotic plants. They used PCA to visualize that each one of these organs had a characteristic metabolite profile and through analysis of the PCA loadings were able to distinguish groups of metabolites which served as characteristic markers for each tissue. As all previous studies on nodule metabolism had been limited to target analysis of a few compound classes, the deployment of broad metabolite profiling has given rise to a data set that requires careful consideration. The authors reported that octadecanoic acid, asparagine, glutamate, homoserine, cysteine, putrescine, mannitol, threonic acid, gluconic acid, glyceric acid-3-phosphate and glycerol-3-phosphate were all relatively enriched in nodules. In addition, a range of metabolites which are known to act as compatible solutes, such as proline, putrescine, ononitol, mannitol and sorbitol, were also found to be elevated. This led the authors to hypothesize that the cells of the nodule are under hypoxic conditions which in turn causes a build up of osmotic stress. However, the technology used in this study was not able to distinguish whether the metabolites were located in the plant cells or the bacteroids. Therefore improvements in resolution are needed before metabolite profiling can contribute further to our understanding of symbiotic nitrogen fixation.

Based upon these recent studies it is possible to speculate that metabolite markers will emerge as key tools to enable the accurate monitoring of physiological and developmental status. Of particular interest is whether it will be possible to identify metabolite markers that are able to predict yield, and whether it might be possible to use this information to develop transgenic strategies for yield enhancement [132].

16.7.2. Plant nutrition

Plant nutrition is a large field where there are now the first indications that metabolite profiling will have an impact. In a landmark study, Hirai *et al.* [57] have reported on a general strategy for integrating transcript and metabolite profiling data in the investigation of the response of plants to nutritional stress. The application that they described in detail was the investigation of sulfur deficiency in *Arabidopsis thaliana* plants. The metabolomics data were primarily generated using FTMS, which, as discussed before, has to be taken with some caution due to the absence of a published validation of the application of this technology to metabolite profiling. Metabolomics and transcriptomics data were integrated and analyzed using a complex statistical technique termed "batch-learning self-organizing maps". This statistical assessment of the data produced groups of metabolites and genes which were regulated by the same mechanism. The authors found that the metabolism of glucosinolates appeared to be under coordinated regulation. Following the observation that three putative sulfotransferase genes clustered together with genes known to be involved in the biosynthesis of glucosinolate, these genes were studied using *in vitro* enzymatic assays, and it was confirmed that the gene products demonstrated desulfoglucosinolate sulphotransferase activity [133]. Further *O*-acetylserine was found to cluster with a range of genes known to be involved in sulfur assimilation. A study of the cluster of anthocyanine biosynthesis genes revealed that a transcriptional factor known to regulate this pathway was also present. The authors concluded that the strategy followed in this study is generally applicable to the elucidation of regulatory metabolites and genes, so long as they cluster together with the downstream genes and metabolites that they regulate.

Further detailed and extensive work on sulfur deficiency has been reported by Nikiforova *et al.* [134, 135, 136], who utilized GC-MS and LC-MS technologies for metabolite profiling. Using both systems, around 6023 non-redundant ion traces were recorded although only 134 metabolites could actually be identified. Evidence was found in the study for the influence of sulfur accumulation on the regulation of nitrogen balance, lipid breakdown, purine metabolism and photorespiration. In addition, evidence for the involvement of auxin in hypo-sulfur stress was also found [136].

The importance of inorganic nitrogen assimilation for plant biochemistry in general and amino acid biosynthesis in particular has been investigated by Urbanczyk-Wochniak and Fernie [137] using GC-MS profiling. Tomato plants were grown in hydroponic culture under three different nitrate supplies (saturated, replete and deficient) and at different light intensities. Leaf samples were taken after 7, 14 and 21 days and in addition the short-term response of leaves to varying nutrient status was investigated. The results of the metabolite profiling revealed quite different profiles for each one of the regimes and time points analyzed. In particular, the authors were able to discern that the nitrate deficiency not only led to a decrease in amino acid contents but that there was also a reduction in organic acids and an increase in sugars, phosphorylated intermediates and some secondary metabolites.

In general terms the data sets generated in the plant nutrition studies discussed here are so complex that it will require considerable time and further experimentation to develop a full understanding of the responses of the metabolic network to nutrient stress [138]. However, the promise of these approaches has already been clearly demonstrated and the number of groups applying these techniques is increasing rapidly.

16.7.3. Abiotic stress

Several groups have now applied metabolite profiling to the study of abiotic stresses. The response of plants to abiotic stresses are complex processes and according to species and genotype differing degrees of tolerance are demonstrated [139]. Underlying many of the response processes are altered requirements for particular metabolites which may act as signals, osmotic protectants or intermediates in energetic processes. Thus the metabolic network shifts considerably upon perception of abiotic stress, and broad metabolite profiling is an ideal tool to investigate stress responses.

At the forefront of the application of metabolite profiling to abiotic stress have been four groups who have reported on cold acclimation and cold/freezing responses. The first report was from Welti *et al.* [140] who used electrospray ionization in tandem with mass spectrometry to profile the membrane lipids of *Arabidopsis* plants subjected to cold and freezing stress. They analyzed both wild type plants and a mutant which is deficient in phospholipase-D and thereby demonstrates a higher tolerance to freezing. They concluded that in wild type plants there is a higher loss of phosphatidylcholine and an increase in phosphatidic acid in comparison to the mutant lines and that these changes may lead to greater destabilization of the membrane bilayer structure. This in turn results in a higher rate of membrane fusion and enhanced cell death in the wild type plants.

Kaplan *et al.* [141] were the first to report on the application of broad metabolite profiling to the study of the induction of acquired thermotolerance in response to heat shock and acquired freezing tolerance in response to cold shock in *Arabidopsis*. These authors used GC-MS metabolite profiling and were able to follow some 81 identified metabolites and 416 unidentified distinct mass spectral tags in their experimentation. In studying both the heat and the cold shock responses the authors were surprised to find that most of the metabolites that changed were similar in both studies. They found elevations in the levels of amino acids derived from pyruvate and oxaloacetate, accumulation of polyamine precursors and increases in the levels of various compatible solutes.

The authors concluded that their results indicated close relationships between heat and cold shock responses, possibly operating through similar underlying mechanisms.

A similar impression was obtained by Gray and Heath in a study of cold acclimation in *Arabidopsis* using metabolic fingerprinting by FTMS. Whilst in principle very powerful, this technology has yet to be validated for quantitative profiling work and therefore the authors restricted themselves to interpreting the overall patterns. They concluded that significant changes in metabolism occurred in *Arabidopsis* plants transferred to low temperature, and that these changes in profile were different to those found in plants which were grown at low temperature. A further study of cold responses to metabolism in *Arabidopsis* has been published by Cook *et al.* [56]. These authors focused on the CBF (CCAAT-Box Binding Factor) pathway which is known to play a key role in the process of cold acclimation. Using GC-MS TOF metabolite profiling, some 434 metabolites were assessed and 75% were found to increase in leaves upon applying low temperature or acclimation. Curiously, 79% were also found to increase in non-acclimated plants following overexpression of one of the CBF transcription factors. These studies were performed in the Wassilewskija-2 (Ws-2) ecotype but a different picture emerged when the less freezing tolerant Cape Verde Islands-1 (Cvi-1) ecotype was analyzed. This ecotype was found to be depleted in the metabolites which responded to low temperature stress in the CBF-overexpressing Ws-2 ecotype indicating that the relative abundance of these metabolites might play an important role in predicting the capability for tolerance. The authors concluded that their data also showed that the metabolome of *Arabidopsis* is reconfigured following low temperature stress. In addition, their work demonstrated that the CBF pathway plays a key role in the acquisition of tolerance [142].

Although the majority of studies have been done to date in the area of cold and freezing stress, it seems equally plausible that metabolite profiling will play a key role in elucidating the mechanisms involved in drought stress and drought tolerance. The only significant study published to date in this area has been that of Rizhsky *et al.* [143], who focused on the response of *Arabidopsis* to a drought and heat stress, alone and in combination. Using broad GC-MS profiling of polar compounds, the authors showed that plants subjected to both stresses responded by accumulating sucrose, maltose and glucose. Interestingly, they observed that proline accumulated in plants subjected to drought but did not increase in plants that received the combination of both stresses indicating different mechanisms at play. Indeed, the authors proposed that heat stress causes enhanced toxicity of proline to cells and that under these conditions sucrose substitutes as an osmoprotectant.

There is one example of the use of metabolite profiling to analyze the consequences of light stress on the metabolome [144]. In this case, both targeted metabolite profiling with HPLC and broad GC-MS and LC-QTOF-MS profiling were applied to the study of a light hyper-responsive high-pigment tomato mutant. Elevated contents of isoprenoids, phenolic compounds and many volatiles were found in fruits. In addition, evidence of the overproduction of a range of metabolites known for their antioxidant or photoprotective properties indicated that the mutant responded to an elevated perception of light stress by shifting metabolism to provide more protective metabolites.

Taken together the use of metabolite profiling to study abiotic stress in plants is in its infancy. Nevertheless the handful of examples that have been published to date indicate that this technology will be very powerful in the elucidation of metabolite markers that are indicative of particular stresses, even at low levels. In addition, the interpretation of metabolic results opens up the possibility of reaching a deeper understanding of the mechanisms behind these complex phenomena, and provides the opportunity to generate new hypotheses that might allow crop plants to be produced with elevated stress tolerance.

16.7.4. Biotic stress

As with the study of abiotic stress, researchers studying different types of biotic stress are just starting to apply metabolite profiling in order to gain deeper insight. As plant defense mechanisms depend to a significant extent on the production of particular defense compounds and in some cases massive functional alterations in infected tissues [145], it is to be anticipated that there are very distinct responses of the metabolic networks to biotic attack that can be detected through metabolite profiling. Of particular importance in the study of biotic stress will be the methodologies for the profiling of volatile compounds. As with biotic stress there is considerable interest to identify metabolite markers which indicate the onset of biotic stress and to understand the mechanisms that underpin the response of the plant to attack. Again, such deeper understanding may lead to hypotheses as to how engineering metabolism might lead to enhanced resistance in valuable crop plants.

Using both directed metabolite profiling data and results from microarray experimentation, Kant *et al.* [146] have studied the response of tomato (*Lycopersicon esculentum*) plants to infestation with spider mites (*Tetranychus urticae*). Tomato plants are fast to respond to attack and defense mechanisms are triggered long before there are any visible phenotypic changes. The results of transcriptional analysis revealed that genes responsible for the biosynthesis of volatile compounds (monoterpenes and diterpenes) were induced on the first day following infestation, and using volatile metabolite profiling an increase in the emission of volatiles could be found four days after initial exposure to spider mites. Further directed study revealed that jasmonic acid signalling is involved in the process of signal transduction which triggers the elevated production of volatiles [147]. One of the consequence of this volatile production is that predatory mites (*Phytoseiulus persimilis*) are attracted to the infested tomato plants and thereby contribute to the overall defence strategy of the tomatoes, a so-called indirect defense response. Whilst the profiling in this study was limited to volatile components, it is interesting to speculate that the enhanced production of terpenoids will also manifest itself in shifts of the complete metabolic network that would be detected with wider metabolite profiling.

Indeed, the role of methyl jasmonate as an elicitor of shifts in the metabolic network has been investigated in some depth using a cell culture system of *Medicago truncatula* [148]. Using GC-MS-based profiling, a range of significant changes in primary metabolites (several amino acids, organic acids, and carbohydrates) was found following elicitation with methyljasmonate. LC-MS measurements revealed the accumulation of triterpene saponins. This work provided further insights into the coordinated regulation of changes in metabolic networks. The authors found that there were correlative changes in the glycine, serine and threonine biosynthetic pathways which they proposed was due to the induction of threonine aldolase activity. Further they found a particularly high correlation between the levels of leucine and isoleucine. In addition, they observed an elevation of beta-alanine which might be indicative of changes in the metabolism of coenzyme A and its esters, an essential component of plant secondary metabolism.

Further evidence that metabolite profiling of volatiles will be a key technology in the study of biotic stress has been provided by the work of Lui *et al.* [149]. This group studied the effect on volatile metabolites of infection of potato tubers with several dry and soft rot pathogens (*Erwinia carotovora* ssp. carotovora, and *Fusarium sambucinum*). The authors used headspace GC-MS and were able to detect some 81 volatile metabolites of which a surprising 58 turned out to be specific for one or a small number of diseases. For example, 2,5-norbornadiene and styrene were unique to inoculation with *Fusarium*, whilst acetic acid ethenyl ester was only found following infection with *Erwinia carotovora* ssp. atroseptica. Based on these results the authors concluded that metabolite profiling may prove to be a valuable tool to control and discriminate diseases in stored potato tubers. However, they also cautioned that the high variability found in their results means that considerable effort would need to be invested into measures for validation and standardization before the results could be commercialized.

As more insight is gained into the metabolic nature of the plant response to pathogens, it is likely that new targeted and broad metabolite profiling methods will need to be developed to cover the secondary metabolites that are found to be of particular importance in individual plant–pathogen interactions. For example, Schmelz *et al.* have recently reported on new methods for profiling phytohormones of relevance for biotic stress, including pathogen-produced mimics of phytohormones such as the phytotoxin coronatine [150, 151].

16.8. Plant functional genomics

The potential of metabolite profiling as a tool for plant functional genomics was identified as the first plant genome project, the sequencing of the *Arabidopsis* genome, was entering its final stages [152]. Whilst much of the world was focusing on microarrays as the key tool for functional genomics, the case for the use of metabolite profiling was also compelling. In particular, it is well known that changes in mRNA do not necessarily lead to changes in protein, changes in protein do not always in turn lead to changes in activity, and a change in activity does not always lead to the expected end result. Thus for studies of gene function in areas where metabolism is at the forefront, it is clearly the best route to seek to directly link a modification in a gene to a change in the metabolic network and not to investigate such functionalities through measuring proxies such as mRNA or protein abundance.

In the past there have been many studies where mutant plant collections have been created and screened for a particular phenotype. Lines which were identified were then subjected to complex genetic analysis, for example map-based cloning, in order to identify the genes that had been affected. The closest to the application of broad metabolite profiling in the past were studies where lipid profiles were analyzed in ethyl methanesulfonate (EMS) mutagenised populations, and following much painstaking work genes were identified which had given rise to an altered balance of fatty acids in plants [153]. With the advent of genomics and reverse genetics the possibility emerged to perform this work in the opposite direction, that is a change in gene function is first engineered and the resulting consequence on a metabolite profile was then determined. The first example which demonstrated the feasibility of this approach with broad metabolite profiling was published by Fiehn *et al.* [154] working at the Max-Planck-Institute for Molecular Plant Physiology. In this study, GC-MS analysis was used to analyze some 326 different compounds in *Arabidopsis* leaves, of which roughly half were known metabolites. The method was used to study four different *Arabidopsis* genotypes: the ecotype Columbia (Col-2) and the *dgd1* mutant (reduced digalactosyldiacylglycerol accumulation) in the Col-2 genetic background, as well as the ecotype C24 and the *sdd1-1* mutant in the C24 genetic background (reduced stomatal density). A univariate statistical appraisal of the profiles identified 41 significant changes in the *sdd1-1* mutant; two of the most significant changes were found in unknown hydrophilic substances. In the *dgd1* mutation a much more pleiotropic pattern was found with a total of 153 significant changes in comparison to the parental Columbia ecotype. This latter observation was particularly interesting as it demonstrated how the absence of one particular enzyme could disrupt the balance in around half of the metabolic network surveyed. Multivariate analysis, in the form of PCA, was also applied to the profiles from the four genotypes and the authors were surprised to find that the "metabolic phenotypes" of the two ecotypes were more divergent, than the mutants were from their respective parental ecotypes. This result, that the differences between ecotypes and sub-species can be greater than those caused by even highly pleiotropic alterations in the function of an individual gene, has been confirmed in a range of subsequent studies and indicates the tremendous latent power in natural genetic diversity.

Whilst mRNA profiling remains today the primary tool used in plant functional genomics [155], metabolite profiling studies are beginning to be integrated into a range of contemporary academic programmes for plant functional genomics [131, 1, 14, 156]. Further it has been demonstrated in yeast that metabolic phenotypes can often be found for genes that on deletion had previously not shown any overt phenotype in terms of growth rate or flux [157].

A detailed experimental approach for the application of metabolite profiling to the functional genomics of secondary metabolism in plant cells has been described [158]. In this approach, targeted metabolite analysis for alkaloids, polyamines and phenylpropanoids was combined with cDNA-amplified fragment length polymorphism-based transcript profiling of jasmonate-elicited tobacco Bright yellow 2 cells. Following elicitation, a major shift in gene expression for around 600 genes was observed which correlated well with changes in metabolite levels and indicated that flux in the metabolic network is significantly altered. A similar approach, but with more precise results, was taken by Achnine *et al.* [159] working with an inducible cell culture system from the model legume, *Medigaco truncatula*. Using targeted metabolite profiling they analyzed 30 different saponins and generated transcript profiles following induction with methyl jasmonate. Using data mining approaches on both data sets, over 300 glucosyltransferases were evaluated to determine if they were potential triterpene glycosyltransferases. Two enzymes were identified and their exact specificities for triterpene saponins determined via *in vitro* studies. In the case of one of the enzymes, it was also able to glycosylate certain isoflavones and the flavonol quercertin; however, the integrated transcript and metabolite profiling data clearly indicated a function for the enzyme in terpenoid biosynthesis.

Tohge *et al.* [58] have published a study that illustrates how metabolite profiling data can be combined with DNA microarray results in order to investigate gene function in transgenic plants. *Arabidopsis thaliana* plants were created which overexpressed the PAP1 gene, which encodes a MYB transcription factor that is implicated in flavonoid biosynthesis. A metabolomics data set was generated using FTMS, although as discussed before (Section 16.3.4) no adequate validation of this technology in the context of metabolite profiling has been published to date. The metabolomics data was enhanced through the addition of results from targeted LC-MS analysis for cyanidin and quercetin derivates. Transcript profiles were generated using DNA microarrays that covered some 22,810 genes and it was found that some 38 genes were induced following ectopic overexpression of PAP1. Whilst some of these genes were already known to be involved in anthocyanin biosynthesis, there were several genes identified which had no assigned function or which encoded putative glycotransferases, acyltransferases, transporters or transcription factors. Further evaluation of two of the genes through *in vitro* enzymatic activity tests enable function to be assigned: flavonoid 3-*O*-glucosyltransferase and anthocyanin 5-*O*-glucosyltransferase.

Despite some very innovative individual academic studies, the systematic application of metabolite profiling as a functional genomics tool in plant research has so far been limited to industry. In the case of metanomics, the company that the authors of this chapter represent, both a loss-of-function and a gain-of-function strategy have been followed in the screening of *Arabidopsis* populations. In the loss-of-function, or knockout, approach, a large *Arabidopsis* T-DNA population [160] has been established using classical insertional mutagenesis to disrupt the endogenous genes. In the gain-of-function approach, a facility has been established which is able to introduce new genes into *Arabidopsis* on an individual basis at the rate of 200 per week. The overexpression *Arabidopsis* populations generated to date include those where each one of the yeast and *Escherichia coli* genes have been individually overexpressed. The results from the metabolite profiling of these populations show that this is an effective and efficient route to link genes to particular metabolic functions, either at the level of pathways or at the level of individual metabolites [161]. Taking a global statistical view of the productivity of such studies, metanomics has observed that a significant alteration in any one metabolite can be induced by altering the activity of 0.1–1.0% of the genes in a genome. From the individual results obtained, only a minority of the genes identified at metanomics could have been associated with the observed alterations in metabolism based upon contemporary gene functional annotation or literature. Therefore, there is currently significant opportunity for new and profound discoveries.

The examples of the use of metabolite profiling in functional genomics are growing rapidly. However, it is also clear that steady state measurements on metabolite levels provide only the first insight into gene function. Ultimately, it will be necessary to understand gene function at the level of cellular dynamics, and in order to do this it will be necessary to integrate data on metabolic flux with data on metabolite levels. Fernie *et al.* [6] have made the first detailed case for the necessity of this development, although due to the technical difficulties associated with measuring metabolic flux in plant systems it remains to be seen how quickly progress will be made.

16.9. Breeding

In the previous sections it has repeatedly been illustrated that metabolite profiling can be used to identify metabolites that serve as markers for particular characteristics.

Therefore, it is to be expected that metabolite profiling has a large potential application in the area of breeding where selection decisions need to be made on which material is advanced in multi-year programmes. If metabolite markers can serve as proxies for phenotypic traits, then, just as is the case for DNA-based markers, they can be utilized earlier in the breeding process to distinguish the desired lines. Such applications are currently restricted through two factors: our limited understanding of which metabolites or patterns can be utilized as markers for particular traits and the fact that the cost per analysis far exceeds a level which could be tolerated in a commercial breeding programme. However, these factors have not held back the initation of academic assessment of this potential and in this section some of the early proof of concepts will be introduced and reviewed.

The first indication that metabolite profiling could be an important tool for breeding came with the publication of a study by Taylor *et al.* [21]. These authors set out to test whether metabolite profiling was able to distinguish between related genotypes of *Arabidopsis*. They selected two background ecotypes Col0 and C24 and created two progeny lines through cross breeding the two ecotypes in both directions (the progeny genotypes are then different only in the inherited mitochondrial and chloroplast DNA). Using GC-MS TOF they were able to analyze some 433 metabolites in leaf samples and the authors used neural networks, PCA and hierachical clustering to investigate the ability to discriminate the genotypes. Using neural networks they were able to discriminate all four lines, however with PCA it was possible to investigate the basis for discrimination. It was found that gluose and fructose were the most important metabolites that were different between the two progeny lines, and that malate and citrate were characteristic for the two parental ecotypes. Whilst this is hardly a breeding application, it was the first indication that metabolite markers could be used to discriminate the products of breeding.

Tomato is currently the crop which has been most studied with respect to the utility of metabolite profiling for breeding. Schauer *et al.* [162] have used GC-MS to investigate the relative metabolic levels of leaves and fruits of *Solanum lycopersicum* and five wild species of tomato that can be crossed with it (*S. pimpinellifolium, S. neorickii, S. chmielewskii, S. habrochaites*, and *S. pennellii*). This study was motivated by the fact that commercially cultivated varieties of *S. lycopersicum* have a relatively narrow genetic base which impacts fruit quality and the ability to withstand biotic and abiotic stress. The authors were able to identify crosses where there were relatively high abundances of nutritionally important metabolites or metabolites linked with stress tolerance. The authors proposed that metabolite profiling could be used as a tool to enable decisions to be made upon what combinations to persue in breeding programmes. These ideas have been extended to the use of tomato introgression lines [163] as has been illustrated by a first study from Overy *et al.* [164]. The latter authors used GC-MS TOF metabolite profiling to create a metabolome fingerprint for six *L. pennellii* introgressions within *L. esculentum* and demonstrated how such metabolite information can be linked to genetic maps for each introgression. This demonstrates the potential to undertake very precise breeding supported by both genetic and metabolic biomarkers.

One area where metabolite profiling might find fast acceptance is in the breeding of trees, due to the very long timelines involved and therefore the huge potential value in using diagnostic markers as proxys for desirable characteristics. Morris *et al.* [165] were the first to report on tree metabolite profiling with the publication of an analysis of differentiating loblolly pine xylem tissue using GC-MS. It was possible to quantify around 60 metabolites in this wood forming tissue, of which a structure could be assigned to approximately half. When the method was applied to six different loblolly pine genotypes, including three high cellulose and three medium cellulose lines, it was possible to separate the two groups following PCA of the data. Similarly, Robinson *et al.* [166] have used GC-MS metabolite profiling to study *Populus tremulaxalba* and two transgenic lines with modified lignin monomer composition. Interestingly the authors were able to distinguish the three lines irrespective of whether they took samples from the cambial zone or from a non-lignifying suspension tissue culture. Further it was surprising to find that carbohydrates, which are not directly associated with the monomers of lignin, enabled differentiation between the lines in PCA analysis. In addition, Terskikh *et al.* [40] have recently shown that ^{13}C NMR spectroscopic metabolite profiling could be used to evaluate seed quality in conifer breeding. Taken together, these studies indicate that metabolite profiling approaches may lead to the development of metabolite markers for application in industrial tree breeding.

More complex aspects of breeding may also be amenable to investigation with metabolite profiling. Kirk *et al.* [167] have studied the phenomena of hybrids using crosses between *Senecio aquaticus* and *Senecio jacobaea*. Hybridization is believed to occur frequently within *Senecio* and it is proposed that this maybe an important mechanism that gives rise to the large diversification of secondary metabolism in this genus. Using NMR spectroscopic profiling the authors observed that F1 hybrids contained differential abundance of many metabolites including lower concentrations of amino acids and flavonoids. They concluded that metabolite profiling is a useful technique for identifying qualitative changes in major metabolites and as such may find application in hybrid breeding approaches.

At this timepoint there is a collection of initial studies on the application of metabolite profiling to different questions in plant breeding. Overall, the results from the first studies have been very promising and there appears now to be little doubt that metabolomics technologies, along with other genomics tools, will find considerable application in breeding. However, the extent to which such approaches are adopted in commercial programmes will depend primarily on the cost/throughput that can be achieved.

16.10. Nutrition and medicine

Plants are consumed by humans in a diverse range of forms and for a wide range of purposes. Whilst the nutritional and medical sciences are not the focus of this chapter, it is worth highlighting that, unlike any other genomics technology, metabolite profiling offers the potential to be a force for holistic developments in the life sciences [168, 169, 170]. As more is understood about the importance of particular metabolites for health, metabolite profiling can be used to investigate the abundance of metabolites in particular dietary or herbal sources. Indeed, through the types of applications already discussed in the area of metabolic engineering and breeding, it may be possible to take steps to ensure an optimization of plant material for human and animal health. Studies may be designed where metabolite profiling contributes both to the development and assessment of the plant material as well as to the investigation of the response of the recipient, for example through profiling of blood or urine. As the metabolites involved in the different materials are the same, direct links between metabolites that are ingested and their occurrence, distribution and effects on human or animal subjects might be identified [172].

Evidence for this view has come from some landmark studies on the metabolic effects of chamomile (*Matricaria recutita* L.) ingestion [173]. Using ^{1}H NMR spectroscopy in conjunction with chemometric methods (discriminant analysis with orthogonal signal correction), the response of human subjects to chamomile tea consumption was evaluated. Urine samples were collected from volunteer subjects over a 6-week period (2 weeks of baseline, 2 weeks of daily chamomile tea consumption, and a 2-week post-treatment period). The analysis revealed that there were clear differences between the samples obtained, despite considerable variation between individuals, before and after chamomile ingestion; in particular, the levels of glycine and hippurate were found to be elevated in urine, whilst creatine was reduced. Further, the samples obtained after the post-treatment period clustered separately indicating that the effects of chamomile consumption were long term. As the same authors had previously demonstrated [174] that ^{1}H NMR spectroscopy and chemometrics can be used to characterize and classify chamomile flowers from three different geographical regions (Egypt, Hungary and Slovakia), this work provides evidence that metabolite profiling may one day be used in an integrated fashion to analyze both diet and health. Similarly, Solanky *et al.* [175] have used ^{1}H NMR-based techniques to profile urine collected from premenopausal women undergoing a controlled dietary invention with soy or miso. The aim of this work was to identify the biochemical effects of diet rich in soy isoflavones. The authors concluded that the dietary intake of conjugated and unconjugated isoflavones had significant effects on several metabolic pathways associated with osmolyte fluctuation and energy metabolism in the human subjects. The changes were more significant following consumption of the miso diet (unconjugated soy isoflavone) suggesting that the exact chemical composition of the isoflavones present in soy-based foods may lead to important differences in the biological efficacy *in vivo*.

Interestingly, the potential for metabolite profiling to form a bridge between modern western medicine and traditional Chinese medicine has now been discussed in the literature. Schaneberg *et al.* [176] have analyzed Ma Huang (*Ephedra sinica*), one of the oldest medicinal herbs used in Chinese medicine. With a simple HPLC metabolite profiling method, these authors were able to show that they could distinguish between *Ephedra* species originating from Eurasia, North America or South America. Further, Chiang *et al.* [177] have shown that metabolite profiling can be used to characterize the natural products found in *Bidens pilosa*, an ingredient used in various folk medicines and herbal teas. Indeed, in a recent review article, *et al.* [178] have argued that metabolomics is the "ultimate phenotyping" and have made the case that metabolite profiling has the potential to revolutionise natural product chemistry and advance the use of scientifically based herbal medicines. In particular, these authors have highlighted that reductionist approaches, where analysis is quickly focused on single components, have traditionally been followed in the discovery and development of new medicines. However, metabolite profiling offers the opportunity to conduct studies that take into account the potential synergistic effects of multiple components [179].

16.11. Outlook to systems biology

There is widespread anticipation that metabolite profiling will play an important role in systems biology. However, consensus has not yet been achieved as to what systems biology actually is and there is much hype associated with this subject. A purist definition of systems biology would include the mathematical modelling of cellular processes, whereby metabolite profiling might play an important role in providing the data for the determination of the parameters in the mathematical models. Gutierrez *et al.* [180], in a visionary article, have proclaimed that the goal of systems biology is "to generate a model of the plant as a whole that describes processes across all layers of biological organization (molecular, cellular, physiological, organismal, and ecological)". A more pragmatic view of systems biology is the process of creating hypotheses and knowledge based on two or more "levels" of biology (e.g. metabolism and proteomics or phenotype and mRNA). In the latter case, the term "systems biology" is often applied when the individual levels are investigated using modern genomics tools with wide coverage.

Throughout this chapter, examples have been provided where metabolite profiling has been combined with other data types, particularly results from transcript profiling, in order to generate new hypotheses. For example, the study of secondary metabolism [158, 58, 159], the investigation of biotic stress in tomatoes [146], the analysis of the consequences of dhurin metabolic engineering [89] or the study of nutritional stress [57, 135, 136]. These studies can be regarded the first steps towards systems biology. However, there is some way to go before any of the visions of plant systems biology are achieved. In order for progress to be made, there must be an increase in the amount of inter-disciplinary projects, and plant scientists will have to work closely with experts in computation, statistics, data mining and visualization and chemistry. Concurrently there must also be an evolution in how science funding bodies approach their work in order to foster multi-disciplinary projects. For an assessment of the opportunities and challenges associated with developing systems biology in plant science, it is worth studying the results of the US Department of Energy Workshop on Plant System Biology [181], or reading the position papers of the European Plant Science Organisation [182], www.epsoweb.org.

16.12. Summary and outlook

These are exciting times to be associated with plant metabolite profiling. After a relatively slow start over the last 10–20 years, where applications focused on crop protection and metabolic engineering, the last few years have seen a rapid increase in the number of groups developing and applying metabolite profiling technologies. Consequently, the range of applications that have been recently explored and published has widened substantially. The value of metabolite profiling in plant functional genomics has now been rigorously demonstrated through a range of studies in primary and secondary metabolism. Proof of concepts for the utility of metabolite profiling in the study of complex biological phenomena such as development, yield and stress have been obtained and the potential of the technology to identify important metabolite biomarkers has been demonstrated. Further, there are now good indications that plant metabolite profiling data will be used in breeding and in conjunction with nutritional research. It seems likely that metabolite profiling will become a routine technology in the tool box of the plant scientist and that the use of metabolite profiling will foster inter-disciplinary exploration of the nature and utility of plants.

References

[1] Bino, R.J., Hall, R.D., Fiehn, O., Kopka, J., Saito, K., Draper, J., Nikolau, B.J., Mendes, P., Roessner-Tunali, U., Beale, M.H., Trethewey, R.N., Lange, B.M., Wurtele, E.S., Sumner, L.W. (2004) Potential of metabolomics as a functional genomics tool. *Trends Plant Sci.* 9:418–425.

[2] Schwab, W. (2003) Metabolome diversity: Too few genes, too many metabolites? *Phytochem.* 62:837–849.

[3] Verpoorte, R. (2000) Plant secondary metabolism. In: R. Verpoorte and A.W. Alfermann (eds) Metabolic Engineering of Plant Secondary Metabolism. Kluwer, Dordrecht, The Netherlands, pp. 1–29.

[4] Wink, M. (1988) Plant breeding: Importance of plant secondary metabolites for protection against pathogens and herbivores. *Theor. Appl. Genet.* 75:225–233.

[5] Baker, D.D., Alvi, K.A. (2004) Small-molecule natural products: New structures, new activities. *Curr. Opin Biotech.* 15:576–583.

[6] Fernie, A.R., Geigenberger, P., Stitt, M. (2005) Flux an important, but neglected, component of functional genomics. *Curr. Opin. Plant Biol.* 8:174–182.

[7] Schwender, J., Ohlrogge, J., Schacher-Hill, Y. (2004) Understanding flux in plant metabolic networks. *Curr. Opin. Plant Biol.* 7:309–317.

[8] Sweetlove, L.J., Fernie, A.R. (2005) Regulation of metabolic networks: Understanding metabolic complexity in the systems biology era. *New Phytol.* 168:9–24.

[9] Zhao, J., Davis, L.C., Verpoorte, R. (2005) Elicitor signal transduction leading to production of plant secondary metabolites. *Biotechol. Adv.* 23:283–333.

[10] Buckhout, T.J., Thimm, O. (2003) Insights into metabolism obtained from microarray analysis. *Curr. Opin. Plant Biol.* 6:288–296.

[11] Stitt, M., Fernie, A.R. (2003) From measurements of metabolites to etabolomics: an "on the fly" perspective illustrated by recent studies of carbon–nitrogen interactions. *Crr. Opin. Biotechnol.* 14:136–144.

[12] Kopka, J., Schauer, N., Krueger, S., Birkemeyer, C., Usadel, B., Bergmuller, E., Dormann, P., Weckwerth, W., Gibon, Y., Stitt, M., Willmitzer, L., Fernie, A.R., Steinhauser, D. (2005) GMD@CSB.DB: The Golm Metabolome Database. *Bioinformatics* 21:1635–1638.

[13] Ratcliffe, R.G., Sachar-Hill, Y. (2005) Revealing metabolic phenotypes in plants: Inputs from NMR analysis. Biological Reviews of the Cambridge Philosophical Society 80:27–43.

[14] Sumner, L.W., Mendes, P., Dixon, R.A. (2003) Plant metabolomics: Large-scale phytochemistry in the functional genomics area. *Phytochem.* 62:817–836.

[15] Horning, E.C., Horning, M.G. (1971) Metabolic profiles: Gas-phase ethods for analysis of metabolites, *Clin. Chem.* 17:802–809.

[16] Niwa, T. (1986), Metabolic profiling with gas chromatography-mass spectrometry and its application to clinical medicine. *J. Chromatogr.* 20:313–345.

[17] Sauter, H., Lauer, M., Fritsch, H. (1991) Metabolic profiling of plants – A new diagnostic technique. In: D.R. Baker, J.G. Fenyes, W.K. Moberg (eds) Synthesis and Chemistry of Agrochemicals II. American Chemical Society, Washington DC, pp. 288–299.

[18] Roessner, U., Wagner, C., Kopka, J., Trethewey, R.N., Willmitzer, L. (2000) Simultaneous analysis of metabolites in potato tuber by gas chromatography-mass spectrometry. *Plant J.* 23:131–142.

[19] Santos, F.J., Galceran, M.T. (2003) Modern developments in gas chromatography-mass spectrometry-based environmental analysis. *J. Chromatogr.* A 1000:125–151.

[20] Fiehn, O. (2003) Metabolic networks of *Cucurbita maxima* phloem. *Phytochem.* 62:875–886.

[21] Taylor, J., King, R.D., Altmann, T., Fiehn, O. (2002) Application of metabolomics to plant genotype discrimination using statistics and machine learning. *Plant J.* 18:241–248.

[22] Dallüge, J., Beens, J., and Brinkman, U.A.T (2003) Comprehensive two-dimensional gas chromatography: a powerful and versatile analytical tool. *J. Chromatogr. A.* 1000:69–108.

[23] Adahchour, M., Brandt, M., Baier, H.U., Vreuls, R.J., Batenburg, A.M., Brinkman, U.A. (2005) Comprehensive two-dimensional gas chromatography coupled to a rapid-scanning quadrupole mass spectrometer: Principles and applications. *J. Chromatogr. A.* 1067:245–254.

[24] Zrostlikova, J., Hajslova, J., Cajka, T. (2003) Evaluation of two-dimensional gas chromatography-time-of-flight mass spectrometry for the determination of multiple pesticide residues in fruit. *J. Chromatogr.* A. 1019:173–186.

[25] Shellie, R., Marriott, P., Morrison, P. (2004) Comprehensive two-dimensional gas chromatography with flame ionization and time-of-flight mass spectrometry detection: Qualitative and quantitative analysis of West Australian sandalwood oil. *J. Chromatogr. Sci.* 42:417–422.

[26] Zhu, S., Lu, X., Dong, L., Xing, J., Su, X., Kong, H., Xu, G., Wu, C. (2005b) Quantitative determination of compounds in tobacco essential oils by comprehensive two-dimensional gas chromatography coupled to time-of-flight mass spectrometry. *J. Chromatogr. A.* 1086:107–114.

[27] Ryan, D., Shellie, R., Tranchida, P., Casilli, A., Mondello, L., Marriott, P. (2004) Analysis of roasted coffee bean volatiles by using comprehensive two-dimensional gas chromatography-time-of-flight mass spectrometry. *J. Chromatogr. A.* 1054:57–65.

[28] Di, X., Shellie, R.A., Marriott, P.J., Huie, C.W. (2004) Application of headspace solid-phase microextraction (HS-SPME) and comprehensive two-dimensional gas chromatography (GC × GC) for the chemical profiling of volatile oils in complex herbal mixtures. *J. Sep. Sci.* 27:451–458.

[29] Welthagen, W., Shellie, R., Ristow, M., Spranger, J., Zimmermann, R., Fiehn, O. (2005) Comprehensive two dimensional gas chromatography – time of flight mass spectrometry, GCxGC-TOF for high resolution metabolomics: Biomarker discovery on spleen tissue extracts of obese NZO compared to lean C57BL/6 mice. *Metabolomics J.* 1:57–65.

[30] Shellie, R.A., Welthagen, W., Zrostlikova, J., Spranger, J., Ristow, M., Fiehn, O., Zimmermann, R. (2005) Statistical methods for comparing comprehensive two-dimensional gas chromatography-time-of-flight mass spectrometry results: Metabolomic analysis of mouse tissue extracts. *J. Chromatogr.* A. 1086:83–90.

[31] Schauer, N., Steinhauser, D., Strelkov, S., Schomburg, D., Allison, G., Moritz, T., Lundgren, K., Roessner-Tunali, U., Forbes, M.G., Willmitzer, L., Fernie, A.R., Kopka, J. (2005a) GC-MS libraries for the rapid identification of metabolites in complex biological samples. *FEBS Lett.* 579:1332–1337.

[32] Wilson, I.D., Plumb, R., Granger, J., Major, H., Williams, R., Lenz, E.A. (2005a) HPLC-MS-based methods for the study of metabonomics. *J. Chromatogr. B.* 817:67–76.

[33] Huhman, D.V., Sumner, L.W. (2002) Metabolic profiling of saponins in Medicago sativa and Medicago truncatula using HPLC coupled to an electrospray ion-trap mass spectrometer. *Phytochem.* 59:347–360.

[34] Lange, B.M., Ketchum, R.E.B., Croteau, R.B. (2001) Isoprenoid biosynthesis. Metabolite profiling of peppermint oil gland secretory cells and application to herbicide target analysis. *Plant Physiol.* 127:305–314.

[35] Niessen, W.M. (2003) Progress in liquid chromatography-mass spectrometry instrumentation and its impact on high-throughput screening. *J. Chromatogr. A.* 1000:413–436.

[36] Hayen, H., Karst, U. (2003) Strategies for the liquid chromatographic-mass spectrometric analysis of non-polar compounds. *J. Chromatogr. A* 1000:549–565.

[37] Roepenack-Lahaye, E. v., Degenkolb, T., Zerjeski, M., Franz, M., Roth, U., Wessjohann, L., Schmidt, J., Scheel, D., Clemens, S. (2004) Profiling of *Arabidopsis* secondary metabolites by capillary liquid chromatography coupled to electrospray ionization quadrupole time-of-flight mass spectrometry. *Plant Physiol.* 134:548–559.

[38] Plumb, R., Castro-Perez, J., Granger, J., Beattie, I., Joncour, K., Wright, A. (2004) Ultra-performance liquid chromatography coupled to quadrupole-orthogonal time-of-flight mass spectrometry. *Rapid Commun. Mass Spectrom.* 18:2331–2337.

[39] Wilson, I.D., Nicholson, J.K., Castro-Perez, J., Granger, J.H., Johnson, K.A., Smith, B.W., Plumb, R.S. (2005b) High resolution "Ultra performance" liquid chromatography coupled to

oa-TOF mass spectrometry as a tool for differential metabolic pathway profiling in functional genomic studies. *J. Proteome Res.* 4:591–598.

[40] Terskikh, V.V., Feurtado, J.A., Borchardt, S., Giblin, M., Abrams, S.R., Kermode, A.R. (2005) In vivo 13C NMR metabolite profiling: potential for understanding and assessing conifer seed quality. *J. Exp. Bot.* 56:2253–2265.

[41] Choi, Y.H., Sertic, S., Kim, H.K., Wilson, E.G., Michopoulos, F., Lefeber, A.W., Erkelens, C., Prat Kricun, S.D., Verpoorte, R. (2005) Classification of Ilex species based on metabolomic fingerprinting using nuclear magnetic resonance and multivariate data analysis. *J. Agric. Food Chem.* 53:1237–1245.

[42] Defernez, M., Colquhoun, I.J. (2003) Factors affecting the robustness of metabolite fingerprinting using 1H NMR spectra. *Phytochemistry* 62:1009–1017.

[43] Ward, J.L., Harris, C., Lewis, J., Beale, M.H. (2003) Assessment of 1H NMR spectroscopy and multivariate analysis as a technique for metabolite fingerprinting of Arabidopsis thaliana. *Phytochem.* 62:949–957.

[44] Bailey, N.J., Oven, M., Holmes, E., Nicholson, J.K., Zenk, M.H. (2003) Metabolomic analysis of the consequences of cadmium exposure in *Silene cucubalus* cell cultures via H-1 NMR spectroscopy and chemometrics. *Phytochem.* 62:851–858.

[45] Skordi, E., Wilson, I.D., Lindon, J.C., Nicholson, J.K. (2004) Characterization and quantification of metabolites of racemic ketoprofen excreted in urine following oral administration to man by 1H-NMR spectroscopy, directly coupled HPLC-MS and HPLC-NMR, and circular dichroism. *Xenobiotica* 34:1075–1089.

[46] Bailey, N.J., Stanley, P.D., Hadfield, S.T., Lindon, J.C., Nicholson, J.K. (2000) Mass spectrometrically detected directly coupled high performance liquid chromatography/nuclear magnetic resonance spectroscopy/mass spectrometry for the identification of xenobiotic metabolites in maize plants. *Rapid Commun. Mass Spectrom.* 14:679–684.

[47] Mesnard, F., Ratcliffe, R.G. (2005) NMR analysis of plant nitrogen metabolism. *Photosynth. Res.* 83:163–180.

[48] Krishnan, P., Kruger, N.J., Ratcliffe, R.G. (2005) Metabolite fingerprinting and profiling in plants using NMR. *J. Exp. Bot.* 56:255–265.

[49] Ratcliffe, R.G., Shachar-Hill, Y. (2001) Probing plant metabolism with NMR. *Annu. Rev. Plant Physiol. Plant Mol. Biol.* 52:499–526.

[50] Bligny, R., Douce, R. (2001) NMR and plant metabolism. *Curr. Opin. Plant Biol.* 4:191–196.

[51] Soga, T., Ueno, Y., Naraoka, H., Ohashi, Y., Tomita, M. and Nishioka, T. (2002) Simultaneous determination of anionic intermediates for *Bacillus subtilis* metabolic pathway by capillary electrophoresis electrospray ionization mass spectrometry. *Anal. Chem.* 74: 2233–2239.

[52] Sato, S., Soga, T., Nishioka, T., Tomita, M. (2004) Simultaneous determination of the main metabolites in rice leaves using capillary electrophoresis mass spectrometry and capillary electrophoresis diode array detection. *Plant J.* 40:151–163.

[53] Brown, S.C., Kruppa, G., Dasseux, J.L. (2005) Metabolomics applications of FT-ICR mass spectrometry. *Mass Spectrom. Rev.* 24:223–231.

[54] Zhang, J., McCombie, G., Guenat, C., Knochenmuss, R. (2005) FT-ICR mass spectrometry in the drug discovery process. *Drug Discovery Today* 10:635–642.

[55] Aharoni, A., de Vos, R.C., Verhoeven, H.A., Maliepaard, C.A., Kruppa, G., Bino, R., Goodenowe, D.B. (2002) Nontargeted metabolome analysis by use of Fourier Transform Ion Cyclotron Mass Spectrometry. *Omics* 6:217–234.

[56] Cook, D., Fowler, S., Fiehn, O., Thomashow, M.F. (2004) A prominent role for the CBF cold response pathway in configuring the low-temperature metabolome of *Arabidopsis. Proc. Natl. Acad. Sci. U.S.A.* 101:15243–15248.

[57] Hirai, M.Y., Yano, M., Goodenowe, D.B., Kanaya, S., Kimura, T., Awazuhara, M., Arita, M., Fujiwara, T., Saito, K. (2004) Integration of transcriptomics and metabolomics for understanding of global responses to nutritional stresses in *Arabidopsis thaliana. Proc. Natl. Acad. Sci. U.S.A.* 101:10205–10210.
[58] Tohge, T., Nishiyama, Y., Hirai, M.Y., Yano, M., Nakajima, J., Awazuhara, M., Inoue, E., Takahashi, H., Goodenowe, D.B., Kitayama, M., Noji, M., Yamazaki, M., Saito, K. (2005) Functional genomics by integrated analysis of metabolome and transcriptome of *Arabidopsis* plants over-expressing an MYB transcription factor. *Plant J.* 42:218–235.
[59] Mungur, R., Glass, A.D., Goodenow, D.B., Lightfoot, D.A. (2005) Metabolite fingerprinting in transgenic nicotiana tabacum altered by the escherichia coli glutamate dehydrogenase gene. *J. Biomed. Biotechnol.* 2005:198–214.
[60] Annesley, T.M. (2003) Ion suppression in mass spectrometry. *Clin. Chem.* 47:1041–1044.
[61] Matuszewski, B.K., Constanzer, M.L., Chavez-Eng, C.M. (2003) Strategies for the assessment of matrix effect in quantitative bioanalytical methods based on HPLC-MS/MS. *Anal. Chem.* 75:3019–3030.
[62] Schrader, W., Klein, H.W. (2004) Liquid chromatography/Fourier transform ion cyclotron resonance mass spectrometry (LC-FTICR MS): an early overview. *Anal. Bioanal. Chem.* 379:1013–1024.
[63] Dunn, W.B., Overy, S., Quick, W.P. (2005) Evaluation of automated electrospray-TOF mass spectrometryfor metabolic fingerprinting of the plant metabolome. *Metabolomics J.* 2:137–148.
[64] Johnson, H.E., Broadhurst, D., Kell, D.B., Theodorou, M.K., Merry, R.J., Griffith, G.W. (2004) High-throughput metabolic fingerprinting of legume silage fermentations via Fourier transform infrared spectroscopy and chemometrics. *Appl. Environ Microbiol.* 70:1583–1592.
[65] Johnson, H.E., Broadhurst, D., Goodacre, R., Smith, A.R. (2003) Metabolic fingerprinting of salt-stressed tomatoes. *Phytochem.* 62:919–928.
[66] Gidman, E., Goodacre, R., Emmett, B., Smith, A.R., Gwynn-Jones, D. (2003) Investigating plant-plant interference by metabolic fingerprinting. *Phytochem.* 63:705–710.
[67] Lindon, J.C., Nicholson, J.K., Holmes, E., Keun, H.C., Craig, A., Pearce, J.T., Bruce, S.J., Hardy, N., Sansone, S.A., Antti, H., Jonsson, P., Daykin, C., Navarange, M., Beger, R.D., Verheij, E.R., Amberg, A., Baunsgaard, D., Cantor, G.H., Lehman-McKeeman, L., Earll, M., Wold, S., Johansson, E., Haselden, J.N., Kramer, K., Thomas, C., Lindberg, J., Schuppe-Koistinen, I., Wilson, I.D., Reily, M.D., Robertson, D.G., Senn, H., Krotzky, A., Kochhar, S., Powell, J., van der Ouderaa, F., Plumb, R., Schaefer, H., Spraul, M. (2005) Standard Metabolic Reporting Structures working group. Summary recommendations for standardization and reporting of metabolic analyzes. *Nat. Biotechnol.* 23:833–838.
[68] Jenkins, H., Hardy, N., Beckmann, M., Draper, J., Smith, A.R., Taylor, J., Fiehn, O., Goodacre, R., Bino, R.J., Hall, R., Kopka, J., Lane, G.A., Lange, B.M., Liu, J.R., Mendes, P., Nikolau, B.J., Oliver, S.G., Paton, N.W., Rhee, S., Roessner-Tunali, U., Saito, K., Smedsgaard, J., Sumner, L.W., Wang, T., Walsh, S., Wurtele, E.S., Kell, D.B. (2004) A proposed framework for the description of plant metabolomics experiments and their results. *Nat. Biotechnol.* 22:1601–1606.
[69] Jenkins, H., Johnson, H., Kular, B., Wang, T., Hardy, N. (2005) Toward supportive data collection tools for plant metabolomics. *Plant Physiol.* 138:67–77.
[70] Grossmann, K. (2005) What it takes to get a herbicide's mode of action. Physionomics, a classical approach in a new complexion. *Pest Manag. Sci.* 61:423–431.
[71] Ott, K.H., Aranibar, N., Singh, B.J., Stockton, G.W. (2003) Metabonomics classifies pathways affected by bioactive compounds. Artificial neural network classification of NMR spectra of plant extracts. *Phytochem.* 62:971–985.

[72] Raghavan, C., Ong, E.K., Dalling, M.J., Stevenson, T.W. (2005) Effect of herbicidal application of 2,4-dichlorophenoxyacetic acid in Arabidopsis. *Funct. Integr. Genomics* 5:4–17.
[73] Lechelt-Kunze, C., Meissner, R.C., Drewes, M., Tietjen, K. (2003) Flufenacet herbicide treatment phenocopies the fiddlehead mutant in *Arabidopsis thaliana. Pest Manag. Sci.* 59:847–856.
[74] Dixon, R.A. (2005) Engineering of plant natural product pathways. *Curr. Opin. Plant Biol.* 8:329–336.
[75] Trethewey, R.N. (2004) Metabolite profiling as an aid to metabolic engineering in plants. *Curr. Opin. Plant Biol.* 7:196–201.
[76] Regierer, B., Fernie, A.R., Springer, F., Perez-Melis, A., Leisse, A., Koehl, K., Willmitzer, L., Giegenberger, P., Kossmann, J. (2002) Starch content and yield increase as a result of altering adenylate pools in transgenic plants. *Nat. Biotechnol.* 20:1256–1260.
[77] Roessner, U., Willmitzer, L., Fernie, A.R. (2001b) High-resolution metabolic phenotyping of genetically and environmentally diverse potato tuber systems. Identification of phenocopies. *Plant Physiol.* 127:749–764.
[78] Bohmert, K., Balbo, I., Kopka, J., Mittendorf, V., Nawrath, C., Poirier, Y., Tischendorf, G., Trethewey, R.N., Willmitzer, L. (2000) Transgenic *Arabidopsis* plants can accumulate polyhydroxybutyrate to up to 4% of their fresh weight. *Planta* 211:841–845.
[79] Beyer, P., Al-Babili, S., Ye, X., Lucca, P., Schaub, P., Welsch, R., Potrykus, I.J. (2002) Golden Rice: Introducing the beta-carotene biosynthesis pathway into rice endosperm by genetic engineering to defeat vitamin A deficiency. *J. Nutr.* 132:506–510.
[80] Houmiel, K.L., Slater, S., Broyles, D., Casagrande, L., Colburn, S., Gonzalez, K., Mitsky, T.A., Reiser, S.E., Shah, D., Taylor, N.B., Tran, M., Valentin, H.E., Gruys, K.J. (1999) Poly(betahydroxybutyrate) production in oilseed leukoplasts of brassica napus. *Planta* 209:547–550.
[81] Kuiper, H.A., Kok, E.J., Engel, K-H. (2003) Exploitation of molecular profiling techniques for GM food safety assessment. *Curr. Opin. Biotech.* 14:238–243.
[82] Roessner, U., Luedemann, A., Brust, D., Fiehn, O., Linke, T., Willmitzer, L., Fernie, A.R. (2001a) Metabolic profiling allows comprehensive phenotyping of genetically or environmentally modified plant systems. *Plant Cell* 13:11–29.
[83] Trethewey, R.N., Reismeier, J.W., Willmitzer, L., Stitt, M., Geigenberger, P. (1999a) Tuber specific expression of a yeast invertase and a bacterial glucokinase in potato leads to an activation of sucrose phosphate synthase and the creation of a futile cycle. *Planta* 208:227–238.
[84] Trethewey, R.N., Geigenberger, P., Riedel, K., Hajirezaei, M.-R., Sonnewald, U., Stitt, M., Riesmeier, J.W., Willmitzer, L. (1998) Combined expression of glucokinase and invertase in potato tubers leads to a dramatic reduction in starch accumulation and a stimulation of glycolysis. *Plant J.* 15:109–118.
[85] Roessner, U., Willmitzer, L., Fernie, A.R. (2002) Metabolic profiling and biochemical phenotyping of plant systems. *Plant Cell Rep.* 21:189–196.
[86] Fernie, A.R., Willmitzer, L., Trethewey, R.N. (2002) Sucrose To Starch: A transition in molecular plant physiology. *Trends Plant Sci.* 7:35–41.
[87] Morino, K., Matsuda, F., Miyazawa, H., Sukegawa, A., Miyagawa, H., Wakasa, K. (2005) Metabolic profiling of tryptophan-overproducing rice calli that express a feedback-insensitive alpha subunit of anthranilate synthase. *Plant Cell Physiol.* 46:514–521.
[88] Matsuda, F., Yamada, T., Miyazawa, H., Miyagawa, H., Wakasa, K. (2005) Characterization of tryptophan-overproducing potato transgenic for a mutant rice anthranilate synthase alphasubunit gene (OASA1D). *Planta* 222:535–545.
[89] Kristensen, C., Morant, M., Olsen, C.E., Ekstrom, C.T., Galbraith, D.W., Moller, B.L., Bak, S. (2005) Metabolic engineering of dhurrin in transgenic *Arabidopsis* plants with marginal inadvertent effects on the metabolome and transcriptome. *Proc. Natl. Acad. Sci. U.S.A.* 102:1779–1784.

[90] Catchpole, G.S., Beckmann, M., Enot, D.P., Mondhe, M., Zywicki, B., Taylor, J., Hardy, N., Smith, A., King, R.D., Kell, D.B., Fiehn, O., Draper, J. (2005) Hierarchical metabolomics demonstrates substantial compositional similarity between genetically modified and conventional potato crops. *Proc. Natl. Acad. Sci. U.S.A.* 102:14458–14462.

[91] Carrari, F., Urbanczyk-Wochniak, E., Willmitzer, L., Fernie, A.R. (2003) Engineering central metabolism in crop species: Learning the system. *Metabolic Engineering* 5:191–200.

[92] Verpoorte, R., Memelink, J. (2002) Engineering secondary metabolite production in plants. *Curr. Opin. Biotechnol.* 13:181–187.

[93] Dixon, R.A., Achnine, L., Kota, P., Liu, C.J., Reddy, M.S.S., Wang, L.J. (2002) The phenylpropanoid pathway and plant defence – a genomics perspective. *Mol. Plant Path.* 3:371–390.

[94] Anterola, A.M., Lewis, N.G. (2002) Trends in lignin modification: A comprehensive analysis of the effects of genetic manipulations/mutations on lignification and vascular integrity. *Phytochem.* 61:221–294.

[95] Humphreys, J.M., Chapple, C. (2002) Rewriting the lignin roadmap. *Curr. Opin. Plant Biol.* 5:224–229.

[96] Guo, D., Chen, F., Inoue, K., Blount, J.W., Dixon, R.A. (2001) Downregulation of caffeic acid 3-*O*-methyltransferase and caffeoyl CoA 3-*O*-methyltransferase in transgenic alfalfa. impacts on lignin structure and implications for the biosynthesis of G and S lignin. *Plant Cell* 13:73–88.

[97] Meyermans, H., Morreel, K., Lapierre, C., Pollet, B., De Bruyn, A., Busson, R., Herdewijn, P., Devreese, B., Van Beeumen, J., Marita, J.M., Ralph, J., Chen, C., Burggraeve, B., Van Montagu, M., Messens, E., Boerjan, W. (2000) Modifications in lignin and accumulation of phenolic glucosides in poplar xylem upon down-regulation of caffeoyl-coenzyme A *O*-methyltransferase, an enzyme involved in lignin biosynthesis. *J. Biol. Chem.* 275:36899–36909.

[98] Chen, F., Duran, A.L., Blount, J.W., Sumner, L.W., Dixon, R.A. (2003) Profiling phenolic metabolites in transgenic alfalfa modified in lignin biosynthesis. *Phytochem.* 64:1013–1021.

[99] Sirtori, C.R., Arnoldi, A., Johnson, S.K. (2005) Phytoestrogens: End of a tale? *Ann. Med.* 37:423–438.

[100] Dixon, R.A. (2004) Phytoestrogens. *Annu. Rev. Plant. Biol.* 55:225–261.

[101] Kachlicki, P., Marczak, L., Kerhoas, L., Einhorn, J., Stobiecki, M. (2005) Profiling isoflavone conjugates in root extracts of lupine species with LC/ESI/MSn systems. *J. Mass Spectrom.* 40:1088–10103.

[102] Wu, Q., Wang, M., Sciarappa, W.J., Simon, J.E. (2004) LC/UV/ESI-MS analysis of isoflavones in Edamame and Tofu soybeans. *J. Agric. Food Chem.* 52:2763–2769.

[103] Zhu, D., Hettiarachchy, N.S., Horax, R., Chen, P. (2005a) Isoflavone contents in germinated soybean seeds. *Plant Foods Hum. Nutr.* 60:147–151.

[104] Gu, L., Laly, M., Chang, H.C., Prior, R.L., Fang, N., Ronis, M.J., Badger, T.M. (2005) Isoflavone conjugates are underestimated in tissues using enzymatic hydrolysis. *J. Agric. Food Chem.* 53:6858–6863.

[105] Liu, C.J., Blount, J.W., Steele, C.L., Dixon, R.A. (2002) Bottlenecks for metabolic engineering of isoflavone glycoconjugates in *Arabidopsis*. *Proc. Natl. Acad. Sci. U.S.A.* 99:14578–14583.

[106] Jung, W., Yu, O., Lau, S.M., O'Keefe, D.P., Odell, J., Fader, G., McGonigle, B. (2000) Identification and expression of isoflavone synthase, the key enzyme for biosynthesis of isoflavones in legumes. *Nat. Biotech.* 18:208–212.

[107] Weichert, H., Kohlmann, M., Wasternack, C., Feussner, I. (2000) Metabolic profiling of oxylipins upon sorbitol treatment in barley leaves. *Biochem. Soc. Trans.* 28:861–862.

[108] Weichert, H., Kolbe, A., Kraus, A., Wasternack, C., Feussner, I. (2002) Metabolic profiling of oxylipins in germinating cucumber seedlings lipoxygenase-dependent degradation of triacylglycerols and biosynthesis of volatile aldehydes. *Planta* 215:612–619.

[109] Sessà, R.A., Bennett, M.H., Lewis, M.J., Mansfield, J.W., Beale, M.H. (2000) Metabolite profiling of sesquiterpene lactones from Lactuca species. Major latex components are novel oxalate and sulfate conjugates of lactucin and its derivatives. *J. Biol. Chem.* 275:26877–26884.

[110] Ketchum, R.E.B., Rithner, C.D., Qiu, D., Kim, Y.S., Williams, R.M., Croteau, R.B. (2003) Taxus metabolomics: Methyl jasmonate preferentially induces production of taxoids oxygenated at C-13 in Taxus x media cell cultures. *Phytochem.* 62:901–909.

[111] Madhusudanan, K.P., Chattopadhyay, S.K., Tripathi, V., Sashidhara, K.V., Kumar, S. (2002) MS/MS profiling of taxoids from the needles of Taxus wallichiana. *Phytochem. Anal.* 13:18–30.

[112] Yamazaki, Y., Urano, A., Sudo, H., Kitajima, M., Takayama, H., Yamazaki, M., Aimi, N., Saito, K. (2003a) Metabolite profiling of alkaloids and strictosidine synthase activity in camptothecin producing plants. *Phytochem.* 62:461–470.

[113] Yamazaki, Y., Nakajima, J., Yamanashi, M., Sugiyama, M., Makita, Y., Springob, K., Awazuhara, M., Saito, K. (2003b) Metabolomics and differential gene expression in anthocyanin chemo-varietal forms of Perilla frutescens. *Phytochem.* 62:987–995.

[114] Huhman, D.V., Berhow, M.A, Sumner, L.W. (2005) Quantification of saponins in aerial and subterranean tissues of Medicago truncatula. *J. Agric. Food Chem.* 53:1914–1920.

[115] Verdonk, J.C., de Vos, C.H.R., Verhoeven, H.A., Haring, M.A., van Tunen, A.J., Schuurink, R.C. (2003) Regulation of floral scent production in petunia revealed by targeted metabolomics. *Phytochem.* 62:997–1008.

[116] Flamini, G., Cioni, P.L., Morelli, I. (2003) Differences in the fragrances of pollen, leaves, and floral parts of garland (*Chrysanthemum coronarium*) and composition of the essential oils from flowerheads and leaves. *J. Agric. Food Chem.* 51:2267–2271.

[117] Brown, K. (2002) Plant genetics. Something to sniff at: unbottling floral scent. *Science* 296:2327–2329.

[118] Parr, A.J., Mellon, F.A., Colquhoun, I.J., Davies, H.V. (2005) Dihydrocaffeoyl polyamines (kukoamine and allies) in potato (Solanum tuberosum) tubers detected during metabolite profiling. *J. Agric. Food Chem.* 53:5461–5466.

[119] Birkemeyer, C., Kolasa, A., Kopka, J. (2003) Comprehensive chemical derivatization for gas chromatography-mass spectrometry-based multi-targeted profiling of the major phytohormones. *J. Chromatogr. A.* 993:89–102.

[120] Wang, H.K., Sakurai, N., Shih, C.Y., Lee, K.H. (2005a) LC/TIS-MS fingerprint profiling of Cimicifuga species and analysis of 23-Epi-26-deoxyactein in Cimicifuga racemosa commercial products. *J. Agric. Food Chem.* 53:1379–1386.

[121] Demirci, B., Goppel, M., Demirci, F., Franz, G. (2004) HPLC profiling and quantification of active principles in leaves of Hedera helix L. *Pharmazie* 59:770–774.

[122] Bailey, N.J., Sampson, J., Hylands, P.J., Nicholson, J.K, Holmes, E. (2002) Multi-component metabolic classification of commercial feverfew preparations via high-field 1H-NMR spectroscopy and chemometrics. *Planta Med.* 68:734–738.

[123] Woo, S.S., Song, J.S., Lee, J.Y., In, D.S., Chung, H.J., Liu, J.R., Choi, D.W. (2004) Selection of high ginsenoside producing ginseng hairy root lines using targeted metabolic. *Phytochem.* 65:2751–2761.

[124] Oksman-Caldentey, K.M., Inze, D. (2004) Plant cell factories in the post-genomic era: New ways to produce designer secondary metabolites. *Trends Plant Sci.* 9:433–440.

[125] Walker, T.S., Bais, H.P., Halligan, K.M., Stermitz, F.R., Vivanco, J.M. (2003) Metabolic profiling of root exudates of *Arabidopsis thaliana*. *J. Agric. Food Chem.* 51:2548–2554.

[126] Narasimhan, K., Basheer, C., Bajic, V.B., Swarup, S. (2003) Enhancement of plant-microbe interactions using a rhizosphere metabolomics-driven approach and its application in the removal of polychlorinated biphenyls. *Plant Physiol.* 132:146–153.

[127] Urbanczyk-Wochniak, E., Baxter, C., Kolbe, A., Kopka, J., Sweetlove, L.J., Fernie, A.R. (2005b) Profiling of diurnal patterns of metabolite and transcript abundance in potato (Solanum tuberosum) leaves. *Planta* 221:891–903.

[128] Jeong, M.L., Jiang, H., Chen, H.S., Tsai, C.J., Harding, S.A. (2004) Metabolic profiling of the sink-to-source transition in developing leaves of quaking aspen. *Plant Physiol.* 136:3364–3375.

[129] Tarpley, L., Duran, A.L., Kebrom, T.H., Sumner, L.W. (2005) Biomarker metabolites capturing the metabolite variance present in a rice plant developmental period. *B.M.C. Plant Biol.* 5:8–20.

[130] Diaz, C., Purdy, S., Christ, A., Morot-Gaudry, J.F., Wingler, A., Masclaux-Daubresse, C. (2005) Characterization of markers to determine the extent and variability of leaf senescence in *Arabidopsis*. A metabolic profiling approach. *Plant Physiol.* 138:898–908.

[131] Desbrosses, G.G., Kopka, J., Udvardi, M.K. (2005) Lotus japonicus metabolic profiling. Development of gas chromatography-mass spectrometry resources for the study of plant-microbe interactions. *Plant Physiol.* 137:1302–1318.

[132] Sinclair, T.R., Purcell, L.C., Sneller, C.H. (2004) Crop transformation and the challenge to increase yield potential. *Trends Plant Sci.* 9:70–75.

[133] Hirai, M.Y., Klein, M., Fujikawa, Y., Yano, M., Goodenowe, D.B., Yamazaki, Y., Kanaya, S., Nakamura, Y., Kitayama, M., Suzaki, H., Sakurai, N., Shibata, D., Tokuhisa, J., Reichelt, M., Gershonzon, J., Paperbrock, J., Saito, K. (2005) Elucidation of gene-to-gene and metabolite-to-gene networks in *Arabidopsis* by integration of *J. Biol. Chem.* 280:25590–25595.

[134] Nikiforova, V.J., Gakiere, B., Kempa, S., Adamik, M., Willmitzer, L., Hesse, H., Hoefgen, R. (2004) Towards dissecting nutrient metabolism in plants: a systems biology case study on sulphur metabolism. *J. Exp. Bot.* 55:1861–1870.

[135] Nikiforova, V.J., Kopka, J., Tolstikov, V., Fiehn, O., Hopkins, L., Hawkesford, M.J., Hesse, H., Hoefgen, R. (2005a) Systems rebalancing of metabolism in response to sulfur deprivation, as revealed by metabolome analysis of *Arabidopsis* plants. *Plant Physiol.* 138:304–318.

[136] Nikiforova, V.J., Daub, C.O., Hesse, H., Willmitzer, L., Hoefgen, R. (2005b) Integrative gene-metabolite network with implemented causality deciphers informational fluxes of sulphur stress response. *J. Exp. Bot.* 56:1887–1896.

[137] Urbanczyk-Wochniak, E., Fernie, A.R. (2005a) Metabolic profiling reveals altered nitrogen nutrient regimes have diverse effects on the metabolism of hydroponically-grown tomato (Solanum lycopersicum) plants. *J. Exp. Bot.* 56:309–321.

[138] Hirai, M.Y., Saito, K. (2004) Post-genomics approaches for the elucidation of plant adaptive mechanisms to sulphur deficiency. *J. Exp. Bot.* 55:1871–1879.

[139] Wang, W., Vinocur, B., Altman, A. (2003) Plant responses to drought, salinity and extreme temperatures: Towards genetic engineering for stress tolerance. *Planta* 218:1–14.

[140] Welti, R., Li, W., Li, M., Sang, Y., Biesiada, H., Zhou, H.E., Rajashekar, C.B., Williams, T.D., Wang, X. (2002) Profiling membrane lipids in plant stress responses. Role of phospholipase D alpha in freezing-induced lipid changes in *Arabidopsis*. *J. Biol. Chem.* 277: 31994–32002.

[141] Kaplan, F., Kopka, J., Haskell, D.W., Zhao, W., Schiller, K.C., Gatzke, N., Sung, D.Y., Guy, C.L. (2004) Exploring the temperature-stress metabolome of *Arabidopsis*. *Plant Physiol.* 136:4159–4168.

[142] Browse, J., Lange, B.M. (2004) Counting the cost of a cold-blooded life: metabolomics of cold acclimation. *Proc. Natl. Acad. Sci. U.S.A.* 101:14996–14997.

[143] Rizhsky, L., Liang, H., Shuman, J., Shulaev, V., Davletova, S., Mittler, R. (2004) When defense pathways collide. The response of *Arabidopsis* to a combination of drought and heat stress. *Plant Physiol.* 134:1683–1696.

[144] Bino, R.J., Ric de Vos, C.H., Lieberman, M., Hall, R.D., Bovy, A., Jonker, H.H., Tikunov, Y., Lommen, A., Moco, S., Levin, I. (2005) The light-hyperresponsive high pigment-2dg mutation of tomato: Alterations in the fruit metabolome. *New Phytol.* 166:427–438.

[145] Gatehouse, J.A. (2002) Plant resistance towards insect herbivores: A dynamic interaction. *New Phytol.* 156:145–169.

[146] Kant, M.R., Ament, K., Sabelis, M.W., Haring, M.A., Schuurink, R.C. (2004) Differential timing of spider mite-induced direct and indirect defenses in tomato plants. *Plant Physiol.* 135:483–495.

[147] Ament, K., Kant, M.R., Sabelis, M.W., Haring, M.A., Schuurink, R.C. (2004) Jasmonic acid is a key regulator of spider mite-induced volatile terpenoid and methyl salicylate emission in tomato. *Plant Physiol.* 135:2025–2037.

[148] Broeckling, C.D., Huhman, D.V., Farag, M.A., Smith, J.T., May, G.D., Mendes, P., Dixon, R.A., Sumner, L.W. (2005) Metabolic profiling of Medicago truncatula cell cultures reveals the effects of biotic and abiotic elicitors on metabolism. *J. Exp. Bot.* 56:323–336.

[149] Lui, L.H., Vikram, A., Abu-Nadal, Y., Kushalappal, A.C., Raghavan, G.S.V., Al-Mughrabi, K. (2005) Volatile Metabolic Profiling for Discrimination of Potato Tubers Inoculated With Dry and Soft Rot Pathogens. *Am. J. Potato Res.* 82:1–8.

[150] Schmelz, E.A., Engelberth, J., Alborn, H.T., O'Donnell, P., Sammons, M., Toshima, H., Tumlinson, J.H. 3rd. (2003) Simultaneous analysis of phytohormones, phytotoxins, and volatile organic compounds in plants. *Proc. Natl. Acad. Sci. U.S.A.* 100:10552–10557.

[151] Schmelz, E.A., Engelberth, J., Tumlinson, J.H., Block, A., Alborn, H.T. (2004) The use of vapor phase extraction in metabolic profiling of phytohormones and other metabolites. *Plant J.* 39:790–808.

[152] Trethewey, R.N., Krotzky, A.J., Willmitzer, L. (1999b) Metabolic profiling: A rosetta stone for genomics. *Cur. Opin. Plant Biol.* 2:83–85.

[153] Somerville, C., Browse, J. (1996) Dissecting desaturation: Plants prove advantageous. *Trends Cell Biol.* 6:148–153.

[154] Fiehn, O., Kopka, J., Dörmann, P., Altmann, T., Trethewey, R.N., Willmitzer, L. (2000) Metabolite profiling for plant functional genomics. *Nat. Biotechnol.* 18:1157–1161.

[155] Borevitz. J.O., Ecker, J.R. (2004) Plant genomics: the third wave. *Annu. Rev. Genomics Hum. Genet.* 5:443–477.

[156] Fiehn, O. (2002) Metabolomics-the link between genotypes and phenotypes. *Plant Mol. Biol.* 48:155–171.

[157] Raamsdonk, L.M., Teusink, B., Broadhurst, D., Zhang, N., Hayes, A., Walsh, M.C., Berden, J.A., Brindle, K.M., Kell, D.B., Rowland, J.J., Westerhoff, H.V., van Dam, K., Oliver, S.G. (2001) A functional genomics strategy that uses metabolome data to reveal the phenotype of silent mutations. *Nat. Biotechnol.* 19:45–50.

[158] Goossens, A., Hakkinen, S.T., Laakso, I., Seppanen-Laakso, T., Biondi, S., De Sutter, V., Lammertyn, F., Nuutila, A.M., Soderlund, H., Zabeau, M., Inze, D., Oksman-Caldentey, K.M. (2003) A functional genomics approach toward the understanding of secondary metabolism in plant cells. *Proc. Natl. Acad. Sci. U.S.A.* 100:8595–8600.

[159] Achnine, L., Huhman, D.V., Farag, M.A., Sumner, L.W., Blount, J.W., Dixon, R.A. (2005) Genomics-based selection and functional characterization of triterpene glycosyltransferases from the model legume Medicago truncatula. *Plant J.* 41:875–887.

[160] Azprioz-Leehan, R., Feldmann, K.A. (1997) T-DNA insertion mutagenesis in *Arabidopsis* going back and forth. *Trends in Genetics* 13:152–156.

[161] Fernie, A.R., Trethewey, R.N., Krotzky, A.J., Willmitzer, L. (2004) Metabolite profiling: From diagnostics to systems biology. *Nat. Rev. Mol. Cell. Biol.* 5:763–769.

[162] Schauer, N., Zamir, D., Fernie, A.R. (2005b) Metabolic profiling of leaves and fruit of wild species tomato: a survey of the Solanum lycopersicum complex. *J. Exp. Bot.* 56: 297–307.

[163] Fridman, E., Carrari, F., Liu, Y.S., Fernie, A.R., Zamir, D. (2004) Zooming in on a quantitative trait for tomato yield using interspecific introgressions. *Science* 305:1786–1789.

[164] Overy, S.A., Walker, H.J., Malone, S., Howard, T.P., Baxter, C.J., Sweetlove, L.J., Hill, S.A., Quick, W.P. (2005) Application of metabolite profiling to the identification of traits in a population of tomato introgression lines. *J. Exp. Bot.* 56:287–296.

[165] Morris, C.R., Scott, J.T., Chang, H.M., Sederoff, R.R., O'Malley, D., Kadla, J.F. (2004) Metabolic profiling: A new tool in the study of wood formation. *J. Agric. Food Chem.* 52:1427–1434.

[166] Robinson, A.R., Gheneim, R., Kozak, R.A., Ellis, D.D., Mansfield, S.D. (2005) The potential of metabolite profiling as a selection tool for genotype discrimination in Populus. *J. Exp. Bot.* 56:2807–2819.

[167] Kirk, H., Choi, Y.H., Kim, H.K., Verpoorte, R., van der Meijden, E. (2005) Comparing metabolomes: The chemical consequences of hybridization in plants. *New Phytol.* 167:613–622.

[168] Gibney, M.J., Walsh, M., Brennan, L., Roche, H.M., German, B., van Ommen, B. (2005) Metabolomics in human nutrition: Opportunities and challenges. *Am. J. Clin. Nutr.* 82:497–503.

[169] German, B., Schiffrin, E.J., Reniero, R., Mollet, B., Pfeifer, A., Neeser, J.R. (1999) The development of functional foods: Lessons from the gut. *Trends Biotechnol.* 17:492–499.

[170] German, J.B., Bauman, D.E., Burrin, D.G., Failla, M.L., Freake, H.C., King, J.C., Klein, S., Milner, J.A., Pelto, G.H., Rasmussen, K.M., Zeisel, S.H. (2004) Metabolomics in the opening decade of the 21st century: Building the roads to individualized health. *J. Nutrition* 134:2729–2732.

[171] German, J.B., Watkins, S.M., Fay, L.B. (2005) Metabolomics in Practice: Emerging Knowledge to Guide Future Dietetic Advice toward Individualized Health. *J. Am. Dietetic Assoc.* 105:1425–1432.

[172] Zeisel, S.H., Freake, H.C., Bauman, D.E., Bier, D.M., Burrin, D.G., German, J.B., Klein, S., Marquis, G.S., Milner, J.A., Pelto, G.H., Rasmussen, K.M. (2005) The nutritional phenotype in the age of metabolomics. *J. Nutr.* 135:1613–1616.

[173] Wang, Y., Tang, H., Nicholson, J.K., Hylands, P.J., Sampson, J., Holmes, E. (2005c) A metabonomic strategy for the detection of the metabolic effects of chamomile (*Matricaria recutita* L.) ingestion. *J. Agric. Food Chem.* 53:191–196.

[174] Wang, Y., Tang, H., Nicholson, J.K., Hylands, P.J., Sampson, J., Whitcombe, I., Stewart, C.G., Caiger, S., Oru, I., Holmes, E. (2004) Metabolomic strategy for the classification and quality control of phytomedicine: A case study of chamomile flower (*Matricaria recutita* L.). *Planta Med.* 70:250–255.

[175] Solanky, K.S., Bailey, N.J., Beckwith-Hall, B.M., Bingham, S., Davis, A., Holmes, E., Nicholson, J.K. (2005) Cassidy A. Biofluid in nutrition research – metabolic effects of dietary isoflavones in humans. *J. Nutr. Biochem.* 16:236–244.

[176] Schaneberg, B.T., Crockett, S., Bedir, E., Khan, I.A. (2003) The role of chemical fingerprinting: Application to *Ephedra. Phytochem.* 62:911–918.

[177] Chiang, Y.M., Chuang, D.Y., Wang, S.Y., Kuo, Y.H., Tsai, P.W., Shyur, L.F. (2004) Metabolite profiling and chemopreventive bioactivity of plant extracts from Bidens pilosa. *J. Ethnopharmacol.* 95:409–419.

[178] Wang, M., Lamers, R.J., Korthout, H.A., van Nesselrooij, J.H., Witkamp, R.F., van der Heijden, R., Voshol, P.J., Havekes, L.M., Verpoorte, R., van der Greef, J. (2005b) Metabolomics in the context of systems biology: Bridging traditional Chinese medicine and molecular pharmacology. *Phytother. Res.* 19:173.

[179] Verpoorte, R., Choi, Y.H., Kim, H.K. (2005) Ethnopharmacology and systems biology: A perfect holistic match. *J. Ethnopharmacol.* 100:53–56.
[180] Gutierrez, R.A., Shasha, D.E., Coruzzi, G.M. (2005) Systems biology for the virtual plant. *Plant Physiol.* 138:550–554.
[181] Minorsky, P.V. (2003) Achieving the in Silico Plant. Systems Biology and the Future of Plant Biological Research. *Plant Phys.* 132:404–409.
[182] EPSO (2005) European plant science: A field of opportunities. *J. Exp. Bot.* 56:1699–1709.

The Handbook of Metabonomics and Metabolomics
John C. Lindon, Jeremy K. Nicholson and Elaine Holmes (Editors)

Chapter 17

In vivo NMR Applications of Metabonomics

Dieter Leibfritz, Wolfgang Dreher and Wieland Willker

Universität Bremen, Fachbereich 2 (Biologie/Chemie) D – 28334 Bremen, Germany

17.1. Introduction

As the metabolite composition within living tissue and its waste management reflects the physiological or pathophysiological state of the total organism, a fast and reliable analysis of metabolites within a specimen or the intact living system will supplement medical diagnosis as well as basic research in various biochemical aspects. NMR spectroscopy is capable of recording many metabolites either *in vivo* or within *ex vivo* samples without preceding isolation of individual chemical compounds. Higher magnetic field strengths and other hardware improvements reveal about 20–30 metabolites *in vivo* in humans or animals and more than hundred metabolites in *ex vivo* specimens like biopsies, tissue extracts and body fluids. The rather moderate sensitivity of NMR spectroscopy is compensated by its high dispersion allowing the recognition of individual metabolites even within complex mixtures of components. The latter may comprise hydrophilic and lipophilic compounds or molecules with low and high molecular weights. This diversity is not overcome by most other analytical methods.

In order to recognize an individual metabolite within such a complex mixture, assignment techniques are required which unambiguously identify the molecular structure without additional manipulations such as separation and/or derivatization. A common procedure to assign a particular component within a mixture is the addition of authentic reference compounds ("spiking"), but this cannot be used for heterogeneous samples or *in vivo*. Therefore, multi-pulse and/or multi-dimensional NMR tools had to be established for a specific assignment of chemical compounds in biological samples. Although developed already for high resolution NMR analysis of homogeneous and pure solutions, they had to be adapted to the specific demands and challenges of physiological specimens or living tissue.

Besides the complexity of their compound composition, biological samples vary enormously with respect to other physical parameters, such as morphological heterogeneity (i.e. extracellular/intracellular compartments; liquid and solid/semi-solid compartments; hydrophilic and lipophilic substructures; phase transitions), viscosity (low and high molecular weight components), solution properties (homogeneous and heterogeneous solution), different nuclear spin relaxivities, different diffusion behaviour and motional artefacts. Therefore, each class of samples will require dedicated assignment techniques, which deal with the texture properties and molecular specificities characterizing the biological material. As body fluids contain many low molecular components with long relaxation times, unrestricted mobility and homogeneous distribution, they allow and require other pulse sequences compared with the *in vivo* situation characterized by an enormous heterogeneity, short relaxation times, motional artefacts and the need of volume-dependent recording of spectra. In the following sections dedicated assignment techniques oriented to the sample demands will be discussed, in particular the body fluid situation versus *in vivo* demands. Besides the *in situ* assignment of individual metabolites the overall spectrum profile might be characterized by pattern recognition methods as described elsewhere in this book.

17.2. *In situ* NMR Assignments of body fluid metabolites

17.2.1. Urine

The large number of components and the variability of their composition are favourably tackled with pattern recognition methods for NMR analysis. However, there are exceptional situations, where one or more metabolites appear unexpectedly in case of intoxications, inborn errors [1] and so on. Metabolite assignment without separation techniques will require a detailed analysis with two-dimensional (2D) heteronuclear-resolved NMR spectroscopy which is a prerequisite.

The analysis of a urine sample from a patient suffering from 3-OH-3-methylglutaryl-CoA lyase deficiency will be used for demonstration. The deficiency of this enzyme, which catalyses the final step of leucine degradation, will increase the concentration of several metabolites in the preceding reactions (Figure 17.1). With improved magnetic field strength and methods, a much more complete assignment was possible compared to former NMR spectroscopic studies [2]. The patient was also treated with the antiepileptic drug Levetiracetam. Figure 17.2 shows the one-dimensional (1D) proton NMR spectrum of the urine with several prominent unusual signals. These signals are the characteristic metabolites of the disease,

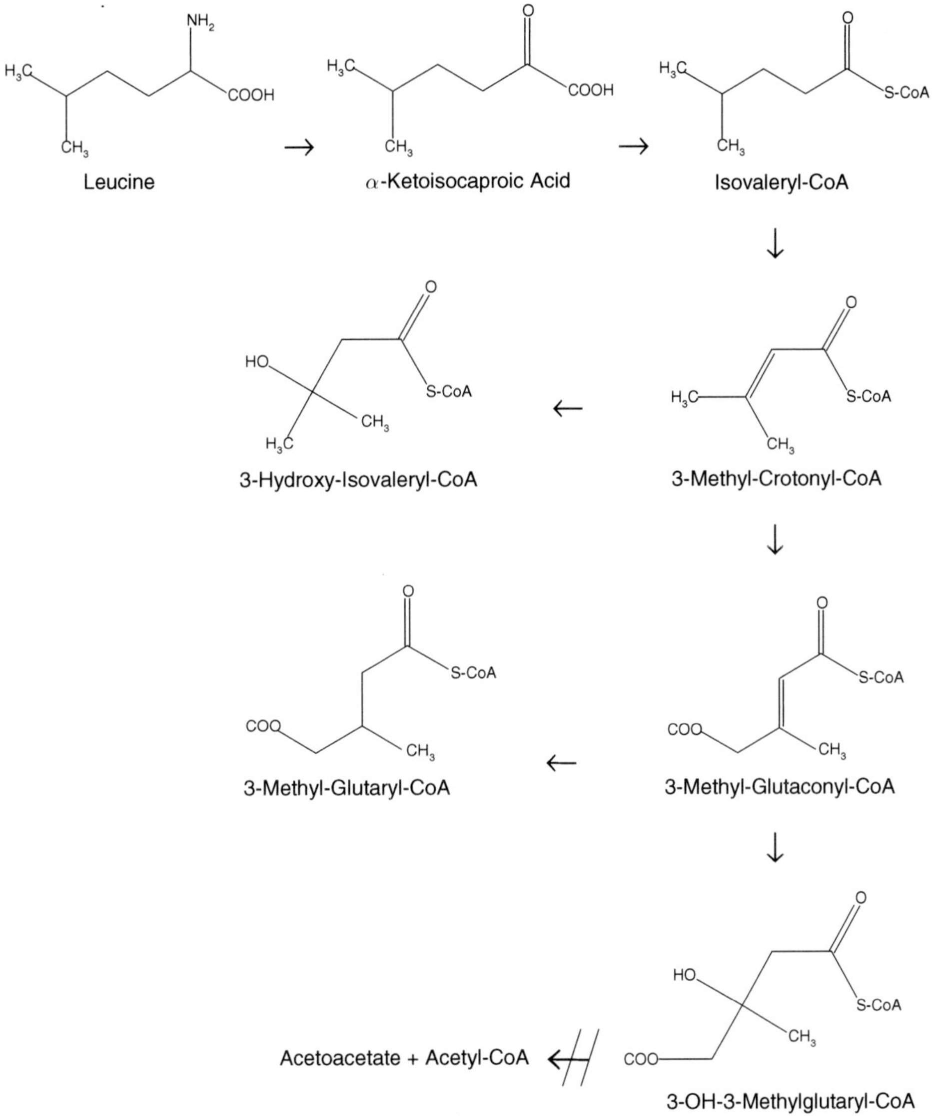

Leucine → α-Ketoisocaproic Acid → Isovaleryl-CoA

↓

3-Hydroxy-Isovaleryl-CoA ← 3-Methyl-Crotonyl-CoA

↓

3-Methyl-Glutaryl-CoA ← 3-Methyl-Glutaconyl-CoA

↓

Acetoacetate + Acetyl-CoA ⇷ 3-OH-3-Methylglutaryl-CoA

Figure 17.1. Leucine catabolism

3-OH-isovalerate (A), 3-methyl-glutaconylate (D) and 3-OH-methyl-glutarate (E). Additionally one can see signals of Levetiracetam.

The 2D heteronuclear single quantum coherence spectroscopy (HSQC) spectrum (Figure 17.3) reveals additional signals from two other metabolites of much lower concentration: 3-methyl-crotonylate and 3-methyl-glutarate (B, C). Several

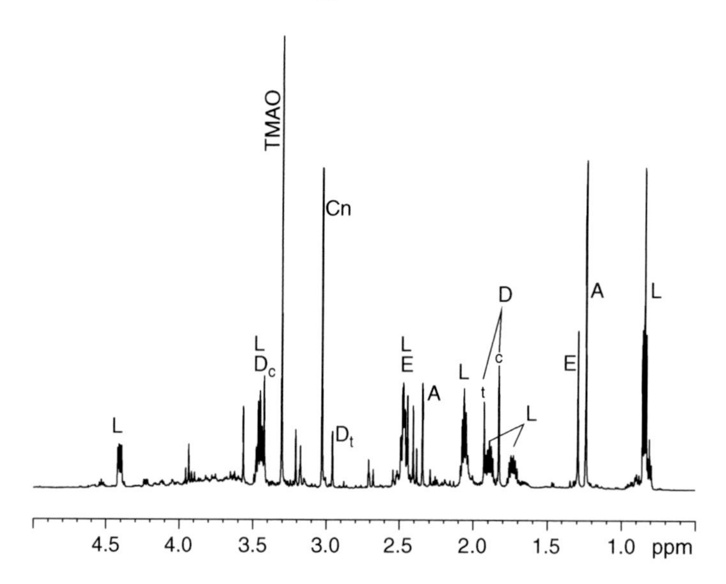

Figure 17.2. High field region of a 1D ^{1}H urine NMR spectrum from a patient suffering from 3-OH-3methylglutaryl-CoA lyase deficiency. The major metabolites are labelled: A (3-OH-isovalerate), D (3-methyl-glutaconylate (*c* (*cis*), *t* (*trans*))), E (3-OH-methyl-glutarate). Drug: L (Levetiracetam) TMAO (trimethylamine-N-oxide), Cn (Creatinine).

superimposed signals within the 1D spectrum are now also separated. Levetiracetam shows a minor second set of signals, which belongs to the free acid. The assignment is further corroborated by a HSQC-TOCSY spectrum (Figure 17.4) showing the correlations of the double bond signals of 3-methyl-crotonylate and 3-methyl-glutaconylate. The assignments are given in Table 17.1.

17.2.2. Plasma and CSF

Different to urine, the composition and the concentration of metabolites of plasma and CSF are maintained in a steady state by the healthy organism using homeostasis. Any detected deviation likely reflects a pathophysiological state. The composition of blood plasma and CSF resemble each other, but differ mainly among the concentration of the metabolites. Plasma contains large amounts of lipoproteins, which are present only at very low concentrations in CSF. The NMR analysis of blood plasma is often hampered because of its heterogeneous mixture of (protein-bound) lipids and water-soluble metabolites. For this reason it is recommended to separate lipids and water-soluble metabolites before NMR analysis (see below).

Of the water-soluble metabolites, glucose and lactate dominate the spectra. They are present in much higher concentrations than all other metabolites. Therefore, their signal sometimes exceeds the dynamic range of the A/D converter and prevents

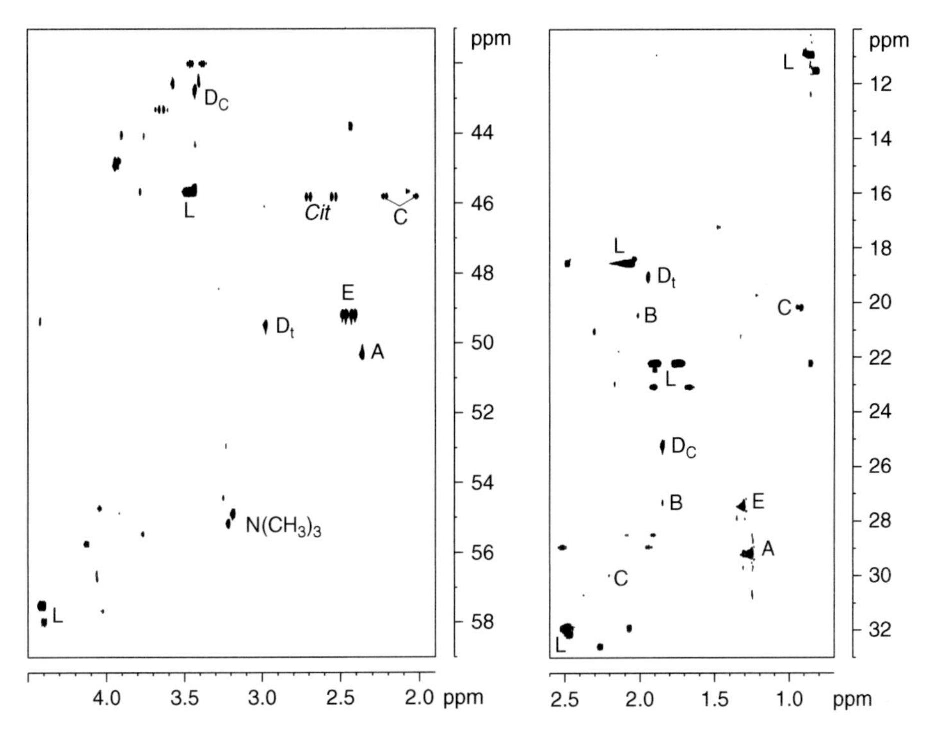

Figure 17.3. High field regions of the 2D ^{1}H–^{13}C HSQC urine NMR spectrum from a patient suffering from 3OH-3-methylglutaryl-CoA lyase deficiency. The metabolites are labelled: A (3-OH-isovalerate), B (3-methyl-crotonylate), C (3-methyl-glutarate), D (3-methyl-glutaconylate (*c* (*cis*), *t* (*trans*)), E (3-OH-methyl-glutarate). Drug: L (Levetiracetam).

the detection of metabolites with lower concentration. In 1D ^{1}H NMR spectra, one cannot easily analyse the chemical shift range between 3 and 4 ppm, because of the large glucose signals. Again the 2D HSQC spectrum offers improved resolution and signal separation for this reason. As an example we compare here one high-field region of plasma and CSF (Figure 17.5). Characteristic differences between plasma and CSF are the absence of proline in CSF and the absence of *N*-acetylglutamate (NAG) in plasma. The following water-soluble compounds have been found and assigned in HSQC spectra:

Plasma: acetate, alanine, aspartate, betaine, citrate, creatine, creatinine, ethanolamine, formate, fructose, glucose, glutamine, glutamate, glycine, histidine, α-hydroxybutyrate, β-hydroxybutyrate, 3-hydroxy-methyl-glutarate, 3-hydroxy-isovalerate, isoleucine, lactate, leucine, lysine, phenylalanine, proline, pyroglutamate, succinate, taurine, threonine, tyrosine, tryptophan, valine.

CSF: acetate, *N*-acetylglutamate, alanine, arginine, aspartate, betaine, citrate, creatine, creatinine, ethanolamine, formate, fructose, glucose, glutamine, glycine,

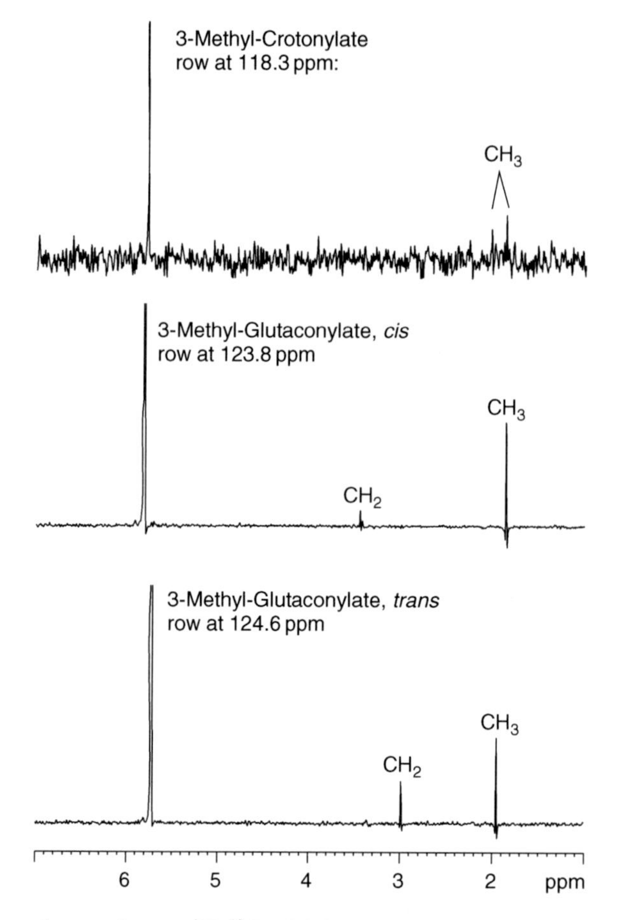

Figure 17.4. Extracted rows from a ^1H–^{13}C HSQC-TOCSY urine NMR spectrum from a patient suffering from 3-OH-3-methylglutaryl-CoA lyase deficiency. The three double bond signals are shown. The top row represents 3-methyl-crotonylate, which is at rather low concentration. The CH_3 correlations of this metabolite are very weak, because they represent ^4J couplings (1.3 Hz).

histidine, α-hydroxybutyrate, β-hydroxybutyrate, 3-hydroxy-methyl-glutarate, 3-hydroxyisobutyrate, myo-inositol, scyllo-inositol, 3-hydroxy-isovalerate, isoleucine, lactate, leucine, lysine, methionine, phenylalanine, 1,2-propanediol, pyroglutamate, succinate, taurine, threonine, TMAO, tyrosine, tryptophan, valine.

17.2.3. Comparison of the lipid composition of blood plasma and CSF

The assignment of lipids in NMR spectra is complicated by the fact that many very similar molecules give rise to the signals. For a detailed assignment very high resolved 2D HSQC NMR spectra have to be used [3, 4]. In absolute amounts, lipids are only present in very low concentrations in CSF (about 1% of the plasma concentration). With respect to their relative concentration ratios one observes some general rules:

(i) Free fatty acids have a much higher relative concentration in CSF than in plasma (30% vs 1–2%) (cf. figure 17.6).

(ii) The compositions of unsaturated fatty acids are different, that is based on setting the fatty acid 18:1 content as 100% in plasma, the relative concentrations of 18:2 are 39% (plasma) and 7% (CSF). For higher unsaturated fatty acids, the values are 38% (plasma) and 22% (CSF); (Figure 17.7).

(iii) Cholesterol esters are lower concentrated in CSF. The ratio is about 1:2 in CSF versus plasma.

Table 17.1
Carbon and proton chemical shifts of metabolites of the leucine catabolism and of 2-(2-oxopyrrolin-1-yl)butanoic acid amide (Levetiracetam). The carbon shift is followed by the attached proton shift where present.

3-OH-Isovalerate (A)	D_c (*cis*): (*cis:trans* ratio is about 3:2)
C1 181.5 –	C1 175.9 –
C2 50.4 2.35 (singlet)	C2 123.8 5.78 (doublet, J = 1.3 Hz)
C3 71.0 –	C3 147.6 –
C4 29.2 1.25 (singlet)	C4 42.8 3.43 (doublet, J < 1 Hz)
	C5 180.4 –
3-Methyl-crotonylate (B)	C3′ 25.1 1.83 (doublet, J = 1.3 Hz)
C1 –	
C2 118.3 5.76 (septet, 1.3 Hz)	3-OH-3-methyl-glutarate (E)
C3 –	C1,5 180.8 –
C4 20.5 2.01 (doublet)	C2,4 49.3 2.46, 2.40 (AB, J = 13.8 Hz)
C3′ 27.4 1.84 (doublet)	C3 71.4 –
	C3′ 27.4 1.30
3-Methyl-glutarate (C)	
C1,5 183.3 –	2-(2-oxopyrrolin-1-yl)butanoic acid amid (Levetiracetam):
C2,4 45.9 2.21, 2.00 (ABX)	C1 175.9 –
C3 30.1 2.20	C2 57.6 4.40 Dd, J_1 = 5.4 Hz J2 = 10.4 Hz
C3′ 20.3 0.91 (doublet, 7.7 Hz)	C3 22.2 1.89, 1.74 multiplet ($ABMX_3$)
	C4 10.9 0.851 J = 7.3 Hz
3-Methyl-glutaconylate	C2′ 180.3 C3′ 32.0 2.48 C4′ 18.5 2.06 C5′ 45.7 3.46
(D_t (*trans*))	(The free acid is also observed as minor metabolite (approx. 15%)
C1 177.1 –	The metabolite ratios are: A:B:C:D_c:D_t:E = 100:5:5:50:34:56
C2 124.6 5.70 (quartet, J = 1.3 Hz)	
C3 145.7 –	
C4 49.5 2.96 (doublet, J = 1 Hz)	
C5 180.9 –	
C3′ 19.0 1.94 (doublet, J = 1.3 Hz)	

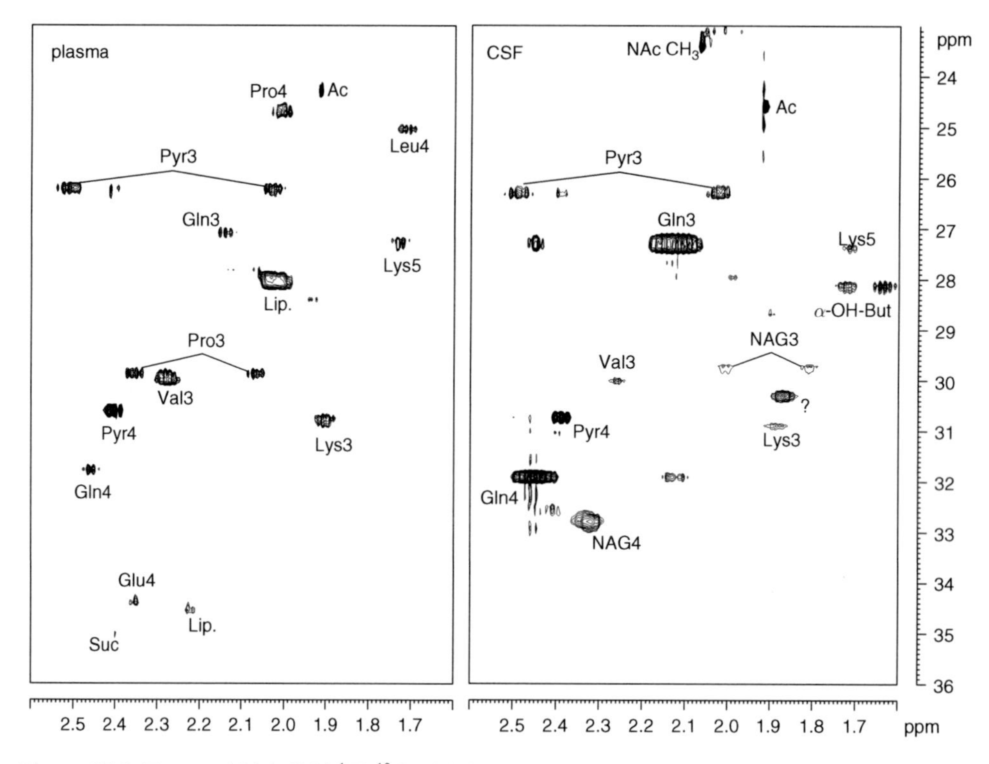

Figure 17.5. Extracted high field ^{1}H–^{13}C HSQC NMR spectral regions from blood plasma and CSF. The plasma sample is from a healthy volunteer, the CSF sample is a mixture from several individuals. Ac (acetate), Gln (glutamine), Glu (glutamate), Leu (leucine), Lip (lipid residues), Lys (lysine), NAG (*N*-acetylglutamate), Pyr (pyroglutamate), Pro (proline), Suc (succinate), Val (valine), α-OH-But (α-hydroxybutyrate).

17.2.4. Creatinine

Creatinine is often used as a reference to calculate relative concentrations or calculate chemical shifts. A peculiar behaviour of creatinine is observed in D_2O solutions. The CH_2 group of creatinine undergoes an H/D exchange. If lyophilized samples of biofluids containing creatinine are dissolved in D_2O, one observes a fast decay of the signal intensity of the CH_2 group in the 1D ^{1}H NMR spectrum while the

O
H_3C—N NH
NH
creatinine

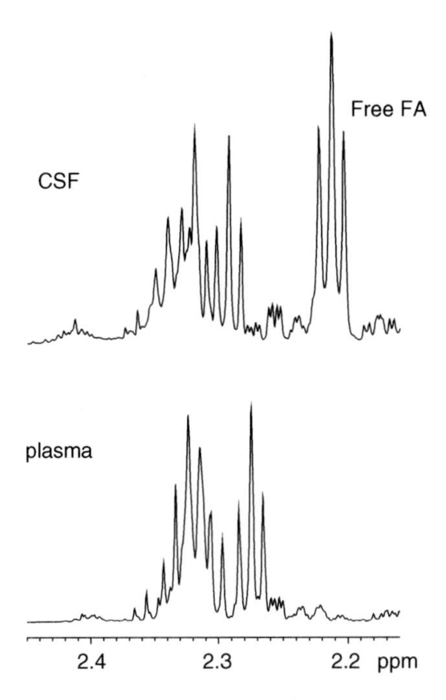

Figure 17.6. A selected region (F_α protons) of 800 MHz ^{1}H NMR spectra of lipid extracts from CSF and blood plasma. A high concentration of free fatty acids is observed in CSF. FA: free fatty acids.

CH_3 group is not affected. This exchange does not happen with creatine. In HSQC spectra the signal of the CH_2 group shows a characteristic tilted triplet pattern, see Figure 17.8. The signal is separated in F1 with the C–D coupling constant (23 Hz) and in the F2 dimension with the H–D coupling constant (2 Hz).

This can be explained by an exchange of one of the two protons of the CH_2 group with deuterium. The spin quantum number of deuterium is 1 and gives rise to a triplet. A similar pattern is known from deuterated methanol (CD_2HOD), which gives a quintet. We have carried out a few tests which indicate that the exchange is faster with increasing pH. It is note worthy that the signal does not disappear completely, but remains as a triplet, suggesting that the deuteration is not complete.

Recently we have found two other (as yet unidentified) signals in blood plasma which represent partially deuterated CH_2 groups. This phenomenon should be kept in mind, especially since creatinine is sometimes used as a concentration standard.

17.2.5. Sample extraction and preparation methods for body fluids

17.2.5.1. Plasma

Blood of 10 ml is collected in a heparinized centrifuge tube and centrifuged at 4000 rpm for 15 min. 3 ml of Plasma is taken from the supernatant and subsequently

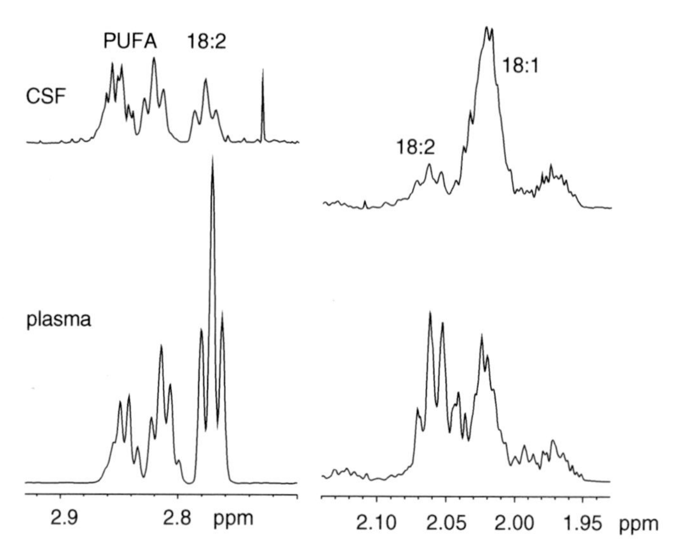

Figure 17.7. PUFA (polyunsaturated fatty acids) $-CH_2=CH_2-CH_2-CH_2=CH_2-$ (left) and MOFA (monounsaturated fatty acids) $-CH_2=CH_2-CH_2-$ regions in 800 MHz ^1H NMR spectra of lipid extracts from CSF and blood plasma.

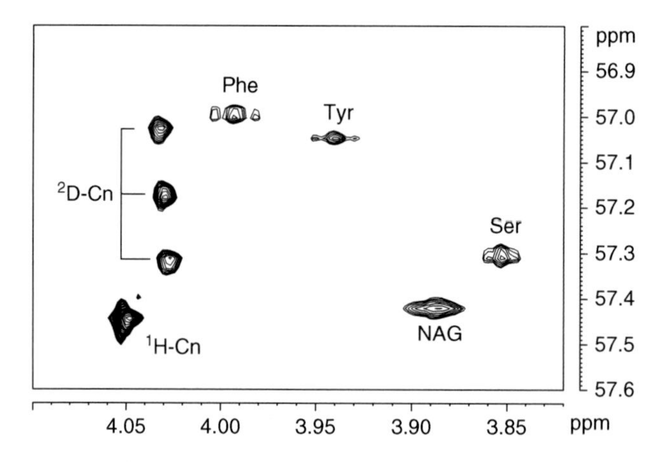

Figure 17.8. Expanded ^1H–^{13}C HSQC NMR spectral region showing the CH_2 group of creatinine in CSF. The characteristic tilted triplet pattern of the mono-deuterated CH_2 group is seen. The same signal structure appears also in spectra of urine and blood plasma.

lyophilized. The lyophilized plasma is transferred into a glass centrifuge tube and is mixed with 2 ml $CHCl_3$ and 2 ml MeOH. The mixture is carefully shaken until the suspension is homogeneous (1–2 min). The mixture is then centrifuged at 4000 rpm for 15 min. The supernatant is then transferred into a clean tube. The remaining pellet is suspended in 1 ml $CHCl_3$ and 1 ml MeOH, shaken and centrifuged again. The supernatants are combined and 1 ml of H_2O is added. After careful shaking the mixture is centrifuged at 4000 rpm for 5 min. The upper H_2O phase is collected into the tube with the solid residue. The $CHCl_3$/MeOH phase is cooled for 15 min at minus 20 °C to allow for a better phase separation. The residual water is removed and the $CHCl_3$/MeOH solvent is evaporated in a stream of dry nitrogen. The lipids are redissolved in 0.6 ml $CDCl_3$/MeOD (2:1) and put in the NMR tube. It is recommended to store lipids in solution, because they are susceptible to oxidation.

The solid residue with the 1 ml of water is carefully stirred up and shaken until the suspension is homogeneous. Then the mixture is centrifuged at 4000 rpm for 15 min. The aqueous solution is transported into a lyophilization flask. The solid residue is stirred up once again with another 1 ml of H_2O, shaken and centrifuged. The combined water phases are lyophilized. The water phase contains considerable amounts of methanol which vaporizes in the beginning (good cooling is recommended). The water phase contains small amounts of ethanol also, which is commonly used as a stabilizer for $CDCl_3$. To remove it completely, the lyophilization should last at least 24 h. The water soluble phase is redissolved in 0.6 ml of D_2O and put in the NMR tube. Then the pH has to be adjusted to 7 using a few drops of diluted DCl and NaOD.

17.2.5.2. Urine

The concentration of metabolites and salt in urine is very variable. Normally 1 ml is sufficient for a 1D NMR spectrum. The urine is lyophilized, redissolved in D_2O and centrifuged. Then the pH has to be adjusted to 7. For a 2D NMR analysis the highest possible amount of sample should be used. Up to about 10 ml urine can be concentrated to 1 ml without salt problems. For higher concentrations the lyophilized urine is extracted several times with MeOH. This procedure may be used to extract up to about 200 ml urine. The MeOH phases are combined and then MeOH removed using evaporation in a stream of dry nitrogen. The residue is redissolved in D_2O and centrifuged. No spectral differences are found in the NMR spectra of urine using either water extraction or MeOH extraction. The MeOH extraction is advantageous as the salt content is significantly lower.

17.2.5.3. CSF

The metabolites in CSF are similar to those in blood plasma, but of lower concentrations and especially lipids are of very low concentration (about a factor of 100 lower). Due to this low concentration of the lipids it is not necessary to remove them prior to measurement. The CSF sample is lyophilized, redissolved in D_2O and centrifuged. An NMR analysis of CSF lipids is normally impossible as 50 ml or more are required for extraction.

17.2.6. NMR acquisition parameters for body fluids

All experiments discussed in this chapter were performed on Bruker DRX 600 and 800 MHz spectrometers at 300 K using a 5 mm H,C,N inverse triple resonance probe with actively shielded field gradient coils. Gradients were shaped by a waveform generator and amplified by a Bruker Acustar amplifier. Sinusoidal z-gradients of 1 ms duration and a recovery time of 100 ms were used for the echo/antiecho gradient selection. Low power adiabatic composite pulse decoupling with WURST [5] has been used for ^{13}C-decoupling. The 1D ^{1}H NMR spectra of urine and CSF can be obtained using normal acquisition with presaturation. Blood plasma still contains some amount of protein which has to be suppressed using the Carr-Purcell-Meiboom-Gill (CPMG) echo-sequence [6–8]. A total CPMG echo-train of 300 ms has been used with an echo time of 200 μs, which has been optimized experimentally. For the 2D NMR analysis a sensitivity-improved HSQC [9, 10] with a relaxation delay of 1 s was used. Normally two spectra are acquired: first an overview spectrum with 1 k data points in F1 (22 Hz) and a high number of scans per increment (typically 32 to 64) and then secondly a high resolution spectrum using 4 k data points in F1 (3 Hz). The latter spectrum is required primarily to unravel the crowded region at 60–75 ppm. To further support the spectral assignments, a HSQC-TOCSY spectrum is acquired with a resolution of 5 Hz in F1. Sometimes also HMBC spectra or homonuclear ^{1}H-TOCSY spectra can be used to corroborate the assignments. All spectra are pure absorption mode spectra and were processed with a $\pi/2$-shifted squared sine bell in F2(^{1}H) and a $\pi/4$-shifted squared sine bell in F1(^{13}C).

17.3. *In vivo* NMR Spectroscopy

17.3.1. Introduction

Pulse sequences used for analytical NMR spectroscopy differ considerably from those applied for localized *in vivo* ^{1}H NMR spectroscopy. While 2D NMR techniques are widely used for high resolution NMR spectroscopy, most *in vivo* spectra are measured by localized NMR spectroscopy or by spectroscopic imaging with only 1D spectral resolution. Therefore, the very basic pulse sequences only are used if sequences with 2D spectral resolution are acquired *in vivo*, in particular localized variants of 2D correlation spectroscopy [11, 12] or 2D J-resolved NMR spectroscopy [13].

Differences arise from various reasons:

(i) The rather long minimum total measurement time T_{min} of 2D MRS experiments limits their *in vivo* applications because of the reasonability for patients or animals and increasing costs. This also prevents measurements of fast metabolic changes, for example in studies of ischaemia or hypoxia. Furthermore, motion will induce artefacts, particularly in the f_1 dimension. Phase cycling schemes to select certain coherence pathways are limited, as only few accumulations are feasible.

(ii) The inherently low signal-to-noise ratio (SNR) per unit measurement time (SNR_t) of *in vivo* MRS restricts the use of those 2D sequences for which the SNR_t is additionally reduced, for example, because the signal is only partially refocused and/or it is split into different components.

(iii) *In vivo* 2D MRS sequences are of interest for an improved signal resolution, signal assignment and quantification, while 2D or 3D sequences intended for the characterization of molecular structures are only of minor interest, if any.

As early as 1983, Balaban *et al.* [14] used ^{31}P 2D exchange spectroscopy to measure the reaction rate of the creatine kinase reaction by observing the chemical exchange between adenosine triphosphate (ATP) and phosphocreatine (PCr) in the leg and in the brain of rats. Later, most *in vivo* 2D MRS studies applied localized variants of 2D chemical shift correlation spectroscopy (COSY) and 2D J-resolved spectroscopy.

17.3.2. 2D chemical shift correlated spectroscopy

The basic COSY pulse sequence, $90°\text{-}t_1\text{–}90°\text{-}t_2$, consists of only two RF pulses. In a series of N_1 measurements, the delay time t_1 is incremented, and the signal is observed during t_2. A 2D Fourier transformation (FT) of the 2D data set S (t_1, t_2) yields the 2D correlation spectrum $S(f_1, f_2)$. Whilst uncoupled spins and J-coupled spins show up as diagonal peaks ($f_1 = f_2$), cross peaks indicate the coherence transfer between J-coupled spins. By this means COSY spectra improve the signal separation and assignment, particularly by measuring cross peaks of J-coupled spins.

Haase *et al.* [15] supplemented 1D localized COSY measurements with water presaturation and combined it with the localization method LOCUS [16] to saturate spins of unwanted regions prior to RF excitation. In other *in vivo* COSY measurements, the basic COSY sequence consisted of two rectangular, non-selective 90° pulses with additional saturation pulses to suppress the intense water and lipid signals. Localization was achieved by use of the spatial B_1 inhomogeneity of a surface coil [17].

Two slice-selective 90° pulses with orthogonal slice orientation ("spatially localized COSY", SLO-COSY) [18] achieve even better spatial localization. This approach towards localized 2D MRS was also used for localized 2D exchange spectroscopy (SLO-NOESY). Localization in the third direction can be achieved by adding a third RF pulse. Two examples are displayed in Figures 17.9a, b. In the L-COSY variant used by Thomas *et al.*, [19] a refocusing 180° is added between the

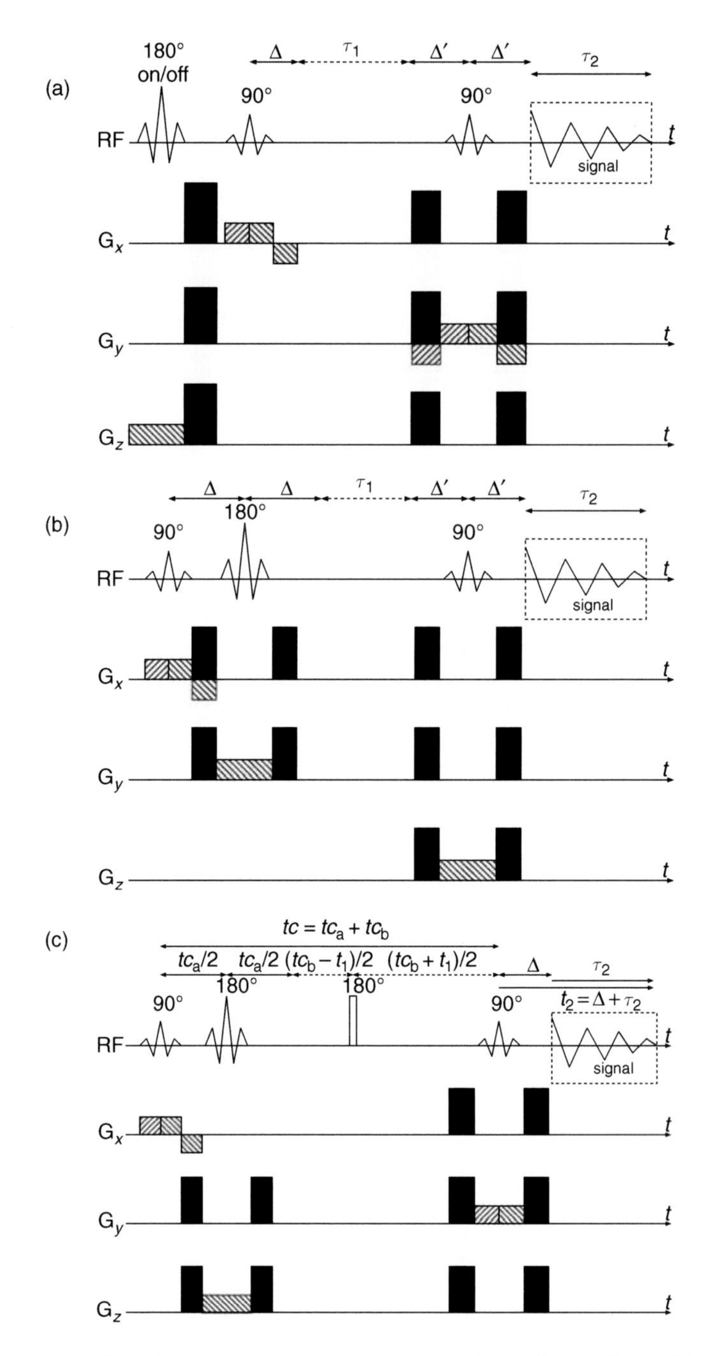

Figure 17.9. Examples for pulse sequences for (a) a two-step scheme for localized COSY with ISIS in one spatial direction and two spatially selective 90° pulses, (b) single-shot localized COSY (L-COSY, cf. [19]), (c) localized COSY with constant time chemical shift encoding (CT-COSY, cf. [27]).

two 90° pulses (cf. Figure 17.9a). Alternatively, the localization method ISIS [20] can be used in one direction which is equivalent to the add/subtract scheme proposed for 1D localization as NOBLE [21]. However, it is advantageous to achieve full localization within each scan and to avoid add/subtract schemes as they further increase the minimum total measurement time T_{min}. For typical repetition times TR of 2–6 s and 128–256 t_1 increment steps, T_{min} is about 4.5–25.6 min which may be acceptable for many *in vivo* applications. However, the rather long T_{min} may result in motion artefacts and restricts the use of phase cycling to suppress unwanted coherence pathways, in addition to the spoiler gradients.

Another way to reduce the f_1 range and thus the necessary number of encoding steps N_1 consists of using spin-echo correlated spectroscopy (SECSY) [22, 23] or Soft COSY [24, 25]. In the SECSY sequence, data sampling starts not immediately after the second 90° pulse, that is at $t = t_1$, but at the echo maximum $t = 2 \cdot t_1$. Thus, all of the diagonal peaks of standard COSY spectra appear at $f_1 = 0$ Hz, while the f_1 values of cross peaks of standard COSY spectra values are given by the chemical shift difference between the coupled spin partners. Therefore, F_1 can be reduced to the largest chemical shift difference between J coupled spins, reducing N_1 and thus also T_{min} for a given spectral resolution in f_1. In Soft COSY, one or several chemical shift-selective RF pulses are used to reduce the spectral ranges F_1 and/or F_2. A reduced spectral range F_1 allows one to use a smaller number of t_1 encoding steps for a given spectral resolution and shortens T_{min}.

To optimize the SNR_t of cross peaks of certain J-coupled spins of interest, CT-COSY, the constant time (CT) chemical shift encoding variant of COSY can be used [26]. The pulse sequence with 3D localization is given in Fig. 17.9c. The delay t_c between the two 90° pulses is kept constant. In a series of measurements, the chemical shift information is encoded, that is t_1 is varied by changing the position of a refocusing 180° pulse within t_c. The delay t_c can be chosen to maximize the transfer function and thus the intensity of a cross peak of interest. In this way, the results can be improved for specific metabolites. However, the cross peak intensity may be decreased for other metabolite signals. Figure 17.10 shows examples for localized CT-COSY spectra measured on a F98 glioma in rat brain *in vivo* at 4.7 T [27]. It is demonstrated that different t_c values of 51 and 85 ms lead to different results which are, for example, optimal for the cross peaks of taurine (Tau) and myo-inositol (Ins), respectively. Note that the optimum t_c values depend on the J-coupling pattern and thus on the field strength B_0, particularly for strongly coupled spins. The CT-COSY experiments are of increasing interest at higher B_0 as shorter t_c times can be realized for a given spectral resolution (in ppm) leading to better SNR_t.

The 2D COSY spectra corroborate the unambiguous assignment of metabolite signals in ^{1}H MRS [28, 29] and the characterization and separation of broader signals assigned to macromolecules such as proteins [30, 31]. *In vivo* applications have been successfully demonstrated on animals and humans, for healthy and diseased tissue

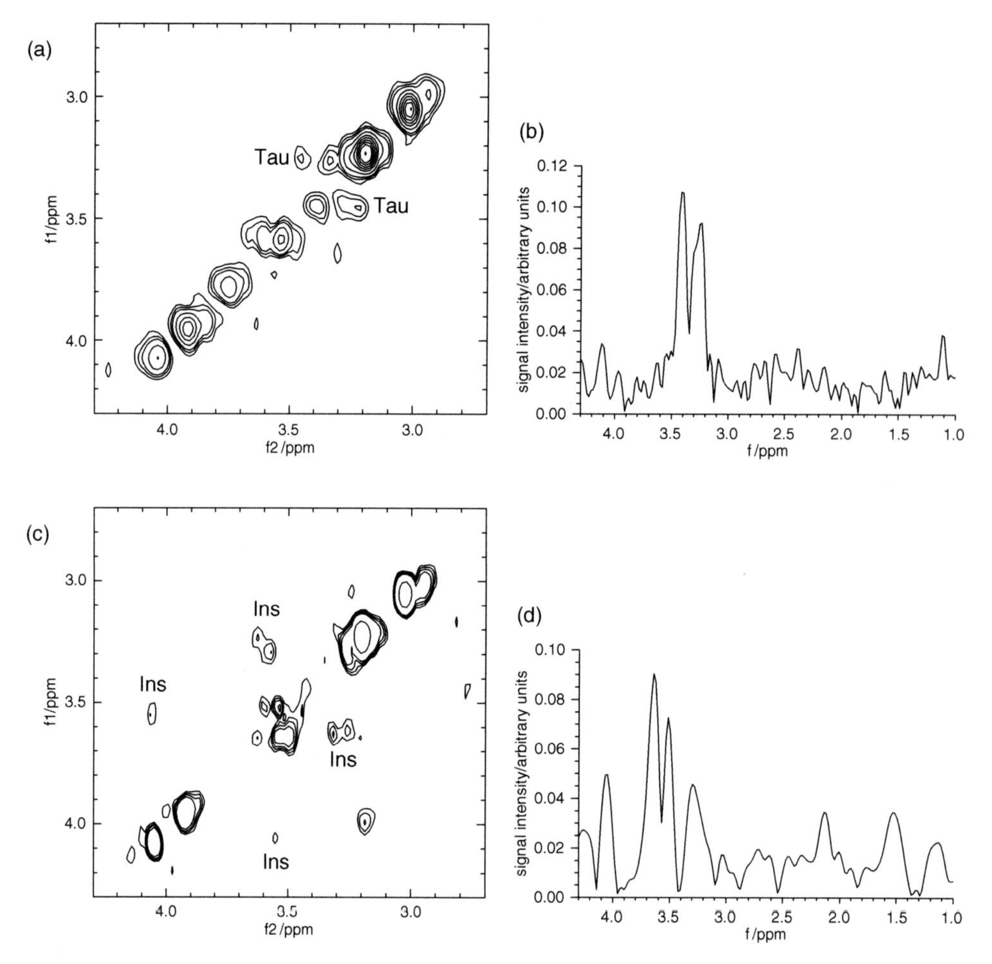

Figure 17.10. Examples of localized CT-COSY spectra measured at 4.7 T *in vivo* on rat brain with a F98 glioma. Parameters: voxel size $4 \times 5 \times 5\,mm^3$; 128 t_1-steps with $-17\,ms \le t_1 \le +17\,ms$; spectral width $F_1 = F_2 = 1880\,Hz$; 512 complex data points in t_2; TR = 2.0 s, two accumulations; apodization in t_1 and t_2 by shifted sine-bell functions; zero filling to 512×512 complex data points; 2D-FFT, magnitude calculation. (a,b) $t_c = 51$ ms, (c,d) $t_c = 85$ ms, (a,c) 2D CT-COSY spectrum, (b) cross-section spectrum at $f_2 = 3.44$ ppm, (d) projection spectrum onto the f_1-axis with $3.50\,ppm \le f_2 \le 3.66\,ppm$.

[19, 23, 32–44]. Various studies have been performed to observe metabolic changes in tumours, for example in brain tumours (glioma), breast cancer or prostate cancer. Other applications comprise ischemia, hepatic encephalopathy or depression.

The improved signal separation in 2D COSY spectra proved to be of particular importance in all cases when it was impossible to separate and quantify certain metabolites of interest in 1D spectra. In 1989, a study on superfused brain slices of rat [45] used the SUPERCOSY [46] sequence to separate and assign signals.

The SUPERCOSY sequence was chosen to remove the antiphase character of the cross peaks. Peres *et al.* [47] demonstrated that variations in cerebral glucose after intravenous injection of ^{13}C-enriched glucose can be detected by the cross peaks in 2D SUPERCOSY spectra when the sequence parameters were optimized to the detection of the cross peaks of glucose.

However, as a large number of metabolites can be separately detected and absolutely quantified by 1D MRS with ultrashort echo time (TE) [48–51] at high magnetic fields, one should consider for each application whether 2D MRS or ultrashort TE 1D MRS measurements are more appropriate for the metabolic characterization of tissue *in vivo*. However, in spite of the great potential of ultrashort TE 1D MRS, 2D MRS studies may be more appropriate in cases when the limited shim quality and/or the large signal overlap at lower B_0 do not allow a sufficient signal separation in 1D spectra, even if prior knowledge is used in sophisticated quantification procedures such as the LCModel [52] or AMARES [53].

An inherent difficulty of 2D MRS is signal quantification. While the dependence of the peak intensities on the sequence parameters and the J-coupling pattern may be simulated, for example by using the GAMMA library [54], there is, as compared to short TE 1D MRS, an additional dependence on T_2 and T_2^* by which quantification is hampered. Thus, for absolute quantification either T_2 and T_2^* have to be known or T_2 and T_2^* should be at least constant in order to attribute signal intensity changes unambiguously to changes in metabolite concentrations.

Many other variants of COSY can also be used for *in vivo* applications as discussed in the textbook by de Graaf [55]. In particular, purged COSY (P.COSY) [56], in which the diagonal peaks are suppressed or even completely eliminated, may be helpful to observe cross peaks close to the diagonal. Total correlation spectroscopy (TOCSY) [57] may be of interest because of the absorption-like line shape for both diagonal and cross peaks and the appearance of multiple-relay cross peaks. However, the latter aspect is of limited use for most molecules observable by 2D *in vivo* MRS contrary to many high resolution NMR studies. Furthermore, the higher RF power deposition caused by the spin-lock pulse or the composite pulse schemes limit the applicability for *in vivo* studies.

17.3.3. 2D J-resolved NMR spectroscopy in vivo

Originally, 2D J-resolved NMR spectroscopy was designed to separate the chemical shift and the line splitting caused by J-coupling into two orthogonal frequency directions. The basic 2D-J pulse sequence [13], $90°\text{-}t_1/2\text{-}180°\text{-}t_1/2\text{-}t_2$, consists of two RF pulses where the delay t_1 and thus the TE is incremented in a series of N_1 measurements. The signal is observed during t_2. The 2D Fourier transformation of the 2D data set $s(t_1, t_2)$ yields the 2D J-resolved spectrum $S(f_1, f_2)$. A 45° tilt of the 2D spectrum is performed to account for the fact that J-modulation influences the signal phase both during t_1 and t_2. Thus all multiplet components of a J-coupled spin can be observed along f_1 at their individual chemical shift in f_2. Uncoupled spins appear at $f_1 = 0\,\text{Hz}$. Signals of J-coupled spins may be evaluated by cross-section spectra at $f_1 \neq 0\,\text{Hz}$ in which signals of uncoupled spins will be suppressed. Of particular use is the projection spectrum onto the f_2 axis as it reflects a spectrum with "effective homonuclear decoupling" displaying signals of both uncoupled and J-coupled spins as single lines without any line splitting. There are various reasons why this pulse sequence is of great use for *in vivo* ^{1}H MRS measurements. First, peak separation and signal assignment are improved compared to spectra with one dimension as explained above. Second, SNR is higher than for COSY measurements because refocusing is done by a 180° pulse and thus coherence transfer to cross peaks is avoided. Third, 2D J-resolved MRS measurements allow short minimum total measurement times T_{min}, as only a narrow frequency range F_1 (less than 50 Hz) and thus a small number of experiments N_1 (typically 16–32) is required for an adequate resolution along f_1. Fourth, 3D spatial localization is easily achieved by adding another 180° pulse and using orthogonal slice orientation of each of the three pulses. Fifth, the implementation of an imaging system is straightforward as a 2D-J MRS measurement is essentially a series of PRESS measurements [58, 59], a basic single voxel MRS method provided by all major manufacturers.

As mentioned above 2D J-resolved measurements were used for the assignments of resonances appearing in *in vivo* ^{1}H spectra of healthy or diseased tissue [28, 23]. Several authors have demonstrated the use of 2D-J MRS for in vivo studies, either on animals [23, 60, 61, 27] or on humans [37, 62–67]. As the SNR of 2D J-resolved measurements is always a critical issue, several approaches have been considered to increase the SNR. A simple way is to start the data acquisition not at the echo maximum, but immediately after the last spoiler gradients bracketing the last 180° pulse (cf. Figure 17.11a). Thus, an asymmetric echo-like signal is detected where the symmetry with respect to the echo maximum at $t_1 = \text{TE}$ increases with increasing TE. This approach corresponds to the so-called FOCSY-J sequence [68, 61, 27] known from analytical 2D J-resolved NMR and allows a better SNR and improved spectral lineshape in magnitude spectra. An example of a localized FOCSY-J spectrum measured at 4.7 T on a rat brain with an implanted F98 glioma is displayed in Figure 17.12. Some typical features of a ^{1}H glioma spectrum are shown: increased total choline, decreased *N*-acetylaspartate (NAA), increased lactate (Lac) and alanine (Ala). In particular, both doublets of Lac and Ala are well detected at 1.33 and 1.48 ppm, respectively. If a 2D J-resolved spectrum is not intended for optimum spectral resolution in f_1, but primarily for projection spectra with effective homonuclear decoupling, both the experimental parameters and the data processing can be adjusted in the so-called CT-PRESS sequence [69] shown in Figure 17.11b. Voxel selection is achieved by a PRESS [58, 59] module with $\text{TE} = t_{c1}$. Chemical shift encoding is performed by shifting the position of an

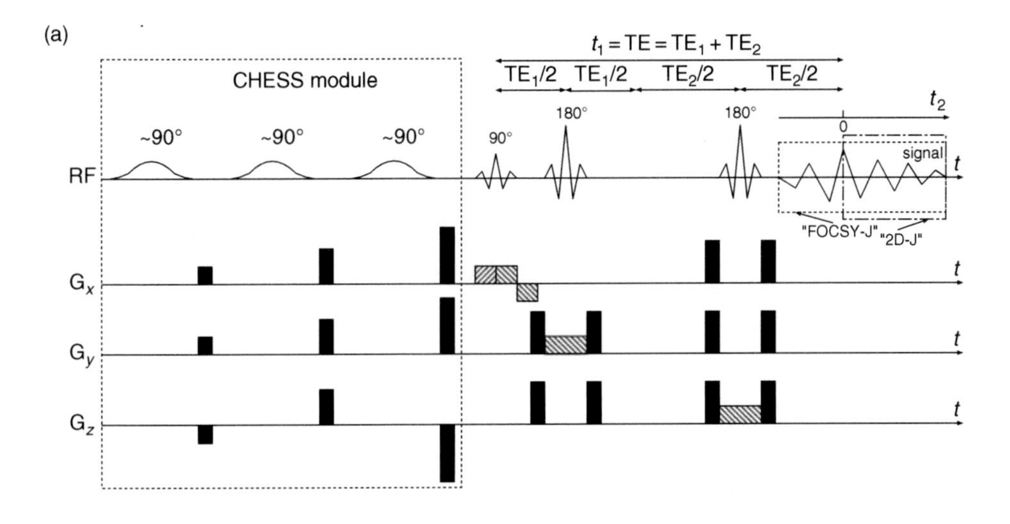

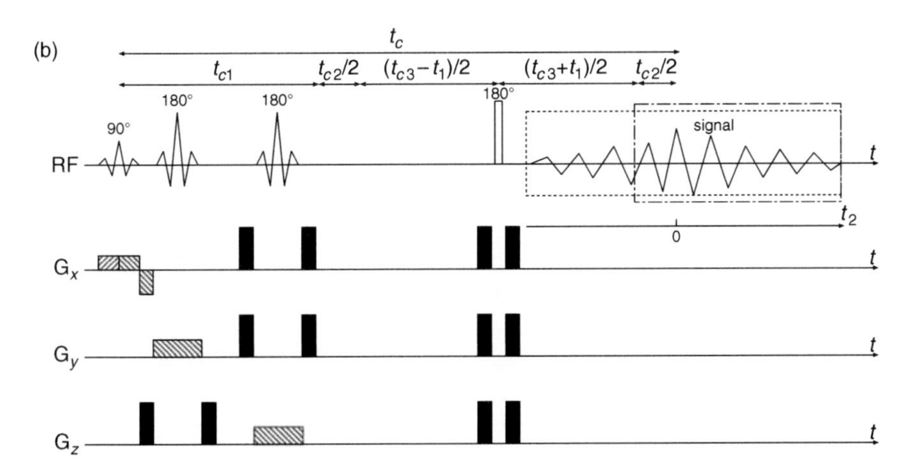

Figure 17.11. Examples of pulse sequences for (a) localized 2D J-resolved MRS with signal detection of the right half of the echo ("2D-J") or of all available data points of the echo ("FOCSY-J"). Water suppression is achieved by three chemical shift selective saturation pulses, (b) localized CT-PRESS aiming at 1D projection spectra with effective homonuclear decoupling (the means of water suppression is not shown).

additional spatially non-selective 180° pulse in a series of measurements within the interval t_{c3}. By means of the delay t_{c2} a value $t_c = t_{c1} + t_{c2} + t_{c3}$ can be chosen to attain the optimum transfer function for a certain J-coupled spin in a metabolite of interest. The use of an additional 180° pulse is advantageous compared to changing the position of the second spatially selective 180° pulse to avoid artefacts in the 2D spectrum at $(f_x/2, f_x)$, which would require phase cycling. Whilst in the original

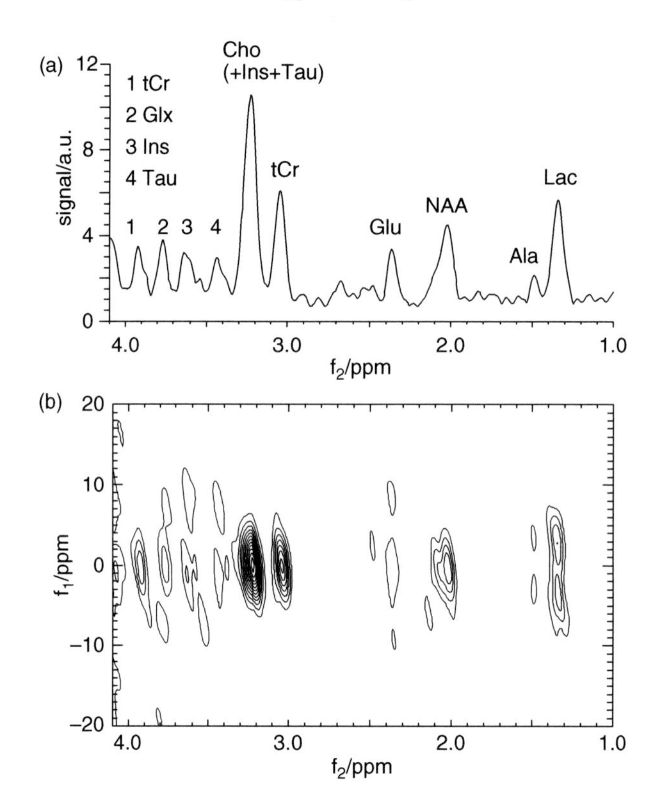

Figure 17.12. Example of a localized FOCSY-J spectrum measured at 4.7 T *in vivo* on rat brain with F98 glioma. Parameters: voxel size: $4 \times 4 \times 4\,\text{mm}^3$; TR=3.5 s; 16 equidistant t_1 values (17–272 ms); 16 accumulations; data acquisition during $t_2 = 272\,\text{ms}$; data processing by shifted sine-bell apodization, 2D-FFT, magnitude calculation and 45° tilt. (a) projection spectrum onto the f_2 axis yielding effective homonuclear decoupling, (b) 2D J-resolved spectrum.

CT-PRESS sequence, data acquisition was started at a constant time after RF excitation, one may also start the acquisition immediately after the last spoiler gradient, similar to FOCSY-J, if the data are appropriately reordered. Figure 17.13 shows a CT-PRESS measurement performed on a healthy rat brain at 4.7 T. Whilst the 2D spectrum (Figure 17.13a) shows that only diagonal peaks appear, Figure 17.13b displays the projection spectrum onto the f_1 axis yielding a 1D spectrum with homonuclear decoupling. Using a t_c of 136 ms yields good results for the CH_2 signal of glutamate at 2.36 ppm as shown in the cross-section and projection spectra, displayed in Figure 17.13c and 17.13d, respectively. Also signals of other J-coupled spins are detected separately as shown by the cross-section spectra yielding the signals of Tau and Ins (cf. Figure 17.13e, f). However, note that the SNR for these signals of J-coupled spins will be higher if different t_c values are used as demonstrated in [69].

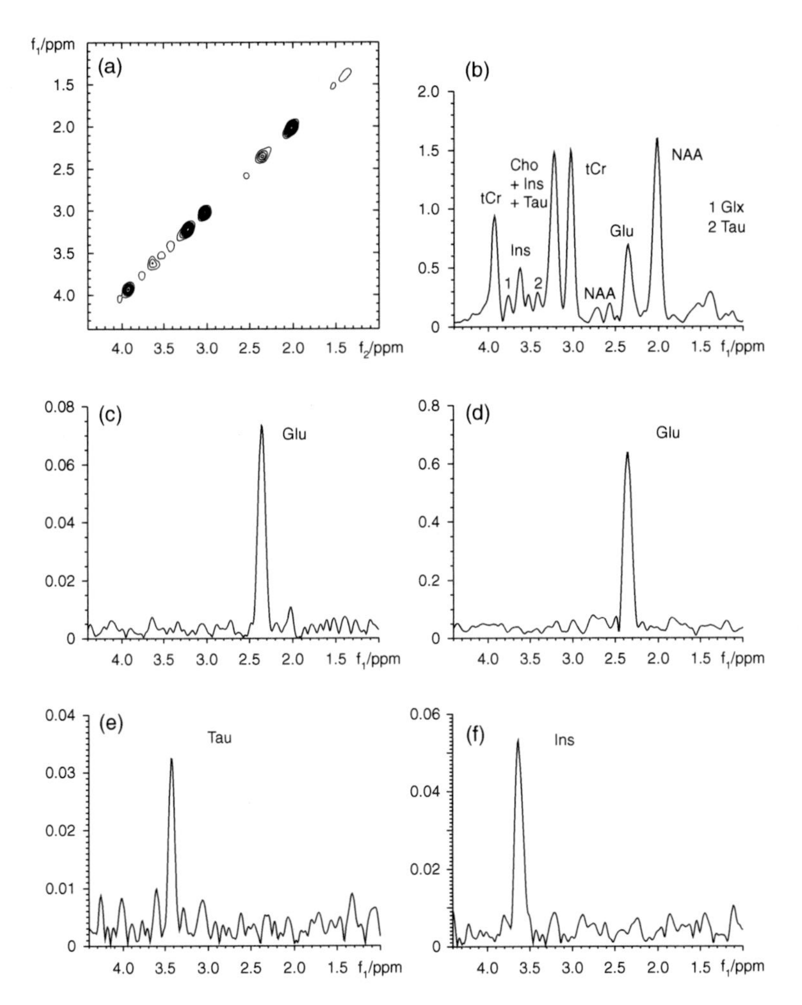

Figure 17.13. Example of a localized CT-PRESS spectrum measured at 4.7 T on healthy rat brain *in vivo*. Parameters: voxel size: $4 \times 4 \times 4\,\text{mm}^3$; TR = 1.1 s; $t_c = 136\,\text{ms}$; $-17\,\text{ms} \leq t_1 \leq +17\,\text{ms}$ and $\Delta t_1 = 266\,\mu\text{s}$; spectral width $F_1 = F_2 = 1880\,\text{Hz}$; 512 complex data points in t_2; (a) contour plot of the 2D spectrum, (b) diagonal spectrum obtained by projecting the 2D data around the diagonal $f_1 = f_2$ with $-11\,\text{Hz} \leq f_1 - f_2 \leq +11\,\text{Hz}$ onto the f_1 axis. (c) cross-section spectrum for Glu ($f_1 = 2.36\,\text{ppm}$), (d) projection spectrum onto the f_1 axis adding all data with $2.40 \geq f_2 \geq 2.32\,\text{ppm}$ yielding Glu, (e) cross-section spectrum for Tau ($f_1 = 3.44\,\text{ppm}$), (f) cross-section spectrum for Ins ($f_1 = 3.63\,\text{ppm}$).

If the experimental setup ensures that only diagonal signals occur and data reconstruction is appropriately modified, undersampling in t_1 can be used to decrease N_1 and thus T_{min}. Then the similarity between CT-PRESS and 2D J-resolved MRS becomes even more obvious, although the differences in both the t_1-range and the data processing remain which ensure the higher SNR of CT-PRESS spectra compared to standard 2D J-resolved spectra.

Applications of CT-PRESS include diffusion-weighted MRS of a large number of metabolites performed on rat brain at 4.7 T [70] or the individual detection of glutamate in human brain performed at 3 T [71]. Hurd *et al.* termed a 2D J-resolved MRS measurement "TE averaged PRESS" [72], whereby simply the cross-section spectrum at $f_1 = 0\,\mathrm{Hz}$ is evaluated.

Recently, an interesting approach has been described to tackle a general problem of *in vivo* 2D MRS, namely the influence of the relaxation times on the peak intensities. A sophisticated data evaluation was proposed to fit the 2D-J MRS data by a model that also contains the T_2 relaxation times [73]. By this means the determination of metabolite concentration is improved and additional information on T_2 values is obtained.

17.3.4. Combination of 2D MRS and spectroscopic imaging

Spectroscopic imaging (SI), which detects simultaneously spectra in an array of voxels, proved to be more appropriate for many medical applications than one or several consecutive single voxel measurements. By using SI sequences, smaller voxels can be measured and the voxel position can be adjusted after the measurement by post-processing to minimize partial volume effects. The simultaneous measurement of all voxels yields a higher SNR than consecutive measurements of single voxels. In particular, the detection of spectra in a large array of voxels is more adequate considering the inhomogeneity of the object to be measured. Thus, more information is acquired and a better characterization of metabolic changes within the spatial transition between healthy and diseased tissue is possible.

Whilst the combination of SI and 2D MRS is highly desirable, the main obstacle is to realize a sufficiently short T_{min} which is acceptable for *in vivo* studies on humans or animals. In some experiments [74–77], one, two or three orthogonal B_0 phase encoding gradients were inserted into a 2D MRS sequence, that is 2D-MRS was combined with classical SI [78, 79], where phase encoding gradients are applied between RF excitation and signal detection for spatial localization. The COSY-SI measurements enable the so-called correlation peak imaging (CPI) which aims at the spatial distribution of cross peaks for metabolites with J-coupled spins [75, 76, 80].

These studies demonstrated the great potential of combined 2D-MRS-SI measurements. However, since the minimum total measurement time $T_{\mathrm{min}} = N_1 \times N_{\mathrm{pe},1} \times N_{\mathrm{pe},2} \times N_{\mathrm{pe},3} \times TR$ is given by N_1, the number of encoding steps for the 2D MRS experiment, and the total number of spatial encodings $N_{\mathrm{pe},1} \times N_{\mathrm{pe},2} \times N_{\mathrm{pe},3}$, the resulting T_{min} is often prohibitively long for typical TR values of 1.5–6 s. For instance, for a COSY-SI measurements with one accumulation, TR=2 s, $N_1 = 128$ and 8×8

or $8 \times 8 \times 8$ voxels, T_{min} would be 4.55 h or 36.4 h, respectively. Some improvements are possible by using reduced k-space sampling [81], shorter TR or smaller N_1 values. Even for the combination of classical SI with 2D J-resolved MRS, with typical $N_1 = 16$–32, T_{min} is often too long.

Therefore, several groups have combined 2D MRS with fast SI methods, which sample the k-space faster than classical SI. These approaches use fast SI methods such as spectroscopic U-FLARE, spectroscopic RARE, echo planar spectroscopic imaging (EPSI) or spiral SI to accelerated k-space sampling and reduce T_{min}. Thus, 2D J-resolved MRS was combined with spiral SI [82, 83] or multi-echo SI [84]. The CT-COSY with its inherent effective homonuclear decoupling in f_1 was combined with EPSI [85], while a sequence based on spectroscopic U-FLARE yielded an array of COSY spectra with effective homonuclear decoupling not only in one, but in both spectral direction [86]. A higher SNR_t was achieved by using spectroscopic RARE with reduced k-space sampling [87]. Presently, these sequences are available only at some sites working on methodological developments and only a broader application will show how useful these approaches will be for specific *in vivo* applications.

17.3.5. Other in vivo applications of 2D MRS methods

Considering the use of localized 2D COSY and 2D J-resolved MRS for *in vivo* studies, the feasibility of a 3D MRS sequence that combines both approaches has been proposed. Thus a localized J-CT-COSY sequence was developed and applied to rat brain *in vivo* at 4.7 T [88]. However, the rather long T_{min} of 35–45 min limits the use *in vivo*. The 2D MRS approaches other than COSY or 2D-J MRS include localized 2D multiple quantum editing sequences described by Crozier *et al.* [89] to detect glutamate/glutamine and lactate in rat brain or heteronuclear 2D MRS (^{13}C-^{1}H) proposed by van Zijl *et al.* to measure glucose and its metabolic products in cat brain *in vivo* [90]. Recently, Thomas *et al.* [91] used a localized version of 2D ^{1}H NMR exchange spectroscopy (EXSY) [92]. In a study performed on human calf muscle at 1.5 T, an exchange cross peak was observed between mobile tissue water at 4.7 ppm and the methyl signal of the total creatine pool at 3.0 ppm supporting earlier studies on magnetization transfer for metabolite signals [93].

17.4. Conclusions

The sample homogeniety and texture properties as well as the metabolite concentrations and mobility within the sample will decide essentially to which extent the repertoire of modern multi-dimensional and multi-nuclear NMR techniques can be recruited to assign and to quantify metabolites in biological sample. As body fluids contain highly mobile low molecular components with long relaxation times, high resolution NMR in combination with multi-pulse sequences is able to detect and assign metabolites with concentrations above micromolar, while *in vivo* and in combination with spatial localization concentrations higher than 0.1–1 mM (depending upon the field strength used) are needed for a reliable metabolite identification. Although the majority of *in vivo* ^{1}H MRS studies are performed with single voxel or spectroscopic imaging sequences with only one spectral dimension, 2D MRS methods are a powerful tool for in vivo studies if measurements with 1D spectral resolution do not yield sufficient signal separation and/or ambiguous signal assignment. Of particular importance are current developments by which 2D MRS methods are combined with fast spectroscopic imaging methods to obtain minimum total measurement times that are short enough for relevant *in vivo* application. In this way both high spectral and high spatial resolution can be obtained. While primarily ^{1}H COSY and 2D J-resolved methods are applied *in vivo*, other 2D methods such as 2D exchange spectroscopy or heteronuclear 2D MRS may gain increasing importance in the future.

References

[1] Moolenaar, S.H., Engelke, U.F.H., Wevers, R.A. Proton nuclear magnetic resonance spectroscopy of body fluids in the field of inborn errors of metabolism. *Ann. Clin. Biochem.* 2003;40:16–24.

[2] Iles, R.A., Jago, J.R., Williams, S.R., Chalmers, R.A. 3-OH-3-Methylglutaryl-CoA lyase deficiency studied using 2-dimensional proton NMR spectroscopy. *FEBS Lett.* 1986;203:49–53.

[3] Willker, W., Leibfritz, D. Assignment of Mono- and Polyunsaturated Fatty Acids in Lipids of Tissue and Bodyfluids. *Magn. Reson. Chem.* 1998;36:79–84.

[4] Willker, W., Leibfritz, D. Lipid oxidation in blood plasma of patients with neurological disorders. *Brain Res. Bull.* 2001;53:437–444.

[5] Kupce, E., Freeman, R. Stretched adiabatic pulses for broadband spin inversion. *J. Magn. Res. A.* 1995;117:246–256.

[6] Hahn, E.L. Spin Echoes. *Phys Rev.* 1950;80:580–594.

[7] Carr, H,Y., Purcell, E.M. Effects of diffusion on the free precession in NMR experiments. *Phys Rev.* 1954;94:630–638.

[8] Meiboom, S., Gill, D. Modified Spin-Echo method for measuring nuclear relaxation times. Rev Sci Instrum. 1958;29:688–691.

[9] Kay, L.E., Keifer, P., Saarinen, T. Pure absorption gradient-enhanced heteronuclear HSQC spectroscopy with improved sensitivity. *J. Am. Chem. Soc.* 1992;114:10663–10665.

[10] Schleucher, J., Schwendinger, M., Sattler, M., Schmidt, P., Schedletzky, O., Glaser, S.J., Srensen, O.W., Griesinger, C. A general enhancement scheme in heteronuclear multidimensional NMR employing pulsed field gradients. *J. Biomol. NMR* 1994;4:301– 306.

[11] Jeener, J. *Ampere Summer School*, Basco Polje, Yugoslavia, 1971.

[12] Aue, W.P., Bartholdi, E., Ernst, RR. Two-dimensional spectroscopy. Application to nuclear magnetic resonance. *J. Chem. Phys.* 1976;64:2229–2246.

[13] Aue, W.P., Karhan, J., Ernst, R.R. Homonuclear and broad band decoupling and two-dimensional J-resresolved NMR spectroscopy. *J. Chem. Phys.* 1976;64:4226–4227.

[14] Balaban, R.S., Kantor, H.L., Ferretti, J.A. In vivo flux between phosphocreatine and adenosine triphosphate determined by two-dimensional phosphorous NMR. *J. Biol. Chem.* 1983;258:12787–12789.

[15] Haase, A., Schuff, N., Norris, D.G., Leibfritz, D. Localised COSY NMR spectroscopy. In: Proceedings: 6th Annual Meeting of the Society of Magnetic Resonance in Medicine, New York, 1987, p.1051.

[16] Haase, A. Localization of unaffected spins in NMR imaging and spectroscopy (LOCUS spectroscopy). *Magn. Reson. Med.* 1986;3:963–969.

[17] Berkowitz, B.A., Wolff, S.D., Balaban, R.S. Detection of metabolites in vivo using 2D proton homonuclear correlated spectroscopy. *J. Magn. Reson.* 1988;79:547–553.

[18] Blackband, S.J., McGovern, K.A., McLennan, I.J. Spatially localized two-dimensional spectroscopy. SLOCOSY and SLO-NOESY. *J. Magn. Reson.* 1988;79:184–189.

[19] Thomas, M.A., Yue, K., Binesh, N., Davanzo, P., Kumar, A., Siegel, B., Frye, M., Curran, J., Lufkin, R., Martin, P., Guze, B. Localized two-dimensional shift correlated MR spectroscopy of human brain. *Magn. Reson. Med.* 2001;46:58–67.

[20] Ordidge, R.J., Conelly, A., Lohman, J.A.B. Image-selected in vivo spectroscopy (ISIS). A new technique for spatially selective NMR spectroscopy. *J. Magn. Reson.* 1986;66:283–294.

[21] Tycko, R., Pines, A. Spatial localization of NMR signals by narrowband inversion. *J. Magn. Reson.* 1984;60:156–160.

[22] Nagayama, K., Wüthrich, K., Ernst, R.R. Two-dimensional spin echo correlated spectroscopy (SECSY) for ^{1}H NMR studies of biological macromolecules. *Biochem. Biophys. Res. Commun.* 1979;90:305–311.

[23] Remy, C., Arus, C., Ziegler, A., Lai, E.S., Moreno, A., Le Fur, Y., Decorps, M. In vivo, ex vivo, and in vitro one-and two-dimensional nuclear magnetic resonance spectroscopy of an intracerebral glioma in rat brain: assignment of resonances. *J. Neurochem.* 1994;62:166–179.

[24] Brüschweiler, R., Madsen, J.C., Griesinger, C., Sorensen, O.W., Ernst, R.R. Two-dimensional NMR spectroscopy with soft pulses. *J. Magn. Reson.* 1987;73:380–385.

[25] Stryjewski, D., Oschkinat, H., Leibfritz, D. Detection of metabolites in body fluids and biological tissue by a 1D soft COSY technique. *Magn. Reson. Med.* 1990;13:158–161.

[26] Girvin, M.E. Increased sensitivity of COSY spectra by use of constant-time t_1 periods (CT COSY). *J. Magn. Reson.* 1994;A108:99–102.

[27] Dreher, W., Busch, E., Röll, S., Hossmann, K.H., Leibfritz, D. F98 glioma in rat brain studied by localized in vivo ^{1}H 2D MR spectroscopy using CT-COSY and FOCSY-J. In: Proceedings 2nd Ann. Sci. Meet. ISMRM, New York, 1996, p. 366.

[28] Behar, K.L., Ogino, T. Assignment of resonances in the ^{1}H spectrum of rat brain by two-dimensional shift correlated and J-resolved NMR spectroscopy. *Magn. Reson. Med.* 1991;17:285–303.

[29] Govindaraju, V., Young, K., Maudsley, A.A. Proton NMR chemical shifts and coupling constants for brain metabolites. *NMR Biomed.* 2000;13:129–153.

[30] Behar, K.L., Ogino, T. Characterization of macromolecule resonances in the ^{1}H NMR spectrum of rat brain. *Magn. Reson. Med.* 1993;30:38–44.

[31] Behar, K.L., Rothman, D.L., Spencer, D.D., Petroff, O.A.C. Analysis of macromolecule resonances in 1H NMR spectra of human brain. *Magn. Reson. Med.* 1994;32:294–302.

[32] Barrere, B., Peres, M., Gillet, B., Mergui, S., Beloeil, J.C., Seylaz, J. 2D COSY ^{1}H NMR: A new tool for studying in situ brain metabolism in the living animal. *FEBS Lett.* 1990;264:198–202.

[33] Peres, M., Bourgeois, D., Roussel, S., Lefur, Y., Devoulon, P., Remy, C., Barrere, B., Decorps, M., Pinard, E., Riche, D., Benabid, A.L., Seylaz, J. Two-dimensional ^{1}H spectroscopic imaging for evaluating the local metabolic response to focal ischemia in the conscious rat. *NMR Biomed.* 1992;5:11–19.

[34] Tang, H.L., Buist, R.J., Rixon, R.H., Whitfield, J.F., Smith, I.C.P. Identification of lactate, threonine and alanine in rat thymus and tumorigenic lymphoid cells using ^{1}H 2D COSY NMR spectroscopy. *NMR Biomed.* 1992;5:69–74.

[35] Brereton, I.M., Galloway, G.J., Rose, S.E., Doddrell, D.M. Localized two-dimensional shift correlated spectroscopy in humans at 2 Tesla. *Magn. Reson. Med.* 1994;32:251–257.

[36] Pierard, C., Satabin, P., Lagarde, D., Barrere, B., Guezennec, C.Y., Menu, J.P., Peres, M. Effects of a vigilance-enhancing drug, modafinil, on rat brain metabolism: A 2D COSY ^{1}H-NMR study. *Brain Res.* 1995;693:251–256.

[37] Kreis, R., Boesch, C. Spatially localized, one- and two-dimensional NMR spectroscopy and in vivo application to human muscle. *J. Magn. Reson. B* 1996;113:103–118.

[38] Thomas, M.A., Binesh, N., Yue, K., DeBruhl, N. Volume-localized two-dimensional correlated magnetic resonance spectroscopy of human breast cancer. *J. Magn. Reson. Imaging* 2001;14:181–186.

[39] Ziegler, A., Gillet, B., Beloeil, J.C., Macher, J.P., Decorps, M., Nedelec, J.F. Localized 2D correlation spectroscopy in human brain at 3 T. *MAGMA* 2002;14:45–49.

[40] Thomas, M.A., Hattori, N., Umeda, M., Sawada, T., Naruse, S. Evaluation of two-dimensional L-COSY and JPRESS using a 3 T MRI scanner: From phantoms to human brain in vivo. *NMR Biomed.* 2003;16:245–251.

[41] Binesh, N., Yue, K., Fairbanks, L., Thomas, M.A. Reproducibility of localized 2D correlated MR spectroscopy. *Magn. Reson. Med.* 2002;48:942–948.

[42] Binesh, N., Kumar, A., Hwang, S., Mintz, J., Thomas, M.A. Neurochemistry of late-life major depression: A pilot two-dimensional MR spectroscopic study. *J. Magn. Reson. Imaging* 2004;20:1039–1045.

[43] Binesh, N., Huda, A., Bugbee, M., Gupta, R., Rasgon, N., Kumar, A., Green, M., Han, S., Thomas, M.A. Adding another spectral dimension to ^{1}H magnetic resonance spectroscopy of hepatic encephalpathy. *J. Magn. Reson. Imaging* 2005;21:398–405.

[44] Thomas, M.A., Wyckoff, N., Yue, K., Binesh, N., Banakar, S., Chung, H.K., Sayre, J., DeBruh, l. N. Two-dimensional MR spectroscopic characterization of breast cancer in vivo. *Technol. Cancer. Res. Treat.* 2005;4:99–106.

[45] Gillet, B., Mergui, S., Beloeil, J.C., Champagnat, J., Fortin, G., Jacquin, T. ^{1}H COSY spectra of superfused brain slices of rat: Ex vivo direct assignment of resonances. *Magn. Reson. Med.* 1989;11:288–294.

[46] Kumar, A., Hosur, R.V., Chandrasekhar, K. A superior pulse scheme for homonuclear two-dimensional correlated spectroscopy. *J. Magn. Reson.* 1984;60:143–148.

[47] Peres, M., Fedeli, O., Barrere, B., Gillet, B., Berenger, G., Seylaz, J., Boloeil, J.C. In vivo identification and monitoring of changes in rat brain glucose by two-dimensional shift-correlated ^{1}H NMR spectroscopy. *Magn. Reson. Med.* 1992;27:356–361.

[48] Tkac, I., Starcuk, Z., Choi, I.Y., Gruetter, R. In vivo ^{1}H NMR spectroscopy of rat brain at 1 ms echo time. *Magn. Reson. Med.* 1999;41:649–656.

[49] Pfeuffer, J., Tkac, I., Provencher, S.W., Gruetter, R. Toward an in vivo neurochemical profile: Quantification of 18 metabolites in short-echo-time ^{1}H NMR spectra of the rat brain. *J. Magn. Reson.* 1999;141:104–120.

[50] Tkac, I., Andersen, P., Adriany, G., Merkle, H., Ugurbil, K., Gruetter, R. In vivo ^{1}H NMR spectroscopy of human brain at 7 T. *Magn. Reson. Med.* 2001;46:451–456.

[51] Geppert, C., Dreher, W., Leibfritz, D. PRESS-based proton single-voxel spectroscopy and spectroscopic imaging with very short echo times using asymmetric RF pulses. *MAGMA* 2003;16:144–148.

[52] Provencher, S.W. Estimation of metabolite concentrations from localized in vivo proton NMR spectra. *Magn. Reson. Med.* 1993;30:672–679.

[53] Mierisova, S., van den Boogaart, A., Tkac, I., Hecke, P.V., Vanhamme, L., Liptaj, T. New approach for quantitation of short echo time in vivo ^{1}H MR spectra of brain using AMARES. *NMR Biomed.* 1998;11: 32–39.

[54] Smith, S.A., Levante, T.O., Meier, B.H., Ernst, R.R. Computer simulation in magnetic resonance. An objektoriented programming approach. *J. Magn. Reson. A* 1994;106:75–105.

[55] de Graaf, R.A. *In vivo NMR Spectroscopy.* New York, John Wiley, 1998.

[56] Marion, D., Bax, A. P.COSY, a sensitive alternative for double-quantum filtered COSY. *J. Magn. Reson.* 1988;80:528–533.

[57] Braunschweiler, L., Ernst, R.R. Coherence transfer by isotropic mixing: Application to proton correlation spectroscopy. *J. Magn. Reson.* 1983;53:521–528.

[58] Gordon, R.E., Ordidge, R.J. Volume selection for high resolution NMR studies. In: Proceeding of the SMRM 3rd Ann. Meet., New York, 1984, pp. 272–273.

[59] Bottomley, P.A. Spatial localization in NMR-spectroscopy in vivo. *Ann. N.Y. Acad. Sci.* 1987;508:333–348.

[60] Dreher, W., Leibfritz, D. On the use of two-dimensional-J NMR measurements for in vivo proton MRS: Measurement of homonuclear decoupled spectra without the need for short echo times. *Magn. Reson. Med.* 1995;34:331–337.

[61] Dreher, W., Leibfritz, D. The use of 2D-J and FOCSY-J proton magnetic resonance spectroscopy for metabolic characterization of rat brain in vivo. In: Proc. 3d Ann. Sci. Meet. SMR, Nice, 1995, p. 1921.

[62] Ryner, L.N., Sorenson, J.A., Thomas, M.A. 3D localized 2D NMR spectroscopy on an MRI scanner. *J. Magn. Reson. B* 1995;107:126–137.

[63] Ryner, L.N., Sorenson, J.A., Thomas, M.A. Localized 2D J-resolved MR spectroscopy: Strong coupling effects in vitro and in vivo. *Magn. Reson. Imaging* 1995;13:853–869.

[64] Thomas, M.A., Ryner, L.N., Mehta, M.P., Turski, P.A., Sorensen, J.A. Localized 2D J-resolved ^{1}H MR spectroscopy of human brain tumors in vivo. *J. Magn. Res. Imaging* 1996;6:453–459.

[65] Kühn, B., Dreher, W., Leibfritz, D., Heller, M. Homonuclear uncoupled ^{1}H spectroscopy of the human brain using weighted accumulation schemes. *Magn. Reson. Imaging.* 1999;17:1193–1201.

[66] Yue, K., Marumoto, A., Binesh, N., Thomas M.A. 2D JPRESS of human prostates using an endorectal receiver coil. *Magn. Reson. Med.* 2002;47:1059–1064.

[67] Adalsteinsson, E., Hurd, R.E., Mayer, D., Sailasuta, N., Sullivan, E.V., Pfefferbaum, A. In vivo 2D J-resolved magnetic resonance spectroscopy of rat brain with a 3-T clinical human scanner. Neuroimage 2004;22:381–386.

[68] Macura, S., Brown, L.R. Improved sensitivity and resolution in two-dimensional homonuclear J-resolved NMR spectroscopy of macromolecules. *J. Magn. Reson.* 1983;53:529–535.

[69] Dreher, W., Leibfritz, D. Detection of homonuclear decoupled in vivo proton NMR spectra using constant time chemical shift encoding: CT-PRESS. *Magn. Reson. Imaging* 1999;17:141–150.

[70] Dreher, W., Busch, E., Leibfritz, D. Changes in apparent diffusion coefficients of metabolites in rat brain after middle cerebral artery occlusion measured by proton magnetic resonance spectroscopy. *Magn. Reson. Med.* 2001;45:383–389.

[71] Mayer, D., Spielman, D.M. Detection of glutamate in the human brain at 3 T using optimized constant time point resolved spectroscopy. *Magn. Reson. Med.* 2005;54:439–442.

[72] Hurd, R., Sailasuta, N., Srinivasan, R., Vigneron, D.B., Pelletier, D., Nelson, S.J. Measurement of brain glutamate using TE-averaged PRESS at 3T. *Magn. Reson. Med.* 2004;51:435–440.

[73] Schulte, R.F., Boesiger, P. Two-dimensional J-resolved spectroscopy fitting. In: Proc. ISMRM, 2005, p. 2471.

[74] Cohen, Y., Chang, L.H., Litt, L., James, T.L. Spatially lokalized COSY spectra from a surface coil using phase-encoding magnetic field gradients. *J. Magn. Reson.* 1989;85:203–208.

[75] Metzler, A., Izquierdo, M., Ziegler, A., Köckenberger, W., Komor, E., von Kienlin, M., Haase, A., Decorps, M. Plant histochemistry by correlation peak imaging. *Proc. Natl. Acad. Sci. U.S.A* 1995;92:11912–11915.

[76] Ziegler, A., Metzler, A., Köckenberger, W., Izquierdo, M., Komor, E., Haase, A., Decorps, M., von Kienlin, M. Correlation-peak imaging. *J. Magn. Reson.* B 1996;112:141–150.

[77] von Kienlin, M., Ziegler, A., Fur, Y.L., Rubin, C., Decorps, M., Remy, C. 2D-spatial/2D-spectral spectroscopic imaging of intracerebral gliomas in rat brain. *Magn. Reson. Med.* 2000;43:211–219.

[78] Brown, T.R., Kincaid, B.M., Ugurbil, K. NMR chemical shift imaging in three dimensions. *Proc. Natl. Acad. Sci. U.S.A* 1982;79:3523–3526.

[79] Maudsley, A.A., Hilal, S.K., Perman, W.H., Simon, H.E. Spatially resolved high resolution spectroscopy by "four-dimensional" NMR. *J. Magn. Reson.* 1983;51:147–152.

[80] Ziegler, A., von Kienlin, M., Decorps, M., Remy, C. High glycolytic activity in rat glioma demonstrated in vivo by correlation peak 1H magnetic resonance imaging. *Cancer Research* 2001;61:5595–5600.

[81] Maudsley, A., Matson, G., Hugg, J., Weiner, M. Reduced phase encoding in spectroscopic imaging. *Magn. Reson. Med.* 1994;31:645–651.

[82] Adalsteinsson, E., Spielman, D.M. Spatially resolved two-dimensional spectroscopy. *Magn. Reson. Med.* 1999;41:8–12.

[83] Hiba, B., Serduc, R., Provent, P., Farion, R., Remy, C., Ziegler, A. 2D J-resolved spiral spectroscopic imaging at 7 T: Application to mobile lipid mapping in a rat glioma. *Magn. Reson. Med.* 2004;52:658–662.

[84] Herigault, G., Zoula, S., Remy, C., Decorps, M., Ziegler, A. Multi-spin-echo J-resolved spectroscopic imaging without water suppression: Application to a rat glioma at 7 T. *MAGMA* 2004;17:140–148.

[85] Mayer, D., Dreher, W., Leibfritz, D. Fast echo planar based correlation-peak imaging: Demonstration on the rat brain in vivo. *Magn. Reson. Med.* 2000;44:23–28.

[86] Mayer, D., Dreher, W., Leibfritz, D. Fast U-FLARE-based correlation-peak imaging with complete effective homonuclear decoupling. *Magn. Reson. Med.* 2003;49:810–816.

[87] Mayer, D., Dreher, W., Leibfritz, D., Spielman, D. Fast correlation peak imaging using spectroscopic RARE and circularly reduced chemical shift encoding. In: Proc. 11th Ann. Sci. Meet. ISMRM, Toronto, 2003, p. 1125.

[88] Dreher, W., Leibfritz, D. Localized 3-dimensional proton MR spectroscopy (J-CT-COSY) used for comprehensive spectroscopic characterization of the rat brain in vivo. In: Proc. 5th Ann. Sci. Meet. ISMRM, Vancouver, 1997, p. 1435.

[89] Crozier, S., Brereton, I.M., Rose, S.E., Field, J., Shannon, G.F., Doddrell, D.M. Application of volume-selected, two-dimensional multiple-quantum editing in vivo to observe cerebral metabolites. *Magn. Reson. Med.* 1990;16:496–502.

[90] van Zijl, P.C.M., Chesnick, A.S., DesPres, D., Moonen, C.T.W., Ruiz-Cabello, J., van Gelderen, P. In vivo proton spectroscopy and spectroscopic imaging of ^{1}H-^{13}C-glucose and its metabolic products. *Magn. Reson. Med.* 1993;30:544–551.

[91] Thomas, M.A., Chung, H.K., Middlekauff, H. Localized two-dimensional ^{1}H magnetic resonance exchange spectroscopy: A preliminary evaluation in human muscle. *Magn. Reson. Med.* 2005;53:495–502.

[92] Jeener, J., Meier, B.H., Bachmann, P., Ernst, R.R. Investigation of exchange processes by two-dimensional NMR spectroscopy. *J. Chem. Phys.* 1979;71:4546–4553.

[93] Leibfritz, D., Dreher, W. Magnetization transfer MRS. *NMR Biomed.* 2001;14:65–76.

The Handbook of Metabonomics and Metabolomics
John C. Lindon, Jeremy K. Nicholson and Elaine Holmes (Editors)

Chapter 18

Applications of Metabonomics Within Environmental Toxicology

Julian L. Griffin[1] and Richard F. Shore[2]

[1]*Department of Biochemistry, University of Cambridge, Cambridge CB2 1QW UK*
[2]*Centre for Ecology and Hydrology, Monks Wood, Abbots Ripton, Huntingdon PE28 2LS, UK*

18.1. Introduction

The application of metabolomic/metabonomic approaches to investigate the effects of environmental contaminants on wild vertebrates has a number of advantages over other 'omic' approaches such as transcriptomics and proteomics. A large practical advantage is that, even though metabolic rates may differ markedly between species, metabolic detoxification pathways remain relatively well conserved. Thus, metabolic biomarkers identified in species used in laboratory studies are generally likely to be transferable to similar wild species, although exceptions may occur. Furthermore, the analytical tools required to profile metabolism in one animal can be used without modification to profile another species. In contrast, transcriptional analysis of a tissue or biofluid for a particular species requires that it has a sequenced (or partly sequenced) genome. Similarly, the majority of proteins identified using proteomic approaches require access to species-specific databases. This translational advantage of metabonomics over these other 'omic' approaches has produced a number of applications of NMR spectroscopy–based metabonomics in environmental toxicology, such as examining cadmium and arsenic toxicity in the bank vole (*Clethrionomys glareolus*) [1, 2], an animal with no sequenced genome.

Unlike other '-omic' technologies, NMR and mass spectrometry based metabonomics/metabolomics [3–6] has the enviable properties of providing a global profiling tool that is cheap on a per sample basis and amenable to high sample throughput [7–10]. While the purchase of a spectrometer/superconducting magnet is a large initial outlay, subsequent analysis of samples costs less than a US $1 a sample, and typically takes ~10 min for a one-dimensional spectrum. This compares favourably to DNA microarrays and proteomic approaches with costs in the region of $200 and $60 respectively, along with significant sample preparation times. A principal benefit of large-sample throughput is that it allows the generation of data matrices which are large enough to model both the variation associated with a toxic insult or disorder and the innate variation associated with the biological system, while minimizing false positives normally associated with such global multivariate analyses. This is particularly appropriate in environmental toxicology studies where target compounds are often dosed at environmentally realistic concentrations that are well below estimated LD_{50} concentrations and so may elicit physiological changes that are relatively small in comparison to normal intra-species biochemical and physiological variation. Variation in both basal metabolism and in response to a given toxicology insult or pathophysiological stimuli is likely to be greater in naturally outbred wild species than inbred and genetically predetermined laboratory animals.

Interfacing biofluid NMR spectroscopy and mass spectrometry based metabonomics with current toxicology approaches has a number of other benefits. The collection of urine and blood plasma is minimally invasive, and volumes of sample required for analysis are usually small enough to allow multiple sampling over time from the same individuals for rats and larger animals. In fact, recent advances in NMR technology mean that it is now possible to obtain reasonable ^1H NMR spectra from as little as ~5 μL of blood plasma or cerebrospinal fluid, thus allowing multiple sampling even in species as small as the laboratory mouse [11] (Figure 18.1). Thus, it is possible to investigate the impact an environmental toxin has on an animal of rodent or larger size in a largely minimally invasive nature. In this manner, wild species can be studied in the field, without the need to kill and take terminal samples from individual animals. This is a highly desirable feature for environmental toxicology where a main aim may be to relate physiological impact to subsequent reproductive success, survival and population dynamics.

It is not always possible to directly apply to environmental studies standard procedures for metabonomic analysis of biofluids or tissues that are used, for example, in the drug safety assessment industry. This is particularly the case when studies involve non-laboratory animals. For example, the common shrew (*Sorex araneus*) is a widely distributed and common small mammal in the UK and is particularly prone to exposure to soil contaminants. This is partly because it directly ingests soil that is adhering to or is contained within (for example in the gut) its invertebrate prey. The soil-dwelling invertebrates that shrews feed on also bioaccumulate some soil contaminants. However, shrews are too small to be orally exposed using gavage or other artificial techniques and so experimental studies

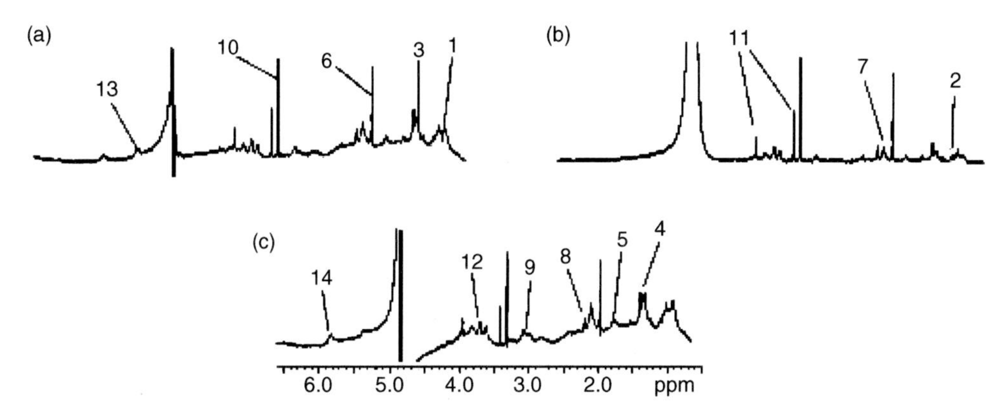

Figure 18.1. Biofluid NMR spectra obtained using a microprobe. The 600 MHz ^{1}H NMR spectra of rat blood plasma using a microlitre probe with 128 scans. The spectra were acquired with three different solvent presaturation sequences: (a) a sequence based on the first increment of the 2D NOESY sequence; (b) a CPMG pulse sequence using continuous wave presaturation and a 40 ms total spin echo time; and (c) a WATERGATE pulse sequence. Key: 1, CH_3CH_2 lipid moieties; 2, leucine, valine and isoleucine; 3, CH_2CH_2 lipid moieties; 4, lactate; 5, $COCH_2CH_2$ lipid moieties; 6, acetate; 7, $CH_2C{=}C$ lipid moieties; 8, glycoprotein $NHCOCH_3$; 9, albumin lysyl ε-CH_2; 10, choline and phosphocholine; 11, betaine; 12, glucose; 13, C=CH lipid moieties; 14, protein amide groups.

on shrews that involve oral exposure require the contaminants to be incorporated either into drinking water or into the diet. In any case, realistic exposure scenarios dictate that shrews are fed prey, such as earthworms, which have been maintained in soil that contains the contaminant of interest. Such experimental designs can greatly complicate modelling of the bioavailability of a given contaminant, and also necessitate the consideration of how the invertebrate prey have modified, metabolized and altered a given xenobiotic prior to it being ingested by the shrew.

Applications of metabonomics to environmental toxicology have focused on wild vertebrates and invertebrates, and plants. The results from animal-based experiments are discussed in this chapter while plant-based metabolomics is covered in Chapter 16.

18.2. Examining the impact of environmental toxins on mammalian species using laboratory models

The metabolic effects of a number of environmentally relevant xenobiotics have been investigated in the laboratory rat and mouse. In particular, the effects of a range of toxic metals and semi-metals, such as mercury, cadmium and arsenic, have been investigated [7, 12]. In many of these studies, specific biomarkers of toxicity have been identified. For example, increases in urinary creatine concentrations have been shown to correlate with testicular damage and taurine excretion correlates with hepatotoxicity [7, 12]. However, a common feature is the increased production of TCA cycle intermediates in the urine. Connor and colleagues [13] have since demonstrated that these intermediaries are associated with the weight loss that accompanies a given toxicological insult, rather than as a direct biochemical result of the initial lesion.

Metabonomics using high resolution ^{1}H NMR spectroscopy of biofluids coupled with pattern recognition has been successfully used to distinguish organ-specific and even sub-organ-specific toxicity from environmental contaminants. For example, Azmi and co-workers [14] used NMR techniques to differentiate the biological effects caused by α-napthylisothiocyanate (ANIT)-induced hepatotoxicity in the rat from those induced by exposure to 1-naphthylisocyanate (NI) and 1-naphthylamine (NA), two products of the metabolism of ANIT. Similar approaches have been applied to environmental contaminants such as polycyclic aromatic hydrocarbons (PAHs) [15], which are produced by the incomplete combustions of petroleum products. In the case of ANIT, NI and NA, all three toxicants produced perturbations in a similar set of urinary metabolites, and the use of principal components analysis (PCA) of the temporal progression resulted in the identification of a rapid initial glycosuria associated with ANIT toxicity; this also occurred in animals dosed with NI but not in those given NA (Figure 18.2). However, longer term perturbations in urinary excretion of succinate, lactate and acetate were common to all three toxicants. The most marked perturbations were induced by ANIT exposure; NI caused more marked effects than NA. This analysis indicated that the effects of ANIT, NI and NA toxicity were distinct, and demonstrated that metabonomics may be useful as a tool for following mechanisms of toxicity in a series of related compounds. Many organic environmental pollutants, such as PAHs and polychlorinated biphenyls (PCBs), are similarly comprised of a series of multiple compounds and metabonomics is likely to be a powerful means of determining specifically which compounds are likely to pose the greatest toxic hazard to wildlife.

NMR spectroscopy has also been applied to investigate the effects of inhalation exposure to contaminants and toxins. 1-nitronaphthalene (1-NN) is a common environmental contaminant in airborne particulates in urban areas and is formed by nitration of PAHs during the combustion of diesel [16, 17]. It is a contributor to the mutagenic activity of airborne particles [18] and an intermediate in the production of a range of chemicals [19]. Both the lungs and the liver of rodents are damaged by exposure to 1-NN, although the lung appears to be the more sensitive target. Single doses of 1-NN given intraperitoneally at 100 mg/kg [20, 21] or orally at ~80–300 mg/kg [22] produce a severe respiratory distress syndrome (RDS), chromodacryorrhea and bronchiolar epithelial injury. By using high resolution ^{1}H NMR spectroscopy, Azmi and co-workers [23] were able to characterize the metabolic

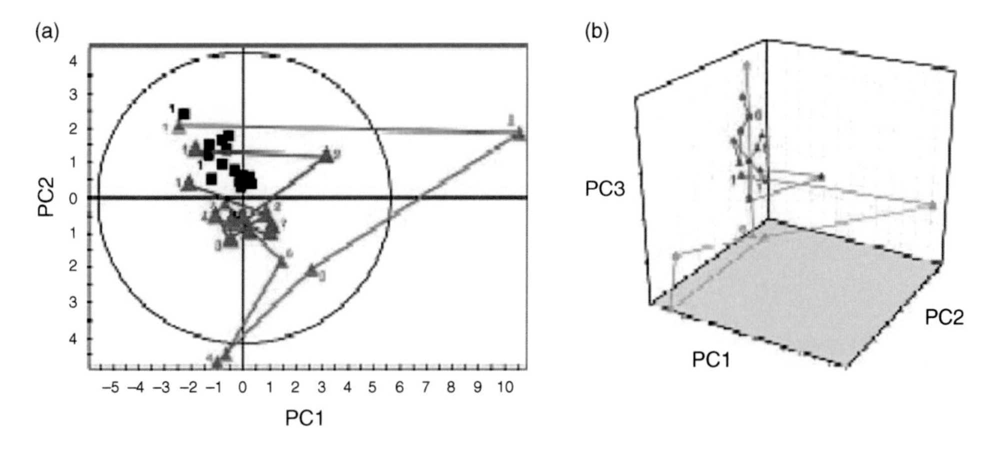

Figure 18.2. A metabonomic investigation into the toxicity of ANIT and its metabolites NI and NA. While all three toxicants produced perturbations in a similar set of urinary metabolites, PCA of the temporal progression identified that the rapid initial glycosuria associated with ANIT toxicity was also present with NI but not NA dosing. (a) PCA scores plot mapping the mean-centered average position of single-pulse urine ^{1}H NMR spectra obtained predose and up to 96 h from individuals following treatment with ANIT (▲), NI, (▲), (NA, ▲) and control data (■). (b) PCA metabolic three-dimensional mean trajectory (PC1 vs PC2 vs PC3) of this data. PC1 and PC2 account for 68.9%, and PC1, PC2 and PC3 account for 77.1% of the data variance [14].

fingerprint, in the form of patterns of high concentration endogenous metabolites, of 1-NN-induced lung toxicity in a range of biofluids and tissues. In addition to examining commonly used biofluids such as urine and blood plasma, bronchoalveolar lavage fluid (BALF) was analysed. In this study, a single dose of 1-NN (75 mg/kg) was administered orally to Sprague-Dawley rats. The BALF and lung tissue were obtained 24 h after dosing from these animals and matched control rats post-mortem. High resolution ^{1}H NMR spectroscopy of BALF samples indicated that 1-NN caused increases in concentrations of choline, amino acids (leucine, isoleucine and alanine) and lactate together with decreased concentrations of succinate, citrate, creatine, creatinine and glucose. In addition, the intact lung weights were higher in the 1-NN treated group, consistent with pulmonary oedema. The NMR-detected perturbations indicated that 1-NN induces a perturbation in energy metabolism in lung and liver tissue, as well as surfactant production and osmolyte levels in the lungs (Figure 18.3). This study demonstrates the advantages of using NMR techniques in that they provide holistic approaches to investigating mechanisms of lung toxicity. This is likely to be of value in evaluating disease progression or the effects of therapeutic intervention in pulmonary conditions such as surfactant disorders or asthma.

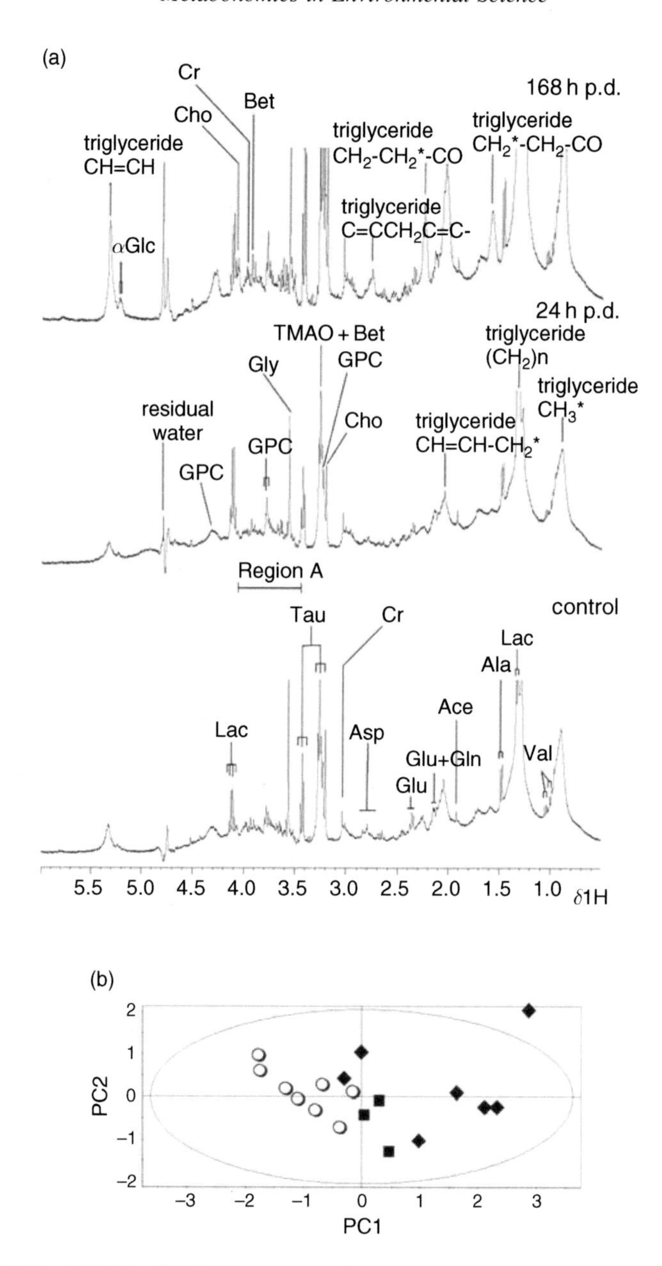

Figure 18.3. (a) The 400 MHz ^{1}H NMR spectra of intact lung from control animals and rats dosed with 1-nitronaphthalene (75 mg/kg). ala, alanine; ace, acetate; Key: Bet, betaine; αGlc, αGlucose; Cr, Creatine; Crn, Creatinine; Gln, glutanine; Gly, glycine; GPC, glycerophosphocholine; Glu, glutamate; Lac, lactate, 2-OG, 2-oxoglutarate; Tau, taurine, TMAO, trimethylamine-N-oxide, Val, valine; p.d., post dose; Region A, Glucose and amino acid CH protons. (b) PCA scores plot of mean-centered post-OSC data obtained from ^{1}H MAS NMR spectra of lung collected 24 h and 168 h post-dose compared with control animals. PCs 1 and 2 account for 80% of the data variance.

18.3. Examining the impact of environmental toxins on wild mammal species

Despite the potential for using metabonomic techniques to investigate the effects of contaminants on wild vertebrates, the number of studies conducted to date are limited, even though the risk of such species being exposed to agrochemicals, pesticides and pollutants is high [24]. Griffin and colleagues assessed the baseline metabolic status of various small mammal species that occur commonly in the UK and Europe. They characterized the metabolite profiles in the urine, kidney and liver tissue of bank voles, wood mice (*Apodemus sylvaticus*), and white-toothed shrews (*Crocidura suaveolens*), and compared them with those obtained from the laboratory rat [1, 2, 25]. The profiles associated with the urine of each small mammal were highly discriminatory in separating the wild species from the laboratory animals (Figure 18.4). In particular, the bank voles excreted large amounts of aromatic amino acids and similar phenolic compounds in their urine. One possible explanation for this is that these phenolic compounds are produced by gut microflora, in a similar manner to hippurate production in rats [26], the large amounts of phenolic compounds resulting from the specialized caecal digestion employed by bank voles to digest their largely herbivorous diet. Given the suggested central importance of gut microflora in terms of xenobiotic metabolism [27], this may have a profound impact on susceptibility to toxins. Gut microflora are responsible for a wide range of metabolic modifications, and increasing the relative proportion of these microbes will increase the range and concentrations of xenobiotics the animal is exposed to. Variation in diet also affects biofluid and tissue composition, although little is known about the influence of changing diet on wild small mammal metabolism. However, the differences detected between bank voles and other (wild and laboratory) rodents in their urinary metabolic profile was not attributable to diet as all the individuals that were analysed had been maintained on a standard laboratory diet.

Rodent species with different trophic strategies have also been found to be profoundly different in terms of their liver and renal tissue compositions (Figure 18.5). Griffin and co-workers [1] examined the normal metabolic composition of these tissues and noted that, compared with the laboratory rat, all the wild species examined had a high lipid content of renal and liver tissue (Figure 18.5). The lipid content of liver and kidney tissues from wood mice were particularly high. This finding suggests that these species may be more vulnerable than laboratory rats to lipophilic xenobiotics [1].

Experimental contaminant studies have also been conducted directly on wild species to examine how susceptible they may be to toxic insults compared with laboratory species. Metabonomic-based ^{1}H MAS NMR spectroscopy has been used to examine cadmium and arsenic toxicity in the bank vole [2]. Following

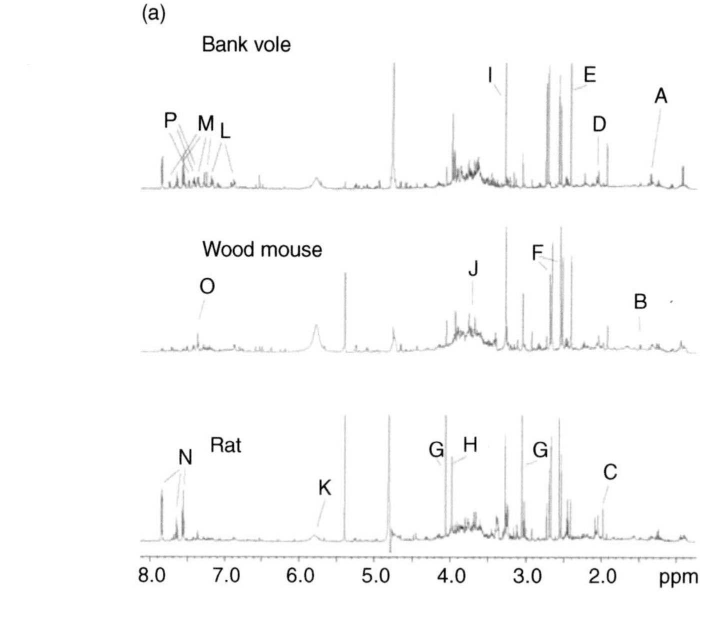

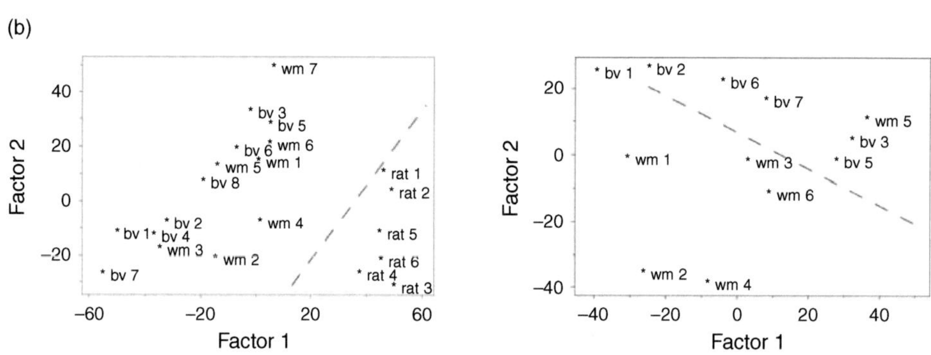

Figure 18.4. (a) The 600 MHz ^{1}H NMR spectra (a) from bank vole wood mouse and rat urine. Bank vole urine contained a variety of aromatic compounds, while rat urine could be distinguished by the high relative concentration of hippurate. Key: A, lactate; B, alanine; C, acetate; D, glutamate; E, succinate; F, citrate; G, creatinine; H, creatine; I, Trimethyl amine oxide (TMAO); J, glucose and sugar containing region; K, urea; L, tyrosine; M, tryptophan; N, hippurate; O, urocanate; P, phenyalanine. (b) Principal component maps of analysis of bank vole, wood mice and rat urine. The left scores plot displays the separation of rat urine from the two wild mammal groups along factor 1 due to hippurate concentration. The right scores plot displays the separation of wood mice and bank vole urine once the rat data has been removed. Separation is almost complete across the diagonal of principal components 1 and 2. Key: bv, bank vole; wm, wood mouse; numbers signify different spectra.

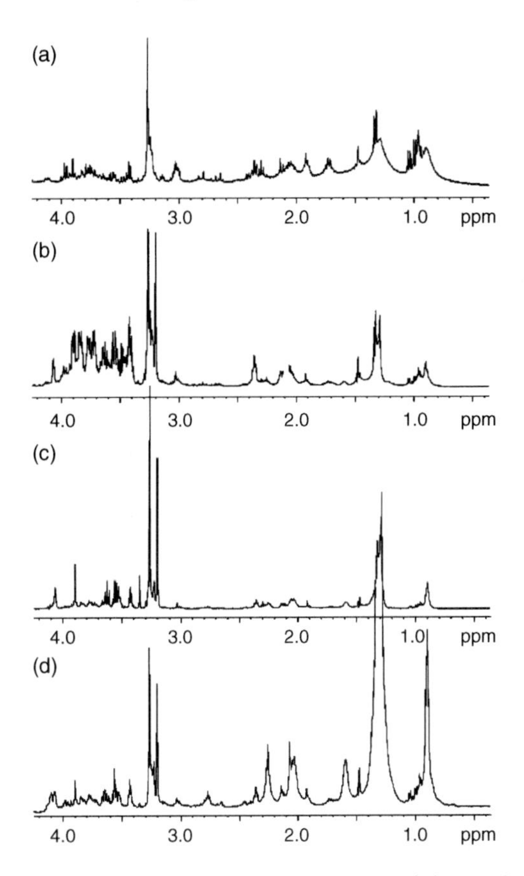

Figure 18.5. ^{1}H MAS NMR spectra from the outer renal cortex of tissue taken from (a) rat, (b) bank vole, (c) shrew and (d) wood mouse kidneys. Increased lipid triglyceride content (corresponding to the broad resonances in the spectra such as at 1.3 and 1.0 ppm. chemical shifts) was found in the wild mammals compared with the laboratory rat. As well as lipids smaller metabolites such as amino acids and sugars could be detected in all animals.

acute exposure of bank voles to cadmium chloride, biochemical changes in lipid and glutamate metabolism that preceded classical nephrotoxicity were detected. Furthermore, these changes occurred after chronic dosing, at levels of exposure that were environmentally realistic and that resulted in a mean renal cadmium concentration (8.4 μg/g dry wt) that was nearly two orders of magnitude below the WHO critical organ concentration (200 μg/g wet wt) [28]. These early stage effects of cadmium on the biochemistry of renal tissue may reflect adaptation mechanisms to the toxic insult or the preliminary stages of the toxicological cascade, and demonstrated the potential sensitivity of the technique. Intriguingly, bank voles also appear susceptible to arsenic toxicity. Diffusion-weighted ^{1}H MAS NMR spectroscopy was used to follow the effects of arsenic-induced hemorrhage via changes in the diffusion properties of water [2]. To monitor toxicity changes in these studies using transcriptomics or proteomics would be vastly complicated by the lack of a sequenced genome for the animal and species-related differences in protein structures.

18.4. Use of high resolution Magic Angle Spinning NMR spectroscopy in studies on the effects of environmental contaminants on mammals

The analysis of biofluids by metabonomic-based techniques offers a minimally invasive procedure for investigating drug toxicity, but it is often necessary to confirm organ toxicity that has been suggested by assessment of systemic metabolism. While histology is still the gold standard for confirming organ and tissue toxicity, NMR spectroscopy is increasingly being used to investigate the biochemical changes within a tissue that accompany a given toxic insult. This approach can also be used in tissues which metabolically are poorly represented in biofluids, for example cerebral tissues which have a small impact on urinary and blood plasma profiles; CSF itself is a relatively dilute solution of metabolites. Solution state spectroscopy, relying on extracting metabolites from the tissue, is already extensively used for this purpose, but this approach is now being augmented by techniques such as high resolution Magic Angle Spinning (HRMAS) NMR spectroscopy that can investigate biochemical changes in tissues directly. These approaches avoid artefacts associated with how soluble a metabolite is, as well as potentially extracting information concerning the metabolic environment of the cells within the tissue (see Chapter 4 for more details of this technique).

Furthermore, ^{1}H MAS NMR spectroscopy can demonstrate when a biofluid biomarker does not originate in a given organ. Nicholson and colleagues (1989) demonstrated that acute exposure of male rats to cadmium chloride resulted in creatinuria following testes-specific toxicity. Thus, it seemed reasonable that similar creatinuria detected in a chronic exposure study of male rats to cadmium chloride resulted from testicular damage [25]. However, using HRMAS ^{1}H NMR spectroscopy, no biochemical changes were detected in testicular tissue, and in particular there was no decrease in tissue creatine content or a change in redox potential in the tissue, known to precede cadmium-induced testicular toxicity. Instead, the most likely explanation for the creatinuria was breakdown of muscle tissue in order to supply glutamine to renal tissue and prevent renal tubular acidosis in the tissue. A major drawback with using HRMAS ^{1}H NMR spectroscopy during routine toxicology studies is that the approach is rather labour intensive and not amenable to automation.

18.5. Examining the impact of environmental toxins on fish

To date there have been few metabonomic studies of fish, despite these organisms being susceptible to environmental contaminants and potentially important as bioindicators of environmental pollution. However, the few studies which have been carried out have shown the general approach of metabolic profiling to be highly discriminatory in assessing fish physiology. Solanky and colleagues [29] have used a combination of ^{1}H NMR spectroscopy of blood plasma and PCA to distinguish Atlantic salmon (*Salmo salar*) exposed to the Gram-negative bacterium *Aeromonas salmonicida* which causes furunculosis. The classification of infected salmon was caused by perturbations in the concentration of lipoproteins, choline-containing residues, carbohydrates, glycerol, trimethylamine-*N*-oxide and betaine. The success of this approach suggests that metabonomics could be used to monitor disease status in salmon as well as a better understanding of the host–pathogen relationship which at present is poorly understood in this fish.

Stentiford and colleagues [30] have used a combined histology, proteomic and metabonomic approach for investigating hepatic cancer in the flat fish dab (*Limanda limanda*). Following histopathology characterization of the tumours, a combination of SELDI-based proteomics and FT-ICR Mass Spectrometry–based metabolomics was used to examine protein and metabolite profiles distinct to liver tumours. In particular, the use of histopathology allowed the exclusion of livers infected by parasite infections, a common problem when examining many wild species. This pilot study suggests that a combination of proteomics and metabonomics could be used to discriminate fish liver tumours.

18.6. Examining the impact of environmental toxins on invertebrates

The relative ease in which metabolic biomarkers can be examined in a range of species has allowed the simultaneous comparison of several different species exposed to a common physiological or environmental stimulus or lesion. This is particularly important for invertebrate animals where few animals have a sequenced genome.

Bundy and colleagues [31] have used the translational power of metabonomics to investigate differences in baseline metabolism between three species of earthworm of the genus *Eisenia*. The three species investigated were *Eisenia fetida, Eisenia andrei and Eisenia venetia*. The two latter species are morphologically indistinguishable, appear to occupy the same ecological niche and only differ genetically. A combination of high resolution ^{1}H NMR spectroscopy, PCA and hierarchical cluster analysis successfully distinguished the three species. This is a particularly important area in invertebrate biology, as soil invertebrates, particularly those with limited dispersive capability, have been used to understand the impacts of evolutionary adaptation and isolation. Limited ability to disperse (~10 m per year) confines populations to small localized areas within which there may be strong selection pressures on sedentary species to adapt to various ecological niches [31]. To date, earthworm taxonomy has largely relied on morphological characteristics, but such methods may have relatively poor power, as is the case with the two *Eisenia* species. The development of a method capable of deriving a metabolic phenotype, or metabotype [32], is highly desirable. Perhaps one of the most surprising results from the study conducted by Bundy and colleagues was that the metabolic profile of coelomic fluid was, in fact, more discriminatory than the analysis of tissue extracts, although the latter did contain some highly discriminatory amino acids.

This metabonomic approach to chemotaxonomy is not confined to animals. Indeed, Bundy and co-workers [33] have used metabonomics to investigate different bacillus cereus strains to separate the microbes according to ecotype. Unlike genomic data, metabonomic analysis using ^{1}H NMR spectroscopy successfully separated the strains into laboratory strains and clinical strains. This approach could be widely applicable to general microbiology in order to identify different functional and physiological ecotypes on bacteria.

To assess the utility of metabonomics for following toxicant exposure in earthworms (*Eisenia veneta*), Bundy and colleagues [34, 35] exposed worms to three different model xenobiotics by a standard filter paper contact test. The impact of these xenobiotics on worm metabolism was investigated by ^{1}H NMR spectroscopy of tissue extracts. PCA demonstrated that perturbations had occurred in the biochemical profiles for all the toxic compounds. Exposure to 2-fluoro-4-methylaniline caused a decrease in 2-hexyl-5-ethyl-3-furansulfonate, as identified using a combination of high-performance liquid chromatography (HPLC)-Fourier transform mass spectrometry and ^{1}H and ^{13}C NMR spectroscopy. An increase in inosine monophosphate was also observed (Figure 18.6). Exposure to 4-fluoroaniline produced a decrease in maltose concentrations, while 3,5-difluoroaniline exerted the same effect as 2-fluoro-4-methylaniline but to a lesser extent. The identification of discrete biomarkers for each xenobiotic indicates that this approach could be used for identifying which toxicants were affecting a given population of worms, as well as identifying a general toxic lesion induced by the exposure.

The impact of metal contamination on three worm species, *Lumbricus rubellus*, *Lumbricus terrestris* and *Eisenia andrei,* has also been investigated using NMR-based metabonomics [36]. In this study, the impact of an environmentally relevant concentration gradient of metal contamination was investigated in the two native lumbricid species and in the *Eiseneia* species that had been introduced to the site. Metabolites identified by ^{1}H NMR spectroscopy and PCA demonstrated that the response to the toxicant was greater in *L. rubellus* than in *E. andrei.* This response was characterized by a profound increase in maltose as well perturbations in histidine

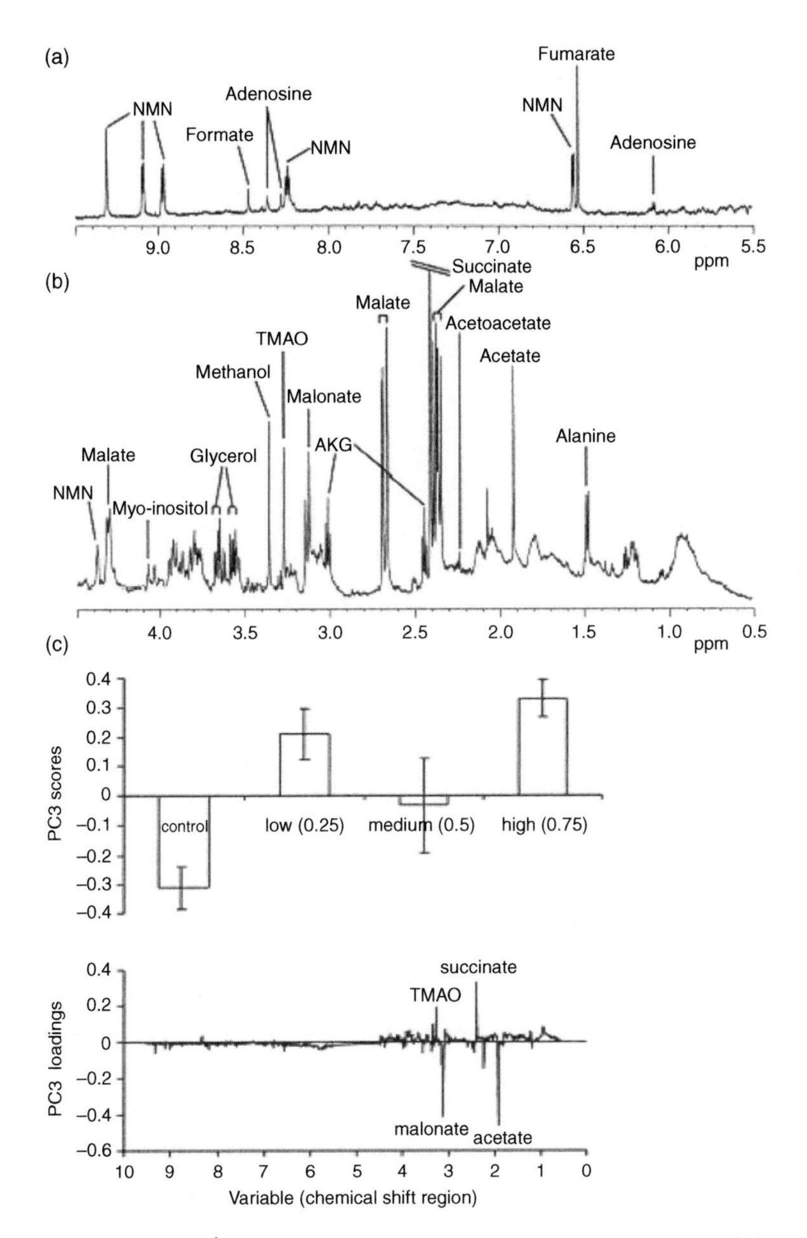

Figure 18.6. High resolution ^{1}H NMR spectra acquired at 600 MHz of coelomic fluid taken from the earthworm *Eisenia veneta*. (a) Aromatic region. NMN: N-methyl nicotinamide α (b) Aliphatic region. AKG: k-ketoglutarate; TMAO: trimethylamine-N-oxide. (c) Principal components analysis of the NMR spectra clearly separated worms that were dosed with 2-fluoro-4-nitrophenol from the control group with this separation being along PC3. Error bars represent the standard error of the mean for these PC scores. The corresponding loadings plot is also shown to demonstrate the importance of malonate, acetate and succinate in this classification. Taken from Reference [34].

and 1-methylhistidine. This study demonstrated the utility of NMR spectroscopy in detecting previously unknown potential biomarkers of toxic insult.

Metabonomics has also been used to determine the effects of contaminants in aquatic invertebrates. Viant and colleagues (2002, 2003) have been using a range of NMR-based approaches to investigate the effects of sublethal doses of environmental contaminants on red abalone (*Haliotis rufescens*), an important wild and farmed shellfish species along the Pacific coast. These workers were able to monitor the effects of exposure to copper (Cu^{2+}) on respiration using ^{31}P NMR spectroscopy [37]. In a follow-up study, they investigated the impact of withering syndrome using NMR-based metabonomics [38]. By examining a range of tissues, the researchers were able to classify abalone as healthy, stunted or diseased according to their metabolic profiles, and identify potential biomarkers associated with the disease. These studies demonstrate the versatility of NMR techniques for investigating the effects of contaminants, other environmental stressors, and for biomonitoring.

18.7. Future directions and conclusions

Although still in its infancy, environmental metabonomics has proven to be a versatile approach for monitoring the impact of xenobiotics within the environment. It has been applied to both vertebrates and invertebrates, and to microbial communities. The approach does not necessitate a sequenced genome, and results are readily transferable from one species to another. However, while many of the approaches used to monitor toxic lesions in laboratory animals can equally be applied to non-laboratory animals, it is clear that this latter group may behave very differently in terms of their responses to a given toxic lesion. Currently, there are a number of large-scale collaborative studies that are assessing and comparing metabonomics with other 'omic' approaches for use in environmental studies. It is likely that metabonomics will become an increasingly used environmental toxicology tool in the future.

References

[1] Griffin, J.L., Walker, L.A., Troke, J., Osborn, D., Shore, R.F. and Nicholson, J.K. (2000) The initial pathogenesis of cadmium induced renal toxicity. *FEBS Lett.*, 478, 147–150.

[2] Griffin, J.L., Walker, L., Shore, R.F. and Nicholson, J.K. (2001) High-resolution magic angle spinning 1H-NMR spectroscopy studies on the renal biochemistry in the bank vole (Clethrionomys glareolus) and the effects of arsenic (As3+) toxicity. *Xenobiotica*, 3, 377–385.

[3] Nicholson, J.K., Lindon, J.C. and Holmes, E. (1999) 'Metabonomics': Understanding the metabolic responses of living systems to pathophysiological stimuli via multivariate statistical analysis of biological NMR spectroscopic data. *Xenobiotica*, 29, 1181–1189.

[4] Nicholson, J.K., Connelly, J., Lindon, J.C. and Holmes, E. (2002) Metabonomics: A platform for studying drug toxicity and gene function. *Nat. Rev. Drug. Discov.*, 1, 153–161.

[5] Oliver, S.G. (2002) Functional genomics: Lessons from yeast. *Philos Trans.* R. *Soc. Lond. B. Biol. Sci.*, 357, 17–23.

[6] Nicholson, J.K. and Wilson, I.D. (2003) Opinion: Understanding 'global' systems biology: Metabonomics and the continuum of metabolism. *Nat. Rev. Drug Discov.*, 2, 668–676.

[7] Beckwith-Hall, B.M., Nicholson, J.K., Nicholls, A.W., Foxall, P.J., Lindon, J.C., Connor, S.C., Abdi, M., Connelly, J. and Holmes, E. (1998) Nuclear magnetic resonance spectroscopic and principal components analysis investigations into biochemical effects of three model hepatotoxins. *Chem. Res. Toxicol.*, 11, 260–272.

[8] Holmes, E., Nicholls, A.W., Lindon, J.C., Ramos, S., Spraul, M., Neidig, P., Connor, S.C., Connelly, J., Damment, S.J., Haselden, J. and Nicholson, J.K. (1998) Development of a model for classification of toxin-induced lesions using 1H NMR spectroscopy of urine combined with pattern recognition. *NMR Biomed.*, 11, 235–244.

[9] Nicholls, A.W., Holmes, E., Lindon, J.C., Shockcor, J.P., Farrant, R.D., Haselden, J.N., Damment, S.J., Waterfield, C.J. and Nicholson, J.K. (2001) Metabonomic investigations into hydrazine toxicity in the rat. *Chem. Res. Toxicol.* 14, 975–987.

[10] Harrigan, G. (2002) Metabolic profiling: Pathways in drug discovery. *Drug Discov. Today*, 7, 351–352.

[11] Griffin, J.L., Nicholls, A.W., Keun, H.C., Mortishire-Smith, R.J., Nicholson, J.K. and Kuehn, T. (2002) Metabolic profiling of rodent biological fluids via 1H NMR spectroscopy using a 1 mm microlitre probe. *Analyst*, 127, 582–584.

[12] Nicholson, J.K., Higham, D.P., Timbrell, J.A. and Sadler, P.J. (1989) Quantitative high resolution 1H NMR urinalysis studies on the biochemical effects of cadmium in the rat. *Mol Pharmacol.*, 36, 398–404.

[13] Connor, S.C., Wu, W., Sweatman, B.C., Manini, J., Haselden, J.C., Crowther, D.J and Waterfield, C.J. (2004) Effects of feeding and body weight loss on the 1H-NMR-based urine metabolic profiles of male Wistar Han rats: Implications for biomarker discovery. *Biomarkers*, 9, 156–179.

[14] Azmi, J., Griffin, J.L., Shore, R.F., Holmes, E. and Nicholson, J.K. (2005) Chemometric analysis of biofluids following toxicant induced hepatotoxicity: A metabonomic approach to distinguish the effects of 1-naphthylisothiocyanate from its products. *Xenobiotica.*, 35, 839–852.

[15] Azmi, J.K. (2002) NMR spectroscopic and chemometric studies on the biochemical effects of substituted aromatic compounds. PhD Thesis, Imperial College London.

[16] Brorstrom-Lunden, E. and Lindskog, A. (1985) Characterization of organic compounds on airborne particles. *Environ. Int.*, 11, 183–188.

[17] Draper, M.W. (1986) Quantitation of nitro- and dinitro polycyclic aromatic hydrocarbons in diesel exhaust particulate matter. *Chemosphere*, 15, 437–447.

[18] Gupta, P., Harger, W. and Arey, J. (1996) The contribution of nitro- and methylnitronaphthalenes to the vapor-phase mutagenicity of ambient air samples. *Atmos. Environ.*, 30, 3157–3166.

[19] Halladay, J.S., Sauer, J.M. and Sipes, I.G. (1999) Metabolism and disposition of (^{14}C)1-nitronaphthalene in male Sprague-Dawley rats. *Drug. Metab. Dispos.*, 27, 1456–1465.

[20] Johnson, D.E., Riley MGI, and Cornish, H.H. (1984) Acute target organ toxicity of 1-nitronaphthalene in the rat. *J. Appl. Toxicol.*, 4, 253–257.

[21] Sauer, J.M. and Sipes, I.G. (1995) All-trans-retinol modulation of nitronaphthalene-induced lung and liver injury in male Sprague-Dawley rats. *Proc. West. Pharmacol. Soc.*, 38, 29–31.

[22] Verschoyle, R.D. and Dinsdale, D. (1990) Protection against chemical-induced lung injury by inhibition of pulmonary cytochrome P450. *Environ. Health Perspect.*, 85, 95–100.

[23] Azmi, J., Connelly, J., Holmes, E., Nicholson, J.K., Shore, R.F., Griffin, J.L. (2005) Characterisation of the biochemical effects of 1-nitronaphthalene in the rat using global metabolic profiling by NMR spectroscopy and pattern recognition. *Biomarkers*, 10, 401–416.

[24] Shore, R.F. and Rattner, B.A. (eds) (2001) *Ecotoxicology of Wild Mammals*. John Wiley & Sons, London, p. 730.

[25] Griffin, J.L., Walker, L.A., Shore, R.F. and Nicholson, J.K. (2001) Metabolic profiling of chronic cadmium exposure in the rat. *Chem. Res. Toxicol.*, 14, 1428–1434.

[26] Phipps, A.N., Stewart, J., Wright, B. and Wilson, I.D. (1998) Effect of diet on the urinary excretion of hippuric acid and other dietary-derived aromatics in rat. A complex interaction between diet, gut microflora and substrate specificity. *Xenobiotica*, 28, 527–537.

[27] Nicholson, J.K., Holmes, E. and Wilson, I.D. (2005) Gut microorganisms, mammalian metabolism and personalized health care. *Nature Rev. Microbiol.*, 3, 431–438.

[28] World Health Organisation (1992) Environmental Health Criteria 134: Cadmium, WHO, Geneva.

[29] Solanky, K.S., Burton, I.W., MacKinnon, S.L., Walter, J.A. and Dacanay, A (2005) Metabolic changes in Atlantic salmon exposed to Aeromonas salmonicida detected by 1H-nuclear magnetic resonance spectroscopy of plasma. *Dis. Aquat. Organ.*, 65, 107–114.

[30] Stentiford, G.D., Viant, M.R., Ward, D.G., Johnson, P.J., Martin, A., Wenbin, W., Cooper, H.J. Lyons, B.P. and Feist, S.W. (2005) Liver Tumors in Wild Flatfish: A histopathological, proteomic, and metabolomic Study. *OMICS: A Journal of Integrative Biology*, 9, 281–299.

[31] Bundy, J.G., Sprurgeon, D.J., Svendsen, C., Hankard, P.K., Osborn, D., Lindon, J.C. and Nicholson, J.K. (2002) Earthworm species of the genus *Eisenia* can be phenotypically differentiated by metabolic profiling. *FEBS Lett.*, 521, 115–120.

[32] Gavaghan, C.L., Wilson, I.D. and Nicholson, J.K. (2002) Physiological variation in metabolic phenotyping and functional genomic studies: Use of orthogonal signal correction and PLS-DA. *FEBS Lett.*, 530, 191–196.

[33] Bundy, J.G., Willey, T.L., Castell, R.S., Ellar, D.J. and Brindle, K.M. (2005) Discrimination of pathogenic clinical isolates and laboratory strains of Bacillus cereus by NMR-based metabolomic profiling. *FEMS Microbiol Lett.*, 242, 127–136.

[34] Bundy, J.G., Osborn, D., Weeks, J.M., Lindon, J.C. and Nicholson, J.K. (2001) An NMR-based metabonomic approach to the investigation of coelomic fluid biochemistry in earthworms under toxic stress. *FEBS Lett.*, 500, 31–35.

[35] Bundy, J.G., Lenz, E.M., Bailey, N.J., Gavaghan, C.L., Svendsen, C., Spurgeon, D., Hankard, P.K., Osborn, D., Weeks, J.M., Trauger, S.A., Speir, P., Sanders, I., Lindon, J.C, Nicholson, J.K. and Tang, H. (2002) Metabonomic assessment of toxicity of 4-fluoroaniline, 3,5-difluoroaniline and 2-fluoro-4-methylaniline to the earthworm *Eisenia veneta* (Rosa): Identification of new endogenous biomarkers. *Environ. Toxicol. Chem.*, 21, 1966–1972.

[36] Bundy, J.G., Spurgeon, D.J., Svendsen, C., Hankard, P.K., Weeks, J.M., Osborn, , Lindon, J.C. and Nicholson, J.K. (2004) Environmental metabonomics: Applying combination biomarker analysis in earthworms at a metal contaminated site. *Ecotoxicology*, 13, 797–806.

[37] Viant, M.R., Walton, J.H., TenBrook, P.L. and Tjeerdema, R.S. (2002) Sublethal actions of copper in abalone (Haliotis rufescens) as characterized by in vivo 31P NMR. *Aquat. Toxicol.*, 57, 139–151.

[38] Viant, M.R., Rosenblum, E.S., Tjeerdema, R.S. (2003) NMR-based metabolomics: A powerful approach for characterizing the effects of environmental stressors on organism health. *Environ. Sci. Technol.*, 37, 4982–4989.

The Handbook of Metabonomics and Metabolomics
John C. Lindon, Jeremy K. Nicholson and Elaine Holmes (Editors)

Chapter 19

Global Systems Biology Through Integration of "Omics" Results

John C. Lindon, Elaine Holmes, and Jeremy K. Nicholson

Department of Biomolecular Medicine, Faculty of Medicine, Imperial College London, Sir Alexander Fleming Building, South Kensington, London, SW7 2AZ UK

19.1. Introduction

The data mining and interpretation of multiple complex metabolic data sets over a cohort of samples is a major statistical challenge, but if achieved successfully, this will present a novel approach to the identification of combination biomarkers of a disease or pharmaceutical effect. In this chapter we describe and discuss the various efforts that have been used to combine information from different data sets, initially using different analytical techniques for the same metabolic analyses, and then move forward to discuss the integration of metabolic data from different samples taken from comprehensive metabolic studies. These include, for example, analysis of various tissues, tissue extracts and biofluids, in what have been termed "integrated metabonomic" investigations.

Next, the integration of data and results from different omics studies at different levels of molecular biology, such as transcriptomics, proteomics and metabo(l/n)omics, is considered, and some examples of where this has been achieved, albeit not usually at the data level but at the interpretation level, are presented and reviewed. Finally the difficulties of modelling such complex biological systems are discussed, and methods for achieving a full systems biology integration are proposed.

19.2. Integration and correlation of metabonomic and metabolomic analytical data

The complementarity of NMR and MS methods as structural tools is well known and has been exploited in structure elucidation studies for natural product research, drug metabolite analysis, and other complex mixture analysis problems for many years. Typically, a variety of NMR and mass spectra are examined together, and structural parameters such as chemical shifts and coupling constants (from NMR) and exact molecular mass and fragmentation patterns (from MS) are extracted from the data and compared, often for a single sample, to generate structural assignment information that is consistent with the outputs of both technologies [1, 2]. However, in many metabonomic studies, multiple samples with a wide range of biochemical variation are available for both NMR and MS analysis, and this creates an opportunity for statistical analysis of signal amplitude co-variation across samples and between technologies for peak assignment and hence metabolite identification purposes.

Recently a new technique has been developed, called Statistical TOtal Correlation SpectroscopY (STOCSY), for analysis of NMR spectroscopic data [3]. STOCSY provides a means of extracting one-dimensional (1D) and two-dimensional (2D) spectral reconstructions of biomarker NMR peaks from both patent and latent signals in 1D NMR spectra. It is based on the properties of the correlation matrix $\mathbf{C}$, computed from a set of sample spectra according to:

$$\mathbf{C} = \frac{1}{n-1}\mathbf{X}_1^t\mathbf{X}_2$$

where $\mathbf{X}_1$ and $\mathbf{X}_2$ denotes the auto-scaled experimental matrices of $n \times v1$ and $n \times v2$ respectively, n is the number of spectra (one for each sample) and $v1$ and $v2$ the number of variables in the spectra for each matrix. $\mathbf{C}$ is therefore a matrix of $v1 \times v2$ where each value is a correlation coefficient between two variables of the matrices $\mathbf{X}_1$ and $\mathbf{X}_2$. Because the different resonance intensities from a single molecule will always have the same ratio whatever the nucleus involved, if the spectrometer conditions are kept identical between samples, the relative intensities will be theoretically totally correlated (correlation coefficient: $r = 1$). In real samples of biofluids, r will be always inferior to 1 because of spectral noise or peak overlaps from other molecules. However, in practice, the correlation matrix from a set of spectra containing different amounts of the same molecule shows very high correlations between the variables corresponding to the resonances of the same molecule. In principle, concentrations of other molecules can also be correlated to the initial molecule of interest, and quantitative relationships between molecules can, therefore, be highlighted. For example, molecules in the same biochemical pathway may exhibit a similar or even co-dependent response to a stimulus. In this case the correlation between resonances from different molecules would be high but not usually as strong as for resonances on the same molecule.

Plotting the correlation matrix provides a graphical representation of the multi-sample spectroscopic data set comparable to that of a 2D correlation NMR experiment conducted on one sample containing all the molecules of all the samples. The closest NMR experiment to STOCSY is TOCSY (TOtal Correlation SpectroscopY), in which signals arise from nucleus within a spin system. Although the 2D plot represents the global correlation map, it requires heavy computer resources (3.2 GB of RAM for a set of spectra with 20 k data points). For this reason, a more concise representation is sometimes easier to obtain.

A 1D representation has been achieved through a chemometric framework using Orthogonal Projection on Latent Structure (O-PLS), which was developed by Trygg *et al.* (see Chapter 6). An O-PLS model is used in order to separate the variation in matrices **X** and **Y** into three parts. The first part contains the covariance of **X** and **Y**, the second one contains the specific variation for **X**, the so-called structured noise, and finally the last one contains the residual variance. Furthermore, the orthogonal loading matrices provide the opportunity to interpret the structured noise, and classical chemometrics tools such as Q^2 can be used in order to test the validity of the model and the presence of outliers that can bias the correlation.

However, it is possible to avoid the full computation of the covariance matrix and to be focused on only one column of **Y** corresponding to a peak of interest. In order to improve the interpretability of the O-PLS model, the method described by Cloarec *et al.* has been applied [3]. It consists in combining the back-scaled O-PLS-DA coefficients from an auto-scaled model with the variable weight of the same model in the same plot. For this purpose, each O-PLS coefficient representing the correlation between the variable of **X** and **Y** is first multiplied by the standard deviation of its corresponding variable to get the covariance which is then plotted as a function of its related chemical shift but with a colour code linked to the initial O-PLS coefficients (correlation), highlighting by this means the resonances that are correlated with the initial resonance of interest.

An example of this approach is given in Chapter 1, Figure 1.5. This shows a STOCSY result from a set of ^{1}H NMR spectra of urine from a metabonomic study of a model of insulin resistance based on the administration of a carbohydrate diet to three different mice strains in which a series of metabolites of biological importance could be conclusively assigned and identified by use of the approach. In this case the metabolite was identified as 3-hydroxyphenylpropionic acid [3].

Extending this concept, it has also been possible to apply an analogous approach (but with some significant differences in the algorithms necessary for scaling and normalisation) to the integration of high resolution NMR and ultra-high-pressure liquid chromatography-mass spectrometry (UPLC-MS) data collected on the same samples in order to achieve a more direct means of cross-assigning signals from these two methods [4]. This approach has been termed Statistical HeterospectroscopY (SHY), as it allows signals in one spectroscopic domain (e.g. NMR) to be dispersed in a second analytical spectroscopic domain (e.g. MS) in order to facilitate data co-analysis. This approach is of wide applicability and can be extended beyond NMR and LC-MS to include any spectroscopic, electrochemical, or other multivariate analytical measurements where multiple samples are measured by more than one technology.

SHY is thus a new statistical paradigm for the co-analysis of multi-spectroscopic data sets acquired on multiple samples. This method again operates through the analysis of the intrinsic covariance between signal intensities in the same and related molecules measured by different techniques across cohorts of samples. The potential of SHY has been illustrated using both 600 MHz ^{1}H NMR and UPLC-TOF-MS data obtained from control rat urine samples ($n = 54$) and from a corresponding hydrazine-treated group ($n = 58$). It was shown that direct cross-correlation of chemical shifts from NMR and m/z data from MS is readily achievable for a variety of metabolites, which leads to improved efficiency of molecular biomarker identification. The approach is illustrated in Figure 19.1 which shows results from the toxicity study of hydrazine mentioned above.

In addition to direct molecular structural information, higher level biochemical information can be obtained on metabolic pathway activity and connectivities by examination of different levels of the NMR to MS correlation and anti-correlation matrixes. Thus Figure 19.2 illustrates this by showing metabolite peaks with both lower, but nevertheless significant, correlation coefficients than would be expected if the peaks came from the same molecule and with peaks showing anti-correlations in peak intensity to the target peak. Both of these observations indicate that, although the two technologies have different detection limits, the detected peaks are from molecules in metabolic pathways related to the target molecule. In addition, if an ion is identified clearly in the MS, and it correlates with a poorly resolved or low signal-noise peak in the NMR spectra, this can provide greater confidence for the NMR assignment than that would have been possible by inspection of the NMR data alone. The SHY approach is of general applicability to complex mixture analysis, if two or more independent spectroscopic data sets are available for any sample cohort. Biological applications of SHY thus show promise as a new systems biology tool for biomarker recovery.

Furthermore, it is possible not only to generate meaningful NMR to m/z correlations to assist in structure assignment using SHY but also to derive connectivities between NMR signals and fragmentation patterns from the same molecules and, further, to cross-correlate NMR spectra and mass spectra from molecules that are not structurally identical but that have commonalities in terms of metabolic pathway regulation. Thus, SHY is not only a structural analytical tool but also a potentially

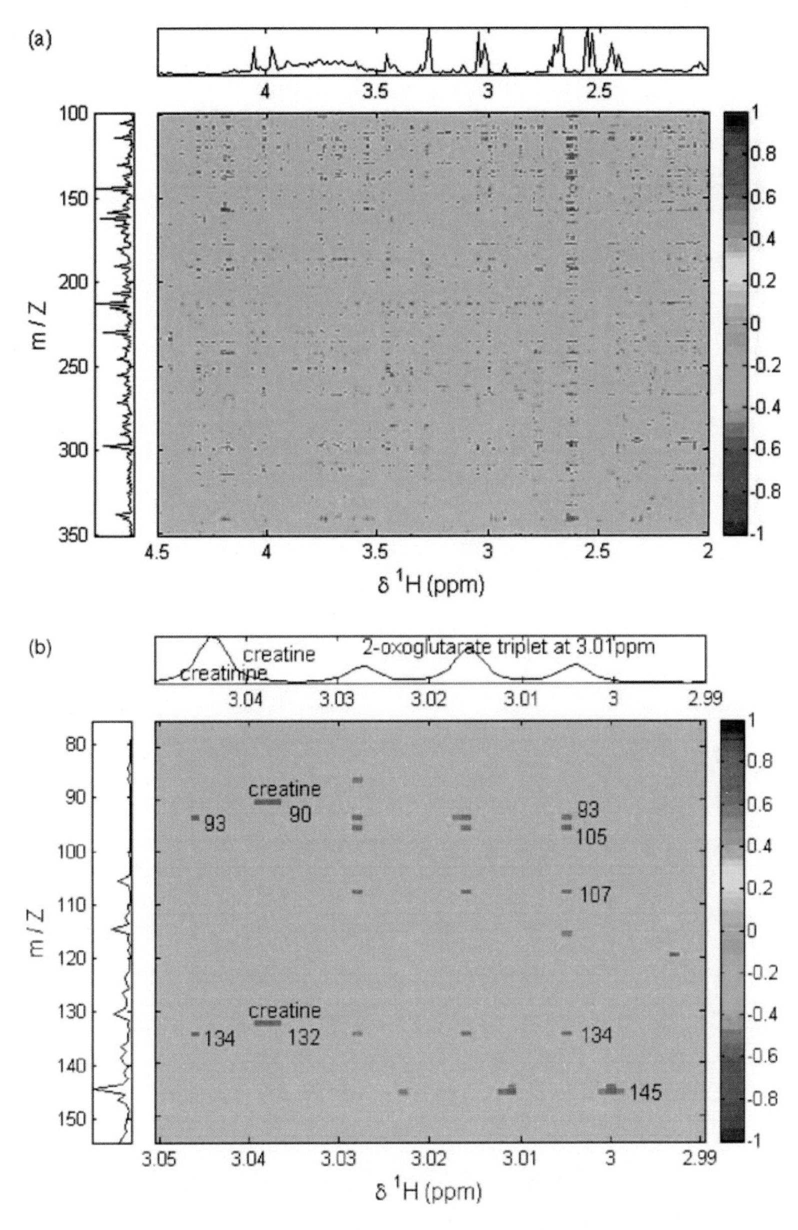

Figure 19.1. (a) NMR-MS correlation for control samples at low resolution, showing part of the aliphatic region of the ^{1}H NMR spectra correlated with the total ion mass spectra in a low molecular weight range (no correlation cut-off applied). (b) NMR-MS correlation for control samples expanded to show a 2-oxoglutarate region of the ^{1}H NMR spectra (cut-off, 0.55). The correlation of the creatine ions with the creatine NMR singlet is clear. There is a negative correlation of unidentified ions at m/z 93 and 134 to the overlapped creatinine peak, probably deriving from the same compound. The ions correlating directly with the 2-oxoglutarate NMR triplet have not been identified but may be related. The ion at m/z 145 correlates with an NMR triplet that overlaps with the peak from 2-oxoglutarate, and it has been proposed as deriving from *N*-acetyllysine [4].

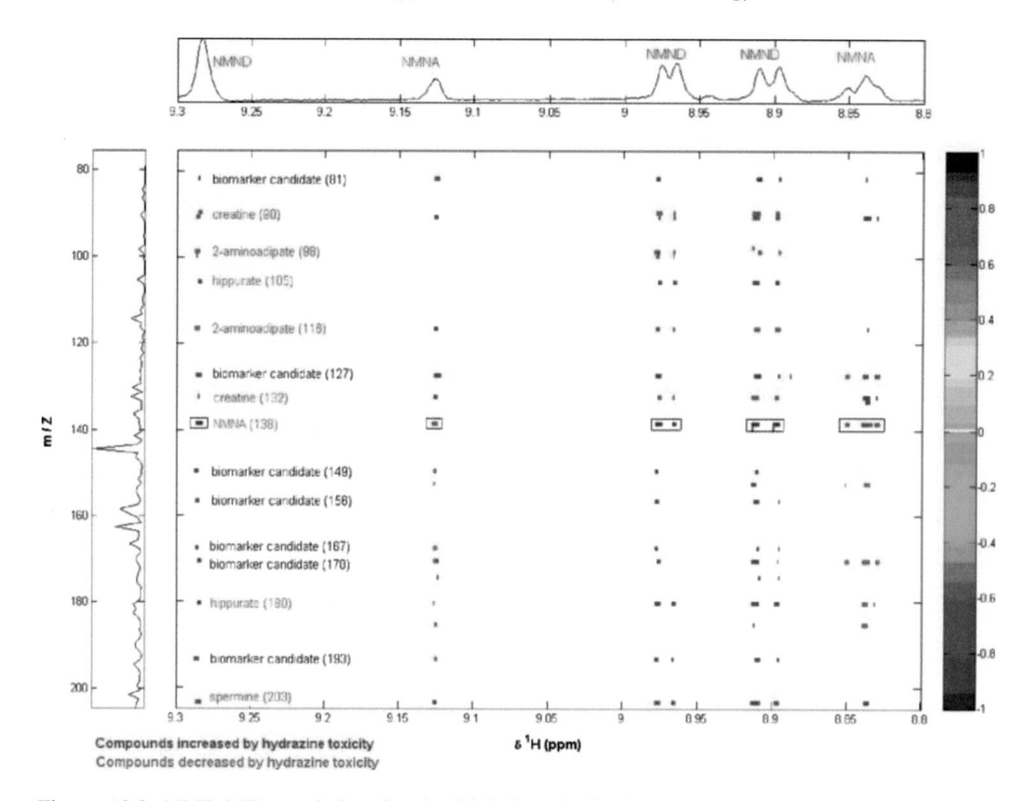

Figure 19.2. NMR-MS correlation for the high-dose hydrazine-treated sample group expanded to show an NMNA/NMND region of the ^{1}H NMR spectrum (cut-off, 0.75). The insets show mean NMR and mass spectra. All identified ions show directions of correlation consistent with known effects of hydrazine. For example, NMNA and NMND are negatively correlated, as they are transaminase-related (inhibited by hydrazine). Also, NMNA correlates positively with itself, as must be the case. These NMNA/NMND correlations are shown boxed. The newly identified spermine ion is shown to correlate positively with toxicity. All unidentified ions are candidate biomarkers. NMND – N-methylnicotinamide; NMNA – N-methylnicotinic acid

powerful bioinformatics tool that can be used in metabonomic or systems biology investigations.

19.3. Integration of metabolomic and metabonomic data from different types of sample

The value of obtaining multiple NMR spectroscopic (or indeed other types of analysis) data sets from various biofluid samples and tissues of the same animals collected at different time points has been demonstrated. This procedure has been termed "integrated metabonomics" [5] and can be used to describe the changes in metabolism in different body compartments affected by exposure to, for example, toxic drugs [6, 7]. Such timed profiles in multiple compartments are themselves characteristic of particular types and mechanisms of pathology and can be used to give a more complete description of the biochemical consequences than can be obtained from one fluid or tissue alone.

The first application of this approach to probe toxic effects of drugs in experimental animals used α-naphthylisothiocyanate (ANIT) as a model hepatotoxicant [6]. Male Han-Wistar rats were dosed with ANIT (150 mg/kg), and plasma, urine and liver samples were collected for NMR and MAS NMR spectroscopy before dosing and at various time points after dosing. Histopathology and plasma clinical chemistry were also performed at all time points. Liver samples were analysed either intact by 600 MHz ^{1}H MAS NMR techniques or using ^{1}H NMR spectroscopy of water-acetonitrile extracts. Typical 600 MHz ^{1}H NMR spectra of urine at predose and various time points following the administration of ANIT and 600 MHz ^{1}H MAS NMR spectra of the left lateral lobe of the liver at various time points following the administration of ANIT are shown in Figure 19.3. These data were then related to sequential ^{1}H NMR measurements in urine and plasma using pattern recognition methods such as principal components analysis (PCA) to show the time-dependent biochemical variations induced by ANIT toxicity. From the eigenvector loadings of the PCA, those regions of the ^{1}H NMR spectra and hence the combinations of endogenous metabolites marking the main phase of the toxic episode were identified. The ANIT-induced biochemical manifestations included a hepatic lipidosis associated with hyperlipidaemia; hyperglycaemia and glycosuria; increased urinary excretion of taurine and creatine; a shift in energy metabolism characterised by increased plasma ketone bodies with reduced urinary excretion of tricarboxylic acid cycle intermediates and raised hepatic bile acids leading to bile aciduria. The integration of metabolic data derived from several sources gave a holistic approach to the study of time-related toxic effects in the intact system and enabled the characterisation of key metabolic effects during the development of, and recovery from, a toxic lesion. A schematic representation of the various metabonomic changes seen in tissues, tissue extracts and biofluids as a function of time is summarised in Figure 19.4. Finally, in cases where tissue MAS NMR data is co-analysed with biofluid NMR data, a more direct link can be made between the biofluid results and the tissue histopathology interpretations.

An integrated metabonomics study using high-resolution ^{1}H NMR spectroscopy has also been applied to investigate the biochemical composition of intact liver tissue (using MAS NMR), liver tissue extracts, and blood plasma samples obtained from control and acetaminophen-treated mice. PCA was used to visualise similarities and differences in biochemical profiles [7]. The time- and dose-dependent biochemical effects of acetaminophen were related to the drug toxicity, as determined using

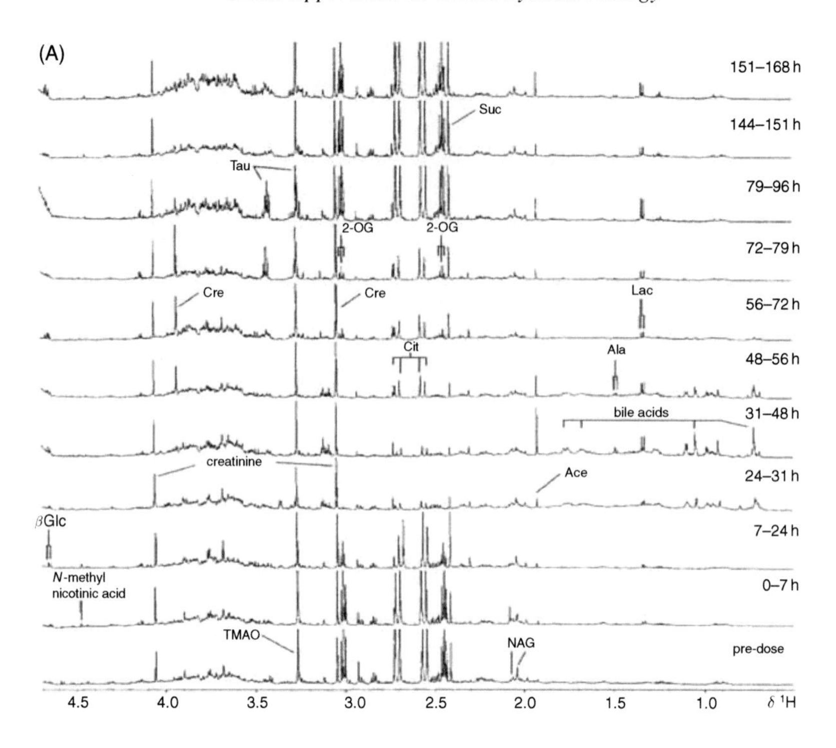

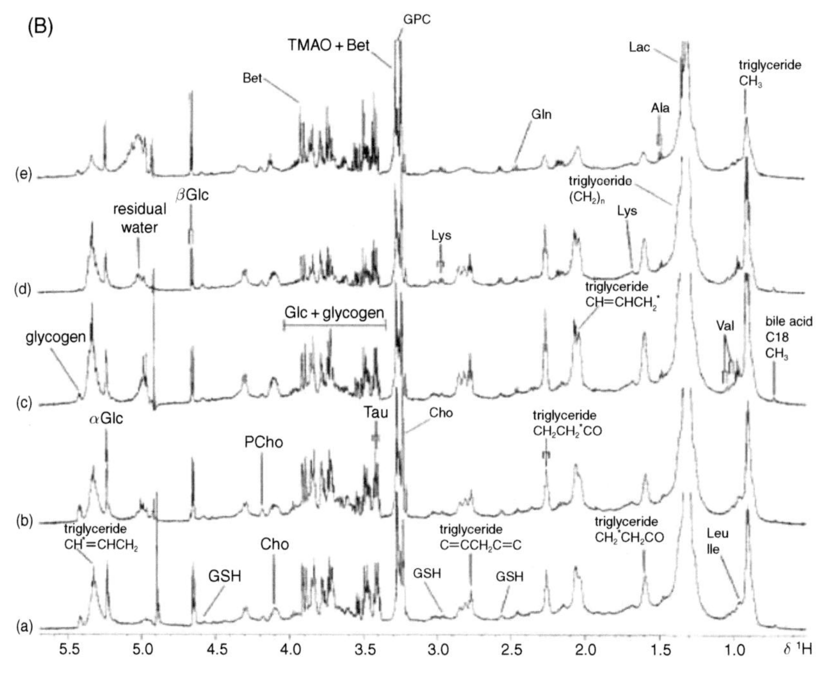

Figure 19.3. (Continued)

histopathology. Metabolic effects in intact liver tissue and lipid-soluble liver tissue extracts from animals treated with the high dose level of acetaminophen included an increase in lipid triglycerides and mono-unsaturated fatty acids together with a decrease in polyunsaturated fatty acids, indicating mitochondrial malfunction with concomitant compensatory increase of peroxisomal activity. In addition, a depletion of phospholipids was observed in treated liver tissue, which suggested an inhibition of enzymes involved in phospholipid synthesis. There was also depletion in the levels of liver glucose and glycogen. In addition, the aqueous soluble liver tissue extracts from high dose animals also revealed an increase in lactate, alanine, and other amino acids, together with a decrease in glucose. Plasma spectra showed increases in glucose, acetate, pyruvate, and lactate. These observations all provided evidence for an increased rate of glycolysis. These findings could indicate a mitochondrial inability to use pyruvate in the citric acid cycle and also reveal the impairment of fatty acid oxidation in liver mitochondria of such treated mice.

A metabonomic analysis has been carried out in order to detect biomarker and metabolite information that might lead to specific protein targets in the toxicological mechanism of bromobenzene-induced centrilobular hepatic necrosis [8]. Male Han-Wistar rats were dosed with bromobenzene (1.5 g/kg) and blood plasma, urine and liver samples were collected for NMR and MAS NMR spectroscopy at various time points post-dose, with histopathology and clinical chemistry being performed in parallel. The liver sample spectra were correlated to sequential ^{1}H NMR measurements in urine and blood plasma using pattern recognition methods (PCA) to show the time-dependent biochemical variations induced by bromobenzene toxicity. In addition to a holistic view of the effect of hepatic toxicity on the biochemistry, a number of putative protein targets of bromobenzene and its metabolites were identified including enzymes of the glutathione cycle, exemplified by the presence of a biomarker, 5-oxoproline, in liver tissue, blood plasma, and urine. This work indicates the importance of metabonomics for helping to resolve the mechanistic complexity of drug toxicity.

Figure 19.3. (Continued) (A) 600 MHz ^{1}H NMR spectra of urine at predose and various time points following the administration of ANIT (150 mg/kg). (B) 600 MHz ^{1}H MAS NMR spectra of the left lateral lobe of the liver at various time points following the administration of ANIT (150 mg/kg): (a) control, (b) 7 h, (c) 24 h, (d) 31 h, and (e) 168 h post-dose. Lorentzian-Gaussian resolution-enhancement was applied to all spectra. Key: Ace, acetate; Ala, alanine; Bet, betaine; Cho, choline; Cit, citrate; Cre, creatine; Glc, glucose; Gln, glutamine; GPC, glycerophosphorylcholine; GSH, glutathione; Ile, isoleucine; Lac, lactate; Leu, leucine; Lys, lysine; NAG, *N*-acetyl glycoproteins; 2-OG, 2-oxoglutarate; PCho, phosphocholine; Suc, succinate; Tau, taurine; TMAO, trimethylamine-*N*-oxide; Val, valine.

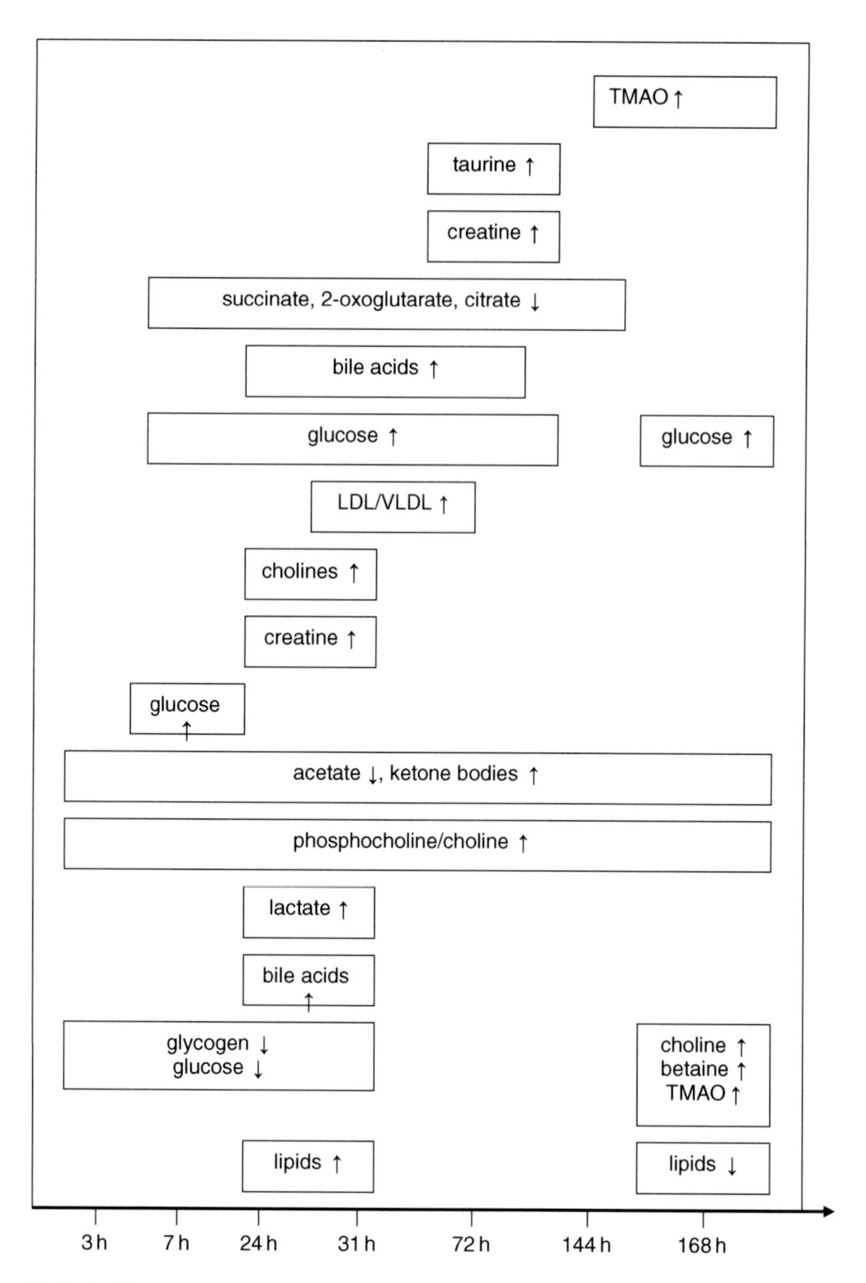

Figure 19.4. An illustration of integrated metabonomics, showing the various biofluids and tissues analysed, and the metabolic changes in each, as a function of time. The toxin α-naphthylisothiocyanate, a model liver toxin, was administered to Han-Wistar rats at time zero [6].

19.4. Integration of data from different omics approaches

A number of studies have attempted to integrate data that has been obtained from various combinations of transcriptomics, proteomics, and metabonomics. Most of these studies have analysed the data from the various "omics" approaches separately and have then tried to discover commonalities and linkages between the altered levels seen in the various experiments.

Genomics and proteomics have thus been combined, together with conventional techniques such as histopathology, to study the effects of paracetamol (acetaminophen), at a range of doses from 0 to 500 mg/kg, on liver in the mouse [9]. Liver was sampled from 15 min post-dose up to 4 h, with changes detected as soon as 15 min post-administration in mitochondrial proteins (apparently preceding most of the gene transcript changes). The bulk of the effects noted by both transcript profiling and proteomics were associated with loss of energy production (a decrease in ATP synthase subunits and β-oxidation pathway proteins). As the authors noted, "transcript profiling and proteomics did not usually detect expression level changes of one mRNA and the corresponding protein, but genes, proteins and pathways identified by transcript profiling and proteomics told a similar story". A subsequent experiment from the same laboratories, using essentially the same study design examined the relationship between the transcript profiles and metabolic profiles [10]. Thus, it has also been possible to integrate data from transcriptomics and metabonomics to find, after acetaminophen administration to mice, common metabolic pathways implicated by both gene expression changes and changes in metabolism. The results of the metabonomic investigation also showed changes consistent with an alteration in energy metabolism (dramatic reductions in hepatic glycogen and glucose and concomitant increases in saturated lipids and reductions in unsaturated and phospholipids). Here the gene changes both preceded and were concurrent with the observed perturbations in metabolism. Taken together, the genomic-proteomic and genomic-metabonomic studies reveal a consistent qualitative picture of cells attempting to respond to the consequences of a global energy failure.

It has also been found that evaluation of both transcriptomic and metabolic changes following administration of the toxin bromobenzene provides a more sensitive approach for detecting the effects of the toxin [11]. This was the conclusion of a study that investigated whether integrated analysis of transcriptomics and metabolomics data could increase the sensitivity of detection and provide new insight in the mechanisms of hepatotoxicity. Metabolite levels in plasma or urine were analysed in relation to changes in hepatic gene expression in rats that received bromobenzene to induce acute hepatic centrilobular necrosis. Multivariate statistical analysis showed that metabolite profiles of blood plasma were largely different from controls when the rats were treated with bromobenzene, also at doses that did not elicit histopathological changes. Changes in levels of genes and metabolites were related to the degree of necrosis, providing putative novel markers of hepatotoxicity. Levels of endogenous metabolites like alanine, lactate, tyrosine and dimethylglycine differed in plasma from treated and control rats. The metabolite profiles of urine were found to be reflective of the exposure levels. This integrated analysis of hepatic transcriptomics and plasma metabolomics was able to more sensitively detect changes related to hepatotoxicity and discover biomarkers. The relationship between gene expression and metabolite levels was explored and additional insight in the role of various biological pathways in bromobenzene-induced hepatic necrosis was obtained. This study also highlighted the fact that the full integration of the methods awaits technical optimisation, especially in the process of identification of metabolites. Nevertheless, corroborating findings from liver transcriptomics and plasma metabolite profiling enabled the generation of new hypotheses concerning cellular mechanisms putatively related to necrosis, such as changes related to apoptosis, glycolysis, and amino acid metabolism. Both liver gene and plasma metabolite markers were discovered to correlate with the degree of hepatocellular necrosis in individual animals.

In a similar fashion, changes in gene expression detected in microarray experiments can lead to the identification of changed enzyme activity, and this can also be achieved by analysis of metabolic perturbations. Orotic acid hepatotoxicity in the rat was also investigated in an integrated reverse functional genomic and metabolic study employing transcriptional and metabonomic analysis combined with multivariate analysis [12]. Perturbed metabolic pathways included those involved with fatty acid, triglyceride and phospholipid biosynthesis, β-oxidation, carbohydrate metabolism, and altered nucleotide and methyl donor metabolism and stress responses whilst transcriptome analysis showed effects on stearoyl-CoA desaturase and other lipid-related transcripts. The relative success of these initial studies suggests that there may indeed be useful synergy to be gained from performing integrated omics. However, these approaches still do not address fully the problems of modelling the complexity of the mammalian superorganism.

Ippolito *et al.* have described combined genomics and metabolomics studies of primary prostate tumours and lymph node and liver metastases from transgenic mice, plus information from cell lines [13]. The gene chip data was used to reconstruct models of cell metabolism in neuroendocrine cancer systems and then these models were used for targeted LC-MS studies for metabolite analysis in tumours and cell lines. It was shown that a distinguishing feature for poor prognosis tumours is a glutamic acid decarboxylase–independent pathway for production of gamma aminobutyric acid (GABA) and a pathway for production of imidazole-4-acetate that involves dopamine (DOPA)-decarboxylase and a membrane-associated amine oxidase, amiloride-binding protein.

Proteomics and metabolomics approaches have also been applied to identify protein and metabolite changes in vessels of apolipoprotein E(−/−) mice on a normal diet [14, 15]. Using 2D gel electrophoresis and mass spectrometry, 79 protein species that were altered during various stages of atherogenesis were identified, including those implicated in immunoglobulin deposition, redox imbalance, and impaired energy metabolism that preceded lesion formation in the mice. Oxidative stress in the vasculature was reflected by the oxidation status of 1-Cys peroxiredoxin and correlated to the extent of lesion formation in 12-month-old apolipoprotein E(−/−) mice. NMR spectroscopy revealed a decline in alanine and a depletion of the adenosine nucleotide pool in tissues from vessels of 10-week-old apolipoprotein E(−/−) mice. Attenuation of lesion formation was associated with alterations of nicotinamide adenine dinucleotide (NADPH)-generating malic enzyme, which provides reducing equivalents for lipid synthesis and glutathione recycling, and successful replenishment of the vascular energy pool. This study provided a comprehensive data set of protein and metabolite changes during atherogenesis and highlighted potential associations of immune-inflammatory responses, oxidative stress, and energy metabolism.

One principal difficulty in integrating data from different omics studies is the need to breakdown the data matrices into systematic variation related to the disease or condition of interest, and thus to separate variation that is unrelated to the biology of interest. By removing extraneous or orthogonal variation, confounding analytical and background factors such as analytical instrument drift or gender variation can be removed, and this can simplify subsequent analyses.

One of the few studies to actually co-analyse data sets from different omics experiments in this fashion used a novel statistical proteomic and metabonomic integration method applied to a human tumour xenograft mouse model of prostate cancer [16]. Parallel 2D differential gel electrophoresis (2D-DIGE) proteomic and ^{1}H NMR metabolic profile data were collected on blood plasma from mice implanted with a prostate cancer (PC3) xenograft and from matched control animals. In order to interpret the xenograft-induced differences in plasma profiles, multivariate statistical algorithms including orthogonal projection to latent structure (O-PLS) were applied to generate models characterising the disease profile. Two approaches to integrating metabonomic data matrices were presented based on O-PLS and O2-PLS algorithms to provide a framework for generating models relating to the specific and common sources of variation in the metabolite and protein data matrices that can be directly related to the disease model. Multiple correlations between metabolites and proteins were found and including associations between serotransferrin precursor and both tyrosine and 3-D-hydroxybutyrate. Additionally a correlation between decreased excretion of tyrosine and increased presence of gelsolin was also observed. This approach was postulated to provide enhanced recovery of combination candidate biomarkers across multi-omic platforms, thus enhancing understanding of *in vivo* model systems studied by multiple omic technologies.

In this study, metabonomic analysis of blood plasma was performed using ^{1}H NMR spectroscopy, and O-PLS-DA modelling of the ^{1}H NMR data was used to investigate the differences in metabolic concentrations between samples obtained from PC3-xenograft-implanted mice and the matched controls. Although complete discrimination between control mice and xenograft mice was not observed in the O-PLS score plot (scores were calculated from cross-validation to ensure that over fitting was avoided), there was an underlying difference between the two sample groups. The ^{1}H NMR shifts that had the largest influence on the O-PLS-DA model, that is those that changed the most between the two classes contained resonances from the amino acids valine, isoleucine, glutamine, leucine, lysine, tyrosine, and phenylalanine together with glucose, 3-D-hydroxybutyrate, and acetate.

The proteomic data contained 392 proteins spots that were present on all 10 DIGE gels, and O-PLS-DA was used to analyse changes in protein levels between the control and PC3 animals, showing a consistent difference between the two classes, although as with the NMR data, complete discrimination between the control and disease class was not achieved. Regression coefficients for the DIGE model showed several protein variables of importance for the discriminant model.

In order to investigate the relationships observed between NMR regions and protein spots, O-PLS and O2-PLS methods were used. The O-PLS models were constructed between 2D-DIGE data spots and individual ^{1}H NMR data peaks that showed the highest discriminatory power between control and PC3 animals. This approach provided a means of adding further confidence in correlations between NMR and DIGE variables since cross-validation could be used for the O-PLS models. Separate O-PLS models were built by regressing all NMR variables against a single DIGE variable, as well as in the opposite direction, using all variables in the DIGE data against one particular NMR shift. For example, the O-PLS model generated for the tyrosine resonance at δ 6.9 (down-regulated in PC3 animals) showed associations between this metabolite and several protein spots (Figure 19.5(a)) as visualised using the correlation approach [3].

The regression coefficients provided a means of interpretation of which, and how, variables in one data matrix, for example NMR, relate to the variables in the other data matrix, for example DIGE. In addition to a transparent model, which enables interpretation of the patterns of change, model statistics (see Chapter 6 for more details) such as R^2 (goodness of fit) and Q^2 (goodness of prediction) yield information about how well the data are modelled as well as quantitative information of the proportion of variance that is modelled and possible to predict.

This approach is generally applicable to proteomic and metabonomic data, but could also be used for integration of other types of "omic" data. The method requires that the data have been collected in parallel on samples from the same animals and that the data matrices are as complete as possible. Using the O-PLS modelling

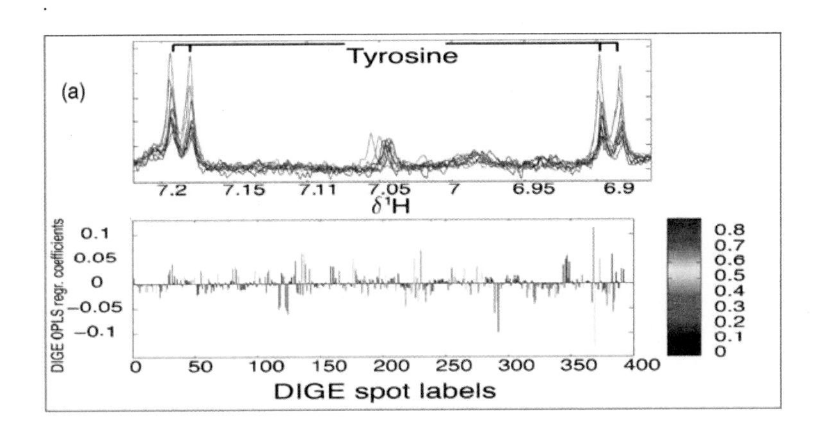

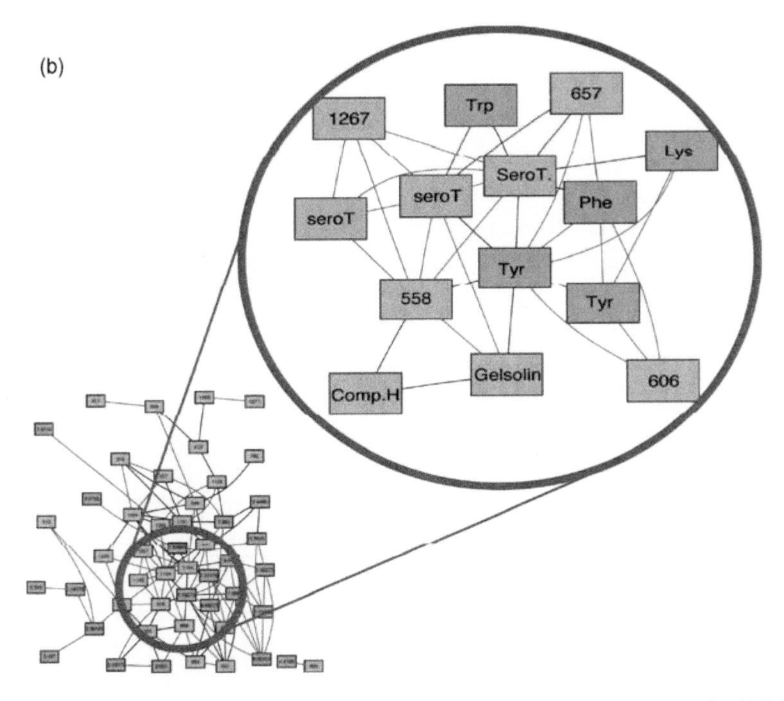

Figure 19.5. (a) The tyrosine aromatic proton resonance region of the NMR spectra (red NMR spectra represent controls) and the corresponding DIGE regression coefficients for the O-PLS models where DIGE data is regressed against the intensity of the 6.9 ppm NMR peak from tyrosine. The colouring of spectra is proportional to the O-PLS-DA regression coefficients. (b) Visualisation of correlations between selected DIGE spots and NMR data points with a correlation to class of >0.77. Edges in the network are present between nodes (NMR/DIGE variables) where the correlation is >0.85. Key: (blue node = NMR variable, red node = DIGE spot, red edge = positive correlation >0.85, blue edge = negative correlation < −0.85).

framework it was possible not only to establish changes in plasma metabolite and protein profiles in response to the PC3 xenograft, but also to establish co-expression patterns between these metabolites and proteins changes in combination in response to disease. O2-PLS was also used for construction of a common model using the two different data sets and this allowed the separation and quantitation of variance that was shared and was predictable from one of the data matrices to the other, but this also allowed identification of variance patterns that are unique in each data matrix.

The analysis showed correlation, and therefore some association, between many proteins and metabolites, as shown in Figure 19.5(b). These correlations may be used to generate hypotheses on biological relationships or pathway activity that can be further tested experimentally *in vivo* or *in vitro*. However, because of the current state of knowledge of the proteome, most of the proteins detected as significant in the current study are, as yet, unidentified. The associations that are revealed by this analysis suggest that extended effort into protein identification may well prove to be a fertile source of new insight into the biology of these implanted tumours. Using the proposed framework for integrating metabonomic and proteomic data [16], no assumptions about network models are made, and the models are not limited by pathway constraints and are therefore open to alternative solutions.

Using the chemometric strategy developed in this study, the data can be separated or modelled in distinct parts relating to metabolites and proteins that co-vary in response to a biological challenge and parts not co-varying (residual and class-orthogonal) with the challenge. It was also possible to separate variance patterns found in proteomic and metabonomic data matrices into parts which are common between the two matrices and parts of the metabolite or protein profiles that vary independently relative to the complementary data matrix.

Where biological sense can be made out of co-varying proteins and metabolites, the results are more robust since the change in protein expression can to some extent validate the change in metabolite levels and *vice versa*. Thus, there is an opportunity to extract and utilise more information from each animal study carried out, and as a consequence, there is the possibility to reduce the number of early studies than that would otherwise be required, which is in alignment with the current emphasis to reduce, replace, and refine animal experimentation. Additionally the appearance of abnormal behaviour in one of the data matrices, for example an animal with abnormal levels of a particular metabolite, can be confirmed as being biologically idiosyncratic when the protein expression profile is also abnormal.

One of the first studies to attempt the combined use of transcriptomics, proteomics, and metabonomics data to gain improved interpretation was a study on the toxicity of hydrazine in the rat [17]. Sprague-Dawley rats were assigned to three treatment groups with 10 animals per group and given a single oral dose of vehicle,

30 or 90 mg/kg hydrazine. RNA was extracted from rat liver 48 h post-dosing and transcribed into cDNA. The abundances of the mRNA were investigated using cDNA microarrays containing 699 rat-specific genes that are involved in toxic responses. In addition, proteins from rat liver samples (48, 120 and 168 h post-dosing) were resolved by 2D-DIGE and proteins with changed expression levels after hydrazine treatment were identified by matrix-assisted laser desorption/ionisation TOF MS peptide mass fingerprinting. In addition, in order to elucidate how the biochemical changes were reflected in biochemical pathways, endogenous metabolites were measured in serum samples collected 48 h post-dosing using 600 MHz ^{1}H NMR spectroscopy. This study showed that a single dose of hydrazine caused gene, protein, and metabolite changes, which could be related to glucose metabolism, lipid metabolism, and oxidative stress. These findings were in agreement with the known effects of hydrazine toxicity but provided potential new biomarkers of hydrazine-induced toxicity.

Most recently, a comprehensive analysis of transcriptomics, proteomics, and NMR-based metabonomics data arising from a toxicity study of methapyriline in the rat has been carried out [18]. Administration of high doses of the histamine antagonist methapyrilene to rats causes periportal liver necrosis but the mechanism of toxicity is ill-defined. Male rats were dosed with methapyrilene for 3 days at 150 mg/kg/day, which was sufficient to induce liver necrosis, or a subtoxic dose of 50 mg/kg/day. Urine was collected over 24 h each day, while blood and liver tissues were obtained at 2 h after the final dose. The resulting data defined the changes that occur in signal transduction and metabolic pathways during methapyrilene hepatotoxicity, revealing modification of expression levels of genes and proteins associated with oxidative stress and a change in energy usage that is reflected in both gene/protein expression patterns and metabolites.

The combination of information from gene, protein, and metabolite levels provided an integrated picture of the response to methapyrilene-induced hepatotoxicity with supporting and validating evidence arising from each biomolecular level. As expected, there were several instances where genes and proteins, either encoded by the same gene or by other genes within the same pathway, were both co-regulated by methapyrilene toxicity, and sometimes this was in concert with an associated metabolic product. This demonstrated the synergy that might be expected from combining these omic approaches. The authors suggested that measurements made across multiple biomolecular levels could potentially yield biomarker combinations that are more specific than combination biomarkers derived from one "omic" platform alone.

However, this work highlighted a number of areas that require continuous refinement for systems biology approaches utilising such multiple "omic" technologies.

First, it was noted that alterations in gene expression, enzyme levels, and protein modifications, while suggesting a potential target of toxic effects, do not imply that function or activity must be altered. This means that from gene and proteomic expression profiling alone, the distinction between cause and effect is generally not possible. Alterations to metabolic profiles do reflect altered function, and so may serve to aid interpretation of corresponding gene expression and proteomic analyses. The correlation between changes in expression of the genome and the proteome on a gene-by-gene basis can be poor, so modulation of pathways at some or all biomolecular levels could indicate a critical control point. In addition, proteomics can also indicate changes in protein modification as well as abundance and any metabolic changes are likely to be rather distant to many signal transduction changes detected by gene expression changes. Furthermore, of the various omics techniques, observed metabolic effects are likely to be more conserved between species, and hence in any integrated systems approaches applied across species, metabolic changes might help to better understand changes in gene expression.

A second issue that was raised is that of experimental design. It could be that looking at time points where toxicity is already well developed mitigates against obtaining a clear understanding of the temporal dynamics of the mechanism, especially as changes at the gene, protein, and metabolite level may proceed at different rates and on different time scales. Analysis of such complex data sets is challenging, and with current tools it also involves a degree of subjectivity. It may therefore be the case in the study under consideration that more subtle relationships present in these data linking the various levels of biomolecular organisation within the cell have been missed.

The third area that was raised might be termed "analytical resolution". The DNA microarray method measures the expression of individual genes and a single expression numerical value can be derived for each. The proteomic and metabonomics approaches used do not yield individual numerical values for every species measured, since the response is measured as the overall change in profile where the variables measured are expressed in terms of differential responses or relative changes. This presents a significant modelling challenge and is one reason why latent variable methods dominate the metabonomics literature.

In the study reported above [18], hundreds of gene expression changes were observed compared to the relatively small number of changes detected by the other two technologies. It may thus be likely that insufficient detail was obtained at each biomolecular level to elaborate fully the mechanism of methapyrilene toxicity. MS-based metabonomics approaches have been shown to be highly complementary to NMR profiling techniques and might have helped. However, it may be necessary to await more sensitive metabonomic and proteomic technologies to enable total resolution of signalling and regulatory pathways in addition to the metabolic pathways.

19.5. Future requirements for modelling hyper-complex biological systems

Currently the bulk of studies in the world of "omics" tend to be in a single discipline (genomics/transcriptomics, proteomics, etc.). This does not necessarily imply a logical approach to the problem under investigation, but often merely reflects the particular expertise and specialism of the groups involved. In any integrated "systems" approach, it would seem beneficial to combine studies in all of the omics to provide an overview of what is going on. Clearly the downside of such a strategy is the requirement for a wide range of technical expertise, and then experimental approaches to combining data from the different omics platforms into a single coherent stream.

For simple systems, the way in which intra-cellular biochemistry can be modelled is now becoming mature, especially for single-celled organisms such as *Saccharomyces*, and a clear direction using conventional metabolic pathways at the metabolite level linked directly to pathways predicted from the genome and confirmed using proteomic studies [19] is being used. Even here, though with a complete knowledge of the genome of a simple organism, such as a pathogen, interpretation of genomic data has not led as simply as expected to new drug targets, and it has been proposed that modelling of metabolic networks is more likely to lead to such new avenues of opportunity for pharmaceutical development. This is particularly true for such simple organisms where the known metabolic pathways are considerably simpler than those for typical mammalian cells with all their secondary interactions. In this case, genes can be mapped to specific proteins/enzymes and hence to metabolic pathways at several levels, that is for the whole genome for a given organism or for a specific part of the genome to compare across organisms (e.g. *Trypanosoma brucei* and *Trypanosoma cruzi*, causing sleeping sickness and Chagas disease, respectively) [20]. Operating at the metabolic level is particularly advantageous when making comparisons across species. However, many metabolites have not been identified, particularly in some species, and hence there remain many gaps in conventional pathways models such as known universally as KEGG (www.genome.ad.jp/kegg) or ExPASY (www.expasy.ch/cgi-bin/search-biochem-index). Also, it has to be remembered that any mathematical model will always be simpler than the real system that it attempts to emulate, and that different models might all be sufficient. Nevertheless, integrated modelling of systems biology at the transcriptomic, proteomic, and metabolic levels is feasible using bioinformatics techniques including rule-based methods such as are now used for predicting pathways for xenobiotic metabolism [21]. However, even at this low level of complexity, many of the feedback and regulatory mechanisms evident in metabolic pathways cannot be addressed. Such metabolic models can be broken down into studies of stoichiometry where metabolic fluxes are calculated to give a view of the system under a particular set of conditions using mathematical optimisation techniques [22] often incorporating stable isotope labelling experiments [23]. This approach has been used to describe all the known reactions in *Escherichia coli* [24]. A second approach attempts to model the reaction kinetics of a system and to use the derived kinetic parameters to predict time profiles [25].

A major problem arises at the next level of complexity – multicellular organisms and superorganisms – because of the need to consider inter-cell transport and other kinetic parameters. Full mathematical treatment of the physicochemical phenomena is necessary. Given the differences in gene and protein expression in different cells, the only coherent and integrated approach is to model the biochemistry at the metabolic level. There has been some considerable thought given to the concept of metabolic networks and their properties. Even for simple organisms, metabolic networks are organised into smaller interconnected modules that combine in a hierarchical manner and, for species such as *E. coli*, these organisations can be recognised as being related to known metabolic functions [26]. A systematic way of defining metabolic networks has been proposed and the properties and scaling of such networks have been evaluated [27]. Giersch [28] has reviewed the techniques used for metabolic modelling, including metabolic control analysis in cells, the effects of oscillations and chaos, and the use of flux analysis, and has provided pointers to software sources. It is possible to conceive of each cell as a node with a set of metabolic pathways within the node as above, but with each node connected to other nodes and then the problem reduces to the modelling of the inter-nodal connections. This could be achieved using conventional mass transfer methods employed in process engineering or by taking advantage of the methods used to monitor transport of drugs across cell membranes or for more generalised pharmacokinetics. The topology and regulatory feedback aspects of such inter-nodal modelling would be complex and would be organism dependent [29]. It is here that the use of non-parametric multivariate statistics will provide a better understanding of the complexities of the situation, in that connections and correlations between metabolic pathways from different cell types and different organs separated in space and with effects displaced in time according to locations and event, capable of being unravelled. One unexpected outcome of such an approach would be the discovery of the natural substrates for a number of promiscuous enzymes involved in xenobiotic metabolism, for example cytochrome P450 families [30] . Additionally, the multivariate statistical approach allows the analysis of combined gene array and metabolic data so that events at the transcriptome level can be related to those seen at the metabolic level in a reverse functional genomics approach as described above.

At the level of interactions between genomes, such as between host and pathogen or symbiotic bacterium, modelling approaches using probabilistic methods are expected to be an efficient way to model interactions between organisms. Bayesian approaches and the concept of maximum entropy allow the construction of a metabolic model at the most sparse and simplest degree of complexity. One recent related example is the determination of the biochemical sources of glucose found in blood plasma using automated Bayesian analysis of the NMR-detected deuterium incorporation patterns after ingestion of D_2O [31]. Hence data that are highly accurate and precise such as metabolite concentrations become hugely valuable given the recognised errors on quantitation at the transcriptomic and proteomic levels. It is possible to envisage an approach whereby both probabilistic and multivariate statistical models can be constructed at the microscale, that is individual genomes, and united using probabilistic criteria to produce models at the macroscale (genome–genome). Thus the probability of a given pathway or the level of a given substance will be assignable based on a combination of the data as measured and the prior information available. For instance, recently it has been possible to model human cytochrome P-450-mediated drug metabolism probabilistically using a machine learning approach [32] and a heuristic approach to prediction of more general xenobiotic metabolism has been described based on a comprehensive library of biotransformations and other reactions and on system-specific transformation probabilities [33]. We conclude that only by this means, and at the metabolic level using the host metabonome or sum of the interacting metabolomes, as the blueprint, will overall system biology models be forthcoming since the genomes of many of the interacting organisms will not be known, and indeed may ultimately be unknowable.

References

[1] J. Shockcor, S. Unger, I.D. Wilson, P.J.D. Foxall, J.K. Nicholson and J.C. Lindon. Combined hyphenation of HPLC, NMR spectroscopy and ion-trap mass spectrometry (HPLC-NMR-MS) with application to the detection and characterisation of xenobiotic and endogenous metabolites in human urine. *Anal. Chem.* 68, 4431–4435 (1996).

[2] K. Albert (Ed.). *On-line LC NMR and Related Techniques*, Wiley: New York, 2002.

[3] O. Cloarec, M.E. Dumas, A. Craig, R.H. Barton, J. Trygg, J. Hudson, C. Blancher, D. Gauguier, J.C. Lindon, E. Holmes and J.K. Nicholson. Statistical total correlation spectroscopy (STOCSY): A new approach for individual biomarker identification from metabonomic NMR datasets. *Anal. Chem.* 77, 1282–1289 (2005).

[4] D.J. Crockford, E. Holmes, J.C. Lindon, R.S. Plumb, S. Zirah, S.J. Bruce, P. Rainville, C.L. Stumpf, and J.K. Nicholson. Statistical HeterospectroscopY (SHY), a new approach to the integrated analysis of NMR and UPLC-MS datasets: Application in metabonomic toxicology studies. *Anal. Chem.*, 78, 363–371 (2006).

[5] J.K. Nicholson, J. Connelly, J.C. Lindon, and E. Holmes. Metabonomics: A platform for studying drug toxicity and gene function. *Nat. Rev. Drug Disc.* 1, 153–162 (2002).

[6] N.J. Waters, E. Holmes, A. Williams, C.J. Waterfield, R.D. Farrant, and J.K. Nicholson. NMR and pattern recognition studies on the time-related metabolic effects of α-naphthylisothiocyanate on liver, urine, and plasma in the rat: An integrative metabonomic approach. *Chem. Res. Toxicol.* 14, 1401–1412 (2001).

[7] M. Coen, E.M. Lenz, J.K. Nicholson, I.D. Wilson, F. Pognan, and J.C. Lindon. An integrated metabonomic investigation of acetaminophen toxicity in the mouse using NMR spectroscopy. Chem. Res. Toxicol. 16, 295–303 (2003).

[8] N.J. Waters, C.J. Waterfield, R.D. Farrant, E. Holmes, and J.K. Nicholson. Integrated metabonomic analysis of bromobenzene-induced hepatotoxicity: Novel induction of 5-oxoprolinosis. *J. Proteome Res.* 5, 1448–1459 (2006).

[9] R. Ruepp, R.P. Tonge, J. Shaw, N. Wallis, and F. Pognan, Genomics and proteomics analysis of acetaminophen toxicity in mouse liver. *Toxicological Sciences* 65, 135–150 (2002).

[10] M. Coen, S.U. Ruepp, J.C. Lindon, J.K. Nicholson, F. Pognan, E.M. Lenz, and I.D. Wilson. Integrated application of transcriptomics and metabonomics yields new insight into the toxicity due to paracetamol in the mouse. *J. Pharmaceut. Biomed. Anal.* 35, 93–105 (2004).

[11] W.H.M. Heijne, R.J.A.N. Lamers, P.J. van Bladeren, J.P. Groten, J.H.J. van Nesselrooij, and B. Van Ommen. Profiles of metabolites and gene expression in rats with chemically induced hepatic necrosis. *Toxicol. Pathol.* 33, 425–433 (2005).

[12] J.L. Griffin, S.A. Bonney, C. Mann, A.M. Hebbachi, G.F. Gibbons, J.K. Nicholson, C.C. Shoulders, and J. Scott. An integrated reverse functional genomic and metabolic approach to understanding orotic acid-induced fatty liver. *Physiol. Genomics* 17, 140–149 (2004).

[13] J.E. Ippolito, J. Xu, S.J. Jain, K. Moulder, S. Mennerick, J.R. Crowly, R.R. Townsend, and J.I. Gordon. An integrated functional genomics and metabolomics approach for defining poor prognosis in human neuroendocrine cancers. *Proc. Nat. Acad. Sci.* U.S.A 102, 9901–9906 (2005).

[14] M. Mayr, Y.L. Chung, U. Mayr, X.K. Lin, L. Ly, H. Troy, S. Fredericks, Y.H. Hu, J.R. Griffiths, and Q.B. Xu. Proteomic and metabolomic analyses of atherosclerotic vessels from apolipoprotein E-deficient mice reveal alterations in inflammation, oxidative stress, and energy metabolism. *Artioscler. Thromb. Vasc. Biol.* 25, 2135–2142 (2005).

[15] M. Mayr, U. Mayr, Y.L. Chung, X. Lin, J.R. Griffiths, and Q.B. Xu. Vascular proteomics: Linking proteomic and metabolomic changes. *Proteomics* 4, 3751–3761 (2004).

[16] M. Rantalainen, O. Cloarec, O. Beckonert, I.D. Wilson, D. Jackson, R. Rowlinson, S. Rayner, R. Tonge, R. Wilkinson, J.D. Mills, J. Trygg, J.K. Nicholson, and E. Holmes. Statistically integrated metabonomic-proteomic studies on a human prostate cancer xenograft model in mice. *J. Proteome Res.*, in press.

[17] T.G. Klenø, B. Kiehr, D. Baunsgaard, and U.G. Sidelmann. Combination of 'omics' data to investigate the mechanism(s) of hydrazine-induced hepatotoxicity in rats and to identify potential biomarkers. *Biomarkers* 9, 116–138 (2004).

[18] A. Craig, J. Sidaway, E. Holmes, T. Orton, D. Jackson, R. Rowlinson, J. Nickson, R. Tonge, I. Wilson, and J. Nicholson. Systems toxicology: Integrated genomic, proteomic and metabonomic analysis of methapyrilene induced hepatotoxicity in the rat. *J. Proteome Res.*, 5, 1586–1601 (2006).

[19] T. Ideker, V. Thorsson, J.A. Ranish, R. Christmas, J. Buhler, J.K. Eng, R. Bumgarner, D.R. Goodlett, R. Aebersold, L. Hood. Integrated genomic and proteomic analyses of a systematically perturbed metabolic network. *Science* 292 (5518), 929–934 (2001).

[20] A.H. Fairlamb, Metabolic pathway analysis in trypanosomes and malaria parasites. *Phil. Trans. R. Soc. Lond.* B 357, 101–107 (2003).

[21] N. Greene, P.N. Judson, J.J. Langowski, and C.A. Marchant. Knowledge-based expert systems for toxicity and metabolism prediction: DEREK SAR and METEOR SAR. *QSAR Environ. Res.* 10, 299–314 (1999).

[22] A. Varma and B.O. Palsson. Metabolic flux balancing: Basic concepts, scientific and practical use. Bio-Technology 12, 994–998 (1994).

[23] T. Szyperski, ^{13}C NMR, MS and metabolic flux balancing in biotechnology research. *Q. Rev. Biophys.* 31, 41–106 (1998).

[24] C.H. Schilling, J.S. Edwards, and B.O. Palsson. Towards metabolic phenomics: Analysis of genomic data using flux balances. *Biotechnol. Prog.* 15, 288–295 (1999).

[25] S. Vaseghi, A. Baumeister, M. Rizzi, and M. Reuss. In vivo dynamics of the pentose phosphate pathway in *Saccharomyces cerevisiae. Metab. Eng.* 1, 128–140 (1999).

[26] E. Ravasz, A.L. Somera, D.A. Mongru, Z.N. Oltvai, and A.-L. Barabasi. Hierarchical organization of modularity in metabolic networks. *Science* 297, 1551–1555 (2002).

[27] A. Giuliani, J.P. Zbilut, F. Conti, C. Manetti, and A. Miccheli. Invariant features of metabolic networks: A data analysis application on scaling properties of biochemical pathways. *Physica A* 337, 157–170 (2004).

[28] C. Giersch. Mathematical modelling of metabolism. *Curr. OP. Plant Biol.* 3, 249–253 (2000).

[29] H. Jeong, B. Tombor, R. Albert, Z.N. Oltvai, and A.-L. Barabási. The large-scale organization of metabolic networks. *Nature* 407, 651–654 (2000).

[30] R. Mohan, and R.A. Heyman. Orphan nuclear receptor modulators. *Curr. Top. Med. Chem.* 3(14) 1637–1647 (2003).

[31] M. Merritt, G.L. Bretthorst, S.C. Burgess, A.D. Sherry, and C.R. Malloy. Sources of plasma glucose by automated Bayesian analysis of H-2 NMR spectra. *Magn. Reson. Med.*, 50, 659–663 (2003).

[32] D. Korolev, K.V. Balakin, Y. Nikolsky, E. Kirillov, Y.A. Ivanenkov, N.P. Savchuk, A.A. Ivashchenko, and T. Nikolskaya. Modelling of human cytochrome P450-mediated drug metabolism using unsupervised machine learning approach. *J. Med. Chem.* 46, 3631–3643 (2003).

[33] O.G. Mekenyan, S.D. Dimitrov, T.S. Pavlov, and G.D. Veith. A systematic approach to simulating metabolism in computational toxicology. I. The TIMES heuristic modelling framework. *Curr. Pharmaceut. Design* 10, 1273–1293 (2004).

Index

1-Naphthylamine, 520
1-Naphthylisocyanate, 520
2-Dimensional, 66–8, 137, 380–2
3-Hydroxyisovalerate, 493
3-Hydroxy-methyl-glutarate, 493, 494
3-Methyl-crotonylate, 491, 492
3-Methyl-glutarate, 491

Absorption, 37, 137, 138, 241, 384, 386, 428, 430, 500, 505
Acetate, 37, 80, 158, 163, 254, 295, 298, 377, 427, 429, 433, 435, 520, 541, 546
Adenosine triphosphate, 501
Adiabatic, 500
Agrochemical Research:
  herbicide, 453–4
Alanine, 55, 251, 257, 331, 338, 377, 378, 419, 432, 493, 506, 521, 541, 544, 545
Alignment, 101, 184
Allostatic load, 300–301, 304–308
α-napthylisothiocyanate, 264, 520, 539
AMARES, 505
Angina pectoris, 332
Angiography, 24, 332, 333, 334, 337, 339, 340, 341
Antiepileptic, 490
Arabidopsis, 211, 448, 455, 457, 459, 461, 463, 464, 465, 466, 469, 470, 471, 472
Artefacts, 490, 501, 503, 507, 526
Arthritis, 422–4
Artificial neural networks, 213–15
Aspartate, 37, 424, 424, 463, 493
Assignment, 490–500
Atherosclerosis, 10, 333
Automation, 69–76, 403–404

Behavioral phenotype, 291, 292, 304, 307, 316
Betaine, 158, 387, 427, 493
Biochemical pathway identification, 42–5
Biofluids, 10–11, 17–18, 25–6, 496, 518, 526
Biomarkers, 20, 267–9, 300–301, 308–10
Biopsies, 216, 244, 356, 365, 431, 433, 434, 489
Blood plasma, 494–6, 526, 541, 543, 545, 546
Body fluids, 375, 387, 395, 397, 490, 497, 500
Brain tumour, 24, 214, 216, 504
Breast Cancer, 355–8
Bronchoalveolar lavage fluid (BALF), 521

Cancer, 345, 361, 363, 366–72, 436
Cancer Metabolomics, 345, 366–72
Capillary electrophoresis, 11, 36, 150, 234, 402, 450
Cardiovascular disease, 327–41
Carr-Purcell-Meiboom-Gill, 500
Cerebrospinal fluid, 389
Chemical shift, 501–505
Chemometrics, 19–20, 171
Cholesterol, 38, 300, 309, 331, 332, 334, 336, 337, 495
Citrate, 65, 81, 82, 84, 269, 423, 434, 493, 521
Classification, 212–13, 217–21
Clinical, 244, 279–85
CLOUDS, 23, 219, 220
Coherence, 95, 122–7, 386
Coherence averaging, 122–7
Coherence averaging by MAS, 122
Consortium for metabonomic toxicology, 23, 79, 219
Coronary artery disease, 24, 339
Coronary heart disease, 331–3, 334–8
Correlation peak imaging (CPI), 510
Correlation spectroscopy, 92–3, 137, 383–4
COSY, 92–3, 137–8, 384, 501
CPMG, 90, 134, 500
Creatine, 55, 104, 303, 425, 522

Creatinine, 104, 496–7
Crop Improvement, 340
CSF, 78, 396, 492–4, 499
CT-COSY, 503, 511

Database, 227, 263, 387–9
Data pre-processing, 96–106
Decoupling, 91, 246, 385, 500, 506, 511
Dereplication, 38
Diabetes, 418–22
Diagnosis, 24–5
Diagnostics, 331–4
Diagnostic tests, 232, 327, 328, 340, 435
Differential gel electrophoresis, 545
Dimension Reduction, 19, 203, 221
Dipolar coupling, 124, 125, 126, 127
Disease Diagnosis, 24–5, 282–5
Disease risk, 292–303
Disease Treatment, 282–5
DOE, 172, 176
Drug development, 26, 242, 265, 328
Drug discovery, 140, 244, 279, 280
Drug metabolising enzymes, 6
Drug safety testing, 22, 26, 518
Dynamic Sampling, 177

Echo planar spectroscopic imaging (EPSI), 511
Echo-sequence, 500
Echo time (TE), 137, 385, 500, 505
Echo/antiecho, 500
Editing sequences, 511
Environmental toxicology, 517
Esters, 468, 499
Ethanol, 253, 400, 429, 499
Ethanolamine, 140, 493
Evaporation, 79, 151, 499
Exchange spectroscopy (EXSY), 501, 511, 512
Experimental Design, 171, 172, 176, 177, 182, 364, 365, 452, 519, 550
Extracellular, 4, 5, 79, 332, 390
Extract, 172–3
Extraction, 151, 497–500

Fatty acid, 158, 178, 256, 269, 270, 282, 292, 304, 336, 367, 419, 420, 422, 446, 469, 541, 544
Fish metabolomics, 219, 282, 395
FOCSY-J, 506, 508
Format, 233
Formate, 80, 158, 403, 423, 493
Fourier transformation, 60, 386, 501, 505
Fructose, 456, 472, 493

Gas chromatography, 150–6, 446–8
Gene expression, 1, 7, 9, 10, 25, 45, 234, 235, 236, 237, 310, 329, 330, 346, 350, 362, 413, 470, 543, 544, 549, 550
Genetic Programming, 215–17
Genetically modified animals, 416–18
Genomics, 351, 469–71
Glioma, 140, 362, 503, 504, 506
Global metabolite profiling, 149, 150, 167
Global systems biology, 4, 414, 533
Glucose, 104, 256, 388, 419
Glutamate, 140, 256, 270, 302, 417, 424, 427, 463, 493, 511, 525
Glutamine, 425, 427, 493, 511, 526, 546
Glycine, 45, 388, 416, 419, 428, 463, 468, 474, 493
Gradient, 89–90, 128–31
Gut microflora, 4, 22, 26, 37, 162, 252, 255, 290, 317, 415, 523

Heparinized, 497
Hepatic encephalopathy, 504
Heteronuclear, 94–6, 125, 386
Hierarchical cluster analysis, 205–207
High Resolution Magic Angle Spinning (HRMAS), 113, 526
High Resolution Magic Angle Spinning Proton Magnetic Resonance Spectroscopy (HRMAS ^{1}H MRS), 365
High throughput, 3, 46, 56, 70, 76, 139, 150, 164, 167, 234, 235, 241, 245, 452
Histidine, 81, 493, 494, 528
HMBC, 96, 500
Homeostasis, 56, 303, 492
Homonuclear, 92–4, 124
H-PCA, 196
HSQC, 95, 386
HSQC-TOCSY, 492, 500
Human brain, 139, 283, 361, 365, 366, 367, 425, 510
Huntington's Disease, 302, 317, 338, 413
Hydrophilic, 469, 489, 490
Hydroxybutyrate, 270, 493, 494, 545, 546
Hypertension, 312, 338

Hyphenation, 150, 156, 402
Hypoxia, 501

Inborn error, 375, 389–93
Inborn errors of metabolism, 375–7, 389–95
Infections, 6, 290, 434
Intact tissue specimen, 364
Integration of omics, 533
Interpretation, 12, 386–7
Invertebrate metabolomics, 527
Ischemia, 337, 338, 504
ISIS, 503
Isoleucine, 426, 429, 463, 468, 493, 494, 521, 546
Isotropic susceptibility heterogeneity, 119–21
In vivo, 489, 55–504, 505–11

J-coupled, 95, 384, 501, 503, 506, 507, 508, 510
J-coupling, 64, 92, 503, 505
J-modulation, 505
J-resolved NMR spectroscopy, 505–10

k-Nearest Neighbour, 212–13
k-space, 176, 511
k-means, 212

Lactate, 55, 81, 285, 419, 429, 434
LCModel, 505
Leucine, 393, 394, 426, 463, 468, 490, 493, 496, 521
Levetiracetam, 490, 491
Lipids, 16, 135, 140, 174, 211, 352, 422, 434, 494, 499, 543
Lipophilic, 40, 489, 490, 523
Lipoproteins, 12, 69, 211, 300, 302, 303, 318, 336, 339, 403, 416, 419, 420, 428, 431, 432, 492, 527
Liquid chromatography, 156–7, 164–6, 448
Liver disease, 163, 283, 430
Lyophilization, 80, 499
Lysine, 429, 493, 494, 546

Macromolecules, 13, 55
Magic angle spinning, 113, 122–7, 526
Magic Angle spinning NMR spectroscopy, 113, 526
MaGiCAD study, 339, 340, 341
Magnetic Resonance Spectroscopy (MRS), 363–4
Mammalian-microbe interactions, 7
Mass spectrometry (MS), 14–17, 149, 156, 164, 166, 446, 448
  capillary electrophoresis mass spectrometry (CE-MS), 150
  gas chromatography mass spectrometry (GC-MS), 150–6
  liquid chromatography mass spectrometry (LC-MS), 446–8
Mechanisms of toxicity, 24, 257, 258, 266, 269, 520
Metabolic networks, 48–51
Metabolic profiling, 301, 443
Metabolic screening, 259–67
Metabolite analysis, 246, 350, 456, 470, 534, 544
Metabolite Biomarkers, 182, 462, 476
Metabolite identification, 11, 512, 536
Metabolites, 37
Metabolome, 36–7, 443–5
Metabonomics, 1–3, 10–11, 21–6, 55, 149, 157–63, 166–7, 171, 241, 279, 282–5, 289
Methanol, 163, 497, 499
Methylglutaryl-CoA, 490
MGED, 234, 235, 236, 237
Microbiome, 5, 6
Model, 180–1, 193, 194–5, 233, 236, 387–8, 416–18, 519–22, 551–3
Molecular fingerprinting, 329–31
Motion averaging, 65, 116–19
MSI, 236–7
Multi-dimensional, 69, 87, 107, 130, 154, 175, 489
Multidimensional scaling, 207–209
Multivariate Design, 176–7
Myo-inositol, 140, 141, 389, 426, 494, 503

N-Acetylaspartate, 114, 427, 506
N-Acetylglutamate, 493
Natural remedies, 415–16
Neurological conditions, 424–7
NMR, 11–14, 56, 86, 87, 92, 113, 449–50, 489, 490, 500, 505, 526
NMR spectroscopy, 11–14, 55, 113, 375, 380–6, 403, 500, 505–10, 526
Non-linear methods, 201

Nuclear magnetic resonance (NMR), 11, 55, 113, 172, 201, 243, 289, 375, 446
Nutrition:
  nutrigenomics, 235

Omics data integration, 45–6
Oncology, 229, 235, 352–3
OPLS, 182, 193
Organ transplants, 430–5

P.COSY, 505
Parasitic infection, 6, 22, 290, 292, 293, 297, 298, 300, 314, 317, 435
Partial least squares, 181–2
Pattern recognition, 171, 192, 204, 207, 209, 211, 213, 214, 221, 282, 283, 284, 328, 365, 387, 417, 420, 425, 430, 435, 490, 520, 541
PCA, 195–6
Pellet, 499
Personalised medicine, 327, 328
Pharmaceuticals, 241, 279, 282–5
Pharmacometabonomics, 25
Phase cycling, 71, 89, 130, 134, 501, 503, 507
Phenylalanine, 162, 493, 494, 546
Phosphocreatine, 501
Physiological variation, 174, 518
Plant abiotic stress:
  cold, 465, 467
  drought, 466
  freezing, 465, 466
Plant biotic stress:
  plant–pathogen interactions, 468
Plant breeding:
  hybrids, 473
Plant development, 461
Plant Genomics:
  overexpression, 459, 460, 466, 471
  plant functional genomics, 469–72
  T-DNA, 471
Plant metabolic engineering, 458
Plant nutrition, 464–5
Plasma, 78, 388–9, 492–6, 497–9
PLS, 18, 19, 98, 99, 140, 171, 173, 181–2
Polycyclic aromatic hydrocarbons (PAHs), 520
Polymorphisms, 6, 25, 328, 329
Population studies, 289, 301
Presaturation, 87, 134, 135
PRESS, 506, 508–10
Principal components analysis, 18, 84, 152, 204, 415, 520, 539
Probabilistic neural networks, 217–21
Profiling, 301, 327, 331–4, 443, 445–53
Proline, 162, 463, 466, 493
Pronostic tests, 328, 337, 341
Prostate Cancer (PC), 353–5
Protein, 4, 7, 10, 44, 329–30, 360, 548
Proteomics, 1, 280, 351, 543, 548
PSI, 234, 235, 236
Psychosocial phenotype, 291, 292, 303, 304, 315
Pyroglutamate, 493, 494

Rat brain, 139, 503, 506, 510, 511
Regression, 22, 23, 26, 50, 140, 174, 202, 214, 215, 332, 370, 546
Relaxation, 10, 59, 90
Renal diseases, 427–30
Renal failure, 427, 428

S. Japonicum, 299, 315, 317
S. mansoni, 294, 295, 297, 299, 313, 317, 434, 435
Sample handling, 76
Schistosomiasis, 293, 294, 434
Screening, 259–67, 352
Secondary metabolites:
  flavonoids, 459, 470, 473
  lignin, 458
SECSY, 503
Self-organizing Maps, 464
Seminal fluids, 10, 24, 92, 384, 435, 436
Serum, 12, 23, 24, 78, 388–9, 403, 436
Signal-to-noise ratio, 13, 401, 501
SIMCA, 180, 219, 436
Spatial localization, 501, 506, 510, 512
Spectroscopic imaging, 510–11
Spectroscopic RARE, 511
Spectroscopic U-FLARE, 511
Spheroids, 11
Standard, 20–1, 227, 234, 237, 452–3
Standardisation, 20–1, 227
Statistical total correlation spectroscopy, 19, 258, 534

Strategy, 40, 173, 178, 242, 283, 290, 315, 318, 328, 339, 366, 373, 430, 435, 458, 463, 464, 468, 471, 523, 548, 551
Stress, 303–304, 465, 467–8
Structural elucidation, 37, 38, 69, 450
Study design, 173–4, 182–3, 192, 252–5
Succinate, 80, 158, 299, 426, 429, 434, 493, 520, 521
SUPERCOSY, 504, 505
Supervised, 18
System biology, 231, 476, 551–3

Taurine, 81, 98, 99, 158, 417, 494, 503, 522
Threonine, 377, 389, 429, 468, 493, 494
Tissues, 1, 3, 419
TOCSY total correlation spectroscopy, 384–5
Toxicity, 257–70
Transcriptomics, 1, 3, 45, 235, 242, 244, 425, 464, 517, 543, 544, 549
Trees, 41, 206, 207, 212, 215, 473
Tryptophane, 160, 162, 298, 434, 456, 457, 493, 494
Tyrosine, 140, 162, 389, 457, 493, 494, 544, 545, 546

Unsupervised, 205–12
Urine, 78, 387–8, 394, 490–2, 499

Valine, 162, 389, 401, 419, 424, 426, 429, 493, 494, 546
Visualisation, 11, 18, 201, 202, 203, 207, 209, 211, 212, 213, 217, 220

Wild mammals, 523–6
WURST, 500

Yield, 461–3